RECENT DEVELOPMENTS IN STATISTICS

RECENT DEVELOPMENTS IN STATISTICS

Proceedings of the European Meeting of Statisticians
Grenoble, 6-11 September, 1976

Edited by

J. R. BARRA
F. BRODEAU
G ROMIER
B. VAN CUTSEM
University of Grenoble

1977

NORTH-HOLLAND PUBLISHING COMPANY
AMSTERDAM • NEW YORK • OXFORD

North-Holland ISBN: 0 7204 0751 6

Published by:
NORTH-HOLLAND PUBLISHING COMPANY
AMSTERDAM • OXFORD • NEW YORK

Distributors for the U.S.A. and Canada:
Elsevier/North-Holland, Inc.
52 Vanderbilt Avenue
New York, N.Y. 10017

PRINTED IN THE NETHERLANDS

FOREWORD

In 1976, the European Meeting of Statisticians was held at the University of Grenoble (France) from September 6-10, and was sponsored by the Institute of Mathematical Statistics and the Bernoulli Society for Mathematical Statistics and Probability. This conference brought together more than 600 statisticians, from many countries, who came here to listen to about 150 talks.

The subjects dealt with were those of classical statistics : theory of estimation and testing hypotheses, non-parametric problems, statistical inference in the classical case and for stochastic processes, experimental designs, multivariate analysis, stochastic approximation, asymptotic expansions, time series, linear models, etc. Moreover, numerous applications of statistics were presented concerning the medical, social and teaching sciences, demography, genetics, physics, etc. Finally, you will notice the great number of papers devoted to data analysis and to the relations between computer science and statistics. Nearly all the invited talks, as well as the most representative ones among the contributed papers, have been combined in this volume and arranged around those principal themes, hoping that this would facilitate the use of this book.

On the occasion of publishing these Proceedings, we would like to thank particularly the Organizing Committee of the European Meetings of Statisticians who agreed to the 1976 Meeting being held at our University, as well as the French authorities who contributed their financial support. Besides, we have the pleasure of saluting the birth of the Bernoulli Society, as this Meeting was the first to be held under its sponsorship. We also feel greatly obliged to Professor H. CRAMER who honoured our Meeting with his presence, as well as to the professors GANI, DURBIN, VAN ZWET and BARNDORFF-NIELSEN for their contributions to the organization of the Meeting and more specifically to the scientific program. Last but not least our thanks are due to the North-Holland Publishing Company, to whom we owe the present publication of the Proceedings.

The Editors - February 1977

CONTENTS

Recent Developments in Statistics
J.R.Barra et al., editors

L'HERITAGE DE BERNOULLI

Grenoble 6 Septembre 1976

H. CRAMER, Suède

Devant cette assemblée, qui s'est réunie sous le patronage de la Bernoulli Society et de l'Institute of Mathematical Statistics, il m'a semblé naturel de dire quelques mots sur L'HERITAGE DE BERNOULLI.

Il va sans dire que c'est de Jacques Bernoulli que je vais parler, et de ce qu'il nous a laissé en notre qualité de statisticiens et probabilistes. Je vais donc passer sous silence les contributions de quelques autres membres de cette famille célèbre, ainsi que les travaux de Jacques Bernoulli lui-même appartenant à d'autres domaines scientifiques.

Quand Jacques Bernoulli mourut en 1705, à l'âge de seulement 51 ans, son grand traité de la théorie des probabilités, auquel il avait travaillé depuis une vingtaine d'années, restait toujours inachevé. Son manuscrit, écrit en latin, fut publié huit ans après sa mort, tel qu'il l'avait laissé, sous le titre ARS CONJECTANDI. Autant que je sache, il n'existe qu'une seule traduction complète de cet ouvrage en une langue moderne, à savoir une édition allemande de l'année 1899, qui d'ailleurs semble être assez difficilement accessible.

L'Ars conjectandi consiste en quatre parties, dont les trois premières n'ont guère un grand intérêt pour nous, même si elles contiennent une étude générale de l'analyse combinatoire, et son application à un grand nombre de questions concernant les jeux de hasard. C'est la quatrième partie, intitulée "Applications de la théorie précédente aux questions sociales, morales et économiques", qui contient les idées fondamentalement nouvelles de l'auteur. Malheureusement c'est aussi cette partie qu'il a due laisser inachevée. Telle que nous l'avons, elle contient des principes pour ces applications, mais pas les applications elle-mêmes.

Je veux essayer de rappeler très brièvement quelques parties essentielles de son raisonnement, en me servant autant que possible de ses propres mots, exprimés dans une terminologie un peu modernisée.

En discutant le caractère de nécessité des événements futurs, il prend une attitude parfaitement déterministe. Le temps qu'il fera demain est parfaitement déterminé par l'état présent de l'atmosphère, de même que les éclipses qui auront lieu l'année prochaine sont déterminées par les positions actuelles des corps célestes. Le résultat d'un coup de dé est déterminé par l'état de mouvement du dé et par sa distance de la table au moment où il sort de la main.

Et pourtant, Bernoulli fait observer, nous tenons à regarder le temps du demain et le résultat d'un coup de dé comme des événements fortuits, dépendant du hasard, tandis que les éclipses sont des événements certains, qu'on peut prédire par des calculs. C'est dans notre ignorance qu'il trouve la raison de cette contradiction apparente : l'état actuel de l'atmosphère, qui en fait est exactement déterminé, ne nous est pas connu avec une précision suffisante pour nous permettre de prédire le temps du demain avec certitude. Et la même situation se présente pour le dé. - Ce n'est pas sans intérêt de trouver de telles pensées si clairement exprimées déjà à cette époque.

Pour un événement de ce caractère fortuit, Bernoulli admet qu'il existe une probabilité, qu'il envisage comme un degré de certitude, ayant une valeur entre zéro et un.

Un événement qui a une probabilité très voisine de l'unité doit être regardé comme moralement certain, tandis qu'une probabilité très petite correspond à un événement moralement impossible. Un détail amusant dans ce contexte est que Bernoulli pense que les gouvernements devraient fixer une limite supérieure officielle pour les probabilités des événements qu'il serait légitime de regarder comme pratiquement impossibles. - Peut-être ça serait-il quelque chose pour nos politiciens contemporains ?

L'art d'estimer les probabilités des événements est ce qu'il appelle Ars conjectandi sive stochastice. Je crois que c'est ici la première fois qu'on rencontre dans cette matière le mot stochastique, si bien connu aujourd'hui.

Il suppose que, pour un événement fortuit, il existe toujours un ensemble de cas possibles, dont les uns sont favorables, les autres défavorables pour que l'événement ait lieu. Si l'on connaît le nombre et les poids relatifs de tous ces cas, on peut trouver la probabilité de l'événement. Et il rappelle que c'est en fait cette méthode qu'on applique dans le domaine des simples jeux de hasard.

Mais, dit Bernoulli, comment serait-il possible de connaître toutes les maladies qui pourraient causer la mort d'un individu donné, ou toutes les variations dans l'atmosphère qui vont déterminer l'évolution du temps ?
Puisque ces événements - et bien d'autres analogues - dépendent de causes infiniment nombreuses et infiniment compliquées, il lui semble impossible de déterminer leurs probabilités par une énumération de tous les cas possibles.

Mais c'est sur ce point qu'une idée nouvelle et féconde fait son entrée dans son raisonnement. Quand une probabilité inconnue ne peut pas être déterminée a priori, par une énumération des cas possibles, il dit qu'on peut souvent la trouver au moins approximativement a posteriori, c'est-à-dire en faisant usage des observations d'un grand nombre d'événements analogues. Si, par exemple, on a observé que, parmi 400 hommes du même âge et constitution que Titius, 300 sont décédés avant dix ans, on peut conclure que le nombre des cas où Titius va mourir avant dix ans est à peu près trois fois plus grand que le nombre des cas où il survivra ; sa probabilité de mourir est donc approximativement 3/4. De même si l'on a observé le temps pendant un grand nombre de jours, on peut former des conclusions p. ex. sur la probabilité d'avoir une pluie demain, et si l'on a noté les résultats d'une longue série de parties du jeu de paume entre deux joueurs Pierre et Paul, on pourra estimer la probabilité que Pierre va gagner la partie suivante.

Il s'agit donc ici d'une méthode générale d'employer des observations statistiques pour déterminer les probabilités des événements. Bernoulli fait observer que la méthode n'est pas nouvelle - en fait, il dit, c'est ainsi qu'on agit dans la vie ordinaire. Sans doute cela est vrai, mais c'est à lui qu'on doit l'introduction systématique des observations statistiques comme un instrument scientifique, intimement lié à la théorie des probabilités.

Cependant il dit que, pour donner une confirmation scientifique de cette méthode déjà appliquée par tant de gens pratiques, il a trouvé nécessaire de se poser d'abord la question fondamentale suivante : Est-il possible, en faisant un assez grand nombre d'observations, d'être moralement certain d'obtenir une approximation aussi bonne que l'on veut de la probabilité cherchée ?

C'est la réponse affirmative à cette question qui constitue le célèbre théorème qu'il annonce comme le résultat principal de ses études.

"Voilà donc", il dit, "ce problème que je me suis déterminé de publier ici, après l'avoir retenu pendant vingt années. Sa nouveauté, sa grande utilité et sa difficulté sont destinées à donner une importance plus élevée à toute cette doctrine."

Et puis il procède à énoncer et à démontrer ce théorème classique qui aujourd'hui fait partie de tous nos traités de la théorie des probabilités.

Il faut admettre que Bernoulli n'exagère point en parlant de la nouveauté de son théorème. En effet, c'est ici la première fois qu'on envisage une distribution de probabilité comme un objet mathématique dépendant d'un paramètre, et qu'on étudie son comportement lorsque ce paramètre tend vers l'infini.

Dans cet ouvrage posthume de Jacques Bernoulli, on trouve donc à la fois une première indication d'un programme pour la statistique mathématique, et la première application de l'analyse infinitésimale à l'étude d'une distribution de probabilité. - Ce n'est pas là un pauvre héritage !

Eh bien : qu'est-ce qu'on a fait - qu'est-ce que nous avons fait - pour développer cet héritage ? Jetons un coup d'oeil rapide, et nécessairement très incomplet, sur ce qui s'est passé en notre domaine après la publication de l'Ars conjectandi.

Pendant tout ce temps de plus de deux cent soixante ans, il y a eu une série continue de travaux qui ont pris leur point de départ directement du théorème original de Bernoulli, tout en le précisant et le généralisant dans une multitude de directions différentes. Ces travaux nous ont donné ce qu'on appelle les théorèmes-limites de la théorie des probabilités.

La première contribution importante à cette série est dûe à De Moivre. Bernoulli avait démontré son théorème par des moyens tout à fait élémentaires, en étudiant la croissance et la décroissance des termes du développement binomial. De Moivre, ayant à sa disposition la formule asymptotique de Stirling, pouvait montrer que la probabilité considérée par Bernoulli est, à la limite, donnée par la distribution normale qui, plus tard, devrait être connue sous les noms de Gauss et de Laplace.

On sait bien que le théorème original de Bernoulli, ainsi que sa forme précisée dûe à De Moivre, peut s'exprimer comme une propriété d'une somme de variables aléatoires indépendantes d'une structure particulièrement simple.
Laplace, dans son grand ouvrage "Théorie analytique des probabilités", a étudié une somme analogue, formée de variables aléatoires beaucoup plus générales, mais toujours mutuellement indépendantes. Pour une telle somme dûment normée, il a énoncé le théorème qu'elle a la même distribution limite exponentielle qui avait été rencontrée par De Moivre dans le simple cas de Bernoulli.

C'est ce théorème pour qui le nom "théorème-limite central de la théorie des probabilités" a été proposé par Georges Polya. C'est un théorème d'une importance capitale, qui contient les résultats antérieurs de Bernoulli et De Moivre comme des cas très particuliers. Mais malheureusement la démonstration donnée par Laplace à son grand théorème n'était point satisfaisante.
Bien que plusieurs des mathématiciens du XIXème siècle se soient efforcés de combler la lacune, c'était seulement Tchebychev qui savait indiquer un chemin qui devait conduire au but. Mais une démonstration vraiment complète et rigoureuse ne fut donnée que dans les premières années de notre siècle, par Liapounov.

La série de ces théorèmes-limites dont le premier était le théorème de Bernoulli s'est prolongée jusqu'à nos jours. Les conditions données par Liapounov pour la validité du théorème central furent généralisées par Lindeberg, et enfin Feller a donné des conditions à la fois nécessaires et suffisantes. On a aussi étudié le passage à la limite de la distribution considérée par ces auteurs plus en détail, en obtenant des développements asymptotiques, dont la distribution normale forme le premier terme, et en considérant ce qu'on appelle les "grandes déviations" qui échappent au théorème de Laplace. Lévy et Khintchine ont apporté des généralisations importantes, en étudiant des cas où il y a une distribution limité appartenant à une classe plus générale que la loi normale, à savoir la classe des distributions infiniment divisibles. D'intéressantes recherches, qui ne sont pas encore terminées, ont été consacrées à l'étude des propriétés limites de sommes de variables aléatoires interdépendantes. Et finalement, toutes ces recherches ont été généralisées aux variables aléatoires multi-dimensionnelles.

Pour tous ces travaux cherchant à généraliser et à approfondir le théorème original de Bernoulli, il était nécessaire d'avoirs recours à des méthodes analytiques nouvelles et puissantes. Lagrange, Laplace et Cauchy avaient employé une première forme de l'outil analytique connu aujourd'hui sous le nom de "fonctions caractéristiques". C'était aussi de cette méthode que s'était servi Liapounov pour sa démonstration du théorème limite central. Mais une théorie détaillée des fonctions caractéristiques ne fut donnée que plus tard, dans les travaux de Paul Lévy, qui a mis en évidence leur haute valeur. Son théorème de continuité pour les fonctions caractéristiques a eu une importance fondamentale pour les recherches sur les théorèmes-limites de la théorie des probabilités. - Une autre méthode importante pour l'étude de ces problèmes est la méthode des moments, employée par exemple par Tchebychev. D'ailleurs il y a des relations intéressantes entre ces deux méthodes.

Les vingt années de paix entre les deux guerres mondiales ont été pour nos domaines une période d'un développement énorme. D'une part, les statisticiens anglais et américains - Fisher, Neyman, Pearson et bien d'autres - ont crée pendant cette époque une nouvelle science, la statistique mathématique, qui s'occupe des relations entre les observations statistiques et les modèles probabilistiques dont on veut se servir pour l'analyse des observations, et pour des prédictions. Ils ont introduit la théorie de l'estimation statistique, la théorie des tests pour des hypothèses statistiques, et des méthodes nouvelles d'échantillonnage.

L'objet principal de cette science est de trouver des méthodes pour déduire des conclusions valides en partant de données statistiques. C'est là précisément le problème que s'était posé Bernoulli dans le cas le plus élémentaire. Nous savons tous combien le développement vigoureux de cette science a continuellement accéléré jusqu'à l'heure actuelle, et qu'elle a trouvé un nombre toujours croissant de domaines d'application.

Il y a ici, comme on sait, deux points de vue opposés. On peut, comme Bernoulli, considérer une probabilité - ou quelque autre paramètre associé à une distribution - comme une constante fixée, quoique inconnue, et se proposer de trouver quelque information sur sa valeur au moyen des observations statistiques. Or, on peut aussi la regarder comme une variable aléatoire, assujettie à une distribution a priori et appliquer un raisonnement du type de Bayes. C'est pourtant toujours la même théorie des probabilités qui fournit l'instrument mathématique dont on se sert. Et sans doute il y a, parmi le nombre immense des champs d'application, pour chacune des deux manières de voir, des cas où elle est convenable.

D'autre part, pendant ces mêmes vingt années entre les guerres, les mathématiciens ont de leur côté développé la théorie des probabilités comme une branche des mathématiques pures. J'ai déjà mentionné quelques-uns des travaux se rattachant aux théorèmes-limites pour les sommes de variables aléatoires. Toute cette ligne de recherche peut être considérée comme un développement naturel du théorème original de Bernoulli. Mais les mathématiciens ont introduit aussi des idées d'un ordre différent, indiquant une ère nouvelle où les relations avec l'héritage de Bernoulli, bien que sans doute encore présentes, n'auront plus le même caractère d'évidence. La théorie des probabilités classique, telle qu'on la connaissait il y a un demi-siècle, a été complètement transformée par l'introduction de ces nouvelles idées.

D'abord ce n'était qu'un développement tout naturel, quand certaines applications statistiques ont conduit à l'étude de suites de variables aléatoires plus générales que de simples sommes, par exemple des suites obtenues en observant les valeurs d'une variable aléatoire à des points de temps équidistants. Encore un pas en avant s'imposait quand, dans les applications, on rencontrait des fonctions du temps telles que leur valeur à tout instant semblait dépendre plus ou moins du hasard - des fonctions aléatoires, comme on dit maintenant. Un exemple célèbre est le mouvement Brownien, étudié en 1906 par Einstein. Se basant sur des arguments physiques, il avait obtenu certains résultats, mais c'était évident que les méthodes de la théorie classique des probabilités ne suffisaient pas pour un traitement rigoureux d'un tel problème.

C'est Norbert Wiener qui, dans une note publiée en 1923, a d'abord attaqué cette difficulté avec succès. Il y a introduit une distribution de probabilité dans un espace fonctionnel, ce qui était alors une innovation remarquable. De cette manière il a pu donner un modèle mathématique du mouvement Brownien où l'on se sert de fonctions aléatoires du temps qui sont continues mais n'ont pas une dérivée. C'était la première fois qu'on avait donné une théorie rigoureuse d'un processus stochastique à temps continu.

C'était encore seulement un cas très particulier de cette idée nouvelle, mais quelques années plus tard Kolmogorov a publié ses grands travaux, où toutes ces questions sont traitées d'une manière générale et profonde.
En montrant l'identité formelle de la probabilité avec la notion purement mathématique de mesure, il a donné un fondement nouveau et un nouvel élan aux recherches dans ce domaine. En particulier, la notion d'une distribution de probabilité se présente chez lui sous une forme infiniment plus générale qu'auparavant, comme une distribution dans un espace tout à fait arbitraire. La distribution dans le "differentiel space" considérée par Wiener apparaît comme un cas de ce qu'on appelle maintenant un processus de Markov, c'est-à-dire un processus tel que la loi de probabilité de son développement futur est bien déterminée par sa valeur au moment actuel.
L'étude de cette classe de processus a été commencée par Kolmogorov en 1931. Il a démontré que leurs distributions satisfont à des équations fonctionnelles qui, sous certaines conditions de continuité, se réduisent à des équations différentielles. On sait bien qu'il y a maintenant une littérature très étendue sur les processus de Markov.

Une autre classe importante de processus stochastiques, les processus stationnaires, a été introduite en 1934 par Khintchine. Pour ces processus on suppose que leurs lois de probabilité satisfont à une condition d'invariance dans le temps. L'entière préhistoire est alors relevante pour la prédiction de leur développement futur. Leur théorie comprend une analyse spectrale qui a conduit à des résultats d'une haute importance pour un grand nombre d'applications.

La théorie des processus stochastiques, dont je n'ai pu rappeller ici que ces deux exemples, constitue aujourd'hui une des parties les plus importantes et les plus caractéristiques de la théorie des probabilités. Elle emploie les méthodes les plus avancées de la mathématique moderne, mais une partie de ses racines se trouvent dans les théorèmes-limites classiques donnés par

Bernoulli et par ses successeurs.

Nous avons donc devant nos yeux ces deux grandes lignes de recherche scientifique : la ligne statistique et la ligne probabilistique. Un parcours du programme de ce congrès met en évidence combien le rapprochement des points de vue et des méthodes de travail de ces deux lignes a été utile pour le progrès de la science. Leurs différentes manières d'attaquer les problèmes se sont mutuellement fertilisées. La pénétration des méthodes statistiques au moyen de l'analyse mathématique a permis la construction d'une théorie statistique logiquement rigoureuse, au même temps que les applications statistiques ont posé de nouveaux problèmes aux mathématiciens, et ont suggéré des méthodes nouvelles pour leur résolution.

L'histoire de chacune de ces deux lignes peut être suivie à travers des siècles passés jusqu'à ces premières indications dans l'Ars conjectandi que j'ai rappelées en commençant. C'est en développant ces indications, et en y ajoutant continuellement des idées nouvelles, qu'on a pu édifier nos branches de la sciences contemporaine, basées sur l'héritage de Bernoulli.

PART ONE: Invited Papers

Session on Asymptotic expansions
organized by J. PFANZAGL and W.R. VAN ZWET

Session on The analysis of survival data in medical investigations
organized by P. ARMITAGE

Session on Computers and statistics
organized by H. CAUSSINUS

Session on Data analysis and applications to human, social and nature sciences
organized by G. ROMIER

Session on Sampling in genetics
organized by G. MALECOT

Session on Statistical problems in control theory
organized by M. METIVIER

Session on Approach to statistics by children from 11 to 15
organized by P. HENNEQUIN

Session on Stochastic approximation
organized by L. SCHMETTERER

Session on Statistical inference and stochastic processes in demography
organized by J. HOEM and N. KEIDING

Recent Developments in Statistics
J.R.Barra et al., editors

Invited Paper

ASYMPTOTIC EXPANSIONS FOR BAYES PROCEDURES

Helmut Strasser [1)]

1. Introduction

a) The general case

Let $(\Omega,\mathfrak{A})$ be a measurable space and $P_\vartheta|\mathfrak{A}, \vartheta\in\Theta\subseteq\mathbb{R}$, a family of probability measures. Assume that the family $\{P_\vartheta\}_{\vartheta\in\Theta}$ is dominated by a σ-finite measure $\mu|\mathfrak{A}$ and let $h_\vartheta := \frac{dP_\vartheta}{d\mu}, \vartheta\in\Theta$. $\mathfrak{B}_\Theta$ denotes the Borel-σ-algebra of Θ.

Many statistical procedures are based on the likelihood functions

$$\vartheta \mapsto \prod_{i=1}^{n} h_\vartheta(\omega_i), \vartheta\in\Theta, \underline{\omega}=(\omega_1,\ldots,\omega_n)\in\Omega^n,$$

(e.g. maximum likelihood estimates). Bayes procedures are based on averages of likelihood functions, i.e. on set functions of the form

$$B \mapsto \int_B \prod_{i=1}^{n} h_\vartheta(\omega_i)p(\vartheta)d\vartheta, \quad B\in\mathfrak{B}_\Theta, \underline{\omega}\in\Omega^n,$$

where $p:\Theta\to\mathbb{R}^+$ is a weight function. If such a set function is a finite measure then the normalized set function is called "posterior" distribution with respect to the (not necessarily finite) "prior" measure defined by p.

The classical theory of asymptotic optimality ("first order efficiency") shows that among optimum procedures there are both Bayes procedures and other ones which seemingly are of non-Bayesian type,

1) H.Strasser, Mathematisches Institut, Justus-Liebig-Universität, Arndtstraße 2, D-6300 Gießen, Federal Republic of Germany.

e.g. maximum likelihood estimates. For practical purposes procedures of non-Bayesian type are preferred since for those the problem of specifying a certain weight function does not occur. However, choosing the maximum likelihood method is of the same arbitrariness as choosing a certain weight function, at least as long as these procedures are equivalent from the mathematical point of view.

Roughly spoken, the classical theory identifies sequences of statistical procedures if they differ by a term of order $o(n^0)$. Recently, the theory has been refined by replacing $o(n^0)$ by $o(n^{-1})$. The basic concepts concerning problems of estimation are as follows: A sequence of estimates $S_n : \Omega^n \to \Theta$ is called approximately median unbiased of order $o(n^{-1})$ if for every $\vartheta \in \Theta$

$$P_\vartheta^n\{S_n \geq \vartheta\} \geq \frac{1}{2} - o(n^{-1})$$

$$P_\vartheta^n\{S_n \leq \vartheta\} \geq \frac{1}{2} - o(n^{-1}).$$

A sequence of estimates $T_n : \Omega^n \to \Theta$ which is approximately median unbiased of order $o(n^{-1})$ is called asymptotically optimal of order $o(n^{-1})$ if

$$P_\vartheta^n\{\vartheta - tn^{-1/2} < T_n < \vartheta + tn^{-1/2}\} \geq$$

$$\geq P_\vartheta^n\{\vartheta - tn^{-1/2} < S_n < \vartheta + tn^{-1/2}\} - o(n^{-1})$$

for every $\vartheta \in \Theta$, $t>0$, and any sequence (S_n) which is approximately median unbiased of order $o(n^{-1})$. It turns out that maximum likelihood estimates are optimal of order $o(n^{-1})$ if they are improved by a bias correction of magnitude n^{-1} (Pfanzagl, 1975, Theorem 7).

The present paper deals with the role of Bayes procedures in the framework of the asymptotic theory of order $o(n^{-1})$. It is shown that Bayes estimates are optimal of order $o(n^{-1})$ if the weight functions are chosen in such a way that the bias of order $o(n^{-1})$ disappears. In some cases this can be achieved by a particularly simple choice of p (confer 1.b). (In this respect the results differ from the results for error terms of order $o(n^0)$ or $O(n^{-1/2})$ where optimality of Bayes estimates hold independently of the weight functions.)

For testing problems things are more complicated. The main result in this direction states that the equivalence classes of asymptotically optimal tests contain tests based on quantiles of posterior distributions. Such tests turn out to be of a particularly simple form if one has to reject relevant alternatives at least with probability 1/2.

b) The case of a single location parameter

Let $\mathcal{B}$ be the Borel-σ-algebra of $\mathbb{R}$. Assume that $P|\mathcal{B}$ is a probability measure with continuous and positive Lebesgue density h. Let $P_\vartheta|\mathcal{B}$ be the probability measure with Lebesgue density $h(.-\vartheta)$, $\vartheta\in\mathbb{R}$. For $B\in\mathcal{B}$ define

$$R_{n,\underline{\omega}}(B) = \frac{\int_B \prod_{i=1}^n h(\omega_i-\vartheta)\,d\vartheta}{\int_{\mathbb{R}} \prod_{i=1}^n h(\omega_i-\vartheta)\,d\vartheta}, \quad \underline{\omega}\in\mathbb{R}^n, \quad n\in\mathbb{N}.$$

Since h is continuous and positive there exists a unique solution $\mu_n^{(\alpha)}(\underline{\omega})$ of the equation

$$R_{n,\underline{\omega}}(\{\sigma\leq\mu_n^{(\alpha)}(\underline{\omega})\}) = \alpha$$

for every $\alpha\in(0,1)$ and every $\underline{\omega}\in\mathbb{R}^n$, $n\in\mathbb{N}$. It is easy to see that $\underline{\omega}\mapsto\mu_n^{(\alpha)}(\underline{\omega})$ is $\mathcal{B}^n$-measurable. As is well known (Welch & Peers, 1963, p.325) the functions $\mu_n^{(\alpha)}$ satisfy

$$P_\vartheta^n\{\mu_n^{(\alpha)}\geq\vartheta\} = \alpha, \vartheta\in\mathbb{R}.$$

In particular, $\mu_n^{(1/2)}$ is a median unbiased estimator of ϑ. For arbitrary $\alpha\in(0,1)$, $\mu_n^{(\alpha)}$ may be used for testing $H=\{\tau\leq\vartheta\}$ against $K=\{\tau>\vartheta\}$ at level α.

For finite sample size $n\in\mathbb{N}$ the functions $\mu_n^{(\alpha)}$ have some properties which are desirable from the decision theoretic point of view. $\mu_n^{(\alpha)}$ is the optimum translation equivariant estimator of ϑ with respect to the loss function

$$L(\vartheta,d) = \begin{cases} (1-\alpha)(d-\vartheta) & \text{if } d \geq \vartheta \\ \alpha(\vartheta-d) & \text{if } d \leq \vartheta . \end{cases}$$

From the theorem of Hunt and Stein it follows that $\mu_n^{(\alpha)}$ is a minimax estimator of ϑ. Fox and Rubin proved that it is even an admissible estimator of ϑ. (These facts can be found in Zacks,1971,Theorems7.4.2,7.5.1,8.4.3.)

The main result of the present paper implies for the case of a single location parameter that $(\mu_n^{(1/2)})$ is asymptotically optimal of order $o(n^{-1})$ (in fact, this is first proved in Strasser, 1975). This implies - as far as median unbiased estimation is concerned - that the asymptotic theory of order $o(n^{-1})$ does not distinguish the maximum likelihood method. On the contrary, for reasons of finite sample properties the maximum likelihood method should be replaced by Bayesian methods.

The same is true for testing $H=\{\tau\leq\vartheta\}$ against $K=\{\tau>\vartheta\}$ at level α as long as the error bound is of order $o(n^{-1/2})$. In this case the sequence $(\mu_n^{(\alpha)})$ is asymptotically optimal, uniformly for alternatives $\vartheta+sn^{-1/2}$ with $|s|\leq \log n$.

If the error bound $o(n^{-1/2})$ is replaced by $o(n^{-1})$ then in general asymptotically optimal tests do not exist (Pfanzagl, 1974 b). Combining the tests based on $(\mu_n^{(\alpha)})$ with a test against moderate and large deviations (confer Pfanzagl, 1974 b, Lemmas 4 and 5) it is possible to compute the deficiency of $(\mu_n^{(\alpha)})$ for rejecting relevant alternatives with power $\beta\in(0,1)$. It turns out that $(\mu_n^{(\alpha)})$ has deficiency zero for $\beta=1/2$.

In Pfanzagl, 1974 b, Theorem 4, it is shown that the tests based on maximum likelihood estimates have deficiency zero for $\beta=1-\alpha$. These tests are constructed by adjusting maximum likelihood estimates in such a way that they define level-α-tests (of order $o(n^{-1})$). Essentially the same can be done with any quantile $\mu_n^{(\gamma)}$ of posterior distributions. One obtains deficiency zero for $\beta=1-\alpha$ iff $\gamma=1/2$. Thus, medians of posterior distributions provide the same tests as maximum likelihood estimates.

2. Notations

Let $(\Omega,\mathcal{A})$ be a measurable space and $P_\vartheta|\mathcal{A}$, $\vartheta\in\Theta$, a family of p-measures. $\Theta\subseteq\mathbb{R}$ is assumed to be an open interval. $\mathcal{B}^1$ denotes the Borel-σ-algebra of $\mathbb{R}^1$ and $\mathcal{B}_\Theta$ denotes the Borel-σ-algebra of Θ. The n-fold product of $(\Omega,\mathcal{A})$ is denoted by $(\Omega^n,\mathcal{A}^n)$ and a single element of Ω^n by $\underline{\omega} = (\omega_1,\dots,\omega_n)$.

Given a dominating measure $\mu|\mathcal{A}$, let $h_\vartheta=\frac{dP_\vartheta}{d\mu}$ and $\ell(\cdot,\vartheta)=\log h_\vartheta$. Let further

$$\ell_j(\omega,\vartheta) = \frac{d^j}{d\vartheta^j}\,\ell(\omega,\vartheta),\quad j\in\mathbb{N},$$

and

$$x_j^{(n)}(\underline{\omega},\vartheta) = n^{-1/2}\sum_{i=1}^{n}\left(\ell_j(\omega_i,\vartheta) - P_\vartheta(\ell_j(\cdot,\vartheta))\right).$$

If $s\in\mathbb{R}$, let

$$y_j^{(n)}(\underline{\omega},\vartheta,s) = n^{-1/2}\sum_{i=1}^{n}\left(\ell_j(\omega_i,\vartheta) - P_{\vartheta+sn^{-1/2}}(\ell_j(\cdot,\vartheta))\right).$$

Denote

$$L_{ijkm}(\vartheta) = P_\vartheta(\ell_1(\cdot,\vartheta)^i\,\ell_2(\cdot,\vartheta)^j\,\ell_3(\cdot,\vartheta)^k\,\ell_4(\cdot,\vartheta)^m).$$

Instead of L_{ijo} or L_{ioo} we shall write L_{ij} or L_i, respectively. For notational convenience we introduce

$$D = \frac{1}{4}\,L_2^{-3}(L_2(L_{o2}-L_2^2)-L_{11}^2)$$

and

$$V_1 = (L_{11}, L_3)^T$$

$$V_2 = (L_{101}, L_2^2, L_{02}, L_4, L_{21})^T$$

$$V_3 = (L_{11}^2 L_2^{-1}, L_{11} L_3 L_2^{-1}, L_3^2 L_2^{-1})^T.$$

We shall use $N_\alpha = \Phi^{-1}(\alpha)$, $\alpha \in (0,1)$, where

$$\varphi(t) = (2\pi)^{-1/2} \exp(-\tfrac{1}{2}t^2), \quad t \in \mathbb{R},$$

$$\Phi(t) = \int_{-\infty}^{t} \varphi(s)\, ds .$$

If G is a distribution function on $\mathbb{R}$ then G(B), $B \in \mathcal{B}^1$, denotes the value of the Borel measure of B defined by G.

3. Regularity conditions

Let $\lambda | \mathcal{B}_\Theta$ be an absolutely continuous Borel measure with positive and continuous Lebesgue density p. Denote $\Lambda = \log p$.

(j) For every $\delta > 0$ and every compact $K \subseteq \Theta$

$$\inf_{\vartheta \in K} \lambda \{\sigma \in \Theta : |\sigma - \vartheta| < \delta\} > 0 .$$

(jj) Λ is twice differentiable and the second derivative satisfies for every compact $K \subseteq \Theta$

$$|\Lambda''(\sigma) - \Lambda''(\tau)| \leq c_K |\sigma - \tau| \quad \text{if } |\sigma - \tau| < e_K, \ \sigma, \tau \in K.$$

(jjj) For every compact $K \subseteq \Theta$

$$\sup_{\vartheta \in K} \int_\Theta \exp(P_\vartheta(\ell_\sigma))\lambda(d\sigma) < \infty.$$

Obviously, (jj) implies (j). Condition (jjj) is needed to obtain consistency of posterior distributions when λ is not necessarily a finite measure. In Strasser, 1975, examples are given where (jjj) holds for $p = L_2^{1/2}$.

The following regularity conditions deal with log-likelihood functions.

(i) $\vartheta \mapsto P_\vartheta$ is continuous with respect to the supremum metric.

(ii) For every $\omega\in\Omega$, $\vartheta\mapsto \ell(\omega,\vartheta)$ is continuous on $\bar{\Theta}$.

(iii) For every $\vartheta\in\Theta$ there exists a neighbourhood W_ϑ of ϑ such that

$$\sup_{\tau\in W_\vartheta} P_\tau(\sup_{\sigma\in W_\vartheta} |\ell_\sigma|^3)<\infty.$$

(iv) For every $\vartheta\in\bar{\Theta}$ there exists a neighbourhood U_ϑ of ϑ such that for every neighbourhood U of ϑ, $U\subseteq U_\vartheta$, and every compact $K\subseteq\Theta$

$$\sup_{\tau\in K} P_\tau(|\sup_{\sigma\in U} \ell_\sigma|^3)<\infty.$$

(v) $\vartheta\mapsto\ell(\omega,\vartheta)$ is four times differentiable on Θ for every $\omega\in\Omega$ and $L_1(\vartheta)=0$.

(vi) For every $\vartheta\in\Theta$ there exists a neighbourhood U_ϑ of ϑ such that

$$\inf_{\tau\in U_\vartheta} L_2(\tau)>0 .$$

(vii) For every $\vartheta\in\Theta$ there exists a neighbourhood U_ϑ of ϑ such that

$$\sup_{\tau,\delta\in U_\vartheta} P_\tau(|\ell_i(\delta)|^{k_i})<\infty,\quad 1\le i\le 4 ,$$

where $k_1=5$, $k_2=5$, $k_3=3$, $k_4=3$.

(viii) For every $\vartheta\in\Theta$ there exists a neighbourhood U_ϑ of ϑ and a function $m(.,\vartheta)$ such that

(a) $|\ell_4(\omega,\tau)-\ell_4(\omega,\delta)|\le|\tau-\delta|\, m(\omega,\vartheta),\ \omega\in\Omega,\ (\tau,\delta)\in\Theta^2.$

(b) $\sup_{\tau\in U_\vartheta} P_\tau(|m(.,\vartheta)|^3)<\infty.$

(ix) For every $\vartheta\in\Theta$ there exists a neighbourhood U_ϑ and a function $k(.,\vartheta)$ such that

(a) $\left|\frac{h(\omega,\delta)}{h(\omega,\tau)}-1\right| = |\delta-\tau|k(\omega,\vartheta),\ \omega\in\Omega,\ (\delta,\tau)\in\Theta^2.$

(b) $\sup_{\tau\in U_\vartheta} P_\tau(|k(.,\tau)|^3)<\infty.$

Let $\tau\in\Theta$, $\vartheta\in\Theta$. Let $Q_{\tau,\vartheta}|\mathcal{B}^3$ be the p-measure which is induced by P_τ and

$$(\ell_1(\vartheta)-P_\tau(\ell_1(\vartheta)),\ \ell_2(\vartheta)-P_\tau(\ell_2(\vartheta)),\ \ell_3(\vartheta)-P_\tau(\ell_3(\vartheta))).$$

Let $Q^n_{\tau,\vartheta}$ be the n-fold product measure of $Q_{\tau,\vartheta}$ and let $Q^{(n)}_{\tau,\vartheta}$ be the distribution of the mapping $\mathbb{R}^{3n} \rightarrow \mathbb{R}^3$ defined by

$$[(\xi_{1i},\xi_{2i},\xi_{3i})]_{1\le i\le n} \mapsto n^{-1/2}\sum_{i=1}^{n}(\xi_{1i},\xi_{2i},\xi_{3i})$$

under $Q^n_{\tau,\vartheta}$.

Lemma 3 in Pfanzagl, 1974 a, gives an Edgeworth expansion of order $o(n^{-1})$ for the distribution of certain polynomials over $\mathbb{R}^3$ under $Q^{(n)}_{\tau,\vartheta}$. This result uses additional conditions.

(x) For every $\vartheta \in \Theta$ there exists a neighbourhood U_ϑ such that

$$\overline{\lim_{\|u\|\to\infty}} \sup_{\tau,\delta\in U_\vartheta} \left| \int \exp\left[i\sum_{j=1}^{3} u_j\xi_j\right] Q_{\tau,\delta}(d\underline{\xi})\right| < 1.$$

(xi) The covariance matrix of $Q_{\vartheta,\vartheta}$ is positive definite for every $\vartheta \in \Theta$.

(xii) For every $\vartheta \in \Theta$ there exists a neighbourhood U_ϑ such that

$$\lim_{a\to\infty} \sup_{\tau,\delta\in U_\vartheta} \int \|\underline{\xi}\|^4 \, 1_{\{\|\underline{\xi}\|>a\}} \, Q_{\tau,\delta}(d\underline{\xi}) = 0.$$

Discussions of conditions (x)-(xii) can be found in Pfanzagl,1974 a,b.

A loss function is a bounded, lower semicontinuous function $W:\mathbb{R}\to\mathbb{R}^+$ such that $W(t)=W(-t)$, $t\in\mathbb{R}$, $W(0)=0$, and $W(t_1)\le W(t_2)$ if $0\le t_1\le t_2$. Consider the following conditions:

Type (A): There exists $c>0$ such that

$$W(t) = \begin{cases} 0 & \text{if } -c\le t\le c \\ 1 & \text{if } t<-c \text{ or } t>c. \end{cases}$$

Type (B): W is continuous and satisfies

(i) $W(t)=0$ iff $t=0$,

(ii) W is twice differentiable on $\mathbb{R}$,

(iii) W'' satisfies a Lipschitz condition on $\mathbb{R}$ and vanishes at infinity.

Moreover

$$V_\vartheta(\sigma) = (2\pi)^{-1/2} \int W(\sigma - \tau L_2(\vartheta)^{-1/2}) \exp(-\frac{\tau^2}{2}) d\tau, \; \sigma \in \mathbb{R}$$

satisfies

(iv) V_ϑ attains its infimum exactly at zero,

(v) $V''_\vartheta(0) > 0$ for every $\vartheta \in \Theta$.

4. The basic expansion

Definition 1. Let $\lambda | \mathcal{B}_\Theta$ be a Borel measure. If for certain $n \in \mathbb{N}, \underline{\omega} \in \Omega^n$, the set function

$$R_{n,\underline{\omega}}(B) = \frac{\int_B \exp(\sum_{i=1}^n \ell(\omega_i, \sigma)) \lambda(d\sigma)}{\int_{\mathbb{R}} \exp(\sum_{i=1}^n \ell(\omega_i, \sigma)) \lambda(d\sigma)}, \; B \in \mathcal{B}_\Theta,$$

is well defined, then it is called posterior distribution for the sample $\underline{\omega}$, the sample size n and the prior distribution λ.

Assume that conditions (i)-(iv), (j) and (jjj) are satisfied. Then Theorem 1 in Strasser, 1975, implies that $R_{n,\underline{\omega}}$ is well defined on sets $M_n(\vartheta) \in \mathcal{A}^n$ satisfying

$$\sup_{\vartheta \in K} P^n_\vartheta(M_n(\vartheta)') = o(n^{-1})$$

for every compact $K \subseteq \Theta$.

If $B \in \mathcal{B}$ define

$$F^\vartheta_{n,\underline{\omega}}(B) = R_{n,\underline{\omega}}\left\{(nL_2(\vartheta))^{1/2}(\sigma - \vartheta) - x_1^{(n)}(\underline{\omega}, \vartheta) L_2^{-1/2}(\vartheta) \in B\right\}.$$

The following theorem gives an eypansion of $F^\vartheta_{n,\underline{\omega}}$ in terms of

$$G^\vartheta_{n,\underline{\omega}}(B) = \frac{\int_B P_n(\underline{\omega}, \vartheta)(s)\, \Phi(ds)}{\int_{\mathbb{R}} P_n(\underline{\omega}, \vartheta)(s)\, \Phi(ds)}, \; B \in \mathcal{B},$$

where $P_n(\underline{\omega},\vartheta)$ are polynomials in $s\in\mathbb{R}$:

$$P_n(\underline{\omega},\vartheta)(s) = 1 + n^{-1/2}\sum_{i=1}^{3}\gamma_{1i}s^i + n^{-1}\sum_{j=1}^{6}\gamma_{2j}s^j.$$

The coefficients γ_{ij} depend on $x_k=x_k^{(n)}(\underline{\omega},\vartheta)$, $1\leq k\leq 3$, and on $L_{ijkm}=L_{ijkm}(\vartheta)$. We give these coefficients in a coded form in order to avoid complicated expressions.

$$\gamma_{11} = a_{11} + b_{11}$$
$$\gamma_{12} = a_{12}$$
$$\gamma_{13} = a_{13}$$
$$\gamma_{21} = a_{21} + b_{21}$$
$$\gamma_{22} = a_{22} + b_{22} + \tfrac{1}{2}a_{11}^2 + a_{11}b_{11} + \tfrac{1}{2}b_{11}^2$$
$$\gamma_{23} = a_{23} + a_{11}a_{12} + a_{12}b_{11}$$
$$\gamma_{24} = a_{24} + \tfrac{1}{2}a_{12}^2 + a_{11}a_{13} + a_{13}b_{11}$$
$$\gamma_{25} = a_{12}a_{13}$$
$$\gamma_{26} = \tfrac{1}{2}a_{13}^2$$

$$a_{11} = x_1x_2L_2^{-3/2} + \tfrac{1}{2}x_1^2L_{oo1}L_2^{-5/2}$$
$$a_{12} = \tfrac{1}{2}x_2L_2^{-1} + \tfrac{1}{2}x_1L_{oo1}L_2^{-2}$$
$$a_{13} = \tfrac{1}{6}L_{oo1}L_2^{-3/2}$$
$$a_{21} = \tfrac{1}{2}x_1^2x_3L_2^{-5/2} + \tfrac{1}{6}x_1^3L_{ooo1}L_2^{-7/2}$$
$$a_{22} = \tfrac{1}{2}x_1x_3L_2^{-2} + \tfrac{1}{4}x_1^2L_{ooo1}L_2^{-3}$$
$$a_{23} = \tfrac{1}{6}x_3L_2^{-3/2} + \tfrac{1}{6}x_1L_{ooo1}L_2^{-5/2}$$
$$a_{24} = \tfrac{1}{24}L_{ooo1}L_2^{-2}$$
$$b_{11} = \Lambda' L_2^{-1/2}$$
$$b_{21} = x_1\Lambda'' L_2^{-3/2}$$
$$b_{22} = \tfrac{1}{2}\Lambda'' L_2^{-1}$$

<u>Theorem 1.</u> Assume that conditions (i)-(ix), (jj) and (jjj) are satisfied. Let $K\subseteq\Theta$ be compact. Then

$$P_\tau^n\{\|F_{n,\underline{\omega}}^\vartheta - G_{n,\underline{\omega}}^\vartheta\| \geq c(\log n)^a n^{-3/2}\} = o(n^{-1})$$

uniformly for $\vartheta \in K$, $|\tau - \vartheta| \leq (\log n)n^{-1/2}$ (where $a < \infty$, $c < \infty$ are to be chosen appropriately).

Proof. Strasser, 1975, Theorem 2. □

Remark 1. The proof of Theorem 1 shows that

$$P_\tau^n\{\sup_{|s| \leq \log n} |F_{n,\underline{\omega}}^{\vartheta\prime}(s) - G_{n,\underline{\omega}}^{\vartheta\prime}(s)| \geq c(\log n)^a n^{-3/2}\} = o(n^{-1})$$

uniformly for $\vartheta \in K$, $|\tau - \vartheta| \leq (\log n)n^{-1/2}$.

5. Quantiles of posterior distributions

Definition 2. Let $\xi \in (0,1)$. A sequence of $\mathcal{A}^n$-measurable functions $\mu_n : \Omega^n \to \Theta$ is called a sequence of (approximate) ξ-quantiles of the posterior distributions $R_{n,\underline{\omega}}$ if for every compact $K \subseteq \Theta$

$$\sup_{\vartheta \in K} P_\vartheta^n\{|R_{n,\underline{\omega}}(\sigma \leq \mu_n) - \xi| \geq c_K(\log n)^a n^{-3/2}\} = o(n^{-1}).$$

Theorem 2 gives an expansion of ξ-quantiles for arbitrary regular weight functions p. The expansion is given as a polynomial in $x_1^{(n)}$, $x_2^{(n)}$, $x_3^{(n)}$. It is convenient to have a name for expansions of such a type.

Definition 3. Let $(T_{n,\vartheta})$ be a sequence of $\mathcal{A}^n$-measurable functions. A sequence $(\tilde{C}_{n,\vartheta})$ of polynomials in $x_1^{(n)}$, $x_2^{(n)}$, $x_3^{(n)}$ of the form

$$\begin{aligned}\tilde{C}_{n,} &= \tilde{a}_o + n^{-1/2}\tilde{a}_1 + n^{-1}\tilde{a}_2 \\ &+ x_1 \\ &+ n^{-1/2}\Big(\sum_i \tilde{b}_i x_i + \sum_{j,k} \tilde{b}_{jk} x_j x_k\Big) \\ &+ n^{-1}\Big(\sum_i \tilde{c}_i x_i + \sum_{j,k} \tilde{c}_{jk} x_j x_k + \sum_{p,q,r} \tilde{c}_{pqr} x_p x_q x_r\Big)\end{aligned}$$

is called x-polynomial of $(T_{n,\vartheta})$ if for every compact $K \subseteq \Theta$

$$P_\tau^n\{|T_{n,\vartheta} - \tilde{C}_{n,\vartheta}| \geq c(\log n)^a n^{-3/2}\} = o(n^{-1})$$

uniformly for $\vartheta \in K$, $|\tau - \vartheta| \leq (\log n) n^{-1/2}$.

Theorem 2. Assume that conditions (i)-(ix), (jj) and (jjj) are satisfied. Let $z \in \mathbb{R}$ and let (μ_n) be a sequence of approximate $\Phi(z)$-quantiles of the posterior distributions. Then the x-polynomial of

$$n^{1/2} L_2(\vartheta)(\mu_n - \vartheta)$$

is given by

$$\tilde{a}_o = zL_2^{1/2}$$

$$\tilde{a}_1 = \Lambda' + \frac{1}{3}L_{oo1}L_2^{-1} + \frac{1}{6}z^2 L_{oo1}L_2^{-1}$$

$$\tilde{a}_2 = z(\frac{1}{2}\Lambda'' L_2^{-1/2} + \frac{1}{8}L_{ooo1}L_2^{-3/2} + \frac{1}{2}\Lambda' L_{oo1}L_2^{-3/2} + \frac{19}{72}L_{oo1}^2 L_2^{-5/2})$$

$$+ z^3(\frac{1}{24}L_{ooo1}L_2^{-3/2} + \frac{5}{72}L_{oo1}^2 L_2^{-5/2})$$

$$\tilde{b}_1 = \frac{1}{2}zL_{oo1}L_2^{-3/2}$$

$$\tilde{b}_2 = \frac{1}{2}zL_2^{-1/2}$$

$$\tilde{b}_{11} = \frac{1}{2}L_{oo1}L_2^{-2}$$

$$\tilde{b}_{12} = L_2^{-1}$$

$$\tilde{c}_1 = \Lambda'' L_2^{-1} + \frac{1}{3}L_{ooo1}L_2^{-2} + \Lambda' L_{oo1}L_2^{-2} + \frac{2}{3}L_{oo1}^2 L_2^{-3}$$

$$+ z^2(\frac{1}{6}L_{ooo1}L_2^{-2} + \frac{1}{3}L_{oo1}^2 L_2^{-3})$$

$$\tilde{c}_2 = \Lambda' L_2^{-1} + \frac{2}{3}L_{oo1}L_2^{-2} + \frac{1}{3}z^2 L_{oo1}L_2^{-2}$$

$$\tilde{c}_3 = \frac{1}{3}L_2^{-1} + \frac{1}{6}z^2 L_2^{-1}$$

$$\tilde{c}_{11} = z(\frac{1}{4}L_{ooo1}L_2^{-5/2} + \frac{3}{8}L_{oo1}^2 L_2^{-7/2} + \frac{1}{4}L_{oo1}^2 L_2^{-7/2})$$

$$\tilde{c}_{12} = z(\frac{3}{4}L_{oo1}L_2^{-5/2} + \frac{1}{2}L_{oo1}L_2^{-5/2})$$

$$\tilde{c}_{13} = \frac{1}{2}zL_2^{-3/2}$$

$$\tilde{c}_{22} = \frac{3}{8}zL_2^{-3/2}$$

$$\tilde{c}_{111} = \tfrac{1}{6}L_{0001}L_2^{-3} + \tfrac{1}{2}L_{001}^2L_2^{-4}$$

$$\tilde{c}_{112} = \tfrac{3}{2}L_{001}L_2^{-3}$$

$$\tilde{c}_{113} = \tfrac{1}{2}L_2^{-2}$$

$$\tilde{c}_{122} = L_2^{-2}$$

Proof. From Theorem 1 we obtain that for $\xi=\Phi(z)$

$$P_\tau^n\{|G_{n,\underline{\omega}}((nL_2)^{1/2}(\mu_n-\vartheta)-x_1L_2^{-1/2})-\xi|\geq c(\log n)^a n^{-3/2}\}$$
$$= o(n^{-1})$$

uniformly for $\vartheta\in K$, $|\tau-\vartheta|\leq(\log n)n^{-1/2}$. Standard expansion techniques show that

$$\hat{\mu}_n = \vartheta + (nL_2)^{-1/2}\Big[N_\xi + x_1L_2^{-1/2} + \frac{\xi - G_{n,\underline{\omega}}(N_\xi)}{G'_{n,\underline{\omega}}(N_\xi)} + \frac{N_\xi}{2}\Big(\frac{\xi - G_{n,\underline{\omega}}(N_\xi)}{G'_{n,\underline{\omega}}(N_\xi)}\Big)^2\Big]$$

satisfy

$$P_\tau^n\{n^{1/2}|\hat{\mu}_n - \mu_n|\geq c(\log n)^a n^{-3/2}\} = o(n^{-1})$$

uniformly for $\vartheta\in K$, $|\tau-\vartheta|\leq(\log n)n^{-1/2}$. (The proof of this fact follows the pattern of Lemma 4 in Strasser, 1975.) Now the assertion is proved by elementary computations. □

It follows from Theorem 2 how any $\Phi(z)$-quantile, $z\in\mathbb{R}$, can be used for testing the hypothesis $\{\tau<\vartheta\}$ against the alternative $\{\tau>\vartheta\}$. The critical regions

$$\{\mu_n \geq \vartheta + (nL_2(\vartheta))^{-1/2}(z - N_\alpha)\}$$

define an asymptotically most powerful sequence of approximate level-α-tests of order $O(n^{-1/2})$. Indeed, we have

$$P^n_{\vartheta+sn^{-1/2}}\{(nL_2(\vartheta))^{1/2}(\mu_n-\vartheta)\geq z-N_\alpha\}$$

$$= P^n_{\vartheta+sn^{-1/2}}\{(L_2(\vartheta)^{-1/2}x_1^{(n)}(.,\vartheta)\geq -N_\alpha-sL_2(\vartheta)^{1/2}\}$$

$$= \Phi(N_\alpha + sL_2(\vartheta)^{1/2}) + o(n^{-1/2})$$

uniformly for $\vartheta\in K$, $|s|\leq \log n$. In order to compare the relative merits of $\Phi(z)$-quantiles for different $z\in\mathbb{R}$ we have to compute the power functions of level-α-tests of order $o(n^{-1})$ based on $\Phi(z)$-quantiles.

Lemma 1. Assume that conditions (i)-(xii), (jj) and (jjj) are satisfied. Let $z\in\mathbb{R}$ and let (μ_n) be a sequence of approximate $\Phi(z)$-quantiles of the posterior distributions. Then the critical regions

$$C_n^\vartheta(\alpha,z) = \{\mu_n\geq\vartheta + (nL_2(\vartheta))^{-1/2}r_n(\vartheta,z,\alpha)\}$$

satisfy

$$P^n_\vartheta(C_n^\vartheta(\alpha,z)) = \alpha + o(n^{-1})$$

uniformly on every compact $K\subseteq\Theta$ iff

$$r_n(\vartheta,z,\alpha) = z - N_\alpha - n^{-1/2}L_2^{-3/2}Q_1^*(\vartheta,z,\alpha)$$
$$- n^{-1}L_2^{-2}Q_2^*(\vartheta,z,\alpha)$$

where

$$Q_1^*(\vartheta,z,\alpha) = (1,z^2,N_\alpha z,N_\alpha^2)\begin{pmatrix}1 & \frac{1}{2} & -\Lambda' \\ \frac{1}{2} & \frac{1}{6} & 0 \\ -1 & -\frac{1}{2} & 0 \\ \frac{1}{2} & \frac{1}{3} & 0\end{pmatrix}\begin{pmatrix}V_1\\ L_2\end{pmatrix}$$

and

$$Q_2^*(\vartheta, z, \alpha) = (z, z^3, N_\alpha, N_\alpha z^2, N_\alpha^2 z, N_\alpha^3).$$

$$\begin{pmatrix} \frac{1}{2} & \frac{1}{8} & \frac{1}{2} & \frac{1}{8} & 1 & -\frac{5}{2} & -2 & -\frac{25}{72} & \frac{3}{2}\Lambda' & \frac{1}{2}\Lambda' & -\frac{1}{2}\Lambda'' \\ \frac{1}{6} & 0 & \frac{1}{8} & \frac{1}{24} & \frac{1}{4} & -\frac{5}{8} & -\frac{5}{12} & -\frac{5}{72} & 0 & 0 & 0 \\ -1 & -\frac{1}{8} & -1 & -\frac{11}{24} & -\frac{5}{2} & 4 & 4 & \frac{70}{72} & -2\Lambda' & -\Lambda' & \Lambda'' \\ -\frac{1}{2} & -\frac{1}{8} & -\frac{3}{8} & -\frac{1}{6} & -1 & \frac{15}{8} & \frac{5}{3} & \frac{1}{3} & 0 & 0 & 0 \\ \frac{1}{2} & \frac{1}{4} & \frac{1}{4} & \frac{1}{4} & \frac{5}{4} & -\frac{7}{4} & -\frac{25}{12} & -\frac{13}{24} & 0 & 0 & 0 \\ -\frac{1}{6} & -\frac{1}{8} & 0 & -\frac{1}{8} & -\frac{1}{2} & \frac{1}{2} & \frac{10}{12} & \frac{5}{18} & 0 & 0 & 0 \end{pmatrix} \begin{pmatrix} V_2 \\ V_3 \\ V_1 \\ L_2 \end{pmatrix}$$

<u>Proof.</u> Lemma 3 in Pfanzagl, 1974 a, implies

$$P_\vartheta^n\{(nL_2)^{1/2}(\mu_n - \vartheta) \geq u + (z - N_\alpha)\} =$$

$$= \Phi(N_\alpha - u + n^{-1/2}L_2^{-1/2}\tilde{a}_1 + n^{-1}L_2^{-1}\tilde{a}_2) -$$

$$- n^{-1/2}L_2^{-3/2}\varphi(N_\alpha - u + n^{-1/2}L_2^{-1/2}\tilde{a}_1) .$$

$$\cdot [r_{10} - r_{11}L_2^{1/2}(N_\alpha - u + n^{-1/2}L_2^{-1/2}\tilde{a}_1) +$$

$$+ r_{12}L_2(N_\alpha - u + n^{-1/2}L_2^{-1/2}\tilde{a}_1)^2] -$$

$$- n^{-1}L_2^{-5/2}\varphi(N_\alpha - u) .$$

$$\cdot [r_{20} - r_{21}L_2^{1/2}(N_\alpha - u) + r_{22}L_2(N_\alpha - u)^2 -$$

$$- r_{23}L_2^{3/2}(N_\alpha - u)^3 + r_{24}L_2^2(N_\alpha - u)^4 - r_{25}L_2^{5/2}(N_\alpha - u)^5] ,$$

where

$$\tilde{a}_1 = (L_2^{-1}, z^2L_2^{-1}) \cdot \begin{pmatrix} -1 & -\frac{1}{3} & \Lambda' \\ -\frac{1}{2} & -\frac{1}{6} & 0 \end{pmatrix} \begin{pmatrix} V_1 \\ L_2 \end{pmatrix}$$

$$\tilde{a}_2 = (zL_2^{-3/2}, z^3L_2^{-3/2}) .$$

$$\cdot \begin{pmatrix} -\frac{1}{2} & 0 & -\frac{3}{8} & -\frac{1}{8} & -\frac{3}{4} & \frac{19}{8} & \frac{19}{12} & \frac{19}{72} & -\frac{3}{3}\Lambda' & -\frac{1}{2}\Lambda' & \frac{1}{2}\Lambda'' \\ -\frac{1}{6} & 0 & -\frac{1}{8} & -\frac{1}{24} & -\frac{1}{4} & \frac{5}{8} & \frac{5}{12} & \frac{5}{72} & 0 & 0 & 0 \end{pmatrix} \begin{pmatrix} V_2 \\ V_3 \\ V_1 \\ L_2 \end{pmatrix}$$

$$r_{10} = (0,\tfrac{1}{6})V_1$$

$$r_{11} = zL_2^{-1/2}(1,\tfrac{1}{2})V_1$$

$$r_{12} = L_2^{-1}(\tfrac{1}{2},\tfrac{1}{3})V_1$$

$$r_{20} = zL_2^{1/2}(0,\tfrac{1}{8},\tfrac{1}{8},0,\tfrac{1}{4},-\tfrac{1}{8},-\tfrac{1}{4},0)(V_2^T,V_3^T)^T$$

$$r_{21} = (1,z^2).$$

$$\cdot \begin{pmatrix} 1 & \frac{1}{8} & 1 & \frac{11}{24} & \frac{5}{2} & -4 & -\frac{23}{6} & -\frac{21}{24} & 2\Lambda' & \Lambda' & -\Lambda'' \\ \frac{1}{2} & \frac{1}{8} & \frac{3}{8} & \frac{1}{6} & 1 & -\frac{7}{8} & -\frac{2}{3} & -\frac{1}{12} & 0 & 0 & 0 \end{pmatrix} \begin{pmatrix} V_2 \\ V_3 \\ V_1 \\ L_2 \end{pmatrix}$$

$$r_{22} = zL_2^{-1/2}(\tfrac{1}{2},\tfrac{1}{4},\tfrac{1}{4},\tfrac{1}{4},\tfrac{5}{4},-\tfrac{1}{4},-\tfrac{1}{2},-\tfrac{1}{8})(V_2^T,V_3^T)^T$$

$$r_{23} = (L_2^{-1},z^2L_2^{-1}).$$

$$\cdot \begin{pmatrix} \frac{1}{6} & \frac{1}{8} & 0 & \frac{1}{8} & \frac{1}{2} & 0 & -\frac{1}{4} & -\frac{1}{9} \\ 0 & 0 & 0 & 0 & 0 & -\frac{1}{2} & -\frac{1}{2} & -\frac{1}{8} \end{pmatrix} \begin{pmatrix} V_2 \\ V_3 \end{pmatrix}$$

$$r_{24} = zL_2^{-3/2}(-\tfrac{1}{2},-\tfrac{7}{12},-\tfrac{1}{6})V_3$$

$$r_{25} = L_2^{-2}(-\tfrac{1}{8},-\tfrac{1}{6},-\tfrac{1}{18})V_3 \quad .$$

Now an application of Lemma 7 in Pfanzagl, 1973, proves the assertion. ▯

It can easily be seen that by an appropriate choice of sequences $\Lambda_n' = (\Lambda_n')_o + n^{-1/2}(\Lambda_n')_1$ it can be achieved that $r_n(\vartheta,z,\alpha) = 0$. It has to be checked, however, in every particular case whether the weight functions $p_n = \exp(\Lambda_n)$ satisfy conditions (j)-(jjj) uniformly for $n \in \mathbb{N}$.

<u>Example 1.</u> Consicer the case $z = N_\alpha$. In this case we have

$$Q_1^*(\vartheta,z,\alpha) = (1,\tfrac{1}{2},-\Lambda')(V_1^T,L_2)^T$$

$$Q_2^*(\vartheta,z,\alpha) =$$

$$= N_\alpha(-\tfrac{1}{2},0,-\tfrac{1}{2},-\tfrac{1}{3},-\tfrac{3}{2},\tfrac{3}{2},2,\tfrac{5}{8},-\tfrac{1}{2}\Lambda',-\tfrac{1}{2}\Lambda',\tfrac{1}{2}\Lambda'')(V_2^T,V_3^T,V_1^T,L_2)^T.$$

The choice of

$$p_n(\vartheta)=\exp(\Lambda_n(\vartheta))= L_2(\vartheta)^{1/2}\exp[-\tfrac{1}{12}n^{-1/2}N_\alpha L_3(\vartheta)L_2(\vartheta)^{-3/2}]$$

implies $r_n(\vartheta,z,\alpha) = 0$. (In a location parameter model this is the constant weight function.) This result has been obtained by Welch and Peers, 1963. If medians of posterior distributions are considered then $N_\alpha = 0$ and $p(\vartheta) = L_2(\vartheta)^{1/2}$ is the desired weight function.

<u>Theorem 3.</u> Assume that conditions (i)-(xii), (jj) and (jjj) are satisfied. Let $z\in\mathbb{R}$ and let (μ_n) be a sequence of $\Phi(z)$-quantiles of the posterior distributions for $p = L_2^{1/2}$. Then the power function of the critical regions

$$C_n(\alpha,z) = \{(nL_2(\vartheta))^{1/2}(\mu_n-\vartheta) \geq r_n(\vartheta,z,\alpha)\}$$

differs at $\tau=\vartheta+sn^{-1/2}$ from the envelope power function by

$$-n^{-1}\,\tfrac{s}{2}DL_2^{3/2}\,(s - \frac{z-2N_\alpha}{L_2^{1/2}})^2 + o(n^{-1})$$

uniformly for $\vartheta\in K$ and $|s| \leq \log n$ for every compact $K\subseteq\Theta$.

<u>Proof.</u> Let $(T_{n,\vartheta})$ be a sequence of $\mathcal{A}^n$-measurable functions. A sequence $(C_{n,\vartheta})$ of polynomials in $y_1^{(n)}$, $y_2^{(n)}$, $y_3^{(n)}$ of the form

$$C_{n,\vartheta} = a_o + n^{-1/2} + n^{-1}a_2$$
$$+ y_1$$
$$+ n^{-1/2}(\sum_i b_i y_i + \sum_{j,k} b_{jk}y_jy_k)$$
$$+ n^{-1}(\sum_i c_i y_i + \sum_{j,k} c_{jk}y_jy_k + \sum_{p,q,r} c_{pqr}y_py_qy_r)$$

is called y-polynomial of $(T_{n,\vartheta})$ if for every compact $K \subseteq \Theta$

$$P_\tau^n\{|T_{n,\vartheta} - C_{n,\vartheta}| \geq c(\log n)^a n^{-3/2}\} = o(n^{-1})$$

uniformly for $\vartheta \in K$, $|\tau - \vartheta| \leq (\log n)n^{-1/2}$.

In the following we put $\tau = \vartheta + sn^{-1/2}$, $|s| \leq \log n$. Then the y-polynomial is obtained from the x-polynomial substituting

$$x_1 = y_1 + sL_2 + n^{-1/2}s^2(\tfrac{1}{2}L_{11} + \tfrac{1}{2}L_3) + $$
$$+ n^{-1}s^3(\tfrac{1}{6}L_{1o1} + \tfrac{1}{2}L_{21} + \tfrac{1}{6}L_4) + O((\log n)^2 n^{-3/2})$$
$$x_2 = y_2 + sL_{11} + n^{-1/2}s^2(\tfrac{1}{2}L_{o2} + \tfrac{1}{2}L_{21}) + O((\log n)^{3/2} n^{-1})$$
$$x_3 = y_3 + sL_{1o1} + O((\log n)n^{-1/2}).$$

We have to compute the power function of the critical regions

$$\{(nL_2)^{1/2}(\mu_n - \vartheta) \geq r_n(\vartheta, z, \alpha)\} \quad .$$

Replacing $(nL_2)^{1/2}(\mu_n - \vartheta)$ by the y-polynomial we obtain an equivalent sequence of critical regions

$$\{y_1 + n^{-1/2}[\ldots] + n^{-1}[\ldots] \geq$$
$$\geq - a_o + L_2^{1/2}(z - N_\alpha)$$
$$+ n^{-1/2}(-a_1 - L_2^{-1}Q_1^*)$$
$$+ n^{-1}(-a_2 - L_2^{-3/2}Q_2^*)\} \quad .$$

The expansion of the power function of these critical regions is obtained from Lemma 3 in Pfanzagl, 1974 a. We give a brief outline of the relevant computations.

Define

$$u = -a_o + (z - N_\alpha)L_2^{1/2}$$
$$+ n^{-1/2}(-a_1 - L_2^{-1}Q_1^*)$$
$$+ n^{-1}(-a_2 - L_2^{-3/2}Q_2^*).$$

Then Lemma 3 in Pfanzagl, 1974 a, yields

$$P^n_{\vartheta+sn^{1/2}}\{y_1 + n^{-1/2}[\ldots] + n^{-1}[\ldots] \geq u\} =$$
$$= 1 - \Phi(\sigma_{11}^{-1/2}u)$$
$$- \varphi(\sigma_{11}^{-1/2}u)\,[n^{-1/2}\sigma_{11}^{-3/2}R_1(u,\vartheta) + n^{-1}\sigma_{11}^{-5/2}R_2(u,\vartheta)]$$
$$+ o(n^{-1}),$$

where

$$\sigma_{11} = L_2 + n^{-1/2}sL_3 + \tfrac{1}{2}n^{-1}s^2(L_{21} + L_4 - 2L_2^2) + o(n^{-1})$$
$$R_1(u,\vartheta) = \sum_{j=0}^{2} r_{1j}(\vartheta)u^j$$
$$R_2(u,\vartheta) = \sum_{j=0}^{5} r_{2j}(\vartheta)u^j \ .$$

Tedious, but elementary computations show that

$$r_{1o} = r_{1o}^o + n^{-1/2}r_{1o}^1 + o(n^{-1/2})$$
$$= (0,\tfrac{1}{6})V_1 + n^{-1/2}s(0,-\tfrac{1}{2},0,\tfrac{1}{6},0)V_2 + o(n^{-1/2})$$
$$r_{11} = r_{11}^o + n^{-1/2}r_{11}^1 + o(n^{-1/2})$$
$$= (s,zL_2^{-1/2})\begin{pmatrix}1 & 1\\ 1 & \tfrac{1}{2}\end{pmatrix}V_1$$
$$+ n^{-1/2}(szL_2^{-1/2},s^2)\begin{pmatrix}0 & -\tfrac{1}{2} & 0 & 0 & -\tfrac{1}{2} & 0 & \tfrac{3}{2} & \tfrac{1}{2}\\ 0 & -1 & 0 & 0 & -1 & 0 & 2 & 1\end{pmatrix}\begin{pmatrix}V_2\\ V_3\end{pmatrix}$$
$$+ o(n^{-1/2})$$

$$r_{12} = r_{12}^{o} + n^{-1/2} r_{12}^{1} + o(n^{-1/2})$$
$$= L_2^{-1}(\tfrac{1}{2},\tfrac{1}{3})V_1$$
$$+ n^{-1/2} s L_2^{-1}(0,-\tfrac{1}{2},0,-\tfrac{1}{6},-1,0,\tfrac{3}{2},\tfrac{2}{3})(V_2^T,V_3^T)^T$$
$$+ o(n^{-1/2})$$

$$r_{2o} = (zL_2^{1/2}, sL_2).$$

$$\cdot\begin{pmatrix} 0 & \frac{1}{8} & \frac{1}{8} & 0 & \frac{1}{4} & -\frac{1}{8} & -\frac{1}{4} & 0 \\ 0 & \frac{1}{2} & 0 & 0 & \frac{1}{2} & 0 & -\frac{1}{2} & 0 \end{pmatrix}\begin{pmatrix} V_2 \\ V_3 \end{pmatrix} + o(1)$$

$$r_{21} = (1, z^2, s^2L_2, szL_2^{1/2}).$$

$$\cdot\begin{pmatrix} 0 & \frac{1}{8} & 0 & -\frac{1}{24} & 0 & 0 & \frac{1}{6} & \frac{1}{8} \\ \frac{1}{2} & \frac{1}{8} & \frac{3}{8} & \frac{1}{6} & 1 & -\frac{7}{8} & -\frac{2}{3} & -\frac{1}{12} \\ \frac{1}{2} & \frac{1}{2} & \frac{1}{2} & \frac{1}{2} & \frac{5}{2} & -\frac{1}{2} & -1 & 0 \\ 1 & \frac{1}{2} & 1 & \frac{1}{2} & 3 & -2 & -2 & -\frac{1}{4} \end{pmatrix}\begin{pmatrix} V_2 \\ V_3 \end{pmatrix} + o(1)$$

$$r_{22} = (zL_2^{-1/2}, s).$$

$$\cdot\begin{pmatrix} \frac{1}{2} & \frac{1}{4} & \frac{1}{4} & \frac{1}{4} & \frac{5}{4} & -\frac{1}{4} & -\frac{1}{2} & -\frac{1}{8} \\ \frac{1}{2} & \frac{1}{2} & \frac{1}{2} & \frac{1}{2} & \frac{5}{2} & -\frac{1}{2} & -\frac{3}{2} & -\frac{1}{2} \end{pmatrix}\begin{pmatrix} V_2 \\ V_3 \end{pmatrix} + o(1)$$

$$r_{23} = (L_2^{-1}, z^2L_2^{-1}, szL_2^{-1/2}, s^2).$$

$$\cdot\begin{pmatrix} \frac{1}{6} & \frac{1}{8} & 0 & \frac{1}{8} & \frac{1}{2} & 0 & -\frac{1}{4} & -\frac{1}{9} \\ 0 & 0 & 0 & 0 & 0 & -\frac{1}{2} & -\frac{1}{2} & -\frac{1}{8} \\ 0 & 0 & 0 & 0 & 0 & -1 & -\frac{3}{2} & -\frac{1}{2} \\ 0 & 0 & 0 & 0 & 0 & -\frac{1}{2} & -1 & -\frac{1}{2} \end{pmatrix}\begin{pmatrix} V_2 \\ V_3 \end{pmatrix} + o)1)$$

$$r_{24} = (zL_2^{-3/2}, sL_2^{-1}) \begin{pmatrix} -\frac{1}{2} & -\frac{7}{12} & -\frac{1}{6} \\ -\frac{1}{2} & -\frac{5}{6} & -\frac{1}{3} \end{pmatrix} \cdot V_3 + o(1)$$

$$r_{25} = L_2^{-2}(-\tfrac{1}{8}, -\tfrac{1}{6}, -\tfrac{1}{18})V_3 + o)1) .$$

The explicite form of u is given by

$$u = - L_2^{1/2}N_\alpha - sL_2 - n^{-1/2}A_1 - n^{-1}A_2$$

where

$$A_1 = (L_2^{-1}, N_\alpha zL_2^{-1}, N_\alpha^2 L_2^{-1}, szL_2^{-1/2}) \begin{pmatrix} 0 & \frac{1}{6} \\ -1 & -\frac{1}{2} \\ \frac{1}{2} & \frac{1}{3} \\ -1 & -\frac{1}{2} \end{pmatrix} \cdot V_1$$

$$A_2 =$$

$$= (zL_2^{-3/2}, N_\alpha L_2^{-3/2}, N_\alpha z^2 L_2^{-3/2}, N_\alpha^2 zL_2^{-3/2}, N_\alpha^3 L_2^{-3/2}, sL_2^{-1}, sz^2L_2^{-1}, s^2zL_2^{-1/2}) .$$

$$\begin{pmatrix} 0 & \frac{1}{8} & \frac{1}{8} & 0 & \frac{1}{4} & -\frac{1}{8} & -\frac{5}{12} & -\frac{1}{12} \\ 0 & -\frac{1}{8} & 0 & \frac{1}{24} & 0 & 0 & 0 & -\frac{1}{36} \\ -\frac{1}{2} & -\frac{1}{8} & -\frac{3}{8} & -\frac{1}{6} & -1 & \frac{15}{8} & \frac{5}{3} & \frac{1}{3} \\ \frac{1}{2} & \frac{1}{4} & \frac{1}{4} & \frac{1}{4} & \frac{5}{4} & -\frac{7}{4} & -\frac{25}{12} & -\frac{13}{24} \\ -\frac{1}{6} & -\frac{1}{8} & 0 & -\frac{1}{8} & -\frac{1}{2} & \frac{1}{2} & \frac{5}{6} & \frac{5}{18} \\ 0 & 0 & 0 & \frac{1}{6} & \frac{1}{2} & 0 & -\frac{2}{3} & -\frac{1}{3} \\ -\frac{1}{2} & 0 & -\frac{1}{2} & -\frac{1}{6} & -1 & 2 & \frac{5}{3} & \frac{1}{3} \\ -\frac{1}{2} & 0 & -\frac{1}{2} & -\frac{1}{4} & -\frac{5}{4} & \frac{3}{2} & \frac{3}{2} & \frac{3}{8} \end{pmatrix} \begin{pmatrix} V_2 \\ V_3 \end{pmatrix}$$

These expressions are used to obtain

$$1 - \Phi(\sigma_{11}^{-1/2}(- N_\alpha L_2^{1/2} - sL_2 - n^{-1/2}A_1 - n^{-1}A_2)) =$$

$$= \Phi(N_\alpha + sL_2^{1/2}) +$$

$$+ \varphi(N_\alpha + sL_2^{1/2})(n^{-1/2}B_1 + n^{-1}(B_2 - \tfrac{1}{2}B_1^2(N_\alpha + sL_2^{1/2})))$$

$$+ o(n^{-1}) ,$$

$$\varphi(\sigma_{11}^{-1/2}(- N_\alpha L_2^{1/2} - sL_2 - n^{-1/2}A_1)) =$$

$$= \varphi(N_\alpha + sL_2^{1/2})(1 - n^{-1/2}B_1(N_\alpha + sL_2^{1/2})) + o(n^{-1/2}) ,$$

$$\sigma_{11}^{-3/2} = L_2^{-3/2}(1 - \tfrac{3}{2}n^{-1/2}sL_3L_2^{-1}) + o(n^{-1/2}) ,$$

$$R_1(- N_\alpha L_2^{1/2} - sL_2 - n^{-1/2}A_1,\vartheta) =$$

$$= r_{1o}^{o} - r_{11}^{o}(N_\alpha L_2^{1/2} + sL_2) + r_{12}^{o}(N_\alpha L_2^{1/2} + sL_2)^2 +$$
$$+ n^{-1/2}[r_{1o}^{1} - r_{11}^{1}(N_\alpha L_2^{1/2} + sL_2) - r_{11}^{o}A_1$$
$$+ r_{12}^{1}(N_\alpha L_2^{1/2} + sL_2)^2 + 2r_{12}^{o}A_1(N_\alpha L_2^{1/2} + sL_2) + o(n^{-1/2}) ,$$

with

$$B_1 = A_1L_2^{-1/2} - \tfrac{1}{2}sN_\alpha L_3L_2^{-1} - \tfrac{1}{2}s^2L_3L_2^{-1/2}$$
$$B_2 = A_2L_2^{-1/2} - \tfrac{1}{2}sA_1L_3L_2^{-3/2} - \tfrac{1}{2}s^2EL_2^{-1}(N_\alpha + sL_2^{1/2})$$
$$E = \tfrac{1}{2}L_{21} + \tfrac{1}{2}L_4 - L_2^2 - \tfrac{3}{4}L_3^2L_2^{-1}.$$

Computing these expressions explicitely and ordering them gives the power functions in the following form:

$$P^n_{\vartheta+sn^{-1/2}}\{(nL_2(\vartheta))^{1/2}(\mu_n - \vartheta) \geq r_n(\vartheta,z,\alpha)\} =$$

$$= \Phi(N_\alpha + sL_2^{1/2}) +$$

$$+ n^{-1/2}\ \varphi(N_\alpha + sL_2^{1/2})\ \left[(sN_\alpha L_2^{-1}, s^2L_2^{-1/2}) \begin{pmatrix} 0 & -\frac{1}{6} \\ \frac{1}{2} & \frac{1}{6} \end{pmatrix} V_1\right] +$$

$$+ n^{-1}\ \varphi(N_\alpha + sL_2^{1/2}).$$

$$[(sL_2^{-3/2}, s^3L_2^{-1/2}, s^2zL_2^{-1}, sz^2L_2^{-3/2}, s^2N_\alpha L_2^{-1}, sN_\alpha^2L_2^{-3/2}, szN_\alpha L_2^{-3/2}).$$

$$\begin{pmatrix} 0 & \frac{1}{8} & 0 & -\frac{1}{24} & 0 & 0 & 0 & \frac{1}{36} \\ \frac{1}{6} & \frac{1}{8} & 0 & \frac{1}{24} & \frac{1}{4} & 0 & -\frac{1}{12} & -\frac{1}{72} \\ 0 & -\frac{1}{4} & \frac{1}{4} & 0 & 0 & -\frac{1}{4} & 0 & 0 \\ 0 & \frac{1}{8} & -\frac{1}{8} & 0 & 0 & \frac{1}{8} & 0 & 0 \\ 0 & \frac{3}{8} & -\frac{1}{2} & -\frac{1}{24} & -\frac{1}{4} & \frac{1}{4} & \frac{1}{6} & \frac{1}{24} \\ 0 & \frac{3}{8} & -\frac{1}{2} & \frac{1}{24} & 0 & \frac{1}{2} & 0 & -\frac{1}{18} \\ 0 & -\frac{1}{2} & \frac{1}{2} & 0 & 0 & -\frac{1}{2} & 0 & 0 \end{pmatrix} \begin{pmatrix} V_2 \\ V_3 \end{pmatrix}$$

$$+ \ (s^5 L_2^{-3/2}, s^2 N_\alpha^3 L_2^{-1}, s^3 N_\alpha^2 L_2^{-1/2}, s^4 N_\alpha) \cdot \begin{pmatrix} -\frac{1}{8} & -\frac{1}{12} & -\frac{1}{72} \\ 0 & 0 & -\frac{1}{72} \\ 0 & \frac{1}{12} & \frac{1}{72} \\ -\frac{1}{8} & 0 & \frac{1}{72} \end{pmatrix} \quad V_3 \Big] \ .$$

Comparing this expansion with the envelope power function (Pfanzagl, 1974 b, Theorem 2, Remark) proves the assertion. □

The statistical interpretation of Theorem 4 in terms of deficiency can be obtained from Lemmas 4 and 5 in Pfanzagl, 1974 b.

6. Some Bayes estimates for general loss functions

Definition 4. Let W be a loss function. A sequence (μ_n) of $\mathcal{A}^n$-measurable functions $\mu_n : \Omega^n \to \overline{\Theta}$ is called a sequence of Bayes estimates for W and the posterior distribution $R_{n,\underline{\omega}}$ if for every compact $K \subseteq \Theta$

$$\inf_{\vartheta \in K} P_\vartheta^n \Big\{ \int W(n^{1/2}(\mu_n - \sigma)) R_{n,\underline{\omega}}(d\sigma) =$$

$$= \inf_{t \in \Theta} \int W(n^{1/2}(t-\sigma)) R_{n,\underline{\omega}}(d\sigma) \Big\} \geq 1 - o(n^{-1}).$$

Since W is assumed to be bounded and lower semicontinuous, there exist Bayes estimates. We are concerned with loss function of type (A) and type (B). For notational convenience we define

$$\int W'(-tL_2^{-1/2})t^k \exp(-\tfrac{1}{2}t^2)dt = w_k,\ k\in\mathbb{N}.$$

It follows that

$$\int W''(-tL_2^{-1/2})\exp(-\tfrac{1}{2}t^2)dt = -w_1,$$

$$\int W''(-tL_2^{-1/2})t^2\exp(-\tfrac{1}{2}t^2)dt = -w_2+2w_1.$$

<u>Theorem 4.</u> Assume that conditions (i)-(ix), (jj) and (jjj) are satisfied. Let (μ_n) be a sequence of Bayes estimates for loss functions of type (A) or type (B). Then the x-polynomial of

$$n^{1/2}L_2(\vartheta)(\mu_n - \vartheta)$$

is as follows: For both type (A) and type (B)

$$\tilde{a}_o = 0 \qquad \tilde{a}_2 = 0$$

$$\tilde{b}_i = 0 \qquad \tilde{c}_{jk} = 0$$

$$\tilde{b}_{jk} = \text{as given in Theorem 2}$$

$$\tilde{c}_{pqr} = \text{as given in Theorem 2.}$$

The remaining coefficients are for type (A)

$$\tilde{a}_1 = \tfrac{1}{6}c^2 L_{oo1} + \Lambda'$$

$$\tilde{c}_1 = \Lambda'' L_2^{-1} + \Lambda' L_{oo1} L_2^{-2} + \tfrac{1}{6}c^2 L_{ooo1} L_2^{-1} + \tfrac{1}{6}c^2 L_{oo1}^2 L_2^{-2}$$

$$\tilde{c}_2 = \Lambda' L_2^{-1} + \tfrac{1}{6}c^2 L_{oo1} L_2^{-1}$$

$$c_3 = \tfrac{1}{6}c^2$$

and for type (B)

$$\tilde{a}_1 = \Lambda' + \frac{1}{6}\frac{w_3}{w_1} L_{oo1}L_2^{-1}$$

$$\tilde{c}_1 = \Lambda'' L_2^{-1} + \Lambda' L_{oo1}L_2^{-2} + \frac{1}{6}\frac{w_3}{w_1} L_{ooo1}L_2^{-2} +$$

$$+ \left(2\frac{w_3}{w_1} + \frac{w_5}{w_1} - \frac{w_3^2}{w_1^2}\right)\frac{1}{12}L_{oo1}^2 L_2^{-3}$$

$$\tilde{c}_2 = \Lambda' L_2^{-1} + \left(2\frac{w_3}{w_1} + \frac{w_5}{w_1} - \frac{w_3^2}{w_1^2}\right)\frac{1}{12}L_{oo1}L_2^{-2}$$

$$\tilde{c}_3 = \frac{1}{6}\frac{w_3}{w_1} L_2^{-1}$$

Proof.

Loss of type (A).

Define

$$\Delta^{\vartheta}_{n,\underline{\omega}}(s) = F^{\vartheta}_{n,\underline{\omega}}(s + cL_1^{1/2}) - F^{\vartheta}_{n,\underline{\omega}}(s - cL_2^{1/2}), \; s\in\mathbb{R}.$$

The Bayes estimates μ_n are characterized by maximizing

$$\Delta^{\vartheta}_{n,\underline{\omega}}((nL_2)^{1/2}(\mu_n(\underline{\omega})-\vartheta)-x_1L_2^{1/2})$$

for some $\underline{\omega}\in\Omega^n$. From Theorem 2 we obtain that

$$P^n_\tau\{|(nL_2)^{1/2}(\mu_n-\vartheta)-x_1L_2^{-1/2}| \geq c(\log n)n^{-1/2}\} = o(n^{-1})$$

uniformly for $\vartheta\in K$, $|\tau-\vartheta| \leq (\log n)n^{-1/2}$. Define $\tilde{\Delta}^{\vartheta}_{n,\underline{\omega}}$ similar to $\Delta^{\vartheta}_{n,\underline{\omega}}$ with $F^{\vartheta}_{n,\underline{\omega}}$ replaced by $G^{\vartheta}_{n,\underline{\omega}}$. From Remark 1 it follows that

$$P^n_\tau\{|(\tilde{\Delta}^{\vartheta}_{n,\underline{\omega}})'((nL_2)^{1/2}(\mu_n-\vartheta)-x_1L_2^{-1/2})| \geq c(\log n)^a n^{-3/2}\}$$

$$= o(n^{-1})$$

uniformly for $\vartheta\in K$, $|\tau-\vartheta| \leq (\log n)n^{-1/2}$. Now standard expansion techniques yield

$$P^n_\tau\{|(nL_2)^{1/2}(\mu_n-\vartheta)-x_1L_2^{-1/2}+\frac{\tilde{\Delta}'_n(0)}{\tilde{\Delta}''_n(0)}| \geq c(\log n)^a n^{-3/2}\}$$

$$= o(n^{-1})$$

uniformly for $\vartheta \in K, |\tau-\vartheta| \leqslant (\log n)n^{-1/2}$. The x-polynomial of (μ_n) follows by elementary computations.

Loss of type (B):

Similar to Theorem 2 in Strasser, 1976 b, it follows that

$$P_\tau^n\{|(nL_2)^{1/2}(\mu_n-\vartheta)-x_1L_2^{-1/2}| \geqslant c(\log n)n^{-1/2}\} = o(n^{-1})$$

uniformly for $\vartheta \in K, |\tau-\vartheta| \leqslant (\log n)n^{-1/2}$. Moreover we have

$$P_\tau^n\{|\int W'(n^{1/2}(\mu_n-\sigma)-x_1L_2^{-1}-tL_2^{-1/2})G_{n,\underline{\omega}}^{\vartheta}(dt)| \geqslant$$

$$\geqslant c(\log n)^a n^{-3/2}\} = o(n^{-1})$$

uniformly for $\vartheta \in K, |\tau-\vartheta| \leqslant (\log n)n^{-1/2}$. Standard expansion techniques yield

$$P_\tau^n\{|n^{1/2}(\mu_n-\vartheta)-x_1L_2^{-1} + \frac{\int W'(-tL_2^{-1/2})F_{n,\underline{\omega}}^{\vartheta}(dt)}{\int W''(-tL_2^{-1/2})F_{n,\underline{\omega}}^{\vartheta}(dt)}| \geqslant c(\log n)^a n^{-3/2}\}$$

$$= o(n^{-1})$$

uniformly for $\vartheta \in K, |\tau-\vartheta| \leqslant (\log n)n^{-1/2}$. The x-polynomial follows by elementary computations. □

Theorem 5. Assume that conditions (i)-(xii), (jj) and (jjj) are satisfied. Let (μ_n) be a sequence of Bayes estimates for loss functions of type (A) or type (B).

(1) If (μ_n) is approximately median unbiased of order $o(n^{-1})$ then (μ_n) is asymptotically optimal of order $o(n^{-1})$.

(2) (μ_n) is approximately median unbiased of order $o(n^{-1})$ iff

$$\Lambda' = L_2^{-1}(\tfrac{1}{6}L_3(1 + \gamma L_2) + \tfrac{1}{2}\gamma L_{11}L_2)$$

where

$$\gamma = c^2 \text{ for loss of type (A)}$$

$$\gamma = \frac{w_3}{w_1} \text{ for loss of type (B).}$$

Proof. The proof contains no new ideas at all. It is quite analogous to the proof of Theorem 3 of the present paper or the proof of Theorem 4 in Strasser, 1975. □

In Theorem 5, Λ' defines $p = L_2^{1/2}$ iff $\gamma = 2L_2^{-1}$.

REFERENCES

Pfanzagl, J. (1973). Asymptotic expansions related to minimum contrast estimators. Ann. Statist. 1 993-1026.

Pfanzagl, J. (1974 a). Asymptotically optimum estimation and test procedures. Proceedings of the Prague Conference on Asymptotic Methods of Statistics, Prague, 201-272.

Pfanzagl, J. (1974 b). Nonexistence of tests with deficiency zero. Preprints in Statistics 8. University of Cologne.

Pfanzagl, J. (1975). On asymptotically complete classes. Statistical Inference and Related Topics, vol. 2, Academic Press, New York, 1-43.

Strasser, H. (1976 a). Asymptotic properties of posterior distributions. Z. Wahrscheinlichkeitstheorie verw. Geb. 35 269-282.

Strasser, H. (1976 b). A note on efficiency. Metrika 23 91-100.

Strasser, H. (1975). Admissible representation of asymptotically optimal estimators. Submitted to Ann. Statist.

Peers, H.W. and B.L.Welch. (1963). On formulae for confidence points based on integrals of weighted likelihoods. J.Roy.Statist.Soc. Ser. B 26 318-329.

Zacks, S. (1971). The theory of statistical inference. John Wiley & Sons, New York.

Recent Developments in Statistics
J.R.Barra et al., editors

Invited Paper

APPLICATIONS OF ASYMPTOTIC EXPANSIONS TO PARAMETRIC INFERENCE

R.Michel
University of Cologne

1. INTRODUCTION

Let $(X,\mathcal{A})$ be a measurable space and $P_\theta|\mathcal{A}$, $\theta \in \Theta$, where Θ is an open subset of $\mathbb{R}$, a family of probability measures. By $P_\theta^n|\mathcal{A}^n$ we denote the n-fold independent product of identical components $P_\theta|\mathcal{A}$. In this paper we consider test-sequences for the hypothesis θ_o (where $\theta_o \in \Theta$ is fixed) against alternatives $\theta > \theta_o$; especially we present results on the power functions for local (contiguous) alternatives $\theta_o + tn^{-1/2}$, $t > 0$, of asymptotically similar tests which are based on test statistics with a certain asymptotic expansion.

In order to put down the asymptotic expansion of our tests in a concise form we write in the following for $\mathcal{A}^n$-measurable functions $g_n(\cdot,\theta)$ and $h_n(\cdot,\theta)$:

$$g_n(\cdot,\theta) \sim h_n(\cdot,\theta) \qquad \text{iff}$$

$$\sup_{\theta \in U_o} P^n\{(x_1,\ldots,x_n)\colon n|g_n(x_1,\ldots,x_n,\theta_o) - h_n(x_1,\ldots,x_n,{}_o)| > (\log n)^\alpha\} = o(n^{-1/2})$$

for some neighborhood U_o of θ_o and some positive α.

We consider sequences of $\mathcal{A}^n$-measurable test statistics $F_n\colon X^n \times \Theta \to \mathbb{R}$, $n \in \mathbb{N}$, which admit the asymptotic expansion

$$(1) \qquad F_n(x_1,\ldots,x_n,\theta) \sim n^{-1/2}\sum_{i=1}^n f_o(x_i,\theta) + n^{-1/2}Q(n^{-1/2}\sum_{i=1}^n f_o(x_i,\theta),\ldots,n^{-1/2}\sum_{i=1}^n f_m(x_i,\theta),\theta)$$

where $f_j: X \times \Theta \to \mathbb{R}$, $j = 0,\dots,m$, and $Q: \mathbb{R}^{m+1} \times \Theta \to \mathbb{R}$ satisfy certain conditions, which ensure that the distribution functions of $F_n(\cdot,\theta_o)$ under $P^n_{\theta_o}$ and $P^n_{\theta_o+tn^{-1/2}}$, $0 < t < \log n$, admit asymptotic expansions up to terms of order $o(n^{-1/2})$.

Using this asymptotic expansion for the distribution function of $F_n(\cdot,\theta_o)$ we define a sequence of tests $C_{n,\alpha}(\theta_o)$, $n \in \mathbb{N}$, based on $F_n(\cdot,\theta_o)$ which is asymptotically similar of level $\alpha + o(n^{-1/2})$, i.e.

$$\lim_{n\to\infty} n^{1/2}|P^n_{\theta_o}(C_{n,\alpha}(\theta_o)) - \alpha| = 0 .$$

These tests have the form

$$(2) \qquad C_{n,\alpha}(\theta_o) = \{(x_1,\dots,x_n): F_n(x_1,\dots,x_n,\theta_o) > c_{n,\alpha}(\theta_o)\}$$

with appropriately chosen cut-off point $c_{n,\alpha}(\theta_o)$.

Furthermore, an asymptotic expansion up to terms of order $o(n^{-1/2})$ of $t \to P_{\theta_o+tn^{-1/2}}(C_{n,\alpha}(\theta_o))$ is given.

From the fact that the most powerful test for P^n against $P^n_{\theta_o+tn^{-1/2}}$ admits an asymptotic expansion (1) (see section 6) we obtain an upper bound (asymptotic up to terms of order $o(n^{-1/2})$) for the power functions of the considered tests. The main result then states that if in the expansion of the power of a test in our class (i.e., a test of the form (2)) the n^o-term agrees with the n^o-term of the expansion of the power function of the most powerful test (i.e., if a test in our class is "first order efficient"), then the $n^{-1/2}$-term in the expansion of its power is uniquely determined and equal to the $n^{-1/2}$-term of the expansion of the power function of the most powerful test (i.e., then the test is necessarily "second order efficient"). Thus for power comparisons between the tests, which are considered here, a more accurate approximation for the power is necessary if we restrict our attention to (first order) efficient tests. Furthermore, the present result suggests, first, that no other test can have substantially greater power than an efficient test and, secondly, that the choice of Q in the expansion (1) is not essential for the power of an efficient test.

Most of the results which are presented in this paper for the one-dimensional case are proved in Pfanzagl [7] for the general (and more complicated) case of testing a one-dimensional parameter in the presence of nuisance parameters. In the latter case the nuisance parameter is "replaced" by an asymptotic studentization procedure.

As the paper of Pfanzagl gives a rigorous treatment of the regularity conditions and of the proofs of the theorems we shall restrict ourselves to try to point out the general ideas of the subject by using more or less heuristic arguments. Furthermore, we do not state regularity conditions explicitely. (The statement of results on asymptotic expansions often is rather lengthy and complex because of the regularity conditions and the necessary notations.)

The upper bound (up to terms of order $o(n^{-1/2})$) for the power functions in the one-dimensional case has been given in Pfanzagl ([5], Theorem 3(i), page 1000). (The approach here is different, as the result directly follows from the considerations of the tests of the form (2).) In the same paper of Pfanzagl it is also shown that a certain test based on the maximum likelihood estimator is second order efficient (Theorem 3(ii), page 1000). This result now may be obtained from the main result of this paper directly.

The test statistic $F_n(x_1,\dots,x_n,\theta) = n^{1/2}(\hat{\theta}_n(x_1,\dots,x_n) - \theta)$, where $\hat{\theta}_n$ is a maximum likelihood estimator, admits under certain regularity conditions an expansion (1) and it is seen that in this expansion the function $f_0(\cdot,\theta)$ is equal to $c(\theta) \frac{\partial}{\partial\theta} \log p(\cdot,\theta)$, which turns out to be equivalent to efficiency of the test (see section 7). For the efficient version of the $C(\alpha)$-tests see Chibisov [3]. In this paper methods similar to those of Pfanzagl [7] are employed and the results are analogous to those of Pfanzagl.

For further references see the papers of Pfanzagl, where an extensive survey of the pertaining literature is given.

2. SOME NOTATIONS

Given a σ-finite measure $\mu|\mathcal{A}$ dominating $P_\theta|\mathcal{A}$, $\theta \in \Theta$, denote by $p(\cdot,\theta)$ a positive density of $P_\theta|\mathcal{A}$ with respect to $\mu|\mathcal{A}$, and let

$$l^{(j)}(\cdot,\theta) = \frac{\partial^j}{\partial\theta^j} \log p(\cdot,\theta), \qquad j = 0,1,2.$$

Let, furthermore,

(3)
$$L_k(\theta) = E_\theta l^{(1)}(\cdot,\theta)^k \quad , k = 2,3, \text{ and}$$
$$L_{11}(\theta) = E_\theta l^{(1)}(\cdot,\theta) l^{(2)}(\cdot,\theta) \ .$$

Given the functions $f_j(\cdot,\theta)$, $j = 0,\dots,m$, of the expansion (1) define

(4)
$$\sigma(\theta) = E_\theta f_0(\cdot,\theta)^2$$

and

(5)
$$\Sigma(\theta) = (E_\theta f_i(\cdot,\theta) f_j(\cdot,\theta))_{i,j=0,\dots,m} \ .$$

(The assumptions on $f_j(\cdot,\theta)$, $j = 0,\dots,m$, include among others that $E_\theta f_j(\cdot,\theta) = 0$, $j = 0,\dots,m$.)

Let $\varphi_{\Sigma(\theta)}$ denote the Lebesgue-density of the normal distribution with mean vector zero and covariance matrix $\Sigma(\theta)$. Let, furthermore, $\varphi(u) = \frac{1}{\sqrt{2\pi}} \exp[-\frac{1}{2}u^2]$, $\Phi(u) = \int_{-\infty}^{u} \varphi(r)dr$, and $N_\alpha = \Phi^{-1}(\alpha)$, $\alpha \in (0,1)$.

For a function $R: \mathbb{R}^{m+1} \to \mathbb{R}$ and $a \in \mathbb{R}^{m+1}$ define

$$(6) \qquad H_{\Sigma(\theta)}[R](v_o,a) = \varphi(\sigma^{-1/2}(v_o-a_o))^{-1}\int\varphi_{\Sigma(\theta)}(v-a)R(v)dv_1\ldots dv_n$$

and

$$(7) \qquad d[R](u,\theta) = \frac{1}{6}(\sigma^{-1}(\theta)u^2 - 1)\sigma^{-3/2}(\theta)E_\theta f_o(\cdot,\theta)^3 + H_{\Sigma(\theta)}[R](u,0) .$$

3. AN EDGEWORTH-EXPANSION

The main device in deriving the present results is a theorem which states that the distribution functions of a statistic $F_n(\cdot,\theta)$ with asymptotic expansion (1) admit an asymptotic expansion (Edgeworth-expansion) up to terms of order $o(n^{-1/2})$:

Assume that $F_n(\cdot,\theta)$ *admits an asymptotic expansion* (1) *and that the functions* $f_j(\cdot,\theta)$, $j = 0,\ldots,m$, *and* Q *in this expansion fulfill certain regularity conditions (see Pfanzagl [6], Theorem 5.8).*

Then uniformly for θ *in a neighborhood of* θ_o *and uniformly for* $u \in \mathbb{R}$

$$(8) \qquad P_\theta^n\{(x_1,\ldots,x_n): F_n(x_1,\ldots,x_n,\theta) < u\} = \Phi(\sigma^{-1/2}(\theta)u - n^{-1/2}\, d[Q(\cdot,\theta)](u,\theta)) + o(n^{-1/2}) .$$

We remark that the regularity conditions needed for deriving this result include among moment conditions a "continuity condition" on the distribution of the vector $f(\cdot,\theta) = (f_o(\cdot,\theta),\ldots,f_m(\cdot,\theta))$. This assumption is related to Cramér's Condition C. The basic idea in the proof is to use results of Bikjalis [1] for obtaining an Edgeworth-expansion for the vector $n^{-1/2} \sum_{i=1}^{n} f(x_i,\theta)$, to make in this expansion a suitable transformation in $\mathbb{R}^{m+1}$ (with first component equal to

$$n^{-1/2} \sum_{i=1}^{n} f_o(x_i,\theta) + n^{-1/2} Q(n^{-1/2} \sum_{i=1}^{n} f_o(x_i,\theta),\dots,n^{-1/2} \sum_{i=1}^{n} f_m(x_i,\theta))$$

and then to make the marginal distribution with respect to the first component. For details see Pfanzagl ([6], Lemmas 1 and 2 and the Corollary, pages 236 - 245). For a different method see also the paper of Chibisov [2]. In Michel [4] the $n^{-1/2}$- and n^{-1}-terms of the Edgeworth-expansion for the distribution of asymptotic maximum likelihood estimators of vector parameters are given explicitely.

4. ASYMPTOTICALLY SIMILAR TESTS

Starting from the expansion in (8) we now are able to define a sequence of asymptotically similar tests $C_{n,\alpha}(\theta_o)$, $n \in \mathbb{N}$, of level $\alpha + o(n^{-1/2})$, i.e. a sequence of tests $C_{n,\alpha}(\theta_o)$, $n \in \mathbb{N}$, fulfilling $\lim_{n\to\infty} n^{1/2}|P^n_{\theta_o}(C_{n,\alpha}(\theta_o)) - \alpha| = 0$, which are based on the test-statistic $F_n(\cdot,\theta_o)$. It is reasonable to investigate tests of the form

$$(9) \qquad C_{n,\alpha}(\theta_o) = \{(x_1,\dots,x_n): F_n(x_1,\dots,x_n,\theta_o) > a + n^{-1/2} b\}$$

where a and b may depend on θ_o and α.

Using (8) we then obtain

$$P^n_{\theta_o}(C_{n,\alpha}(\theta_o)) = \Phi(-\sigma^{-1/2}(\theta_o)(a + n^{-1/2} b) + n^{-1/2} d[Q(\cdot,\theta_o)](a + n^{-1/2} b, \theta_o))) + o(n^{-1/2})$$

$$= \Phi(-\sigma^{-1/2}(\theta_o)a + n^{-1/2}(d[Q(\cdot,\theta_o)](a,\theta_o) - \sigma^{-1/2}(\theta_o)b)) + o(n^{-1/2}).$$

Hence, $P^n_{\theta_o}(C_{n,\alpha}(\theta_o)) = \alpha + o(n^{-1/2})$ is fulfilled, if

$$(10) \qquad a = -\sigma^{-1/2}(\theta_o)N_\alpha$$

and

$$(11) \qquad b = \sigma^{1/2}(\theta_o)d[Q(\cdot,\theta_o)](-\sigma^{1/2}(\theta_o)N_\alpha,\theta_o) .$$

This gives the following result:

Assume that $F_n(\cdot,\theta)$ *admits an asymptotic expansion (1). Then (under certain regularity conditions) the sequence of tests defined by*

$$(12) \qquad C_{n,\alpha}(\theta_o) = \{(x_1,\ldots,x_n): F_n(x_1,\ldots,x_n,\theta_o) > -\sigma^{1/2}(\theta_o)N_\alpha + n^{-1/2}\sigma^{1/2}(\theta_o)d[Q(\cdot,\theta_o)](-\sigma^{1/2}(\theta_o)N_\alpha,\theta_o)$$

is asymptotically similar of level $\alpha + o(n^{-1/2})$.

5. ASYMPTOTIC EXPANSIONS OF THE POWER

The power function $t \to P^n_{\theta_o+tn^{-1/2}}(C_{n,\alpha}(\theta_o))$ of the test (12) may also be obtained from the Edgeworth-expansion in (8). Let $\tau = \theta + tn^{-1/2}$, $0 < t < \log n$, and write

$$(13) \qquad F_n(x_1,\ldots,x_n,\theta) \sim n^{-1/2}\sum_{i=1}^n f_o(x_i,\theta) + n^{-1/2}Q(n^{-1/2}\sum_{i=1}^n f_o(x_i,\theta),\ldots,n^{-1/2}\sum_{i=1}^n f_m(x_i,\theta),\theta)$$

$$= n^{-1/2}\sum_{i=1}^n (f_o(x_i,\theta) - E_\tau f_o(\cdot,\theta)) + n^{1/2}E_\tau f_o(\cdot,\tau)$$

$$+ n^{-1/2}Q(n^{-1/2}\sum_{i=1}^n (f_o(x_i,\theta) - E_\tau f_o(\cdot,\theta))$$

$$+ n^{1/2}E_\tau f_o(\cdot,\theta),\ldots,n^{-1/2}\sum_{i=1}^n (f_m(x_i,\theta) - E_\tau f_m(\cdot,\theta))$$

$$+ n^{1/2}E_\tau f_m(\cdot,\theta),\theta) .$$

Using $E_\theta f_0(\cdot,\theta) = 0$ we obtain by a Taylor expansion,

(14) $$n^{1/2} E_\tau f_0(\cdot,\theta) = r_n(t,\theta) + O(n^{-1}(\log n)^3)$$

where

(15) $$r_n(t,\theta) = tE_\theta f_0(\cdot,\theta)l^{(1)}(\cdot,\theta) + \tfrac{1}{2}t^2 n^{-1/2}[E_\theta f_0(\cdot,\theta)l^{(2)}(\cdot,\theta) + E_\theta f_0(\cdot,\theta)l^{(1)}(\cdot,\theta)^2]$$

and, for $j = 1,\ldots,m$,

(16) $$n^{1/2} E_\tau f_j(\cdot,\theta) = tE_\theta f_j(\cdot,\theta)l^{(1)}(\cdot,\theta) + O(n^{-1/2}(\log n)^2).$$

Hence, (13) - (16) imply

(17) $$F_n(x_1,\ldots,x_n,\theta) - r_n(t,\theta) \sim \hat{F}_n(x_1,\ldots,x_n,\theta),$$

where

(18) $$\hat{F}_n(x_1,\ldots,x_n,\theta) = n^{-1/2}\sum_{i=1}^{n}(f_0(x_i,\theta) - E_\tau f_0(\cdot,\theta)) + n^{-1/2}\,Q(n^{-1/2}\sum_{i=1}^{n}(f_0(x_i,\theta) - E_\tau f_0(\cdot,\theta)) + tE_\theta f_0(\cdot,\theta)l^{(1)}(\cdot,\theta), \ldots, n^{-1/2}\sum_{i=1}^{n}(f_m(x_i,\theta) - E_\tau f_m(\cdot,\theta) + tE_\theta f_m(\cdot,\theta)l^{(1)}(\cdot,\theta),\theta).$$

Furthermore, we use

$$\sigma(\tau) := E_\tau(f_0(\cdot,\theta) - E_\tau f_0(\cdot,\theta))^2 = \sigma(\theta) + n^{-1/2}\,tE_\theta f_0(\cdot,\theta)^2\, l^{(1)}(\cdot,\theta) + O(n^{-1}(\log n)^2)$$

which implies

(19) $$\sigma^{-1/2}(\tau) = \sigma^{-1/2}(\theta)(1 - \tfrac{1}{2}n^{-1/2}t\sigma^{-1}(\theta)E_\theta f_0(\cdot,\theta)^2\, l^{(1)}(\cdot,\theta)) + O(n^{-1}(\log n)^2)\ .$$

Applying (8) to $\hat{F}_n(\cdot,\theta)$ and using (17), (19), and (12) we therefore obtain:

Assume that $F_n(\cdot,\theta)$ *admits an asymptotic expansion (1). Then (under certain regularity conditions) the power* $t \to P^n_{\theta_o+tn^{-1/2}}(C_{n,\alpha}(\theta_o))$, $0 < t < \log n$, *of the sequence of tests given by (12) has the asymptotic expansion*

$$(20)\qquad P^n_{\theta_o+tn^{-1/2}}(C_{n,\alpha}(\theta_o))$$

$$= \Phi(N_\alpha + t\sigma^{-1/2}(\theta_o)E_{\theta_o} f_o(\cdot,\theta_o)l^{(1)}(\cdot,\theta_o) + n^{-1/2}s(\cdot,\theta_o)$$

$$+ n^{-1/2}[H_{\Sigma(\theta_o)}[Q(\cdot,\theta_o)](-N_\alpha\sigma^{1/2}(\theta_o),tE_{\theta_o} l^{(1)}(\cdot,\theta_o)f(\cdot,\theta_o))$$

$$- H_{\Sigma(\theta_o)}[Q(\cdot,\theta_o)](-N_\alpha\sigma^{1/2}(\theta_o),0)]) + o(n^{-1/2}) ,$$

where

$$(21)\qquad s(\alpha,\theta) = \tfrac{1}{2}t^2\sigma^{-1/2}(\theta)(E_\theta f_o(\cdot,\theta)l^{(1)}(\cdot,\theta)^2 + E_\theta f_o(\cdot,\theta)l^{(2)}(\cdot,\theta))$$

$$- \tfrac{1}{2}t\sigma^{-1}(\theta)(N_\alpha + t\sigma^{-1/2}(\theta)E_\theta f_o(\cdot,\theta)l^{(1)}(\cdot,\theta))E_\theta f_o(\cdot,\theta)^2 l^{(1)}(\cdot,\theta)$$

$$+ \tfrac{1}{6}t\sigma^{-2}(\theta)(2N_\alpha + t\sigma^{-1/2}(\theta)E_\theta f_o(\cdot,\theta)l^{(1)}(\cdot,\theta))E_\theta f_o(\cdot,\theta)^3 E_\theta f_o(\cdot,\theta)l^{(1)}(\cdot,\theta).$$

6. ASYMPTOTIC UPPER BOUND FOR THE POWER

At first we look at the behavior of the power for the special case

$$(22)\qquad f_o(\cdot,\theta) = c(\theta)l^{(1)}(\cdot,\theta) , \qquad c(\theta) > 0 .$$

By straightforward calculations we obtain that in the expansion of the power-function the terms, which depend on Q, cancel each other out. Furthermore, we have in this case

$$(23)\qquad N_\alpha + t\sigma^{1/2}(\theta)E_\theta f_o(\cdot,\theta)l^{(1)}(\cdot,\theta) = N_\alpha + tL_2^{1/2}(\theta)$$

and (see (21))

(24) $$s(\alpha,\theta) = -\tfrac{1}{6}tN_\alpha L_2^{-1}(\theta)L_3(\theta) + \tfrac{1}{6}t^2L_2^{-1/2}(\theta)(L_3(\theta)+3L_{11}(\theta)).$$

We now show that the most powerful test based on

(25) $$\prod_{i=1}^{n} \frac{p(x_i,\theta_o+tn^{-1/2})}{p(x_i,\theta_o)}$$

is for $t > 0$ up to $o(n^{-1/2})$ equivalent to the test based on

(26) $$n^{-1/2}\sum_{i=1}^{n} l^{(1)}(x_i,\theta_o) + n^{-1/2}\,\tfrac{1}{2}tn^{-1/2}\sum_{i=1}^{n}(l^{(2)}(x_i,\theta_o) - E_{\theta_o} l^{(2)}(\cdot,\theta_o)).$$

This equivalence immediately follows from a Taylor expansion of

$$t^{-1}\sum_{i=1}^{n}(l(x_i,\theta_o+tn^{-1/2}) - l(x_i,\theta_o)) .$$

Consider now a test $C_{n,\alpha}(\theta_o)$ (see (12)) which is asymptotically similar of level $\alpha + o(n^{-1/2})$ and choose the cut-off point in (12) such that the test based on (25) is asymptotically similar of level $\alpha + o(n^{-1/2})$ (use (26) to achieve this). By adding an appropriate term of order $o(n^{-1/2})$ to the cut-off point (call this test $\varphi^*_{n,\alpha}(\theta_o)$) we then obtain

$$P^n_{\theta_o}(C_{n,\alpha}(\theta_o)) = \alpha + o(n^{-1/2}) \leq P^n_\theta(\varphi^*_{n,\alpha}(\theta_o)) .$$

Therefore,

$$P^n_{\theta_o+tn^{-1/2}}(C_{n,\alpha}(\theta_o)) \leq P^n_{\theta+tn^{-1/2}}(\varphi^*_{n,\alpha}(\theta_o)) .$$

As $\varphi^*_{n,\alpha}(\theta_o)$ is equivalent to a test based on (26) we obtain from (20), (23), (24) and the fact that $f_o(\cdot,\theta)$ in (26) is equal to $l^{(1)}(\cdot,\theta)$:

Let $C_{n,\alpha}(\theta_o)$ be a test given according to (12). Then uniformly for $0 < t < \log n$,

$$P^n_{\theta_o+tn^{-1/2}}(C_{n,\alpha}(\theta_o)) \tag{27}$$
$$\leq \Phi(N_\alpha + tL_2^{1/2}(\theta_o) + n^{-1/2}[tN_\alpha h_1(\theta_o) + t^2h_2(\theta_o)]) + o(n^{-1/2})$$

where

$$h_1(\theta) = -\frac{1}{6}L_2^{-1}(\theta)L_3(\theta) \tag{28}$$
$$h_2(\theta) = \frac{1}{6}L_2^{-1/2}(\theta)(L_3(\theta) + 3L_{11}(\theta)) .$$

7. FIRST ORDER EFFICIENT TESTS

According to the results of section 6 we call a test sequence $C_{n,\alpha}(\theta_o)$ *first order efficient* (or asymptotically efficient) if uniformly for $0 < t < \log n$

$$P^n_{\theta_o+tn^{-1/2}}(C_{n,\alpha}(\theta_o)) = \Phi(N_\alpha + tL_2^{1/2}(\theta_o)) + o(n^o) . \tag{29}$$

We now show that this is the case iff the function $f_o(\cdot,\theta)$ in (1) is P_θ - a.e. proportional to $l^{(1)}(\cdot,\theta)$.

First order efficiency implies by (20) and (27) that

$$\sigma^{-1/2}(\theta_o)E_{\theta_o}f_o(\cdot,\theta_o)l^{(1)}(\cdot,\theta_o) = L_2^{1/2}(\theta_o) .$$

Therefore equality holds in the Cauchy-Schwarz inequality

$$E_{\theta_o}f_o(\cdot,\theta_o)l^{(1)}(\cdot,\theta_o) \leq \sigma^{1/2}(\theta_o)L_2^{1/2}(\theta_o) .$$

This gives the assertion.

8. FIRST ORDER EFFICIENCY IMPLIES SECOND ORDER EFFICIENCY

According to section 6 we call a test sequence $C_{n,\alpha}(\theta_o)$ *second order efficient* if the expansion (20) of its power agrees with (27) up to terms of order $o(n^{-1/2})$. By the results of sections 6 and 7 we now easily obtain the announced main result (see Theorem 7.5 in Pfanzagl [7]):

Assume that a test sequence $C_{n,\alpha}(\theta_o)$ *given according to (12) is first order efficient. Then (under appropriate regularity conditions) it is necessarily second order efficient.*

9. AN EXAMPLE

In the introduction we remarked that the result of section 8 holds true for the test based on the maximum likelihood estimator (as it admits an expansion of the form (1)). The following example shows that the result is not true in general for the maximum likelihood estimator. Here the differentiability conditions which ensure the expansion (1) for the maximum likelihood estimator are not fulfilled.

Assume that $P_\theta|\mathcal{B}$ has Lebesgue-density

$$p(x,\theta) = \frac{1}{2}\exp[-|x-\theta|] .$$

Then the distribution function of the maximum likelihood estimator $\hat{\theta}_n(x_1,\ldots,x_n) = \underset{1\leq i\leq n}{\text{med}}\; x_i$ admits the following asymptotic expansion

$$P_\theta^n\{(x_1,\ldots,x_n):\ n^{1/2}(\hat{\theta}_n(x_1,\ldots,x_n)-\theta) < t\}$$

$$= \Phi(t-\tfrac{1}{2}n^{-1/2}\ t^2 \text{ sign } t) + o(n^{-1/2}) .$$

Hence the test

$$C_{n,1/2}(\theta) = \{(x_1,\ldots,x_n):\ \hat{\theta}_n(x_1,\ldots,x_n) > \theta\}$$

is asymptotically similar of level $\frac{1}{2}+o(n^{-1/2})$ and its power for $t > 0$ is given by

$$P^n_{\theta+tn^{-1/2}}(C_{n,1/2}(\theta)) = \Phi(t-\tfrac{1}{2}n^{-1/2}\ t^2) + o(n^{-1/2}).$$

The power of the most powerful test ($\alpha = 1/2$) can be obtained by some algebraic calculations and is given by

$$\Phi(t - \frac{1}{6}n^{-1/2}\ t^2) + o(n^{-1/2}), \qquad t > 0 .$$

Hence $C_{n,1/2}(\theta)$ is first order efficient, but not second order efficient.

REFERENCES

[1] Bikjalis, A. (1968). Asymptotic expansions for the densities and distributions of sums of independent and identically distributed random vectors. Litovsk. Mat. Sb. 8, 405 - 422. (In Russian).

[2] Chibisov, D.M. (1972). An asymptotic expansion for the distribution of a statistic admitting an asymptotic expansion. Theor. Prob. Appl. 17, 620 - 630.

[3] Chibisov, D.M. (1973). Asymptotic expansions for Neyman's C(α) tests. Proceedings of the Second Japan-USSR Symposium on Probability Theory. (Ed. by Maruyama, G., and Prokhorov, Yu.V.). Lecture Notes in Mathematics 330, Springer Verlag, 16 - 45.

[4] Michel, R. (1975). An asymptotic expansion for the distribution of asymptotic maximum likelihood estimators of vector parameters. J. Multivar. Anal. 5, 67 - 82.

[5] Pfanzagl, J. (1973). Asymptotic expansions related to minimum contrast estimators. Ann. Statist. 1, 993 - 1026.

[6] Pfanzagl, J. (1973). Asymptotically optimum estimation and test procedures. Proceedings of the Prague Symposium on Asymptotic Statistics. (Ed. by J.Hájek). Vol. 1, 201 - 272.

[7] Pfanzagl, J. (1976). First order efficiency implies second order efficiency. To appear in Hájek's Memorial Volume on Asymptotic Statistics.

Recent Developments in Statistics
J.R.Barra et al., editors

Invited Paper

ANALYSIS OF SURVIVAL DATA IN HETEROGENEOUS POPULATIONS

David P. Byar*
National Cancer Institute, U.S.A.

A method is given for identifying prognostic variables based on marginal death rates. A non-parametric method for comparing survival curves while adjusting for prognostic variables is presented. Selection and fitting of parametric survival models incorporating covariate information is discussed and methods for comparing survival distributions using these models are presented. Finally a method for detecting significant treatment-covariate interactions using parametric models is presented and the concept of treatment of choice is introduced.

INTRODUCTION

In many medical investigations the comparison of survival curves is of paramount interest. Typically we will be interested in comparing heterogeneous populations since for most diseases we know that there are important prognostic variables (covariates) which may influence the survival time distribution. In studies of cancer for example these variables might include the patient's age, the stage of the disease at diagnosis, the histological appearance of the tumor, his performance status, and perhaps the levels of certain laboratory measurements. In randomized studies we may ignore the effects of these covariates in looking for an overall best treatment, but when treatment comparisons are made using retrospective data it is essential to adjust for covariates in order to remove potential biases. Even in randomized studies we may desire to increase the precision of our estimates of the treatment effects by methods of covariate adjustment or to identify special subgroups of patients in which treatment effects vary. Also in randomized studies where appreciable differences can be demonstrated in the distributions of important covariates, adjustment procedures are often desirable in interpreting the results of a trial.

---

*Clinical and Diagnostic Trials Section, National Cancer Institute, Landow Building, Room C-509, Bethesda, Maryland 20014 U.S.A.

IDENTIFYING PROGNOSTIC VARIABLES

Quite often in medical studies information on a large number of variables is recorded for each patient. Of particular importance in comparing survival distributions are those variables measured before treatment was begun. Often these variables will have been measured because they are known to affect the outcome as a result of the analysis of the data collected in previous studies. In addition other variables may be included because it is suspected that they may be related to the outcome. Our problem in analyzing the data is to determine which variables in the data at hand can be demonstrated to affect the outcome, in this case survival. A very useful technique for this purpose is to examine each variable marginally, pooling together patients from all treatment groups, to see if it affects survival. The variables under consideration may be either categorical or continuous but we have generally chosen to treat all variables as categorical at this stage of the analysis by means of grouping the data for continuous variables into adjacent categories. We then compute the death rate defined as the number of deaths divided by the total observation time for patients in that category whether the observations are censored or uncensored. The death rate can be recognized as the maximum likelihood estimate of the parameter for a simple exponential distribution.

Table 1. Marginal death rates for pretreatment variables

Variable	Value	Number of Patients	Number of Deaths	Total months observation	Death rate per 100 patient-mos	Chi-square*	Degrees of freedom
1. Previous antibiotic treatment	No	98	91	1152	7.90	6.881 (6.879)	2 (1)
	Yes	70	63	504	12.50		
	Unknown	19	17	213	7.98		
2. Dysphonia	No	152	136	1661	8.19	8.845 (8.692)	2 (1)
	Yes	33	33	190	17.37		
	Unknown	2	2	18	11.11		
3. Age	< 55	35	31	381	8.14	1.160	4
	55-59	43	41	474	8.65		
	60-64	46	43	458	9.39		
	65-69	47	43	419	10.26		
	≥ 70	16	13	137	9.49		
4. Performance rating (%)	20-60	22	21	137	15.33	11.145	2
	70-80	75	70	619	11.31		
	90-100 & unknown	90	80	1113	7.19		

*Computed for uncensored data since censoring was <10%.

An example of such an analysis is presented in Table 1 for four variables in a study of treatment for inoperable lung cancer. The first two variables are categorical and both include a special category for "unknown". Unknown values are quite common in medical studies because of the nature of the data collection process. If the decision is made to use either of these variables we must further decide what to do with the unknown values. For the first variable the death rate in the unknown category is similar to that in the category designated "no" and we may choose to combine these two categories assuming that if a patient received no previous antibiotic treatment either the box on the form was ignored or marked as no. In general how to treat unknown values will depend on the variable being studied and our best guess as to why the information is unknown as well as our estimate of the effects of treating unknown values in various ways. The third and fourth variables are continuous but have been treated as categorical variables in this analysis. Performance ratings were originally given to the nearest 10 percent but have been previously grouped in adjacent categories for this analysis. By inspection we see that variables 1, 2 and 4 appear to be related to survival as judged by some variation in the death rates for the various categories but the third variable, age, does not appear to affect the death rate appreciably. Under the assumption of exponential survival in each category we may perform an approximate test for the homogeneity of the death rates, as suggested by Mitchell Gail. Let $\hat{\lambda}_i$ represent the estimated death rate in the i'th category with the variance of $\hat{\lambda}_i$ estimated as $\hat{\sigma}_i^2$ and let $\hat{\lambda}$ represent a weighted least squares estimate of the common death rate. Then asymptotically

$$X^2 = \sum_{i=1}^{k} \frac{(\hat{\lambda}_i - \hat{\lambda})^2}{\hat{\sigma}_i^2} \tag{1}$$

is distributed as a chi-square with k-1 degrees of freedom if there are k categories. Formulae for the estimates are given in Appendix I where both censored and uncensored cases are treated. Values of the statistic are given in Table 1 in the right-hand column along with the degrees of freedom. The values in parentheses were obtained by pooling the unknown with the no's for the first two variables. The advantage of pooling similar categories for categorical variables or adjacent categories for continuous variables is illustrated with the fourth variable. Originally there were ten categories and the chi-square value was 14.042 with nine degrees of freedom (not significant) but after pooling into three categories the chi-square value is 11.145 with two degrees of freedom ($p < .004$).

Undoubtedly there are many other ways of selecting variables for inclusion in adjusted survival analyses, but the method just illustrated has worked well in practice. It does not guarantee that when multivariate methods are used all variables will be retained in the final analysis but it will tend to select those variables most likely to be significant in a multivariate context. This would certainly be true if the variables were all independent. If two variables have prognostic significance and are highly correlated, this method will tend to select both, but in a multivariate analysis perhaps only one of the two would be needed for prediction. However if the prognostic significance of a variable depends on the level of some other variable (a significant interaction is present) this method is not likely to detect such an effect and other techniques would be required.

Armitage and Gehan (1974) have recently provided an excellent review on the subject of identifying prognostic factors for a disease. Zelen (1975) has discussed the importance of prognostic factors in planning therapeutic trials. Most of the observations he makes are equally applicable to the analysis of survival data in heterogeneous populations.

NONPARAMETRIC METHODS FOR COMPARING SURVIVAL DISTRIBUTIONS

Once the important covariates have been selected perhaps the simplest nonparametric method would be to divide the data into various strata on the basis of these covariates and then compare survival directly within each stratum. Unless the data are quite abundant this approach is likely to lead to small numbers of patients in the various strata, especially if there are many important covariates, resulting in poor power in the survival comparisons. In this situation the overall results may be difficult or impossible to interpret. A frequent remedy in this sort of situation is to use some method which combines estimates of the treatment effects within the various strata into a summary statistic. Such methods are of course most appropriate if the effect being studied is consistent across the various strata. But even failing this condition, such a method may provide a summary test of overall difference in the treatments which in effect adjusts for differences in the distribution of prognostically important covariates between the two or more treatment groups.

Although many nonparametric tests for comparing survival distributions may be used in this way we will illustrate this method with the widely used Mantel-Haenszel statistic as presented by Mantel (1966). Assume that the data have been divided into strata and that we arrange the data in each stratum at each point in time at which a response occurs as a two-fold table:

	Dead	Alive	At Risk
Treatment I	A_{ij}	B_{ij}	N_{1ij}
Treatment II	C_{ij}	D_{ij}	N_{2ij}
Total	M_{1ij}	M_{2ij}	T_{ij}

where $i=1,2,\cdots,k$ (k=the number of strata) and $j=1,2,\cdots,t_i$ (t_i=the number of unique death times in the i'th stratum). Conditional on the marginal totals, A_{ij} is hypergeometric with expectation

$$E(A_{ij}) = N_{1ij}M_{1ij}/T_{ij} \tag{2}$$

and variance

$$V(A_{ij}) = N_{1ij}N_{2ij}M_{1ij}M_{2ij}/(T_{ij}^2(T_{ij}-1)). \tag{3}$$

The Mantel-Haenszel statistic is computed as

$$\chi^2 = \left[\left|\sum_{i=1}^{k}\sum_{j=1}^{t_i} A_{ij} - \sum_{i=1}^{k}\sum_{j=1}^{t_i} E(A_{ij})\right| - 1/2\right]^2 \Big/ \sum_{i=1}^{k}\sum_{j=1}^{t_i} V(A_{ij}) \tag{4}$$

and under the null hypothesis of no treatment difference is distributed as a chi-square variable with 1 degree of freedom. Right censored patients are treated as being at risk for all deaths in their stratum preceding their censoring times, but not for those following them.

This method may be used even when the numbers of patients appearing in the separate tables are quite small since it is the sum of the A_{ij} which must approach normality for the validity of the chi-square approximation rather than the separate A_{ij}. Of course the use of too many strata might produce some 2×2 tables in which one or the other treatment was not represented, in effect removing the patients who did appear from the analysis. In some instances this might be desirable on the grounds that comparisons should not be made between treatments if in fact patients on one of the treatments are not available for comparison in certain strata.

The method just described is useful for comparing survival between two treatment groups while adjusting for covariates when we do not wish to make any assumption about interrelationships between the covariates or about the distributional form of the survival curve. It can also be recommended because of its simplicity. We may study indirectly the effects of the various covariates by systematically eliminating them from the analysis, that is, by reducing the number of strata, but we are unable to estimate formally the effects of the covariates on survival. If we are interested in assessing the effects of the covariates when they are allowed to act jointly, we will need a parametric or semi-parametric model.

GENERAL COMMENTS ON FITTING PARAMETRIC SURVIVAL MODELS

Before making some comments on selecting a particular parametric survival model we might consider in general how the parameters of these models may be estimated. Assuming that we have chosen a survival model for continuous time with probability density function $f(t,\underset{\sim}{x},\underset{\sim}{\theta})$ where t represents the time of death, $\underset{\sim}{x}$ is a vector of the covariates, and $\underset{\sim}{\theta}$ a vector of parameters to be estimated, then the likelihood for a sample, possibly including censored observations, is given by

$$L = \prod_{i=1}^{n} f(t_i, \underset{\sim}{x}_i, \underset{\sim}{\theta})^{z_i} \quad (1 - F(t_i, \underset{\sim}{x}_i, \underset{\sim}{\theta}))^{1-z_i} \tag{5}$$

where F( ) is the cumulative distribution function of f( ), and $z_i=1$ for uncensored observations and 0 for censored ones. The t_i represent death times for uncensored observations and follow-up times for censored ones. Frequently we may estimate $\underset{\sim}{\theta}$ by maximum likelihood simply by obtaining the derivatives of the likelihood or the log likelihood with respect to θ_j, $j=0,1,2,\cdots,k$ (k=the number of covariates studied) and setting the derivatives equal to 0. The solution to the set of k+1 simultaneous and usually non-linear equations so obtained will be the maximum likelihood estimate of $\underset{\sim}{\theta}$ provided that we have not achieved a local maximum and that the solution is not a saddle point. The latter possibility can be excluded by demonstrating that the eigenvalues of the matrix of second partial derivatives with respect to the parameters are all negative. The covariate x_0 may be taken identically equal to 1 so that θ_0 may represent an intercept. An estimate of the asymptotic covariance matrix $\sum$ of $\hat{\underset{\sim}{\theta}}$ is usually obtained as the negative inverse of the matrix of expectations of the second partial derivatives

$$\sum = -\left[E(d^2 \ell n L / d\theta_j d\theta_\ell) \right]^{-1} . \tag{6}$$

A specific example using the exponential distribution and some comments relating to the asymptotic covariance matrix will be found in Appendix II.

It is usually advisable in fitting a model to consider as candidates all covariates which are known or suspected to influence survival. Some selection procedures should then be used to determine the final set. Greenberg et al. (1974) have suggested a step-down procedure for this purpose and Krall et al. (1975) have suggested a step-up procedure.

SELECTING A PARAMETRIC SURVIVAL MODEL

In the last few years a great deal of attention has been devoted to constructing survival models incorporating covariate information suitable for use with censored observations. The first problem for the data analyst is to choose which among the many models proposed is most appropriate for the data at hand. The most important characteristics of the models which have been proposed are the assumptions about the hazard and the manner in which the covariates are related to the parameter(s) of the distribution.

The simplest assumption about the hazard is that it is constant over time which implies that the survival function is exponential. If the hazard changes in time other distributions may be suggested such as the Weibull, log-normal, gamma, or Gompertz distributions. Prentice (1974) has recently proposed a log-gamma model which is of sufficient generality to include the log-normal, exponential and Weibull distributions as special cases. A model selection parameter is estimated which allows one to judge which of the special cases best fits the data at hand. Cox (1972) has suggested a method which incorporates many features of the parametric models but which introduces a general underlying hazard function assumed to be factorable into a time-dependent portion and a non-time-dependent portion incorporating the covariate information. In effect this allows for an arbitrary response time distribution. Brown (1975) has suggested a novel approach allowing for a time-dependent hazard by means of fitting the survival distribution in pieces using dummy variables for separate intervals of time. Using likelihood ratio tests he distinguishes between changes in the underlying survival distribution in time and changes in the importance of the covariates in time.

The relationship of the covariates to the parameter(s) of the model may be illustrated by the simple exponential in which various authors have proposed that a linear combination of the covariates may be related directly to the hazard, to its reciprocal (the expected survival time), or to the log hazard. Myers et al. (1973) and Byar and Mantel (1975) have pointed out some interesting interrelationships between several of these models.

There has been little work on formal procedures for comparing models from different families that can be applied to these complex situations. However, Cox (1962) has suggested a procedure based on the log likelihood ratio for two competing models. He has proved that the log likelihood ratio minus its expectation and divided by its standard deviation, where both expectation and variance are computed under the assumption that one of the models is true, is asymptotically normally distributed. We have been exploring this approach using simulation estimators for the mean and variance of the log likelihood ratio. Lindsey (1974) has suggested a general method of comparing statistical models by means of separate comparisons with some base model. His procedures are based on likelihood inference and he does not discuss any tests of significance. At present it appears that the best guide to choosing a survival distribution would be to divide the observations into groups with approximately comparable risk and plot the estimated actuarial survival function versus that obtained by a parametric fit. In other words, we might choose various models which are possibly appropriate and then see how well they correspond to the data at hand by graphical inspection. A possible alternative suggested by Gehan (1969,1975) is to plot the sample hazard function versus time and then choose a distribution based on knowledge of the behaviour of the hazard functions of different distributions.

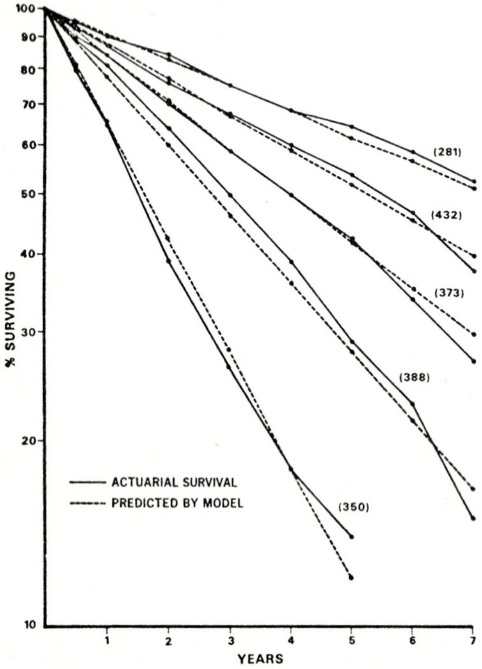

Fig. 1. Comparison of actuarial survival curves for five risk groups of patients with cancer of the prostate versus predicted curves based on the additive exponential model using eight covariates.

Sometimes a knowledge of the disease under study may provide some hints as to an appropriate survival model. Metcalf (1974) has shown that for a wide variety of cancers, survival for patients of comparable risk follows an exponential distribution. An example illustrating the fit of the exponential survival model using eight covariates to a large group of patients with prostatic cancer (Byar et al., 1974) is shown in Figure 1. After the parameters of the model were estimated the patients were divided into five risk groups based on their fitted hazards. In each of the five risk groups we see that the survival functions predicted by the model (dotted lines) agreed quite well with the actuarial curves (solid lines) estimating the observed survival for patients in each of the five groups.

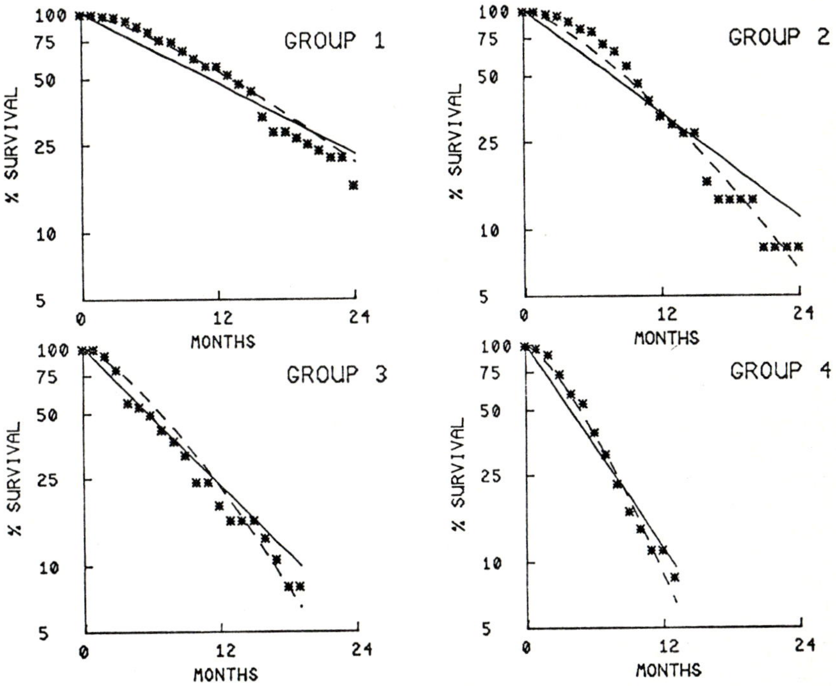

Fig. 2. Comparison of exponential (——) and Weibull (---) models with actuarial curves (* * *) for four risk groups of patients with inoperable bronchogenic carcinoma.

The same exponential model, this time incorporating three covariates, was used in a study of patients with advanced lung cancer. In order to avoid overlapping curves, each of four risk groups is plotted separately (Figure 2). In this example there appears to be some systematic deviation from the predicted exponential curves in each of the four risk groups characterized by better observed survival in the early part of the curves and worse survival in the later parts of the curves. This pattern suggests that the data might better be fit by a Weibull model which can incorporate an increasing (or decreasing) hazard as a function of time. The predictions of the Weibull model (dashed lines) appear to be in closer agreement with the observed survival curves. In this instance we could perform a likelihood ratio test comparing the fit of the two models since the exponential is a special case of the Weibull. In fitting the Weibull we assumed the same shape parameter for all risk groups, thus minus twice the log likelihood ratio for the exponential versus the Weibull fit is asymptotically distributed as a chi-square with 1 degree of freedom. The observed value of this statistic was 24.12 indicating that the Weibull model fits the data significantly better than the exponential. In addition to graphical tests such as those just described, numerical tests of goodness of fit such as those proposed by Feigl and Zelen (1965) or Greenberg et al. (1974) are sometimes appropriate.

COMPARING TREATMENTS AFTER FITTING A PARAMETRIC MODEL

There are several approaches to comparing treatments after fitting a parametric survival model incorporating covariate information. One simple approach is to regard the goal of fitting the parametric model as simply that of determining which covariates, when they are all allowed to act together, significantly affect the survival distribution and of estimating the magnitude of these effects. Having done this we may then cross-classify the patients on the basis of these covariates and apply a nonparametric method such as the one discussed in an earlier section. Another form of this hybrid approach would be to form groups of comparable risk based on the parametric analysis such as the groups illustrated in Figures 1 and 2 and use a summary risk score to stratify patients in a nonparametric analysis. For example, in the case of the exponential model, risk groups can be determined on the basis of the computed hazard for each patient based on fitting with all treatments combined. If the distribution has a time-dependent hazard then the hazard would have to be computed at some specific point in time. Another alternative is to introduce dummy variables representing treatment along with the other covariates. The model may then be fitted with and without these dummy variables and the importance of treatment effects may be tested using likelihood ratio procedures. Breslow (1974) has illustrated this approach. Prentice (1973) using the technique of structural probability has illustrated yet another approach in which a dummy variable is introduced to represent treatment effects, but inferences about the importance of treatment are based on the value of the estimated coefficient for treatment itself since he is able to obtain the exact distribution of the regression coefficient. This method, while methodologically elegant, may not be suitable in practice in some instances because of the assumptions introduced to handle censoring and because of computational difficulty.

TESTING FOR TREATMENT-COVARIATE INTERACTIONS: THE CONCEPT OF TREATMENT OF CHOICE

All the methods for comparing treatment while adjusting for covariate values discussed in the previous section are based on the assumption that there are no important treatment-covariate interactions. The question of interest in the analyses just suggested was "Is there any best overall treatment?" In a heterogeneous population the possibility exists that no one treatment may be best for all patients. One treatment may be better for some kinds of patients and another for other kinds. This possibility, however, can only exist in the presence of significant treatment-covariate interactions. In order to test for such interactions we must estimate the parameters of our model separately for each treatment group as well as for all treatments combined.

Table 2. Regression coefficients for the additive exponential model obtained for pooled and separate treatments

Parameter	Variable	Symbol	Both treatments combined*	Treatment P*	Treatment E*	Ratio (P/E)
θ_0	Intercept	--	4.36	7.35	2.08	3.53
θ_1	Hemoglobin	HG	6.28	12.40	3.26	3.80
θ_2	Performance rating	PF	15.55	6.70	22.64	0.30
θ_3	History of cardiovascular disease	HX	8.27	5.36	11.15	0.48
θ_4	Stage-grade category	SG	9.51	12.30	7.93	1.55
θ_5	Standardized weight	WT	3.99	1.19	4.88	0.39
θ_6	Age at diagnosis	AG	6.02	1.17	9.89	0.12
θ_7	Size of lesion	SZ	15.21	23.28	9.46	2.46

* Regression coefficients expressed in deaths per 1000 patient-months

Byar and Corle (1976) have performed such an analysis (Table 2) using data from a randomized clinical trial of treatment of prostatic cancer. The exponential model with the hazard a linear function of the covariates was used in this analysis. The seven covariates chosen for the final fit were obtained using a step-down procedure when data from both treatments were combined. Note that for all these covariates the regression coefficients when estimated separately for the two treatments differ to some extent suggesting the presence of significant treatment-covariate interactions. The details of a likelihood ratio procedure for detecting significant treatment-covariate interactions are presented in Appendix III. For these data the test for an overall best treatment was not significant.

If we have detected significant treatment-covariate interactions, then the question of interest is "Which treatment is best for which kind of patient?" This new question embodies the concept of treatment of choice, that is, how do we use the information in the covariate vectors to select the appropriate treatment for different kinds of patients? In the case of the exponential model the fitted hazard for each of the various treatments may be computed for each individual patient and the patient may be assigned to the treatment associated with the smallest estimated hazard. Of course, random variation must be taken into account when assessing whether the hazard associated with some specific treatment is significantly lower than that for the others. An approximate test based on the distribution of the range in normal samples is presented in Appendix IV. In interpreting the overall results of an analysis it is probably advisable to devote most attention to those cells arising in the cross-classification of categorical covariates in which the range tests are significant (or in which some other test has been used to screen out effects most likely due to random variation). The values of the covariates associated with these cells may then be tabulated according to the optimal treatment in these cells. Such tabulations may then be examined for patterns of meaning with respect to making treatment recommendations on the basis of the analysis.

Table 3. Covariate values for patients in cells with significant range tests

Optimal Treatment	Number of patients	Values of covariates[1] HG	PF	HX	SG	WT	AG	SZ	Hazard for Treatment P[2]	Hazard for Treatment E[2]	T_3[3]
P	1	0	0	1	0	0	2	0	15.0	33.0	3.07
	11	0	0	1	0	1	1	0	15.8	28.0	3.34
	6	0	0	1	0	1	2	0	17.0	37.9	3.64
	1	0	0	1	0	2	2	0	18.9	42.8	3.60
	4	0	1	1	0	1	1	0	22.5	50.6	2.78
	1	0	1	1	0	1	2	0	23.7	60.5	3.37
	1	0	1	1	0	2	2	0	25.6	65.4	3.52
E	41	0	0	0	0	0	0	0	7.3	2.1	3.06
	22	0	0	0	1	0	0	0	19.6	10.0	3.02
	5	0	0	0	1	0	0	1	42.9	19.5	2.87
	4	1	0	0	0	0	0	0	19.7	5.3	2.98
	4	1	0	0	1	0	0	0	32.0	13.3	3.61
	3	1	0	0	1	0	0	1	55.3	22.7	3.59
	5	1	0	0	1	1	0	0	33.9	18.2	3.22
	3	1	0	0	1	1	0	1	57.2	27.6	3.35
	1	2	0	0	1	1	0	0	46.3	21.4	2.82

[1] See table 2 for meaning of symbols HG, etc.

[2] Hazards expressed in deaths per 1000 patient-months.

[3] Value for range test. See equation A.14 of appendix IV.

An example of such a tabulation is presented in Table 3. Here we might tentatively conclude that patients with values 1 or greater for variables representing initial hemoglobin, combined stage-grade, or size of the tumor should receive treatment E while those with non-zero values for initial performance rating, history of cardiovascular disease, weight, or age should receive treatment P. Of course, if we truly believe the model, for any given covariate vector we could determine the optimal treatment. In situations where the hazard is time-dependent the strategy we have employed for determining optimal treatments would require choosing some point in time for defining optimality unless the hazards happen to remain ordered in time for all possible combinations of covariates.

Possibly the greatest value of the kind of analysis just described is heuristic - it suggests relationships which might not have been apparent by more conventional methods. The proof of any conclusions drawn, as in most statistical applications, should depend on future experiments designed specifically to test those conclusions.

We have attempted in this paper to present some general approaches to the problem of comparing survival distributions in heterogeneous populations. Many variations of these methods and more detailed applications have appeared in the literature, but in an article of this length a thorough review of the literature has not been possible. Nevertheless, the material presented here and the references cited should provide a good beginning in approaching problems of this sort.

APPENDICES

I. CHI-SQUARE TEST FOR HOMOGENEITY OF DEATH RATES IN k CATEGORIES

Assume there are n_i patients and d_i deaths in the i'th category, $i=1,2,\cdots,k$, and that for each patient there is an observed follow-up time t_{ij}, $j=1,2,\cdots,n_i$, and a potential follow-up time t^*_{ij} (the maximum time the patient could have been followed if he had not died or been censored earlier).

Assuming simple exponential survival in each category the maximum likelihood estimate of the death rate λ_i is given by

$$\hat{\lambda}_i = d_i/T_i \quad (A.1)$$

where $T_i = \sum_{j=1}^{n_i} t_{ij}$. The asymptotic variance is estimated by

$$\hat{\sigma}_i^2 = \hat{var}(\hat{\lambda}_i) \cong 1/(-E\frac{d^2 \ell n L}{d\lambda^2}) = \hat{\lambda}_i^2/(n_i - \sum_{j=1}^{n_i} \exp(-\hat{\lambda}_i t^*_{ij})). \quad (A.2)$$

If censoring is absent or slight we may take

$$\hat{\sigma}_i^2 \cong \frac{\tilde{\lambda}_i^2}{d_i} = d_i/T_i^2 \quad (A.3)$$

where the approximation sign in eqs. A.2 and A.3 signifies the assumption of negligible censoring in eq. A.3 as well as the fact that the true variance depends on the unknown parameter λ_i which has been replaced by its estimate in both eqs. A.2 and A.3.

The weighted least squares estimate of the common death rate for the k categories, $\hat{\lambda}$, is obtained as

$$\hat{\lambda} = \left(\sum_{i=1}^{k} \hat{\lambda}_i/\hat{\sigma}_i^2\right)/\sum_{i=1}^{k} (1/\hat{\sigma}_i^2) \tag{A.4}$$

which for the uncensored case simplifies to

$$\hat{\lambda} = \sum_{i=1}^{k} T_i/\sum_{i=1}^{k} (T_i^2/n_i). \tag{A.5}$$

The test statistic is computed as

$$X^2 = \sum_{i=1}^{k} (\hat{\lambda}_i-\hat{\lambda})^2/\hat{\sigma}_i^2 \tag{A.6}$$

which for the case of negligible or no censoring reduces to

$$X^2 = \sum_{i=1}^{k} (1-\hat{\lambda}/\hat{\lambda}_i)^2 d_i. \tag{A.7}$$

The test statistic is asymptotically distributed as a chi-square with k-1 degrees of freedom under the null hypothesis of homogeneous death rates in the categories. (This test was proposed by Dr. Mitchell Gail).

II. ESTIMATION OF $\underset{\sim}{\theta}$ AND Σ FOR THE EXPONENTIAL MODEL

Assume that $f(t,\underset{\sim}{x},\underset{\sim}{\theta})$ for the i'th patient is given by $\lambda_i \exp(-\lambda_i t)$ where $\lambda_i = g(\underset{\sim}{x}_i,\underset{\sim}{\theta})$. From equation 5 we have

$$L = \prod_{i=1}^{n} \exp\,(z_i \ln\lambda_i - \lambda_i t_i). \tag{A.8}$$

At this point we may introduce the function $g(\underset{\sim}{x}_i,\underset{\sim}{\theta})$ for the λ_i. Three functions which have been studied are $\lambda_i = \underset{\sim}{x}_i'\underset{\sim}{\theta}$, $1/\lambda_i = \underset{\sim}{x}'_i\underset{\sim}{\theta}$, and $\ln\lambda_i = \underset{\sim}{x}_i'\underset{\sim}{\theta}$ where $\underset{\sim}{x}_i$ and $\underset{\sim}{\theta}$ are column vectors. The first of these relationships may be termed <u>additive</u> since the $\hat{\theta}_j$ estimate the additive effect on the hazard (death rate) associated with unit increases in the x_j. This model, while perhaps easiest to interpret, may occasionally be difficult to fit if negative values for any λ_i are encountered during iteration since then the likelihood becomes undefined. The second relationship may be termed for convenience the <u>reciprocal</u> model and relates $\underset{\sim}{x}_i'\underset{\sim}{\theta}$

to the expected survival time. This model, unlike the two others, belongs to the exponential family so that fitting by weighted least squares will be essentially equivalent to fitting by maximum likelihood. The third model may be termed multiplicative since here the effect of a unit change in x_j, say, multiplies the hazard by the factor $\exp(\theta_j x_j)$. For this model the likelihood is always defined.

For the sake of concreteness we will use the additive model $\lambda_i = \underset{\sim}{x}_i{}'\underset{\sim}{\theta}$ to illustrate the estimation procedure.

Taking the derivatives of $\ell n L$ with respect to the θ_j and setting them equal to 0 we obtain the k+1 non-linear equations

$$\frac{d\ell nL}{d\theta_j} = \sum_{i=1}^{n} (z_i x_{ij}/(\underset{\sim}{x}_i{}'\underset{\sim}{\theta}) - \sum_{i=1}^{n} x_{ij} t_i) = 0 . \tag{A.9}$$

These equations may be solved iteratively, for example by the vector analog of the Newton-Raphson technique, to obtain $\hat{\underset{\sim}{\theta}}$. We obtain $\hat{\sum}$ using equation 6 where

$$d^2 \ell nL/d\theta_j d\theta_\ell = - \sum_{i=1}^{n} z_i x_{ij} x_{i\ell}/(\underset{\sim}{x}_i{}'\underset{\sim}{\theta})^2 . \tag{A.10}$$

Here to obtain the expected value of A.10 we need only replace the z_i by $E(z_i) = (1 - \exp(-x_i{}'\hat{\underset{\sim}{\theta}}\, t_i^*))$ where t_i^* represents the potential follow-up time for the i'th patient. If we prefer to estimate $\sum$ conditional on the observed pattern of censoring we need not replace the z_i by $E(z_i)$.

With the additive model the random variable t_i does not appear in the second partial derivative of the log likelihood. If t_i had appeared we would have had to take the expectations with respect to t_i as well. Evaluation of this expectation requires care because of the nature of the likelihood function which under censoring is composed of mixed continuous (truncated) and discrete (point-mass) distributions. In general for a censored likelihood as given in equation 5

$$E(t_i) = \int_0^{t_i^*} t\left[f(t,\underset{\sim}{x}_i,\underset{\sim}{\theta})\right] dt + t_i^* \ (1-F(t_i,\underset{\sim}{x}_i,\underset{\sim}{\theta})). \tag{A.11}$$

(I am grateful to Max Myers for clarifying this relationship to me.)

Had we used the reciprocal or multiplicative models then estimation of $\sum$ would have required taking expectation with respect to t. For some models the expectation with respect to t may not be readily evaluable because of difficulty in performing the integrations. This is true, for example, of Weibull models where the expectation involves an integral which does not exist in closed form.

In such instances the data analyst must be satisfied with using the observed values of the second partial derivatives of the log likelihood rather than their expectations in estimating $\sum$.

III. LIKELIHOOD RATIO TESTS FOR TREATMENT-COVARIATE INTERACTIONS

We assume that k covariates have been selected and two treatments are under study. Let $\theta_0=\alpha$ and let $\underset{\sim}{\beta}$ represent the vector of remaining θ_j, j=1,2,···,k. A sample likelihood will be written as $L(\hat{\alpha},\hat{\underset{\sim}{\beta}})$ where absence of subscripts for $\hat{\alpha}$ and $\hat{\underset{\sim}{\beta}}$ denotes fitting for both treatments combined and subscripts denote separate fitting for the individual treatments. The first test is

$$T_1 = L(\hat{\alpha},\underset{\sim}{\beta})/L(\hat{\alpha}_1,\hat{\alpha}_2,\hat{\underset{\sim}{\beta}}) \qquad (A.12)$$

and asymptotically $-2\ln T_1$ is distributed as chi-square with 1 degree of freedom, testing whether there is an overall best treatment when adjusting for covariates.

The test for treatment-covariate interactions is given by

$$T_2 = L(\hat{\alpha}_1,\hat{\alpha}_2,\hat{\underset{\sim}{\beta}})/L(\hat{\alpha}_1,\hat{\alpha}_2,\hat{\underset{\sim}{\beta}}_1,\hat{\underset{\sim}{\beta}}_2) \qquad (A.13)$$

where $-2\ln T_2$ is asymptotically distributed as chi-square with k degrees of freedom. This test determines whether the data are better fit with separate regression coefficients for the covariates for each treatment in addition to separate intercepts.

IV. A RANGE TEST FOR DETECTING SIGNIFICANT DIFFERENCES IN THE DEATH RATES (HAZARDS) FOR SPECIFIC COVARIATE VECTORS

Assume we have estimated $\underset{\sim}{\theta}$ separately for each of 2 treatments, denoting the estimates by subscripts, $\hat{\underset{\sim}{\theta}}_1$ and $\hat{\underset{\sim}{\theta}}_2$. Assuming the additive exponential model, where $\lambda_i = \underset{\sim}{x}_i'\underset{\sim}{\theta}$ represents the hazard (death rate) for patients with covariate vector $\underset{\sim}{x}_i$, we wish to test whether the death rate is significantly different for the two separate estimates of $\underset{\sim}{\theta}$. Treating $\underset{\sim}{x}_i$ as a vector of constants, our two estimates of the death rate $\hat{\lambda}_{1i}$ and $\hat{\lambda}_{2i}$ are asymptotically normal because of the asymptotic multivariate normality of $\hat{\underset{\sim}{\theta}}_1$ and $\hat{\underset{\sim}{\theta}}_2$. Under the null hypothesis of no treatment covariate interactions, $\hat{\lambda}_{1i}$ and $\hat{\lambda}_{2i}$ may be taken to represent random observations from a normal distribution with mean $\underset{\sim}{x}_i'\underset{\sim}{\theta}$ and variance $\underset{\sim}{x}_i'(2\hat{\sum})\underset{\sim}{x}_i$ where absence of subscripts for $\hat{\theta}$ and $\hat{\sum}$ indicates estimation for the 2 treatments pooled together. We compute

$$T_3 = (\underset{\sim}{x}_i'\hat{\underset{\sim}{\theta}}_1 - \underset{\sim}{x}_i'\hat{\underset{\sim}{\theta}}_2)/(\underset{\sim}{x}_i'(2\hat{\sum})\underset{\sim}{x}_i)^{\frac{1}{2}}. \qquad (A.14)$$

If T_3 exceeds the critical value for the studentized range for a sample of size 2, we conclude that the difference between the two estimated death rates is too large to be assigned to random variation alone. If there are $m > 2$ treatments we use $\max(\underset{\sim}{x}_i'\hat{\underset{\sim}{\theta}}_j) - \min(\underset{\sim}{x}_i'\hat{\underset{\sim}{\theta}}_j)$, $j=1,2,\cdots,m$, in the numerator of T_3 and $(\underset{\sim}{x}_i'(m\hat{\Sigma})\underset{\sim}{x}_i)^{\frac{1}{2}}$ in the denominator to adjust the variance estimate for the sample size used in obtaining the $\hat{\underset{\sim}{\theta}}_j$. In this case T_3 is referred to critical values for samples of size m. This test is only approximate but has been useful in identifying the $\underset{\sim}{x}_i$ associated with the most important treatment-covariate interactions.

REFERENCES

Armitage, P. and Gehan, E.A. (1974). Statistical methods for the identification and use of prognostic factors. Int. J. Cancer 13, 16-36.

Breslow, N. (1974). Covariance analysis of censored survival data. Biometrics 30, 89-99.

Brown, C.C. (1975). On the use of indicator variables for studying the time-dependence of parameters in a response-time model. Biometrics 31, 863-872.

Byar, D.P. and Corle, D.K. (1976). Selecting optimum treatment in clinical trials using covariate information. (To appear in J. Chron. Dis.)

Byar, D.P., Huse, R., Bailar III, J.C. and Veterans Administration Cooperative Urological Research Group (1974). An exponential model relating censored survival data and concomitant information for prostatic cancer patients. J. Natl. Cancer Inst. 52, 321-326.

Byar, D.P. and Mantel, N. (1975). Some interrelationships among the regression coefficient estimates arising in a class of models appropriate to response-time data. Biometrics 31, 943-947.

Cox, D.R. (1962). Further results on tests of separate families of hypotheses. J. Royal Stat. Soc., Series B 24, 406-424.

Cox, D.R. (1972). Regression models and life-tables. J. Royal Stat. Soc., Series B 34, 187-220.

Feigl, P. and Zelen, M. (1965). Estimation of exponential survival probabilities with concomitant information. Biometrics 21, 826-838.

Gehan, E.A. (1969). Estimating survival functions from the life table. J. Chron. Dis. 21, 629-644.

Gehan, E.A. (1975). Statistical methods for survival time studies. Cancer Therapy: Prognostic Factors and Criteria of Response, ed. M.J. Staquet, (Raven Press, New York), 7-35.

Greenberg, R.A., Bayard, S. and Byar, D.P. (1974). Selecting concomitant variables using a likelihood ratio step-down procedure and a method of testing goodness of fit in an exponential survival model. Biometrics 30, 601-608.

Krall, J.M., Uthoff, V.A. and Harley, J.B. (1975). A step-up procedure for selecting variables associated with survival. Biometrics 31, 49-57.

Lindsey, J.K. (1974). Construction and comparison of statistical models. J. Royal Stat. Soc., Series B 36, 418-425.

Mantel, N. (1966). Evaluation of survival data and two new rank order statistics arising in its consideration. Cancer Chemother. Rep. 50, 163-170.

Metcalf, W. (1974). Analysis of cancer survival as an exponential phenomenon. Surg. Gynecol. Obstet. 138, 731-740.

Myers, M.H., Hankey, B.F. and Mantel, N. (1973). A logistic-exponential model for use with response-time data involving regressor variables. Biometrics 29, 257-269.

Prentice, R.L. (1974). A log gamma model and its maximum likelihood estimation. Biometrika 61, 539-544.

Prentice, R.L. (1973). Exponential survivals with censoring and explanatory variables. Biometrika 60, 279-288.

Zelen, M. (1975). Importance of prognostic factors in planning therapeutic trials. Cancer Therapy: Prognostic Factors and Criteria of Response, ed. M.J. Staquet, (Raven Press, New York), 1-6.

Recent Developments in Statistics
J.R.Barra et al., editors

Invited Paper

METHODS OF ANALYSIS OF SURVIVAL DATA IN ANIMAL EXPERIMENTS WHICH LEAD TO AN UNDERSTANDING OF THE DATA AS WELL AS AVOIDING BIASSED CONCLUSIONS

Peter N. Lee
Tobacco Research Council,
Glen House, Stag Place,
London SW1E 5AG
(Great Britain)

A commonly occurring experimental situation in carcinogenesis is one in which animals are given repeated treatments until they either develop a tumour or die tumourless. To compare the tumour producing effect of different treatments adequately it is necessary to take into account possible differences in non-tumour mortality. A desirable method of analysis should be unbiassed, statistically efficient and allow a simple dose-response relationship. To achieve these properties a mathematical model for time-to-tumour is needed and for such a model to fit the data adequately it should be based on biological considerations. Evidence from human data suggests that some cancers might arise from a multistage process. Under certain simplifying assumptions a multistage process predicts a Weibull distribution for time-to-tumour, a distribution dependent on three parameters, one only of which, b, is likely to depend on treatment given.

Methods for estimation of the parameters of the Weibull distribution are discussed, and evidence given of its superiority over the log-normal distribution in fitting data from a large scale mouse skin painting experiment. Under certain circumstances the parameter b should vary according to a simple power law of dose and data using the pure carcinogen benzpyrene is presented in which not only is time-to-tumour Weibull distributed but b is almost exactly proportional to dose squared. Data in which smoke condensate is applied also exhibits these properties but only over a limited range of dose levels. A method of estimating the value and significance of between-treatment contrasts is discussed, analogous to multiple regression methods in concept and application. The effect of age of the animal and of cessation of treatment at a given time is also discussed and experimental data is shown which conforms with the multistage hypotheses.

It is concluded that the use of Weibull distributions based on multistage models can give a good fit to observed data, insight into the mechanisms involved and allow an adequate representation of the tumorigenic effect of a particular treatment by a single parameter.

TYPE OF EXPERIMENT CONSIDERED

In this paper we shall limit ourselves to the discussion of one particular type of experimental situation. This is where a number of groups of animals receive different treatments and, for each animal within each group, we observe: either the time at which the animal develops a tumour of a given type or the time at which the animal dies without such a tumour.

A typical experiment might be one in which the backs of mice in a particular treatment group are painted regularly with a particular dose level of a given tobacco smoke condensate and the time to skin tumour of a given size is observed. It is also possible to include in our definition experiments on internal tumours discoverable only at necropsy provided that the tumour is very fast growing or causes death rapidly (or both). Time to death can then be taken as time to tumour with little loss of accuracy.

PROPERTIES OF A GOOD STATISTICAL METHOD OF ANALYSIS

For a method of statistical analysis of such data to be of the greatest practical use it should have a number of properties.

Firstly, it should allow the tumour experience of the different groups to be compared in such a way that differences in non-tumour mortality between the groups do not bias the comparison. A measure of tumorigenicity not likely to be satisfactory for two reasons is the percentage of animals getting tumours by a given point of time. Firstly, a treatment resulting in half the original animals getting tumours, with the remaining half dying early due to treatment toxicity, is clearly very much more tumorigenic than a second treatment which results in half of the animals having tumours, but with the remaining half still alive and tumourless, despite the two groups having equal percentages of animals with tumours. Secondly, a percentage measure will fail to discriminate in circumstances where 100% of the animals in both groups being compared have tumours.

Apart from being unbiassed, a second feature of our method of analysis is that it should be efficient, making use of all the available data, and not just data up to a given time point or in a given time period.

It would also be helpful for our analysis to be able to quantify the tumour experience in a group in such a way that, at least under favourable circumstances, it should allow a simple dose-response relationship.

The most logical way to proceed towards developing an analysis with the above properties appears to be to fit to the observed distribution of time-to-tumour in the groups a mathematical expression which depends on a limited number of parameters. Assuming this expression can be shown to fit the data adequately, it would then be valid to compare the treatment groups by making the appropriate tests on the fitted values of these parameters in the individual groups.

However, to see the situation simply as fitting a mathematical expression to the data is to ignore biological considerations completely. One is far more likely to be able to find a suitable mathematical expression if it is a reasonable biological model of the situation.

THE WEIBULL DISTRIBUTION; DERIVATION AND DEFINITION

Armitage and Doll (1954) observed that in humans the age-specific incidence rate of many types of cancer is proportional to a power of age (or time from first exposure) and showed that this result would be expected under a multistage model. This model makes the following simplifying assumptions:

(i) that there are a large and constant number (N) of cells at risk,

(ii) that all these cells start in an identical state,

(iii) that at least one of them has to go through a fixed number (k) of stages or transformations before a tumour appears,

(iv) that any cell in a given stage has the same very small, but constant, probability per unit time of commencing the transformation into the next stage (kinetic rate-constants b_1, $b_2 \ldots\ldots b_k$) and

(v) that each transformation takes a constant time (w_1, $w_2 \ldots\ldots w_k$).

Under these assumptions it can be shown that the age-specific incidence rate I(t) at time t (often called the hazard function) among a homogeneous population will be given by

$$I(t) = bk\ (t-w)^{k-1}$$

where k is the fixed number of stages, w is the sum of the times of the constant transformations and b is proportional to the product of the kinetic rate-constants for each stage and the number of cells at risk. This is a particular case of the Weibull distribution with cumulative density function

$$G(t) = 1 - \exp(-b(t-w)^k)$$

Though this model is clearly an over-simplification it may be expected that, if cancer mechanisms in animals and man are of this type, incidence rate data from laboratory animals, which are often inbred and presumably therefore more homogeneous, may follow a Weibull distribution even more clearly than for humans.

We assume that, in the absence of treatment, the kinetic rate-constant for each stage is non-zero (untreated control mice do get skin tumours occasionally) and that the effect of continuous application of a carcinogen is to alter at least one of these rate-constants so that the parameter b, and hence the incidence rate, will be increased. The parameters k and w will not be altered.

Thus if the model is applicable to our animal situation, one should be able to compare the results in different experimental groups by considering the value of b in each group as an index of carcinogenicity, having fitted a common k and w.

A METHOD FOR FITTING THE WEIBULL DISTRIBUTION

Peto and Lee (1973) have described a maximum likelihood method for estimating these parameters based on earlier work by Pike (1966). The contribution to the log-likelihood of an animal in group i dying tumourless at time t_1 is given by

$$\log_e(1-G(t_1)) = -b_i(t_1-w)^k$$

and that of an animal dying with a tumour at time t_2 is

$$\log_e(dG(t_2)) = \log_e b_i + \log_e k + (k-1)\log_e(t_2-w) - b_i(t_2-w)^k$$

Maximization of the log-likelihood function with respect to b_i, k and w simultaneously is not straightforward but can be achieved satisfactorily by a modified Newton-Raphson iterative procedure.

In practice, however, if one is interested mainly in between-treatment comparison, i.e. in the relative magnitude of the b_i's, then it is reasonably satisfactory, and much simpler computationally, to fix k and w values from previous experience and compute the b_i by the formula

$$b_i = s_i/v_i$$

where s_i is the number of animals bearing tumour in the ith group and $v_i = \Sigma(t-w)^k$, the summation being over both the times of tumour and of death without tumour. Pike (1966) has demonstrated that the ratio of b's between different groups is virtually independent of the actual values of k and w chosen provided the k,w pair is not too far from the best fitted values.

COMPARISON WITH THE LOG-NORMAL DISTRIBUTION

Paige, in Day (1967), when analysing data from the largest mouse-skin painting experiment ever carried out at the time, assumed that the incidence rate followed a log-normal distribution with time. Subsequent analysis by Peto, Lee and Paige (1972) showed that the Weibull distributions with a k,w pair common to all groups proved a significantly better fit to the data than did the log-normal.

DOSE-RESPONSE RELATIONSHIPS

In the multistage hypothesis, if a carcinogen has an effect directly proportional to dose on each of c kinetic rate-constants and if the dose applied is sufficiently large for the background rate-constants to be neglected for those c stages of the cancer process, it follows that the age specific tumour incidence rate will be proportional to dose to the power of c.

Lee and O'Neill (1971) analysed an experiment in which the strong pure chemical carcinogen 3,4-benzpyrene was painted continuously at 6, 12, 24 and 48 micrograms per week on four groups of 300 mice. They found that not only could skin tumour incidence rates be well described by a Weibull distribution with k and w common to all four groups, but that b was proportional to dose squared. This is demonstrated graphically in Figure 1 using the fact that

$$G(t) = 1 - \exp(-b_i(t-w)^k)$$

implies

$$\log_e \log_e (1/(1-G(t))) = \log_e b + k \log_e (t-w)$$

and estimating G(t) non-parametrically by the actuarially simulated number of tumours with a zero mortality standard population (Lee, 1970). As can be seen, the plots of $\log_e \log_e (1/(1-G(t))$ against $\log_e (t-17.70)$ form approximately parallel equidistant straight lines. The slope of the lines, 2.95, estimates k and is not significantly different from an integer value as suggested by the model. The average distance between the lines in a vertical direction is almost exactly twice the logarithm of the ratio of successive doses implying a value of c of 2.

The results are therefore consistent with a multistage hypothesis in which benzpyrene affects two out of three of the stages of the process.

Davies et al (1974) have described another dose-response study in which two cigarette smoke condensates dissolved in either of two solvents were each tested at seven equal intervals of log dose; 65, 84, 108, 139, 180, 232 and 300 mg/week. Here also the pattern of tumour incidence rate with time followed a Weibull distribution with a k,w pair common to all 28 groups, but the relationship of b to dose was far more erratic (Table 1). Between 84 and 108 mg the response did increase approximately proportionately to the second power of dose. However above that dose there was a marked flattening off in response. One possible explanation of this phenomenon is that the toxicity of the condensate at high dose levels reduce the number of cells at risk contrary to the hypothesis underlying the Weibull distribution. A greater number of arimals did in fact show areas of epidermal cell necrosis at the higher dose levels. There was no obvious reason why the incidence at 65 mg should have been higher than that at 84 mg.

So far the effect of one carcinogen has been considered but the model extends easily to the multi-carcinogen situation. If two carcinogens affect the same stage of the cancer process, their joint response, as measured by the parameter b,

FIGURE 1

Fit of Weibull distribution to BP data of Lee and O'Neill (1971).

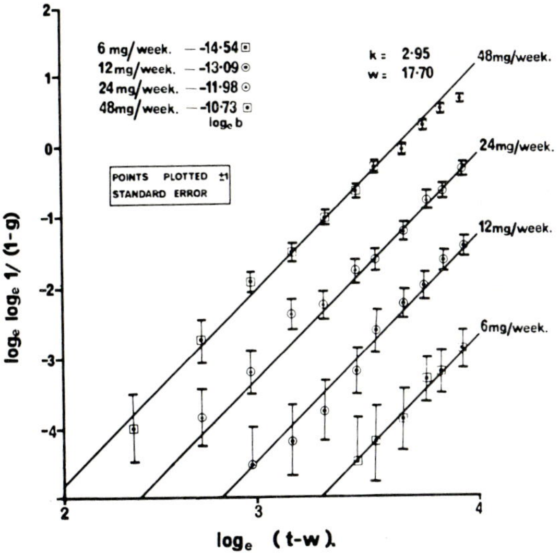

Table 1

RELATIONSHIP OF b TO DOSE IN SMOKE CONDENSATE DATA OF DAVIES ET AL (1974)

Dose level (mg/week)	Number of mice bearing tumours (out of 204)	Relative b values*
65	35	1.00
84	29	0.81
108	45	1.48
139	70	2.54
180	103	6.78
232	105	7.63
300	119	10.69

* averaged over results from both solvents and both condensates.

will be an additive function of the two doses, whereas if they each strongly affect different stages their joint action will be a multiplicative function. If one or both carcinogens affect more than one stage then the joint dose-response, though more complex, can be predicted similarly.

TREATMENT EFFECTS; ESTIMATION AND SIGNIFICANCE TESTING

If k and w are known or have been estimated from the data then the log-likelihood for an r group experiment is given by

$$L = \sum_{i=1}^{r} s_i \log_e b_i - \sum_{i=1}^{r} b_i v_i$$

Suppose that we wish the parameter b for each group to depend on certain explanatory variables (dose, carcinogen, method of application, etc.) $x_0, x_1, \ldots\ldots x_p$, where x_{iu} is the value of the uth variable in the ith group. Then following Cox (1972) we relate b_i to the x_{iu} by the expression:

$$\log_e b_i = \sum_{u=0}^{p} c_u x_{iu} \qquad \text{or}$$

$$b_i = \exp \sum_{u=0}^{p} c_u x_{iu}$$

where the c_u are coefficients to be estimated. The log-likelihood function then becomes:

$$L = \sum_{i=1}^{r} s_i \sum_{u=0}^{p} c_u x_{iu} - \sum_{i=1}^{r} v_i b_i$$

As Peto and Lee (1973) showed, the maximization of L with respect to the c_i by a Newton-Raphson search converges rapidly.

In standard multiple regression techniques to compare two hypotheses, in one of which the regressor variables form a subset of the regressor variables in the other, one tests differences in sums of squares for the two maximum likelihood solutions. To compare two such hypotheses with Weibull methods, one compares the two maximum log-likelihoods, taking twice the difference as being distributed approximately as chi-squared with degrees of freedom equal to the difference in the number of regressor variables provided the total number of tumour bearing animals in the experiment is large. All the standard techniques of multiple regression have their analogues and Peto and Lee give examples in their paper.

OTHER INFERENCES FROM MULTISTAGE HYPOTHESES

a) Effect of age and treatment duration

The fourth assumption of the multistage hypotheses underlying the Weibull distribution implies that, if treatment is continuous, the kinetic rate-constants for each stage do not depend on the age of the animal. It follows that, provided the carcinogen affects the first stage of the process strongly enough for it to be a reasonable approximation to assume all observed tumours to have arisen because of this, the age-specific incidence rates will depend wholly on duration of treatment and not at all on age *per se*.

In a large experiment carried out by Peto et al (1975) 3,4-benzpyrene was applied to the skin of mice in four groups of increasing size starting at 10, 25, 40 and 55 weeks old respectively. As can be seen from Figure 2, the percentage of tumourless mice when plotted against age differs markedly between the four groups. However, when plotted against treatment duration the four groups are virtually identical, as expected from multistage assumptions. The data from this experiment is also summarised in Table 2.

FIGURE 2

Percentages of tumourless mice against (a) age, (b) duration of exposure to benzo(a)pyrene; from Peto et al (1975).

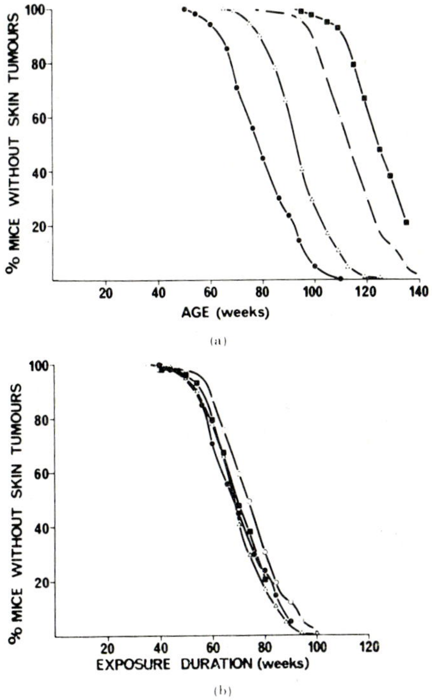

KEY

● = GROUP 1
Δ = GROUP 2
O = GROUP 3
■ = GROUP 4

Table 2

RELATIONSHIP OF CROP OF 10MM EPITHELIAL TUMOURS TO REPEATED APPLICATIONS OF BENZO(A)PYRENE TO THE SKIN; FROM PETO ET AL (1975)

	Group 1	Group 2	Group 3	Group 4
Age (weeks) at start of regular BP	10	25	40	55
No. of animals allocated randomly (at age 5 weeks) to the four groups	140	170	220	420
No. which developed a 10 mm epithelial tumour	110	128	119	134
Life-table estimate of the duration (in weeks) of BP treatment by which, in the absence of other causes of death,				
(a) 1% of mice	44	46	44	44
(b) 33% of mice	64	64	68	66
(c) 67% of mice	74	74	78	78
would have got a 10 mm epithelial tumour.				
(The mice were examined fortnightly)				

Having shown that the relationship in the four groups between tumour incidence and duration is not significantly different, Peto et al combined the results in the four groups to illustrate the overall fit to the Weibull distribution. This is shown in Figure 3 and it is obvious that over the 100-fold range from 0.25% per fortnight up to the massive incidence rate of 25% per fortnight observed after 90 weeks of regular BP administration, the points do approximately fit a straight line. As for Lee and O'Neill's (1971) data the best fitting whole number to the slope is three.

b) Effect of stopping painting

Continuous carcinogenesis experiments are able to show how many stages of the cancer process a carcinogen affects, but not which stages. The position is different when the carcinogens are applied for a period and then stopped. If only early stages of the process have been affected by the carcinogen, little immediate change in incidence rate rise may be seen as tumours appearing soon after treatment is stopped will tend to come from those cells which have already passed through these stages during the period of treatment. On the other hand, if the final stage of the process is affected, the drop off in rate will become apparent very rapidly.

If one assumes that the effect of stopping treatment is simply to cause the kinetic rate-constants that are affected by the carcinogen to revert to the background values, it is possible to work out the expected form of the relationship between incidence rate and age after stopping. For example, in a multistage hypothesis with 3 stages where treatment, causing kinetic rate-constants α_1, α_2 and α_3, is applied up to time S and then stopped, causing reversion to background kinetic rate-constants β_1, β_2 and β_3, the incidence rate at time T ($>S+w$) is given by the expression:

FIGURE 3

Incidence rates of 10mm epithelial tumours at successive fortnightly chartings against duration of BP application, on a log/log scale from 28 weeks. The points are statistically independent and 90% confidence intervals are indicated; from Peto et al (1975).

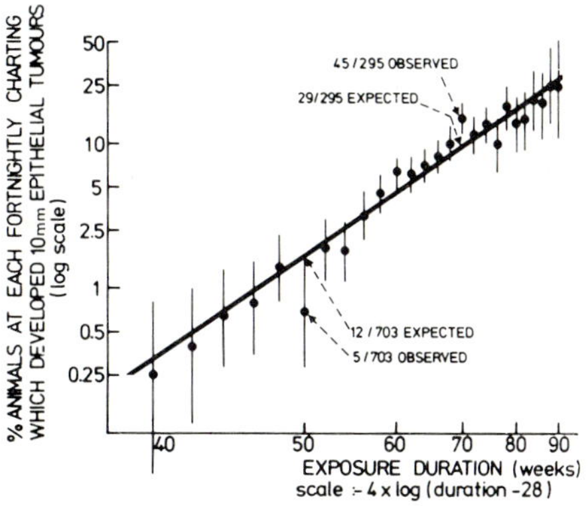

$$I = N\left[\frac{\alpha_1\alpha_2\beta_3S^2}{2} + \alpha_1\beta_2\beta_3S(T-S-w) + \frac{\beta_1\beta_2\beta_3(T-S-w)^2}{2}\right]$$

By comparing maximum likelihood solutions for the various cases where different α_j are fixed to be equal to the β_j it is possible theoretically to test which stages the carcinogen affects. The Tobacco Research Council (1975) describe an experiment in which mice were painted with whole smoke condensate, a chemical sub-fraction of it or with benzpyrene for various periods of time and then treatment was stopped after varying periods of time. For all three treatments a fairly good agreement was obtained between the observed incidence rate pattern and that predicted by the formula, provided one assumed that at least two stages of the cancer process, one early and one late, were affected.

ADEQUACY OF THE MODEL

The experiments we have described have all agreed well with the predictions of the multistage model and it is perhaps not unreasonable to infer that its assumptions approximate reality at least over the range of responses that have been observed. Although the model is not very specific, inasmuch as it does not show what the actual stages are in biological terms, it not only provides insight into the mechanisms involved but also suggests a quantitative method of comparing the mouse-skin tumorigenic effect of different smoke condensates by a single parameter.

REFERENCES

Armitage, P. and Doll, R., (1954), Br.J.Cancer, 8, 1.
Cox, D.R., (1972), J.Roy.Statist.Soc. Ser.B, 34, 187.
Davies, R.F., Lee, P.N. and Rothwell, K. (1974), Br.J.Cancer, 30, 146.
Day, T.D., (1967), Br.J.Cancer, 21, 56.
Lee, P.N., (1970), Biometrics, 26, 777.
Lee, P.N. and O'Neill, J.A., (1971), Br.J.Cancer, 25, 759.
Peto. R. and Lee, P.N., (1973), Biometrics, 29, 457.
Peto, R., Lee, P.N. and Paige, W.S., (1972), Br.J.Cancer, 26, 258.
Peto, R., Roe, F.J.C., Lee, P.N., Levy, L. and Clack, J., (1975), Br.J.Cancer, 32, 411.
Pike, M.C., (1966), Biometrics, 22, 142.
Tobacco Research Council, (1975), Review of Activities, 1970-74, London, 28.

Recent Developments in Statistics
J.R.Barra et al., editors

Invited Paper

INTELLIGENT PROGRAMS, THE NEXT STAGE IN STATISTICAL COMPUTING

J.A. Nelder
Rothamsted Experimental Station
Harpenden, Herts, England

Current computer programs for statistical computing are largely unintelligent in that they do not use the data to check if the assumptions underlying the statistical procedure being used are approximately satisfied. Regression analysis is used to illustrate what procedures exist to check various possible kinds of distortion caused by faults both in the data and in the model, and the effects that can arise from the interaction between different faults occurring together. Some general alternatives for the implementation of intelligent programs are briefly discussed.

0. INTRODUCTION - THE PRESENT

Present-day statistical programs are primarily concerned to make available to users at various levels of sophistication the standard procedures of statistics, such as regression or principal component analysis, in a way that makes it easy for users to approach the computer. To achieve this has meant developing and implementing problem-oriented languages (GENSTAT, SAS, SPSS, etc.) that make specification of problems easier and hide from the user the details of internal storage allocation etc., details vital for a working program but tending to obscure the statistics.

Current programs are largely unintelligent, so that, for example, a regression procedure will accept any data set of the right shape (a set of variates all of the same length), and having elements of the right mode (reals). Upon that data set it will perform certain calculations and will print out the results, such as regression coefficients, standard errors, and P-values for test statistics. These results are intelligible if the model underlying the analysis is appropriate to the data. If (as happened to me recently) a number 2.000000, alleged to be on a paper tape containing some data, is read into the computer as 2000000, then the values for quantities such as standard errors will be wrong. Most current programs will not give the user any explicit warning that anything is wrong, so the uncritical user will not know if the results are wrong or not. An unintelligent program does what it is told and does not answer back.

The amount of uncritical use of standard procedures is enormous and, in my view, is tending to bring the subject of statistics into disrepute. This paper will discuss what we might do to improve matters, by using the results of recent work on the detection of false assumptions about data and models to begin the construction of less unintelligent programs. The subject is a vast one, and to sharpen the focus we shall consider regression analysis, i.e. involving data on continuous quantities for which the classical least-squares model with a linear systematic component and independent equal-variance errors is thought to be appropriate. The two main questions are 'What could we do?' and 'How should we do it?', i.e. what statistical tools are available, and how we should present them to the user.

1. THE FUTURE- WHAT COULD WE DO?

Ideally a regression analysis should be protected against the following sources of distortion:

(i) Faults in the data (gross errors, by mismeasurement, faults in transcription, etc.)

(ii) Faults in the model - systematic part.

- (a) Wrong choice of x-variates.
- (b) Wrong scale for x-variates (not producing linearity).
- (c) Wrong scale for y-variate (not producing linearity).

(iii) Faults in the model - random part.

- (a) Wrong scale for y-variate (not producing right variance/mean relation).
- (b) Wrong shape of distribution for errors (e.g. long tails).
- (c) Lack of independence of errors (e.g. auto-correlation).

We consider these sources in turn, indicating techniques that have been developed that seem likely to be specially useful in detecting each source of distortion.

1.1. Faults in the data

We are concerned with faulty data items called variously outliers, rogue observations, gross errors etc. Their essential characteristic is that they are wrong, e.g. result from a misreading of a scale, a faulty recording instrument, or an error in transcription (e.g. 415 for 145). With data sets of any appreciable size it is most unusual for there to be no faulty items, and survey analysts habitually regard the validation (or data-cleaning) stage as a most necessary part of the process of analysis. Such attitudes seem much less strong when it comes to dealing with experimental results, but validation is still important (one fault in 200 data can easily lead to a quite false inference about, say, the equality of two slopes).

Much useful validation can be achieved with very crude tools, e.g. the inspection of the minimum, mean and maximum of each variate. Prior information can often set limits on the ranges. More complex tests can involve impossible combinations of variate values (the case of the male teenaged widow). Note that such tests are difficult to automate because of the dependence on external knowledge outside the data values themselves. There is another kind of test which depends upon the internal consistency of the data; such consistency has to be expressed in statistical terms, and involves assumptions about the form of the relevant model. In regression analysis the model predicts a set of fitted values $\hat{\underset{\sim}{Y}}$ for the data $\underset{\sim}{y}$, say; tests for outliers consider the set of residuals $\underset{\sim}{y} - \hat{\underset{\sim}{Y}}$, and in particular the size of the largest in absolute value in relation to the rest. The problem, of course, is that the fitted values depend on parameters that must be estimated from the data themselves, and hence may be distorted by the 'bad' points. As an example of this consider the following set of artificial data with x = 1, 2...12 and $y = x^2$ except that for the last point the x and y values have been transposed so that x = 144 and y = 12. If we now fit a quadratic to these data we find that the residuals vary from 16.0 and -10.9 at one extreme of absolute size to .899 and -.050 at the other, with the smallest absolute residual, by an order of magnitude, belonging to the bad point! The pattern of residuals taken as a whole is peculiar but any statistic based on the largest would clearly be wholly misleading.

1.1.1 Robust methods of estimation

It is to cope with this kind of problem that robust methods of estimation of the parameters, and hence fitted values have been developed, mainly by Tukey and others of the Princeton school. (See, e.g. Andrews et al (1972) and Andrews (1974).)

In effect the methods weight the points in a fit by a function depending on the current estimate of the residual, so that points with larger residuals receive less weight. The process of fitting is necessarily iterative, and does not present any great practical difficulty given a program that can do iterative weighted regression. The main interpretive difficulty arises from the uncertainty about the model, both in its systematic parts (e.g. assuming linearity when it is false) and in its random part (e.g. assuming the variance independent of the mean when it is not), and the interaction between this uncertainty and the possible presence of data faults. We return to this point later. Another, practical, difficulty concerns the starting values for the iterative fit. The example discussed above shows that the least-squares solution may be quite unsatisfactory in providing a starting configuration. Andrews (1974) has proposed an alternative based on medians; there is more work to be done here in balancing ease of computation against effectiveness.

To allow suitable flexibility in using robust estimation technique any regression program must allow iterative fitting with weights that can be functions of the data and the current set of fitted values.

1.2 Faults in the model - systematic part

1.2.1 Wrong choice of x-variates

Of the three types of fault listed in Section 1 under this heading, the first, wrong choice of x-variates, is at once the most fundamental and the least tractable for the designer of intelligent programs, for, clearly, the omission of an important x-variate by the user will often be undetectable using the internal evidence of the data that are provided. If the omitted variate has values in the sample well distributed over its range and is uncorrelated with the x-variates that are included, then its effect on the variation of the y-variate is unlikely to be distinguishable from random error. Only if its values are strongly clumped and its regression coefficient appreciable, may the sub-populations so defined be detectable. The alternative models are then essentially of the mixture kind where the components have to be estimated from the data. The problem here is that, to put it crudely, mixture models can explain anything; the chance of some kind of apparent clumping appearing by accident in sampling homogeneous populations is quite high, and with multivariate data there are virtually certain to be planes which divide the data most convincingly into clusters (Day (1969)). Perhaps the most useful action that an intelligent program could take would be to discourage the use of regression procedures on large data sets (which are almost certain to be heterogeneous), especially when possible classifying variables exist but have been ignored.

1.2.2 Wrong scale for x-variates

Techniques are well-developed for a single x-variate. Box and Tidwell (1965) describe how a power θ may be found from the data when the regression is linear on x^{θ} rather than x. Residuals plotted against the x-values will give qualitative information on curvature, indicating if a $\theta > 1$ or < 1 is required. or perhaps a more complicated transformation. More information is required about the behaviour of these techniques when many x-variates are involved, and particularly about the interaction of transforms of individual x-variables and the presence or absence of interaction terms involving 2 or more xs. For we can generalise the linear relation

$$Y = \beta_1 x_1 + \ldots + \beta_k x_k$$

either in the direction

$$Y = \beta_1 f_1(x_1) + \ldots + \beta_k f_k(x_k) \tag{1}$$

or in the direction

$$Y = \beta_1(x_2, \ldots, x_k)\, x_1 + \ldots \beta_k(x_1, \ldots, x_{k-1}) x_k \tag{2}$$

If, as a first approximation, the βs are taken as linear functions of the remaining xs, this is equivalent to including some of the product terms $x_i x_j$ in the regression. Both types of extension, exemplified by equations (1) and (2), could be investigated by ranking the points by values of one of the xs, dividing them accordingly into two or more groups, and testing for homogeneity of regression within the groups.

1.2.3 Wrong scale for y-variate

This section is concerned with the possibility that the relationship of the expected value of y is not linear in the xs, i.e. so that although there exists a linear predictor

$$\eta = \Sigma \beta_i x_i$$

nevertheless $E(y) = \mu \neq \eta$. If, however, there exists a link function (Nelder and Wedderburn (1972))

$$\mu = f(\eta)$$

for some function f(.), not the identity function, then the model becomes one of the class of generalised linear models and may be fitted by the method described by Nelder and Wedderburn. The program GLIM (Nelder (1973)) allows a limited set of link functions with Normal errors, and so allows some exploration of the effect of changing the assumption about the link. Note that changing the link is logically distinct from the use of a transformation of the data, where it is assumed that

$$E(g(y)) = \eta = \Sigma \beta_i x_i$$

This will be further discussed in Section 1.3.1. Techniques for detecting removable non-additivity, i.e. the need for a non-identity link, in 2-way tables and other orthogonal structures have been described by Anscombe and Tukey (1963). Again more general techniques are still needed, but meanwhile empirical tests involving alternative link functions can be informative.

1.3 Faults in the model - random part

1.3.1 Wrong scale for y-variate

The standard regression model assumes that the variance of the errors is constant, and so, in particular, independent of the mean. We can generalise this assumption by thinking in terms of models where the variance V is some function of the mean (possibly involving an unknown scale factor). Thus with gamma errors we have $V \propto \mu^2$. By the introduction of quasi-likelihood Wedderburn (1974) extended the models based on errors belonging to an exponential family to those with an arbitrary variance-mean relationship. A non-constant variance function can be dealt with either by the introduction of a weight function in the fitting (which necessarily becomes iterative) or by a transformation of the ys to give an approximately equal variance on the transformed scale. Thus for $V \propto \mu^2$ we have the ln transform producing equal variance to a first-order approximation. There is, of course, no guarantee that the transformation of the data which produces the required structure of constant variance will also produce an identity link function (Section 1.2.3). It is the great advantage of the approach via generalised linear models (Nelder and Wedderburn (1972)) that two quite separate transformations can be defined, one on the data to produce the right variance structure and one on the linear predictor to produce additivity. The program GLIM is designed to allow a variety of combinations of these two transformations to be applied in the fitting of models to data, and so give the user the opportunity to explore alternatives to the standard model.

1.3.2 Wrong shape of error distribution

We are concerned here with problems other than those expressible by variance/mean relationships, e.g. the existence of long tails where no transformation will produce approximate Normality. The use of a weight function is still possible, but its form depends on having sufficient data to establish the shape of the distribution. Establishing the shape of such a distribution is greatly complicated by artefacts produced (a) by faults in the systematic part of the model and (b) by faults in the data. Both of these can mimic the effect of long tails. See Anscombe and Tukey (1963) for a discussion relating mainly to two-way tables.

1.3.3 Non-independent errors

Regression procedures are frequently applied to data in the form of time-series, when there are a priori reasons to expect some auto-correlations in the error structure. The Durbin-Watson statistic was developed to test for such autocorrelations. Other methods include the use of lagged variables on the right-hand side of the equation. More generally the class of models can be expanded to encompass the ARIMA models associated with Box and Jenkins. Beale and Seeley (unpublished) point out that the fitting of some auto-regressive models can be done quite easily with minor extensions to a good-quality regression program. Such alternatives would only be considered when the data were ordered in time. There are other types of correlation structure, often describable as the sum of independent error terms indexed by differing combinations of classifying factors. Such structures are widely found in models for the analysis of designed experiments, but occur much more generally, a particularly common example being the occurrence of two variance components, one between and one within groups. The correct method of fitting would use a weight function suitably combining estimates of the two variance components. The joint estimation of means and variances in a multivariate normal distribution is not, however, a completely solved problem. The effect of ignoring the variance structure will be loss of efficiency in estimating the means.

1.4 The interaction between data faults and model faults

The techniques discussed briefly above for the detection of faults in data and in models are generally satisfactory when only one source of distortion is present. Thus if we have a satisfactory model we can detect data faults, or if we know the form of the systematic part of the model, we know how to check that the variance assumption is correct. However, in practice we do not know which sources of distortion are present, and we certainly cannot guarantee that there will be only one. The possible presence of multiple sources of distortion introduces new and very difficult problems of interpretation. An interesting example is provided by the data set first published by Brownlee (1965), analysed in considerable detail by Daniel and Wood (1971), and again by Andrews (1974). The data set has 21 points and gives the output y of a chemical plant as related to three input variables x_1, x_2, and x_3. Daniel and Wood, after a multi-stage analysis, concluded that there was no justification for transforming y and that a model using terms in x_1, x_2, and x_1^2 was adequate if 4 points (nos. 1,3,4,21) were treated as abnormal. They gave a possible explanation why the yields from these points might be unreliable, the explanation depending on the fact that the yields were derived serially in time. Andrews used a robust estimation procedure to show that the same points as those noted by Daniel and Wood were highlighted as discrepant by his method. I investigated the data using another robust estimation procedure and also obtained discrepant points, but they were the points (2,3,4,21) instead of (1,3,4,21). I also found that the inclusion of an extra term in the regression equation, which did not itself improve the fit, changed the discrepant points to the original four. Aitken (personal communication) then proposed that the use of lny in place of y not only simplified the regression relationship (by removing a quadratic term), but also, as judged by plots of the residuals against the fitted values and the $\underline{x}$s, gave no indication of there being any discrepant points at all!

Several lessons can be learnt from this data set. First the discovery, by use of a robust regression technique, of a set of apparently discrepant points, should not obscure the possibility that other sets of points if discovered may give almost as good fits; thus it may not be enough to have one solution, other near-optima are relevant. Secondly the interpretation of a point as discrepant depends crucially on the variance assumption. The use by Aitken of lny implies that, approximately, $\text{var}(y) \sim \mu^2$ and hence that a wider scatter is to be expected for large μ. In this example this means that point 1 (found discrepant by Daniel and Wood, and by Andrews) and point 2 (found discrepant by me), which have large predicted values, must be inconsistent if the variance is assumed constant, but become entirely consistent if the variance is assumed constant on the log scale. The third lesson, and perhaps the most important for designers of intelligent programs, is that any interpretation of sources of distortion must be presented to the user in a tentative way, it being made very clear that there may be other alternative explanations which the user should be encouraged to seek out for himself.

2. HOW SHOULD WE IMPLEMENT INTELLIGENT PROGRAMS?

We have a number of useful tools for looking at the relevance of regression models to data, and these could be applied in ways ranging from the wholly automatic to the wholly user-controlled. We now discuss some possibilities and indicate some of their advantages and disadvantages.

2.1 The role of external information

The value of some procedures depends entirely on external information, e.g. tests for serial correlation require the data to be ordered in space or time, and it would be pointless to apply them when this was not so. It could be valuable, therefore, if a program required the user to specify certain information external to the data themselves, such as

(a) Are the data ordered in space or time?
(b) Are there any possibly relevant groupings of the data not indicated in the data matrix?
(c) What type of observation do the variates consist of, e.g. measurement, count, proportion, scaled assessment, group no. etc.?

The answers to these, and other such questions, could be used to guide the selection of procedures used to check the assumptions and to print warnings and comments about the model used and possible dangers in its use. There is, of course, no guarantee that such a program would be popular with users, some of whom would undoubtedly much rather not think about their data, and who want just a black box delivering the analysis which makes their data respectable.

Given suitable external information there remains a variety of choice in the subsequent behaviour of the program, and we look at three possibilities.

2.2.1 Standard checks by default

The program does certain standard checks, with conventional limits on acceptability, and prints warnings if any test fails, possibly accompanied by further information, for example a plot. The checks could be guided by the external information discussed in 2.1, and might include tests for outliers, for linearity, for variance/mean relations. The advantage of such a procedure is that it would give considerable built-in protection for the user, of a kind which does not exist at present. The disadvantages include extra computing, often not needed by the experienced user, and the fact that the set of tests is closed and may not always be relevant. The needs of the experienced user are perhaps better met by the next possibility.

2.2.2 Standard checks by request

The same set of checks are involved as in 2.2.1 but are now carried out if and only if requested by the user. Advantages include better control by the user, and less computing where the tests are known *a priori* to be unnecessary. The disadvantages again include the fact that the set of tests is closed, and now additionally that there is no longer any built-in protection for the user.

Two points should be made about both 2.2.1 and 2.2.2. First both procedures assume that we know how to define the test procedures in considerable generality, i.e. for balanced and unbalanced data, and for quantitative and qualitative x-variates. Secondly the balance of cost of CPU time and peripheral use is changing, and is tending to make CPU time relatively cheaper. This meansthat the marginal costs of extra computing, given that the data have been input, is declining, so that the cost of the checks will become relatively less.

To those who rate the disadvantage of having a closed set of tests as particularly important, there is a third possible form of implementation.

2.2.3 User control of a good basic tool-kit

Given a suitable set of facilities for fitting regression models, including weighted regression and the saving of residuals and fitted values, together with facilities for creating derived variates, plotting and looping over sets of instructions, the user is himself equipeed to program checking procedures himself, adapting them to his particular needs, and adding new ones as they appear in the literature. The process will be easier and more convenient if the procedures can be stored as macros or subprograms easily callable in the program. The user will be further helped if the manual that accompanies the packages provides standard macros for well-established tests with examples of their use. The advantage of this method is its open-endedness; it is adaptable. Its disadvantages are that it gives no built-in protection, and that it requires more detailed knowledge of the programming language and the more advanced features of the system concerned. It also requires the system itself to have sufficiently general facilities in its control language. These are by no means always available.

3. DISCUSSION

It will be clear from the questions that have been raised that the implementation of intelligent programs is likely to be full of difficulties. These difficulties are partly methodological - much remains to be done on efficient procedures for detecting possible faults in data and models, particularly where more than one kind of distortion may be present. Other difficulties involve implementation, the possible adaptation and improvement of existing systems, and how to make the facilities of intelligent programs interesting and attractive to a wide range of users. Finally there are difficulties of education, how statisticians can and should communicate with the ever-growing body of users who use and misuse the techniques of statistics with the aid of computing tools often developed by people with little or no statistical training. I believe that a potent aid to such communication could be the use by statisticians and non-statistical research workers of the same programs. The development of intelligent programs could be a way of bringing about this common use.

REFERENCES

Andrews, D.F. (1974). A robust method for multiple linear regression. Technometrics, 16, 523-531.

Andrews, D.F. et al. (1972). Robust estimates of location: survey and advances. Princeton University Press.

Anscombe, F.J. and Tukey, J.W. (1963). The examination and analysis of residuals. Technometrics, 5, 141-160.

Box, G.E.P. and Tidwell, P.W. (1965). Transformation of the independent variables. Technometrics, 4, 531-550.

Brownlee, K.A. (1965). Statistical theory and methodology in science and engineering (2nd Ed.), New York: Wiley.

Daniel, C. and Wood, F.S. (1971). Fitting equations to data. New York: Wiley

Day, N.E. (1969). Estimating the components of a mixture of normal distributions. Biometrika, 56, 463-474.

Nelder, J.A. (1973). GLIM (Generalised Linear Interactive Modelling) User's Manual. Oxford: Numerical Algorithms Group, 57 pp.

Nelder, J.A. and Wedderburn, R.W.M. (1972) Generalised linear models. Journal of the Royal Statistical Society, Series A, 135, 370-384.

Wedderburn, R.W.M. (1974) Quasi-likelihood functions, generalised linear models, and the Gauss-Newton method. Biometrika, 61, 439-447.

Recent Developments in Statistics
J.R.Barra et al., editors

Invited Paper

A SOFTWARE PROJECT
SOME NEW IDEAS IN STATISTICAL COMPUTING

Y. Schektman, J. Jockin, P. Pasquier, D. Vielle
Laboratoire de Statistique (1)
Université Paul Sabatier
Toulouse - FRANCE

In this paper we do not carry out any comparative study about the numerous statistical software packages now operating for users. The general problems arising from automation are taken into account, but the main topic concerns reflections and research work on a large statistical software project. The authors mainly unfold their proceedings on how to make statistical model programming easier for those working in research.

1 - INTRODUCTION

Work on real data leads to humility and clarifies the finesse of statistical tools. It especially points out the difficulties encountered not only (i) by the users applying statistical techniques (important size of data, complex computations,...) but also (ii) by those programming statistical models (unfitness of usual programming languages, arbitrary and mysterious system interfaces,...).

Now, numerous software packages [11], [13] , [15] , [19] are operating for users, [7] and [12] are two rather powerful complementary examples. On the contrary few things have been done to make the work of software designers easier (automation of routine problems, high level concepts in languages,...).

Our team works on the project "Software aids in programming and in uses of statistical models" (LAPUMS (2)). This project is divided into four parts,

Part I : batch and interactive subsystems (interfaces with operating system, definition of the resources required for the three other parts,...)

Part II : software aids in programming statistical models

Part III : software aids in uses of statistical models

Part IV : definition of a statistical display terminal in interactive mode (keyboard,...)

We first worked on the last two parts [16] , but very early we needed deep knowledge in computer science which led us to define the first two parts of the project. This project is now in process and about 70 % of the work of the first two parts have been accomplished. It amounts to over 6000 programme cards and 300 pages of scientific directions.

(1) E.R.A. - C.N.R.S. N° 591
(2) The stage of this research work as reached in May 1976, is set out in the report at the end of the contract of the A.T.P. N° 72 99 10 (Centre National de la Recherche Scientifique).

In this paper we deal with the main lines of some aspects of our study, especially how to make the researcher's work easier.

2 - GENERAL PROBLEMS - CHOICES

2.1. Problems arising with automation

* Deontology : a sentence seems to sum up adequately this aspect, "Do not increase the risk of errors by putting tools forward at the expense of thought".
* Correct, intelligent use of statistical models : the model or the statistical summaries must fit the problem and the data structure. The encoding must reflect the perceived nature of the data.
* Reliability and correct use of numerical routines : validity field, rounding errors, ...
* Intelligent and efficient use of the interactive mode.

These important problems lead our studies but we shall not deal with them here.

2.2. Problems to be solved to achieve automation

* Unification of statistical methods : the work carried out under J.P. PAGES [14] is most useful. In appendix, we enclose a short survey of our research in this field.
* Unification of interactive and batch modes.
* Structure and definition of a large statistical software : reliability, adaptability and portability are general features. The specific features are proposed in 2.3.
* Scope of contribution from computer science.

2.3. Choices in the definition of a large statistical software

Confronted with the multiplicity and variety of statistical tools we nevertheless opted for a single subsystem, but we do not think that a universal language is a good thing[2]. We agree on this point with G.N. WILKINSON [11, p. 205] and D.F. HENDRY [11, p. 230]. For instance, we have chosen to define :

* a single subsystem valid for the batch and interactive modes
* a basic language called LOM which is a declarative type language . LOM is characterized by describing "what is to be done" as opposed to "how it is to be done".
* several problem oriented languages (LOMOP). The LOMOP are built around the LOM which is the heart. They are differentiated by the set of specific operators (array operators, M-orthogonal projector,...) which they are provided with. The LOMOP will mainly be used by researchers.

LOMOP = LOM + Specific operators

* numerous users' dialects, more or less powerful, activated and linked to one another by any of the LOMOPs. They are differentiated by objects to be treated, data, structure, statistical terminology, kinds of statistical models, mathematical and algorithmic tools, external aspect (syntax). For example, the syntax proposed by G.N. WILKINSON and C.E. ROGERS [22] is most specific of factorial models for variance analysis.

Relatively to the various lines of research in computer science concerning the means of access to a computer [6, p. 337 - 341] , the LOM mainly possesses the characteristics of the algorithmic and declarative languages constituting a programming system.

3 - DIFFICULTIES ENCOUNTERED BY STATISTICIANS WHEN WRITING A PROGRAMME AND SOLUTIONS

3.1. Difficulties

a) Unfitness of usual programming languages for algorithmic logicfor high level concepts, such as a statistical model programme, or for the statistical needs (array operators, intervals generator, list operators for formal derivation,...).

b) Processing of large amounts of data : linear organization of the working space, use of magnetic diskpacks or tapes and study of a management policy (access method, block length calculus, actual size of a file,...)

c) Addition of new statistical treatments

d) Insertion of an old programme into a larger one

e) Control of the reliability of numerical algorithms

f) Arbitrary and "mysterious" interfaces with operating system

g) Passage from a batch programme to an interactive one

h) A statistic on our FORTRAN programmes reveals that only about 25 % of the instructions concern statistical calculi. The remaining instructions include data management instructions (~ 10 %), edit instructions (~ 50 %) and conversational instructions (~ 15 %).

In conclusion, the researcher must attend to several tasks at a time, many of which do not concern statistical work. This implies a great loss of efficiency and especially of reliability.. These difficulties lead us to define the basic LOM language and the characteristics of the subsystem.

3.2. Solutions

* For {a,b,c,g,h}

Splitting into two parts of the constituents of a programme, so as to write separately and with maximum clarity the instructions required by the sole programmation of the algorithm associatedwith the statistical model.The first part describes the statistical calculi, the second part deals with the data management policy, the interactive guidance, the implementation tree of subroutine code, inserting or deleting instructions into the programme and, last but not least, with the selection of a subroutine from among other subroutines giving the same result. In this latter case,the choice generally depends on the management policy of the working space, this choice being the best with regard to the values of the descriptive parameters of the data.

After having taken into account the values of the descriptive parameters of the data as well as those of the parameters characteristic of the statistical treatment, the subsystem works out, just before running the programme, the synthesis of the informations concerning both parts (of the programme) and updates the tables included in the object code.

It is important to point out that for debugging, only the first part of a programme (statistical calculi) is needed. The second part, required by the frequently large amount of data to be treated where applications are concerned, may be postponed and dealt with by the researcher or any competent person.

* For {a,b,c}

A set of consistent control instructions with structured programming [4], [5]. This choice makes printing, proving, re-reading and alterations of the programme easier. By imposing a clear linear programming, it ensures a better reliability of the programmes and simplifies the future conception of an optimal data management policy of the programme working space.

* For {d}

A LOM instruction, called "command instruction" ensures the linking of the dialects (programme) through an interface language.

The first two syntaxic rules of this instruction are :

< Command Instruction > = < Identifier > : < Interface > FIN

< Interface > = < Input Interface > < Output Interface > < Results list >
"< Dialect Instructions >"

The example quoted in 4 will give an idea of < Input/Output Interface > . For the class of dialects which can be called "not very communicative" (Programme of a statistical model with few parameters), the part < Dialect Instructions > may be empty. On the contrary, for other dialects, this part may be important and submitted to a rather elaborate syntax (processing of surveys [16], factorial models for variance analysis [22], ...)

The programmes of the dialects must be written in one of the LOMOPs. When running a "command instruction", the subsystem will allocate all the memories of its working space to the called programme (dialect) and will carry out all the tasks required by the linking : output on diskpack of the interface data and the specific environment of the calling programme, stacking the state variables, loading and synthesis of the called programme.

* For {a,c,h}

A LOM instruction called "selection instruction" ensures the branching to subroutines (utilitarian or specific).).

The first three syntaxic rules of this instruction are :

< Selection Instruction > = EXECUTER < Calling subroutines list > FIN

< Calling subroutines list > = < Calling subroutines list > ;
< Calling subroutine >

< Calling subroutine > = < Identifier > (< Effective parameters list >)

The subroutines are written in a scientific programming language. At present , the interface is only possible with FORTRAN, but the subsystem is so designed as to eventually permit the interface with other languages (Pascal,...).

* For {e}

The use, in a control instruction, of the LOM predicate "ANOMALIE" conditions the running of a programme as soon as an anomaly has been detected by the subsystem or by utilitarian subroutines written in this view. Research such as in [8] may be very useful.

* For {g}

All the routine problems (assigning and numbering of the logical units, input/output on magnetic, block length calculus,...) and all the facilities mentioned in the preceding paragraphs, are performed by the tasks of the subsystem (synthesizer, librarian, loader, keeper, bailiff,...).

The subsystem, designed according to a topdown conception, detects automatically the mode of computer exploitation chosen by the user, and hence unifies in a single subsystem both interactive and batch modes.

* For {a,h}

The work can be thus divided as shown in the following diagram : number 2 is the complement of number 1 and it gives some dynamic informations.

The object code "O.C.U." is entirely interpreted, contrary to the object code "O.C.S." which is composed of about 90 % of machine instructions.

Note :

As the greater part of the programming code is composed of subroutines, the LOM, together with the "command" and "selection" instructions, has indeed the characteristics of a declarative language organizing modules [3].

Programming levels	Actors	Motivations Productions	Languages in use
1	User	Specific research	Dialects
2	Researcher	Statistical models Dialects	LOMOP (Part I) (Statistical calculi)
2'	Analyst	System Interface	LOMOP (Part II) (Interfaces)
3	Programmer	Subroutines	FORTRAN, ...

Diagram 1

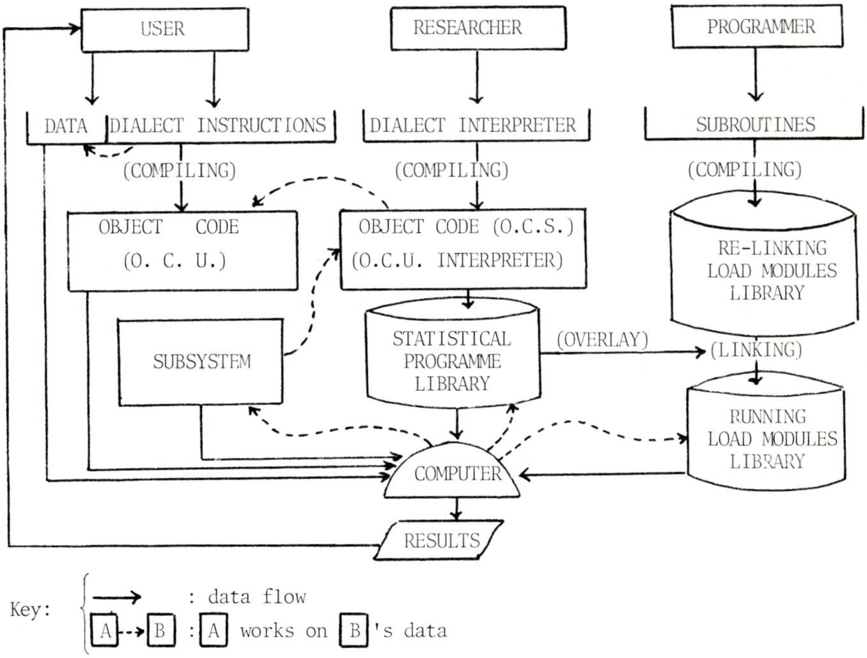

Diagram 2

4 - A DEVELOPED EXAMPLE

4.1. Example

The example submitted here is inspired from an encoding method proposed by M.TENENHAUS [20] during the data analysis conference organized by the "Association Française pour la Cybernétique Economique et Technique" (December 1975). The external appearance of the language is recognized by a syntaxic analyser SLR(1).

```
DEBUT "CODAGE.TENENHAUS"
 $    -----------------

      NB.VARIABLES = NOMBRE DE VARIABLES NOMINALES
      NB.US = NOMBRE D'UNITES STATISTIQUES
      DONNEE = TABLEAU DES INDICATRICES D'UNE VARIABLE NOMINALE
      NB.MODALITES = NOMBRE DE MODALITES DE REPONSE D'UNE VARIABLE NOMINALE
      Q = NB.MODALITES - 1
      M = NOMBRE DE COMPOSANTES PRINCIPALES UTILISEES POUR LE CODAGE
      T = TABLEAU DES CODAGES
      DELTA, Z, CODAGE, TCODAGE, INERT = TABLEAUX DE TRAVAIL $

DECLARATION

      TABLEAU NOM:DONNEE  DIMENSION:(NB.US,NB.MODALITES[NB.VARIABLES])
              NOM:DELTA   DIMENSION:(NB.US,Q[NB.VARIABLES])
              NOM:Z       DIMENSION:(NB.US,M:NB.VARIABLES)
              NOM:CODAGE  DIMENSION:(Q[NB.VARIABLES])
              NOM:T       DIMENSION:(NB.US,NB.VARIABLES)
              NOM:TCODAGE DIMENSION:(NB.US)                DUPLIC:NB.VARIABLES
              NOM:INERT   DIMENSION:(M:NB.VARIABLES,M:NB.VARIABLES) TYPE:DIAGONAL

FINDECLARATION

CALCUL
   POUR  I=1   PAS 1   JUSQUA NB.VARIABLES
   FAIRE EXECUTER CALCULD(NB.US,NB.MODALITES[I],DELTA[I],DONNEE[I])   FIN
   FIN

   POUR  M=1   PAS 1   JUSQUA NB.VARIABLES

   FAIRE CANONIQUE.GENERALISEE.CARROLL :

         ENTREE (DONNEE[I]←DELTA[I], I=1,NB.VARIABLES) ; NB.FACTEURS ← M ;
         SORTIE Z←VARIABLES.AUXILIAIRES ;

       FIN

       REPETER  C1 = 0 ;

                POUR  I=1   PAS 1   JUSQUA NB.VARIABLES

                FAIRE CANONIQUE :

                        ENTREE DONNEE[1]←DELTA[I] ; DONNEE[2]←Z ;
                        SORTIE CODAGE[I]←FORME.LINEAIRE.1 ;
                               PI ← "POURCENTAGE INERTIE EXPLIQUEE FACTEUR 1" ;
                        "NOMBRE DE FACTEURS = 1"

                      FIN

                      C1=C1+PI ; TCODAGE[I]= DELTA[I]*CODAGE[I];

                FIN

                EXECUTER CONCATENER (NB.US,NB.VARIABLES,T,TCODAGE) FIN
```

```
            COMPOSANTES.PRINCIPALES :
              ENTREE  DONNEE ← T ; NB.FACTEURS ← M ;
              SORTIE  INERT ← INERTIE ; IT ← INERTIE.TOTALE ; Z ← FACTEURS ;
              "DISTANCE USUELLE"
            FIN
            Z = Z*INV(INERT) ; C2 = TRACE(INERT)/IT ;
      JUSQUA   (ABS (C1-C2) < .001)    FIN
      SI (C2 > .80) ALORS SORTIR   FIN
 FIN
 RESULTAT NB.FACTEUR
   ECRIRE ("NOMBRE DE FACTEURS" M) ;
 FIN
 RESULTAT CODAGE
   ECRIRE ("CODAGE VARIABLES NOMINALES" T) ;
 FIN
FINCALCUL
FINPROGRAMME
```

4.2. Comments

In this example, only the first part (statistical calculi) of the programme is presented.

Three command instructions define the linking between the programmes (dialects) "CANONIQUE GENERALISEE CARROLL" (which evaluates the outset values transmitted to the calling programme via the "Z" array), and the "CANONIQUE" followed by "COMPOSANTES PRINCIPALES". These three programmes are performed in two nesting loops ending when the convergence criteria is satisfied and, when the quality of the results is deemed adequate or else when "M" reaches its maximum value "NB VARIABLES".

The "Z" array quoted above is implemented in the "DONNEE [2]" array of the "CANONIQUE" programme which, for each value of "I", supplies results in the "CODAGE [I]" array. After alterations (multiplication by "DELTA [I]" and concatenation in the "T" array, the latter are implemented in the "DONNEE" array of the "COMPOSANTES PRINCIPALES" programme.

The "CODAGE TENENHAUS" programme calls upon two subroutines "CONCATENER" (utilitarian) and "CALCULD" (specific to this statistical model).
"INV" (inversion), "TRACE", "ABS" (absolute value) are operators of the LOM.

"DONNEE", "FACTEURS", "INERTIE" are words included in the vocabulary of the "COMPOSANTES PRINCIPALES" dialect.

The content of the non terminal symbol < Dialect Instructions > of the command instruction "COMPOSANTES PRINCIPALES" is limited to "DISTANCE USUELLE" and it is empty for the command instruction "CANONIQUE GENERALISEE CARROLL".

In the < Output Interface > of the command instruction "CANONIQUE" we find a sentence of this dialect "POURCENTAGE INERTIE EXPLIQUEE FACTEUR 1". This possibility is provided for in the syntax of the LOM, but we do not think, at least in this example, that its use is justified, considering the amount of work required by the writing of a recognition module (Part II of the programme). This points out one of the aspects of this type of research, in which only practice can validate or invalidate certain decisions.

Relatively to usual languages, the LOM offers extra possibilities for the array declarations. All these possibilities are present in the above example :

a) In the model "CANONIQUE GENERALISEE CARROLL" the number of the sets of variables treated ("NB VARIABLES") is unknown at the programming stage. Should we wish to avail ourselves of the possibilities due to an independent management of the arrays, corresponding to each set of variables, it will then be necessary to plan a dynamic duplication of all the arrays when running the programme.

For example, "NB VARIABLES" arrays "DONNEE [1]", "DONNEE [2]", ... have "NB US" lines and "NB MODALITES [1]", "NB MODALITES [2]", ..., "NB MODALITES[NB VARIABLES]" columns respectively. Or the "NB VARIABLES" arrays "TCODAGE [1]", "TCODAGE [2]", ... have "NB US" lines and one column.

b) The progressive trials performed upon an increasing number "M" ∊ [1,NB VARIABLES] of factors result in arrays whose sizes vary until we obtain an adequate precision, as the programme is running.

For example, the "Z" array has a number "M" of columns which is variable but which cannot exceed the value "NB VARIABLES".

Lastly, it can be seen from this example that the LOM has no reading instruction : this function, among others, is secured by the "bailiff" task of the subsystem in the "CODAGE TENENHAUS" programme or by the LOM compiler in the "APPEL CODAGE TENENHAUS" programme submitted below. As to the editing, only one standard writing instruction on the printer is authorized. A more sophisticated editing of results or the use of working files on magnetic units will necessarily have to be carried out in subroutines.

The programme below commands the calling of the "CODAGE TENENHAUS" programme. The values of the data descriptive parameters are affected directly, whereas in the command instructions, quoted above, these values are transmitted via the array descriptors. Among the two possible results, NB FACTEURS and CODAGE (in the "CODAGE TENENHAUS" programme), only CODAGE will be printed.

```
DEBUT "APPEL.CODAGE.TENENHAUS"
  $   -----------------------   $

CALCUL

   CODAGE.TENENHAUS :

     ENTREE  NB.VARIABLES ← 4 ; NB.US ← 500 ;
             NB.MODALITES ← 8, 6, 4, 9 ;
             DONNEE ← FICHIER  JOCKIN00134, PASQUIER0048,
                               SCHEKTMAN00135, VIELLE00179 ;

     RESULTAT CODAGE ;

   FIN

FINCALCUL

FINPROGRAMME
```

5 - SCOPE OF THE COMPUTER SCIENCE CONTRIBUTION

To write subsystem tasks and routines we need the operating system procedures library and the processor FORTRAN routine library.

To define the syntax and the semantics of LOM we use the DELTA meta-compiler of the Institut de Recherche en Informatique et en Automatisme LABORIA/IRIA-FRANCE).

Below, we give two simplified diagrams which throw some light on the main operations which a programme undergoes.

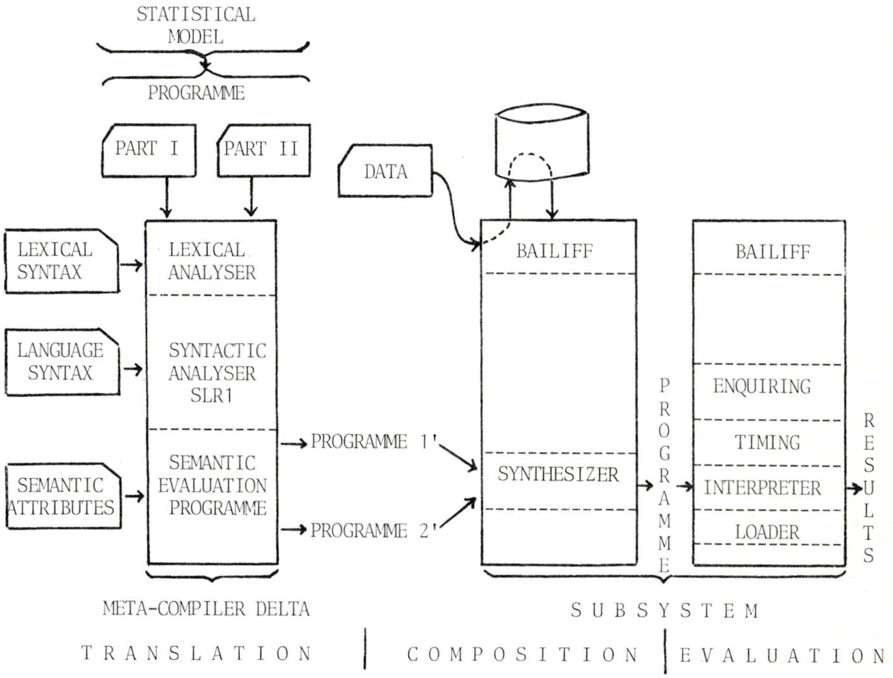

DIAGRAM : PROGRAMME FLOW

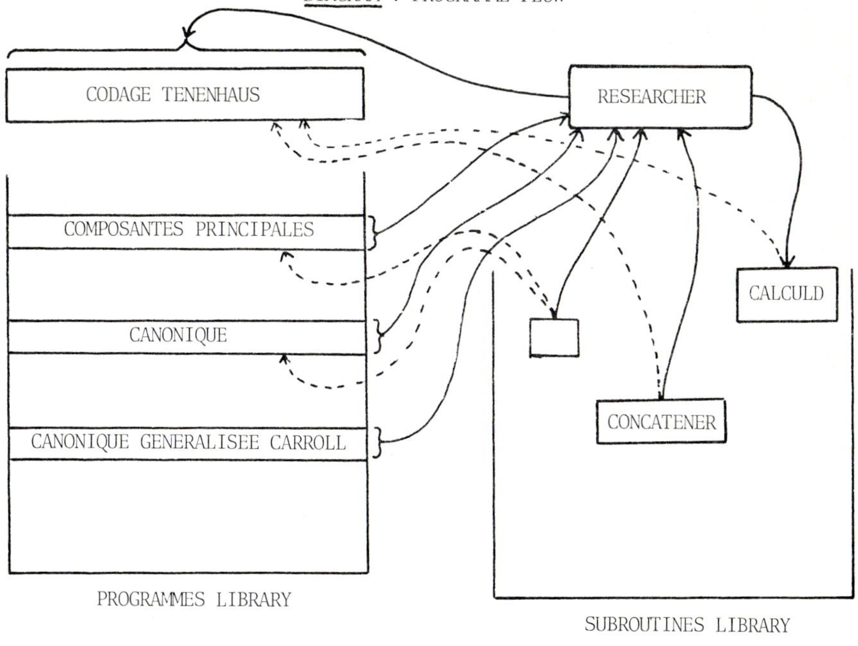

DIAGRAM: HOW TO RAISE THE ABSTRACTION LEVEL OF THE PROGRAMMES (DIALECTS) LIBRARY

6 - CONCLUSIONS

6.1. Limitations

Computing speed : since we only have partial measures at our disposal we cannot yet assess the slowing down rate due to our subsystem. However, if we consider the technological evolution and the various notions being questioned in computer science, is running speed a major problem for the years to come ? Besides, this loss of efficiency may be greatly made up for by the gain in programming time and a better reliability.

Portability : it is necessarily limited by the interfaces of our subsystem with the CII Siris 8 system and the FORTRAN processor. On the other hand, we have purposely overlooked portability of running speed by programming in machine language the most frequently used instruction sequences.

Adaptability : the existence of command instructions, the possibility to build up LOMOPs from the LOM, and the facilities offered by the subsystem either to programme or reprogramme are many positive points in favour of adaptability. Few things can be planned to adapt dialects, but what is proposed above will most probably be sufficient.

Programming constraints : so as to set up a homogenous subroutine library[9], compatible with the subsystem and having an acceptable rate of portability, the programmer must abide by programming norms when writing subroutines.

6.2. Possible improvements

More flexibility : Provided that there is no ambiguity, the automatic correction of spelling or agreement mistakes [2], (in the sentences written in a dialect) would increase the researcher's confidence when debugging the interfaces of command instructions. It may also be desirable to explicit the interfaces with the subroutines as it is done for the programmes.

Programming aid in interactive mode : e.g., where interfaces with the subroutines are concerned, the subsystem could ask explicitly for the effective values of the formal parameters, the subsystem could indicate the implicit values of some parameters, ... but we encountered a much larger field where our choices have not yet been made.

Extendibility : why not use the mecanisms of syntaxic and semantic extension, incorporated in the extendible language [1] , [6] , to build up the LOMOPs automatically from the LOM? Although the association of a programming aid subsystem to extendibility mecanisms is a need for the future, this possibility goes beyond the fields of our abilities and above all of our preoccupations. The latter are mainly aimed at defining tools answering our needs. In this aim, the DELTA metacompiler is for us a valuable tool.

Simplification or suppression of declarations : without sufficient practice, we dared not enter this course on our first attempt at a semantic definition of the LOM. Later on, having deepened our knowledge in this subject, we contemplated assessing the contributions and the feasability of this improvement. This type of approach seems sensible for, with DELTA, the semantics of the LOM are defined in a modular declarative way through attributes[10].

6.3. A hope and a duty

Considering the numerous and vast problems arising from automation in statistics, it may be wise to develop even further the coordination between the different teams of statisticians. It is necessary to develop this coordination in order to define research directions, to find means to inflect everyone's objectives, even to limit them momentarily in view of more homogeneity, and to distribute the efforts to be more efficient and rational.

Considering the fast development of the data basis systems, parallel to that of mass core storages likely to keep a practically unlimited number of information, statisticians must assume a scientific responsability endeavouring to ensure a rational utilisation of this data in accordance with certain ethical values.

ACKNOWLEDGEMENT

Our grateful thanks are due,

to Mr M. GALINIER, Professor in computing science at Paul Sabatier University (Toulouse - France) for advices he gave us,

to Mr B. LHORO and also to his team (LABORIA - IRIA) for the courses they gave us and the services rendered by the DELTA meta-compiler they have designed,

to Mr G. PERENNOU, Director of "Centre Interuniversitaire de Calcul de Toulouse (C.I.C.T.)" for the facilities he always granted us,

to Mr J.P. GALLOU, system engineer at C.I.C.T. and also to Mr G. MOREL engineer at C.C.E.T.T. (Rennes - France) for a great deal of help in solving the problems of interfaces with the CII - SIRIS - 8 system.

to Mr BABITS (C.N.E.S. - Toulouse) and Mrs COUSTEL (Mirail University - Toulouse) who proof read this paper.

REFERENCES

[1] CHERBONNEAU, B. (1976). Le langage extensible LET : définition et exemples d'application - Thèse de 3ème cycle, Toulouse.

[2] CHIARAMELLA, Y., COURTIN, J., GRANDJEAN, E., VEILLON, G. (1976). Utilisation de techniques de recherche en parallèle au contrôle de données ambigües. Application aux sciences humaines. Congrès AFCET, Panorama de la nouveauté informatique en France - Tome 1, p. 203-213.

[3] DEREMER, F., KRON, H. (1975). Programming in the large versus. Programming in the small. Proceedings of international conference on reliable software.

[4] DIJKSTRA, E.W. (1972). The Humble programmer - Communication of ACM, Vol. 15, N° 10, p. 859.

[5] GALINIER, M. and collaborators (1975). Programmation structurée - Ecole d'été AFCET - RABAT.

[6] JORRAND, P. (1975). Contribution au développement des langages extensibles. Thèse d'Etat, Grenoble.

[7] GLIM (by NELDER, J.A. (1975). Rothamsted Experimental Station, Harpenden, Herts. NAGLIM G55/G33 : REL 2

[8] LA PORTE M. et VIGNES J. (1973). Méthode numérique de détection de la singularité d'une matrice. Institut de programmation. PARIS

[9] LARMOUTH, J. (1973). Serious FORTRAN - Part 1. Software - Practice and Experience, Vol. 3, p. 87-107.

[10] LORHO, B. (1974). De la définition à la traduction des langages de programmation : Méthode des attributs sémantiques. Thèse d'Etat, Toulouse.

[11] MILTON, R.C., NELDER, J.A. (1969). Statistical computation. Academic Press . New-York - London.

[12] NELDER, J.A. and members of the Rothamsted Statistics Department (1973). Genstat reference manual. Inter-University/Research councils series, report n° 3, second edition. Edinburgh : Program Library Unit. Edinburgh Regional Computing Center.

[13] NELDER, J.A. (1974). A User's Guide to the evaluation of statistical packages and systems, International statistical review, vol. 42, N° 3.

[14] PAGES, J.P., CAILLIEZ, F. (1976). Introduction à l'analyse des données - S.M.A.S.H.

[15] POTTIER, F., ROUZAUD, A., JULIEN - LAFERRIERE, B. (1973). Etude comparative de systèmes de programmation d'analyse statistique pour les sciences de l'homme et de la vie - Informatique et Sciences Humaines, N° 18.

[16] SCHEKTMAN, Y., PASQUIER, P. (1972). Un nouvel outil à la disposition du médecin du travail : le langage de dépouillement d'enquête (LADEN), revue médecine du travail, Tome 1, N° 2.

[17] SCHEKTMAN, Y. (1976). Rapport scientifique - ATP-CNRS N° 729910. 200 p.

[18] SCHEKTMAN, Y., JOCKIN, J., PASQUIER, P., VIELLE, D. (1976). Logiciel d'aide à la programmation et à l'utilisation des modèles statistiques - Congrès AFCET, Panorama de la nouveauté informatique en France - Tome 1, p. 191 à 201.

[19] SCHUCANY, W.R., MINTON P.D. (1972). A survey of statistical packages - computing surveys, vol. 4, N° 2.

[20] TENENHAUS, M. (1976). Analyse en composantes principales d'un ensemble de variables nominales ou numériques. C.E.S.A. 78350 JOUY-EN-JOSAS

[21] TUCKER, A. (1975). Very high-level language design : A view point. Computer languages, Vol. 1, p. 3-16.

[22] WILKINSON, G.N., ROGERS, C.E. (1973). Symbolic description of factorial models for variance analysis. Applied Statistics, Vol. 22, N° 3, J.R.S.S. serie C.

APPENDIX : STRUCTURAL HYPOTHESES ON FACTORS IN LINEAR DATA ANALYSIS

As we point out in 2.2. the development of statistical computing is closely linked to the development of statistical methodology. If the unification of statistical techniques makes calculus automation easier, similarly the existence of an adequate informatic tool enables us to take up more easily fairly complex analyses. Below, we give some indications about a methodological research approach illustrating these two points.

1 - RESEARCH CONTEXT

Many authors have written about structural hypotheses on factors and proposed different models. Below, we list some principal aspects of the research concerning this topic.

a) THURSTONE L.L.

Hypothesis : simple structure

One reference : (1954) "An analytical method for simple structure" Psychometrika, vol. 19, p. 173-182.

b) CARROLL J.R., CRAWFORD C., FERGUSSON G.A., HORST P., KAISER H.F., ...

Hypothesis : analytical criterion (Varimax, quartimax, oblimax, parcimax,...)

References : many papers published in "Psychometrika" concerning orthogonal and oblique rotations.

c) LAWLEY D.N.

Hypothesis : some factor loadings are zero

Reference : (1958) "Estimation in factor analysis under various initial assumptions" Brit. J. Stat. Psych., Vol. 11, 1-12.

d) RAO C.R.

Hypothesis : some factors are uncorrelated with instrumental variables

Reference : (1964) "The use and interpretation of principal component analysis in applied research". Sankhya, serie A, 26, p. 329-358.

e) BURNABY T.P.

Hypothesis : it concerns a particular case of d) (discriminant analysis)

Reference : "growth-invariant discriminant functions and generalized distances" Biometrics - vol. 22 - N° 1 , p. 96-110.

2 - GENERAL STRUCTURE HYPOTHESES ON FACTORS

2.1. Notations

$F = \mathbb{R}^n$ is the variable space.
θ is the weight scalar product defined on $F \times F$.
$x^j \in F$ $(j=1,\dots,p)$ are main variables, we assume $\bar{x}^j = 0$ for all j.
$y^k \in F$ $(k=1,\dots,q)$ are instrumental variables, we assume $\bar{y}^k = 0$ for all k.
$\mathcal{X}$ is the subspace of F generated by $\{x^j \;/\; j=1,\dots,p\}$.
SF_K is the subspace of F generated by $\{y^k \;/\; k \in K\}$, where K is a subset of $\{1,2,\dots,q\}$.
A^ℓ $(\ell = 1,\dots,p)$ are the principal factors built up from $\{x^j \;/\; j=1,\dots,p\}$.

2.2. Basic hypotheses

We deal with the following basic hypotheses

(i) Linearity : $A^\ell \in SF_K \quad \forall \ell \in L$

where L is a given subset of $\{1,2,\dots,p\}$ such that $\text{Card}(L) \leq \dim(SF_K)$

(ii) "Interior" : $A^\ell = \Sigma \{a_k y^k \;/\; k \in K\}$, where $a_k \geq 0 \quad \forall k \in K$

(iii) Minimal correlation : $|\rho(A^\ell, y^k)|$ is greater than a given constant for all $\ell \in L$

(iv) Uncorrelation : A^ℓ is θ-orthogonal to SF_K for all $\ell \in L$

Note : for linearity and "interior" hypotheses, we suppose $SF_K \subset \mathcal{X}$. If $y^k \notin \mathcal{X}$, then we suggest to use $\hat{y}^k$, the orthogonal projection of y^k on $\mathcal{X}$.

2.3. Definitions

Let K_r, K_s be subsets of $\{1,2,\dots,q\}$. The constraint imposed upon A^ℓ by a number η of linear (or interior) hypotheses is called :

η-balanced if $\text{Card}(L) = \text{Card}(K)$
η-separable if for all integers $r,s=1,\dots,\eta$ $(r \neq s)$ $SF_{K_r} \cap SF_{K_s} = 0_F$ holds

η-complete if $\Sigma\ \{SF_{K_r}\ /\ r=1,\ldots,\eta\} \equiv F$

Notes : - uncorrelation (resp. minimal correlation) hypotheses can be treated as linear (resp. interior) hypotheses.

- it is possible to define and take into account indirect constraints . For example, a linear indirect constraint would be: y^k belongs to a certain subspace generated by some A^ℓ.

3 - GLOBAL OPTIMALITY IN PRINCIPAL COMPONENT ANALYSIS (PCA)

3.1. Notations :

$E = \mathbb{R}^p$ is the statistical units space (or subjects space) - we assume $p \leq n$.
$\{e_1,e_2,\ldots,e_p\}$ is the canonical base of E.
$x_i = \Sigma\ \{x_i^j\ e_j\ /\ j=1,\ldots,p\}$ is the ith statistical unit vector.
M is a scalar product defined on E × E.
V is the quadratic form of inertia defined on $E^* \times E^*$.
$\{t_j\ /\ t_j \in E,\ \|t_j\|_M = 1,\ j=1,\ldots,p\}$ is the base of E such that
$\Sigma\ \{A_i^j\ t_j\ /\ j=1,\ldots,p\} = x_i$, where $\{A_i^j\ /\ i=1,\ldots,n\}$ are the coordinates of A^j in the canonical base of F.

3.2. Global criteria :

First, in view of taking into account the most general hypotheses we have to set out global criteria.

a) First group criteria : we introduce the following criteria

(1) $\Pi\{M \circ V \circ M(u_j,u_j)\ /\ j=1,\ldots,p\}$ is a minimum

(2) $\Sigma\{[M \circ V \circ M(u_j,u_j)]^2\ /\ j=1,\ldots,p\}$ is a maximum

(3) $\Sigma\{M \circ V \circ M(u_j,u_j) \times M \circ V \circ M(u_k,u_k)\ /\ j=1,\ldots,p-1\ ;\ k=j+1,\ldots,p\}$ is a minimum

and we introduce the condition

(4) $M(u_j,u_k) = \delta_{jk}\ \forall j,k=1,\ldots,p$

Proposition : *t_j $(j=1,\ldots,p)$ are solutions of (1) [resp. (2), (3)] under (4).*

b) Second group criteria : we keep the same criteria but we introduce the new condition

(5) $M(u_j,u_j) = 1,\ M \circ V \circ M(u_j,u_k) = 0\ \forall j,\ k=1,\ldots,p\ (j \neq k)$

Proposition : *t_j $(j=1,\ldots,p)$ are solutions of (1) [resp. (2), (3)] under (5).*

The proofs of both propositions can be found in [17].

Our criteria seem better [1] than those set out by

(i) OKAMOTO M. (1968) in "Multivariate Analysis" KRISNAIAH Academic Press p. 673-685.

(ii) DENIAU C., OPPENHEIM G. (1972) in "Publication I.S.U.P." Paris p. 27-42.

4 - P.C.A. UNDER STRUCTURE HYPOTHESES ON FACTORS (P.C.A.H.)

To define the P.C.A.H. we have resolved to use the first group of criteria. Indeed, the factors are generally correlated under structure hypotheses. This choice agrees with some ideas of THURSTONE L.L., THOMSON G.H. [2] and REUCHLIN M. [3].

4.1. Notations - Definitions

SE_H $(H=1,\ldots,g)$ is a subspace of E with dimension p_H.

${}_HM$ (resp. ${}_HV$) is the restriction of M (resp. V) to SE_H (resp. SE_H^*).

$\forall j=1,\ldots,p_H \qquad {}_Hu_j \in SE_H$.

Let us define

(i) Strong PCAH (SPCAH) : we call SPCAH the determination of t_ℓ $(\ell=1,\ldots,p)$ such that (1) [resp. (2), (3)] while (4) hold under some given structure hypotheses on A^ℓ (defined in 2)

(ii) Weak and context sensitive PCAH (WSPCAH) : we call WSPCAH the determination of t_ℓ $(\ell=1,\ldots,p)$ such that (1) [resp. (2) , (3)] while $M({}_Hu_j , {}_Hu_k) = \delta_{jk}$ for all $j,k = 1,\ldots,p_H$ and $H=1,\ldots,g$ hold under some given structure hypotheses on A^ℓ.

(iii) Weak and context free PCAH (WFPCAH) : we call WFPCAH the determination of t_ℓ $(\ell=1,\ldots,p)$ such that $\Sigma\{[{}_HM \circ {}_HV \circ {}_HM ({}_Hu_\ell , {}_Hu_\ell)]^2 / \ell= 1,\ldots,p_H ; H=1,\ldots,g\}$ is a maximum while $M({}_Hu_j,{}_Hu_k)=\delta_{jk}$ for all $j,k = 1,\ldots,p_H$ and $H=1,\ldots,g$ hold under some given structure hypotheses on A^ℓ.
Other equivalent criteria can be built from (1) and (3).

4.2. Some properties on linear constraints

Proposition : *a WFPCAH under linear constraints can be solved by means of g independent PCA.*

Proposition : *for a SPCAH there exists $SE_K \subset E$ such that $A^\ell \in SF_K \iff t_\ell \in SE_K$*

Proposition : *if the constraint imposed upon A^ℓ is linear , 1-balanced, then there exists $SE_M \subset E$ such that $A^m \notin SF_L \iff t_m \in SE_M$ $\forall m \notin L$.*

---

(1) HAIT J.R., SCHEKTMAN Y. : A propos de l'analyse en composantes principales sous contraintes d'orthogonalité.
Publication du Laboratoire de Statistique, TOULOUSE (N°76-06)

(2) Analyse factorielle des aptitudes humaines - P.U.F. p. 280

(3) Méthodes d'analyse factorielle à l'usage des psychologues - P.U.F. p. 243

Proposition : *if the constraint imposed upon A^{ℓ} is linear , g-complete separate balanced, then there exists SE_{K_r} $(r=1,\dots,g)$ included in E such that for all ℓ and SF_{K_r}*

$$A^{\ell} \in SF_{K_r} \iff t_{\ell} \in SE_{K_r}$$

Proposition : *under linear constraints, if SE_{K_r} is M-orthogonal to SE_{K_s} for all r,s $(r\neq s)$ and $\oplus\{SE_{K_r} / r=1,\dots,g\} \equiv F$ then the notions of SPCAH, WSPCAH and WFPCAH are equivalent.*

Proposition : *if SE_{K_r} is M-orthogonal to SE_{K_s} for all r,s $(r\neq s)$, then a SPCAH under a linear constraint, η-separate balanced, can be solved by means of η+1 independent PCA.*

Proposition : *if SE_{K_r} is M-orthogonal to SE_{K_s} , for all r,s $(r\neq s)$, and $\oplus\{SE_{K_r} / r=1,\dots,g\} \equiv F$, then a WSPCAH under a linear constraint, g-complete separate balanced, can be solved by means of g independent PCA.*

Other properties on linear constraints and properties on interior (or minimal correlation) constraints can be found in [17].

4.3. Algorithm for linear or uncorrelation hypotheses

The main lines are :

a) Determination of the SE_{K_r} $(r=1,\dots,\eta)$.

b) Compatibility of constraints (existence of a solution), especially in SPCAH.

c) For $r=1,\dots,\eta$, construction of an orthonormal vector system which generates SE_{K_r} and which is used for the beginning of the iterative process quoted in d).

d) For $r=1,\dots,\eta$ determination of the ${}_r t_j \in SE_{K_r}$. This is done by an iterative process, for each iteration of which, the algorithm calculates the best angle of an elementary rotation in a two dimensional subspace included in SE_{K_r} .

NOTES :

- For interior (or minimal correlation) constraints, this algorithm is adapted without difficulties.
- Without constraint this algorithm is the Jacobi's method for finding eigenvectors.: Hence we have given in 3.2. a statistical interpretation of Jacobi's algorithm for a positive matrix.
- For linear or uncorrelation hypotheses it can be proved that PCAH can always be solved by finding the eigenvectors of a square matrix. Its dimension is equal to $\dim(SE_r)$. Relatively to RAO's calculi, this latter matrix is therefore of a smaller dimension and moreover is obtained by simpler calculi.

5 - CONCLUSION

Our research generalizes and unifies the studies listed in [1] By comparison with RAO's results, the formal framework used to define the PCAH (i) leads to a more complete approach in the determination of factors under structural hypotheses and (ii) leads to simpler calculi.

Besides, the criteria used, based upon an optimal reproduction of the inertia of the cloud of statistical units (or subjects), constitute a firm point of departure to determine factors under structural hypotheses.

One can say that the PCAH constitutes a unique sound framework compared to the very numerous criteria of oblique rotations.

We consider that criteria like these of the second group would allow us to define the uncorrelated factors under structural hypotheses.

Obviously, the PCAH makes it possible to define correspondence analysis, descriptive discriminant analysis and CARROLL generalized canonical analysis (1) under structural hypotheses.

In a nutshell, this method appears like a linear descriptive tool of structured multidimensional data. Besides, the general framework defined by this method makes calculus automation easier.

(1) The results of our research in this topic can be found in [17].

SESSION : STATISTICAL COMPUTING

DISCUSSION ON THE INVITED PAPERS

Discussants :

- T. CALIŃSKI, Department of mathematical and statistical methods - Academy of agriculture - 60637 , POZNAN - POLAND.

- R. TOMASSONE, I.N.R.A., Laboratoire de Biométrie C.N.R.Z. F 78350 JOUY-EN-JOSAS - FRANCE

1 - CONTRIBUTION OF R. TOMASSONE

Before discussing the topic "computing and statistics" it is first important to examine if there is some need to particularize this sort of computing compared with numerical analysis in its broad sense. Of course, some people have gone on ahead ; statistical softwares exist, and their existence seems to meet some requirements. This pragmatic approach is a concrete proof, even if it is not perfectly defined, of a need deeply felt by statisticians as well as by users. But it is time now to have a more complete discussion with non-statisticians and non-users of statistics, especially with numerists and specialists of language theory. In fact, from the statistician's point of view -i.e. the statistician involved daily in computing- four directions seem important :

a) Direct influence of information processing on statistical softwares :

at this point, the softwares' portability and the difficulties presented by operating systems cannot be neglected. What the statisticians, as everybody using computers, want is to obtain a sort of superficiality : not to learn too much in computers but use them in an optimal manner ; for example, how to use a batch system and/or a conversational system indistinctly (BUHLER, 1976) ? It is evident that softwares having this clarity have the best chance to be adopted.

b) Simultaneous use of different languages :

when dealing with concrete statistical problems several stages are always present; well informed users may now find programs adapted to them : survey programs for a first analysis, statistical programs like anova, regression, factor analysis, clustering techniques for a more detailed analysis, optimization or simulation programs to complete the whole. Programs for each stage are often written in different languages, and the links between them come down to a file process. But this process is not an easy one, particularly on small computers ; this fact being magnified by the frequent recycling of the data which is common in statistical work. Until now, no satifactory solutions exist.

c) Technical comparisons of softwares :

too many statisticians, having had a poor education in numerical analysis, ignore the importance of the numerical precision of computations. Recent papers show that this problem is not so simple as it seems at a first glance, even for the computations of basic statistics like means or covariances (LONGLEY, 1967 ; HAWKINS, 1974). The software evaluation is an essential need, especially due to the proliferation of softwares, uses of which don't always correspond to a scientific value but to a good marketing service (YATES, 1975 ; WILLMOTT, 1976). In this evaluation it is necessary to take into consideration that statisticians may be forced to use small computers even if networks are a solution for the near future (BRYCE and al., 1976).

d) Towards the user :

even if it is evident that computers have simplified the statistician's task, some significant improvements may be found. A special case, at the core of statistical thinking, is the simple writing of a model. To write an anova model, as a statistician may write it in a methodological paper, some solutions exist but they are seldom used (WILKINSON and ROGERS, 1973 ; NELDER and al. , 1975). The problem of intelligent programs is quite different because the definition of what they are is not easy;but the statistician's responsibility towards the user is important. In particular, the most widely used programs are the regression ones ; can the statistical program libraries ignore the existence of robust techniques, especially when solutions, even partial ones, exist (ANDREWS, 1974) ? At last, it seems that computers have more to provide to statistical knowledge ; firstly in the teaching of statistics (ROSENBRANKS, 1974 ; von MAYDELL and SMILLIE, 1975), then also in the learning of essential statistical structures which may be more clearly illustrated by the practice of a "statistical language" (NELDER, 1974).

In conclusion, we think that it is time now to merge information processing with statistics by a better integration of it in statistical thinking. This integration is one of the most important factors for a better use of statistics.

References .

ANDREWS, D.F. (1974) . Alternative calculations for regression and analysis of variance problems, in Applied Statistics, Gupta ed. North Holland, 1-7.

BRYCE, G.R., FRANCIS, J., HEIBERGER, R.M. (1976) . Statistical Software Package Evaluation in the U.S.A. in COMPSTAT 1976, Gordesch et Naeve ed. Physica Verlag, 327-334.

BUHLER, R. (1976) . Designing statistical software for use in both batch and interactive environments, in COMPSTAT 1976, Gordesch et Naeve ed. Physica Verlag, 335-342.

HAWKINS, D.M. (1974) . Computing mean vectors and dispersion matrices in MANOVA algorithm AS 72, Applied Statistics 23, 234-237.

LONGLEY, J.W. (1967) . An appraisal of least squares for the electronic computer from the point of view of the user, JASA, 62, 819-841.

von MAYDELL, U.M. and SMILLIE, K.W. (1975) . Two methods of using APL in the teaching of probability and statistics, in APL 75, Pise , 244-248.

NELDER, J.A. (1974) . A user's guide to the evaluation of statistical packages and systems, Int. Stat. Rev. , 42, 3, 291-298.

NELDER, J.A. and coll. (1975) . GENSTAT, Univ. of Edinburgh, report 3.

ROSENBRANKS, B. (1974) . APL as a notational language in statistics, in COMPSTAT 74, Bruckmann ed. Physica, Verlag 507-515.

WILKINSON, G.N. and ROGERS, C.E. (1973) . Symbolic description of factorial models for analysis of variance. Applied Statistics 22, 392-399.

WILLMOTT, A.J. (1976) . Design of computer programs for survey analysis. Biometrics, 32, 700.

YATES, F. (1975) The design of computer program for survey analysis ; a contrast between the Rothamsted General Survey Program (RGSP) and SPSS Biometrics, 31, 573-584.

2 - ANSWER OF Y. SCHEKTMAN

Mr TOMASSONE's four preoccupations are part and parcel of the essential elements which led us to define and then to design the LAPUMS project.

The tools of the LAPUMS answering the first two preoccupations are : the tasks of the subsystem which automatically performs a great deal of routine acts (interfaces with the operating system, passage from the batch to the interactive mode, ...) and the "command" instruction of the LOM language which performs the connection of the calling dialect or programme with the called dialect.

As far as the numerical reliability of statistical calculi is concerned, we do not think it is now possible to give a clear, complete answer : in the LAPUMS, this third preoccupation is partly satisfied with the use of the "ANOMALIE" predicate and the activation (as required) of subsystem tasks performing certain numerical controls (see, e.g [8]) or altering the implicit limits of some parameters (the number of loops,...). This latter possibility obviously requires putting in "COMMON" these parameters between the subroutines under consideration and the subsystem.

Mr TOMASSONE's fourth preoccupation must be kept in mind by future designers of dialects. In the third part of our project, we shall deal with this task, which will be made easier by the use of the LOM language when programming and the use of the Delta metacompiler when defining the syntax and the semantics of those dialects.

3 - CONTRIBUTION OF T. CALIŃSKI

3.1. Comments on "Intelligent programs, the next stage in statistical computing"

Ideas and suggestions of this paper emerge from the growing feeling shared among many academical and professional statisticians that we are facing a serious danger of misuse of statistical methods to which an easy access has been granted by means of computers. A vast collection of statistical programs, packages and systems is available and offered to the disposal of various users by numerous computing units and centres. Even if we could say that all the statistical methods and procedures embodied into the statistical programs etc. have been designed by experienced statisticians, there is no guarantee that the methods will always be used with proper understanding of their aims and scopes. In particular nobody can prevent the user of applying a statistical program based on a model which is partially or completely inappropriate for the data and/or the purpose of the statistical analysis employed.

To avoid this kind of misuse of a statistical programme, it is necessary to implement it with a variety of additional, more or less sophisticated, procedures able to check assumptions and at least produce sound warnings against any false understanding of the output. This means to make the program more intelligent. But how, if ever, is it possible to provide intelligent programs protecting against any misuse of them ?

Three possible solutions are suggested in the paper and the Author is aware that none of them is fully satisfactory. It is hoped that a thorough discussion may help in finding a proper solution on the way to the next stage in statistical computing.

Some points for discussion have already been put forward in the paper.

"Can we define a suitable set of tests for, e.g. regression ?" is one of the questions.

If we take the question literally, as it is asked in its broad generality, the answer must be negative. There exists hardly any procedure for testing distortion of a possible kind in all circumstances. Most of the techniques mentioned as relevant for searching distortions of some specified sources are applicable in rather simple layouts only. Take also the known tests for normality. They are in fact applicable for samples composed of independent random variables. But beyond some simple cases, the experimental residuals, as obtainable from observations, are usually correlated. And in most cases there is no unique transformation into uncorrelated variables to which the Shapiro-Wilk or any other exact test could be applied.

The next question that arises is "How should we communicate the results to the user ?"

Certainly proper warnings of any distortions found by the tests employed, are what at least the user expects to get from an intelligent program. But he would be also interested to know how to overcome the defiances of assumptions when they appear. Therefore, not only tests but also suitable transformations or refinements are to be implemented into intelligent programs.

Finally, it is not a surprise that the Author asks "Should we attempt to replace classical methods altogether by robust techniques ?"

It is tempting to reply affirmatively to this question. But to do so does not help very much. One has then immediately to ask another question "Which robust techniques do you recommend for practical use ?" And as they have abounded so much recently it would be very difficult to suggest the right technique for each of the situations occuring in practical applications. Some promising general methods have been suggested (see, e.g. the review given by Huber, The Annals of Mathematical Statistics, 1972, Vol. 43, No 4, 1041-1067), but not much experience has yet been gathered to adventure their wide application in statistical packages.

As for the present, until the various tests and robust techniques are generalized sufficiently, so that their use in different situations becomes quite practical and comprehensive, nothing better can be suggested as the protection (A) indicated in the paper. And this also is the best way of teaching the user to explore his data with care and profoundness. Also of preparing him to the next stage, when the protection (C) will be fully implemented into intelligent statistical programs. It is doubtful if introducing the protection (B) would ever be a wise solution, as this would only teach the user to leave everything to the computer and to devolve the responsibility for the analysis on the statistician who stands behind the supposed intelligent program or system.

3.2. Comments on "Some new ideas in statistical programming - A software project"

The discussion on intelligent programs brings us right into the problem of further developments of statistically orientated computer languages. If the user is to take full advantage of the intelligent programs, he has to be supported by flexible programming facilities. These may only be provided by a very high-level language designed specifically for applied statistics. The LAPUMS-system described in the paper seems to be aimed at intelligent use of statistical programs.

From what is presented one can judge that LAPUMS is not merely one more statistical package or system, but a generalization of the most advanced achievements in statistical programming. This sounds very promising, but also seems to be a very complex and difficult endeavour.

Without going into technical details one can see the following questions arising in this context :

1. Which of the present general statistical packages and systems will be covered by LAPUMS ?

2. How far will it be possible to ensure that only well defined, optimal and not ambiguous statistical techniques and procedures will be included in the system ?

3. In what extent will the user be free in choosing desirable techniques and how will he be guided in his choice?

4. What kind of facilities will be available to teach the potential users how to take full advantage of the system?

5. How open will the system be for accepting new developments in statistical techniques and numerical solutions that will prove better than those already included ?

6. Is there any prospect for a wide distribution of the system among statis-

ticians that could not only use it but also contribute to its further development and improvement ?

If in the discussion on the paper at least some of the above questions could receive satisfactory answers, the session on "computers and statistics" would be a success.

4 - ANSWER OF Y. SCHEKTMAN

We have now carried out about 70 % of the work required by the first two stages of the LAPUMS project and it may be reasonable to assume that about within one year, the LOM language and its subsystem (at least in its simplified version) will be operational. At this stage, an implementation of all the means (wide diffusion of programmes and user's guide manuals, conferences, ...) likely to foster a close collaboration between all the research groups working on this subject would be most welcome.

The flexibility of the subsystem, the properties of the LOM language and the modular characteristics of the Delta meta-compiler will enable us to consider building up, in a single framework, the dialects and LOMOPs, required by the various points of interest of the applied statistics.

The quality (reliability, intelligence, ...) of the dialects (programmes) will of course depend on the quality of the researchers having carried out the analyses and performed the programmation.

We hope this answers adequately Mr.CALINSKI's questions.

Recent Developments in Statistics
J.R.Barra et al., editors

Invited Paper

THE ANALYSIS OF ASYMMETRY AND ORTHOGONALITY

J.C. Gower
Rothamsted Experimental Station
Harpenden, Herts
England

Multidimensional scaling and ordination methods analyse a symmetric matrix A by approximating it by distances, but when A is square but asymmetric, distance approximations become inappropriate. Amongst several possible methods discussed for analysing asymmetric data, illustrations are given of multidimensional unfolding and of the use and interpretation of the canonical decomposition of skew-symmetric matrices. The latter approach is also examined with two specific models that generate asymmetry, and with respect to analytical canonical decompositions of specific skew-symmetric matrices. The canonical decomposition of orthogonal matrices is interesting in its own right and also as another way of analysing asymmetry. Non-metric extensions of this work are outlined.

1. INTRODUCTION

Visual inspection and interpretation of diagrams is one of the most powerful tools available to statisticians. Most data benefit from a preliminary graphical analysis, if only to examine the plausibility of assumptions underlying a more formal analysis, and graphical plots and charts play an important part in presenting results and illustrating conclusions. Because of the difficulties of visualising the relationships between points in multidimensional space, there has been a special interest amongst multivariate statisticians in finding simple geometrical representations that approximate the relationships in useful ways.

An important class of multivariate methods, is that which presents results in the form of a map, possibly in many dimensions. Given a data-matrix X, either the elements classifying the rows or the elements classifying the columns can be represented on a map. In some methods both row-elements and column-elements are represented on the same map. The familiar concept of distance (usually, but not necessarily, Euclidean distance) is used to interpret relationships. One of the fundamental assumptions of this type of analysis is that relationships are symmetric so that the distance property $d_{ij} = d_{ji}$ is plausible. The additional assumption $d_{ii} = 0$ sometimes proves troublesome but can often be satisfied by some simple preliminary transformations of the data. The matrix D with elements d_{ij} may be derived from X or its elements may be recorded directly.

In the following I give a preliminary account of work on the analysis of square non-symmetric matrices D whose rows and columns are classified by the same elements. Thus although $d_{ij} \neq d_{ji}$ there may be a feeling that distance ideas at least partially explain the elements of D but that the differences between d_{ij} and d_{ji}, representing the asymmetry, may contain additional information that cannot be regarded as "noise".

As an example consider the Rothkopff (1957) morse-code data. This is a square matrix classified by the 36 morse code signals (26 letters and 10 digits). The (i,j)th cell contains the proportion of times (out of 150) that signal i followed by signal j was said to be the same. The relationship is not symmetric but rows and columns are similarly classified so that a map might throw light on the structure of the data-matrix. Shepard (1963), in one of the first applications of non-metric multidimensional scaling, analysed the symmetric matrix $\frac{1}{2}(D+D')$ and obtained a two-dimensional map of the 36 signals.

Its main feature was a trend in number of symbols (dots and dashes) in one direction coupled with a nearly orthogonal trend of number of dots to number of dashes. Thus in the lower left-hand corner occurs the symbol E(·) and in the upper right-hand corner occurs zero (- - - - -). Finer detailed structure also occurs. Successful as this analysis is, it says nothing about the asymmetric nature of the data.

Asymmetric data of this kind is not uncommon. For example in ecology we may count the number of times species i is the nearest neighbour of species j, in archaeology the number of times a coin minted in city i is excavated in city j, in genetics there is the diallele cross experiment, in growth studies the height of variety i when grown surrounded by variety j. Tobler (1975) discusses data giving the number of persons from place i who work/study/holiday etc. in site j. Here the sites form a known geographical map so the interest of the data lies entirely in its asymmetry. A feature of most of these examples is that row and column classifications correspond to different aspects of the same basic entities. Thus for morse code, rows correspond to pre-signals and columns to post-signals; in the growth studies rows correspond to inside plants and columns to outside plants, and so on.

Of course with some of these problems, a plausible model can be set up and its parameters estimated by standard statistical techniques. This is so for example with the diallele cross experiment (Yates (1947)). When such models exist they should be used, but here I am concerned more with general data-analytical techniques applicable when explicit models are not available.

In section 2 four possible methods of analysing asymmetry are outlined and in section 3 a fifth method that seems to be especially interesting is developed more fully. Section 4 examines the performance of the analysis with some explicit models and section 5 briefly sketches how a non-metric form of asymmetry analysis. Finally section 6 discusses the relationship between asymmetry and orthogonality.

2. METHODS OF ANALYSIS

Method 1

Expressing $D = U'\Sigma V$ in its singular value decomposition form where U and V are orthogonal and Σ diagonal, then

$$V'UD = V'\Sigma V$$

$$\text{and } DV'U = U'\Sigma U$$

are both symmetric. Thus the orthogonal matrix $H = V'U$ may be regarded as a measure of asymmetry. In particular, when D is symmetric $V = U$ and $H = I$, so interest focuses on the difference between H and I. The analysis of orthogonal matrices is discussed more fully below in section 6.

Method 2

Express $D = L\Lambda L^{-1}$ where Λ is diagonal (eigenvalues) and L and L^{-1} are pre- and post-vectors of D. Procrustes rotation theory (Schonemann (1970), and Gower (1971), (1975)) shows that $L\Lambda^{\frac{1}{2}}$ and $(L^{-1})'\Lambda^{\frac{1}{2}}$ are rotated to best fit, so the rows of these two matrices may be plotted against common axes. When D is symmetric $L^{-1} = L'$ so the two sets of points coincide. Differences between like points measure asymmetry, and the centroids of pairs of like points represent the symmetrical features of D. Alternatively if the elements of D are regarded as pseudo-distances, in analogy with metric scaling, it should be best first to transform D to have elements $(-\frac{1}{2}d_{ij}^2)$ and also to centre its rows and columns. We should then plot $L\Lambda^{\frac{1}{2}}$ and $(L^{-1})'\Lambda^{\frac{1}{2}}$ derived from the transformed matrix.

This difference in outlook, depending on whether the elements of D are better thought of as coordinates of points or as some kind of distance, is common in much of this and related work. Which is the better must be decided at the outset. A more serious practical difficulty is what to do about complex decompositions which may arise with real non-symmetric matrices.

Method 3

Permute the rows and columns of $D = (\,_{L}\!\!\diagdown^{U})$ such that $\big|\,|U| - |L|\,\big|$ is maximum, where $|U|$ ($|L|$) is the sum of the elements of the upper (lower) triangular matrix. Thus U and L (made symmetric) may be regarded as two distance matrices that give the worst possible interpretation on the asymmetry of D. The two maps derived from U and L may be visually compared or analytically fitted, noting that when D is symmetric $U = L'$. Minimising $\big|\,|U| - |L|\,\big|$ is less successful because many different permutations may give a value of the criterion that is near zero.

Method 4

Superimpose separate row and column maps using established techniques such as multidimensional unfolding or correspondance analysis.

Fig.1 shows a non-metric unfolding of the Rothkopff morse-code data, obtained from the MDSCAL program. Of ten trial solutions obtained, the figure reproduced is that with minimum stress found (.4932). The maximum stress for the remaining solutions was .5147 but despite the small range of stress-values there were some notable differences between the configurations, especially in the positions of letters E and T which correspond to single component morse-code signals. In obtaining these solutions, the diagonal elements giving the confusion between a symbol followed by itself were deleted, to reduce the tendency for pairs of points referring to like symbols to become too closely identified because of the strong natural "confusion" on the diagonal of the data-matrix. The diagram very largely reproduces Shepard's analysis of the symmetry, but does so for the pre- and post-signals separately. This supports the validity of analysing $\frac{1}{2}(D + D')$. However some interesting asymmetric relationships may also be discerned. Perhaps the clearest suggestion is that a two-component signal followed by a three component signal gives more confusion than the reverse. This tendency is more prevelant for pairs of similar signals. Thus M(- -) followed by O(- - -) or G(- - .) has more confusions than the reverse order but M followed by S(· · ·) has little differential confusion. However I(· ·) followed by S has many more confusions than the reverse order. A similar, but less pronounced pattern is seen when comparing three with four component signals. This is not the place for a full discussion of the analysis of this data, but the example suffices to show how unfolding methods can be useful in analysing asymmetry.

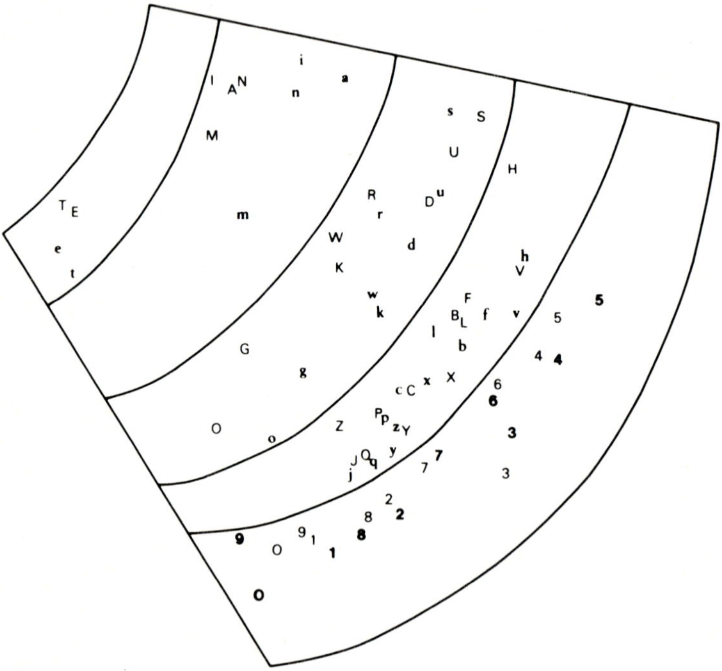

Fig.1. A non-metric unfolding of the Rothkopff Morse-Code Data, excluding self-confusions, obtained from MDSCAL. Upper-case letters refer to post-signals and lower-case letters to pre-signals.

The above methods are being appraised on the analysis of various data sets and it is hoped soon to publish examples of these types of analysis. However a fruitful mathematical approach has been found that has interesting interpretive possibilities. This will be described next.

3. CANONICAL ANALYSIS OF ASYMMETRY

We can ask what rank 1 matrix when added to D makes the result most symmetric. Symmetricity of a matrix A is measured simply as $\sum_{i<j} (a_{ij} - a_{ji})^2$ and this is the quantity that is to be minimised.

It turns out that the question is not meaningful, for any improvement to the symmetricity of D gained by adding a rank 1 matrix, can be equalled by adding a further rank 1 matrix. Thus we have to confine ourselves to asking what rank 2 matrix can best be added to D. A proof of this result is not given here but it is a consequence of the standard canonical decomposition of a skew-symmetric matrix N in the form

$$N = \sum_{i=1}^{[\frac{1}{2}n]} \sigma_i(u_{2i-1}u'_{2i} - u_{2i}u'_{2i-1}) \tag{1}$$

In (1) the vector pairs (u_{2i-1},u_{2i}) are conveniently (but not necessarily) chosen to be orthogonal. This fixes the values of σ_i but the solution is arbitrary to the extent that the vectors may be replaced by any other pair of orthogonal vectors in the same plane. The constraint $u'_{2i}.1 = 0$, where 1 is a vector of n units, gives a unique representation. Alternatively (1) may be written in matrix form

$$N = U\Sigma JU' \tag{2}$$

where $\Sigma = \text{diag}(\sigma_1, \sigma_1, \sigma_2, \sigma_2, \ldots)$ with a final zero if n, the order of N, is odd, and U is orthogonal with its columns equal to the vectors u_i and J is the elementary block-diagonal skew symmetric orthogonal matrix made up of 2 x 2 diagonal blocks $\begin{pmatrix} 0 & 1 \\ -1 & 0 \end{pmatrix}$ with a final unit diagonal element if n is odd. Because J is orthogonal, so is JU' and therefore (2) is the special form taken by the singular value decomposition of N, when it is skew-symmetric. From the Eckart-Young theorem (1935) this implies that the terms of (1) corresponding to the r largest singular values σ_i provide a least-squares fit of rank 2r to N.

Now if D is expressed as the sum of symmetric and skew symmetric matrices D = M + N, then the best (in the least squares sense) rank 2r matrix to add to D is formed from the first r terms in the canonical decomposition of N. Thus the canonical decomposition of N forms the basis for an analysis of the asymmetry of D, and the matrix M forms the basis for an analysis of symmetry.

To get the equivalent of a map from (1) we may plot the n points with coordinates given by the rows of (u_{2i-1},u_{2i}) $(i=1,2,\ldots,[\frac{1}{2}n])$. When G_1 is sufficiently large a 2-dimensional plot will give a good approximation to N, but just as in principal components and similar forms of analysis, more dimensions may be required for a good fit. It is important to realise that because the elements of the skew symmetric matrix N do not form a metric, then any kind of distance interpretation of such diagrams so drawn in unwarranted. To see the appropriate geometrical interpretation, consider the first vector pair $(u_1, u_2) \equiv (a, b)$ (say). Thus the (i,j)th element of N is fitted by $n_{ij} = \sigma_1 (a_i b_j - a_j b_i)$ so that if the point $P_i \equiv (a_i, b_i)$ and $P_j \equiv (a_j, b_j)$ then it is the <u>area</u> of the triangle formed by P_i,P_j and the origin, 0, that is proportional to n_{ij}. A consequence is that $\Delta(OP_iP) = -\Delta(OP_jP_i)$ accounting for the skew-symmetry. Also if $n_{ij} = n_{ik}$ then P_jP_k must be parallel to OP_i. Thus a property of this non-metric space is that all items equally skew-symmetric with item i lie on lines parallel to OP_i. Items with corresponding points on different sides of OP_i have opposite signs in their fitted skew symmetries. Consider next a fixed point P_j, then points lying on a line through the origin will have increasingly large asymmetry with P_i as one proceeds further from the origin. To familiarise oneself with this non-metric space takes time, but I think it will be found just as useful for interpreting skew-symmetry as Euclidean spaces have been found for interpreting symmetry.

In fact the space does have some metric properties. These follow from noting that

$$NN' = \sum_{i=1}^{[\frac{n}{2}]} \sigma_i^2 (u_{2i-1}u'_{2i-1} + u_{2i}u'_{2i}) \tag{3}$$

showing that the points P_i,P_j now plotted with scaling σ_i^2, rather than σ_i as previously, are distant from each other by an amount approximating to the distances between the ith and jth rows (or columns) of N regarded as coordinates. Consequently if the ith and jth items have similar skew-symmetries with the other items, the points P_i and P_j will be close on the map.

Fig.2 shows a skew-symmetric analysis in two dimensions of the Rothkopff Morse-Code data.

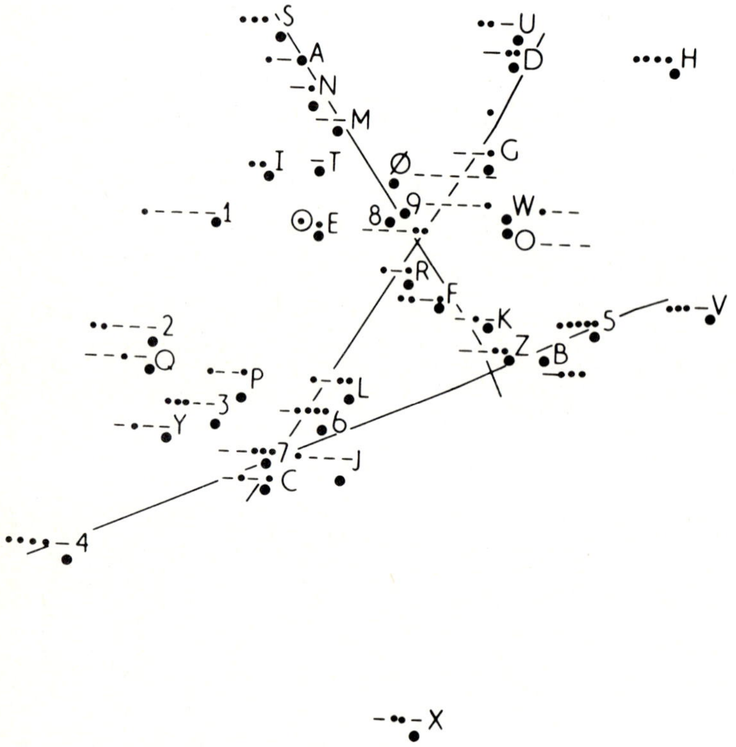

Fig.2. Canonical analysis of skew-symmetry. (Rothkopff Morse-Code)

The fit is not so good as Shepard's analysis of the symmetry, and perhaps some preliminary transformation would improve things. However the figure contains some structure of interest. The one, two and three symbol characters are progressively in bands further from the origin and generally, but not always, the four and five character symbols are even more remote from the origin. This suggests a progressive increase in differential confusion as the morse code signals get more complex, although of course points close together but both remote from the origin also have small differential confusion. The figure also shows the anomolous positions of H(····), V(...-), and X(-..-) and 4(....-) which all have large differential confusions with many other morse code symbols. There are also suggestions of co-linearities that are marked on the figure.

As another example consider table 1 of Jorré & Curnow (1975). This gives (theoretical) first passage times for the transition of one amino acid into another for all possible pairs chosen from 21 amino acids. The supposition is that easy transitions have a bearing on evolutionary developments. Figure 3 shows the analysis of asymmetry to give a nearly linear plot. This implies that the skew-symmetry can be expressed in the form 1u'-u1' so that in this case the distance $u_i - u_j$ is also a measure of asymmetry. This is because all relevant triangles in figure 2 have the same altitudes so that their areas are proportional to the lengths of their basis. Figure 4 shows a combination in one diagram of the analysis of symmetry and of skew symmetry for this set of data. Symmetry has been analysed in two dimensions by what Sibson (1972) terms "local order scaling". The asymmetry, being one-dimensional, is expressed in a third dimension which, remarkably, admits of the contour representation shown in the figure. The overall picture is of a deep trough with MET at the bottom. Transition times between amino acids lying on the same contour are equal in both directions. Those lying on different contours are easier transitions when going "downhill" than when going "uphill". At the very least figure 4 is a useful geometrical representation of a complex two-way table. However as the few codons, also marked on the figure, indicate there is additional structure that suggests other interesting relationships. A further feature of this example is that it illustrates how symmetric and skew-symmetric aspects can be combined in the one figure. This is something that we shall meet again in the following discussion.

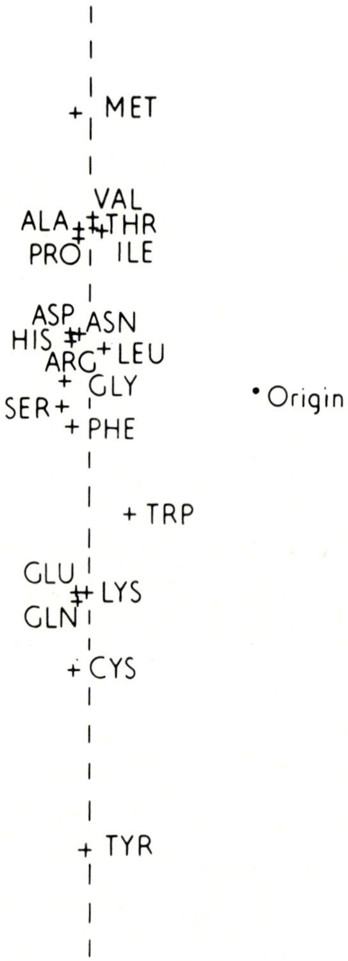

Fig.3. Amino acids - one dimensional asymmetry.

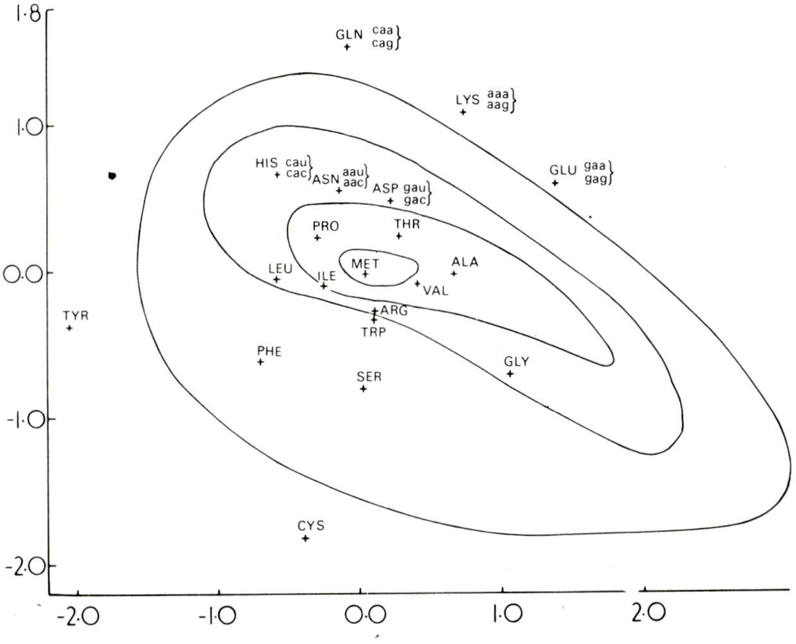

Fig.4. Amino acid first passage times represented in two dimensions by local-order scaling, with superimposed one dimmensional skew-symmetry represented in contour form. Codons bracketed and in lower case letters.

4. EXPLICIT MODELS

Clearly further insight on how asymmetry analysis operates in practice can be obtained by examining explicit models.

(1) Jet-stream model

First we consider a plane flying at constant velocity V between two towns P_i and P_j distance d_{ij} apart. The flight time will be d_{ij}/V, irrespective of the direction of flight. If now there is a jet-stream, velocity v, making an angle θ_{ij} with P_iP_j then the flight times t_{ij} and t_{ji} will be

$$t_{ij} = d_{ij}/(V+v\cos\theta_{ij})$$
$$\text{and } t_{ji} = d_{ij}/(V-v\cos\theta_{ij}). \qquad (4)$$

Thus the symmetric part of the flight time matrix has elements

$$m_{ij} \sim d_{ij}/V \qquad (5)$$

and the skew symmetric part has elements

$$n_{ij} \sim v\, d_{ij} \cos \theta_{ij}/V^2 \qquad (6)$$

where $\frac{v}{V}$ is sufficiently small to ignore $\frac{v^2}{V^2}$.

Given the matrix of flight times for n towns, an analysis of symmetry of (5) will reproduce the correct map of the country. The asymmetry of equation (6) represents the orthogonal projection of the map of the country on to the jet-stream direction, and is hence one dimensional. Consequently an analysis of the asymmetry (6) will give a one dimensional representation expressed algebraically in the form $1u'-u1'$, just as for the amino acids. Now however, rather than express the elements of the vector u as a third dimension, we know from the model that it can be represented as the orthogonal projection of the map onto a suitably chosen direction. To find this direction we have to rotate the direction until the sum of squares of the deviations between the projection and the elements of u is minimum, simultaneously estimating a scaling factor which gives the ratio of the two velocities v/V. The least squares problem of fitting $A(n \times k)$ to an orthogonal projection of $B(n \times p)$ where $p > k$ is non-trivial, but when $k = 1$ and $p = 2$ admits to simple algebraic treatment.

In practise the above method works very well and the effect of the approximation is barely noticeable for $v/V = 60\%$ and even beyond this. Again we have a combination of symmetry and skew-symmetry in one analysis. Given the flight time matrix it is possible to recover the map, the direction of the jet-stream and the velocity ratio v/V. But this has been possible, mainly because we know the model in advance; if we did not, it might be easy to overlook this simple interpretation of the data.

(2) Cyclone model

Next we consider a model similar to the jet-stream model except that the jet-stream is replaced by a cyclonic wind rotating about a point C at constant angular velocity ω. The flight times are now

$$\left.\begin{aligned} t_{ij} &= d_{ij}/(V + \omega h_{ij}) \\ \text{and } t_{ji} &= d_{ij}/(V - \omega h_{ij}) \end{aligned}\right\} \qquad (7)$$

where h_{ij} is the perpendicular distance of C from P_iP_j. Symmetric elements are:-

$$m_{ij} \sim d_{ij}/V \qquad (8)$$

and skew-symmetric elements are:-

$$n_{ij} \sim \omega\, d_{ij} h_{ij}/V^2 \qquad (9)$$

ignoring terms in ω^2/V^2.

Again the geographical map is obtained from (8). Equation (9) may be written

$$n_{ij} \sim \omega\, r_i r_j \sin(\theta_i - \theta_j)/V^2 \qquad (10)$$

where (r_i, θ_i) and (r_j, θ_j) are polar coordinates of P_i and P_j referred to C as origin. Because (10) is proportional to the area of the triangle CP_iP_j, it is clear that an analysis of the skew-symmetric matrix (10) will give points in two dimensions reproducing the areas given by (10). In general the maps given by analysing (8) and (10) will differ, because the vectors, satisfying (10) are not unique. Scaling one axis by a factor λ and the other by λ^{-1}

will not affect the areas; neither will any plane rotation. Areas are therefore invariant to an indefinitely long sequence of scalings followed by rotations, and such sequences of operations do not seem easily to be representable as a finite number of elementary transformations. The map given by (8) will occur somewhere amongst this set of solutions and can perhaps be found by fitting the skew-symmetric map to the basic map under a sequence of scalings and Procrustean rotations. Fortunately this indeterminacy is unimportant as it is more simple to estimate the centre of the cyclone as the point in the plane of the basic map that gives areas of best fit to those given by (10), allowing for a scaling factor that estimates ω/V.

Thus the complete model structure is once again reproducible and this requires a combination of analysis of symmetry and of skew-symmetry.

The jet stream, cyclone models and the amino acids data give three different ways of combining symmetrical and skew-symmetrical aspects in one analysis. In future analyses these possibilities can be borne in mind as useful interpretive devices that also give information on actual model specification. No doubt other similar devices remain to be discovered and further odels should be studied to build up experience and knowledge in this area.

(3) Explicit skew-symmetric decompositions

As a further contribution to understanding we have investigated the explicit singular value decompositions of some skew-symmetric matrices, that seem to be of potential interest. For example if we envisage an ecological situation with n species and define

$$x_{ij} = \begin{cases} 1 & \text{if species } i \text{ dominates species } j \\ 0 & \text{otherwise} \end{cases} \qquad (11)$$

then, if the species can be ordered from the most to the least dominant, then X is a matrix of units on and above the diagonal with zero below the diagonal. The derived skew-symmetric matrix N has +1 above and -1 below the diagonal.

To generalise slightly we may define

$$n_{ij} = \begin{cases} 1 + |i-j|\,a & \text{if } i < j \\ 0 & \text{if } i = j \\ -1-|i-j|\,a & \text{if } i > j \end{cases} \qquad (12)$$

a model that indicates a simple linear difference in growth size. The previous model is the special case of (12) with a = 0.

It turns out that the singular value decomposition of this matrix is given by the following result.

(i) The singular values σ_i are given by $\sigma_i = \cot \frac{1}{2}\theta$ where $\cot \frac{1}{2}\theta$ are the roots of the polynomial

$$\frac{\tan \frac{1}{2} n\,\theta}{\tan \frac{1}{2}\theta} = \frac{n\,p^2}{p^2 - 4} \qquad (13a)$$

where $p = na + 2$.

(ii) The pair of vectors corresponding to the ith singular value have elements:

$$\left.\begin{array}{l} \left.\begin{array}{l} u_{2i-1,j} = r\,[\cos(\tfrac{1}{2}(n+1) - j)\,\theta + c] \\ u_{2i,j} = r\,\sin(\tfrac{1}{2}(n+1) - j)\,\theta \end{array}\right\} j = 1,2,\ldots,n \\ \text{where} \quad r^2\left[n - \dfrac{\sin(n\theta)}{\sin\theta}\right] = 2 \\ \text{and} \quad c = -\dfrac{a}{p}\,\dfrac{\sin\frac{1}{2}n\theta}{\sin\frac{1}{2}\theta} \end{array}\right\} \quad (13b)$$

Geometrically this means that the n points arising from any pair of vectors lie on a circle radius r, centred at (rc,0), which is the origin when a = 0. The points are evenly distributed around part of the circumference, adjacent pairs all subtending the same angle θ at the centre, where $\cot\frac{1}{2}\theta$ is the appropriate singular value. Thus as the singular values get smaller the angles subtended get larger.

I have not yet investigated how fast the singular values fall off or how this fall off is related to the values of a, but it seems likely that it will be rare for (12) to have an adequate approximation in only two dimensions. However as $a \to \infty$, N/a tends to matrix of rank 2 of form $1u'-u1'$ where $u_j = [(4n^2-1)(n-2j+1)/3]^{\frac{1}{2}}$ suggesting that rank 2 solutions are likely to be more appropriate for higher values of a.

A further generalisation is to define:

$$\left.\begin{array}{lll} n_{ij} & = a_k & i > j \\ & = 0 & i = j \\ & = -a_k & 1 < j \end{array}\right\} \quad (14)$$

where k = Max (i,j).

We term this a <u>canopy</u> matrix or a <u>dominance</u> matrix. It arises in studies of forest canopies where species i grows to a height a_i so that if i and j grow together, the dominant species will determine the height of the forest canopy. Little progress has been made with an explicit decomposition of canopy matrices.

Yet another form of generalisation, to block skew-symmetric matrices, has yielded some results. Consider skew-matrices of the form

$$\begin{pmatrix} S_{11} & T_{12} & T_{13} & \cdots & T_{1p} \\ -T'_{12} & S_{22} & T_{23} & \cdots & T_{2p} \\ -T'_{13} & -T'_{23} & S_{33} & \cdots & T_{3p} \\ \vdots & \vdots & \vdots & & \vdots \\ -T'_{1p} & -T'_{2p} & -T'_{3p} & \cdots & S_{pp} \end{pmatrix} \quad (15)$$

where T_{ij} $(n_i \times n_j)$ has all its elements equal to t_{ij} and S_{ii} is skew-symmetric with all elements equal to s_i above the diagonal and $-s_i$ below. When p = 1 we have the matrix (11) with a = 0. For other values of p, plots for the vector pairs give p separate fan-shaped regions, the ith fan being an arc of a circle centred at the origin with n_i points distributed evenly along the circumference, subtending equal angles θ_i at the origin. The angles for successive fans are related to the singular value σ corresponding to the vectors, by:-

$$s_1 \cot \tfrac{1}{2} \theta_1 = s_2 \cot \tfrac{1}{2} \theta_2 = \dots = s_p \cot \tfrac{1}{2} \theta_p = \sigma \qquad (16)$$

When p = 2 and 3 explicit results have been obtained for the characteristic equations, giving the singular values, and for the radii of the fans and for the angles between the centres of every pair of fans (when p = 2, the centres subtend a right angle at the origin). These results all give information of potential value in interpreting skew-symmetric analyses of real data.

5. NON-METRIC ANALYSIS-OF-SKEW-SYMMETRY

Non-metric multidimensional scaling has been valuable in the analysis of symmetry. The basic idea is to fit a distance (usually Euclidean distance) to a symmetric matrix of data which are not necessarily metric, using a goodness-of-fit criterion (e.g. stress, Kruskal (1964)) that is invariant to monotonic transformations of the data.

If the original data are in the form of a skew-symmetric matrix there is no further problem in principle if, instead of using a distance, the term

$$\sum_1^r (u_{2i-1} u'_{2i} - u_{2i} u'_{2i-1})$$

is fitted to give a skew-symmetric 2r dimensional representation. The resulting map admits the geometrical interpretation in terms of areas discussed above. For definiteness we can seek solutions with $u'_{2i-1} u_{2i-1} = u'_{2i} u_{2i}$ and perhaps with $1'u_{2i-1} = 0$. The goodness-of-fit criteria remain valid but in the ranking process due regard must be taken of sign, as fitted values also are signed. To ensure invariance to permutations of the rows and columns of the skew-symmetric data-matrix N, stress should be calculated over all the non-diagonal elements of N or over all positive non-diagonal elements of N. The latter method permits computer programs to store only the elements below the diagonal of N, and their fitted values, but raises the difficulty of whether to associate the negative or positive fitted value with any zero non-diagonal element of N. If these minor precautions are kept in mind, existing multidimensional scaling programs should need little modification for them to include asymmetry analysis. This possibility is being investigated.

6. ANALYSIS OF ORTHOGONALITY

In section 2, method 1 of analysis lead to measuring asymmetry by the difference between a certain orthogonal matrix and the unit matrix. There is a well-known relationship between orthgonal and skew-symmetric matrices, due to Cayley, who showed that if N is skew then

$$H = (I + N)(I - N)^{-1} \qquad (17)$$

is orthogonal. Also if H is orthogonal

$$N = (I + H)(I - H)^{-1} \qquad (18)$$

is skew, subject to certain provisos when H has unit eigenvalues. Writing N in its singular value decomposition form $N = U \Sigma JU'$ and substituting in (17) gives

$$H = U(I + \Sigma J)(I - \Sigma J)^{-1} U' \qquad (19)$$

a minor variant of the standard canonical form for orthogonal matrices:-

$$H = UDU' \qquad (20)$$

where D is composed of 2 x 2 block diagonal matrices corresponding to elementary rotations

$$\begin{pmatrix} \cos\theta & \sin\theta \\ -\sin\theta & \cos\theta \end{pmatrix}$$

through an angle θ in the plane defined by the corresponding pairs of columns of U, and of diagonal values -1 (corresponding to reflections) and +1 (corresponding to identity or null transformations).

Thus to analysis H we can convert it to a skew-symmetric matrix, using (18) paying special attention to the effects of unit eigen values, and proceed as before. Alternatively we can use (20) and write

$$H - I = U(D-I)U'. \qquad (21)$$

D-I is similar to D but the 2 x 2 diagonal matrices become

$$2 \sin \tfrac{1}{2}\theta \begin{pmatrix} -\sin \tfrac{1}{2}\theta & \cos \tfrac{1}{2}\theta \\ -\cos \tfrac{1}{2}\theta & -\sin \tfrac{1}{2}\theta \end{pmatrix}$$

and the diagonal values of -1 and 1 become -2 and zero respectively.

This gives the singular value decomposition of H-I so that the best least squares fit of an orthogonal matrix H_1 to H which minimises $H-H_1$ requires the least squares fit to H-I which is given by the Eckart-Young theorem. The diagonal matrix of singular values is

$$(2 \sin \tfrac{1}{2}\theta_1,\ 2 \sin \tfrac{1}{2}\theta_1,\ 2 \sin \tfrac{1}{2}\theta_2, 2 \sin \tfrac{1}{2}\theta_2,\dots,2 \sin \tfrac{1}{2}\theta_k,\ 2 \sin \tfrac{1}{2}\theta_k,\ 0,0,\dots 0,2,2,\dots,2)$$

showing that the larger angles of rotation contribute most to the least squares fit. Thus for example the best fitting rank 2 matrix to H-I is

$$-(1 - \cos\theta)\ (u_1u_1' + u_2u_2') - \sin\theta\ (u_1u_2'-u_2u_1')$$

where θ is the value of θ_i which maximises $\sin^2 \frac{1}{2}\theta_i$.
Thus this decomposition gives a method for approximating an orthogonal matrix by a unit matrix associated with progressively less important plane rotations. Obvious applications are in interpreting orthogonal matrices of eigenvectors given, for example, by principal components analysis, but I have no applications to report at present.

ACKNOWLEDGEMENTS

This work was initiated in 1975 when I was with CSIRO in Australia. Collaborators working on certain aspects were G. Laslett, P. Milne and D. Ratcliffe. Special thanks go to A.G. Constantine for his contribution to the canonical decomposition aspects which we hope to describe in more detail in a joint publication.

REFERENCES

Eckart, V. & Young, G. (1935). The approximation of one matrix by another of lower rank. Psychometrika, 1, 211-218.

Gower, J.C. (1971). Statistical methods of comparing different multivariate analyses of the same data. In Mathematics in the Archaeological & Historical Sciences. Eds. Hodson, Kendall & Tartu. The Univ. Press. Edinburgh, 138-149.

Gower, J.C. (1976). Procrustes rotational fitting problems. The Mathematical Scientist. Suppl. to Vol.1, No.1.

Jorre, R.P & Curnow, R.N. (1975). The evolution of the genetic code. Biochimie, 57, 1147-1154.

Kruskal, J.B. (1964). Multidimensional scaling by optimising goodness of fit to a non-metric hypothesis. Psychometrika, 29, 1-27.

Rothkopff (1957). A measure of stimulus similarity and errors in some paired-associate learning tasks. J.exp.Psych., 53, 94-101.

Schonemann, P.H. & Carroll, R.M. (1970). Fitting one matrix to another under choice of a central dilation and a rigid motion. Psychometrika, 35, 245-256.

Shepard, R.N. (1962a). The analysis of proximities: multidimensional scaling with an unknown distance function I. Psychometrika, 27, 125-139

Shepard, R.N. (1962b). The analysis of proximities: multidimensional scaling with an unknown distance function II. Psychometrika, 27, 219-146.

Sibson, R. (1972). Order invariant methods for data analysis. J.R.statist.Soc., (B), 34, 311-349.

Tobler, W. (1975). Spatial Interaction Patterns, IIASA research report RR-75-19.

Yates, F. (1947). The analysis of data from all possible reciprocal crosses between a set of parental lines. Heredity, 1, 297-301.

Recent Developments in Statistics
J.R.Barra et al., editors

Invited Paper

OPERATORS RELATED TO A DATA MATRIX

Y. Escoufier
Centre de Recherche en Informatique
et Gestion
Université des Sciences et Techniques
du Languedoc
Montpellier , France

I - INTRODUCTION

The results shown in this paper concern the analysis of data, and especially the comparison of several analyses. They have as a common starting point an accepted party which must be specified before anything else; a data analysis of n subjects is characterized by the couple (D,D_p) where D is the matrix n x n of distances between subjects and D_p a diagonal matrix of weights affected to each of the subjects.

The inevitable consequence of this point of view is that we are led to make a comparison of the data analyses given for the same subjects by comparing the couples (D,D_p) which characterize them ; paragraph 2 deals with the way in which this aim can be achieved. Paragraph 3 shows how this approach allows a new and unifying look at the different methods of multidimensional statistical analysis. Paragraph 4 touches the problem of variable choice while paragraph 5 deals with the joint treatment of several data matrices.

II - THE COMPARISON OF DATA ANALYSIS

II-1 Let us take, for this paragraph, a point of view which could be called "traditional" in which a data analysis of n subjects is defined by the triplet (X,Q,D_p) in which :

X is a matrix, p x n, containing the values taken by p numeric variables on each of the n subjects.

The column X^j of X contains the observations made on the individual j.

Q is a positive definite or semi-definite matrix, p x p, which allows the calculation of the distances between the individuals

$$D_{jk} = \left[(X^j - X^k)'Q(X^j - X^k)\right]^{1/2}$$

D_p is a diagonal positive matrix, n x n, of weights affected to the subjects.

Let us define the matrix W, the elements of which are : $W_{jk} = X^{j'}Q\,X^k$

We have : $$D_{jk} = \left[W_{jj} + W_{kk} - 2\,W_{jk}\right]^{1/2}$$

It is obvious that the knowledge of W allows the calculation of D and that two analyses which lead to the proportional matrices W of proportionality $k > 0$ lead necessarily to the matrices D which have a coefficient of proportionality equal to $k^{1/2}$. This allows us, in this "traditional" point of view to substitute in the comparaison of the couples (D,D_p) those of the couples (W,D_p).

Remarks :

Let $\underset{\sim}{1}$ be the vector, n x 1, of components all equal to 1 and Y the centered matrix p x n associated with X. We can write : $Y = X(I - D_p\underset{\sim}{1}\,\underset{\sim}{1}')$.

We can easily verify that :

a) $Y'Q\,Y = (I - \underset{\sim}{1}\,\underset{\sim}{1}'\,D_p)\,W(I - D_p\,\underset{\sim}{1}\,\underset{\sim}{1}')$

b) The distances calculated from either W or Y'Q Y are the same.

c) Y' Q Y admits the eigenvector $D_p\,\underset{\sim}{1}$ associated with the zero eigenvalue so that for all other eigenvectors U, we have $U'D_p\,\underset{\sim}{1} = 0$.

Because of these remarks, we choose to calculate W and then to work with $(I - \underset{\sim}{1}\,\underset{\sim}{1}'\,D_p)\,W\,(I - D_p\,\underset{\sim}{1}\,\underset{\sim}{1}')$.

II-2 Let us now take a point of view less traditional in which the data are a couple (D^*, D_p) in which :

D_p has the same significance as in II-1.

D^* is a n x n matrix of dissimilarities between subjects obtained either as the result of manipulations of variables, eventually qualitatively, or in a purely subjective way.

Given $\mathcal{D}$ the matrix n x n the elements of which are the squares of those of D^*, the Torgerson method defines a matrix of scalar products W^* by the formula :

$$W^* = -(I - \underset{\sim}{1}\,\underset{\sim}{1}'D_p)\,\mathcal{D}\,(I - D_p\,\underset{\sim}{1}\,\underset{\sim}{1}')$$

which is such that :

$$D^*_{jk} = \left[W^*_{jj} + W^*_{kk} - 2\,W^*_{jk}\right]^{1/2}$$

but nothing assures that there exists a real configuration of n points, accepting as mutual distances the elements of D^*.

Because the obtaining of real configurations of points is the center of all methods of data analysis, we have chosen in this case to substitute for W^* the positive definite matrix W which is its approximation in the sense of least squares. If $\{U_i\ ;\ i = 1, \ldots, n\}$ is the set of eigenvectors (of unit norm) of W^* associated with the eigenvalues $\{\lambda_i\ ;\ i = 1, \ldots, n\}$ we know that :

$$W = \sum_{i \in I} \lambda_i\,U_i\,U_i' \quad \text{where} \quad I = \left\{ i / \lambda_i > 0 \right\}$$

We remark that for all numbers k, the matrix $W^* + kI$ has the same eigenvectors as W^*, the eigenvalues being $\lambda_i + k$, thus by taking $k^* = \max_{\lambda_i < 0} |\lambda_i|$, $W^* + k^*I$ is positive semi-definite and the distances corresponding to it are equal to :

$$\left[D^{*2}_{jk} + 2\,k^*\right]^{1/2} = D^*_{jk}\left[1 + \frac{2\,k^*}{D^{*2}_{jk}}\right]^{1/2}$$

Consequently, working with the matrix $W^* + k^*I$ (a method suggested to me by J.P. Pagès) does not modify the order of distances and thus is a happy alternative to the "additive constant method". Secondly, the eigenvectors associated with the largest eigenvalues of W and $W^* + k^* I$ are the same (those also of W^*). The configurations given by W and $W^* + k^* I$ are thus neighbours and are only differenciated by the fact that the eigenvectors of W have as norm $\lambda^{1/2}$ where those of $W^* + k^* I$ have as norm $(\lambda + k^*)^{1/2}$.

Thus we consider that in this less traditional point of view we can substitute for the couple (D^*, D_p) the couples (W, D_p) or $(W^* + k^* I, D_p)$. No matter which solution is chosen, the study is charácterized by a couple which we call (W, D_p).

II-3 Let us now look at the operator on R^n defined by the matrix WD_p. Our previous choices put WD_p in the rank $k \leqslant n$.

If U_i is an eigenvector of WD_p, $D_p^{1/2} U_i$ is the eigenvector of $D_p^{1/2} W D_p^{1/2}$ for the same eigenvalue. It follows:

a) that the eigenvectors of $W D_p$ are D_p-orthogonal.

b) that if the U_i are chosen as D_p-orthogonal, then we have :

$$\sum_{i=1}^{n} \lambda_i U_i U_i' = W$$

At this point, the operator WD_p appears as characteristic of the couple (W, D_p), and comparing two studies E_1 and E_2 is the same as comparing the operators $W_1 D_1$ and $W_2 D_2$ which are associated with them.

It is known that for the square matrices A and B, Tr(AB) is a scalar product, the corresponding norm being $\left[Tr(A^2)\right]^{1/2}$. If thus two studies E_1 and E_2 lead to the matrices $W_1 D_1$ and $W_2 D_2$, we can define a distance between E_1 and E_2 by :

$$d_1(E_1, E_2) = \left[Tr(W_1 D_1 - W_2 D_2)^2\right]^{1/2}$$

$$= \left[Tr(W_1 D_1)^2 + Tr(W_2 D_2)^2 - 2\, Tr(W_1 D_1 W_2 D_2)\right]^{1/2}$$

In order to make the proportional matrices equivalent, we use as well

$$d_2(E_1, E_2) = \left[Tr\left[W_1 D_1 / \left[Tr(W_1 D_1)^2\right]^{1/2} - W_2 D_2 / \left[tr(W_2 D_2)^2\right]^{1/2}\right]^{1/2}\right]^{1/2}$$

$$= 2\left[1 - 1 Tr(W_1 D_1 W_2 D_2) / \left[Tr(W_1 D_1)^2\, Tr(W_2 D_2)^2\right]^{1/2}\right]^{1/2}$$

To clarify the significances of these distances, we must make the following remarks :

a) Let us represent a study by a point P of $R^{n \times n}$ of coordinates $p_{(i-1)n+j} = (W D_p)_{ij}$. Using in $R^{n \times n}$ the identity metric, if p^1 and p^2 are the points representative of the two studies E_1 and E_2, the distance between p^1 and p^2 in $R^{n \times n}$ is $d_1(E_1, E_2)$.

b) Let us now return to the point of view which we called "traditionnal" and suppose that we have to compare two studies (X, I, D_p) and (Y, I, D_p) given for the same individuals with the same weights and the identity metrics in both cases. If we remark that :

$X D_p X' = S_{11}$		the sample covariance matrix as estimated from the variables defining the rows of X,
$Y D_p Y' = S_{22}$		the sample covariance matrix as estimated from the variables defining the rows of Y,

$Y D_p X' = S_{21}$ and $X D_p Y' = S_{12}$,

then $Tr(W_1 D_p W_2 D_p) = Tr(S_{12} S_{21})$

and $Tr(W_1 D_p W_2 D_p) / \left[Tr(W_1 D_p)^2 \; Tr(W_2 D_p)^2\right]^{1/2} = Tr(S_{12} S_{21}) / \left[Tr(S_{11}^2) Tr(S_{22})^2\right]^{1/2}$

The expressions on the right-hand side of the two preceding equalities are analogues of the coefficients COVV and RV introduced in (2) for two random vectors defined on the same probability space. By extension we thus use :

$$COVV(E_1, E_2) = Tr(W_1 \; D_p \; W_2 \; D_p)$$

$$RV(E_1, E_2) = Tr(W_1 \; D_p \; W_2 \; D_p) / \left[Tr(W_1 D_p)^2 \; Tr(W_2 \; D_p)^2\right]^{1/2}$$

III - MULTIDIMENSIONAL STATISTICAL ANALYSIS

Without going into the details of the demonstrations which the interested reader can find in a recent paper (5), we will now show how the different methods of multidimensional statistical analysis could be presented starting from the coefficient RV.

Note, first, that for every matrix Q, p x p, positive semi-definite, there exists a matrix L, p x t, $t \leq p$, such that $Q = LL'$. This allows us to describe a study on a matrix X, p x n in the form (X, LL', D_p). If we consider a second study (Y, MM', D_p) with Y of dimensions q x n, for the same individuals and the same weights, we can adopt the notation RV(L'X, M'Y). Note also that the matrix L is such that Q=LL' is determined to within a rotation in R^t. Inasmuch as we want to determine L, we are thus forced to introduce the conditions which allow its determination. In the following table, we can find the description of problems the solutions of which correspond with the methods of multidimensional analysis.

Problems	solutions
Find M, p x t, $t \leq p$ which maximises RV(X, M'X) under the constraint : $M' X D_p X M$ diagonal	First t principal components of X
Find M, q x t, $t \leq \inf(p,q)$ which maximises RV(X, M'Y) under the constraint : $M'Y D_p Y M$ diagonal	First t principal components of Y with respect to X
<u>particular case</u> : t = p	$M' = (X D_p Y')(Y D_p Y')^{-1}$ gives the maximum and $(X-M'Y)D_p Y' = 0$ Thus we have the regression of X with respect to Y.
Find L, p x t, and M, q x t which maximise RV(L'X, M'Y) under the constraints $L' X D_p X'L = M' Y D_p Y'M = I_t$	First t couples of canonical variables
<u>Cas particulier</u> : The individuals are divided into k groups. Y^j is the mean vector in the group containing the jth indidual	First t discriminant functions (L = M)

N.B. - Our approach leads to the introduction of all the methods in terms of comparison of distance matrices. This gives sometimes an unusual point of view ; for example the regression of X with respect to Y is interpreted as the research for the metric MM' to be taken for Y, in such a way that the distances in the study (Y, MM', D_p) are as close as possible to the distances in the study (X,I,D_p).

Let us finish this paragraph by noting that for all the studies given under the form of a couple (D, D_p) we have associated a couple (W, D_p). The factorization of W in the form $W = X'X$ allows the presentation of the study in the form (X,I,D_p) and thus the extension of all the classical methods of multidimensional analysis for data (D,D_p).

IV - CHOICE OF VARIABLES

The use of the RV coefficient and in particular the results of the previous paragraph allows us to study much more clearly the problem of the choice of variables by clearly showing that the choice is always accompanied by a choice of the metric to be used. Let us take two given studies (X,LL',D_p) and (Y,MM',D_p) with X, p x n, and Y, q x n, given for the same individuals (possibly Y = X and M = L).

The problem is to find Z , t x n, t $<$q, extracted from Y, which for a metric NN' where N is given or has to be found, realizes the maximum of RV(L'X, N'Z).

Problem 1 : $N = I_t$

We want to use the variables Z as they are. The distance between two individuals is the classical euclidean distance. The study (Z, I_t, D_p) is, from the point of view of the principal components method the closest possible to $(X. LL', D_p)$.

Problem 2 : N positive diagonal

The affecting of weight to the chosen variables is accepted (i.e. the changing of units) but we want to conserve the experimental significance of the variables. We look for both Z and N.

Problem 3 : Any N

Both Z and N have to be found, which means that the principal components analysis on (X, LL', D_p) and (Z, NN', D_p) must be as close as possible. Remark that for a given Z, N is given by the principal components of Z with respect to X.

Problem 4 : $N' = (L' X D_p Z') (Z D_p Z')^{-1}$

Following the results of the previous paragraph, we realize the regression of L'X with respect to Z. The problem is thus to find the best sub-set of Y from the point of view of the regression of L'X.

Problem 5 : $LL' = (X D_p X')^{-1}$, $MM' = (Z D_p Z')^{-1}$

One can see as well (by calculating RV for example), that the variables retained are those that are susceptible to allow the canonical analysis, with X the most satisfying.

Problem 6 : X matrix p x k of the averages of k groups $(LL')^{-1}$ inter-groups covariance matrix for X .

Y matrix q x k of the averages of the same k groups $(NN')^{-1}$ inter-groups covariance matrix for Z extracted from Y. The maximisation of RV(L'X, N'Z) is the same thing as looking for the best sub-set of the variables Y, in the sense where it allows a discriminant analysis, the closest possible to that which allows X.

V - JOINT ANALYSES OF SEVERAL DATA MATRICES

V-1 Let us define $\{E_i ; i \in I\}$ a family of data analyses on the same individuals with the same weights. With $\{E_i ; i \in I\}$ is associated the family of operators $\{W_i D_p ; i \in I\}$. We propose to study the proximities and the differences of E_i.

Consider the matrix C the elements of which are $C_{ij} = COVV(E_i, E_j)$ C is the matrix of scalar products between the operators and, following the remark (a) in II_3, there exists a real configuration of points compatible with C. The canonical factorisation of C gives a visualisation of this configuration of points in which each study is represented by a point. The distances between the points are $(C_{ii} + C_{jj} - 2 C_{ij})^{1/2}$. Particularly if $W_j D_p = k W_i D_p$, the origin and the points p^i and p^j associated with E_i and E_j are colinear and $\vec{Op^j} = k\vec{Op^i}$. The converse is true . Of course practical thought leads to limit the representation to two or three eigenvectors of C associated with the largest eigenvalues. The quality of this approximation is appreciated by the usual tools : rate between the extracted eigenvalues and Tr(C) for example.

Rather than work with C, we can choose to work with R the elements of which are $R_{ij} = RV(E_i, E_j)$. In this case, studies leading to proportional operators are represented by confused points. Moreover, for all $i \in I$, $\|\vec{Op^i}\| = 1$ and $\langle \vec{Op^i}, \vec{Op^j} \rangle = RV(E_i, E_j)$. If the representation by the first two eigenvectors of R keeps, for the projections of the vectors Op^i, the norms neighbour to the unity, then the value of $RV(E_i, E_j)$ is rather equal to the cosine of the angle made by the projections of $\vec{Op^i}$ and $\vec{Op^j}$.

V-2 The coefficients $COVV(E_i, E_j)$ and thus $RV(E_i, E_j)$ are always non-negative. So, the matrices C and R have a first eigenvector the elements of which are always non-negative. Let us take $\{\alpha_i ; i \in I\}$ as the components of this vector. We have the following theorem :

For all $\{\beta_i ; i \in I\}$ such that for all i , $\beta_i \geq 0$ and $\sum_{i \in I} \beta_i^2 = \sum_{i \in I} \alpha_i^2$, we have :

$$\sum_{i \in I} \left[RV\left(\sum_{j \in I} \beta_j W_j D_p\right)\right]^2 \leq \sum_{i \in I} \left[RV\left(\sum_{j \in I} \alpha_j W_j D_p, W_i D_p\right)\right]^2$$

Thus, $W D_p = \sum_{i \in I} \alpha_i W_i D_p$ is an operator which constitutes the best compromise between all the studies in the sense of the criteria used in the theorem. Because it is a positive linear combination of positive semi-definite operators, $W D_p$ is also positive semi-definite. We can thus obtain by a canonical factorisation a visualisation of the objects which constitutes a compromise between all the studies.

Remark : We have as well an analogous theorem for COVV.

V-3 The space of the representations of objects as they are seen by the compromise can be used as reference space into which each one of the initial studies is projected. This possibility allows us to see the way in which each one of the studies differs from the compromise. In the case of a chronological study, it allows us to visualise the evolution of the different objects during the time. Examples of the application of the theory developed in the fifth section are treated in (3). A program exists and the way in which it can be obtained will be given by the author.

REFERENCES

Caillez F. and Pages J.P. (1976), Introduction à l'Analyse des données - SMASH 9, rue Duban, 75016 PARIS.

Escoufier Y. (1973), Le traitement des variables vectorielles, Biometrics 29, p. 751-760.

L'Hermier des Plantes H. (1976), Structuration des tableaux à trois indices de la statistique. Thèse de 3ème cycle. Centre de Recherche en Informatique et Gestion, avenue d'Occitanie, Montpellier.

Pages J.P., A propos des opérateurs d'Y. Escoufier, séminaires IRIA, classification automatique et perception par ordinateur, 1974, p. 261-271.

Robert P. and Escoufier Y., A unifying tool for linear multivariate statistical methods : the RV. coefficients. Applied Statistics (à paraître).

Recent Developments in Statistics
J.R.Barra et al., editors

Invited Paper

APPLICATIONS OF CONVEX ANALYSIS TO MULTIDIMENSIONAL SCALING

Jan de Leeuw
Department of Data Theory
University of Leiden
Leiden, The Netherlands

In this paper we discuss the convergence of an algorithm for metric and nonmetric multidimensional scaling that is very similar to the C-matrix algorithm of Guttman. The paper improves some earlier results in two respects. In the first place the analysis is extended to cover general Minkovski metrics, in the second place a more elementary proof of convergence based on results of Robert is presented.

1: INTRODUCTION

In multidimensional scaling (MDS) problems the data consist of m nonnegative square matrices $\Delta_1, \Delta_2, \ldots, \Delta_m$ of order n, whose elements are interpreted as measures of dissimilarity between the n objects $o_1, o_2, \ldots, o_n$, measured at m replications $r_1, r_2, \ldots, r_m$. Thus δ_{ijk} is the dissimilarity between objects o_i and o_j at replication r_k. In a psychological context the objects are often called stimuli, and the replications are defined by the dissimilarity judgments of different subjects. Moreover we assume that m nonnegative square matrices $W_1, W_2, \ldots, W_m$ of order n are given, whose elements are interpreted as weights, i.e. w_{ijk} indicates the relative importance or precision of measurement δ_{ijk}.

Multidimensional scaling techniques represent the objects $o_1, o_2, \ldots, o_n$ as points $x_1, x_2, \ldots, x_n$ in a metric space $\langle\Omega,d\rangle$ in such a way that the distances $d(x_i,x_j)$ are approximately equal to the dissimilarities δ_{ijk}. We sometimes write d_{ij} for $d(x_i,x_j)$.

In this paper we study representations of $O = \{o_1, o_2, \ldots, o_n\}$ in the space of all p-tuples of real numbers, in which the metric is defined by a norm $||.||$. Thus $d_{ij} = ||x_i - x_j||$. A representation of O is then an n x p matrix X, with row i representing o_i. We also define the notation $d_{ij}(X)$ for the distance between x_i and x_j.

The loss function we use in this paper to evaluate the badness-of-fit of a particular representation X is

$$\sigma(X) = \sum_{k=1}^{m} \sum_{i=1}^{n} \sum_{j=1}^{n} w_{ijk}(\delta_{ijk} - d_{ij}(X))^2.$$

Clearly $\sigma(X) \geq 0$, and $\sigma(X) = 0$ if and only if $d_{ij}(X) = \delta_{ijk}$ for all i,j,k with $w_{ijk} \neq 0$. If $w_{ijk} = 0$ then the value of $\sigma(X)$ does not depend on δ_{ijk}. This provides us with a simple device for handling <u>missing data</u>: if the observation corresponding with the triple i,j,k is missing, then we can choose δ_{ijk} arbitrarily, and set $w_{ijk} = 0$.

The first and most basic MDS problem we study in this paper is the minimization of $\sigma(X)$ over all n x p <u>configuration matrices</u> X. This is usually called <u>metric</u> MDS, to distinguish it from the more general <u>nonmetric</u> problem in which the δ_{ijk} are only partially known. Or, more precisely, in the MDS problem as we have defined it so far the δ_{ijk} have to be either completely known or completely unknown (missing), nonmetric MDS deals with various kinds of intermediate cases. In a later section of the paper we discuss straigthforward extensions of our techniques that cover nonmetric MDS.

2: PREVIOUS WORK

Algorithms for the minimization of $\sigma(X)$ have been proposed earlier by Kruskal (1964 a,b) and Guttman (1968). In fact both Kruskal and Guttman propose algorithms to solve the more general nonmetric scaling problems. In this general nonmetric case there are substantial differences between the two approaches, but if we specialize them to the metric MDS problem they become very similar. A detailed discussion and comparison of the algorithms and the corresponding computer programs is available in Lingoes and Roskam (1973). We only discuss the main ideas, and the major differences between the two approaches in the metric case.

Kruskal proposes a gradient method of the form

$$X \leftarrow X - \alpha\nabla\sigma(X),$$

where $\nabla\sigma(X)$ is the <u>gradient</u> of σ at X, i.e. the n x p matrix of partial derivatives, and where $\alpha > 0$ is a <u>step-size</u>. Guttman on the other hand shows that the stationary equation $\nabla\sigma(X) = 0$ can be rewritten in the form $X - C(X)X = 0$, where $C(X)$ is a square symmetric matrix valued function of X. He proposes the iterative process

$$X \leftarrow C(X)X.$$

By substituting Guttman's formula for the gradient in Kruskal's algorithm we find

$$X \leftarrow X - \alpha(X - C(X)X) = (1 - \alpha)X + \alpha C(X)X.$$

Thus Guttman's algorithm is a special case of Kruskal's with a constant step-size α equal to one. And Kruskal's algorithm can be interpreted as an over- or underrelaxed version of Guttman's algorithm. Interesting geometrical and mechanical interpretations of these algorithms have been discussed by Kruskal and Hart (1965), McGee (1966), and Gleason (1967).

There are two problems with the Kruskal-Guttman approach that specifically interest us. In the first place the distance function $d_{ij}(X)$ is typically not differentiable at all configurations X with $x_i = x_j$. This implies that a gradient method cannot be applied to $\sigma(X)$ without further specifications, it implies that the usual convergence theorems for gradient methods are invalid, and it also implies that local minimum points of $\sigma(X)$ need not satisfy the stationary equations. The second problem is that it has not been shown, for either Kruskal's "heuristic" or for Guttman's "constant" step-size procedure, that the resulting algorithms are indeed convergent. Kruskal (1969, 1971) has proved some partial results, and Guttman (1968) and Lingoes and Roskam (1973) have some heuristic arguments and some empirical results, but there is no complete convergence proof.

Until recently these problems have been ignored, or they have been "dissolved" by transforming the model and, through the model, the loss function. ALSCAL, for example, defines the loss on the squared distances and squared dissimilarities (Takane, Young, De Leeuw, 1976). Classical metric scaling methods apply both squaring and double centering to the dissimilarities, and then define the loss on the scalar products (Torgerson, 1958). These transformations do make the loss functions better behaved in some respects, but they do not really solve the problems with the Kruskal-Guttman approach, they merely transform them away. Moreover using transformations seems less direct, and does not generalize to other distance functions than the usual Euclidean one.

In this paper we derive a simple algorithm for directly minimizing $\sigma(X)$, that can easily proved to be convergent. Although the derivation of the algorithm does not use differentiation or stationary equations, it turns out that the algorithm is identical to Guttman's C-matrix method. One (modest) interpretation of the main result of this paper is that it provides a convergence proof for Guttman's algorithm. Another interpretation is that we show that the C-matrix method should not be interpreted as a gradient method. It is more natural to view it as a minimization method based on an analysis of the convexity properties of the distance function. In fact it may very well be better not to interprete Kruskal's algorithm as a gradient method, but as a relaxed version of the C-matrix method. This interpretation makes it possible, for example, to construct interesting optimal step-size procedures, that do not use heuristic arguments and several arbitrary parameters.

3: PROBLEM REDUCTION: PARTITIONING

In this section we reduce the MDS problem to a more simple form by partitioning the loss function into additive components. For this purpose we define

$$\underline{w}_{ij} = \frac{1}{m}\sum_{k=1}^{m} w_{ijk},$$

$$\underline{\delta}_{ij} = \sum_{k=1}^{m} w_{ijk}\delta_{ijk} \;/\; m\underline{w}_{ij},$$

$$w_{ij} = \tfrac{1}{2}(\underline{w}_{ij} + \underline{w}_{ji}),$$

$$\delta_{ij} = \frac{\underline{w}_{ij}\underline{\delta}_{ij} + \underline{w}_{ji}\underline{\delta}_{ji}}{\underline{w}_{ij} + \underline{w}_{ji}} \quad \text{for } i \neq j,$$

$$w_{ii} = \delta_{ii} = 0.$$

As is customary in the analysis of variance we collect the components of the partitioning in a table.

SOURCE	LOSS COMPONENT
Proper loss	$\sum_{i=1}^{n}\sum_{j=1}^{n} w_{ij}(\delta_{ij} - d_{ij}(X))^2.$
Symmetry	$\sum_{i \neq j}^{n}\sum^{n} \{\underline{w}_{ij}\underline{\delta}_{ij}^2 - w_{ij}\delta_{ij}^2\}.$
Hollowness	$\sum_{i=1}^{n} \underline{w}_{ii}\underline{\delta}_{ii}^2 .$
Individual differences	$\sum_{k=1}^{m}\sum_{i=1}^{n}\sum_{j=1}^{n} w_{ijk}(\delta_{ijk} - \underline{\delta}_{ij})^2.$
Total loss	$\sum_{k=1}^{m}\sum_{i=1}^{n}\sum_{j=1}^{n} w_{ijk}(\delta_{ijk} - d_{ij}(X))^2.$

It is obvious that we minimize the total loss if we minimize the proper loss, and that the proper loss is more simple. In fact in defining the proper loss we can suppose without loss of generality that both the weigths and the dissimilarities are symmetric and hollow. The only assumption we make about the δ_{ij} is that they are nonnegative. The weigths are also assumed to be nonnegative, but we make the additional nondegeneracy assumption of <u>irreducibility</u>: we suppose that there is no partitioning of $\{1,2,\ldots,n\}$ such that $w_{ij} = 0$ whenever i and j belong to different members of the partition. Again this assumption causes no real loss of generality, because if all between-subset weigths are zero the MDS problem separates into a number of smaller problems corresponding with each of the subsets.

4: PROBLEM MANIPULATION: USE OF HOMOGENEITY

As we have shown in the previous section the metric MDS problem can be reformulated without loss of generality as the minimization of

$$\sigma(X) = \sum_{i<j}\sum w_{ij}(\delta_{ij} - d_{ij}(X))^2,$$

over the n x p configuration matrices X. In this section we study a closely related maximization problem, that is in some respects more simple. For the discussion of this alternative problem we need the following definitions.

$$\rho(X) = \sum_{i<j}\sum w_{ij}\delta_{ij}d_{ij}(X),$$

$$\eta^2(X) = \sum_{i<j}\sum w_{ij}d^2_{ij}(X),$$

$$\eta^2_\delta = \sum_{i<j}\sum w_{ij}\delta^2_{ij},$$

and

$$\lambda(X) = \rho(X) / \eta(X)\eta_\delta.$$

Theorem 4.1: For all X we have $0 \leq \lambda(X) \leq 1$. Moreover $\lambda(X) = 1$ if and only if the dissimilarities δ_{ij} and the distances $d_{ij}(X)$ for which $w_{ij} \neq 0$ are proportional.

Proof: This follows directly from the Cauchy-Schwartz inequality applied to $\rho(X)$. //

Theorem 4.2: a) Suppose $\hat{X}$ minimizes $\sigma(X)$. Then $\hat{X}$ also maximizes $\lambda(X)$.
b) Suppose $\hat{X}$ maximizes $\lambda(X)$. Then $\{\rho(\hat{X})/\eta^2(\hat{X})\}.\hat{X}$ minimizes $\sigma(X)$.

Proof: Because $d_{ij}(\beta X) = \beta d_{ij}(X)$ for all X and all $\beta \geq 0$ we can reformulate the MDS problem as the minimization of

$$\sum_i\sum_j w_{ij}(\delta_{ij} - \beta d_{ij}(X))^2$$

over the n x p matrices X and over all $\beta \geq 0$. The minimum over β for fixed X is attained at

$$\hat{\beta} = \rho(X) / \eta^2(X),$$

and the value at the minimum is $\eta^2_\delta(1 - \lambda^2(X))$. The theorem follows from these computations. //

It follows from theorem 4.2 that we can solve the metric MDS problem by finding the configuration matrix that maximizes $\lambda(X)$.

5: THE EUCLIDEAN CASE

Suppose $d_{ij}(X)$ is Euclidean, i.e.

$$d^2_{ij}(X) = \sum_{s=1}^{p} (x_{is} - x_{js})^2.$$

In this case it is convenient to derive some matrix expressions for $\rho(X)$ and $\eta(X)$. Define the matrix valued function $B(X)$ by

$b_{ij}(X) = - w_{ij}\delta_{ij}s_{ij}(X)$ if $i \neq j$,

$b_{ii}(X) = \sum w_{ij}\delta_{ij}s_{ij}(X)$.

Here

$$s_{ij}(X) = \begin{cases} d_{ij}^{-1}(X) & \text{if } d_{ij}(X) \neq 0, \\ 0 & \text{if } d_{ij}(X) = 0. \end{cases}$$

We also define the matrix V by

$v_{ij} = - w_{ij}$ if $i \neq j$,

$v_{ii} = \sum w_{ij}$.

Both $B(X)$ and V are real symmetric matrices with nonpositive off-diagonal and nonnegative diagonal elements, whose rows and columns sum to zero. By a familiar matrix theorem they are consequently both positive semi-definite of rank not exceeding $n - 1$. Because V is irreducible by assumption we have in fact $\text{rank}(V) = n - 1$, and the null space of V is the set of all vectors with constant elements. (Taussky, 1949; also Varga, 1962, sections 1.4 and 1.5). If e is the n-vector with all elements equal to one, then the Moore-Penrose inverse of V is simply

$V^+ = (V + \frac{1}{n} ee')^{-1} - \frac{1}{n} ee'$.

The following results can be verified easily.

<u>Theorem 5.1</u>: a) $\rho(X) = \text{tr } X'B(X)X$.

b) $\eta^2(X) = \text{tr } X'VX$.

We also define, for all pairs of configuration matrices,

$\mu(X,Y) = \text{tr } X'B(Y)Y$.

<u>Theorem 5.2</u>: $\mu(X,Y) \leq \rho(X)$ for all X,Y.

<u>Proof</u>: The Cauchy-Schwartz inequality implies

$$d_{ij}(X) \geq s_{ij}(Y) \sum_{s=1}^{p} (x_{is} - x_{js})(y_{is} - y_{js}).$$

If we multiply both sides with $w_{ij}\delta_{ij}$, sum all inequalities, and simplify, we find the inequality stated in the theorem. //

Using the notation developed in this section we can now define the B-matrix algorithm for Euclidean metric multidimensional scaling as the recursion

$X^{k+1} = V^+B(X^k)X^k$.

The only difference between Guttman's C-matrix and our B-matrix is due to the fact that we have removed the homogeneity from the problem in section 4, this makes B more simple than C.

Theorem 5.3: a) The three sequences $\rho(X^k)$, $\eta(X^k)$, and $\lambda(X^k)$ are bounded and increasing. The limits are ρ_∞, $\eta_\infty = \rho_\infty^{\frac{1}{2}}$, and $\lambda_\infty = \rho_\infty / \eta_\delta \eta_\infty$.

b) The sequence X^k has convergent subsequences. If X_∞ is the limit of a convergent subsequence, then $\lambda(X_\infty) = \lambda_\infty$. Moreover X_∞ is a fixed point, i.e. $X_\infty = V^+B(X_\infty)X_\infty$, and if λ is differentiable at X_∞, then $\nabla\lambda(X_\infty) = 0$.

c) $||X^{k+1} - X^k|| \to 0$.

Proof: From the Cauchy-Schwartz inequality

$\rho(X^k) = \text{tr } X^kVX^{k+1} \leq \eta(X^k)\eta(X^{k+1})$.

From theorem 5.2

$\rho(X^{k+1}) \geq \text{tr } X^{k+1}B(X^k)X^k = \text{tr } X^{k+1}VX^{k+1} = \eta^2(X^{k+1})$.

If we combine these inequalities we obtain

$$\eta(X^k) \leq \frac{\rho(X^k)}{\eta(X^k)} \leq \eta(X^{k+1}),$$

and

$$\rho(X^k) \leq \frac{\eta(X^k)}{\eta(X^{k+1})} .\rho(X^{k+1}) \leq \rho(X^{k+1}).$$

Because $\rho(X^k) \leq \eta_\delta^2$ and $\eta(X^k) \leq \eta_\delta$ it follows that both $\rho(X^k)$ and $\eta(X^k)$ are convergent increasing sequences, with limits, say, ρ_∞ and η_∞. Because $\lambda(X^k) \leq 1$ it follows that $\lambda(X^k)$ is another convergent increasing sequence with limit λ_∞. Moreover $\rho_\infty = \eta_\infty^2$, and $\lambda_\infty = \rho_\infty / \eta_\delta\eta_\infty$. It also follows that the sequence X^k lies in the compact set $\eta(X) \leq \eta_\delta$, and has convergent subsequences. There is equality in the basic chain of inequalities if and only if X is a fixed point. This implies that subsequential limits are fixed points. Finally

$\text{tr } (X^{k+1} - X^k)'V(X^{k+1} - X^k) = \eta^2(X^{k+1}) + \eta^2(X^k) - 2\rho(X^k) \to 2(\eta_\infty^2 - \rho_\infty) = 0$,

which implies part c of the theorem. //

A very similar result appears in Robert (1967). The general convergence theorems of Zangwill (1969) and Meyer (1976) are also relevant. Observe that theorem 5.3 does not say that X^k converges. This follows only if we make some rather arbitrary additional assumptions, for example that there is only a finite number of fixed points, or that one of the subsequential limits is an isolated fixed point. If X^k does not converge, it follows that the set of limit points is a continuum (a result due to Ostrowski, cf Daniel, 1971, section 6.3).

Theorem 5.3 is our basic convergence theorem for metric Euclidean MDS. It is quite satisfactory, and it has been proved by very elementary methods. In fact the proof only uses some elementary properties of sequences, and the Cauchy-Schwartz inequality.

6: NONEUCLIDEAN METRICS

Our method can be generalized to general Minkovski metrics, with the metric defined by a gauge ϕ. We shall use some elementary facts about gauges without proof. The proofs follow easily from the beautiful introduction to gauges and norms in Rockafellar (1970, chapter 15). Introductions to Minkovski geometry are given in Busemann and Kelley (1953), and Busemann (1955).

A gauge is a function $\phi:R^n \rightarrow R$ satisfying

G1: $\phi(x) \geq 0$.

G2: $\phi(x) = 0$ if and only if $x = 0$.

G3: $\phi(\mu x) = \mu\phi(x)$ for all $\mu \geq 0$.

G4: $\phi(x + y) \leq \phi(x) + \phi(y)$.

A gauge is a norm if we can replace G3 by the stronger

G5: $\phi(\mu x) = |\mu|\phi(x)$ for all μ.

A gauge defines a distance function by the rule

$d_{ij}(X) = \phi(x_i - x_j)$.

Unless the gauge is a norm this distance is not necessarily symmetric. With some minor modifications our results are also valid if we replace G2 by the weaker

G6: $\phi(0) = 0$.

In fact most of the results remain valid if we only assume G3 and G4, i.e. for all homogeneous convex functions. Thus gauges are relevant for our problem because they can be used to construct very general distance functions. But they are even more relevant because of the following result.

Theorem 6.1: Both $\rho(X)$ and $\eta(X)$ are gauges on the space of all n x p configuration matrices.

Proof: For both $\rho(X)$ and $\eta(X)$ property G1 is obvious. For $\eta(X)$ property G2 follows from irreducibility, for $\rho(X)$ property G6 is obvious, we could assume G2, but we never need it. Properties G3 and G4 follows from the fact that each $d_{ij}(X)$ is convex and homogeneous on the space of configuration matrices. //

The theorem shows that the metric MDS problem reduces to the maximization of a ratio of two gauges. Problems of that type have been studied by Robert (1967), Boyd (1974), Pham Dinh Tao (1975, 1976). Before we discuss their results and apply them to our problem we state some of the elementary facts about gauges. First define the polar of a gauge as the function $\phi^o:R^n \rightarrow R$ given by

$$\phi^o(y) = \max_x \frac{<x,y>}{\phi(x)} .$$

Here <.,.> denotes inner product.

Fact 6.2: a) The polar of a gauge is a gauge, the polar of a norm is a norm.

b) The polar of the polar of a gauge ϕ is the gauge ϕ.

c) The Euclidean norm $\langle x,x\rangle^{\frac{1}{2}}$ is its own polar.

d) (Hölder's inequality). If ϕ and ϕ^{o} are polar gauges, then $\langle x,y\rangle \leq \phi(x)\phi^{o}(y)$ for all x,y in R^n.

We can study the conditions for equality in Hölder's inequality by introducing subdifferentials. Remember that a subgradient of a function ϕ at a point x is a vector y such that $\phi(z) \geq \phi(x) + \langle y, z - x\rangle$ for all $z \in R^n$. The set of all subgradients of ϕ at a point x is the subdifferential of ϕ at x, and is written as $\partial\phi(x)$. Thus for each x the symbol $\partial\phi(x)$ stands for a subset of R^n, possibly empty. Again we mention some facts about subdifferentials, without proof. The proofs can be found in part V of Rockafellar (1970).

Fact 6.3: a) If ϕ is a finite convex function, then for each $x \in R^n$ the set $\partial\phi(x)$ is nonempty, convex, and compact.

b) If ϕ is differentiable at x with gradient $\nabla\phi(x)$, then $\partial\phi(x) = \{\nabla\phi(x)\}$.

c) The map $x \rightarrow \partial\phi(x)$ is closed, i.e. if $x_i \rightarrow x_\infty$, if $y_i \rightarrow y_\infty$, and if for each i also $y_i \in \partial\phi(x_i)$, then $y_\infty \in \partial\phi(x_\infty)$.

By combining the results of fact 6.2 and fact 6.3 we find the following results.

Fact 6.4: a) Suppose ϕ is a gauge. Then $y \in \partial\phi(x)$ if and only if $\phi(z) \geq \langle y,z\rangle$ for all $z \in R^n$, and $\phi(x) = \langle y,x\rangle$.

b) Suppose ϕ and ϕ^{o} are polar gauges. Then $x \in \partial\phi^{o}(y)$ if and only if $\langle x,y\rangle = \phi^{o}(y)$ and $\phi(x) = 1$. Moreover $y \in \partial\phi(x)$ if and only if $\langle x,y\rangle = \phi(x)$ and $\phi^{o}(y) = 1$.

c) Suppose ϕ and ϕ^{o} are polar gauges. Then $\langle x,y\rangle = \phi(x)\phi^{o}(y)$ if and only if $x \in \phi(x)\partial\phi^{o}(y)$ if and only if $y \in \phi^{o}(y)\partial\phi(x)$.

Now consider the problem of maximizing the ratio

$$\lambda(x) = \frac{\phi(x)}{\psi(x)},$$

with both ϕ and ψ gauges. From the definitions of gauges and their polars we obtain the following result.

Theorem 6.5: Maximizing $\lambda(x)$ over R^n is equivalent to maximizing

$$\xi(x,y) \triangleq \frac{\langle x,y\rangle}{\psi(x)\phi^{o}(y)}$$

over $R^n \times R^n$, and this is equivalent to minimizing

$$\lambda^{o}(x) \triangleq \frac{\phi^{o}(x)}{\psi^{o}(x)}$$

over R^n.

By using Hölder's inequality we can derive the following necessary conditions for an extreme value.

Theorem 6.6: a) If $\hat{x},\hat{y}$ maximizes $\xi(x,y)$ then $\hat{y} \in \phi^o(\hat{y})\partial\phi(\hat{x})$ and $\hat{x} \in \psi(\hat{x})\partial\psi^o(\hat{y})$.

b) If $\hat{x}$ maximizes $\lambda(x)$ or minimizes $\lambda^o(x)$ then $\hat{x} \in \psi(\hat{x})\partial\psi^o.\partial\phi(\hat{x})$.

For maximizing $\lambda(x)$ the following algorithm was proposed by Robert (1967). We start with x^0 such that $\psi(x^0) = 1$. Then define $y^k \in \partial\phi(x^k)$ and $x^{k+1} \in \partial\psi^o(y^k)$.

Theorem 6.7: a) The sequence $\lambda(x^k)$ is increasing and convergent. The sequence $\lambda^o(y^k)$ is decreasing and convergent. Both sequences converge to the same limit $\underline{\lambda}$.

b) All accumulation points of (x^k,y^k) correspond with the same function value $\underline{\lambda} = \underline{\xi}$. Moreover all accumulation points satisfy the necessary conditions of theorem 6.6.

Proof: From our facts about gauges

$\phi^o(y^k) = 1,$

$\psi(x^{k+1}) = 1,$

$\langle x^k,y^k\rangle = \phi(x^k),$

$\langle x^{k+1},y^k\rangle = \psi^o(y^k).$

By applying Hölder's inequality

$$\phi(x^k) = \langle x^k,y^k\rangle \leq \psi^o(y^k)\psi(x^k) = \psi^o(y^k),$$

$$\psi^o(y^k) = \langle x^{k+1},y^k\rangle \leq \phi(x^{k+1})\phi^o(y^k) = \phi(x^{k+1}),$$

and thus

$$\lambda(x^k) \leq 1/\lambda^o(y^k) \leq \lambda(x^{k+1}),$$

which implies part a. Because the subdifferentials are closed and the iterations remain in a compact set we can apply the general convergence theorems of Zangwill (1969) to get b. //

If we compare 6.7 and 5.3 we see that 6.7 has no part c, and is consequently weaker than 5.3. It is possible to prove that $||x^{k+1} - x^k|| \to 0$ in this more general context too, but we need additional assumptions. One of the more natural ones is that ϕ or ψ^o or both are differentiable at all stationary points, other possibilities are discussed by Meyer (1976). A far more important difference between the algorithms of sections 5 and 6 is that in most cases the function ψ^o cannot be computed in closed form. The same thing is true for the subdifferential $\partial\psi^o$. This means that we must compute x^{k+1} by maximizing $\langle x,y^k\rangle$ over $\{x \mid \psi(x) = 1\}$. This is a convex programming problem, which cannot be solved in a finite number of steps in general. Consequently we need a version of theorem 6.7 in which this convex programming problem is truncated after a finite number of steps. Zangwill's general convergence theory shows how this truncating should be done.

The simplifications in section 5 are possible, because the gauge ψ is ellipsoidal in this case. If both ψ and ϕ are ellipsoidal, then Robert has pointed out that the algorithm reduces to the power method for solving a generalized eigenvalue problem. Compare also Pham Dinh Tao (1976). It is of some interest that Guttman already pointed out that his C-matrix method for MDS looked like a sort of generalized power method. The analysis in this paper shows what the exact relationships are.

7: NONMETRIC MDS

In the simplest forms of nonmetric MDS we must minimize

$$\tau(X,\Delta) = \frac{\sum_{i=1}^{n}\sum_{j=1}^{n} w_{ij}(\delta_{ij} - d_{ij}(X))^2}{\sum_{i=1}^{n}\sum_{j=1}^{n} w_{ij}d_{ij}^2(X)},$$

over all n x p configuration matrices X and over all n x n disparity matrices Δ. The disparity matrices must be chosen from a known convex cone Γ, the metric MDS problem is the special case in which Γ is a ray, the additive constant problem is the special case in which Γ is a two-dimensional subspace. We briefly indicate the modifications needed to apply our ideas to nonmetric MDS in this simple form. More complicated partitioned loss functions, with more complicated normalizations, will be discussed in subsequent publications.

By using the homogeneity of the distance function as in section 4 we can show that the nonmetric MDS problem is equivalent to the maximization of

$$\lambda(X,\Delta) = \frac{\rho(X,\Delta)}{\eta(X)\eta(\Delta)},$$

with

$$\rho(X,\Delta) = \sum_{i=1}^{n}\sum_{j=1}^{n} w_{ij}\delta_{ij}d_{ij}(X),$$

$$\eta^2(\Delta) = \sum_{i=1}^{n}\sum_{j=1}^{n} w_{ij}\delta_{ij}^2,$$

and $\eta(X)$ as before. If we define

$$\rho(X) = \max_{\Delta\,\varepsilon\,\Gamma} \frac{\rho(X,\Delta)}{\eta(\Delta)},$$

then $\rho(X)$ is a homogeneous convex function, in fact a gauge. Thus we have a problem of the familiar form, a ratio of gauges must be maximized, and the algorithm of section 6 can be applied.

If the distance function is Euclidean, the analysis of section 5 can be used. The only difference with metric MDS is in the definition of $\rho(X)$, in the nonmetric case we have to compute the optimum Δ for given X in order to compute $\rho(X)$. We can compute the optimum $\hat{\Delta}(X)$ as the unique minimizer of

$$\sum_{i=1}^{n} \sum_{j=1}^{n} w_{ij}(\delta_{ij} - d_{ij}(X))^2$$

over the cone Γ. After solving this regression problem we can normalize the solution such that $\eta(\hat{\Delta}(X)) = 1$, but this is not strictly necessary. The B-matrix algorithm for nonmetric Euclidean MDS is defined as the recursion

$$X^{k+1} = V^{+}B(X^{k},\hat{\Delta}(X^{k}))X^{k},$$

with $B(X,\hat{\Delta}(X))$ defined as $B(X)$, but with $\hat{\delta}_{ij}(X)$ substituted for δ_{ij}. The same inequalities and equations can be derived as in 5.1, 5.2, and the proof of 5.3.

Theorem 7.1: Parts a,b,c of theorem 5.3 are also true for the nonmetric Euclidean B-matrix algorithm.

In the nonmetric case the differences between our B-matrix method and Guttman's C-matrix method are larger than in the metric case. One important reason is that Guttman uses rank images, while our convexity approach forces us to use monotone regression estimates of the δ_{ij}. I have not been able to find a rigorously defined optimization problem in which rank images can be used. This does not mean, of course, that we cannot use rank images in the earlier iterations of an MDS algorithm. In the earlier iterations we can do anything we please. As in TORSCA we can use the semi-nonmetric Young-Householder process, or as in MINISSA we can use rank images. We only have to switch to monotone regression and the B-matrix algorithm if things are getting out of hand (if the loss starts to increase, for example).

REFERENCES

Busemann, H. The geometry of geodesics. New York, Academic Press, 1955.

Busemann, H. and Kelly, P. Projective geometry and projective metrics. New York, Academic Press, 1953.

Daniel, J.W. The approximate minimization of functionals. Englewood Cliffs, Prentice Hall, Inc, 1971.

Gleason, T.C. A general model for nonmetric multidimensional scaling. Michigan mathematical psychology program, 1967, 3.

Guttman, L. A general nonmetric technique for finding the smallest coordinate space for a configuration of points. Psychometrika, 1968, 33, 469 - 506.

Kruskal, J.B. Multidimensional scaling by optimizing goodness-of-fit to a nonmetric hypothesis. Psychometrika, 1964, 29, 1-28.

Kruskal, J.B. Nonmetric multidimensional scaling: a numerical method. Psychometrika, 1964, 29, 115-129.

Kruskal, J.B. A new convergence condition for methods of ascent. Unpublished memo, Bell Labs, Murray Hill, 1967.

Kruskal, J.B. Monotone regression: continuity and differentiability properties. Psychometrika, 1971, 36, 57-62.

Kruskal, J.B. and Hart, R.E. A geometric interpretation of diagnostic data from a digital machine. Bell System Technical J., 1966, 45, 1299-1338.

Lingoes, J.C. and Roskam, E.E. A mathematical and empirical analysis of two multidimensional scaling algorithms. Psychometrika, 1973, 38, monograph supplement.

McGee, V.E. The multidimensional analysis of 'elastic' distances. Brit. J. Math. Statist. Psychol., 1966, 19, 181-196.

Meyer, R.R. Sufficient conditions for the convergence of monotonic mathematical programming algorithms. J. Comp. System Sciences, 1976, 12, 108-121.

Pham Dinh Tao Eléments homoduaux d'une matrice A relatifs à un couple des normes (ϕ,ψ). Applications au calcul de $S_{\phi\psi}(A)$. Séminaire d'analyse numérique, Grenoble, 1975, no 236.

Pham Dinh Tao Calcul du maximum d'une forme quadratique définie positive sur la boule unité de ψ_∞. Séminaire d'analyse numérique, Grenoble, 1976, no 247.

Robert, F. Calcul du rapport maximal de deux normes sur R^n. R.I.R.O., 1967, 1, 97-118.

Rockafellar, R.T. Convex analysis. Princeton, Princeton University Press, 1970.

Takane, Y, Young, F.W., and De Leeuw, J. Nonmetric individual differences multidimensional scaling. Psychometrika, 1976, 41, in press.

Taussky, O. A recurring theorem on determinants. Amer. Math. Monthly, 1949, 56, 672-676.

Torgerson, W. Theory and methods of scaling. New York, Wiley, 1958.

Varga, R.S. Matrix iterative analysis. Englewood Cliffs, Prentice Hall, 1962.

Zangwill, W.I. Nonlinear programming: a unified approach. Englewood Cliffs, Prentice Hall, 1969.

Recent Developments in Statistics
J.R.Barra et al., editors

Invited Paper

KINSHIP IN THE BIRTH AND DEATH PROCESS OF A POPULATION SUBDIVIDED IN FINITE PANMICTIC GROUPS

G. Malécot
Université de Lyon 1 - Mathématiques Appliquées
Boîte Postale 2037 - 69603 VILLEURBANNE - FRANCE

When a population is subdivided in panmictic groups with gene flow ("migration") between them, the dependence of kinship on the migration law is not modified by the overlapping of generations. All asymptotic formulas are easily obtained by a Laplace transform with respect to time.

I - Let us suppose that every group contains a constant number N of diploïd individuals, i.e. of 2N "useful gametes" ; let us suppose that these gametes are (instead of the diploïds) individually and independently replaced, i.e. that the death between t and t+τ [with probability $\mu\tau + o(\tau)$] of a given gamete in group i immediately replaces it by another useful gamete which is a copy (with probability 1-k of no mutation) of another useful gamete sampled in group ℓ (with probability $g_{i\ell}$) : this sampling is with replacement if we suppose that many copies of a "living" gamete are independently possible ; the probability of identity of living gametes sampled at date t in ℓ and j with replacement will be called $\phi_{\ell j}(t)$.

If, at date t+τ, two distinct gametes are sampled in i and j (sampled without replacement if j = i), one only [except with a probability $o(\tau)$] may have replaced between t and t+τ a dead one ; if the probability of identity of living gametes sampled at date t in i and j without replacement is called $\overline{\phi}_{ij}(t)$, we have, by a well known technique :

$$\frac{d\,\overline{\phi}_{ij}(t)}{dt} = -\,2\mu\,\overline{\phi}_{ij}(t) + \mu(1-k)\left[\sum_{\ell} g_{i\ell}\,\phi_{\ell j}(t) + \sum_{m} g_{jm}\,\phi_{im}(t)\right] \tag{1}$$

with (each group of 2N gametes being panmictic) :

$$\overline{\phi}_{ij} - \phi_{ij} = \delta_{ij}(\phi_{ii}-1)/M \qquad ; \qquad M = 2N-1 \tag{2}$$

The groups are supposed indexed by integers i, j, $\in \mathbf{Z}$, in the unidimensional case (migration on the straight line, or on a circle) ; in the h-dimensional case, they are each indexed by a h-uple of integers ($\in \mathbf{Z}^h$).

To insure convergence of formulas, we introduce the Laplace Transform, real, continuous, and bounded by $1/\lambda$ if $\lambda > 0$:

$$\int_0^\infty e^{-\lambda t}\ \overline{\phi}_{ij}(t)\ dt = \overline{\Phi}_{ij}(\lambda)\ ;$$

in the same way for $\phi_{ij}(t)$:

$$(\lambda+2\mu)\ \overline{\Phi}_{ij}(\lambda) - \overline{\phi}_{ij}(o) = \mu(1-k)\left[\sum_\ell g_{i\ell}\ \Phi_{\ell j}(\lambda) + \sum_m g_{jm}\ \Phi_{im}(\lambda)\right] \tag{3}$$

$\overline{\phi}_{ij}(o)$ is replaced by 0 if at t = 0 there is no kinship between distinct gametes, which will be supposed henceforth.

II - We now suppose that migration is homogeneous, i.e. :

$$\phi_{\ell j}(t) = \phi(y,t) \quad ; \quad \Phi_{\ell j}(\lambda) = \Phi(y,\lambda) \quad ; \quad [y = j-\ell]$$

We then use Fourier Transforms :

$$\begin{cases} \sum_y \alpha^y\ \Phi(y,\lambda) = R(\alpha,\lambda) = \sum_y \int_o^\infty e^{-\lambda t}\ \phi(y,t)dt \\ \sum_\ell g_{i\ell}\ \alpha^{\ell-i} = L(\alpha) \qquad [|\alpha| = 1] \end{cases}$$

(they may be shown to exist if $g_{i\ell}$ has an exponential decrease when $|\ell-i|$ increases) :

$$R(\alpha,\lambda) = (\lambda+2\mu)[1/\lambda-\Phi(o,\lambda)]/M\ \{\lambda+2\mu-\mu(1-k)[L(\alpha)+L(1/\alpha)]\} \tag{4}$$

We shall now use the notation :

$$\beta = 2\mu(1-k)/(\lambda+2\mu)\ , \tag{5}$$

$\# 1$ if k small or zero and $\lambda \to 0$.

III - In unidimensional case on infinite lattice, j, i,... $\in \mathbb{Z}$; Fourier inversion gives :

$$\Phi(y,\lambda) = \frac{1}{2\pi i}\int_C \alpha^{y-1}\ R(\alpha,\lambda)\ d\alpha = (1/2\pi)\int_{-\pi}^{+\pi} e^{ivy}\ R(e^{iv},\lambda)dv \tag{6}$$

When migration is bounded ($L(\alpha)$ polynomial in α and $1/\alpha$), residues of all poles of $R(\alpha,\lambda)$ are negligible (when $\beta \to 1$) % the residue of the root $\alpha_1 < 1$ and $\# 1$ of equation

$$(1/2)\ [L(\alpha) + L(1/\alpha)] = 1/\beta$$

The equation (6) gives :

$$\Phi(y,\lambda) \sim \frac{1/\lambda-\Phi(o,\lambda)}{M}\ \alpha_1^{|y|}\ /(1/\alpha_1-\alpha_1)\sigma^2/2 \tag{7}$$

(σ^2 = variance of migration)

a) When $k > 0$:

$$\phi(y\ ,\ +\infty) = \lim_{\lambda \to o} \lambda\ \Phi(y,\lambda) = \phi_o\ \alpha_1^{|y|}$$

(expon. decrease of kinship with distance.)

b) When $k = 0$, the limit is 1 ; $1-\phi(y,t) = P(y,t)$, "panmictic index", has Laplace transform :

$$\Pi(y,\lambda) = 1/\lambda - \phi(y,\lambda) \quad ; \tag{8}$$

The equation (7) implies that $\Pi(o,\lambda) = O(\lambda^{-1/2})$ when $\lambda \to o$; so (tauberian theorem) :

$$P(y,t) = O\left(\frac{1}{\sqrt{t}}\right) \qquad (t \to +\infty)$$

IV - In h-dimensional infinite case, $y = j-i$ is a vector of $\mathbb{Z}^h$ (with h components y_ℓ) ; v is a vector of the dual space, and the last member of (6) is replaced by $(1/2\pi)^h$ times a h-dimensional integral over the "cube" $|v_\ell| \leqq \pi$. Let us now rotate the axis so as to diagonalize the covariance matrix of the symmetrized migration distribution defined in (4) by its real and even characteristic function $\left[L(e^{iv}) + L(e^{-iv})\right]/2 = H(v)$ which then becomes :

$$1 - \sum_{\ell=1}^{h} \sigma_\ell^2\ v_\ell^2/2(1+\varepsilon)\ ,$$

the σ_ℓ^2 being the variances along the new axis, and ε being $O(|v|^\gamma)$ if $|v|$ is the maximum modulus of the v_ℓ s, and if all the moments of order $2+\gamma$ are finite (we shall suppose, for ulterior convenience, $0 < \gamma \leqq 2$).

We now take on each axis a new unit which is the corresponding standard deviation σ_ℓ ; taking account of the jacobian $J = 1/\sigma_1 \ldots \sigma_h = 1/\prod_\ell \sigma_\ell$, we have from (4), (5), and the multidimensional version of (6) (where vy means $\sum_{\ell=1}^{h} v_\ell\ y_\ell$) :

$$M\Phi(y,\lambda) = [1/\lambda - \Phi(o,\lambda)][J/(2\pi)^h] \int \ldots \int_\Delta \frac{e^{ivy}\ d\omega}{1-\beta H(v)} = [1/\lambda - \Phi(o,\lambda)]F(y,\beta) \tag{9}$$

with now the development :

$$H(v) = 1 - \sum_{\ell=1}^{h} v_\ell^2\ (1+\varepsilon)/2 = 1 - v^2\ [1+O(v^\gamma)]/2 \tag{10}$$

(putting for simplicity, $\|v\| = v$, $\sum_\ell v_\ell^2 = v^2$) ; $d\omega$ is the elementary volume (in new units) ; Δ is a "parallelepipedic" domain of center O including a sphere Σ' of center O, of radius A', and included in a sphere Σ'', of center O, of radius A''. If σ_ℓ is independent of ℓ, $\sigma_\ell = \sigma$, we may take $A' = \Pi\sigma$, $A'' = \Pi\sigma\sqrt{h}$.

$F(y,\beta)$ is a shorter notation for the inverse Fourier transform of $\frac{1}{1-\beta\ H(v)}$, which is a continuous function for $\beta < 1$.

For $\beta = 1$, 0 is the unique singularity inside Δ (otherwhere in Δ, $H(v) < 1$, since domain Δ is the "period" of $H(v)$, i.e. the dual of the span of the h-dimensional arithmetic migration distribution if the migration law is supposed aperiodic) ; all further formulas will be related to the infinite part, when $\beta \to 1$, of $F(y,\beta)$ and also of its derivatives for $\beta < 1$ (all integrands being continuous), which we shall write (including F by putting $F^{(o)} = F$) :

$$F^{(p)}(y,\beta) = \frac{\partial^p F^{(o)}}{\partial \beta^p} = p![J/(2\pi)^h] \int\cdots\int_\Delta \frac{H^p(v)\, e^{ivy}\, d\omega}{\left[1-\beta H(v)\right]^{p+1}} \quad (p \geq 0\ ,\ \beta < 1) \qquad (11)$$

The infinite part when $\beta \to 1$ is unchanged when replacing Δ by the sphere Σ of radius A, and even by every concentric sphere of whatever small radius ; let us show how this infinite part and the remainder depend on the first terms (when $v \to 0$) of development (10) ; we may write from (10) :

$$[1-\beta H(v)]^q = \left[1-\beta+\beta(1+\varepsilon)v^2/2\right]^q \qquad (12)$$

and, after a change of notations,

$$a = \frac{2}{\beta} > 0 \quad , \quad D = 2(1-\beta)/\beta = (\lambda/\mu+2k)/(1-k) > 0 \quad , \quad D \to 0 \text{ when } \beta \to 1 \qquad (13)$$

$$\left[1-\beta\ H(v)\right]^{-q} = a^q\ (v^2+D)^{-q}\left[1+O(v^{2+\gamma})/(D+v^2)\right] \qquad (14)$$

Since (for $\gamma \leq 2$) :

$$H^p(v)-1 = O(v^2) = O(v^\gamma) = O[v^{2+\gamma}/(D+v^2)] \quad ;$$

Equation (11) gives :

$$F^{(p)}(y,\beta) = p!\ a^{p+1}\left[J/(2\pi)^h\right] \int\cdots\int_\Delta \frac{1+O[v^{2+\gamma}/(D+v^2)]\ e^{ivy}\ d\omega}{(D+v^2)^{p+1}} \qquad (15)$$

This Fourier inverse may be approximated by "Hankel inverse" if we notice that $(D+v^2)^{-p-1}$ and $v^{2+\gamma}/(D+v^2)^{p+2}$ are isotropic functions inside spheres Σ of radius A (such as Σ' and Σ'' previously defined) : their integral over the domain Δ differs, as to its real part, from their integral inside Σ' or inside Σ'', by a quantity which is bounded by their integral between spheres Σ' and Σ'' for y=o ; the continuity insures that there exists a sphere Σ of radius A ($A' \leq A \leq A''$) such that the integral over Δ is equal to the integral on the domain S inside Σ (its radius A is between $\pi\sigma$ and $\pi\sigma\sqrt{h}$ when all standard deviations σ_ℓ are equal (to σ) ; when the σ_ℓ s are not equal, we may put $A' = \min_\ell \sigma_\ell$ and $A'' = \sqrt{h}\max_\ell \sigma_\ell$; when the σ_ℓ s are nearly equal to a quantity σ ("quasi isotropic case"), A', A", and A are $O(\sigma)$. We now only have to study integrals such as :

$$I_{mq}(r,A) = (1/2\pi)^h \int\cdots\int_S \frac{v^m}{(D+v^2)^q}\, e^{ivy}\, d\omega \qquad (16)$$

which is, putting $\|y\| = r$, a simple integral constructed from the Hankel kernel :

$$\mathcal{H}_h(r,v) = (2\pi)^{-h-2}\, r^{1-h/2}\, v^{h/2}\, J_{h/2-1}(vr)$$

by the formulas :

$$\begin{cases} I_{mq}(r,A) = \int_0^A \mathcal{H}_h(r,v) \dfrac{v^m}{(D+v^2)^q}\, dv \\ \mathcal{H}_1 = (1/\pi)\cos vr \;;\; \mathcal{H}_2 = (1/2\pi)v\, J_o(vr) \;;\; \mathcal{H}_3 = (1/2\pi^2 r)v \sin vr \end{cases} \tag{17}$$

To study the divergence of I_{mq} when $D \to 0$, it may be useful to put $A = +\infty$: I_{mq} is, for $r \neq 0$, semi-convergent (owing to alternating sign) when $A \to +\infty$ if $v^m/(D+v^2)^q$ is a decreasing function tending to 0 at infinity ; when $r = 0$, we may use the formula :

$$\int_0^A \frac{v^m}{(D+v^2)^q}\, dv = (1/2)\, D^{m/2+1/2-q}\, B(m/2+1/2 \;,\; q-m/2-1/2) \tag{18}$$

B being an incomplete beta-function which remains finite (when $A \to \infty$) if $m+1-2q < 0$.

Let us transform (15), replacing the jacobian J by $1/\prod_\ell \sigma_\ell$, and using the majoration $|\mathcal{H}_h(r,v)| \leq \mathcal{H}_h(o,v)$, which gives the formula (19) :

$$|(\prod_\ell \sigma_\ell)F^{(p)}(r,\beta) - p!a^{p+1} I_{o,p+1}(r,\infty)| \leq O[I_{o,p+1}(0,\infty) - I_{o,p+1}(0,A)] + O[I_{2+\gamma,p+2}(0,A_1)]$$

(A_1, finite, may be different from A)

The majorants in second member are independent of r (A depends on p, but this does not matter).

Since $\mathcal{H}_h(o,v) = v^{h-1}$ times a constant, we have :

$$I_{mq}(r,A) \leq I_{mq}(o,A) = \int_0^A \frac{v^{m+h-1}}{(D+v^2)^q}\, dv \text{ times a constant} = O(D^{m/2+h/2-q})$$

So the last majorant in formula (19) is $O[D^{\gamma/2+h/2-p-1}]$; the first majorant in (19) is $O\left[\int_A^\infty \dfrac{\mathcal{H}_h(o,v)}{(D+v^2)^{p+1}}\, dv\right]$, proportional to $\int_A^\infty \dfrac{v^{h-1}}{(D+v^2)^{p+1}}\, dv$, which by (18) is $D^{h/2-p-1}$ times the difference between an incomplete and a complete beta function, difference found to be $O[(D/A^2)^{p+1-h/2}]$; so the first majorant in (19) is $O[A^{h-2p-2}]$, which is interesting when A is large, i.e. when all the σ_ℓ s are large, i.e. when the arithmetic migration distributions are well approximated by continuous distributions.

Equation (19) allows (except for a noticeable exception which will be given further) an approximation of $F^{(p)}(r,\beta)$ (defined by (11)) by means of :

$$I_{o,p+1}(r,\infty) = \int_0^{+\infty} \frac{\mathcal{H}_h(r,v)}{(D+v^2)^{p+1}}\, dv = \frac{(-1)^p}{p!} \frac{d^p}{dD^p}\left[\int_0^{+\infty} \frac{\mathcal{H}_h(r,v)\, dv}{D+v^2}\right]$$

Let us recall that D is > 0 (and → 0 when β → 1) ; all derivations under ∫ are then allowed if there is convergence (for p ≧ 0) and uniform convergence (for p ≧ 1) up to ∞, i.e. when r > 0 [which gives $|\mathcal{H}_h| \leq O(v^{h/2-1/2})$].

$$\left\{\begin{aligned} (-1)^p\ p!\ I_{o,p+1}(r,\infty) &= \frac{d^p}{dD^p}\left(e^{-r\sqrt{D}}/2\sqrt{D}\right) && \text{if } h = 1 \\ &= \frac{1}{2\pi}\frac{d^p}{dD^p}K_o(r\sqrt{D}) && \text{if } h = 2\ (r \neq 0) \\ &= \frac{-1}{4\pi}\frac{d^{p-1}}{dD^{p-1}}\left(e^{-r\sqrt{D}}/2\sqrt{D}\right) && \text{if } h = 3 \text{ and } p \geq 1 \end{aligned}\right. \tag{20}$$

So (19) gives :

$$\left\{\begin{aligned} (\prod_\ell \sigma_\ell)\ F^{(p)}(r,\beta) &= a^{p+1}\ p!\ I_{o,p+1}(r,\infty) + O_1 + O_2 \\ O_1 &= O(A^{h-2p-2}) && \text{(independent of } r > 0) \\ O_2 &= O(D^{\gamma/2+h/2-p-1}) && \text{(independent of A and } r > 0) \end{aligned}\right. \tag{21}$$

a) There is an exception when h = 2, r = 0 , p = 0 :

$$I_{o1}(0,A) = \frac{1}{2\pi}\int_o^A \frac{v}{D+v^2}\,dv \sim \frac{1}{2\pi}\log(A/\sqrt{D}) \quad \text{divergent when } A \to \infty.$$

Instead of (19), we then have (leaving A instead of ∞ in I_{o1}) :

$$|F(0,\beta)-(a/2\pi)\log(A/\sqrt{D})| \leq O\,[I_{2+\gamma,2}(0,\infty)] \tag{19'}$$

$$|\prod_\ell \sigma_\ell F(0,\beta)-(a/2\pi)\mathrm{Log}(A/\sqrt{D})| \leq O\left[\int_o^{A_1}\frac{v^{2+\gamma}\mathcal{H}_2(o,v)}{(D+v^2)^2}\,dv\right] = O(D^{\gamma/2}) \text{ by (17).} \tag{22}$$

(A_1 is between A' and A", but may be different from A)

b) But, for h = 2, r = 0, p ≧ 1, formulas (21) remain valid when putting

$$I_{o2}(0,\infty) = \int_o^\infty \frac{\mathcal{H}_2(o,v)}{(D+v^2)^2}\,dv = (1/2\pi)\int_o^\infty \frac{v\,dv}{(D+v^2)^2} = 1/4\pi D$$

All further derivations % D give formulas (21) for r = 0 as if $I_{o1}(0,\infty)$ was replaced by $(1/2\pi)\ \mathrm{Log}\ (A/\sqrt{D})$ (with A = cte).

V - The coefficient of kinship (probability of identity by descent) $\phi(y,t)$, the "panmictic index" $1-\phi(y,t) = P(y,t)$, may be studied on their Laplace transforms $\Phi(y,\lambda)$, defined by (9), and $\pi(y,\lambda)$, defined by (8), which jointly give :

$$\Phi(y,\lambda)/F(y,\beta) = \Phi(0,\lambda)/F(0,\beta) = 1/\lambda-\Phi(0,\lambda) = \pi(0,\lambda),$$

whence :

$$\lambda\Phi(y,\lambda) = 1-\lambda\pi(y,\lambda) = F(y,\beta)/[M+F(0,\beta)] \tag{23}$$

Equation (9) shows that Φ is a function of k only through β (given by formula (5)) ; it now will be more explicit to write Φ in the form $\Phi(y,k,\lambda)$; the partial derivative :

$$\lambda \frac{\partial\Phi}{\partial(1-k)} = (1+\lambda/2\mu)^{-1} \frac{\partial}{\partial\beta}\left[\frac{F(y,\beta)}{M+F(0,\beta)}\right] \tag{24}$$

will have an interesting significance :

VI - The identity by descent of 2 gametes sampled (with replacement) at date t in ℓ and j ($j-\ell = y$), identity which has probability $\phi(y,k,t)$ (Laplace inverse of $\Phi(y,k,\lambda)$), may be realised through p links, p being a random integer $\in \mathbb{N}$ ($p = 0$ when the <u>same</u> gamete is sampled) ; the probability of identity through exactly p links, "probability of a <u>kinship chain of order p within time t</u>", is called $\varpi_p(y,t)$ (it does not depend on k), and of course :

$$\phi(y,k,t) = \sum_{p=o}^{\infty} (1-k)^p\, \varpi_p(y,t)$$

This series is uniformly convergent for $k \geq 0$, since ϕ is a probability ;
Putting $c_p(y,\lambda) = \int_0^{\infty} e^{-\lambda t}\, \varpi_p(y,t)dt$, we have :

$$\Phi(y,k,\lambda) = \sum_{p=o}^{\infty} (1-k)^p\, c_p(y,\lambda) \qquad (k \geq 0)$$

We deduce another entire series :

$$\frac{\partial\Phi(y,k,\lambda)}{\partial(1-k)} = \sum_{p=1}^{\infty} (1-k)^{p-1}\, p\, c_p(y,\lambda) , \tag{25}$$

convergent when $k > 0$, but which diverges for $k = 0$, then giving a tauberian result for $p\, c_p(y,\lambda)$ and $p\, \varpi_p(y,t)$.
Equations (24) and (25) give :

$$\sum_p (1-k)^{p-1}\, p\, \varpi_p(y,\infty) = \lim_{\lambda\to 0} \lambda \frac{\partial\Phi}{\partial(1-k)} = \frac{\partial}{\partial\beta}\left[\frac{F(y,\beta)}{M+F(0,\beta)}\right]_{\beta=1-k} \tag{25'}$$

The last member is infinite when $k \to 0$, and gives an asymptotic equivalent of $p\, \varpi_p(y,\infty)$.

VII - A curiously related problem is the study of the difference :

$$\psi(y,t) = \phi(y,\infty) - \phi(y,t) \tag{26}$$

When the asymptotic value $\phi(y,\infty)$ ($\neq 1$ when $k > 0$) is given by :

$$\phi(y,\infty) = \lim_{\lambda\to 0} \lambda\, \Phi = F(y,1-k)/[M+F(0,1-k)] \quad \text{by (23)} \tag{27}$$

Because we have supposed $\bar{\phi}(y,0) = 0$, $\phi(y,t)$ is the probability of having a common ancestor over an interval smaller thant t, which is the p. distribution of time spent since the nearest common ancestor ;
So it is an increasing function of t, and (26) is a <u>positive</u> function, the Laplace transform of which is :

$$\int_0^\infty e^{-\lambda t}\,\psi(y,t)dt = \left[\frac{F(y,1-k)}{M+F(0,1-k)} - \frac{F(y,\beta)}{M+F(0,\beta)}\right]/\lambda \qquad (28)$$

When $k > 0$, the abscissa of convergence of this Laplace integral is a negative number, since (28) is regular for $\lambda = 0$ ($\beta = 1-k$), being then the (finite) derivative % β of the bracket, for $\beta = 1-k$; the second member of (28) is even continuous and regular for all values $\beta < 1$ (i.e. $\lambda > -2\mu k$), since the $F^{(p)}(y,\beta)$ given by (11) are then continuous and positive [as mixtures of the probability distributions defined by $H^{\ell}(v)$]. But, when $\beta = 1$, it was seen after (15) when $D \to 0$ that some derivatives are discontinuous, so the abscissa of convergence of (28) is exactly $\lambda = -2\mu k$, and we now put $\lambda+2\mu k = s$ to study, when $s = (\lambda+2\mu)(1-\beta) \to 0$, the derivative % λ of $\lambda\int_0^\infty e^{-\lambda t}\,\psi(y,t)dt$ given by (28) :

$$\int_0^\infty e^{-\lambda t}\psi(y,t)dt - \lambda\int_0^\infty e^{-\lambda t}t\psi(y,t)dt = \frac{(1-k)}{2\mu(1+\lambda/2\mu)^2}\,\frac{\partial}{\partial\beta}\left[\frac{F(y,\beta)}{M+F(0,\beta)}\right] \qquad (29)$$

The last member is infinite when $\beta \to 1$, and gives, by Tauber's theorem, applied when $s \to 0$ to $\int_0^\infty e^{-st}\,[t\,e^{2\mu kt}\,\psi(y,t)]dt$, an equivalent, when $t \to \infty$, of the bracket in the last integrand.

VIII - In bidimensional case ($h = 2$), formulas (21) give :

$$\lambda\Phi = [K_o(r\sqrt{D})+O(1)]/[2\pi M\sigma_1\sigma_2/a + \mathrm{Log}(A/\sqrt{D}) + O(D^{\gamma/2})] \quad (r \neq 0) \qquad (30)$$

$$\lambda\Phi(0,k,\lambda) = 1/[1+M/F(0,\beta)] = \frac{1}{1+[2\pi\sigma_1\sigma_2 M/a\ \mathrm{Log}(A/\sqrt{D})]} + O(D^{\gamma/2}/\mathrm{Log}^2 D) \qquad (30')$$

$$\lambda\pi(0,k,\lambda) = \frac{1}{1+a\ \mathrm{Log}(A/\sqrt{D})/2\pi\sigma_1\sigma_2 M} + O(D^{\gamma/2}/\mathrm{Log}^2\ D)$$

1° When $k = 0$, $D = \lambda/\mu$: $P(0,+\infty) = \lim_{\lambda\to 0}\lambda\pi = 0$: $\phi(y,+\infty) = 1$:
The speed towards genetic homogeneity is given by the fact that when $\lambda \to 0$, $a \sim 2$, $\lambda\pi \sim 2\pi\sigma_1\sigma_2 M/\mathrm{Log}(1/\lambda)$, whence by Tauber's theorem :

$$P(0,t) \underset{(t\to\infty)}{\sim} (2\pi\sigma_1\sigma_2 M)/\mathrm{Log}\ t$$

(slower than in unidimensional case)

When $r \neq 0$, $\lambda\Phi(y,k,\lambda) = [F(r,\beta)/F(0,\beta)]\ \lambda\Phi(0,k,\lambda)$

$$F(0,\beta)-F(r,\beta) = (a/2\pi)\ \mathrm{Log}(Ar) + O(1),$$

which is interesting only if Ar is large (A is of order of the σ_ℓ s measured with units which were the spans of the arithmetic migration laws, r is the distance measured with the σ_ℓ s as units ; so that "large Ar" means the distance of the 2 studied groups is large % the distance between 2 adjacent groups). We then have :

$$\lambda\pi(y,k,\lambda)-\lambda\pi(0,k,\lambda) \underset{(\lambda\to 0)}{\sim} \frac{\text{Log}(Ar)}{\text{Log}(A/\sqrt{D})} \sim 2\ \text{Log}(Ar)/\text{Log}(1/\lambda)$$

$$P(y,t) \underset{(t\to\infty)}{\sim} [2\pi\sigma_1\sigma_2 M+2\ \text{Log}(Ar)]/\text{Log}\ t \quad (Ar\ \text{large}) \tag{31}$$

(the increase of panmictic index with distance is only logarithmic)

2° When $k > 0$ (but small), $D \sim \lambda/\mu + 2k$

$$\phi(y,\infty) = \lim_{\lambda\to 0} (\lambda\Phi) = [K_o(r\ \sqrt{2k})+O_3]/[\text{Log}(A/\ \sqrt{2k})+\pi M\sigma_1\sigma_2+O(1)] \tag{32}$$

a) When $r\sqrt{2k}$ is small, $K_o(r\sqrt{2k}) \sim -\text{Log}(r\sqrt{2k})$

b) For $r = 0$, we replace $K_o(r\sqrt{2k})$ by $\text{Log}(A/\ \sqrt{2k})$

c) When $r\sqrt{2k}$ is large, $K_o(r\sqrt{2k}) \sim \sqrt{\pi/2r\sqrt{2k}}\ e^{-r\sqrt{2k}}$; by formulas (21), O_3 would be $O_1 = O(1)$, as was written in (30). By a more precise calculation, for $r \neq 0$, using the decrease and the alternating sign of $\mathcal{H}_h(r,v)$, would give in (21) instead of O_1+O_2 :

$$O_3 = O\left[r^{-1/2-h/2}\ A^{h/2-2p-5/2}\right]+O\left[r^{1/2-h/2}\ A^{h/2-2p-3/2}\right]$$

So that, if Ar is large, O_3 may be taken $O\left[(Ar)^{-1/2}\right]$; if O_3 may be neglected with respect to $K_o(r\sqrt{2k})$ (this is seen to be valid if r remains small with respect to $\text{Log}(1/k)/6k$, which is practically the case in all experiments, on account of the smallness of mutation rate), the Malécot's 1966 formula of continuous migration is extended (for Ar large) to all laws of finite moments of order $2+\gamma$ (cf. SAWYER, private communication) except the correction $\text{Log}(A/\ \sqrt{2k})$ instead of $\text{Log}(1/\ \sqrt{2k})$.

d) It is interesting that, if $r\sqrt{2k}$ increases among the smallest multiples of unity, the decrease is quasi exponential, as was found by MORTON for many populations.

3° When $k > 0$, the speed of $\phi(y,t)$ towards its limit is given by (29) ; remembering that $2(1-\beta)/\beta$ was called D, we have, when $r\sqrt{D}$ is small so as we replace K_o and K_1 by the first terms of their developments near 0 :

$$\frac{\partial}{\partial\beta}\left[\frac{F(y,\beta)}{M+F(0,\beta)}\right] \sim 4\ [\pi M\sigma_1\sigma_2+\text{Log}(Ar)]/D\ \text{Log}^2 D = C/D\ \text{Log}^2 D \tag{33}$$

(Ar is supposed large)

Equation (29) then gives, when $\lambda = s-2\mu k \to -2\mu k$:

$$2\mu k \int_o^\infty e^{-st}\ [t\ e^{2\mu kt}\ \psi(y,t)]dt \sim C/2\mu\ D\ \text{Log}^2\ D$$

with $D \sim 2(1-\beta) = 2s/(\lambda+2\mu) \sim s/\mu$

$$t\, e^{2\mu kt}\, \psi(y,t) \underset{(t\to\infty)}{\sim} [\pi M\sigma_1\sigma_2 + \mathrm{Log}(Ar)]/\mu k\, \mathrm{Log}^2 t \qquad (34)$$

The decrease of $e^{2\mu kt}\, \psi(y,t)$ is larger than was in (31) for $k = 0$.

4° The probability $\varpi_p(y,\infty)$ of a kinship chain of order p (of whatever length in time) is given by the series (25') when $k \to 0$, i.e. when in (33) $D = 2k/(1-k) \to 0$; Tauber's theorem gives :

$$p\, \varpi_p(y,\infty) \underset{(p\to\infty)}{\sim} 2\, [\pi M\sigma_1\sigma_2 + \mathrm{Log}(Ar)]/\mathrm{Log}^2 p$$

The decrease of $p\, \varpi_p(y,t)$ is very similar to the decrease in (34) which is a general result whatever the dimensionality [compare (25') and (29)] .

N.B. - These results involve those of NAGYLAKI and SAWYER (Theor. Pop. Biol. 10, 1, 1976, p. 70) ; for h = 1, the results of G. MALECOT, 1975 (unpublished), were the following :

$$\lambda\Phi = \left[e^{-r\sqrt{D}} + O(A^{-1}\sqrt{D})\right] / \left[1 + M\sigma\sqrt{D} + O(A^{-1}\sqrt{D})\right]$$

1° When $k = 0$, $D = \lambda/\mu$; $\lambda\pi \underset{(\lambda\to o)}{\sim} [M\sigma + r + O(A^{-1})]\sqrt{\lambda/\mu}$

$$P(y,t) \underset{(t\to\infty)}{\sim} \left[M\sigma + r + O(A^{-1})\right] / \sqrt{\pi\mu t}$$

2° When $k > 0$, $D \sim \lambda/\mu + 2k$;

$$\phi(y,\infty) = \lim_{\lambda\to o} \lambda\Phi = \left[e^{-r\sqrt{2k}} + O(A^{-1}\sqrt{2k})\right] / \left[1 + M\sigma\sqrt{2k} + O(A^{-1}\sqrt{2k})\right]$$

If $r\sqrt{2k}$ is not large and if $M\sigma$ is large, we may neglect $O(A^{-1}\sqrt{2k})$ with respect to the other terms.

3° $$\frac{\partial}{\partial\beta}\left[\frac{F(y,\beta)}{M+F(O,\beta)}\right] = e^{-r\sqrt{D}} \left\{[M\sigma + r + O(A^{-1})]\, D^{-1/2} + O(M^2\sigma^2) + O(M\sigma r)\right\}$$

a) For given $k > 0$, (29), with $D \sim s/\mu$, gives :

$$t\, e^{2\mu kt}\, \psi(y,t) \underset{(t\to+\infty)}{\sim} [M\sigma + r + O(A^{-1})]/4k\mu^{3/2}\sqrt{\pi t}$$

b) When $k \to 0$, (25') gives, (with $D = 2k/(1-k)$:

$$p\, \varpi_p(y,\infty) \underset{(p\to\infty)}{\sim} [M\sigma + r + O(A^{-1})]/\sqrt{2\pi p}$$

Recent Developments in Statistics
J.R.Barra et al., editors

Invited Paper

EXISTENCE, UNIQUENESS, AND ABSOLUTE CONTINUITY FOR STOCHASTIC DIFFERENTIAL EQUATIONS

Jean JACOD*
Université de Rennes

In this paper, we follow two objectives: first, we review a number of stochastic differential equations, and we try to make them enter the same general framework, that of semi-martingales, via the setting of such equations as "martingale problems". Secondly, we present some results concerning two questions: representation of martingales, and comparison between different measures corresponding to stochastic processes. These results are derived for semi-martingales, with of course the idea of applying them to solutions of stochastic differential equations. The domain of application for such results includes detection, filtering and control theory for stochastic processes and, more widely, statistics of stochastic processes, with the meaning given to those words by Liptzer and Shyriaev [22].

Let us be more explicit by describing a fairly classical example, namely a diffusion equation. This equation may be termed as

$$dX_t = a(X_t)dt + dW_t, \tag{1}$$

where a is some function, $W=(W_t)$ is a Wiener process, and $X=(X_t)$ is the unknown process which is to be the solution of equation (1). Actually there may be several kinds of solutions, according to the fact that the underlying probability space is given or not, or the Wiener process W is given or not. We will be interested in one particular type of solutions, called "weak solutions": a measurable space $(\Omega,\mathcal{F})$ is given, as well as an increasing family $(\mathcal{F}_t)$ of sub-σ-algebras of $\mathcal{F}$, and a continuous process X such that each X_t is $\mathcal{F}_t$-measurable. Then a (weak) solution is a probability measure P on $(\Omega,\mathcal{F})$ for which there exists a Wiener process W on $(\Omega,(\mathcal{F}_t),\mathcal{F},P)$ such that (1) holds.

Which problems concerning equation (1) will we be interested in ?

(i) Conditions on coefficient a for having existence and uniqueness.

(ii) If P is a solution, can each martingale on the space $(\Omega,(\mathcal{F}_t),\mathcal{F},P)$ be expressed as a stochastic integral with respect to the Wiener process ?

(iii) Let a and b be two different coefficients, and P_a and P_b be solutions of the corresponding equations; do we have $P_a \ll P_b$, or $P_a \sim P_b$? (of course, the answers to those questions are well known for equation (1)).

---

*I.R.I.S.A., *Lab. associé au C.N.R.S. n° 227, Dép. Math. Inf., Université de Rennes. BP 25 A, 35031 RENNES Cédex.*

One more remark concerning equation (1): if P is a solution, then X is a semi-martingale for P. This will be a common property of all the stochastic differential equations we have in mind. This fact motivates the organization of the paper: in part 1, semi-martingales and their local characteristics are introduced. Part 2 displays many examples of stochastic differential equations, for continuous or discontinuous processes, in relation with part 1; in particular we shall see that local characteristics are linked to coefficients of the corresponding equation. Parts 3 and 4 are concerned respectively with representation properties for martingales, and absolute continuity conditions and obtention of likelihood ratio.

This paper being a review, we try to avoid any technical details and, of course, no proofs are provided. For simplicity we speak only about 1-dimensional processes, but everything would be true in the d-dimensional case. Also for simplicity, the time interval is $[0,1]$.

1 - SEMI-MARTINGALES AND THEIR LOCAL CHARACTERISTICS

1.1 - Some notations. Let Ω be a space equipped with an increasing family $(\mathcal{F}_t)_{0\le t\le 1}$ of σ-algebras. We put $\mathcal{F}=\mathcal{F}_1$, and we define the predictable σ-algebra $\mathcal{P}$ on $\Omega\times[0,1]$ as being the σ-algebra generated by all the sets $A\times]s,t]$, with $s,t\ge 0$, $A\in\mathcal{F}_s$ (cf. Dellacherie [4]).

We also consider a process $X=(X_t)_{0\le t\le 1}$ defined on Ω, such that each X_t is $\mathcal{F}_t$-measurable, and whose paths are right-continuous and left-hand limited. $\Delta X_t = X_t - X_{t-}$ is the size of the jumps of X at time t. If $E=\mathbb{R}\setminus\{0\}$ is equipped with its Borel σ-algebra $\mathcal{E}$, the jumps of X are completely described by the random measure $\mu(\omega;dt,dx)$ on $[0,1]\times E$ which is defined by

$$\mu(\omega;[0,t]\times A) = \sum_{s\le t} 1_A(\Delta X_s) \qquad (t\ge 0,\ A\in\mathcal{E}) \tag{2}$$

(the symbol 1_C means the indicator function of the set C).
Let $\tilde{\Omega} = \Omega\times[0,1]\times E$, with the σ-algebra $\tilde{\mathcal{P}} = \mathcal{P}\otimes\mathcal{E}$.

Let P be a probability measure on $(\Omega,\mathcal{F})$. For each fact concerning $(\Omega,(\mathcal{F}_{t+}),\mathcal{F},P)$-martingales, one may consult for example Meyer [23] (we put $\mathcal{F}_{t+} = \bigcap_{s>t}\mathcal{F}_s$ if $t<1$, and $\mathcal{F}_{1+}=\mathcal{F}_1$). Let us just recall that a right-continuous process M is a local martingale on this space if there exists an increasing sequence (T_n) of stopping times such that $\bigcup_{(n)}\{T_n=1\}=\Omega$ P-a.s., and that each process $(M_{T_n\wedge t})_{0\le t\le 1}$ is a martingale.

1.2 - Semi-martingales. Let P be a probability measure on $(\Omega,\mathcal{F})$. We say that X is a P-semi-martingale if it has a decomposition $X=A+M$, where M is a P-local martingale and A is a process whose paths have bounded variation on $[0,1]$. Of course, such a decomposition is not unique.

The class of semi-martingales covers a fairly wide range of random processes. The simplest non-trivial example is when, under P, X is a process with independent increments (say, homogeneous, for simplicity); then X is a P-semi-martingale. But this example presents another fundamental property: at least when $\mathcal{F}$ is not too big, namely when $\mathcal{F} = \sigma(X_s, s\in[0,1])$, then P is entirely characterized, via Lévy-Khintchine formula, by three terms which are the diffusion coefficient b, the drift coefficient a, and the Lévy measure F. Therefore this example will prove crucial for us, in the sense that we shall try to extend the characterization of P by something looking like the triplet (F,a,b), to each P for which X is a semi-martingale.

So, suppose X is a P-semi-martingale. We define three terms as follows:

(i) The Lévy system ν: this is a positive random measure $\nu(\omega;dt,dx)$ on $[0,1]\times E$. It is characterized by the fact that for each $A\in\mathcal{E}$ lying at a strictly positive distance of 0, the process $(\nu([0,t]\times A))_{0\le t\le 1}$ is predictable, and the process $(\mu([0,t]\times A) - \nu([0,t]\times A))_{0\le t\le 1}$ is a local martingale.

(ii) The drift A: this is a predictable process whose paths are right-continuous and have bounded variation, and such that $A_0 = 0$. It is characterized by the fact that the process $M_t = X_t - X_0 - \sum_{s\le t} 1_{\{|\Delta X_s|>1\}}\Delta X_s - A_t$ is a local martingale.

(iii) The "diffusion part" B: as M is a local martingale, it has a unique decomposition $M = M^c + M^d$, where M^c is a continuous local martingale such that $M_0^c=0$, and M^d is "orthogonal" to each continuous local martingale. Then we put $B = \langle M^c, M^c\rangle$, which is the unique continuous increasing process such that $B_0 = 0$ and $(M^c)^2-B$ is again a local martingale.

The triplet $\mathcal{C}=(\nu,A,B)$ is called the set of P-local characteristics of X. When X is a process with independent increments, associated with (F,a,b) as above, its P-local characteristics are :

$$\nu(\omega;dt,dx) = dt\times F(dx), \quad A_t(\omega) = at, \quad B_t(\omega) = b^2 t \tag{3}$$

The above notion and terminology were introduced by Grigelionis [12,13], under some restrictions. However, the triplet $\mathcal{C}$ does not necessarily characterize the measure P, and the drift A is not intrinsic because the bound 1 imposed on the jumps of X in its definition is of course arbitrary (the same problem arises for processes with independent increments, a different version of the Lévy-Khintchine formula giving different values for the drift, while the other two terms are intrinsic). On the other hand, the Lévy system ν may clearly be defined for more general processes than semi-martingales, and it was introduced under this name for Markov processes by Watanabe [31].

In addition to their simple relationships with coefficients of stochastic differential equations (see part 2 below), local characteristics may have a statistical interest: in the best cases they characterize the measure P, and at the same time they are much easier to apprehend statistically than the process X itself because they result of a "smoothing" of X (by taking something similar to a conditional expectation): this is most apparent in formula (3) for processes with independent increments.

Let us end this part by saying some words about the "initial condition". By definition, the initial condition is a probability measure on $(\Omega,\mathcal{F}_0)$: usually $\mathcal{F}_0=\sigma(X_0)$, so the initial condition is the distribution of X_0.

Now we can set our problem in suitable terms. We say that P belongs to $\mathcal{S}(Q,\mathcal{C})$ if its restriction to $(\Omega,\mathcal{F}_0)$ is Q, and if X is a P-semi-martingale with P-local characteristics $\mathcal{C}$. Very often we start with a probability measure Q on $(\Omega,\mathcal{F}_0)$ and a triplet $\mathcal{C}=(\nu,A,B)$, and we study the set $\mathcal{S}(Q,\mathcal{C})$, or, as it is put sometimes, the "problem $\mathcal{S}(Q,\mathcal{C})$".

2 - EXAMPLES

Below we state some results about existence and uniqueness for problems $\mathcal{S}(Q,\mathcal{C})$. However, existence and uniqueness usually do not require the same assumptions: existence requires regularity conditions on $\mathcal{C}$, and also that the space Ω be sufficiently rich, whereas uniqueness only requires regularity conditions (often more stringent than for existence).

Therefore we follow a general rule. Consider the assumptions:

(A-1): $\mathcal{F}_t = \sigma(X_s, s\leq t)$ for each t.

(A-2): Ω is the set of all functions $\omega:[0,1]\to\mathbb{R}$ which are right-continuous and left limited, $X_t(\omega)=\omega(t)$ and $\mathcal{F}_t=\sigma(X_s,s\leq t)$

(A-2) implies (A-1). Then everywhere in part 2, when a problem $\mathcal{S}(Q,\mathcal{C})$ is said to satisfy existence (resp. uniqueness), it is automatically understood that assumption (A-2) (resp. (A-1)) is in force. Saying that $\mathcal{S}(Q,\mathcal{C})$ satisfies both existence and uniqueness, therefore, means that existence holds under (A-2), while uniqueness holds under (A-1).

2.1 - Diffusion equation. The general (non-homogeneous) diffusion equation is usually written as

$$dX_t = a(t,X_t)dt + b(t,X_t)dW_t, \tag{4}$$

where a and b are functions: $[0,1]\times\mathbb{R}\to\mathbb{R}$, and W is a Wiener process. In fact equation (4) has a symbolic meaning, and a (weak) solution is a probability measure P on $(\Omega,\mathcal{F})$ for which there exists a Wiener process W on $(\Omega,(\mathcal{F}_t),\mathcal{F},P)$ such that

$$X_t = X_0 + \int_0^t a(s,X_s)ds + \int_0^t b(s,X_s)dW_s \tag{5}$$

(the second integral is a stochastic integral). Therefore it is easy to see that a weak solution of (4) with a distribution Q for X_0 is exactly an element of $\mathcal{S}(Q,\mathcal{C})$, with

$$\mathcal{C} = (\nu,A,B) : \nu=0,\ A_t = \int_0^t a(s,X_s)ds,\ B_t = \int_0^t b^2(s,X_s)ds.$$

After the pioneering work of Ito [14], Stroock and Varadhan [28], among a number of people, have given a fairly general condition for both existence and uniqueness for this problem $\mathcal{S}(Q,\mathcal{C})$: a is a measurable bounded function, b is a continuous bounded function satisfying identically $b \neq 0$; moreover if $P \in \mathcal{S}(Q,\mathcal{C})$, the process (Ω,X_t,P) is then Markov.

Let us say that the same equation when X is restricted to belong to some subset of $\mathbb{R}$ (or $\mathbb{R}^d$) has been extensively studied (diffusion processes on a domain, diffusion processes with boundary conditions); all these versions of (4) may be fitted within the same setting of semi-martingales (cf. for example El Karoui [6], Watanabe [32]). Finally some cases of existence without uniqueness have also been studied in detail (Girsanov [11]).

2.2. - Generalized diffusion processes. We consider the equation

$$dX_t = \tilde{a}_t dt + \tilde{b}_t dW_t \tag{6}$$

(with the same interpretation as (5) above) where W is again a Wiener process, while $\tilde{a}$ and $\tilde{b}$ are "non-anticipating functionals", that is random processes which are measurable and adapted to the family $(\mathcal{F}_t)$. A weak solution of (6) is an element of $\mathcal{S}(Q,\mathcal{C})$, with

$$\mathcal{C} = (\nu,A,B) : \nu=0, \quad A_t = \int_0^t \tilde{a}_s ds, \quad B_t = \int_0^t \tilde{b}_s^2 ds.$$

Various conditions on a and b are known, for existence or uniqueness of solutions (see for example Liptzer and Shyriaev [22]), but usually b must satisfy very restrictive conditions. For example existence and uniqueness hold when $\tilde{a}$ is bounded and when $b_t(\omega) = b(t,X_t(\omega))$, with $\tilde{b}$ satisfying the conditions of example 2.1.

2.3 - Continuous martingales. This example does not derive directly from a stochastic differential equation. It is the problem $\mathcal{S}(Q,\mathcal{C})$, where $\mathcal{C} = (0,0,B)$. Quite recently, Ershov [8] and Skorokhod [27] have shown that for a wide class of increasing continuous processes B, the above problem admits at least one solution; unfortunately, conditions are too technical to be given here. When B is such that existence holds for $\mathcal{S}(Q,\mathcal{C})$, then existence holds as well for $\mathcal{S}(Q,\mathcal{C}')$ where $\mathcal{C}' = (0,A,B)$, with $A_t = \int_0^t \tilde{a}_s dB_s$, the process $\tilde{a}$ being predictable and bounded. Conditions for having uniqueness remain to be found.

Of course, examples 2.1, 2.2 and 2.3 are of increasing generality, but they all concern continuous processes (that is, if P is a solution, X is P-a.s. continuous). Now we turn to some examples of discontinuous processes.

2.4 - Point processes. This is a rather trivial example, but it is of much importance for the applications. At first, a remark: if X is P-a.s. the counting process of a point process on [0,1] (that is, if X is a pure jump process and ΔX_t equals 0 or 1), it is certainly a P-semi-martingale. Moreover the P-local characteristics are $\mathcal{C} = (\nu,A,0)$, with

$$\nu(\omega;dt,dx) = dA_t(\omega)\varepsilon_1(dx), \quad A \text{ is increasing} \tag{7}$$

Conversely if $P \in \mathcal{S}(Q,\mathcal{C})$ where $\mathcal{C} = (\nu,A,0)$ satisfies (7), then necessarily X is P-a.s. a counting process (the drift A is then called the "stochastic intensity" of the point process; for example for a Poisson point process $A_t=t$: compare with (3)).

Let $\mathcal{C}$ as above. Then uniqueness for $\mathcal{S}(Q,\mathcal{C})$ holds true in any case (recall that assumption (A-1) is in force), while if A_1 is everywhere finite we have existence (cf. [15], or Kabanov, Liptzer and Shyriaev [18]).

2.5 - Skorokhod equation. This equation may be found in his book [26]:

$$X_t = X_0 + \int_0^t \tilde{a}_s ds + \int_0^t \tilde{b}_s dW_s + \int_0^t\int_E \tilde{f}(s,X_x,x)q(ds,dx). \tag{8}$$

Here, a and b are non-anticipating functionals. f is a function: $\Omega\times[0,1]\times\mathbb{R}\times E\rightarrow\mathbb{R}$, which is $\mathcal{P}\otimes\mathcal{R}\otimes\mathcal{E}$-measurable. W is a Wiener process. Now we describe q: let $p(\omega;dt,dx)$ be a "Poisson random measure", that is a random measure on $[0,1]\times E$ such that the random variables $(p(C_i))_{i\leq n}$ are integer-valued and mutually independent for any family $(C_i)_{i\leq n}$ of pairwise disjoint Borel subsets of $[0,1]\times E$ and that $E[p([0,t]\times A)] = tF(a)$ for some measure F. Then we put $q(\omega;dt,dx) = p(\omega;dt,dx) - dt\,F(dx)$ (a typical example of Poisson random measure is as follows: p is the random measure associated by (2) with a process with independent increments, with Lévy measure F). Of course, the third integral in (8), as well as the second one, are stochastic integrals. Let us remark that in [26], the measurability condition on $\tilde{f}$ is a little less stringent than here, and functionals $\tilde{a}$ and $\tilde{b}$ may depend on the value of the process X: but this greater generality is only apparent.

Then it can be shown that a weak solution of (8) is an element of $\mathcal{S}(Q,\mathcal{C})$, where

$$\mathcal{C} = (\nu,A,B) : \begin{cases} \nu([0,t]\times A) = \int_0^t ds \int_E F(dx) 1_A(\tilde{f}(s,X_{s-}+x,x)) \\ A_t = \int_0^t \tilde{a}_s ds \\ B_t = \int_0^t b_s^2 ds, \end{cases}$$

and Skorokhod gives conditions for having both existence and uniqueness for $\mathcal{S}(Q,\mathcal{C})$.

He also studies a less general equation, for which he gives other (of course weaker) conditions for having existence or uniqueness: more precisely this is the case where

$$\tilde{a}(\omega,s) = a(s,X_x(\omega)),\ \tilde{b}(\omega,s) = b(s,X_s(\omega)),\ \tilde{f}(\omega,s,x,y) = f(s,x,y) \qquad (9)$$

When we have (9) and when uniqueness holds for each initial condition Q, then the process X is Markov.

2.6. - Diffusion processes with jumps. Here again, this example does not derive directly from a stochastic differential equation, but it is a natural generalization, in the Markovian setting, of example 2.1, which allows jumps for the solution. It has been studied recently by Stroock [29] , Lepeltier and Marchal [21]. This is the problem $\mathcal{S}(Q,\mathcal{C})$, with

$$\mathcal{C} = (\nu,A,B) : \begin{cases} \nu(dt,dx) = dt\ N(t,X_{t-};dx) \\ A_t = \int_0^t a(s,X_s)ds, \quad B_t = \int_0^t b^2(s,X_s)ds \end{cases}$$

where a and b are like in example 2.1, and N is a kernel from E into $[0,1]\times\mathbb{R}$. Then if a is measurable and bounded, if b is continuous, bounded, and satisfies $b\neq 0$ identically, and if N is such that $\int f(y)\frac{y^2}{1+y^2}N(\cdot;dy)$ is continuous for each bounded continuous function f, then existence holds for $\mathcal{S}(Q,\mathcal{C})$. We obtain uniqueness if we add the assumption: $\int 1_A(y)\frac{y^2}{1+y^2}N(\cdot;dy)$ is continuous for each $A\in\mathcal{E}$ (other conditions for existence may be found in [21]).

Of course this example is very similar to example 2.5 when (9) holds. But the original formulation, as well as the conditions which are obtained, are quite different.

3 - REPRESENTATION PROPERTIES FOR MARTINGALES.

Let $P\in\mathcal{S}(Q,\mathcal{C})$. In the definition of the "diffusion part" B of $\mathcal{C}$, a continuous local martingale M^c appears, and we may classically define stochastic integrals

$\int_0^t u_s dM_s^c$ of suitable processes u with respect to M^c (see [23]). On the other hand, $(\mu([0,t]\times A) - \nu([0,t]\times A))$ is also a local martingale. By a monotone class argument, it is clear that for any $\tilde{\mathcal{P}}$-measurable function W on $\tilde{\Omega}$ such that the (ordinary) integral $\int_0^t\int_E W(s,x)(\mu-\nu)(ds,dx)$ makes sense for each t, this defines a process which is again a local martingale. So, there is no wonder that we can extend the definition of such stochastic integrals with respect to $(\mu-\nu)$ to a fairly large class of suitable $\tilde{\mathcal{P}}$-measurable functions W, by standard approximation arguments (an example of such a stochastic integral with respect to a particular $(\mu-\nu)$ has already been encountered in example 2.5).

We say that P satisfies the property of representation for martingales (with respect to X) if any P-local martingale N may be written as

$$N_t = E(N_0) + \int_0^t u_s dM_s^c + \int_0^t\int_E W(s,x)(\mu-\nu)(ds,dx) \tag{10}$$

for some suitable u and W (above integrals are evidently stochastic integrals). Let us notice that (10) implies N_0 is constant. Both terms M^c and $(\mu-\nu)$ are completely determined if X is known.

This property of representation for martingales is crucial for many problems in filtering and control theory. So the following theorem has some interest [17]:

THEOREM 1: *Let* $P\in\mathcal{S}(Q,\mathcal{C})$. *Then* P *enjoys the property of representation for martingales if and only if* P *is an extremal point in the convex set* $\mathcal{S}(Q,\mathcal{C})$.

In view of the applications, the most important case is when uniqueness holds for $\mathcal{S}(Q,\mathcal{C})$: then necessarily the unique solution P enjoys the property of representation. This theorem covers many classical results, for example the well-known theorem of Fujisaki, Kallianpur and Kunita [9], or representation properties for point processes (Chou and Meyer [3], [15]), or for processes with independent increments (Galtčuk [10]).

4 - ABSOLUTE CONTINUITY CONDITIONS AND LIKELIHOOD RATIO.

4.1 - The effect of a change of measure on local characteristics. We begin by some theoretical considerations, which will be of primary importance for deriving absolute continuity conditions.

Let $P\in\mathcal{S}(Q,\mathcal{C})$. Let P' be another probability measure on $(\Omega,\mathcal{F})$, such that $P'<<P$. We denote by Q' the initial condition associated with P', that is the restriction of P' to $(\Omega,\mathcal{F}_0)$. Of course $Q'<<Q$, and we have (Van Schuppen and Wong [30], [16]):

THEOREM 2: *There exists a* $\tilde{\mathcal{P}}$*-measurable nonnegative function* Y *on* $\tilde{\Omega}$ *and a predictable process* v *such that* $P' \in \mathcal{S}(Q',\mathcal{C}')$, *where*

$$\mathcal{C}' = (\nu',A',B') : \begin{cases} \nu'(\omega;dt,dx) = Y(\omega,t,x)\nu(\omega;dt,dx) \\ A'_t = A_t + \int_0^t v_s dB_s + \int_0^t \int_{\{|x|\le 1\}} x[Y(s,x)-1]\,\nu(ds,dx) \\ B' = B. \end{cases} \quad (11)$$

In particular, that theorem asserts that X is a P'-semi-martingale. It extends a part of Girsanov's theorem (this theorem concerns diffusion processes) the generalization of the rest of Girsanov's theorem will be given in part 4.3.

<u>4.2 - Absolute continuity conditions</u>. Now we want to find conditions on $\mathcal{C},\mathcal{C}'$, and Q' for having P'<<P or P'∼P, when $P \in \mathcal{S}(Q,\mathcal{C})$ and $P' \in \mathcal{S}(Q',\mathcal{C}')$. Of course we must have Q'<<Q or Q'∼Q respectively, and, in view of theorem 2, it is only natural to consider the problem in case $\mathcal{C}'$ is obtained from $\mathcal{C}$ by formula (11), for some Y and v.

Before giving the results, <u>let us proceed to some comments</u>. Why do we look for conditions on $\mathcal{C}$ and $\mathcal{C}'$? Firstly because local characteristics play a crucial role from the theoretical point of view, especially for filtering and control problems. But also for the next "statistical" reason: one of the main problems is often to find the correct model which describes a stochastic phenomenon. When this phenomenon is governed by a stochastic differential equation, this reduces to finding the correct coefficients of the equation or, finally, the correct local characteristics: therefore the importance, from the statistical point of view, of knowing conditions in terms of $\mathcal{C}$ and $\mathcal{C}'$ and also of possibly deriving the likelihood ratio $\frac{dP'}{dP}$ from $\mathcal{C}$ and $\mathcal{C}'$, when P'<<P.

We add one more remark: suppose we compare $P \in \mathcal{S}(Q,\mathcal{C})$ and $P' \in \mathcal{S}(Q',\mathcal{C}')$ and that we do not have P'<<P nor P<<P'. Then what follows still applies (with some precautions) via the next considerations: the set of all probability measures P for which X is a P-semi-martingale is convex, so there exist Q" and $\mathcal{C}''$ such that $P'' = \frac{1}{2}(P+P') \in \mathcal{S}(Q'',\mathcal{C}'')$ and now we have P<<P" and P'<<P". In other words, any two solutions of two problems are always dominated by a solution of a third "problem of semi-martingales".

Now we come to the statement of the results. <u>We suppose that</u> $P \in \mathcal{S}(Q,\mathcal{C})$ <u>and</u> $P' \in \mathcal{S}(Q',\mathcal{C}')$, <u>with</u> Q'<<Q <u>and</u> $\mathcal{C}'$ <u>obtained from</u> $\mathcal{C}$ <u>by formula (11)</u>. Results will be expressed in terms of Y and v, but if $\mathcal{C}$ and $\mathcal{C}'$ are given, usually it is not difficult to compute Y and v with (11). We define two subsets of Ω:

$$G = \{\int_0^1 v_s^2 dB_s + \int_0^1\int_E [(Y(s,x)-1)^2 1_{\{Y(s,x)\le 2\}} + (Y(s,x)-1)1_{\{Y(s,x)>2\}}]\nu(ds,dx) < \infty\}$$

$$\hat{G} = \{\int_0^1\int_E 1_{\{Y(s,x)=0\}}\nu(ds,dx)=0\}$$

(when Y>0 identically, $\hat{G}=\Omega$). We will also make an essential assumption: X is quasi-left continuous for P, which means that for any increasing sequence (T_n) of stopping times going to T, we have lim $X_{T_n}=X_T$ P-a.s.; this is also equivalent to having $\nu(\omega;\{t\}xE) = 0$ identically, and thus X is also quasi-left continuous for P'. Except possibly for example 2.4, all examples of part 2 satisfy this assumption. At last a condition of local uniqueness for $\mathcal{S}(Q,\mathcal{C})$ or $\mathcal{S}(Q',\mathcal{C}')$ will appear in some places below: we do not want to go into details, let us just say that for example 2.4 and for all the Markovian examples of part 2, local uniqueness is equivalent to uniqueness (see [16]).

Then we have:

THEOREM 3: (a) If P'≪P *we have* P'(G)=1.
(b) *If* P'∼P P(G∩$\hat{G}$)=1 *and* Q'∼Q.

THEOREM 4: Assume that $\mathcal{F}_t=\sigma(X_x,s\le t)$ *for each* t, *and that local uniqueness holds for* P *(resp.* P'*). Then*
(a) *if* P'(G∩$\hat{G}$)=1 *(resp.* P'(G)=1) *we have* P'<<P,
(b) *if* P(G∩$\hat{G}$)=P'(G∩$\hat{G}$)=1 *(resp.* P(G∩$\hat{G}$)=1) *and* Q'∼Q, *we have* P'∼P.

A vast literature has been written on this subject, giving various conditions for the validity of these theorems, mainly for the continuous case. See for instance Kailath and Zakaï [20], Orey [24,25], Liptzer and Shyriaev [22], Ershov [7] when X is continuous, [16] for the general case. When the assumption of quasi-left continuity is dropped, almost nothing is known (however, see Kabanov, Liptzer and Shyriaev [18] for the case of point processes).

4.3 - Expression of the likelihood ratio. When P'<<P, the expression of the likelihood ratio is based upon the "exponential formula" for martingales, due to Doléans-Dade [5].

All the assumptions of part 2.2 are in force. Let

$$T_n = \inf(t: \int_0^t v_s^2 dB_s + \int_0^t\int_E [(Y(s,x)-1)^2 1_{\{Y(s,x)\le 2\}} + (Y(s,x)-1)1_{\{Y(s,x)>2\}}]\nu(ds,dx)\ge n).$$

Thus $G = \bigcup_{(n)} \{T_n=1\}$. We can put

$$N_t = \begin{cases} \int_0^t v_s dM_s^c + \int_0^t\int_E (Y(s,x)-1)(\mu-\nu)(ds,dx) & \text{if } t\le T_n \text{ for some } n \\ 0 & \text{if not.} \end{cases} \quad (12)$$

Here, the stochastic integrals have the same meaning as in (10); in particular M^c is the continuous part of the local martingale intervening in the definition of the drift A. For each n, the process $(N_{T_n\wedge t})_{0\le t\le 1}$ is a P-local martingale, and we put

$$Z_t = \begin{cases} [\exp(N_t - \frac{1}{2}\int_0^t v_s^2 dB_s)] \prod_{s\le t} [(1+\Delta N_s)e^{-\Delta N_s}] & \text{if } t\le T_n \text{ for some } n \\ 0 & \text{if not.} \end{cases} \quad (13)$$

Let us remark that N and Z may be computed when ℓ, ℓ', and the process X are known, because μ and M^c are entirely determined by X. We have

THEOREM 5: Assume $\mathcal{F} = (X_s, s\in[0,1])$. *Then if* $P' \ll P$ *and if* $q = \frac{dQ'}{dQ}$, *we have* $\frac{dP'}{dP} = qZ_1$ *each time one of the following conditions holds true: local uniqueness for P, or local uniqueness for P', or property of representation for martingales for P.*

In fact, one can say more: qZ_t is a version of the likelihood ratio $\frac{dP'_t}{dP_t}$, where P_t (resp. P'_t) is the restriction of P (resp. P') to $(\Omega, \mathcal{F}_{t+})$. This fact is often expressed by the formula

$$qZ_t = E(\frac{dP'}{dp}|\mathcal{F}_{t+}).$$

The assumption of quasi-left continuity for X is not really necessary for the validity of that theorem; however if it does not hold, Z is still defined by (13) but N has a more complicated expression than (12) (see [16]).

Finally, the consideration of Z allows us to give a partial converse to theorem 2, which is the second half of Girsanov's theorem. We suppose that $P\in \mathcal{S}(Q,\ell)$ and that Y and v are given, as well as N and Z by (12) and (13).

THEOREM 6: Assume that $E(Z_1)=1$. *Then the probability measure* $\hat{P}$ *defined by* $\hat{P}=Z_1.P$ *belongs to* $\mathcal{S}(Q,\ell')$, *where* ℓ' *is defined by* (11).

This theorem is an important step to prove theorems 3 and 4. It also provides a theorem of existence for problems $\mathcal{S}(Q,\ell')$ when existence holds for $\mathcal{S}(Q,\ell)$ and ℓ' is given by (11). A number of authors have written about theorems 5 and 6, under various conditions. This includes those quoted in part 2.2, but also several others who worked more specifically on point processes (Segall and Kailath [19], Brémaud [2], Boël, Varaiya and Wong [1]).

One last word to say that the following bibliography, although already long, has no pretension of being complete: in particular we have usually selected a few papers for each author, and many related fields are completely ignored. A comprehensive bibliography for the continuous case may be found in [22], and additional references in [25].

BIBLIOGRAPHY

[1] BOEL R., VARAIYA P., WONG E. Martingales on jump processes. SIAM J. of control, 13, pp 999-1060, 1975.

[2] BREMAUD P. A martingale approach to point processes. PhD Thesis, El. Res. Lab. Berkeley, M-345, 1972.

[3] CHOU C.S., MEYER P.A. La représentation des martingales relative à un processus ponctuel discret. C.R.A.S., A, 278, 1561-1563, 1974.

[4] DELLACHERIE C. Capacités et processus stochastiques. Berlin: Springer 1972.

[5] DOLEANS-DADE C. Quelques applications de la formule de changement de variables pour les semi-martingales. Z. Wahr., 16, 181-194, 1970.

[6] EL KAROUI N. Processus de réflexion dans $\mathbb{R}^n$. Sém. Proba. Strasbourg IX, Lect. Notes in Math. 465. Berlin: Springer 1975.

[7] ERSHOV M.P. On the absolute continuity of measures corresponding to diffusion processes. Th. Proba. Appl. 17; 173-178, 1972.

[8] ERSHOV M.P. The existence of a martingale with a given diffusion functional. Th. Proba. Appl. 19, 633-655, 1974.

[9] FUJISAKI M., KALLIANPUR G., KUNITA K. Stochastic differential equations for the nonlinear filtering problem. Osaka J. Math., 9, 19-40, 1972.

[10] GALTČUK L. The structure of a class of martingales. Proc. School-Seminar on random processes. Vilnius, Acad. Sci. Lit. SSR, I, 7-32, 1975.

[11] GIRSANOV I. An example of non-uniqueness of a solution of Ito's stochastic equation. Th. Proba. Appl., 7, 336-342, 1962.

[12] GRIGELIONIS B. On the absolute continuity of measures corresponding to stochastic processes. Lit. Math. S. 11, 783-794, 1972.

[13] GRIGELIONIS B. On non-linear filtering theory and absolute continuity of measures, corresponding to stochastic processes. 2^{d} Japan-USSR Symp. Lect. Notes in Math. 330, Berlin: Springer, 1973.

[14] ITO K. On stochastic differential equations. Mem. Am. Math. Soc. 4, 1951.

[15] JACOD J. Multivariate point processes: predictable projection, Radon-Nikodym derivatives, representation of martingales. Z. Wahr. 31, 235-253, 1975.

[16] JACOD J., MEMIN J. Caractéristiques locales et conditions de continuité absolue pour les semi-martingales. Z. Wahr. 35, 1-37, 1976.

[17] JACOD J. A general theorem of representation for martingales. Proc. of Symposia in Pure Mathematics (Meeting A.M.S. of Urbana), Vol. 31, 27-43, 1977.

[18] KABANOV I.M., LIPTZER R., SHYRIAEV A. Martingale method in the theory of point processes. Proc. School-Seminar on random processes. Vilnius Acad. Sci. Lit. SSR, II, 296-353, 1975.

[19] KAILATH T., ZAKAI M. Absolute continuity and Radon-Nikodym derivatives for certain measures relative to Wiener measure. Ann. Math. Stat. 42, 130-140, 1971.

[20] KAILATH T., SEGALL A. Radon-Nikodym derivatives with respect to measures induced by discontinuous independent increments processes. Annals of Prob., 3, 449-464, 1975.

[21] LEPELTIER J.P., MARCHAL B. Problème des martingales et équations différentielles stochastiques associées à un opérateur intégro-différentiel. Ann. Inst. H. Poincaré, B, XII, 43-103, 1976.

[22] LIPTZER R., SHYRIAEV A. Statistics of stochastic processes. Moscou: Nauka 1974 (French translation available, English translation in preparation).

[23] MEYER P.A. Un cours sur les intégrales stochastiques. Sém. Proba. Strasbourg X, lect. Notes in Math. 511, Berlin: Springer 1976.

[24] OREY S. Conditions for the absolute continuity of two diffusions. T.A.M.S. 193, 413-426, 1974.

[25] OREY S. Radon-Nikodym derivatives of probability measures: martingale methods. Dept. Found. Math. Sci. Tokyo Univ. of Education, 1974.

[26] SKOROKHOD A.V. Studies in the theory of random processes. Reading: Addison-Wesley 1965.

[27] SKOROKHOD A.V. A remark on square-integrable martingales. Theor. Prob. Appl. 20, 195-199, 1975.

[28] STROOCK D., VARADHAN S.R.S. Diffusion processes with continuous coefficients I,II. Comm. Pure Appl. Math. 22, 345-400 and 479-530, 1969.

[29] STROOCK D. Diffusion processes associated with Lévy generators. Z. Wahr. 32, 209-244, 1974.

[30] VAN SCHUPPEN J.H., WONG E. Transformations of local martingales under a change of law. Ann. Proba. 2, 879-888, 1974.

[31] WATANABE S. On discontinuous additive functionals and Lévy measures of a Markov process. Japan J. Math. 34, 53-79, 1964.

[32] WATANABE S. On stochastic differential equations for multi-dimensional diffusion processes with boundary conditions. J. Math. Kyoto Univ. 11, 169-180, 1971.

Recent Developments in Statistics
J.R.Barra et al., editors

Invited Paper

CHILDREN (AGE 11 TO 15) APPROACHING STATISTICS

C. Comiti and P. Clarou, G. Dumousseau, G. Le Nezet
Research Institutes for the Teaching of Mathematics
Universities of Grenoble, Bordeaux, Rennes

J. Pincemin
National Institute for Pedagogical Research and Documentation, Paris

The subject we are going to talk about will perhaps surprise some of the participants of this Meeting. However, the reason why we chose to deliver this talk precisely at this Meeting is because, in our opinion, the specialists in this field may very well be interested in learning about the research work presently being conducted in France by members of the teaching staff, aiming at making the children very early sensitive to the importance of statistics.

As you may already know, French schools do not teach probability and statistics before the end of the second teaching cycle (age 15 to 18). Thus a great number of pupils - those leaving school earlier - have been confronted only with deterministic problems and never have approached any aleatory phenomena.

In order to fill in this gap, a group of mathematics teachers, of mathematicians in research and of psychologists, all of them sponsored by the I.R.E.M. (Research Institutesfor the Teaching of Mathematics) and the I.N.R.D.P. (National Institute for Pedagogical Research and Documentation) have started a research project in 1973 on the approach of statistics by pupils of the first cycle of secondary school (age 11 to 15).

The age of 11 turned out to be very favourable for a first approach. Indeed, at this age, children's interest centers are broadening. Being more and more receptive to the mass of information they generously receive via the newspapers and TV, they are now being confronted with new problems and may well be interested by data investigation.

Of course, the mathematical tools children of this age dispose of, are far from being sufficient for permitting a complete analysis of the submitted problems. However, to be confronted with such problems already at this age will progressively enlarge their experimentation field in the aleatory domain and will incite them to construct a certain number of new tools.

What are the main goals of the teachers engaged in this research project ? Once the experimentation completed, the children should know

- *how to read and organize data in great number (furnished by the teacher, or collected by the pupils themselves) ;*
- *how to find one or several investigation methods allowing to answer their questions concerning these data ;*
- *how to interprete critically the results of such a treatment ;*
- *how to approach with prudence the decision and forecast problems.*

All this has to be accomplished in the actual framework of the teaching schedule. Do not forget that the pupils of these forms have four one-hour lessons of mathe-

matics per week and that their teachers are obliged to cover the normal teaching programme. Thus we cannot teach these children statistics, but we can propose them activities of statistical order as a work-theme which should be connected with the mathematical context of the school-year.

How far have we got up to date ?

Our work-group has realized fifteen experimentation projects. Out of these, the following four have been chosen to be presented here.

A - Study of the population of fifth form students at a lycee in Valence, realized by the GRENOBLE team.

B - Survey into the most widely read magazines at a school of the Parisian area, realized by the PARIS team.

C - Bordeaux Harbour, an investigation realized by the BORDEAUX team.

D - Registration numbers of French and foreign cars ; a survey realized by the RENNES team.

In the present paper we shall emphasize primarily the notions of statistical ordering as conceived by the children, and set aside the more general notions these studies allowed us to introduce and to work out (such as order, quasi-ordering, barycenter, vectorial, etc.).

Problems encountered during the first phase of our study, i.e. at the stage of collecting or reading the data.

For each experimentation theme, in order to get the work started, the teacher concerned provided the children with :

- *either a subject inciting the children to collect the data themselves (projects A and B) ;*
- *or a collection of data (projects C and D).*

Here are a few examples of problems that were encountered when collecting the data.

TALK Nr. 1 (Philippe CLAROU)

Problems encountered when collecting the data :

The experimentation carried through at the Lycee C. Vernet in Valence was centered around the problems presented by the realisation of a survey and its exploitation.

In the proposed survey, the following characteristics were to be specified : the height, the weight, the month and year of birth, and the sex of each 5th form student of the lycee (age 12).

The experimentation comprised two parts.

1 - The survey which made the students appreciate the problems of collecting data (particularly, the exactitude and the validity conditions of these data).

2 - The treatment of the data which made the students work on numerous, but simple data involving only familiar notions.

Here we just talk about the first part.

Before starting out on their survey, the children, four or five in a group, specified the different methods they planned to apply.

Mainly they proposed :

A - either to question the children when entering their class-rooms, or else have them fill in forms ;

B - or to weigh and to measure each pupil ;

C - or to go and get the information at the school's medical office.

Then they actually carried through their survey and compared the different methods with one another, as well as the results obtained. This led them to the following statements :

- when questioning fifth form students, certain informations are fanciful (some kids cheated especially to the weight) ;
- weighing and measuring each pupil turned out to be too long ; this method had to be abandoned in the framework of the lycee ;
- to go and get the information at the medical office proved more reliable, but this method had also its drawbacks (reading errors and transcription errors, empty files because of absence).

When inquiring themselves, each students had taken pains in writing down the family name as well as the first name of the questioned child. The information given by the medical office however was anonymous and therefore appeared to the children devoid of interest.

During this experience, we observed the following.

- It takes a lot of time to make children proceed with a survey.
- At the beginning, most of the pupils were rather abashed about the fact that such an activity was proposed to them in the framework of the mathematics lesson.
- They of course had no statistical motivation, to start with. Why then collect this information ? What were they to do with it ? All questions we could not answer a priori.
- However, the students were obliged to get organized.

They also became aware of the fact that, when handling measurable data, the conditions in which they were collected were most important and even could influence the very results.

Still at this same stage, i.e. the collection of data, the PARIS group, when working on the most widely read magazines, approached the problem of sampling and of sample surveys.

TALK Nr. 2 (Jacqueline PINCEMIN)

Here fifth form students were asked to proceed with a survey on the reading of magazines in all the sixth and fifth forms of their lycee. At this occasion we wanted to approach tentatively the concepts of "sample" and of "sample survey". Through the influence of the mass-media, the notion of "survey" is closely associated in the children's minds with the notion of "sample survey", and sample surveys are indeed far more frequent than exhaustive surveys.

In order to reach our aim we adopted a method which surely will appear very open to criticism. Anyhow, we asked the children to organize, on the same population, first sample polls and then an exhaustive survey. The children, in work-groups of three or four, were asked to organize and realize a sample poll, without taking any notice about what was going on in the other groups. After that, the same work-groups re-distributed among themselves the total population in order to proceed with the exhaustive survey. Then they analysed the answers obtained, both by the sample polls and the exhaustive survey, expressing the results in percentages.

Finally they tried to determine which was the best poll or the best polls and to find out the reason why certain polls produced particularly poor results and others a bit better ones.

To be sure, our goal was not to teach children how to proceed with sample polls nor how to constitute samples. We just wanted them to find out by themselves why one has recourse to polls, i.e. we wanted them to understand that, starting from a subset of a population, valid information may be gained from it on behalf of the entire population ; and secondly, that the correct choice of the sample is the basic condition of success.

This is what happened during the first part of the experiment, the construction of the sample polls.

The children had decided to limit the survey to the five magazines that were the most widely read ones in their own class. They established a questionary divided into two parts.

Part 1 - One question on regular reading and one question on occasional reading of the five magazines.

Part 2 - 14 questions on personal characteristics, such as : class, sex, age, brothers and sisters, hobbies, etc.

The teacher then explained the children his idea of first making sample polls and then an exhaustive survey on the total population. He also gave them a definition of a sample poll : the problem is to determine a subset of a population such that the informations obtained from this subset be valid for the entire population. To determine this subset, they had to answer the following questions.

- How many pupils were to be questioned ?
- Which pupils were to be questioned ?
- How where they to be chosen ?

Right from the beginning, an important difficulty showed up. Many children found it hard to understand that the different polls had to be made independently one from the other and that they could bear on different and not necessarly disjunct sets. This difficulty appeared in several ways :

- some groups proposed to divide the whole population into 12 about equal parts and to allot one part to each of the 12 work-groups ;
- others simply stated that one must not question too many children, because some had to be left for the other groups ;
- others finally insisted on specifying that it was to be avoided that two groups question the same person ; so, in practice, everyone was recommended to make sure, when proposing the questionary to a child, that he or she had not already filled it in for another group.

All these children bore in mind to put together all those sample polls in order to obtain as complete an information as possible on behalf of the whole population. This became quite clear during a later lesson : when the children had finished analysing their polls, the teacher asked them what could be done with all these results. The idea of adding them up was proposed as often as the one of comparing them.

Secondly, we noticed that very few pupils asked themselves interesting questions about the size of the sample. Only two groups had some interesting ideas as for the question : "How many pupils were to be questioned ?".

1) In a group that had decided to question three boys and three girls of each form, one child asked : "Let's see. If I only question six pupils out of thirty-five, will I really get everybody's opinion ?".

2) In another group, two children discuss : one would like to question only ten pupils, the other fifty. The latter says : "With ten, we won't get an opinion. The ten may read Mickey Magazine."

On the opposite of these two groups, some declared that "the number didn't matter". And, like them, most of the groups focused their discussion on who was to be questioned. As for the size of the sample, some took their decision for practical reasons (time needed for getting through with the sample-poll, convenience for an easy calculation of the percentages), and the others for reasons we ignore completely.

3) As for the choice of pupils to be questioned, the children were more explicit. There were two opinions.

1) Some simply decide to question a fixed number of persons at random. "To question at random", this means to them :

- not to fix beforehand the composition of the sample by taking into account personal characteristics that could influence the reading of the magazines ;
- not to question persons you know, but persons you meet : in the corridors, when leaving school, in the school's dining-hall, during the break.

2) In the other groups, the children discussed the characteristics that had to be taken into account in order to ensure the "representativity" of the sample :

- take a small number of pupils in each form ;
- take boys and girls ;
- take sixth form students as well as fifth form students ;
- form an arrangement on the base of the three preceding decisions.

We observed that the children always took an equal number of pupils in each form, or else an equal number of boys and girls, or else an equal number of sixth form and of fifth form students. As for the choice of the individual to be questioned, they proceeded either by acquaintance or "at random" (at random being defined as above).

During the weeks following this session about the elaboration of the sample-polls, the children realized and analysed the polls as well as the exhaustive survey. Further on we will see how they tried to evaluate each poll.

This type of situation presents the advantage of motivating the students right from the beginning who are thus induced to organize, construct and realize themselves their sample survey. Unluckily, much time is then needed for the organization, the realisation and finally the critical inspection of the data collected, so that it proved difficult to still push on considerably the analysis of the obtained data.

In the second type of situation where the students are directly confronted with numerous and sometimes difficult data furnished by their teacher, the above difficulties are replaced by problems of reading, understanding and organizing these data. The survey on Bordeaux harbour gives you an example of this type of situation.

TALK Nr. 3 (Guy DUMOUSSEAU)

This subject has been experimented on sixth form and fifth form students (age 11 to 13).

The harbour is an important source of activity for Bordeaux. All the children had seen the piers and the boats, some had made an organized visit to the port with their school.

The Direction of the Autonomous Port of Bordeaux has furnished the readings over the past 12 years :

- sharing out of the traffic according to the national colours ;
- traffic of goods.

Thus the data were very abundant and very diversified. The experience with numerous workgroups has shown that a child is not afraid of such a mass of data.

Before starting out it was necessary to understand the significance of the data well, to master the techniques used for its presentation and finally to select a subject to investigate. The general theme was indeed rich enough to contain numerous and open problems (the children had spontaneously asked 22 questions).

The group that chooses a subject (for instance : the importation of cocoa from 1962 to 1973) must extract from the tables the relevant elements. Thus the group establishes the chronological series relating to the selected subject and calculates the totals for each year.

Some groups have shown functions of the set of colours towards the set N of natural numbers according to the years, or else of the product of the set of colours by the set of years towards the set of natural numbers.

Others, when investigating the importations of cocoa, sugar, and cod looked for the application of the product of the set of colours by the set of the investigated goods towards the set of natural numbers.

Some groups worked on rates. For example : the variation of the percentage of the number of French boats as compared with the total number of boats that had entered the port.

We noticed that, for the majority of the pupils, percentages had little meaning only : they had heard talk about them ! Though they had worked on the subject the preceding year, the very notion of it seemed to have faded away... It was necessary to re-examine the concept, first in the group, then in the entire form, during a synthesis.

There were two main motivations :

- the spokesmen of the groups had to explain to the others the reasons why they had chosen to conduct the work a certain way ;
- in the data the percentages were not directly given for certain years ; they had to be calculated.

Moreover, some had noticed that as for the tonnage of goods entered by French boats one could read :

- in 1962, 2,138,521 t, representing 61.16 % ,
- in 1964, 2,402,276 t, representing 56.45 % ,

which posed the problem of relativity.

Thus frequency was introduced. It has been defined as the decimal factor by which the total effectives were to be multiplied in order to obtain the effectives relating to the value of the character.

Problems encountered during the second phase of our study, i.e. at the stage of finding a method for describing the data.

The methods proposed by the students are of two types :

- *calculation of position parameters and first approach of dispersion ;*
- *construction of graphical representations.*

The RENNES group working on the registration of French and foreign cars gives us an example of the work accomplished in the calculation of means and absolute values of the deviations from the mean.

TALK Nr. 4 (Georges LE NEZET)

The subject "Registration of Cars" was studied during the first term of the first year in high-school, i.e. with children aged 11. The project consisted in two phases.

1) The children were provided with data from automobile statistics relating to :

a) the number of registrations of new private cars in France, month by month, from 1970 to 1974 ;

b) the number of registrations of new private cars in the EEC countries, by year and by make, in 1972, 1973 and 1974.

2) The children also proceeded with a survey about their own parents' cars. They then compared the national distributions, by year and by make, to the corresponding distributions obtained from this survey.

The teaching team chose this subject because it was related to the global and personal environment of the child, because it allowed to present the pupils different types of tables, and because it allowed to study an economical phenomenon connected with the "1974-crisis", which could eventually be extended as a subject of interest to other disciplines.

One of the first notions dealt with was the concept of the mean.

There were two different approaches depending on the pupils' knowledge.

- Either going out from a precise question : "How many French cars were registered during the years '70, '71, '72, '73 in one month on an average ?".
- Or going out from a very vague idea. For instance, from the question : "In which month of the year were sold the maximum of cars ?".

When reading the data table, the pupils became aware, first, that the peak was not always in the same month ; this gave them the idea of calculating the mean for each month of the year.

The calculation of the means presented no difficulty as long as it was the mean of a unique statistical series (a well-known problem for them). Difficulties appeared, however, when they were asked to calculate a global mean over the known partial means.

Interpretation.

The children wanted first of all to obtain an approximate idea of the number range without taking into account the seasonal variations. The mean, at the beginning, was not considered a characteristic. Representing this mean on a diagram, however, gave the children a better approach of the notion. The diagram helped the children to separate, first of all, the good months from the bad ones, then establish a classification with the help of the deviation from the mean.

They also approached the notion of dispersion.

As for the graphical representation, selecting the variable to be represented as well as choosing a scale and abiding by it are hard problems to solve. Here are a few examples of representations as proposed by the children.

TALK Nr. 5 (Philippe CLAROU)

The analysis of the data from the project about the population of the fifth forms (exp. A) led us to diverse graphical representations.

In particular, the children proposed to represent the number of pupils per form. They then joined these dots "in order to make a diagram as in geography". The comments of one child made them realize that this did not mean anything at all. We proposed a representation by vertical strokes. Some children who wanted to compare the average height of the boys to the average height of the girls in each class, superposed the corresponding strokes, thus avoiding the problem of scale concordance.

Then the students became interested, not so much in the average height, but the distribution of the heights. They proposed to represent the different heights in diagrams, and to do so, wanted "to round off". During a discussion we agreed on regrouping the heights by 5 cm ; hence the classical histogrammes obtained.

Finally we asked the children to find out whether the children with same weights had same statues.

In a first stage, they constructed a diagram by tracing the lines having as ordinates the given weights and their intersections with the lines having as abscissae the corresponding heights. As this diagram very quickly became unreadable, the teacher suggested to delete the lines of recall, leaving nothing but the intersection dots. Thus a cloud of dots was obtained.

TALK Nr. 6 (Guy DUMOUSSEAU)

Children often find it difficult to choose a scale, particularly when faced with great numbers. They find great numbers really difficult to handle.

In the project concerning Bordeaux Harbour (exp. C), there were boats of 10,000 tons, of 6,000 tons, imported goods weighing 3,497,083 tons, etc.

What do such big numbers mean to a child ? During a discussion pupils made the following remark :

"18°, 20° - one can imagine what this means ; 2,500,000 - one cannot imagine. It's much, for sure !".

Consequently, they mainly apply themselves to arrange these big numbers in some order, without, first, caring for their relative values.

Later on, they become aware of the interest of evaluating the differences between the numbers, and the choice of a regular scale where a unit is perfectly determined by its graphical representation.

This was the case when the children studied the variations in the importations of certain goods in the course of the years, or else when forecasting what was going to happen in 1974 by means of a linear graph method.

TALK Nr. 7 (Georges LE NEZET)

When working on the project "Registration of Cars" (exp. D), we obtained three types of diagrams.

1) Semi-circular graphs ("cheeses") representing the distribution percentages.

These diagrams had been obtained by 12 years old students who had had to overcome several difficulties to realize them :

- calculation of percentages ;
- notion of an angle and its measurement ;
- notion of proportionality.

2) Graphs with dotted lines representing either effectives or distribution percentages or else evolution percentages (increase rates from one year to the other).

Note that the curves representing evolution percentages surprised the students as the represented function, in certain cases, was decreasing whereas the effectives were increasing.

3) Finally, some star-shaped diagrams, the registration effectives being represented on 12 axes (one per month).

Problems encountered during the third phase of our study, i.e. at the stage of interpretation and of forecasting.

The interpretation of the obtained results proved problematic in all experimentation projects, though in different respects.

TALK Nr. 8 (Jacqueline PINCEMIN)

The analysis of the sample polls and of the exhaustive survey being completed, we presented the children with a table indicating the percentages of regular and of occasional readers (corresponding to the first two questions asked), percentages that had been obtained by means of the exhaustive survey on one side, and of each of the 11 polls (*) on the other.

We then asked the students to try to find methods allowing them to determine the best poll.

All the children defined the best poll as the one that, on the whole, came closest to the percentages corresponding to the entire population. They considered several ways of determining it.

1) The graphical process consisting in tracing profiles of the entire population and of each poll was rapidly abandoned. The profiles crossed too often to allow to deduct rapidly simple conclusions.

2) One group had thought of establishing, for each poll, the total of the absolute values of the deviations from the percentages of the entire population, the best poll being the one for which this total was the lowest.

3) Another method consisted in finding, on each line, not only the percentage that came closest to the one of the exhaustive survey, but also the most remote percentage. Then, per column, one would add up, for each poll, the number of closest approaches and substract therefrom the number of remotest approaches, the best poll, then, being the one having the highest surplus of closest approaches over the remotest ones.

(*) One of the 12 sample polls had been lost !

4) Finally, one group first attributed a rank to each "division" on a line, depending on the more or less great distance from the percentage registered for the whole population : the closest distance was ranked 1, etc. up to rank 11. The children then added up the ranks of each poll, the best poll being the one having the lowest total.

With the help of these methods, the children not only determined "the best poll", but also established a classification of the polls. The results differed considerably, depending on the methods employed (though much less with respect to "the best poll"). This gave the teacher an opportunity of starting a discussion :

A certain poll generally appeared to be the best one. Is it possible to find out why ? Is it possible to know whether the sampling was better defined ?" Here, to tell the truth, we failed completely. For the children started discussing again the characteristics they in general had examined during the preparation, but did not want to plunge into a detailed examination of how the samples were composed (as resumed by a table) which would have led them to considering the techniques of selecting the persons to be questioned.

TALK Nr. 9 (Georges LE NEZET)

Starting out from the survey on the registration of cars, the students of the six experimental classes asked the following questions :

- "In which of the six forms the distribution comes closest to the national distribution ?"
- "Do the six forms together come closer ?"

Whenever the results are given in percentages, in form of a table, few conclusions are deducted. To compare two samples with respect to the national distribution amounts to, first of all, count the months when it is "better" and the months when it is "not so good".

For a good number of pupils, the graphical representation is sufficient to make a rough comparison between the different samples. However, several students sensed the necessity of calculating the exact distance between the samples. In the majority of cases, they totalled up the absolute values of the deviations of the distributions from the national distribution.

Moreover, two other ideas were noteworthy :

- the comparison of the lenghts of the curves ;
- the comparison of the areas limited by the curve representing the sample and by the one representing the national distribution.

Our aim being only to make the children responsive to the notion of a sample, we thought it preferable not to insist too much on the interpretation of the obtained results.

The problematical value of forecasting became particularly evident when the children were working on the chronological evolution of a phenomenon. The problems touched hereby are all the more tricky as, at this stage, we do not dispose of the tools needed for introducing the concept of confidence interval.

Let us see how these problems were handled when investigating about Bordeaux harbour.

TALK Nr. 10 (Guy DUMOUSSEAU)

The following projects had been considered by the different groups :

- forecasts of the number of boats to enter the harbour in 1974 ;

- forecasts of the quantities of certain goods to be imported in 1974.

In general the students immediately thought of having recourse to the diagrams they disposed of, envisaging graphical solutions.

In certain cases where the diagram did not show any movement indicating the general tendency, neither increasing nor decreasing, the pupils very quickly used an interval the extremities of which being the minimum and the maximum encountered during the investigated period.

However, in this domain most of the studies show rather slow increasing or decreasing variations. Therefore, all students asserted intuitively that the evolution had to render an account of what preceded and that a finer interval had to be proposed.

A first method, frequently proposed, was to produce the last line of the graph. But this proposal was quickly rejected, as certain pupils had noticed that there were irregular increases and decreases from one year to another, inspite of some general tendency.

How then represent graphically this idea of decreasing or increasing, considered as a long lasting movement of the series ? Finding a solution, therefore, was very slow.

The consideration of simple cases, as pointed out at the beginning where the dots were distributed around a line having as ordinate the mean, as well as the assistance of the teacher,induced the children to represent a tendency line : "Trace a line on the graph such that there is the same number of dots on either side".

But all the children were convinced that there was no reason why, on the diagram, the forecast for 1974 should be given by the intersection of the tendency line with the line having as abscissa 1974.

After discussion whithin the groups, some proposed to trace two parallels to the tendency lines passing through the more important peaks and hollows.

The forecast, then, was given by means of an interval.

Remarks.

1 - Very naturally, the groups did not take into account any dots that were exceptionally far away from the others. When asked why, they replied : "That's too exceptional a case ; one mustn't take it into consideration".

2 - The verification made thanks to the harbour readings for 1974 allowed the children to see for themselves that their forecasts had often proved right, but also, that some results were not within the foretold interval. These latter cases did not upset the children who understood very well that the forecast value was but a guess for a future event which could not be measured precisely before its realisation. Moreover, for certain graphs one should have considered the problem of the pertinence of the linear model used.

Thus, it often has proved possible to make a child sensitive to criticizing proposed models for the resolution of forecast problems.

To conclude, we wish to insist on the fact that the situations proposed to young students must be chosen very carefully. We hope that they then will constitute a favourable basis for a non-axiomatical approach to probability.

Recent Developments in Statistics
J.R.Barra et al., editors

Invited Paper

ALGORITHMS OF PENALIZATION TYPE AND OF DUAL TYPE FOR THE SOLUTION OF STOCHASTIC OPTIMIZATION PROBLEMS WITH STOCHASTIC CONSTRAINTS

J.B. HIRIART-URRUTY

Introduction.- Let f_0, $f_1, \ldots, f_m$ be functions defined on $X = \mathbb{R}^p$ and Q a nonempty closed convex subset of X and C defined as follows

$$C = Q \quad D \text{ where } D = \{x \in X \mid \quad \forall i \in <1, m> f_i(x) \leq 0\}$$

Let us consider the following optimization problem :

$$(\mathcal{P}) \quad \textit{Find } \bar{x} \in C \textit{ such that } f_0(\bar{x}) = \textit{Inf } \{f_0(x) \mid x \in C\}$$

We suppose that we don't know the exact form of the economic function f_0 and of the functions f_j defining the set of the constraints D, but we have access to the noise observations of f_0 and f_j. At each point x of Q, $f_0(x)$ and $f_j(x)$ are observed with a random error $\xi^k(x)$

$$Z^o(x) = f_0(x) + \xi^o(x) \qquad Z^j(x) = f_j(x) + \xi^j(x) \qquad \forall j \in <1,m>$$

The distribution of the random errors (or noises) is unknown but we suppose that at a point x one can make one or several stochastic evaluations $Z^k(x)$. These situations appear frequently, in particular in the system theory ; for other problems we refer the reader to the examples in [5].

In the case where the presence of noise only changes the knowledge of the economic function f_0 whereas the domain of the constraints C is known exactly, different methods have been proposed to solve $(\mathcal{P})$ and almost surely (a.s.) convergence theorems for the sequence (or for a subsequence) of the stochastic iterations towards the solutions of $(\mathcal{P})$ (or the stationary points) have been obtained ([1], [3], [4]). When we can only make the noise observations on the functions f_j, the preceding methods are no longer applicable because, for example, we are not assured that the nth stochastic iteration X_n generated by a stochastic algorithm is in C or not.

In the deterministic case, the problem $(\mathcal{P})$ has been discussed by several others specially using the penalization notion and the theory of duality. However, in such approaches, it is to be remarked that frequently the problem $(\mathcal{P})$ is replaced by a sequence of subproblems of optimization $(\mathcal{P}_n)$ with less constraints, which is solved either exactly or approximately. In the stochastic case such an approach is not acceptable because we cannot be sure if we are close to a solution of $(\mathcal{P}_n)$: in fact, we only have the noise observations $Z^k(x)$ of $f_k(x)$; also the gradients cannot be evaluated exactly. In the case of noise observations on the constraints, we refer to $(\mathcal{P})$ as a *stochastic optimization problem with stochastic constraints*. The solution of such problems have only been considered, to our knowledge,by H.J. Kushner and E. Sanvicente [5] who give a stochastic algorithm analogous to the deterministic one of Arrow and Hurwicz using in this problem the ordinary dual associated with the primal problem $(\mathcal{P})$. In that study it is in particular supposed that f_o is strictly convex, f_j is convex and that one has a bound a priori on the set of the solutions and on the set of the vectors of Kuhn-Tucker of $(\mathcal{P})$.

Considering that to the same primal problem $(\mathcal{P})$ may be associated different dual problems with the specific properties for everyone of them, our approach to the stochastic problem uses a dual problem originally proposed in the case of inequality constraints by R.T. Rockafellar [10] and which with its langrangian has interesting properties for the problems we are treating.

In particular the so called exterior penalty methods are connected with these lagrangians ; this naturally leadsus to consider different families of stochastic algorithms.

In the first family of algorithms we consider the methods of the exterior penalty type : an a.s. convergence theorem of a subsequence of the sequence of stochastic iterations is proved. In the stochastic algorithms of the dual type, we are interested in the convex case, that is to say the case where f_o and f_i are convex. In a first method the stochastic iterations follow the directions of $\nabla_x L_{\rho_n}$ and $\nabla_y L_{s_n}$ where L_{ρ_n} and L_{s_n} are the lagrangians which can be changed at each iteration ; the basic idea of the algorithm is to amount to the usual lagrangian L_o. We prove both a result of a.s. convergence and in $L^1(\Omega, \mathcal{A}, P)$ of the stochastic iterates sequence towards the set S of solutions of $(\mathcal{P})$. In a second stochastic algorithm of the dual type,the iterations follow the directions $\nabla_x L_\rho$

and $\nabla_y L$ of a fixed lagrangian L_ρ and the proof of the theorem of convergence brings out a remarkable property of the lagrangians which L_0 does not have.

These algorithms are capable of variations notably in the random steps and the lagrangians L_{ρ_n} with ρ_n random. We will discuss only the basic algorithms. Some results of this paper have been announced as a Compte Rendu de l'Académie des Sciences de Paris (t. 282, p. 907-910, 1976). For further details and missing proofs of this paper, see the report of the author [12].

I.- STOCHASTIC ALGORITHM OF THE PENALTY TYPE

I.1.- Lagrangians and dual problems.- Recall the construction of the family of perturbed problems starting from the optimization problem $(\mathcal{P})$ and then the associated dual problem. Let f_0 and f_i for $i \in < 1, m >$ be convex functions, Q a closed convex (nonempty) subset of X and $(\mathcal{P})$ the problem of convex optimization :

$$(\mathcal{P}) \quad f_0^* = \operatorname{Inf} \{f_0(x) \mid x \in C\}$$

with $C = Q \cap D$ and $D = \{x \in X \mid \ \forall i \in < 1, m > f_i(x) \leqslant 0\}$.

$U = \mathbb{R}^m$ being the space of perturbations, for all $r \geqslant 0$, let θ_r be the function of perturbation defined on X x U by :

$$\theta_r(x,u) = f_0(x) + r \sum_{i=1}^{m} u_i^2 \quad \text{if} \quad x \in Q \text{ and } f_i(x) \leqslant u_i \quad \forall i \in < 1,m > \ ; \ +\infty \text{ if not}$$

The Lagrangian associated with these perturbations θ_r is given by :

$$L_r(x,y) = \operatorname{Inf} \{\theta_r(x,u) - < u,y > \mid u \in U\}$$

If $r = 0$, this reduces to the usual Lagrangian L_0, that is

$$L_0(x,y) = \begin{cases} f_0(x) + \sum_{i=1}^{m} y_i f_i(x) \text{ if } x \in Q \text{ and } y \in U^+ \text{ (i.e. } y_i \geqslant 0 \ \ \forall i \in <1,m>) \\ -\infty \text{ if } x \in Q \text{ and } y \notin U^+ \ ; \ +\infty \text{ if } x \notin Q \end{cases}$$

For $r > 0$, we obtain ([10]) :

$$L_r(x,y) = f_0(x) + \frac{1}{4r} \sum_{i=1}^{m} \left[(y_i + 2 r f_i(x))^+ \right]^2 - y_i^2 \ ; \ +\infty \text{ if } x \notin Q$$

The problem $(\mathcal{D}_r)$ dual to $(\mathcal{P})$, associated with the perturbations θ_r is given by :

$$(\mathcal{D}_r) \quad -\beta_r = \sup_{y \in U} \left[\inf_{x \in X} L_r(x,y) \right]$$

We discuss the properties of this problem $(\mathcal{D}_r)$ in connection with the stochastic algorithms of the dual type

I.2.- Approximations of gradients.- The stochastic algorithms which we develop requires only the evaluations of the functions f_k ($k \in <0, m>$) in different points. The gradient of f_k at a point x_n of Q will be approached by a divided difference. Let us give two types of approximation of a gradient for a function f_k differentiable in the sense of Gâteaux (G-differentiable) [8]). For a given $\varepsilon_o > 0$, define Q_{ε_o} by :

$$Q_{\varepsilon_o} = \{x + \varepsilon_o e_i \, , \; x \in Q \qquad i \in <1, p>\}$$

where e_i is the unit vector in the direction of the $i^{\underline{th}}$ coordinate. We are thus led to make the following hypothesis on the functions f_k :

(A_1) $\begin{cases} f_k \text{ is G-differentiable over an open set } O \text{ containing } Q_{\varepsilon_o} \text{ and the} \\ \text{G-derivative satisfies the condition :} \\ \exists\, \eta > 0 \;\; \forall x,\, y \in Q_{\varepsilon_o} \;\; \|x-y\| \leq \eta \Longrightarrow \|\nabla f_k(x) - \nabla f_k(y)\| \leq O(1)\,\|x-y\| \end{cases}$

As a first approximation, $\Delta f_k(x_n)$ is approached by a unilateral divided difference, that is to say one considers the vector $\Delta f_k(x_n)$ defined by :

$$(1) \qquad \forall i \in <1, p> \quad [\Delta f_k(x_n)]_i = \frac{f_k(x_n + c_n e_i) - f_k(x_n)}{c_n}$$

The approximation of $\nabla f_k(x_n)$ may be given through the bias of the bilateral divided differences, that is to say considering $\Delta f_k(x_n)$ defined by :

$$(2) \qquad \forall i \in <1, p> \quad [\Delta f_k(x_n)]_i = \frac{f_k(x_n + c_n e_i) - f_k(x_n - c_n e_i)}{2\, c_n}$$

We have then the following propositions :

Proposition 1.- Let f_k be a function satisfying the hypothesis (A_1), $\{c_n\}_{n \geq 1}$ a *sequence of real numbers converging to 0 ; Then for all $x_n \in Q$, the vector $\Delta f_k(x_n)$ defined in (1) or (2) is given in the following form :*

(3) $$\Delta f_k\,(x_n) = \Delta f_k\,(x_n) + \theta_n^k\, c_n \qquad \text{with } ||\theta_n^k|| = O\,(1)$$

<u>*Proposition 2.*</u>*- Let f_k be a function satisfying the hypothesis (A_1), K a compac in Q. For all $x \in Q$, let M (x) be the set of elements $\bar{x}$ satisfying :*

$$\bar{x} \in K \qquad ||x - \bar{x}|| = \mathrm{Inf} \quad \{||x - y|| \mid y \in K\}$$

Then

(4) $$\forall\, x \in Q \quad \forall\, \bar{x} \in M\,(x) \qquad ||\nabla f_k(x)|| \leq O(1)\,(1 + ||x-\bar{x}||)$$

I.3.- *Iterative method*

a) In the family of Lagrangians L_{r_n} defined in $(x_n, y_n) \in Q \times \mathbb{R}^m$ by

$$L_{r_n}\,(x_n, y_n) = f_o\,(x_n) + \frac{1}{4\,r_n} \sum_{j=1}^{m} \left[(y_n^j + 2\,r_n\, f_j\,(x_n))^+\right]^2 - y_n^{j^2}$$

we put $y_n = 0$ for all n ; that is to say the iterations for the dual variable remain unchanged and do not yield a sequence of estimations of a Lagrange multiplier. The iterative method proposed for solving the problem $(\mathcal{P})$ uses a function of type $L_{r_n}\,(x,0) = f_o(x) + r_n \sum_{j=1}^{m} |f_j^+(x)|^2$ or more precisely a function Φ_n the general form of which is :

$$\Phi_n\,(x) = f_o(x) + r_n \sum_{j=1}^{m} p_j\,(f_j\,(x))$$

Briefly, it concerns an iterative method of the penalization type where the penalization functions satisfy the following conditions :

$$(\Pi)\quad \begin{cases} p_j \text{ is an application of } \mathbb{R} \text{ in } \mathbb{R}^+, \text{ convex, differentiable such that} \\ \forall\, u \leq 0 \qquad p_j(u) = \nabla p_j\,(u) = 0 \qquad \forall u > 0 \qquad p_j(u) > 0 \\ \nabla\, p_j \text{ is Lipschitzian on } \mathbb{R} \end{cases}$$

The function $p_j(t) = (t^+)^2$ does satisfy the hypotheses (Π) and in this case : $\Phi_n\,(x) = L_{r_n}\,(x,0)$. Let us consider the penalized problem

$(\mathcal{P}_n)$ *To find $x^* \in Q$ minimizing Φ_n on Q*

The proposed method consists of a coupling of the penalization method and the approached gradient method, that is to say one makes only a step in the direction of the *approached stochastic gradient* of Φ_n.

b) *Algorithm :* starting from the stochastic evaluations of f_o at $X_n + c_n e_i$ and $X_n - c_n e_i$ (we take for example the case of the bilateral approximation). We are going to construct a stochastic evaluation of an approximation of $\nabla f_o (X_n)$. For $i \in \langle 1, p \rangle$, let $Z^o_{2n,i}$ and $Z^o_{2n+1,i}$ be the stochastic evaluations of $f_o (X_n + c_n e_i)$ and $f_o (X_n - c_n e_i)$. We define the noises $\tilde{\xi}^o_{2n,i} = \xi^o (X_n + c_n e_i)$ and $\tilde{\xi}^o_{2n+1,i} = \xi^o (X_n - c_n e_i)$:

$$Z_{2n,i} = f_o (X_n + c_n e_i) + \tilde{\xi}^o_{2n,i} \qquad Z^o_{2n+1,i} = f_o (X_n - c_n e_i) + \tilde{\xi}^o_{2n+1,i}$$

We shall take as a stochastic evaluation of $\nabla f_o (X_n)$ the vector Z^o_n defined by :

$$\forall i \in \langle 1, p \rangle \qquad (Z^o_n)_i = \frac{Z^o_{2n,i} - Z^o_{2n+1,i}}{2 c_n} \qquad \text{that is to say :}$$

$$(Z^o_n)_i = \frac{f_o (X_n + c_n e_i) - f_o (X_n - c_n e_i)}{2 c_n} + \frac{\xi^o_{n,i}}{2 c_n}$$

If f_o satisfies the hypothesis (A_1), the evaluation Z^o_n gives, following Proposition 1, the following development

$$Z^o_n = \nabla f_o (X_n) + \theta^o_n c_n + \frac{\xi^o_n}{2 c_n}$$

where θ^o_n is bounded and ξ^o_n is the vector of random error (of $\mathbb{R}^p$) with components $\xi^o_{n,i}$. In a similar way, if f_j satisfies the hypothesis (A_1), the stochastic evaluation Z^j_n of an approximation of $\nabla f_j (X_n)$ leads to the following decomposition :

$$Z^j_n = \nabla f_j (X_n) + \theta^j_n c_n + \frac{\xi^j_n}{2 c_n} \quad \text{with} \quad \|\theta^j_n\| = O(1)$$

On the other hand, the form of the proposed algorithm leads us to consider the stochastic evaluations of f_j at X_n. For this purpose, k_n being a positive integer, we make k_n stochastic evaluations $\bar{Z}^j_{n,k}$ of $f_j (X_n)$ and we take as a final evaluation for defining the algorithm the mean of these evaluations :

$$\forall k \in \langle 1, k_n \rangle \qquad \bar{Z}^j_{n,k} = f_j(X_n) + \bar{\xi}^j_{n,k} \quad \text{and} \quad \bar{Z}^j_n = \frac{1}{k_n} \sum_{k=1}^{k_n} \bar{Z}^j_{n,k} = f_j(X_n) + \Xi^j_n$$

If $\mathcal{B}_n$ is the σ-algebra generated by the first n stochastic iterations $X_1, \ldots, X_n$, we shall make the (natural) hypothesis that the k_n evaluations are conditionally independant, given $\mathcal{B}_n$. Moreover, we shall make the usual hypotheses on the noises, that is to say :

$$E^{\mathcal{B}_n} \bar{\xi}^j_{n,k} = 0 \quad \text{and} \quad E^{\mathcal{B}_n} |\bar{\xi}^j_{n,k}|^2 \leqslant \sigma^2$$

Consequently, in the decomposition of $\bar{Z}^j_{n,k}$, we have the following relations on the resulting noise Ξ^j_n :

$$E^{\mathcal{B}_n} \Xi^j_n = 0 \qquad \text{and} \qquad E^{\mathcal{B}_n} |\Xi^j_n|^2 \leqslant \frac{\sigma^2}{k_n}$$

In the same manner, we suppose that $E^{\mathcal{B}_n} \xi^j_n = 0$ and $E^{\mathcal{B}_n} ||\xi^j_n||^2 \leqslant \sigma^2$.

Moreover, the existence of several observations of f_j on X_n, $X_n - c_n e_i$, $X_n + c_n e_i$ leads us to make a hypothesis on the conditional correlation of the energies of the noises on those different points. We suppose that

$$E^{\mathcal{B}_n} \left(|\Xi^j_n|^2 \cdot ||\xi^j_n||^2 \right) \leqslant \frac{O(1)}{k_n}$$

This hypothesis is in particular verified if $\xi^j_{n,i}$ and $\bar{\xi}^j_{n,k}$ are conditionally inde-pendent, given $\mathcal{B}_n$ or if

$$E^{\mathcal{B}_n} |\xi^j_{n,i}|^4 \leqslant O(1) \quad \text{and} \quad E^{\mathcal{B}_n} |\bar{\xi}^j_{n,k}|^4 \leqslant O(1)$$

Under the hypotheses of differentiability of f_o and f_j , the penalized function Φ_n is differentiable and :

$$\nabla \Phi_n(x) = \nabla f_o(x) + r_n \sum_{j=1}^{m} \nabla p_j(f_j(x)) \; \nabla f_j(x)$$

A stochastic evaluation of an approximation of $\nabla \Phi_n(X_n)$ is given by :

$$Z^o_n + r_n \sum_{j=1}^{m} \nabla p_j(\bar{Z}^j_n) \; (Z^j_n)$$

The sequence of stochastic iterations is defined recursively by

$$
(5) \quad \begin{cases} X_1 \text{ random in } Q \text{ such that } E(||X_1||^2) < \infty \\ X_{n+1} = P_Q \left\{ X_n - a_n \left[Z_n^o + r_n \sum_{j=1}^{m} \nabla p_j (\overline{Z}_n^j)(Z_n^j) \right] \right\} \end{cases}
$$

Before giving a theorem of a.s. convergence concerning the sequence of stochastic iterations defined by the algorithm (5), we shall give the convergence result of the set of solutions S_n and the optimal value Φ_n^* of the penalized problem $(\mathcal{P}_n)$ towards the set S of solutions and the optimal value f_o^* of the problem $\mathcal{P}$.

I.3.- Recall that a function f_o is said to be inf-compact on Q if

$$\forall \lambda \in \mathbb{R} \quad \{x \in Q \mid f_o(x) \leqslant \lambda\} \quad \text{is compact}$$

Proposition 3.- *Let $f_o, f_1, \ldots, f_m$ be continuous functions on Q, relatively to Q. f_o is supposed to be inf-compact on Q. Let S be the set of solutions of $(\mathcal{P})$, S_n the set of solutions of the penalized problem $(\mathcal{P}_n)$ and Φ_n^* the optimal value of $(\mathcal{P}_n)$. Then : $\lim_{n\to\infty} \Phi_n^* = f_o^*$ and if we denote $e(S_n, S) = \operatorname{Sup}\{d(x,S) \mid x \in S_n\}$, $\lim_{n\to\infty} e(S_n, S) = 0$*

I.5.- Convergence Theorem : *unless explicitely mentioned, let S be the set of solutions of the problem $(\mathcal{P})$ and for $x \in Q$, we say that $\overline{x}$ is a projection of x on S if x has closest point $\overline{x}$ in S. Let $D_j = \{x \in X \mid f_j(x) \leqslant 0\}$. We also use the same notations as defined in (I.3) above.*

I.5.1.- Theorem 1 : Under the following hypotheses :

$(\mathcal{H}_1)$ *f_o is inf-compact on Q and satisfies the hypothesis (A_1)*

$(\mathcal{H}_2)$ $\forall \varepsilon > 0 \quad \operatorname{Inf}\{< \nabla f_o(x), x-\overline{x} > \mid x \in C \; ||x-\overline{x}|| \geqslant \varepsilon\} \geqslant \alpha(\varepsilon) > 0.$
if $\overline{x}$ projection of x on S.

$(\mathcal{H}_3)$ *For $x \in Q$, $x \notin C$ and $y \notin C$, f_o satisfies the inequality :*
$f_o(x) - f_o(y) \leqslant \; < \nabla f_o(x), x-y >$

$(\mathcal{H}_4)$ *$\forall j \in <1, m>$ f_j is a convex function satisfying the hypothesis (A_1)*

$(\mathcal{H}_5)$ *For* $x \in Q$, $f_j(x) \leqslant O(1)\,(1 + d(x, D_j))$

$(\mathcal{H}_6)$ $\forall j \in \langle 1, m \rangle$, *the function of penalization* p_j *satisfies the hypotheses* (Π)

$(\mathcal{H}_7)$ $E^{\mathcal{B}_n}\, \xi_n^o = 0$, $E^{\mathcal{B}_n}\, ||\xi_n^o||^2 \leqslant \sigma^2$;

$\forall j \in \langle 1, m \rangle$ $E^{\mathcal{B}_n}\, \xi_n^j = 0$ $E^{\mathcal{B}_n}\, ||\xi_n^j||^2 \leqslant \sigma^2$

$\forall k \in \langle 1, k_n \rangle$ $E^{\mathcal{B}_n}\, \bar{\xi}_{n,k}^j = 0$ $E^{\mathcal{B}_n}\, |\bar{\xi}_{n,k}^j|^2 \leqslant \sigma^2$

$\forall j \in \langle 1, m \rangle$ $E^{\mathcal{B}_n}(||\xi_n^j||^2 \,.\, |\Xi_n^j|^2) \leqslant \dfrac{O(1)}{k_n}$.

$(\mathcal{H}_8)$ $\{a_n\}_{n\geqslant 1}$, $\{c_n\}_{n\geqslant 1}$, $\{r_n\}_{n\geqslant 1}$ *are sequences of real positive numbers,* $\{k_n\}_{n\geqslant 1}$ *a sequence of positive integers such that :*

$$\sum_{n=1}^{\infty} a_n = +\infty \qquad \sum_{n=1}^{\infty} a_n c_n r_n < +\infty \qquad \sum_{n=1}^{\infty} \frac{a_n^2 r_n^2}{c_n^2} < \infty$$

$$\sum_{n=1}^{\infty} \frac{a_n r_n}{c_n k_n^{1/2}} < +\infty \qquad \lim_{n\to\infty} c_n = 0 \qquad \lim_{n\to\infty} r_n = +\infty$$

Then, there exists a subsequence $\{X_{n_k}\}$ *of the sequence of the stochastic iterations generated by the algorithm defined in (I.3) such that :*

$$\lim_{k\to\infty} d(X_{n_k}, S) = 0 \text{ a.s.}$$

I.5.2.- Comments on the different hypotheses

1. The hypothesis of inf-compactness of f_0 on Q implies that f_0^* is finite and that S is compact. It can be replaced by all other hypothesis of the same type as :

$$\liminf_{\substack{||x||\to\infty \\ x \in Q}} f_0(x) > \inf \{f_0(x) \,/\, x \in Q\}$$

or

$$\liminf_{\substack{||x||\to\infty \\ x \in Q}} ||\nabla f_0(x)|| \;\dot{>}\; 0$$

2. The hypothesis $(\mathcal{H}_3)$ is related to the behaviour of f_0 for $x \notin C$. It is obviously verifiable in the case where f_0 is convex and can be weakened as the proof of the theorem will make evident. It is sufficient that :

$$\forall x \in Q , \quad x \notin C \qquad f_o(x) - f_o(\bar{x}) \leqslant \langle \nabla f_o(x) , x - \bar{x} \rangle$$

Similarly, the convexity of the functions f_j is not necessary ; it is sufficient that

$$\forall x \in Q \quad x \notin C \qquad f_j(x) - f_j(\bar{x}) \leqslant \langle \nabla f_j (x), x - \bar{x} \rangle$$

3. The conditions $(\mathcal{H}_7)$ on the noises are usual in the stochastic approximation processes ([1], [3], [4]).

4. The hypothesis $(\mathcal{H}_8)$ implies the usual conditions in the approximation procedures of type Kiefer-Wolfowitz ([11]), that is to say :

$$\sum_{n=1}^{\infty} a_n c_n < + \infty \quad \text{and} \quad \sum_{n=1}^{\infty} \frac{a_n^2}{c_n^2} < \infty$$

but it involves moreover the conditions on the penalization factor r_n and on the integer k_n.

I.5.3. <u>Proof of Theorem 1</u> : At the beginning, we give some inequalities on the terms used in the proof.

f_o and f_j satisfying the hypothesis (A_1), we have following Proposition 2,

(1) $$\forall k \in \langle 0, m \rangle \qquad ||\nabla f_k(X_n)|| \leqslant O(1)(1 + ||X_n - \bar{X}_n||)$$

Let us consider the expression $\nabla p_j (f_j (X_n))$ for $j \in \langle 1, m \rangle$

If $f_j (X_n) \leqslant 0$, $\nabla p_j (f_j (X_n)) = 0$. Otherwise, ∇p_j being lipschitzian, one has for $f_j (X_n) > 0$: $\nabla p_j (f_j (X_n)) \leqslant O(1)(1 + f_j (X_n))$

Following the hypothesis $(\mathcal{H}_5)$, one has, for $f_j (X_n) > 0$

$$f_j (X_n) \leqslant O(1)(1 + d(X_n, D_j))$$

As $\bar{X}_n \in S \subset D_j$, one concludes that

(2) $$\nabla p_j (f_j (X_n)) \leqslant O(1)(1 + ||X_n - \bar{X}_n||)$$

<u>First point</u> : the vector X_{n+1} is defined through $\tilde{X}_{n+1}$: $X_{n+1} = P_Q (\tilde{X}_{n+1})$

where $\tilde{X}_{n+1} = X_n - a_n \left\{ \nabla f_o (X_n) + \theta_n^o \; c_n + \dfrac{\xi_n^o}{2 c_n} \right\}$

$$- a_n r_n \left\{ \sum_{j=1}^{m} \nabla p_j (f_j(X_n) + \Xi_n^j)(\nabla f_j (X_n) + \theta_n^j c_n + \frac{\xi_n^j}{2c_n} \right\}$$

After decomposition, we write $\widetilde{X}_{n+1}$ in the following form :

$$\widetilde{X}_{n+1} = X_n - a_n \left\{ \nabla f_o (X_n) + r_n \sum_{j=1}^{m} \nabla p_j (f_j (X_n)) \nabla f_j (X_n) \right\}$$

$$- a_n (\theta_n^o c_n + \frac{\xi_n^o}{2 c_n}) - a_n r_n \left\{ \sum_{j=1}^{m} \nabla p_j (f_j(X_n)) \left(\theta_n^j c_n + \frac{\xi_n^j}{2c_n} \right) \right\}$$

$$- a_n r_n \left\{ \sum_{j=1}^{m} \left[\nabla p_j (f_j(X_n) + \Xi_n^j) - \nabla p_j (f_j(X_n)) \right] (\nabla f_j(X_n) + \theta_n^j c_n + \frac{\xi_n^j}{2 c_n} \right\}$$

Let $\overline{X}_n$ be a projection vector of X_n on the set S of the solutions of the problem ($\mathcal{P}$). We propose to evaluate $||X_{n+1} - \overline{X}_n||^2$ conditionally ; we set :

$$U_n = \nabla f_o (X_n) + r_n \sum_{j=1}^{m} \nabla p_j (f_j (X_n)) \nabla f_j (X_n)$$

$$V_n = r_n \sum_{j=1}^{m} \nabla p_j (f_j (X_n)) (\theta_n^j c_n + \frac{\xi_n^j}{2 c_n})$$

$$W_n = r_n \sum_{j=1}^{m} \left[\nabla p_j (f_j(X_n) + \Xi_n^j) - \nabla p_j (f_j(X_n)) \right] (\nabla f_j(X_n) + \theta_n^j c_n + \frac{\xi_n^j}{2c_n})$$

We write :

$$\widetilde{X}_{n+1} - X_n = X_n - \overline{X}_n - a_n (U_n + \theta_n^o c_n + \frac{\xi_n^o}{2c_n}) - a_n (V_n + W_n).$$

After expansion of $||\widetilde{X}_{n+1} - \overline{X}_n||^2$, we get :

$$(3) \qquad ||\widetilde{X}_{n+1} - \overline{X}_n||^2 = ||X_n - \overline{X}_n||^2 - 2 a_n < U_n , X_n - \overline{X}_n >$$

$$- 2 a_n < \theta_n^o c_n + \frac{\xi_n^o}{2c_n} , X_n - \overline{X}_n > - 2 a_n < V_n + W_n , X_n - \overline{X}_n >$$

$$+ a_n^2 ||U_n + \theta_n^o c_n + \frac{\xi_n^o}{2 c_n}||^2 + a_n^2 \left\{ ||V_n + W_n||^2 + 2 (U_n + \theta_n^o c_n + \frac{\xi_n^o}{2c_n})(V_n + W_n) \right\}$$

Let us examine successively the terms of this expansion and let us give the inequalities in terms of conditioned expectations on $\mathcal{B}_n$, generated by X_1, ..., X_n.

1.0.- S being closed, the application $x \to d(x, S)$ is continuous and projections exist. Moreover, the application φ_n defined on $\mathbb{R}^p \times \Omega$ by $\varphi_n (y,\omega) = ||X_n(\omega) - y||$ is a normal convex integrand on ($\mathbb{R}^p \times \Omega$, $\mathcal{A}_p \otimes \mathcal{B}_n$) ([9]).

It follows that the multiapplication K defined by :

$$K(\omega) = \{y \in \mathbb{R}^p ,\ \varphi_n (y, \omega) \leqslant d(X_n(\omega), S)\}$$

is $\mathcal{B}_n$-measurable([9] Corollary 4.4).The constant multiapplication S is $\mathcal{B}_n$-measurable. It follows that the multiapplication M defined by : $M(\omega) = K(\omega) \cap S$ is $\mathcal{B}_n$-measurable ([9], corollary 1.3). M, $\mathcal{B}_n$-measurable, taking compact nonempty values admits measurable selections following theorem of Kuratowski-Ryll-Nardzewski ([9] corollary 1.1). Let $\bar{X}_n$ be such a selection, that is to say one has :

$$||X_n - \bar{X}_n|| = d(X_n, S)$$

1.1.- Since $E^{\mathcal{B}_n} \xi_n^0 = 0$, we have :

$$- 2 a_n E^{\mathcal{B}_n} < \theta_n^0 c_n + \frac{\xi_n^0}{2c_n} , X_n - \bar{X}_n > = - 2 a_n E^{\mathcal{B}_n} < \theta_n^0 c_n , X_n - \bar{X}_n >$$

then :

$$(1.1) \qquad - 2 a_n E^{\mathcal{B}_n} < \theta_n^0 c_n + \frac{\xi_n^0}{2c_n} , X_n - \bar{X}_n > \leqslant O(1)\, a_n c_n ||X_n - \bar{X}_n|| \leqslant O(1)\, a_n c_n (1 + ||X_n - \bar{X}_n||^2)$$

1.2.- Similarly, we have by hypothesis : $\forall j \in <1, m>$ $E^{\mathcal{B}_n} \xi_n^j = 0$

Consequently : $$- 2 a_n E^{\mathcal{B}_n} < V_n, X_n - \bar{X}_n > = - 2 a_n r_n \sum_{j=1}^{m} \nabla p_j (f_j(X_n)) . < E^{\mathcal{B}_n} \theta_n^j c_n , X_n - \bar{X}_n >$$

Following the inequality (2) given in the preliminary, one has :

$$\nabla p_j (f_j (X_n)) \leqslant O(1) (1 + ||X_n - \bar{X}_n||)$$

$$(1.2) \qquad - 2 a_n E^{\mathcal{B}_n} < V_n , X_n - \bar{X}_n > \leqslant O(1)\, a_n r_n c_n ||X_n - \bar{X}_n|| + O(1)\, a_n c_n r_n ||X_n - \bar{X}_n|^2 \leqslant O(1)\, a_n r_n c_n (1 + ||X_n - \bar{X}_n||^2)$$

1.3.- Regarding the term W_n, we have :

$$|W_n| \leqslant r_n \sum_{j=1}^{m} O(1) \, |\Xi_n^j| \, (||\nabla f_j (X_n)|| + O(1)\, c_n + \frac{||\xi_n^j||}{2 c_n})$$

By construction of Ξ_n^j, we have : $\forall j \in \langle 1, m \rangle$, $E^{\mathcal{B}_n} |\Xi_n^j|^2 \leqslant \frac{\sigma^2}{k_n}$

(see iterative method). It follows that $\forall j \in \langle 1, m \rangle$ $E^{\mathcal{B}_n} |\Xi_n^j| \leqslant \frac{\sigma}{k_n^{1/2}}$

Similarly, following an inequality of $(\mathcal{H}_7)$

$$E^{\mathcal{B}_n}\left(|\Xi_n^j| \cdot ||\xi_n^j||\right) \leqslant \left[E^{\mathcal{B}_n} |\Xi_n^j|^2 \cdot E^{\mathcal{B}_n} ||\xi_n^j||^2\right]^{1/2} \leqslant \frac{O(1)}{k_n^{1/2}}$$

It follows then :

$$E^{\mathcal{B}_n} |W_n| \leqslant O(1) \frac{r_n}{k_n^{1/2}} \sum_{j=1}^{m} ||\nabla f_j(X_n)|| + O(1) \frac{r_n c_n}{k_n^{1/2}} + O(1) \frac{r_n}{c_n k_n^{1/2}}$$

Following the inequality (1) in the preliminary, one has :

$$\forall j \in \langle 1, m \rangle \qquad ||\nabla f_j(X_n)|| \leqslant O(1)(1 + ||X_n - \bar{X}_n||)$$

The sequence c_n converges to 0, then :

$$\frac{r_n c_n}{k_n^{1/2}} \leqslant O(1) \frac{r_n}{k_n^{1/2}} \leqslant O(1) \frac{r_n}{k_n^{1/2} c_n}.$$

Consequently :

$$|E^{\mathcal{B}_n} \langle W_n, X_n - \bar{X}_n \rangle| \leqslant O(1) \frac{r_n}{k_n^{1/2} c_n} ||X_n - \bar{X}_n|| + O(1) \frac{r_n}{k_n^{1/2} c_n} ||X_n - \bar{X}_n||^2.$$

and :

$$(1.3) \qquad -2 a_n E^{\mathcal{B}_n} \langle W_n, X_n - \bar{X}_n \rangle \leqslant O(1) \frac{a_n r_n}{k_n^{1/2} c_n} (1 + ||X_n - \bar{X}_n||^2)$$

1.4.- Let us consider the expression $||U_n + \theta_n^o c_n + \frac{\xi_n^o}{2 c_n}||^2$. We have : $E^{\mathcal{B}_n} ||\xi_n^o||^2 \leqslant \sigma^2$; c_n converging to 0, one has :

$$E^{\mathcal{B}_n} ||U_n + \theta_n^o c_n + \frac{\xi_n^o}{2 c_n}||^2 \leqslant O(1) E^{\mathcal{B}_n} ||U_n||^2 + \frac{O(1)}{c_n^2}$$

Following the inequalities (1) and (2) in the preliminaries :

$$E^{\mathcal{B}_n} ||U_n||^2 \leqslant O(1)(1 + r_n^2)(1 + ||X_n - \bar{X}_n||^2)$$

such that :

$$(1.4) \qquad a_n^2 E^{\mathcal{B}_n} ||U_n + \theta_n^o c_n + \frac{\xi_n^o}{2c_n}||^2 \leqslant O(1) a_n^2 + O(1) a_n^2 r_n^2 + O(1) \frac{a_n^2}{c_n^2} + O(1) a_n^2 (1+r_n^2) ||X_n - \bar{X}_n||^2.$$

1.5.- We have :

$$||V_n+W_n||^2 + 2 < U_n + \theta_n^o c_n + \frac{\xi_n^o}{2c_n} , V_n+W_n >$$

$$\leqslant O(1)\ ||V_n||^2 + O(1)\ ||W_n||^2 + O(1)\ ||U_n + \theta_n^o c_n + \frac{\xi_n}{2\ c_n}||^2$$

Being already given an inequality of $E^{\mathcal{B}_n}\ ||U_n + \theta_n^o c_n + \frac{\xi_n^o}{2c_n}||^2$, it remains to give the inequalities of $E^{\mathcal{B}_n}\ ||V_n||^2$ and of $E^{\mathcal{B}_n}\ ||W_n||^2$.

$$||V_n|| \leqslant r_n \sum_{j=1}^m \nabla p_j\ (f_j\ (X_n))\ (O\ (1)\ c_n + \frac{||\xi_n^j||}{2\ c_n})$$

But $\nabla p_j\ (f_j\ (X_n)) \leqslant O(1)\ (1 + ||X_n - \overline{X}_n||)$ and $E^{\mathcal{B}_n}\ ||\xi_n^j||^2 \leqslant \sigma^2$. It follows that : $E^{\mathcal{B}_n}||V_n||^2 \leqslant O(1)\ r_n^2\ c_n^2\ (1 + ||X_n-\overline{X}_n||^2) + O(1)\ \frac{r_n^2}{c_n^2}\ (1 + ||X_n-\overline{X}_n||^2)$

c_n converging to 0, one has : $r_n^2\ c_n^2 \leqslant O\ (1)\ \frac{r_n^2}{c_n^2}$. Thus :

$$a_n^2\ E^{\mathcal{B}_n}\ ||V_n||^2 \leqslant O\ (1)\ \frac{a_n^2\ r_n^2}{c_n^2}\ (1 + ||X_n-\overline{X}_n||^2). \tag{1.5}$$

1.6.- Moreover, we have :

$$||W_n||^2 \leqslant O(1)\ r_n^2 \sum_{j=1}^m\ |\Xi_n^j|^2\ (||\nabla f_j\ (X_n)||^2 + c_n^2 + \frac{||\xi_n^j||^2}{c_n^2})$$

Now : $E^{\mathcal{B}_n}\ |\Xi_n^j|^2 \leqslant \frac{\sigma^2}{k_n}$, $E^{\mathcal{B}_n}\ ||\xi_n^j||^2 \leqslant \sigma^2$, and

$E^{\mathcal{B}_n}\left(|\Xi_n^j|^2\ .\ ||\xi_n^j||^2\right) \leqslant \frac{O(1)}{k_n}$ (see the iterative method).

Thus :

$$E^{\mathcal{B}_n}\ ||W_n||^2 \leqslant O(1)\ \frac{r_n^2}{k_n}\ (1 + ||X_n-\overline{X}_n||^2) + O(1)\ \frac{r_n^2\ c_n^2}{k_n} + O(1)\ \frac{r_n^2}{k_n\ c_n^2} \tag{1.6}$$

1.7.- Following the hypothesis $(\mathcal{H}_8)$, all the numerical terms occurring in the inequalities on the conditional expectation (1.1) to (1.6) are general terms of convergent series by means of the following implications :

$\sum_{n=1}^{\infty} a_n c_n r_n < +\infty$ and $\lim_{n\to\infty} r_n = +\infty$ imply $\sum_{n=1}^{\infty} a_n c_n < +\infty$

$\sum_{n=1}^{\infty} \frac{a_n^2 r_n^2}{c_n^2} < +\infty$ and $\lim_{n\to\infty} r_n = +\infty$ imply $\sum_{n=1}^{\infty} \frac{a_n^2}{c_n^2} < +\infty$

$\sum_{n=1}^{\infty} a_n c_n < +\infty$ and $\sum_{n=1}^{\infty} \frac{a_n^2}{c_n^2} < +\infty$ imply $\sum_{n=1}^{\infty} a_n^2 < +\infty$

$\sum_{n=1}^{\infty} \frac{a_n^2 r_n^2}{c_n^2} < +\infty$ and $\lim_{n\to\infty} c_n = 0$ imply $\sum_{n=1}^{\infty} a_n^2 r_n^2 < +\infty$

Using the fact that $||X_{n+1} - \overline{X}_{n+1}|| \leqslant ||X_{n+1} - \overline{X}_n|| \leqslant ||\widetilde{X}_{n+1} - \overline{X}_n||$ where $\overline{X}_{n+1}$ is chosen as in (1.0) and the succesive inequalities 1.1 to 1.6 one obtains from inequality (3) :

$$(4) \quad E^{\mathcal{B}_{n+1}} ||X_{n+1}-\overline{X}_{n+1}||^2 \leqslant (1+\alpha_n)\, ||X_n-\overline{X}_n||^2 - a_n < \nabla f_0\, X_n),\, X_n-\overline{X}_n >$$

$$- a_n r_n \sum_{j=1}^{m} \nabla p_j\, (f_j(X_n)) < \nabla f_j(X_n),\, X_n-\overline{X}_n > + \delta_n$$

with $\sum_{n=1}^{\infty} \alpha_n < +\infty$ et $\sum_{n=1}^{\infty} \delta_n < +\infty$

Sécond point : Let us consider the expression :

$$\psi\,(X_n) = < \nabla f_0\,(X_n),\, X_n-\overline{X}_n > + r_n \sum_{j=1}^{m} \nabla p_j\,(f_j(X_n)) < \nabla f_j(X_n),\, X_n-\overline{X}_n >$$

Different cases are to be considered according to X_n is in C or not.

<u>1st case</u> : $X_n \in C$. In this case $\psi\,(X_n) = < \nabla f_0\,(X_n),\, X_n - \overline{X}_n >$ and by the hypotheses, one has :

(5) $< \nabla f_0\,(X_n),\, X_n - \overline{X}_n > \;\geqslant 0$. Moreover, by the hypothesis $(\mathcal{H}_2)$:

$d\,(X_n, S) \geqslant \varepsilon$ implies that $< \nabla f_0\,(X_n),\, X_n - \overline{X}_n > \;\geqslant \alpha\,(\varepsilon) > 0$.

Briefly : $\psi\,(X_n)\, 1_{\{X_n \in C\}} \geqslant \alpha\,(\varepsilon)\, 1_{\{X_n \in C\} \cap \{d\,(X_n,S)\geqslant\varepsilon\}}$

<u>2nd case</u> : $X_n \notin C$. The functions f_j being convex, one has the inequality

$$< \nabla f_j\,(X_n),\, X_n - \overline{X}_n > \;\geqslant f_j\,(X_n) - f_j\,(\overline{X}_n)$$

So, $\nabla p_j\,(f_j(X_n)) < \nabla f_j(X_n),\, X_n - \overline{X}_n > \;\geqslant \nabla p_j\,(f_j(X_n))\,(f_j(X_n) - f_j(\overline{X}_n))$

p_j being a convex function, one has also :

$$p_j\,(f_j\,(\overline{X}_n)) \geqslant p_j\,(f_j\,(X_n)) + \nabla p_j\,(f_j\,(X_n))\,(f_j\,(\overline{X}_n) - f_j\,(X_n))$$

Since $p_j\,(f_j\,(\overline{X}_n)) = 0$, one can conclude that

$$\nabla p_j\,(f_j(X_n))\,(f_j\,(X_n) - f_j\,(\overline{X}_n)) \geqslant p_j\,(f_j\,(X_n))$$

and $\psi\,(X_n) \geqslant \,< \nabla f_o\,(X_n),\, X_n - \overline{X}_n > + \,r_n \sum_{j=1}^{m} p_j\,(f_j\,(X_n)$

Since $X_n \notin C$ and since $\overline{X}_n \in C$, by the hypothesis $(\mathcal{H}_3)$, one has :

$$< \nabla f_o\,(X_n),\, X_n - \overline{X}_n > \;\geqslant\; f_o\,(X_n) - f_o\,(\overline{X}_n)$$

Consequently, if $X_n \notin C$: $\psi\,(X_n) \geqslant f_o\,(X_n) - f_o\,(\overline{X}_n) + r_n \sum_{j=1}^{m} p_j\,(f_j\,(X_n))$

Since $f_o\,(\overline{X}_n) = f_o^*$ is the optimal value of the problem $(\mathcal{P})$ and Φ_n denotes the penalized function defined in I.3, we have :

$$(6) \qquad \psi\,(X_n) \;\geqslant\; \Phi_n\,(X_n) - \Phi_n^* + (\Phi_n^* - f_o^*)$$

Let us now consider the term $\Phi_n\,(X_n) - \Phi_n^*$. Following Proposition 3, $\lim_{n\to\infty} e\,(S_n,\, S) = 0$ et $\lim_{n\to\infty} \left[f_o^* - \Phi_n^* \right] = 0$. For a fixed $\varepsilon > 0$, let $n_o\,(\varepsilon)$ be such

that : $\quad \forall\, n \geqslant n_o\,(\varepsilon),\; e\,(S_n,\, S) \leqslant \frac{\varepsilon}{2}$ and $f_o^* - \Phi_n^* \leqslant \varepsilon$

One has : $\forall x \in Q \qquad \Phi_{n_o(\varepsilon)}\,(x) - \Phi^*_{n_o(\varepsilon)} \geqslant 0$ and $\Phi_{n_o(\varepsilon)}$ is inf-compact on Q.

Consequently : $\quad \Phi_{n_o(\varepsilon)}\,(x) - \Phi^*_{n_o(\varepsilon)} \geqslant \beta_{n_o(\varepsilon)} > 0$ if $d\,(x,\, S_{n_o(\varepsilon)}) \geqslant \frac{\varepsilon}{2}$.

Let x be such that $d\,(x,\, S) \geqslant \varepsilon$; then : $\forall\, n \geqslant n_o\,(\varepsilon)$, $d\,(x,\, S_n) \geqslant \frac{\varepsilon}{2}$.

Hence it follows that for all $\quad n \geqslant n_o\,(\varepsilon),\;\; \Phi_n\,(x) - \Phi_n^* \geqslant 0$

and

$$\Phi_n(x) - \Phi_n^* \geqslant \Phi^*_{n_o(\varepsilon)}\,(x) - \Phi^*_{n_o(\varepsilon)} + \Phi^*_{n_o(\varepsilon)} - \Phi_n^*$$

$$\geqslant \beta_{n_o(\varepsilon)} + (\Phi^*_{n_o(\varepsilon)} - f_o^*) \geqslant \beta_{n_o(\varepsilon)} - \varepsilon$$

Briefly, for $n \geqslant n_o(\varepsilon)$: $\Phi_n\,(X_n) - \Phi_n^* \geqslant (\beta_{n_o(\varepsilon)} - \varepsilon)\, 1_{\{d(X_n,S)\geqslant\varepsilon\}}$

Consequently, if $X_n \not\in C$, we have :

$$\psi(X_n) \geq \beta_{n_0}(\varepsilon)\, 1_{\{d(X_n,S)\geq\varepsilon\}} - \varepsilon + \Phi_n^* - f_0^* .$$

Summing up the two cases, we have then for $n \geq n_0(\varepsilon)$:

$$\psi(X_n) \geq \alpha(\varepsilon)\, 1_{\{X_n \in C\} \cap \{d(X_n,S)\geq\varepsilon\}} + \beta_{n_0}(\varepsilon)\, 1_{\{X_n \not\in C\} \cap \{d(X_n,S)\geq\varepsilon\})}$$
$$- (\varepsilon + f_0^* - \Phi_n^*)\, 1_{\{X_n \not\in C\}}$$

Let $\gamma(\varepsilon) = \mathrm{Min}\,(\alpha(\varepsilon), \beta_{n_0}(\varepsilon)), \gamma(\varepsilon) > 0$:

$$(7) \qquad \psi(X_n) \geq \gamma(\varepsilon)\, 1_{\{d(X_n,S)\geq\varepsilon\}} - (\varepsilon + f_0^* - \Phi_n^*).$$

The inequality (4) becomes by the estimation (7) : $\forall \varepsilon > 0$, $\exists\, n_0(\varepsilon)$ such that for all $n \geq n_0(\varepsilon)$:

$$E^{\mathcal{B}_{n+1}} \|X_{n+1}-\overline{X}_{n+1}\|^2 \leq (1+\alpha_n)\|X_n-\overline{X}_n\|^2 - a_n\,\gamma(\varepsilon)\, 1_{\{d(X_n,S)\geq\varepsilon\}}$$
$$+ a_n(\varepsilon + f_0^* - \Phi_n^*) + \delta_n$$

Taking the expectation in this inequality and summing from $n_0(\varepsilon)$, one obtains :

$$(8) \qquad \sum_{n>n_0}^{N} a_n \prod_{m>n}^{N-1}(1+\alpha_n) \left[\gamma(\varepsilon)\, P\{d(X_n,S) \geq \varepsilon\} - (\varepsilon + f_0^* + \Phi_n^*)\right] < \infty$$

The series of the general term a_n being divergent, one deduces from the inequality (8)

$$\forall \varepsilon > 0 \qquad \liminf_{n\to\infty} \left[P\{d(X_n, S) \geq \varepsilon\} - (\varepsilon + f_0^* - \Phi_n^*)\right] \leq 0$$

But $\lim_{n\to\infty}(f_0^* - \Phi_n^*) = 0$; then : $\forall \varepsilon > 0 \quad \liminf_{n\to\infty} P\{d(X_n,S) \geq \varepsilon\} \leq \varepsilon$

Let $\{\varepsilon_k\}_{k\geq 1}$ be the general term of a convergent series. It follows from above that there exists n_k such that: $P\left\{d(X_{n_k}, S) \geq \varepsilon_k\right\} \leq \varepsilon_k + \frac{1}{2^k}$. This implies $\sum_{k=1}^{\infty} P\{d(X_{n_k}, S) \geq \varepsilon_k\} < +\infty$. We obtain using the Borel-Cantelli lemma that : $\lim_{k\to\infty} d(X_{n_k}, S) = 0$. Let us remark that using inequalities (5) and (6), we have :

$$(9) \qquad E^{\mathcal{B}_n} \|X_{n+1}-\overline{X}_{n+1}\|^2 \leq (1+\alpha_n)\|X_n-\overline{X}_n\|^2 + a_n(f_0^* - \Phi_n^*)$$

Without further details on the general term of the series $\{a_n(f_0^* - \Phi_n^*)\}$, one cannot prove the a.s. convergence of the whole sequence $\|X_n - \overline{X}_n\|^2$.

I.5.4 - Remarks

a) The penalization method which we have studied uses differentiable penalization functions p_j which lead to

$$\lim_{n\to\infty} \Phi_n^* = f_o^* \quad \text{and} \quad \lim_{n\to\infty} e\,(S_n, S) = 0 \text{ (Proposition 3)}$$

We do not have results on the speed of convergence of Φ_n^* towards f_o^*. If moreover we assume the hypotheses of Theorem 1 we suppose that

$$\sum_{n=1}^{\infty} a_n \,(f_o^* - \Phi_n^*) < +\infty .$$

One can easily show, using the inequality (9), that the whole sequence $\{d\,(X_n, S)\}_{n\geq 1}$ converges a.s. towards 0.

b) If we want to have, in the convex case, an exact penalization theorem, that is to say : $\forall n \geq n_o$, $\Phi_n^* = f_o^*$ and $S_n = S$, it is necessary, as it has been remarked by D.P. Bertsekas ([2]), that the functions p_j should satisfy

$$\lim_{t\to 0^+} \frac{p_j(t)}{t} > 0$$

For our type of proof, the hypothesis of differentiability has been introduced to control the effects of noises on the constraints.

II - STOCHASTIC ALGORITHM OF DUAL TYPE : $1^{\underline{st}}$ method

II.1.- Properties of the augmented Lagrangians L_r : Let θ_r be the family of perturbations defined in paragraph I.1, we consider the perturbed problem

$$h_r(u) = \text{Inf}\ \{\theta_r\ (x,\ u)\ |\ x \in X\}$$

$h_r(o) = f_o^*$ and the problem $(\mathcal{P})$ is said to be *stable* if f_o^* is finite and h_r is continuous at $0 \in U$ ([6] p. 395). In the sequel, f_o is a convex function, inf-compact on Q and the functions f_j, $j \in <1,\ m>$ are convex. Under these conditions, we have the following properties ([10]).

Proposition 4 : If the Slater hypothesis is satisfied, that is to say if there exists $x_o \in Q$ such that $\forall j \in <1,\ m>$ $f_j\ (x_o) < 0$. Then

a) relatively to the perturbations θ_r $(r \geqslant 0)$ $(\mathcal{P})$ is stable

b) $\mathcal{D}_r$ and $\mathcal{D}_o$ have the same solutions and optimal values.

c) Relatively to the lagrangians L_r $(r \geqslant 0)$, one has the same Kuhn-Tucker vectors and the same saddle-points. Thus $(\bar{x},\ \bar{y})$ is a saddle-point of L_r if and only if $\bar{x}$ is a solution of $(\mathcal{P})$. $(\bar{x} \in S)$ and $\bar{y}$ is a Kuhn-Tucker vector $(\bar{y} \in \mathcal{Y})$.

In the sequel, we suppose that the Slater hypothesis is satisfied : thus S and $\mathcal{Y}$ are nonempty convex compact sets. The following property is a special property of the lagrangians L_r for $r > 0$.

Proposition 5 ([10]). If f_o is inf-compact on Q and if the Slater hypothesis is satisfied, one has for all $y \in \mathcal{Y}$ and $r > 0$

$$\bar{x} \in S \iff \bar{x} \text{ minimizes the function } L_r\ (r,\ y) \text{ on } Q$$

II.2.- Iterative method

a) We know that the exact formula of $\nabla_y L_{s_n}$ on $(X_n,\ Y_n)$ is given by the vector whose components are: :

$$\forall j \in <1,\ m> \quad [\nabla_y L_{s_n}\ (X_n,\ Y_n)]_j = \text{Max}\ \{f_j\ (X_n),\ -\frac{Y_n^j}{2s_n}\} \qquad ([10])$$

Since one has only a stochastic evaluation of $f_j\ (X_n)$ at $X_n = x$, one takes as a vector of observation of the gradient $\nabla_y L_{s_n}\ (X_n,\ Y_n)$ the vector $\tilde{D}_y L_{s_n}\ (X_n,\ Y_n)$

defined as follows :

$$\forall j \in \langle 1, m \rangle \left[\tilde{D}_y L_{s_n} (X_n, Y_n)\right]_j = \text{Max} \left\{ Z_n^j , \frac{Y_n^j}{2s_n} \right\}$$

where Z_n^j is the result of a single stochastic evaluation of $f_j (X_n)$. Let us decompose Z_n^j by defining the variable of error $\bar{\xi}_n^j$, that is $Z_n^j = f_j (X_n) + \bar{\xi}_n^j$.

It is to be remarked that in the present construction $\tilde{D}_y L_{s_n} (X_n, Y_n)$ is not, in general, an unbiased estimation of $\nabla_y L_{s_n} (X_n, Y_n)$. In fact, if $E^{\mathcal{B}_n} \bar{\xi}_n^j = 0$, one only has : $\tilde{D}_y L_{s_n} (X_n, Y_n) = \nabla_y L_{s_n} (X_n, Y_n) + \zeta_n$ where $E^{\mathcal{B}_n} \zeta_n \geq 0$ in $\mathbb{R}^m$.

We define the sequence $\{Y_n\}_{n \geq 1}$ of *stochastic iterations on the dual variable* in the following manner:

(3.1) $$\begin{cases} Y_1 \text{ random in } (\mathbb{R}^m)^+ \text{ such that } E(||Y_1||^2) < +\infty \\ \text{Having } (X_n, Y_n),\ Y_{n+1} \text{ is to be determined by the formula} \\ \forall i \in \langle 1, m \rangle \quad Y_{n+1}^i = Y_n^i + a_n \text{ Max} \left\{ f_i (X_n) + \bar{\xi}_i^n , - \frac{Y_n^i}{2s_n} \right\} \end{cases}$$

We take s_n as a general term of a sequence of positive real numbers such that : $s_n \geq \frac{a_n}{2}$. So, $\{Y_n\}_{n \geq 1}$ is a sequence of stochastic iterates of $(\mathbb{R}^m)^+$.
Indeed, one has :

(3.1)' $$\forall i \in \langle 1, m \rangle \quad Y_{n+1}^i = Y_n^i \left(1 - \frac{a_n}{2s_n}\right) + a_n \left[f_i (X_n) + \bar{\xi}_i^n + \frac{Y_n^i}{2s_n} \right]^+$$

Let us remark that in the particular case where $s_n = \frac{a_n}{2}$, the formulation (3.1)' takes the following simplified form :

$$\forall i \in \langle 1, m \rangle \quad Y_{n+1}^i = \left[Y_n^i + a_n (f_i (X_n) + \bar{\xi}_n^i) \right]^+$$

Of course, one can also fix s_n and set : $s_n = s$ with $2\, s \geq \sup_n a_n$.

b) The sequence of *stochastic iterates on the primal variable,* that is to say the sequence of estimates of the solution of the primal problem $(\mathcal{P})$ is built as in the method of the penalization type with the addition of the dual variables. The exact gradient of L_{ρ_n} with respect to x at (X_n, Y_n) is given by :

$$\nabla_x L_{\rho_n} (X_n, Y_n) = \nabla f_o (X_n) + \sum_{i=1}^{m} \left[Y_n^i + 2 \rho_n f_i (X_n) \right]^+ \nabla f_i (X_n)$$

We are going to take an approximation of this gradient using the divided differences approximations of $\nabla f_o (X_n)$ and of $\nabla f_i (X_n)$. We define $D_x L_{\rho_n} (X_n, Y_n)$ to be the vector :

$$\forall i \in \langle 1, p \rangle \quad \left[D_x L_{\rho_n} (X_n, Y_n) \right]_i = \frac{f_o (X_n + c_n e_i) - f_o (X_n - c_n e_i)}{2 c_n}$$

$$+ \sum_{j=1}^{m} \left[Y_n^j + 2 \rho_n f_j (X_n) \right]^+ \left[\frac{f_j (X_n + c_n e_i) - f_j (X_n - c_n e_i)}{2 c_n} \right]$$

Starting from the observations of $f_o (X_n + c_n e_i)$, $f_o (X_n - c_n e_i)$, $f_j (X_n)$, $f_j (X_n + c_n e_i)$, $f_j (X_n - c_n e_i)$, one constructs a vector $\widetilde{D}_x L_{\rho_n} (X_n, Y_n)$ which is a stochastic evaluation of $D_x L_{\rho_n} (X_n, Y_n)$ denoting :

$$\forall i \in \langle 1, p \rangle \quad \widetilde{D}_x L_{\rho_n} (X_n, Y_n)_i = \frac{f_o (X_n + c_n e_i) - f_o (X_n - c_n e_i) + \xi^o_{n,i}}{2 c_n}$$

$$+ \sum_{j=1}^{m} \left[Y_n^j + 2 \rho_n (f_j (X_n) + \bar{\xi}^j_n) \right]^+ \left[\frac{f_j (X_n + c_n e_i) - f_j (X_n - c_n e_i) + \xi^j_{n,i}}{2 c_n} \right]$$

Recall that in this formulation, for $k \in \langle 0, m \rangle$, $\xi^k_{n,i}$ denotes the sum of noises in the observations of f_k at $X_n + c_n e_i$ and $X_n - c_n e_i$ and that $\bar{\xi}^j_n$ represents the noise in the observation of f_j at X_n. The sequence $\{X_n\}_{n \geq 1}$ is defined as follows:

$$(3.2) \quad \begin{cases} X_1 \text{ random in } Q \text{ such that } E(\|X_1\|^2) < +\infty \\ X_{n+1} = P_Q (X_n - a_n \widetilde{D}_x L_{\rho_n} (X_n, Y_n)). \end{cases}$$

II.3.- *Convergence Theorem* : We recall that f_o has been supposed to be inf-compact on Q and that one supposes that Slater's hypothesis is satisfied. $\mathcal{B}_n$ denotes the σ-algebra generated by $(X_1, \ldots, X_n ; Y_1, \ldots, Y_n)$.

II.3.1. Theorem 2

(I_1) *f_o and f_j for $j \in \langle 1, m \rangle$ are convex functions satisfying (A_1) and :*

$$\forall j \in \langle 1, m \rangle \qquad \liminf_{\substack{\|x\| \to \infty \\ x \in Q}} f_j(x) > -\infty$$

$\forall x \in Q \; ; \; f_j(x) \leqslant O(1) \; (1 + d\,(x, D_j))$

(I_2) $\quad E^{\mathcal{B}_n} \xi_n^o = 0 \, , \quad E^{\mathcal{B}_n} \|\xi_n^o\|^2 \leqslant \sigma^2$

$\forall j \in \langle 1, m \rangle \quad E^{\mathcal{B}_n} \xi_n^j = 0$ and $E^{\mathcal{B}_n} \|\xi_n^j\|^2 \leqslant \sigma^2$

$E^{\mathcal{B}_n} \bar{\xi}_n^j = 0 \, , \quad E^{\mathcal{B}_n} (|\bar{\xi}_n^j|^2 \, . \, \|\xi_n^j\|^2) \leqslant \tau^2$

(I_3) $\quad \{a_n\}_{n\geq 1}$, $\{c_n\}_{n\geq 1}$, $\{\rho_n\}_{n\geq 1}$, $\{s_n\}_{n\geq 1}$ *are sequences of real positive numbers such that :*

$$\sum_{n=1}^{\infty} a_n c_n < \infty \qquad \sum_{n=1}^{\infty} \frac{a_n^2}{c_n^2} < \infty \qquad \sum_{n=1}^{\infty} a_n = +\infty$$

$$\sum_{n=1}^{\infty} \frac{a_n \rho_n}{c_n} < \infty \qquad \lim_{n\to\infty} c_n = 0 \qquad 0 < \rho_n \leqslant \bar{\rho} < \infty$$

$s_n > 0$ *and* $2\, s_n \geqslant a_n$

(I_4) $\quad \bar{x}$ *being a projection of* x *on* S, *for all vector* $\bar{y}$ *of Kuhn-Tucker, we suppose that* $x \neq \bar{x}$ *implies* $L_o\,(x, \bar{y}) > L_o\,(\bar{x}, \bar{y})$.

Under these hypotheses, the sequence of stochastic iterates $\{X_n, Y_n\}$ *defined by (3.1) and (3.2) is such that :* $d\,(X_n, S) \to 0$ *a.s. and in* $L^1\,(\Omega, \mathcal{A}, P)$

II.3.2.- Remarks

1. In this method, one puts $k_n = 1$, that is a single stochastic evaluation of $f_j\,(X_n)$.
2. The hypothesis (I_3) constitutes the conditions on a_n and c_n for the method introduced by Kiefer-Wolfowitz [11] with an additional assumption on the sequence $\{\rho_n\}$. The condition $\sum_{n=1}^{\infty} \frac{a_n \rho_n}{c_n} < +\infty$ in particular implies that : $\liminf_{n\to\infty} \rho_n = 0$. This is a difficulty in the consideration of Lagrangians L_{ρ_n} ; it will be avoided in the second stochastic algorithm of dual type where ρ_n will be fixed but where we will need several stochastic evaluations of $f_j\,(X_n)$.
3. If $\bar{y}$ is a Kuhn-Tucker vector , one knows that : $\inf_{x\in Q} L_o\,(x, \bar{y}) = f_o^*$

   If $\tilde{x}$ minimizes $L_o\,(., \bar{y})$ on Q and if $\forall j \in \langle 1, m \rangle \quad f_j\,(\tilde{x}) \leqslant 0 \quad \bar{y}_j . f_j(\tilde{x}) = 0$

then $\tilde{x} \in S$. However if the set of minimizing points of $L_0(., \bar{y})$ is reduced to a single point $\tilde{x}$ (if f_0 is strictly convex for instance) then $\tilde{x}$ is the solution of $(\mathcal{P})$ et (I_4) is verified.

4. As we will see in the proof, we obtain also the a.s. convergence of $d(X_n, S)^2 + d(Y_n, \mathcal{Y})^2$ but we cannot deduce any convergence result for the sequence $\{Y_n\}$. With a different stochastic algorithm, the convergence result described in Theorem 2 is of the same order as that of H.J. Kushner and E. Sanvicente ([5]).

II.3.3.- Proof of Theorem 2. The proof of the theorem is divided in three points. The first two ones concern inequalities of $E^{\mathcal{B}_n} ||Y_{n+1} - \bar{Y}_{n+1}||^2$ and of $E^{\mathcal{B}_n} ||X_{n+1} - \bar{X}_{n+1}||^2$; the a.s. convergence result p.s. is proved in the third point. To begin with, we give two lemmas :

Lemma A : *Let* $\{U_n\}_{n\geq 1}$ *be a sequence of positive random variables adapted to the increasing sequence of* σ*-algebras* $\{\mathcal{B}_n\}_{n\geq 1}$ *such that :*

$$E^{\mathcal{B}_n} U_{n+1} \leq (1+\alpha_n) U_n + \beta_n$$

where $\alpha_n \geq 0$, $\beta_n \geq 0$ *and* $\sum_{n=1}^{\infty} \alpha_n < +\infty$ $\sum_{n=1}^{\infty} \beta_n < +\infty$

Then $\lim_{n\to\infty} U_n$ *exists a.s.*

The proof is mentioned in ([7], p. 33). One will say that a sequence satisfying the assumptions of Lemma A is "almost" a positive supermartingale

Lemma B : *Let* $\{U_n\}_{n\geq 1}$ *be a sequence of random variables such that :*

$$\forall \varepsilon > 0 \qquad \sum_{n=1}^{\infty} a_n P \left\{ ||U_n|| \geq \varepsilon \right\} < +\infty$$

where $a_n \geq 0$ *and* $\sum_{n=1}^{\infty} a_n = +\infty$. *Then, there exists a subsequence* $\left\{U_{n_k}\right\}_{k\geq 1}$ *such that :* $\lim_{k\to\infty} U_{n_k} = 0$ *a.s.*

As a preliminary inequality, we have :

$$f_j(X_n) \leqslant O(1)\,(1 + d(X_n, D_j))\ \text{(by the hypothesis } (I_1))$$

Moreover : $\liminf_{\substack{||x||\to\infty \\ x\in Q}} f_j(x) > -\infty$. So, $|f_j(X_n)| \leqslant O(1)\,(1 + ||X_n - \overline{X}_n||)$

1st point : Let $(\overline{X}_n, \overline{Y}_n)$ be the projection of (X_n, Y_n) on the set of the saddle-points of the lagrangian L_0 (or equivalently of L_{ρ_n}). According to the structure of this set, $\overline{X}_n$ is the projection of X_n on the set S of the solutions of $(\mathcal{P})$ and $\overline{Y}_n$ is the projection of Y_n on the set $\mathcal{Y}$ of vectors of Kuhn-Tucker for the problem $(\mathcal{P})$. $\overline{Y}_n$ belongs to $(\mathbb{R}^m)^+$; the function $x \to x^+$ is lipschitzian with coefficient 1. By the formulation (3.1) of the sequence $\{Y_n\}_{n\geqslant 1}$ it follows that :

$$\forall i \in <1, m>\ |Y^i_{n+1} - \overline{Y}^i_n|^2 = \left[\left(1 - \frac{a_n}{2s_n}\right) Y^i_n - a_n \left[f_i(X_n) + \overline{\xi}^i_n + \frac{Y^i_n}{2s_n}\right]^+ - \overline{Y}^i_n\right]^2$$

$$\leqslant |Y^i_n + a_n (f_i(X_n) + \overline{\xi}^i_n) - \overline{Y}^i_n|^2$$

Let $f_\bullet(x) = (f_1(x), \ldots, f_m(x))$ and $\overline{\xi}_n$ be the vector $(\overline{\xi}^i_n)_{i\in<1,m>}$. The preceding inequality then implies

$$||Y_{n+1} - \overline{Y}_n||^2 \leqslant ||Y_n - \overline{Y}_n||^2 + 2\,a_n < f_\bullet(X_n) + \overline{\xi}_n\,,\ Y_n - \overline{Y}_n > +\ a_n^2\,||f_\bullet(X_n) + \overline{\xi}_n||^2$$

Let us take the conditional expectation in relation to $\mathcal{B}_n$, σ-algebra generated by $(X_1, \ldots, X_n)$ and $(Y_1, \ldots, Y_n)$. It follows from the hypotheses that $E^{\mathcal{B}_n}\,||\overline{\xi}_n||^2 \leqslant O(1)$. Moreover, following the inequalities indicated in the preliminaries concerning $|f_j(X_n)|^2$, one has :

$$||f(X_n)||^2 \leqslant O(1)\,(1 + ||X_n - \overline{X}_n||^2)$$

From which it follows

$$(1)\quad E^{\mathcal{B}_n}\,||Y_{n+1} - \overline{Y}_n||^2 \leqslant ||Y_n - \overline{Y}_n||^2 + 2\,a_n < f_\bullet(X_n), Y_n - \overline{Y}_n >$$

$$+\ O(1)\,a_n^2 + O(1)\,a_n^2\,||X_n - \overline{X}_n||^2.$$

On $(\mathbb{R}^m)^+$, the application $L_o(x, .)$ is differentiable and $\nabla_y L_o(x,y) = f_.(x)$. Moreover, if we denote by $\overline{Y}_{n+1}$ the projection of Y_{n+1} on $\mathcal{Y}$, one has : $||Y_{n+1} - \overline{Y}_{n+1}|| \leqslant ||Y_{n+1} - \overline{Y}_n||$, such that the inequality (1) is reduced to :

$$(2) \qquad E^{\mathcal{B}_n} ||Y_{n+1}-\overline{Y}_{n+1}||^2 \leqslant ||Y_n-\overline{Y}_n||^2 + 2\, a_n < \nabla_y L_o(X_n,Y_n), Y_n-\overline{Y}_n > + O(1)\, a_n^2 + O(1)\, a_n^2 ||X_n-\overline{X}_n||^2$$

2nd point : Let us take $\widetilde{X}_{n+1} = X_n - a_n \widetilde{D}_x L_{\rho_n}(X_n, Y_n)$

In the developed expression of $\widetilde{X}_{n+1}$, let us bring out $\nabla_x L_{\rho_n}(X^n, Y^n)$, that is :

$$\nabla f_o(X_n) + \sum_{j=1}^m \left[Y_n^j + 2\rho_n f_j(X_n)\right]^+ \nabla f_j(X_n)$$

Thus :
$$\widetilde{X}_{n+1} = X_n - a_n \nabla_x L_{\rho_n}(X_n, Y_n) - a_n\left(\theta_n^o c_n + \frac{\xi_n^o}{2 c_n}\right) - a_n \sum_{j=1}^m \left[Y_n^j + 2\rho_n f_j(X_n)\right]^+ \left(\theta_n^j c_n + \frac{\xi_n^j}{2 c_n}\right) - a_n \sum_{j=1}^m \zeta_n^j \left(\nabla f_j(X_n) + \theta_n^j c_n + \frac{\xi_n^j}{2 c_n}\right)$$

where we have set $\zeta_n^j = \left[Y_n^j + 2\rho_n (f_j(X_n) + \overline{\xi}_n^j)\right]^+ - \left[Y_n^j + 2\rho_n f_j(X_n)\right]^+$

The arguments we shall develop for having an inequality of $||X_{n+1} - \overline{X}_n||^2$ in the conditional expectation are of the same type as those of the proof of Theorem 1, except that in addition the variable Y_n of the stochastic iteration on the dual variables enters into the formulation of $\widetilde{X}_{n+1}$. Let us set : $U_n = \theta_n^o c_n + \frac{\xi_n^o}{2 c_n}$,

$$V_n = \sum_{j=1}^m \left[Y_n^j + 2\rho_n f_j(X_n)\right]^+ \left(\theta_n^j c_n + \frac{\xi_n^j}{2 c_n}\right)$$

$$W_n = \sum_{j=1}^m \zeta_n^j \left(\nabla f_j(X_n) + \theta_n^j c_n + \frac{\xi_n^j}{2 c_n}\right)$$

So :

$$(3) \quad ||X_{n+1}-\overline{X}_n||^2 = ||X_n-\overline{X}_n||^2 - 2\, a_n < X_n - \overline{X}_n, \nabla_x L_{\rho_n}(X_n, Y_n) > - 2\, a_n < X_n-\overline{X}_n, U_n + V_n + W_n > + a_n^2 ||\nabla_x L_{\rho_n}(X_n,Y_n)||^2 + a_n^2 ||U_n+V_n+W_n||^2 + 2\, a_n^2 < \nabla_x L_{\rho_n}(X_n,Y_n), U_n + V_n + W_n >$$

1.1. θ_n^o is a bounded vector and $E^{\mathcal{B}_n} \xi_n^o = 0$, thus :

$$- 2 a_n E^{\mathcal{B}_n} < X_n - \overline{X}_n , U_n > \leqslant 0(1) a_n c_n (1 + ||X_n - \overline{X}_n||^2)$$

1.2. $\forall j \in < 1, m >$, $E^{\mathcal{B}_n} \xi_n^j = 0$; the sequence $\{\rho_n\}_{n\geqslant 1}$ is bounded ; the set $\mathcal{Y}$ of Kuhn-Tucker vectors is bounded. Otherwise, one has :

$$|f_j (X_n)| \leqslant 0(1) (1 + ||X_n - \overline{X}_n||)$$

θ_n^j is a bounded vector. Consequently :

$$||E^{\mathcal{B}_n} V_n|| \leqslant 0(1) c_n (1 + ||X_n - \overline{X}_n|| + ||Y_n - \overline{Y}_n||)$$

and :

$$- 2 a_n E^{\mathcal{B}_n} < X_n - \overline{X}_n , V_n > \leqslant 0(1) a_n c_n (1 + ||X_n - \overline{X}_n|| + ||Y_n - \overline{Y}_n||) ||X_n - \overline{X}_n||$$

$$\leqslant 0(1) a_n c_n (1 + ||X_n - \overline{X}_n||^2 + ||Y_n - \overline{Y}_n||^2)$$

1.3. Concerning W_n, we remark that : $|\zeta_n^j| \leqslant 2 \rho_n |\overline{\xi}_n^j|$ and

$$||W_n|| \leqslant 0(1) \rho_n \sum_{j=1}^{m} |\overline{\xi}_n^j| (||\nabla f_j (X_n)|| + c_n + \frac{||\xi_n^j||}{c_n})$$

Now $E^{\mathcal{B}_n} ||\xi_n^j||^2 \leqslant \sigma^2$ and $E^{\mathcal{B}_n} |\overline{\xi}_n^j|^2 \leqslant \sigma^2$. Moreover, we have : $||\nabla f_j (X_n)||^2 \leqslant 0(1) (1 + ||X_n - \overline{X}_n||^2)$. One deduces :

$$- 2 a_n E^{\mathcal{B}_n} < X_n - \overline{X}_n , W_n > \leqslant 0(1) a_n \rho_n + 0(1) a_n \rho_n c_n + 0(1) a_n \frac{\rho_n}{c_n}$$

$$+ 0(1) a_n \rho_n ||X_n - \overline{X}_n||^2 .$$

The sequences $\{\rho_n\}_{n\geqslant 1}$ and $\{c_n\}_{n\geqslant 1}$ are bounded ; finally it follows :

$$- 2 a_n E^{\mathcal{B}_n} < X_n - \overline{X}_n , W_n > \leqslant 0(1) a_n \frac{\rho_n}{c_n} (1 + ||X_n - \overline{X}_n||^2).$$

1.4. The term $||\nabla_x L_{\rho_n} (X_n, Y_n)||^2$ is bounded in the following manner :

$$||\nabla_x L_{\rho_n} (X_n , Y_n)||^2 \leqslant 0(1) (||\nabla f_o (X_n)||^2 + \sum_{j=1}^{m} (|Y_n^j + 2\rho_n f_j (X_n)|^2 + ||\nabla f_j (X_n)||^2)$$

$$\leqslant 0(1) (1 + ||X_n - \overline{X}_n||^2 + ||Y_n - \overline{Y}_n||^2)$$

1.5. Using the hypothesis $E^{\mathcal{B}_n} \|\xi_n^o\|^2 \leqslant \sigma^2$, one immediately sees that $E^{\mathcal{B}_n} \|U_n\|^2 \leqslant \frac{O(1)}{c_n^2}$

Similarly : $E^{\mathcal{B}_n} \|V_n\|^2 \leqslant \frac{O(1)}{c_n^2} (1 + \|X_n - \overline{X}_n\|^2 + \|Y_n - \overline{Y}_n\|^2)$.

For the term $\|W_n\|^2$, we obtain :

$$\|W_n\|^2 \leqslant O(1)\, \rho_n^2 \sum_{j=1}^{m} |\overline{\xi}_n^j|^2 (\|\nabla f_j(X_n)\|^2 + c_n^2) + \frac{|\overline{\xi}_n^j|^2 \cdot \|\xi_n^j\|^2}{c_n^2}$$

$E^{\mathcal{B}_n}\left(|\xi_n^j|^2 \cdot \|\xi_n^j\|^2\right)$ is bounded, it follows :

$$a_n^2\, E^{\mathcal{B}_n} \|W_n\|^2 \leqslant O(1) \frac{a_n^2}{c_n^2} (1 + \|X_n - \overline{X}_n\|^2).$$

Substituting the successive inequalities 1.1 to 1.5 in the inequality (3), one obtains :

$$(4) \quad E^{\mathcal{B}_n} \|X_{n+1} - \overline{X}_{n+1}\|^2 \leqslant (1 + \alpha_n^{(1)}) \|X_n - \overline{X}_n\|^2 + \beta_n^{(1)} \|Y_n - \overline{Y}_n\|^2 + \gamma_n^{(1)}$$
$$- 2\, a_n < \nabla_x L_{\rho_n} (X_n, Y_n), X_n - \overline{X}_n >$$

In this expansion, $\alpha_n^{(1)}$, $\beta_n^{(1)}$, $\gamma_n^{(1)}$ are positive and general terms of convergent series (hypothesis (I_3)). $\{X_n, Y_n\}$ is a sequence of $Q \times (\mathbb{R}^m)^+$ and for $(x, y) \in Q \times (\mathbb{R}^m)^+$, one has : $\nabla_x L_o (x, y) = \nabla f_o (x) + \sum_{i=1}^{m} y_i \nabla f_i(x)$

Let us give a bound for $D_n = < \nabla_x L_o (X_n, Y_n) - \nabla_x L_{\rho_n} (X_n, Y_n), X_n - \overline{X}_n >$

$$D_n = \sum_{i=1}^{m} \left\{ Y_n^i - \left[Y_n^i + 2\, \rho_n\, f_i (X_n) \right]^+ \right\} < \nabla f_i (X_n), X_n - \overline{X}_n >$$

f_i being convex, we have : $< \nabla f_i (X_n), X_n - \overline{X}_n > \;\geqslant f_i (X_n) - f_i(\overline{X}_n)$. It follows that if $f_i (X_n) \geqslant 0$, $< \nabla f_i (X_n), X_n - \overline{X}_n > \;\geqslant 0$ and $D_n \leqslant 0$. Let us examine the case where $f_i (X_n) < 0$

$$\left| Y_n^i - \left[Y_n^i + 2\, \rho_n\, f_i (X_n) \right]^+ \right| \leqslant - 2\, \rho_n\, f_i (X_n)$$

Using the hypothesis (I_1), $f_i (X_n) > - M > - \infty$. Moreover :

$$|< \nabla f_i (X_n), X_n - \overline{X}_n >| \leqslant O(1)\, (1 + \|X_n - \overline{X}_n\|^2).$$

So $\left\{Y_n^i - \left[Y_n^i + 2\,\rho_n\, f_i(X_n)\right]^+\right\} < \nabla f_i(X_n),\, X_n-\overline{X}_n > \leqslant \rho_n\, O(1)\,(1 + ||X_n-\overline{X}_n||^2)$

The series $\{a_n\, \rho_n\}_{n\geqslant 1}$ being convergent, the inequality (4) may be put in the following form :

$$(5)\quad E^{\mathcal{B}_n}\, ||X_{n+1}-\overline{X}_{n+1}||^2 \leqslant (1 + \alpha_n^{(1)})\, ||X_n-\overline{X}_n||^2 + \beta_n^{(1)}\, ||Y_n-\overline{Y}_n||^2 + \delta_n^{(1)}$$

$$- 2\, a_n < \nabla_x L_o\,(X_n, Y_n),\, X_n - \overline{X}_n > \quad \text{with} \sum_{n=1}^{\infty} \delta_n^{(1)} < +\infty$$

3nd point

a) Let $(\overline{X}_n, \overline{Y}_n)$ be the projection of (X_n, Y_n) on the set of saddle-points of L_o

and $\Delta\,(X_n, Y_n) = ||X_n - \overline{X}_n||^2 + ||Y_n - \overline{Y}_n||^2$

From inequalities (2) and (5), it follows :

$$(6)\quad E^{\mathcal{B}_n}\, \Delta\,(X_{n+1}, Y_{n+1}) \leqslant (1+ \alpha_n)\, \Delta(X_n,Y_n) - 2\, a_n < \nabla_x L_o\,(X_n,Y_n),\, X_n-\overline{X}_n >$$

$$+ 2\, a_n < \nabla_y L_o\,(X_n, Y_n),\, Y_n - \overline{Y}_n > + \delta_n \quad \text{with} \sum_{n=1}^{\infty} \alpha_n < +\infty \text{ and } \sum_{n=1}^{\infty} \delta_n < +\infty$$

Let A be the operator defined on $Q \times (\mathbb{R}^m)^+$ by $A\,(X,Y) = \left\{\nabla_x L_o\,(X,Y),\, - \nabla_y L_o\,(X,Y)\right\}$

L_o being a convex-concave function, A is monotone and

$$< \nabla_x L_o\,(X_n,Y_n),\, X_n-\overline{X}_n > - < \nabla_y L_o\,(X_n,Y_n),\, Y_n - \overline{Y}_n > \geqslant 0$$

The inequality (6) shows that $\{\Delta(X_n,Y_n)\}$ is "almost" a positive supermartingale. From lemma A, $\Delta\,(X_n, Y_n)$ converges a.s. towards a random variable Δ_∞. Taking the expectation in (6), one obtains :

$$E\,\Delta\,(X_{n+1}, Y_{n+1}) \leqslant (1 + \alpha_n)\, E\,\Delta\,(X_n, Y_n) + \delta_n$$

As $\sum_{n=1}^{\infty} \alpha_n < +\infty$, $\sum_{n=1}^{\infty} \delta_n < +\infty$ and $E\,\Delta\,(X_1, Y_1) < +\infty$, one deduces that :

$$(7)\qquad \operatorname{Sup}_n\; E\,\Delta\,(X_n,Y_n) < +\infty$$

b) $(\overline{X}_n, \overline{Y}_n)$ is a saddle point of L_o ; thus :

$$- < \nabla_x L_o\,(X_n,Y_n),\, X_n - \overline{X}_n > + < \nabla_y L_o\,(X_n, Y_n),\, Y_n - \overline{Y}_n >$$

$$\leqslant L_o\,(\overline{X}_n, Y_n) - L_o\,(X_n, \overline{Y}_n) \leqslant L_o\,(\overline{X}_n, \overline{Y}_n) - L_o\,(X_n, \overline{Y}_n)$$

Following the hypothesis (I_4), one has : $\forall \varepsilon > 0, \ \exists \alpha(\varepsilon) > 0$ such that :

$$||X_n - \overline{X}_n|| = d(X_n, S) \geqslant \varepsilon \Longrightarrow L_o(\overline{X}_n, \overline{Y}_n) - L_o(X_n, \overline{Y}_n) \leqslant -\alpha(\varepsilon)$$

The inequality (6) may be written :

$$E^{\mathcal{B}_n} \Delta(X_{n+1}, Y_{n+1}) \leqslant (1+\alpha_n) \Delta(X_n, Y_n) - a_n \alpha(\varepsilon) 1_{\{d(X_n,S) \geqslant \varepsilon\}} + \delta_n$$

Consequently $\forall \varepsilon > 0 \quad \forall N \quad \sum_{n=1}^{N} a_n \prod_{m>n}^{N-1} (1+\alpha_m) P\{d(X_n, S) \geqslant \varepsilon\} < +\infty$

The series of general term a_n is divergent ; thus there exists, using Lemma B, a subsequence $\{X_{n_k}\}$ such that : $\lim_{k\to\infty} d(X_{n_k}, S) = 0$ a.s.

c) Let $\varepsilon_1 < \varepsilon_2$ be two strictly positive real numbers, let us define the random variables μ_k $(k \geqslant 1)$ successively by :

$$\mu_1 = \text{Inf}\{n \in \mathbb{N} \quad d(X_n, S) \leqslant \varepsilon_1\}$$
$$\mu_2 = \text{Inf}\{n \mid n > \mu_1 \quad d(X_n, S) \geqslant \varepsilon_2\}$$
$$\vdots$$
$$\mu_{2k} = \text{Inf}\{n \mid n > \mu_{2k-1} \ d(X_n, S) \geqslant \varepsilon_2\}$$
$$\mu_{2k+1} = \text{Inf}\{n \mid n > \mu_{2k} \ d(X_n, S) \leqslant \varepsilon_1\}$$ with the condition $\text{Inf}\,\emptyset = +\infty$.

The random variables $\{\mu_k\}_{k\in\mathbb{N}}$ associated with the sequence $\{d(X_n, S)\}_{n\geqslant 1}$ define stopping times. If $\mu_{2k_o} = +\infty$, then $\mu_p = +\infty$ for all $p > 2k_o$. According to the convergence result of a subsequence $d(X_{n_k}, S)$ towards 0, if μ_{2k} is finite, then necessarily μ_{2k+1} is finite. Let us denote by $\beta_{\varepsilon_1,\varepsilon_2}$ the following random variable :

$$\beta_{\varepsilon_1,\varepsilon_2} = \text{Inf}\{k \in \mathbb{N} \mid \mu_{2k} < +\infty\}$$

$\beta_{\varepsilon_1,\varepsilon_2}(\omega)$ represents the *number of successive times that the sequence* $\{d(X_n(\omega),S)\}$ *crosses from* $\leqslant \varepsilon_1$ *to* $\geqslant \varepsilon_2$. ([7] p. 26). In the second point of the proof, we have seen :

$$||X_{n+1}-X_n|| \leqslant ||\widetilde{X}_{n+1}-X_n|| = a_n \quad ||\nabla_x L_{\rho_n}(X_n,Y_n) + U_n + V_n + W_n||$$ and

$$||\nabla_x L_{\rho_n}(X_n,Y_n) + U_n + V_n + W_n||^2 \leqslant O(1)\,(1 + ||X_n-\overline{X}_n||^2 + ||Y_n-\overline{Y}_n||^2)$$

If $\{\Gamma_n\}$ is defined by $\Gamma_n = \dfrac{\Delta_n}{\prod_{i=1}^{n-1}(1+\alpha_i)} + \sum_{i=n}^{\infty} \dfrac{\delta_i}{\prod_{j=1}^{i}(1+\alpha_j)}$

the inequality (6) implies the following one: $E^{\mathcal{B}_n}\,\Gamma_{n+1} \leqslant \Gamma_n$. $\{\Gamma_n\}$ is also a positive supermartingale and $\operatorname{Sup}_n \Gamma_n$ is a.s. finite ([7] Proposition II.2.7). Consequently $\operatorname{Sup}_n \Delta_n$ is a.s. finite and $X_{n+1} - X_n = a_n R_n$ with $\operatorname{Sup}_n \|R_n\| < +\infty$ a.s.

Let $A_{\varepsilon_1,\varepsilon_2} = \left\{\omega \mid \beta_{\varepsilon_1,\varepsilon_2}(\omega) = +\infty\right\} = \left\{\omega \mid \forall k \;\; \mu_{2k}(\omega) < +\infty\right\}$, that is to say $A_{\varepsilon_1,\varepsilon_2}$ is the set of ω for which $d(X_n(\omega), S) \geqslant \varepsilon_2$ an infinitely number of times. We will show that $P(A_{\varepsilon_1,\varepsilon_2}) = 0$. By construction of μ_{2k} and μ_{2k+1}, we have

$$1_{A_{\varepsilon_1,\varepsilon_2}} \cdot \|X_{\mu_{2k}} - X_{\mu_{2k+1}}\| \geqslant (\varepsilon_2 - \varepsilon_1)\, 1_{A_{\varepsilon_1,\varepsilon_2}}$$

$$(\varepsilon_2 - \varepsilon_1) \cdot 1_{A_{\varepsilon_1,\varepsilon_2}} \leqslant \Big(\sum_{n=\mu_{2k}}^{\mu_{2k+1}-1} a_n \|R_n\|\Big)\, 1_{A_{\varepsilon_1,\varepsilon_2}}$$

As $S = \operatorname{Sup}_n \|R_n\|$ is a.s. finite, for all $K > 0$, we have :

$$(8) \qquad (\varepsilon_2 - \varepsilon_1)\, 1_{A_{\varepsilon_1,\varepsilon_2}} \cdot 1_{\{S\leqslant K\}} \leqslant \sum_{n=\mu_{2k}}^{\mu_{2k+1}-1} a_n K \cdot 1_{A_{\varepsilon_1,\varepsilon_2}} \cdot 1_{\{S\leqslant K\}} .$$

Moreover, we know that :

$$A_n(X_n,Y_n) = \langle \nabla_x L_o(X_n,Y_n), X_n-\overline{X}_n \rangle - \langle \nabla_y L_o(X_n,Y_n), Y_n-\overline{Y}_n \rangle$$

$$\geqslant \alpha(\varepsilon_1)\, 1_{\{d(X_n,S) \geqslant \varepsilon_1\}}$$

and that : (9) $\displaystyle\sum_{n=1}^{\infty} a_n A_n(X_n, Y_n) < +\infty$ a.s.

So, for all n included between μ_{2k} and $\mu_{2k+1}-1$, $d(X_n, S) \geqslant \alpha(\varepsilon_1)$.

Consequently :

$$1_{A_{\varepsilon_1,\varepsilon_2}} \cdot \sum_{k=1}^{+\infty} \sum_{n=\mu_{2k}}^{\mu_{2k+1}-1} a_n\, \alpha(\varepsilon_1) \leqslant \sum_{n=1}^{\infty} a_n A_n(X_n, Y_n)$$

Following the inequalities (8) and (9), we have then :

$$1_{A_{\varepsilon_1,\varepsilon_2}} \cdot 1_{\{S\leqslant K\}} \sum_{k=1}^{+\infty} \frac{\varepsilon_2-\varepsilon_1}{K} \leqslant \left[\sum_{n=1}^{\infty} a_n A_n(X_n, Y_n)\right] 1_{\{S\leqslant K\}} < +\infty$$

this implies $\forall K$ $\quad P\left\{A_{\varepsilon_1,\varepsilon_2} \cap (S \leqslant K)\right\} = 0$ thus : $P\left\{A_{\varepsilon_1,\varepsilon_2}\right\} = 0.$

It follows that the number of upcrossings of $[\varepsilon_1, \varepsilon_2]$ $\beta_{\varepsilon_1,\varepsilon_2}$ is a.s. finite, that is $d\,(X_n, S)$ converges a.s. ([7] , Lemma II.2.10). Having the a.s. convergence of a subsequence towards 0, one has the a.s. convergence of the whole sequence $d\,(X_n, S) = ||X_n - \overline{X}_n||$ towards 0.

Furthermore,

$$\sup_n E\,(||X_n - \overline{X}_n||^2) \leq \sup_n E\,\Delta\,(X_n, Y_n) < +\infty$$

The sequence $\{d\,(X_n, S)\}$ is then equi-integrable and

$$d\,(X_n, S) \xrightarrow[n\to\infty]{} 0 \text{ in } L^1\,(\Omega, \mathcal{A}, P)$$

III - STOCHASTIC ALGORITHM OF DUAL TYPE : 2nd method

In this second stochastic method of the dual type, one uses the *mean of several stochastic evaluations of* $f_j(X_n)$ as in the stochastic algorithm of the penalization type. Unlike the preceding algorithm the lagrangian used is the augmented lagrangian L_ρ where ρ is a strictly positive number *invariant* for all successive iterations.

III.1. Iterative method .- Let k_n be a positive integer and $\{Z^j_{n,k}\}_{k \in <0,k_n>}$ k_n stochastic evaluations of $f_j(X_n)$; $Z^j_{n,k} = f_j(X_n) + \xi^j_{n,k}$

$\mathcal{B}_n$ is the σ-algebra generated by $(X_1, \ldots, X_n, Y_1, \ldots, Y_n)$; we will suppose that $E^{\mathcal{B}_n}\, \xi^j_{n,k} = 0$. The evaluations $Z^j_{n,k}$ are conditionally independent given $(X_1, \ldots, X_n ; Y_1, \ldots, Y_n)$. The mean stochastic evaluation $\bar{Z}^j_n$ of $f_j(X_n)$ will be :

$$\bar{Z}^j_n = \frac{1}{k_n} \sum_{k=1}^{k_n} Z^j_{n,k}$$

Let us write moreover $\bar{Z}^j_n = f_j(X_n) + \Xi^j_n$ with $\Xi^j_n = \sum_{k=1}^{k_n} \xi^j_{n,k}$

a) The sequence of *stochastic iterates on the dual variable* is defined as follows :

$$(4.1)\quad \begin{cases} Y_1 & \text{random in } (\mathbb{R}^m)^+ \text{ such that } E(||Y_1||^2) < +\infty \\ Y_{n+1} & \text{determined from } (X_n, Y_n) \text{ by :} \\ \forall i \in <1, m> & Y^i_{n+1} = Y^i_n + a_n \operatorname{Max}\left\{f_i(X_n) + \Xi^i_n \,,\, -\dfrac{Y^i_n}{2\rho}\right\} \end{cases}$$

ρ is a positive constant such that : $2\rho \geqslant \sup_n a_n$.

b) Let us define the sequence of stochastic iterates on the primal variable

$$(4.2)\quad \begin{cases} X_1 & \text{random in } Q \text{ such that } E(||X_1||^2) < +\infty \\ X_{n+1} = P_Q(X_n - a_n \tilde{D}_x L_\rho(X_n, Y_n)) \end{cases}$$

In this formula, $\tilde{D}_x L_\rho(X_n, Y_n)$ is as in the first stochastic algorithm with ρ instead of ρ_n and Ξ^j_n instead of ξ^j_n.

III.2. *Convergence Theorem*

III.2.1. *Theorem 3*

$(\mathcal{J}_1)$ *Let* f_o *and* f_j $(j \in <1, m>)$ *be convex functions satisfying the hypothesis* (I_1) *of Theorem 2*

$(\mathcal{J}_2)$ $E^{\mathcal{B}_n}\ \xi_n^o = 0$, $\quad E^{\mathcal{B}_n}\ ||\xi_n^o||^2 \leqslant \sigma^2$

$E^{\mathcal{B}_n}\ \xi_n^j = 0$, $\quad E^{\mathcal{B}_n}\ ||\xi_n^j||^2 \leqslant \sigma^2$

$\forall j \in <1, m>$, $\forall k \in <1, k_n> E^{\mathcal{B}_n}\ \overline{\xi}^j_{n,k} = 0 \qquad E^{\mathcal{B}_n}\ |\overline{\xi}^j_{n,k}|^2 \leqslant \sigma^2$

$E^{\mathcal{B}_n}\ (|\Xi_n^j|^2 \cdot ||\xi_n^j||^2) \leqslant \dfrac{O(1)}{k_n}$

$(\mathcal{J}_3)$ $\{a_n\}_{n\geqslant 1}$, $\{c_n\}_{n\geqslant 1}$ *are sequences of real positive numbers,* $\{k_n\}_{n\geqslant 1}$ *a sequence of positive integers such that :*

$$\sum_{n=1}^{\infty} a_n = +\infty \qquad \sum_{n=1}^{\infty} a_n c_n < +\infty \qquad \sum_{n=1}^{\infty} \frac{a_n^2}{c_n^2} < \infty$$

$$\lim_{n\to\infty} c_n = 0 \qquad \sum_{n=1}^{\infty} \frac{a_n}{c_n k_n^{1/2}} < +\infty$$

Under these assumptions, the sequence $\{X_n\}_{n\geqslant 1}$ *defined in (4.1) and (4.2) is such that :* $\lim\limits_{n\to\infty} d\ (X_n, S) = 0$ *a.s. and in* $L^1\ (\Omega, \mathcal{A}, P)$

III.2.2. Sketch of the proof

1st point : Y_{n+1} may be written as :

$$\forall i \in <1, m> \quad Y^i_{n+1} = (1 - \frac{a_n}{2})\ Y^i_n + \frac{a_n}{2} \left[Y^i_n + 2\rho\ (f_i\ (X_n) + \Xi^i_n)\right]^+$$

Let $\tau^i_n = \dfrac{1}{2\rho} \left\{ \left[Y^i_n + 2\rho\ (f_i(X_n) + \Xi^i_n)\right]^+ - \left[Y^i_n + 2\rho\ f_i\ (X_n)\right]^+ \right.$

Since $\left[Y^i_n + 2\rho\ f_i\ (X_n)\right]^+ = Y^i_n + 2\rho \left[\nabla_y L_\rho\ (X_n, Y_n)\right]_i$, we have :

$$Y_{n+1} = Y_n + a_n \nabla_y L_\rho\ (X_n, Y_n) + a_n \tau_n .$$

Let $\overline{Y}_n$ be the projection of Y_n on $\mathcal{Y}$; then :

(1) $||Y_{n+1}-\overline{Y}_n||^2 = ||Y_n-\overline{Y}_n||^2 + 2\ a_n < \nabla_y L_\rho\ (X_n, Y_n), Y_n - \overline{Y}_n >$

$$+ 2 a_n < \tau_n , Y_n - \bar{Y}_n > + a_n^2 ||\tau_n + \nabla_y L_\rho (X_n, Y_n)||^2$$

According to the definition of τ_n, $|\tau_n^i| \leqslant |\Xi_n^i|$ and consequently

$$E^{\mathcal{B}_n} ||\tau_n|| \leqslant E^{\mathcal{B}_n} ||\Xi_n^i|| \leqslant O(1) \frac{\sigma}{k_n^{1/2}}$$

Then : $2 a_n E^{\mathcal{B}_n} < \tau_n , Y_n - \bar{Y}_n > \leqslant O(1) \frac{a_n}{k_n^{1/2}} ||Y_n - \bar{Y}_n||$

Otherwise, $\left[\nabla_y L_\rho (X_n, Y_n)\right]_i = \dfrac{\left[Y_n^i + 2 \rho f_i (X_n)\right]^+ - Y_n^i}{2 \rho}$; this implies

$$\left|\left[\nabla_y L_\rho (X_n, Y_n)\right]_i\right|^2 \leqslant |f_i (X_n)|^2 \leqslant O(1) (1 + ||X_n - \bar{X}_n||^2)$$

It follows from the inequality (1) :

$$(1') \quad E^{\mathcal{B}_n} ||Y_{n+1} - \bar{Y}_{n+1}||^2 \leqslant ||Y_n - \bar{Y}_n||^2 + 2 a_n < \nabla_y L_\rho (X_n, Y_n), Y_n - \bar{Y}_n >$$

$$+ O(1) \frac{a_n}{k_n^{1/2}} ||Y_n - \bar{Y}_n|| + O(1) a_n^2 (1 + ||X_n - \bar{X}_n||^2) + O(1) \frac{a_n^2}{k_n} .$$

Following the assumptions of the theorems, $\left\{\dfrac{a_n}{k_n^{1/2}}\right\}_{n\geqslant 1}$ $\left\{a_n^2\right\}_{n\geqslant 1}$ are general terms of convergent series. Briefly :

$$(2) \quad E^{\mathcal{B}_n} ||Y_{n+1} - \bar{Y}_{n+1}||^2 \leqslant (1+\alpha_n^{(1)}) ||Y_n - \bar{Y}_n||^2 + 2 a_n < \nabla_y L_\rho (X_n, Y_n), Y_n - \bar{Y}_n >$$

$$+ \beta_n^{(1)} ||X_n - \bar{X}_n||^2 + \gamma_n^{(1)} \text{ with } \sum_n \alpha_n^{(1)} < + \infty \quad \sum_n \beta_n^{(1)} < + \infty \quad \sum_n \gamma_n^{(1)} < + \infty .$$

<u>2nd *point*</u> : we refer the reader to the notations of the proof of the preceding theorem (second point). T_n^j is ζ_n^j with ρ instead of ρ_n and Ξ_n^j instead of $\bar{\xi}_n^j$ and $W_n = \sum_{j=1}^m T_n^j (\nabla f_j (X_n) + \theta_n^j c_n + \dfrac{\xi_n^j}{2 c_n})$. If $\bar{X}_n$ is the projection of X_n on the solution set S of $(\mathcal{P})$, we have :

$$(3) \quad ||X_{n+1} - \bar{X}_{n+1}||^2 \leqslant ||X_n - \bar{X}_n - a_n \nabla_x L_\rho (X_n, Y_n) - a_n (U_n + V_n + W_n)||^2$$

We use the same lines of arguments as those developed in the second point of the proof of Theorem 2 (inequalities (1.1) and (1.2))

1.3. For an inequality on $E^{\mathcal{B}_n} < X_n - \bar{X}_n , W_n >$, we remark that

$|T_n^j| \leqslant 2 \rho \ |\Xi_n^j|$ and

$$||W_n|| \leqslant O(1) \sum_{j=1}^{m} |\Xi_n^j| \left(||\nabla f_j (X_n)|| + c_n + \frac{||\xi_n^j||}{2 c_n} \right)$$

Since $E^{\mathcal{B}_n} |\Xi_n^j| \leqslant \frac{O(1)}{k_n^{1/2}}$, $E^{\mathcal{B}_n} \left(|\Xi_n^j| \cdot ||\xi_n^j|| \right) \leqslant \frac{O(1)}{k_n}$.

We have the following inequality :

$$- 2 a_n E^{\mathcal{B}_n} < X_n - \bar{X}_n, W_n > \leqslant O(1) \frac{a_n}{k_n^{1/2}} (1 + ||X_n - \bar{X}_n||) + O(1) \frac{a_n c_n}{k_n^{1/2}} + O(1) \frac{a_n}{c_n k_n^{1/2}}$$

The sequence $\{c_n\}$ is bounded, so :

$$- 2 a_n E^{\mathcal{B}_n} < X_n - \bar{X}_n, W_n > \leqslant O(1) \frac{a_n}{c_n k_n^{1/2}} (1 + ||X_n - \bar{X}_n||^2).$$

1.4. Bounding $||\nabla_x L_\rho (X_n, Y_n)||^2$, $E^{\mathcal{B}_n} ||U_n||^2$, $E^{\mathcal{B}_n} ||V_n||^2$, we have

$$||\nabla_x L_\rho (X_n, Y_n)||^2 \leqslant O (1) (1 + ||X_n - \bar{X}_n||^2 + ||Y_n - \bar{Y}_n||^2).$$

$$E^{\mathcal{B}_n} ||U_n||^2 \leqslant \frac{O(1)}{c_n^2} , \quad E^{\mathcal{B}_n} ||V_n||^2 \leqslant \frac{O(1)}{c_n^2} (1 + ||X_n - \bar{X}_n||^2 + ||Y_n - \bar{Y}_n||^2)$$

1.5. Concerning $||W_n||^2$, one has :

$$||W_n||^2 \leqslant O(1) \sum_{j=1}^{m} |\Xi_n^j|^2 (||\nabla f_j (X_n)||^2 + c_n^2 + \frac{||\xi_n^j||^2}{c_n^2})$$

$$E^{\mathcal{B}_n} ||W_n||^2 \leqslant O(1) \frac{a_n^2}{c_n^2} (1 + ||X_n - \bar{X}_n||^2)$$

By the convergence of $\left\{ \frac{a_n}{c_n k_n^{1/2}} \right\}$, $\left\{ \frac{a_n^2}{c_n^2} \right\}$, we deduce from the inequality (3)

$$(4) \quad E^{\mathcal{B}_n} ||X_{n+1} - \bar{X}_{n+1}||^2 \leqslant (1 + \alpha_n^{(2)}) ||X_n - \bar{X}_n||^2 - 2 a_n < \nabla_x L_\rho (X_n, Y_n), X_n - \bar{X}_n >$$

$$+ \beta_n^{(2)} ||Y_n - \bar{Y}_n||^2 + \gamma_n^{(2)} \quad \text{with } \sum_n \alpha_n^{(2)} < + \infty \ \sum_n \beta_n^{(2)} < + \infty \text{ et } \sum_n \gamma_n^{(2)} < + \infty$$

3rd point

a) Combining the previous inequalities, one obtains :

(5) $E^{\mathcal{B}_{n+1}} \Delta(X_{n+1}, Y_{n+1}) \leqslant (1+\alpha_n)\, \Delta(X_n, Y_n) - 2\, a_n < \nabla_x L_\rho (X_n, Y_n), X_n - \overline{X}_n >$

$+ 2\, a_n < \nabla_y L_\rho (X_n, Y_n), Y_n - \overline{Y}_n > + \delta_n$ with $\sum_{n=1}^{\infty} \alpha_n < \infty$ et $\sum_{n=1}^{\infty} \delta_n < \infty$

b) According to the fundamental property of L_ρ with a *fixed* ρ (Proposition 5), one has :

$(x \in S \iff)\ d(x, S) = 0 \iff L_\rho (x, \overline{y}) = \inf L_\rho (x, \overline{y}) \quad \forall\, \overline{y} \in \mathcal{Y}$

that is to say : $\forall \varepsilon > 0$, if $d(X_n, S) \geqslant \varepsilon$, there exists $\alpha(\varepsilon) > 0$ such that :

$$< \nabla_x L_\rho (X_n, Y_n), X_n - \overline{X}_n > - < \nabla_y L_\rho (X_n, Y_n), Y_n - \overline{Y}_n > \leqslant L_\rho (\overline{X}_n, \overline{Y}_n) - L_\rho (X_n, \overline{Y}_n) \leqslant - \alpha(\varepsilon)$$

The inequality (5) implies that :

(6) $\forall \varepsilon > 0,\ E^{\mathcal{B}_{n+1}} \Delta(X_{n+1}, Y_{n+1}) \leqslant (1+\alpha_n)\, \Delta(X_n, Y_n) - 2\, a_n\, \alpha(\varepsilon)\, 1_{\{d(X_n, S) \geqslant \varepsilon\}} + \delta_n$

Consequently, $\{\Delta(X_n, Y_n)\}_{n \geqslant 1}$ is "almost" a supermartingale and $\Delta(X_n, Y_n)$ converges a.s. Moreover, following the inequality (6), we have :

$$\forall \varepsilon > 0 \qquad \sum_i a_i\, P\, \{d(X_i, S) \geqslant \varepsilon\} < + \infty$$

Thus, according to lemma B, there exists a subsequence of $\{d(X_n, S)\}_{n \geqslant 1}$ which converges a.s. towards 0.

The proof of the convergence of the whole sequence is similar to the one developed in the first stochastic approximation method of dual type.

REFERENCES

[1] *Annales de l'Université de Clermont n° 58, Mathématiques, 12ème fascicule (1976).*

[2] *D.P. BERTSEKAS, Necessary and sufficient conditions for a penalty method to be exact. Mathematical Programming, Vol. 9 (1975), n° 1.*

[3] *T. GAVIN, Stochastic approximation type methods for unconstrained and constrained optimization problems, Ph. D. Thesis, Brown University (1974).*

[4] *H.J. KUSHNER, Stochastic approximation algorithms for constrained optimization problems. The Annals of Statistics (1974). Vol. 2, n° 4, p. 713-723.*

[5] *H.J. KUSHNER, E. SANVICENTE, Stochastic approximation methods for constrained systems with observation noise on the systems and constraints. Stochastic Control Symposium, Budapest (preprints).*

[6] *P.J. LAURENT, Approximation et Optimisation, Paris Hermann (1972).*

[7] *J. NEVEU, Martingales à temps discret. Paris, Masson et Cie (1972).*

[8] *J.M. ORTEGA, W.C. RHEINBOLDT. Iterative solution of non linear equations in several variables, Academic Press, New-York and London (1970).*

[9] *R.T. ROCKAFELLAR, Measurable dependence of convex sets and functions on parameters, J. Math. Anal. Appl. 28, p. 4-25 (1969).*

[10] *R.T. ROCKAFELLAR, A dual approach to solving non linear programming problems by unconstrained optimization, Mathematical Programming 5 (1973). p. 354-373.*

[11] *L. SCHMETTERER, L'approximation stochastique. Cours de D.E.A. de Mathématiques Appliquées, 2ème édition, Université de Clermont (1972).*

[12] *J.B. HIRIART-URRUTY, Algorithmes stochastiques de type pénalisation et de type dual pour la résolution de problèmes d'optimisation stochastique avec contraintes stochastiques. Rapport de Recherche, Département de Mathématiques Appliquées, Université de Clermont (1975).*

J.B. HIRIART-URRUTY
Université de Clermont
Complexe Scientifique des Cézeaux
Département de Mathématiques
Appliquées
Boîte Postale n° 45

63170 AUBIERE (FRANCE)

Recent Developments in Statistics
J.R.Barra et al., editors

Invited Paper

POINT PROCESS MODELS FOR LINE TRANSECT EXPERIMENTS

Tore Schweder
University of Tromsø
Tromsø, Norway

The models allow the animals to move freely about and do not assume that they are independently (Poisson) distributed over the space, or are sighted independently of each other. The basic parameter, the sighting efficiency, is identified in a first order point process model for the experiment. Unbiased and asymptotically normally distributed estimators for the population density are briefly discussed, assuming stationarity and mixing conditions.

KEY WORDS

population density estimate, point process, asymptotic normality.

INTRODUCTION

The line transect method is one of the more frequently used census methods for animal populations. It is used for a variety of species: whales, grouse and other birds, deer and other ungulates and also for dead objects like deer carcasses or water fowl nests. The experiment is conducted by an observer linearly traversing the field populated by the species in question, counting the number of animals seen and recording the position of the animals relative to the observer when sighted. The statistical problem with line transect data is to estimate the sighting efficiency or the width of the strip effictively observed. Seber (1973) gives a brief review of the line transect litterature. He notes that the basic assumptions underlying the current statistical analysis of the line transect method are those stated below. See also Burnham and Anderson (1976) for the most recent account.

(1) The animals are not moving, except possibly when they are

flushed. They are distributed over the area as a homogeneous Poisson process.

(2) The sighting of one animal is independent of the sighting of another.

(3) No animal is counted more than once.

(4) When an animal is sighted, its position relative to the observer is recorded without error.

(5) The response behaviour of the population as a whole does not substantially change in the course of running the transect.

(6) The individuals are homogeneous with regard to their response behaviour, regardless of position in the field and also sex, age etc.

(7) An animal located on the line transect path will be sighted with probability one.

We will basically keep assumptions 3-7, but relax assumpions 1 and 2, which are practically never satisfied. Instead of these strong independence assumptions, we will assume homogeneity of the following character.

At reference time zero the animals are homogeneously distributed over the area. More precisely, we shall assume that their positions may be represented by a first order stationary stochastic point process on the two dimensional Euclidean space $\mathbb{R}^2$. Theory of point processes is found in Daley and Vere-Jones (1972). This point process may be overdispersed as will be the case for gregarious animals like whales or underdispersed as will be the case for territorial animals like certain birds. Moreover, the animals are allowed to move around. The movements are however assumed to be stocastically identical for different individuals and independent of the position at a particular time. And the movement is assumed to be stationary in time. We furthermore assume that the sighting of one animal depends only on the locations of this animal relative to the observer, and not on how the other animals move about. The sighting of two animals need however not be stochastically independent.

Based on general assumptions of the process of sighting, we will derive a point process model for the line transect observations: the time T_j and the position relative to the observer $\underline{X}_j \in \mathbb{R}^2$ of the j-th sighted animal (vectors of $\mathbb{R}^2$ are underlined). More precisely, we will derive the first order structure, or the first order moment measure of this point process of observations.

By linking up the observation process with the population process by the sampling rule, the relevant parameter called the sighting efficiency is identified and the problem of estimating this parameter and the population density is briefly discussed.

Under stationarity and mixing conditions the asymptotic behaviour of certain estimators is derived.

To avoid border effects, we have defined the space of the population to be the whole two dimensional Euclidean space. We furthermore assume that the observer has been moving at unit velocity for an infinitely long time.

The above population model is intended for non flocking animals populating a fairly homogeneous habitat. If the habitat is inhomogeneous, stratification may be necessary. If the animals are flocking, one may represent each flock by its initial position, its flocksize and its displacement process, and proceed more or less as below.

CONTINUOUS LINE TRANSECT MODELS

Let $\underline{Z}_{it} \in \mathbb{R}^2$ denote the location of the i-th animal at time t. We shall assume that the population is characterized by the following.

(i) At time t=0, the population is scattered as a stochastic point process $\{\underline{Z}_{i0}\}_i$ over $\mathbb{R}^2$, with first order moment measure $\lambda d\underline{z}$ and without multiple points.

(ii) For each animal i, the location process $\{\underline{Z}_{it}\}_t$ is continuous with probability one and has stationary increments in the sense that $\{\underline{Z}_{i\ t+u} - \underline{Z}_{iu}\}_t$ is the same process as $\{\underline{Z}_{it} - \underline{Z}_{i0}\}_t$. The displacement process is assumed independent of the locations $\{\underline{Z}_{i0}\}_i$ at time zero, and is stochastically the same for various i's.

The parameter λ is the population density which is to be estimated by the line transect experiment. By the above assumptions the locations of the population at any time t, $\{\underline{Z}_{it}\}_i$ is a first order stationary point process with density λ.

For the sampling process, assume that the observer moves at constant velocity along the vector $\underline{e}$ so that his position at time t is $t\underline{e}$. For the sighting process we shall assume the existence of a non negative hazard function Q such that the probability of sighting an animal at position $\underline{Z}$ relative to the observer in a time period of length dt is of the order

$$Q(\underline{Z})dt$$

More precisely, let

$$\rho_i(t) = \begin{cases} 1 & \text{if the i-th animal was sighted at time t or earlier} \\ 0 & \text{if not yet sighted at time t.} \end{cases}$$

The sighting mechanism is assumed to be:

(iii) Given the paths of all the animals $\{\{\underline{Z}_{jt}\}_t\}_j$, the probability that the i-th animal is not yet sighted at time t is

$$P(\rho_i(t) = 0 \mid \{\{\underline{Z}_{ju}\}_u\}_j) = \exp(-\int_{-\infty}^{t} Q(\underline{Z}_{iu}-u\underline{e})du). \tag{1}$$

We shall assume, as Seber, that no animal is counted more than once. This may be reasonable if the observer moves faster than the animals, if he is able to keep track of where the sighted animals go or if he is able to recognize a previously sighted animal. Under this assumption the i-th animal is sighted at most once, if $\rho_i(\infty)=1$, and this sighting is made at time

$$T = \int_{-\infty}^{\infty} t \; \rho_i(dt)$$

and at relative position

$$\underline{X} = \int_{-\infty}^{\infty} (\underline{Z}_{it}-t\underline{e}) \; \rho_i(dt).$$

The observation process

$$\{(T_k, \underline{X}_k)\} \text{ on } \mathbb{R} \times \mathbb{R}^2$$

may therefore be represented by

$$\{(T_k, \underline{X}_k)\} = \{(\int_{-\infty}^{\infty} t\rho_i(dt), \int_{-\infty}^{\infty} (\underline{Z}_{it} - t\underline{e})\, \rho_i(dt)) \mid \rho_i(\infty) = 1\} \qquad (2)$$

By relating the observation process to the population process in this way we have:

<u>Model 1</u>

The observation process $\{(T_k, \underline{X}_k)\}$ is a point process on $\mathbb{R} \times \mathbb{R}^2$ with first order moment measure

$$\mu(dt, d\underline{x}) = \lambda\, Q(\underline{x})\; E[\exp(-\int_{-\infty}^{0} Q(\underline{x}+\underline{Z}_u - u\underline{e})du) \mid \underline{Z}_0 = \underline{0}]dtd\underline{x}$$

$$= \lambda\, \nu\, q_1(\underline{x})\; dtd\underline{x}$$

where ν is the velocity of the observer, i.e. the length of $\underline{e}$, and $\{\underline{Z}_u\}$ is the path of a typical animal.

The quantity $\int_{\mathbb{R}^2} q_1(\underline{x})d\underline{x}$ will be called the *sighting efficiency*. It indicates the expected number of sightings per unit population density and per unit distance travelled by the observer. If the sighting efficiency was known, there would be no problem with estimating the population density λ. The sighting efficiency needs usually to be estimated from the data. This problem is briefly discussed below.

By assuming more specific models for the displacement process $\{\underline{Z}_u\}$ and the hazard function Q, one obtains more specific models for the function $q_1(\underline{x})$, which after division by the sighting efficiency is the marginal density of observed relative distances $\underline{X}_k$. It may be an interesting excercise to see what kind of distribution $\underline{X}_k$ will have for various models for $\{\underline{Z}_u\}$ and Q, but for the purpose of estimating the sighting efficiency and finally the population density, this may not be of too much value. More specific displacement and sighting models will usually be hard to check in any other way than to confront the deduced distribution q_1 with observed $\underline{X}_k$'s, and it is usually better just to fit a suitable model to these observations.

We will however in model 2 and 3 consider two simplifications of the model.

To demonstrate that model 1 follows from the above assumptions, we give the following

proof.

Let ξ be a suitable real function. By definition and relation (2), the first order moment measure μ of the observation process must satisfy

$$\int \xi(t,\underline{z})\mu(dt, d\underline{z}) = E \sum_k \xi(T_k, \underline{X}_k)$$

$$= E \sum_i \xi(\int t\rho_i(dt), \int(\underline{Z}_{it}-t\underline{e})\rho_i(dt))\rho_i(\infty)$$

Since ρ_i either is a one point distribution or has no mass,

$$\int \xi(t,\underline{z})\mu(dt,d\underline{z}) = E \sum_i \int \xi(t, \underline{Z}_{it}-t\underline{e})\rho_i(dt).$$

By conditioning on the paths of all the animals, we have by (1),

$$\int \xi(t,\underline{z})\mu(dt,d\underline{z}) = E \sum_i E\,[\int \xi(t,\underline{Z}_{it}-t\underline{e})\rho_i(dt) \mid \{\{\underline{Z}_{ju}\}_u\}_j]$$

$$= E \sum_i E \int \xi(t,\underline{Z}_{it}-t\underline{e})\; E[\rho_i(dt) \mid \{\underline{Z}_{iu}\}_u]$$

$$= E \sum_i E\,[\int \xi(t,\underline{Z}_{it}-t\underline{e})\Omega(\underline{Z}_{it}-t\underline{e}) \exp(-\int_{-\infty}^{t}\Omega(\underline{Z}_{iu}-u\underline{e})du)dt \mid \underline{Z}_{i0}]$$

Since the displacement process $\{\underline{Z}_{iu}-\underline{Z}_{i0}\}_u$ is assumed independent of $\{\underline{Z}_{j0}\}_j$ and this is a first order stationary point process with density λ, we have

$$\int \xi(t,\underline{z})\mu(dt,d\underline{z}) = E \int_{\mathbb{R}^2}\int_{\mathbb{R}} \xi(t,\underline{z}+\underline{Z}_{it}-\underline{Z}_{i0}-t\underline{e})\Omega(\underline{z}+\underline{Z}_{it}-\underline{Z}_{i0}-t\underline{e}) \exp(\int_{-\infty}^{t}\Omega(\underline{z}+\underline{Z}_{iu}-Z_{i0}-u\underline{e})du)dt\lambda d\underline{z}$$

By conditioning on the displacement process, changing the order of integration and using the transformation

$$\underline{z} + \underline{Z}_{it} - \underline{Z}_{i0} - t\underline{e} \to \underline{z},$$

we end up with

$$\int \xi(t,\underline{z})\mu(dt,d\underline{z}) = \int \xi(t,\underline{z})\Omega(\underline{z}) \; E[\exp(\int_{-\infty}^{0} \Omega(\underline{z}+\underline{Z}_u-u\underline{e})du) \mid \underline{Z}_0=\underline{0}] \; \lambda d\underline{z}dt.$$

Finally, by varying ξ over a rich enough family of functions, it is seen that μ must have the form of model 1.

If the animals have a tendency to move in a certain direction $\underline{d} \in \mathbb{R}^2$ such that

$$E[\underline{Z}_t | \underline{Z}_0=\underline{0}] = t\underline{d},$$

the first moment measure of the observation process may preferably be written

$$\mu(dt,d\underline{z}) = \lambda dt \; Q(\underline{z}) \; E[\exp(-\int_{-\infty}^{0} \Omega((\underline{Z}_u-u\underline{d})-u(\underline{e}-\underline{d}))du) | \underline{Z}_0=\underline{0}] \; d\underline{z}$$

and we have the situation where there is no drift tendency among the animals, but where the observer moves along the direction $\underline{e}-\underline{d}$ at constant velocity. We may therefore generally assume that

$$E(\underline{Z}_u \mid \underline{Z}_0 = \underline{0}) = \underline{0}$$

provided the drift $\underline{d}$ of the animals is known.

Typically, the hazard function $Q(\underline{x})$ of sighting an animal at relative position $\underline{x}$ is very small when $\underline{x}$ is of some size. Also, the velocity of the observer is typically larger than that of the animals. In that case we may use the following approximation to model 1.

<u>Model 2</u>

When the animals are not moving, the observation process $\{(T_k, \underline{X}_k)\}$ has first order moment measure

$$\mu(dt,d\underline{x}) = \lambda \; Q(\underline{x})\exp(-\int_{\infty}^{0} \Omega(\underline{x}-u\underline{e})du) \; dtd\underline{x} = \lambda \; \nu \; q_2(\underline{x}) \; dtd\underline{x}$$

In Cartesian coordinates, letting the observer move with velocity along the x-axis, model 2 reads

$$\mu(dt,dx,dy) = \lambda\ Q(x,y)\ \exp(-\int_{-\infty}^{0} Q(x-u\nu,y)du)dtdxdy$$

and in polar coordinates (r,φ) with φ being the angle,

$$\mu(dt,dr,d\varphi) = \lambda\ Q(r\cos\varphi, r\sin\varphi)$$

$$\exp(-\int_{r}^{\infty} Q(\sqrt{r'^2-r^2\sin^2\varphi},\ r\sin\varphi)\nu^{-1}\ (1-(\frac{r}{r'})^2\sin^2\varphi)^{-\frac{1}{2}}dr')$$

$$dt\ rdr\ d\varphi$$

provided $Q(x,y) = 0$ for $x<0$, i.e. no observations are made behind the observer.

If the sighting hazard furthermore depends only on the distance r ahead to the animal and not on the angle, model 3 is appropriate. This model was developed informally by Hayne (1949).

Model 3

If the animals do not move and the sighting hazard, denoted $G(r)$, depends only on the distance r ahead to the animal, the observation process has first order moment measure

$$\mu(dt,dr,d\varphi) = \lambda\ \cos\varphi\ rG(r)\ \exp(-\nu^{-1}\int_{r}^{\infty} G(r')dr')dtdrd\varphi \qquad (3)$$

$$= \lambda\nu\ \cos\varphi\ rg(r)\ dtdrd\varphi \qquad |\varphi| \leqslant \frac{\pi}{2}.$$

This is obtained from model 2 by taking

$$Q(r\cos\varphi, r\sin\varphi) = \cos\varphi\ G(r) \qquad |\varphi| \leqslant \frac{\pi}{2}$$

to make

$$Q(x,y)dx = G(r)\ dr.$$

In this model, the first order moment measure (3) factors, and the angle φ_i and the distance R_i of a sighting is therefore (first order) independent.

It may be of interest to note that model 3 may be obtained in a different way. We still assume the animals to be immobile. Let the i-th animal be located at position (X_i, Y_i) in usual coordinates, and let it be characterized by a critical distance R_i such that when the observer comes at distance R_i the animal is flushed and thus sighted. Assume that the critical distances are (first order) independent of the locations, and distributed with probability density $g(r)$, The population process

$$\{(X_i, Y_i, R_i)\} \text{ on } \mathbb{R}^2 \times \mathbb{R}^+$$

has then first order moment measure

$$\lambda\, g(r)\,dx\,dy\,dr.$$

Since all animals sitting closer to the observer's path than their critical distance are sighted, and since such an animal (X_i, Y_i, R_i) with $|Y_i| \leqslant R_i$ is observed at position $X_i - \sqrt{R_i^2 - Y_i^2}$ and at angle Arcsin (Y_i/R_i), the observation process may be represented:

$$\{(T_k, R_k, \varphi_k)\} = \{(\nu^{-1}(X_i - \sqrt{R_i^2 - Y_i^2}\,),\ R_i,\ \text{Arcin}\frac{Y_i}{R_i}) \;\Big|\; |Y_i| \leqslant R_i\}$$

With I being the indicator function, the first order moment measure must, for suitable ξ, satisfy,

$$\int_{-\infty}^{\infty}\int_{0}^{\infty}\int_{-\frac{\pi}{2}}^{\frac{\pi}{2}} \xi(t,r,\varphi)\,\mu(dt,dr,d\varphi)$$

$$= E\sum_i \xi(\nu^{-1}(X_i - \sqrt{R_i^2 - Y_i^2}),\ R_i,\ \text{Arcsin}\,\frac{Y_i}{R_i})\ I(|Y_i| < R_i)$$

$$= \int_{0}^{\infty}\int_{-\infty}^{\infty}\int_{-\infty}^{\infty} \xi(\nu^{-1}(x - \sqrt{r^2 - y^2}),\ r,\ \text{Arcsin}\,\frac{y}{r} I(|y| < r)\ \lambda g(r)\,dx\,dy\,dr$$

$$= \int_{-\infty}^{\infty}\int_{0}^{\infty}\int_{-\frac{\pi}{2}}^{\frac{\pi}{2}} \xi(t,r,\varphi)\ \lambda\nu\cos\varphi\ r g(r)\,d\varphi\,dr\,dt$$

and model 3 is obtained by varying ξ over a rich enough family of functions.

Assuming that circles are distributed randomly on the plane such that the circle centers follow a homogeneous Poisson process and the radii are independent and identically distributed with density $g(r)$, the distribution of the radii of those circles intersected by a randomly chosen straight line, is by a standard argument in geometric probability found to have density proportional to

$$r\ g(r),$$

see Kendall and Moran (1963).

This corresponds to our moment measure (3), which is reasonable because model 3 is a model for the experiment of dropping a straight line randomly on a plane where circles are distributed randomly. We have however not made the strong independence assumption that the process

$$\{(X_i, Y_i, R_i)\} \qquad \text{is a Poisson process on } \mathbb{R}^2 \times \mathbb{R}.$$

We have only assumed that the first order structure of the process is characterized by the moment measure

$$\lambda dxdy\ g(r)dr.$$

The strong independence assumptions above, or assuming that animals at time 0 are Poisson distributed, that they move independently of one another and that they are sighted independently, does not change the first order structure of the observation process stated in the models above. It does however imply that the observation processes are Poisson processes, Schweder (1974). That is, for model 1, that the number $N(\tau)$ of sighted animals in a time period of length τ is Poisson distributed with mean

$$\lambda \nu \tau \int_{\mathbb{R}^2} q_1(\underline{x})d\underline{x}$$

and that the observed relative positions of the sighted animals, $\underline{X}_1, \ldots, \underline{X}_{N(\tau)}$ are stochastically independent and identically distributed with probability density

$$q_1(\underline{x})(\int_{\mathbb{R}^2} q_1(\underline{x})d\underline{x})^{-1}.$$

DISCRETE SIGHTING MODELS

In the preceding section it was assumed that each animal is continuously available for sighting. This is however not the case for whales which only are available for visual sighting when they are up spouting. This may also be the case for certain birds, which only may be observed while they are singing.

Parallel to model one, the i-th whale is characterized by

$\underline{Z}_i^o$, its position at time zero

$\underline{Z}_{ij}$, its position relative to Z_i^o when spouting for the j-th time

U_{ij}, the time of its j-th spout.

Model 4

Suppose the position $\{\underline{Z}_i^o\}$ at reference time 0 is a point process on $\mathbb{R}^2$ with first order moment measure $\lambda d\underline{z}$. Suppose the spouting history $\{(\underline{Z}_{ij},U_{ij})\}_j$ for each i is a point process on $\mathbb{R}^2 \times \mathbb{R}$ with first order moment measure $\delta\, f(\underline{z};u)d\underline{z}du$, δ being the spouting intensity and $f(\cdot;u)$ the probability density of $\underline{Z}_{ij}$ given $U_{ij}=u$. The spouting histories are assumed stochastically identical for all i, and independent of $\{\underline{Z}_i^o\}$. Suppose also that a spout at position $\underline{x}$ relative to the boat is observed with probability $Q(\underline{x})$, independent of the sighting of other spouts.

Then the observation process

$$\{(T_k,\underline{X}_k)\}$$

where T_k is the time and $\underline{X}_k$ is the position of the spout relative to the boat when the k-th whale first is sighted, is a point process on $\mathbb{R} \times \mathbb{R}^2$ with first order moment measure

$$\lambda\ \delta\ Q(\underline{x})\ E[\prod_{j<0} (1-Q(\underline{Z}_j-U_j\underline{e}))\ |\ \underline{Z}^o = \underline{0}]\ dtd\underline{x}$$

$$= \lambda\nu\delta q_4(\underline{x})\ dtd\underline{x},$$

ν being the speed of the boat.

The proof of this model is parallel to the proof of model 1, see Schweder (1974).

Since the scouts in the crow's nest in the observation boat usually are able to distinguish the spouts of one whale from the spouts of another, by noticing the size of the whale, its colour and its direction of movement, model 4 may be appropriate.

One may however also take record of all the spouts seen, and we have

Model 5

Under the assumptions of model 4, the observation process

$$\{(\overline{T}_k, \underline{\overline{X}}_k)\},$$

where $\overline{T}_k$ is the time and $\underline{\overline{X}}_k$ the position of the k-th spout observed relative to the boat, is a point process on $\mathbb{R} \times \mathbb{R}^2$ with first order moment measure

$$\lambda\delta \; Q(\underline{x}) \; dt \; d\underline{x}.$$

ON THE ASSUMPTION THAT ALL THE ANIMALS SITTING RIGHT ON THE OBSERVERS PATH ARE SIGHTED

When the animals are not moving until flushed, Seber's assumption 7 mentioned in the introduction, is a natural starting point for estimation purposes.

Considering model 2 and using Cartesian coordinates, an animal on the observer's path, the x-axis, is sighted with probability one if the sighting hazard along the x-axis integrates to inifinity,

$$\int_{-\infty}^{\infty} Q(x,0)dx = \infty.$$

But

$$\int_{\infty}^{\infty} q_2(x,y)dx$$

$$= \nu^{-1} \int_{-\infty}^{\infty} Q(x,y)\exp(-\int_{-\infty}^{0} Q(x-u\nu,y)du)dx$$

$$= 1 - \exp(-\int_{-\infty}^{\infty} Q(x,y)dx)$$

The assumption is therefore

$$\int_{-\infty}^{\infty} q_2(x,0)dx = 1$$

which of course means that close to the x-axis, the expected number of sightings per unit transect is the same as the expected number of animals close to the x-axis per unit distance.

The corresponding assumption for model 1,

$$\int_{-\infty}^{\infty} q_1(x,0)\,dx = 1$$

has the same interpretation, which has somewhat the taste of an ergodic hypothesis: Along the x-axis, averaging over time is the same as averaging along the x-axis at fixed time.

For model 3 it is seen that the assumption is

$$\int_0^{\infty} G(r)\,dr = \infty,$$

which makes

$$\phi(r) = \nu^{-1} G(r) \exp(-\nu^{-1}\int_r^{\infty} G(r')\,dr'), \qquad r>0$$

a proper probability density. This corresponds to assuming that all the critical distances are strictly positive.

SIGHTING EFFICIENCY

The parameter

$$\theta^{-1} = \int_{\mathbb{R}^2} q(\underline{x})\,d\underline{x} \tag{4}$$

is previously coined the sighting efficiency.

The expected number of sightings resulting from running a line transect of length τ is according to model 1 and 2,

$$\lambda\tau\theta^{-1}$$

which in fact is equal to the expected number of animals found by a complete count of a strip of length τ and width θ^{-1}, at time 0.

The corresponding parameter

$$\theta^{-1} = \int_{\mathbb{R}^2} q_4(\underline{x})\,d\underline{x}$$

of model 4 indicates the width of a strip of length τ in which an equal expected number of spouts will be seen in one unit of time, as whales sighted by going along the line transect a length τ.

As argued above, we will assume that "all animals along the path of the observer will be sighted". By defining

$$g(y) = \int_{-\infty}^{\infty} q(x,y)\,dx$$

(using Cartesian coordinates), this assumption reads

$$g(0) = 1.$$

Consequently, the marginal distribution of the observed y-coordinates $Y_1, Y_2, \ldots$ is given by the probability density

$$f_1(y) = \theta g(y) \qquad y \in \mathbb{R}$$

and our parameter of interest, the inverse sighting efficiency, is therefore the value of this density at zero:

$$\theta = f_1(0)$$

Kelker (1945) seems to be the first explicitly aware of this being the parameter for use as correction factor. In terms of the density g of model 3, we find

$$\theta^{-1} = \int_{-\frac{\pi}{2}}^{\frac{\pi}{2}} \cos\theta d\theta \cdot \int_0^{\infty} r\ g(r)dr = 2\int_0^{\infty} r\ g(r)dr.$$

The observed radial distances have however density

$$g_1(r) = (\int_0^{\infty} r'\ g(r')dr)^{-1} r\ g(r),$$

and hence

$$\theta^{-1} = 2(\int_0^{\infty} r^{-1}\ g_1(r)dr)^{-1}, \qquad (5)$$

i.e. for model 3, the sighting efficiency is twice the harmonic mean of the distribution of observed radial distances.

ESTIMATION OF THE INVERSE SIGHTING EFFICIENCY AND THE POPULATION DENSITY λ

Letting $N(\tau)$ be the number of animals sighted by running the line transect a distance τ, we have for model 1

$$EN(\tau) = \lambda\frac{1}{\theta}\tau,$$

and the obvious estimator for λ is

$$\hat{\lambda}_\tau = \frac{1}{\tau} N(\tau)\hat{\theta}_\tau$$

where $\hat{\theta}_\tau$ is an estimate of the inverse sighting efficiency.

In model 1, θ is the value at zero of the marginal density $\theta g(y)$ of observed y-cordinates.

Natural assumptions on the detection function $g(y)$ are

(i) symmetry: $g(y) = g(-y)$

(ii) decreasing sighting efficiency: $g(y)$ is decreasing as $|y|$ is increasing.

By (i) the signs of the observed y-coordinates may be disregarded. The density of the observed perpendicular distances $|Y_i|$ is

$$f(y) = 2\theta g(y) \qquad y>0$$

and the estimation of θ may be approached non parametrically or through a parametric model for $f(y)$.

There is a rich choice of nice parametric densities on $[0,\infty)$ which are finite at 0 and non increasing. In the context of line transect experiments the following have been mentioned: the uniform, Amman and Baldwin (1960)-woodpeckers; the exponential, Gate et al (1968) - ruffed grouse; $f(y) = (a+1)(aw)^{-1}(1-y^a w^{-a})$ $0 \leq y \leq w$, Eberhardt (1968); the left tail of the two parameter gamma distribution $f(y) = \frac{\gamma}{\beta}\int_y^\infty \gamma^\beta \Gamma(\beta)^{-1} z^{\beta-1} e^{-\gamma z} dz$, Sen et at (1974). One may also try the folded normal, Järvinen and Väisänen (1975), which I found to fit some data for gerenuk (a long necked gazelle) very well, or also $f(y) = \frac{1}{\sigma}\Gamma(\frac{1}{\gamma}+1)^{-1}\exp(-y^\gamma\sigma^{-\gamma})$.

When fitting a parametric model it is not important that the model fits well at the right tail, it is the density at zero we want to estimate! Estimators should therefore be robust versus perturbations of the model at the right tail. This may be accomplished by grouping the data, with a large group to the right.

If Θ is estimated by the maximum likelihood method, using a parametric model for f(y) and assuming that $Y_1,\dots, Y_n$ are independent, the population density estimator $\hat{\lambda}_\tau$ is the maximum likelihood estimator of λ assuming the observation process is Poisson. This is seen by writing the likelihood function of the observations $N(\tau)$, $\{(T_i,Y_i) \mid i = 1,\dots, N(\tau)\}$ in the form

$$\prod_{i=1}^{N(\tau)} \lambda g(Y_i) \exp(-\tau\lambda \int_{-\infty}^{\infty} g(y)dy) = ((\frac{\lambda}{2\theta})^{N(\tau)} \exp(-\tau\frac{\lambda}{2\theta}))(\prod_{i=1}^{N(\tau)} f(Y_i)).$$

In this case, $\hat{\theta}$ based on n observations is normally $\sqrt{n}$ equivalent to an average $\frac{1}{n}\Sigma\, h(Y_i)$, and

$$\hat{\lambda}_\tau = \frac{1}{\tau} N(\tau)\ \hat{\theta}_\tau \approx \frac{1}{\tau}\sum_{i=1}^{N(\tau)} h(Y_i)$$

is asymptotically normally distributed.

In the next section we prove that under stationarity and mixing conditions, this result may also hold when the observationprocess is not Poisson.

As mentioned by Burnham and Anderson (1976), a simple non parametric estimator is provided by

$$\hat{\theta} = \frac{1}{2} \max_{i=1,\dots,k} (t_i^{-1} F_n(t_i)) \tag{6}$$

where $0<t_1<t_2<\dots<t_k<\infty$ is a suitable grid, and F_n is the cumulative distribution function of the observed perpendicular distances. If $g(y) = 1$ (<1) for $y \leqslant y_0$ ($y > y_0$), an optimal choice of t_1 is y_0. With such a choice of the grid, the results of the next section imply asymptotic normality under mixing conditions. The estimator (6) is what Barlow et al (1972) called an isotonic density estimate. They advocate to take t_i as the $i\cdot n^{-0.2}$ empirical quantile, $i=1,\dots,n^{0.2}$ This is however only sensible for extremely large samples.

Many of the existing estimation methods for line transect data are based on the radial distances R_i rather than the perpendicular distances $|Y_i|$. One of the reasons being that it is easier to measure the radial distances without error than the perpendicular ones, Robinette et al (1974).

With the exception of a particular case treated by Gates (1969) and certainly Hayne (1949) who basically developed model 3 above, the relation between the inverse sighting efficiency θ and the distribution of observed radial distances R_i has not been well understood. When the assumptions of model 3 are met, the sighting efficiency θ^{-1} is by (5) twice the harmonic mean in the distribution of observed distances, and Hayne's estimate

$$\hat{\theta} = 2\,\frac{1}{n}\sum_{i=1}^{n} R_i^{-1}$$

is thus unbiased.

This estimator may however be very unstable if small R's are observed with positive probability. In that case it may be preferable to trim away the small R's or to fit a parametric model to the distribution of R's. The two parametric γ-model,

$$h(r) = \gamma^{\beta}\,\Gamma(\beta)^{-1}\,r^{\beta-1}\,e^{-\gamma r} \qquad r>0, \qquad \beta>1$$

may provide a good fit. In this model

$$\theta^{-1} = 2\frac{1}{\gamma}(\beta-1).$$

As Hayne informally argued, one may check the validity of model 3 by testing whether the observed angles φ_i follow a cosin distribution, or equivalently if $\sin|\varphi_i| = \left|\frac{Y_i}{R_i}\right|$ follow a uniform distribution, and whether the angles and radial distances are independent.

ON THE ASYMPTOTIC BEHAVIOUR OF THE POPULATION DENSITY ESTIMATE

In our treatment, we have abandoned the strong independence assumptions, which imply the number of observed animals to be Poisson distributed and the observed relative positions independent and identically distributed. Our models 1,...,5 are only first order models. To investigate the asymptotics, one needs assumptions on the dependence structure of the point process of observation times and relative positions $\{(T_i,\underline{X}_i)\}$, represented by the (random) counting measure V on $\mathbb{R} \times \mathbb{R}^2$:

$$V(A \times B) = \#\{(T_i,\underline{X}_i) \mid T_i \in A,\ \underline{X}_i \in B\}$$

We will say that V is stationary in time if for all $t \in R$.

$$\{(T_i+t\ ,\underline{X}_i)\} \text{ is the same process as } \{(T_i,\underline{X}_i)\}.$$

In that case the k-th order cumulant measure C^k defined by

$$C^k(A_1,\ B_1,\ldots,\ A_k,B_k) = \text{cum}(V(A_1 \times B_1),\ldots,\ V(A_k \times B_k)),$$

$A_i \in \mathbb{R}$, $B_i \in \mathbb{R}^2$, cum being the ordinary cumulant operator, is of the form

$$C^k(dt_1,d\underline{x}_1\ldots,\ dt_k,\ d\underline{x}_k) = C_k(d\underline{x}_1,\ldots,d\underline{x}_k;$$
$$d(t_1-t_k),\ldots,\ d(t_{k-1} - t_k))dt_k$$

A basis for hope that the population density estimate $\hat{\lambda}$ for large line transect experiments follows a normal distribution is provided by the following theorem and corollary.

Theorem

If V is time stationary mixing with respect to the real function h on $\mathbb{R}^2$ such that for $k = 2,3,\ldots.$

$$\int_{\mathbb{R}^{k-1}} \int_{\mathbb{R}^{2k}} \prod_{i=1}^{k} |h(\underline{x}_i)| \quad |C_k(d\underline{x}_1,\ldots,d\underline{x};\quad du_1,\ldots du_{k-1})$$

is finite, then

$$\sqrt{\tau}\left(\frac{1}{\tau}\sum_{i=1}^{N(\tau)} h(\underline{X}_i) - \lambda\int_{\mathbb{R}^2} h(\underline{x})q_1(x)d\underline{x}\right)$$

converge, as $\tau\to\infty$, in distribution to a normal distribution centered at zero and with variance

$$\sigma^2 = \int_{\mathbb{R}}\int_{\mathbb{R}^4} h(\underline{x}_1)h(\underline{x}_2)\ C_2(d\underline{x}_1,d\underline{x}_2;\ du) \tag{7}$$

proof We need to show that all cumulants of

$$\tau^{-\frac{1}{2}}\sum_{i=1}^{N(\tau)} h(\underline{X}_i) = \tau^{-\frac{1}{2}}\int_0^\tau\int_{\mathbb{R}^2} h(\underline{x})\ V(dt,d\underline{x}) \tag{8}$$

of order higher than 2 vanish in the limit, that the second cumulant converges to (7) and that the first one equals

$$\tau^{\frac{1}{2}}\lambda\int_{\mathbb{R}^2} h(\underline{x})\ q_1(\underline{x})d\underline{x}.$$

The first cumulant is the expectation, so by model 1,

$$E\ \tau^{-\frac{1}{2}}\int_0^\tau\int_{\mathbb{R}^2} h(\underline{x})\ V(dt,d\underline{x}) = \tau^{\frac{1}{2}}\lambda\int_{\mathbb{R}^2} h(\underline{x})\ q_1(\underline{x})d\underline{x}.$$

The k-th cumulant of (8) is now

$$\tau^{-\frac{k}{2}}\underbrace{\int_0^\tau\cdots\int_0^\tau}_{k}\int_{\mathbb{R}^{2k}}\prod_{i=1}^{k} h(\underline{x}_i)\ C^k(dt_1,d\underline{x}_1,\ \ldots,\ dt_k,d\underline{x}_k)$$

$$= \tau^{-\frac{k}{2}}\int_0^\tau [\underbrace{\int_{-t_k}^{\tau-t_k}\cdots\int_{-t_k}^{\tau-t_k}}_{k-1}\int_{\mathbb{R}^{2k}}\prod_{i=1}^{k} h(\underline{x}_i)\ C_k(d\underline{x},\ldots,d\underline{x}_k\ ;$$

$$du_1,\ldots,du_{k-1})]dt_k.$$

By the mixing condition, this converges to zero for k=3,... and to σ^2 for k=2, and the proof is completed.

Often it is possible to prove that

$$N(\tau)\ (\hat{\theta}_\tau - N(\tau)^{-1}\sum_{i=1}^{N(\tau)} h(\underline{X}_i)) \overset{P}{\to} 0 \tag{9}$$

as $N(\tau) \overset{P}{\to} \infty$. For $\hat{\theta}_\tau$ not to be assumptotically biased, one will have

$$E\ h(\underline{X}_i) = \theta$$

which by the definition (4) of θ, implies

$$\int_{\mathbb{R}^2} h(\underline{x})\ q_1(\underline{x})d\underline{x} = 1. \tag{10}$$

Since $N(\tau) \overset{P}{\to} \infty$ and $\frac{1}{\tau}N(\tau) \overset{P}{\to} \lambda\frac{1}{\theta}$ as $\tau\to\infty$, (9) implies that $\hat{\theta}_\tau$ is $\sqrt{\tau}$

equivalent to the average appearing in (9). We therefore have by the theorem the

Corollary

If under the stationarity and mixing conditions of the theorem, $\hat{\theta}_\tau$ is equal or $\sqrt{N(\tau)}$ equivalent to the average

$$\frac{1}{N(\tau)} \sum_{i=1}^{N(\tau)} h(X_i)$$

with h-satifying (10), then

$$\hat{\lambda}_\tau = \frac{1}{\tau} N(\tau) \hat{\theta}_\tau$$

is asymptotically normally distributed with center at λ and variance given by (7).

Let us finally, as an example, show how to use these results to prove that the population density estimate based on the non-parametric estimator (6) is asymptotically normal under the stationarity and mixing conditions above relative to the function

$$h_k(x,y) = I(|y|<t_k),$$

given in Cartesian coordinates. Now, for $j \geq 1$, we have

$$\frac{1}{2}\frac{1}{\tau} N(\tau) \frac{1}{t_j} F_n(t_j) = \frac{1}{2}\frac{1}{t_j}\frac{1}{\tau}\int_0^\tau \int_{\mathbb{R}^2} I(|y| \leq t_j)\ V(dt,dx,dy). \tag{11}$$

The expectation of this quantity is for j=1

$$\frac{1}{2t_1} \int_{-\infty}^{\infty} \int_{-t_1}^{t_1} \lambda q_1(x,y)\,dxdy = \lambda,$$

when $t_1 = y_0$ as defined below (6). For $j>1$, the expectation for (11) is less than λ.

Since the process is mixing relative to $h_j(x,y) = I(|y| \leq t_j)$ whenever it is mixing with respect to h_k $j \leq k$, the theorem implies that (11) asymptotically is normally distributed around its expectation and with variance of the order $\frac{1}{\tau}$. Consequently, the maximum of (11) over $j=1,\ldots,k$ will asymptotically be (11) with j=1, and thus the asserted normality of

$$\hat{\lambda}_\tau = \frac{1}{2}N(\tau) \max_{j=1,\ldots,k} \frac{1}{t_j}F_n(t_j)$$

is proved.

REFERENCES

Amman, G.A. and Baldwin, P.H. (1960). A comparison of methods for censusing woodpeckers in spruce-fir forests of Colorado. Ecology 41, 699-706.

Barlow, R.E., Bartholomew, D.J., Bremner, J.M. and Brunk, H.D.(1972). Statistical Interference Under Order Restrictions, Wiley, New York.

Burnham K.P. and Anderson D.R. (June 1976). Mathematical models for nonparametric interference from line transect data. Biometrics 32, no.2, 325-336.

Daley, D.J. and Vere-Jones, D. (1972). A summary of the theory of point processes, in Stochastic Point Processes; Ed. by P.E. Lewis, Wiley, New York.

Eberhardt, L.L. (1968). A preliminary appraisal of line transects. J. Wildl. Manage. 32, 82-88.

Gates, C.E., Marshall, W.H. and Olson, D.P. (1968). Line transect methods of estimating grouse population densities, Biometrics 24, no. 1, 132-145.

Gates, C.E. (1969). Simulation study of estimators for the line transect sampling method, Biometrics 25, no. 2, 317-328.

Hayne, D.W. (1949). An examination of the strip census method for estimating animal populations. J. Wildl. Manage. 13, 145-157.

Järvinen, O. and Väisänen R.A. (1975). Estimating relative density of breeding birds by the line transect method. OIKOS 26, 316-322.

Kelker, G.H. (1945). Measurement and interpretation of forces, that determine populations of managed deer. Ph.D. thesis. Univ. of Michigan, Ann Arbor. 422 pp.

Kendal, M.G. and Moran, P.A.P. (1963). Geometric Probability. Griffin, London.

Robinette, W.L., Loveless, C.M. and Jones, D.A. (1974). Field tests of strip census methods. J. Wildl. Manage. 38, no. 1, 81-96.

Seber, G.A.F. (1973). The estimation of animal Abundance and Related Parameters, Griffin, London.

Sen, Tourigny and Smith (1974). On the line transect sampling method. Biometrics 30, 329-340.

Schweder, T. (1974). Transformation of point processes:applications to animal sighting and catch problems, with special emphasis on whales. Ph. D. thesis. University of California, Berkely. 183 pp.

Recent Developments in Statistics
J.R.Barra et al., editors

Invited Paper

STUDY OF A CLASS OF DENSITY ESTIMATORS

by D. Bosq (University of Lille I - France)

(Summary)*

The purpose of this paper is the construction and study of a class of density estimators which contains many usual estimators.

It deals with estimators of the form

$$\hat{f}_n(t) = \frac{1}{n}\sum_{i=1}^{n} K_{r(n)}(X_i,t) \quad ; \quad t \in E, \quad n \in N^* \tag{1}$$

where E is the observation space ; $X_1,\dots,X_n$ is a sample from the distribution of which we want to estimate the density and (K_r) a family of numerical functions defined on $E\times E$.

This class of estimators contains the estimators obtained by the "convolution kernel method" that is, estimators of the form

$$\hat{f}_n^{(1)}(t) = \frac{r(n)^s}{n}\sum_{i=1}^{n} K[r(n)(X_i-t)] \quad ; \quad t \in \mathbb{R}^s \quad , \quad n \in \mathbb{N}^*$$

where K is defined on R^s.

Likewise for the estimator

$$\hat{f}_n^{(2)}(t) = \sum_{j=0}^{r(n)}\Big[\sum_{i=1}^{n}\frac{1}{n}\, e_j(X_i)\Big]e_j(t) \quad ; \quad t \in E, \quad n \in \mathbb{N}^*$$

where $(e_j, j \in N^*)$ is an orthonormal basis of $L^2(\mu)$ ("orthogonal functions method").

In part one we attempt to justify the method of estimation used. We proceed as follows : under the assumption that the density to be estimated belongs to $L^2(\mu)$, we start by looking for a minimum covariance operator unbiased estimator.

Now, the existence of an unbiased estimator (U.E.) for the density based on one

---

* A detailed version of this paper was submitted to the "Annals of Statistics".

observation signifies that the identity of $L^2(\mu)$ is an integral operator ; whence we infer that a necessary and sufficient condition for the existence of such an estimator is that μ be purely atomic. This result is more precise than those generally given.

It is then possible to construct a density estimator, either in estimating unbiasedly a function of the density (which must be an integral operator), or in restricting the family $\mathcal{D}$ of possible densities in order to obtain an U.E. (the space generated by $\mathcal{D}$ must thus be "selfreproducing").

For the convergence of a sequence of such estimators it is necessary for the associated sequence of integral operators to converge to identity (or again for the associated sequence of reproducing kernel spaces to cover $L^2(\mu)$).

In part two, assumptions being made on (K_r), we give conditions on $(r(n), n \in N^*)$ for the convergence of the estimator. Thus, we obtain necessary and sufficient conditions for pointwise and uniform convergence according to various types. Lower and upper bounds for rate of convergence are also given.

Finally we indicate applications to specific estimators, that is, estimators constructed with Fourier series, Hermite functions or Haar basis.
These asymptotic results were obtained in collaboration with J. BLEUEZ.

The proofs of some of the necessary conditions (in particular necessary conditions for uniform convergence) use methods introduced by GEFFROY and BERTRAND-RETALI.

On the other hand, let us point out that FÖLDES and REVESZ have studied a completely similar class of density estimators : they have obtained sufficient conditions for uniform convergence when $E = R$.

It seems that the results of the present paper are the most general ones obtained up to now.

Recent Developments in Statistics
J.R.Barra et al., editors

Invited Paper

CONFORMITY OF INFERENCE PATTERNS

A.P. Dawid
Department of Statistics and Computer Science
University College
London WC1E 6BT, England

In memory of Allan Birnbaum, 1923-1976.

We reconsider the rôle which general principles of inference have to play in the interpretation of statistical data. Such principles are viewed as testing-grounds for proposed inference patterns, rather than as rules for deriving correct inferences. The interplay between various principles is discussed, and their application to some familiar inference patterns. It is shown that a careful consideration of the scope of these principles allows a wider class of inference patterns than commonly supposed. Nevertheless some difficulties remain.

1. INTRODUCTION

Among R.A. Fisher's many valuable contributions to statistical inference was the insight that valid inference should be based on a sufficient statistic, and related to a frame of reference that conditions on the observed value of an ancillary statistic. However, while these precepts may have seemed clear-cut at first, it gradually became apparent that difficulties could arise in the choice of ancillary statistic (Basu, 1959), and that Fisher's guidelines were ambiguous. In a classic paper, Birnbaum (1962) recast Fisher's ideas in an unambiguous way, and found they implied that inference should depend only on the observed likelihood function. Birnbaum himself, together with many others, was unhappy with this conclusion, and his result is often passed over, perhaps out of a feeling that the structure he imposed was inappropriate.

In this paper we set up a framework, very similar to Birnbaum's, in which to introduce various principles of inference, and investigate the relationships between, and implications of, these principles. It is hoped that this treatment, which to some extent covers old ground, will help to clarify the content and scope of the principles.

2. EXPERIMENTS

Let Θ be an unknown quantity of interest. Various experiments may be performed to gain information about Θ. Let ξ denote such an experiment. For the purpose of this paper we shall suppose that any unknown aspects of the structure of ξ are completely determined by the value of Θ. A typical ξ might be described in some detail. We take, as part of this description, the specification of what variable X is to be recorded in the experiment, and write $\xi:X$, or $\xi:X\epsilon\mathcal{X}$ where $\mathcal{X}$ is the space of possible values for X. In most cases of statistical interest, description of ξ will also include (but may go beyond) specification of a distribution P_θ for X when $\Theta=\theta$. Then we write $X\sim P_\theta$.

Suppose $\xi:X\epsilon\mathcal{X}$, and consider a statistic $T=t(X)$, where $t:\mathcal{X}\to\mathcal{Y}$. Then we can define a <u>marginal experiment</u> ξ^T, and (for $t_0\epsilon t(\mathcal{X})$) a <u>conditional experiment</u> $\xi_{T=t_0}$. Both have the same experimental set-up as ξ, but ξ^T records only the value of T, while $\xi_{T=t_0}$ records X, but only if $T=t_0$. If $X\sim P_\theta$, then we take the derived probability specifications in ξ^T and $\xi_{T=t_0}$ to be the usual marginal and conditional distributions based on P_θ. Similarly, we may construct from ξ hybrid experiments of the form $\xi^T_{S=s_0}$ (observe T, but only if $S=s_0$).

We may also compound experiments sequentially. Thus let $H=(\eta_\lambda:\lambda\epsilon\Lambda)$ be a class of experiments, and let $\xi:X\epsilon\Lambda$. Then $\xi\otimes H$ is the experiment which performs ξ and then, if $X=\lambda,\eta_\lambda$. The outcomes of both ξ and η_λ are recorded. If the distribution for X in ξ is completely specified, independently of Θ, we get a <u>mixture</u> experiment. The converse case in which the distributions in η_λ are all independent of Θ corresponds to <u>post-randomisation</u>.

3. INFERENCE PATTERNS

Any sufficiently formal general method of inference should be capable of application in a wide variety of performable experimental situations, and comparison between inferences made in different experiments is an essential aspect of several of the principles of inference below. We shall therefore need to specify the family Ξ of experiments to which the method may be applied. This family is to contain the experiment actually performed, together with others which will serve as a basis of comparison. For the moment we shall not be too specific about the choice of Ξ, but reconsider this question in §5.

Now consider such a method of inference about a quantity Θ. After performing any $\xi\epsilon\Xi$, and observing the outcome X=x, this method produces an inference statement about Θ. We impose no restrictions on the <u>nature</u> of this inference. Typical instances might be, for example, "$\Theta=0$", "Θ is estimated as 17.2", "$2\leqslant\Theta\leqslant7$ with confidence 95%", or "uncertainty about Θ is expressed by the Normal distribution with mean 12 and variance 26". However, we shall ignore all semantic problems of interpretation of such inferences, and allow the range of possible inference to be an arbitrary unstructured set $\mathcal{Z}$. An inference pattern (i.p.) I is then a function having range $\mathcal{Z}$ and domain $\mathcal{S}=\{(\xi,x):\xi\varepsilon\Xi$ and $x\varepsilon\mathcal{X}$ (where $\xi:X\varepsilon\mathcal{X}$)$\}$, with the interpretation that $I(\xi,x)$ is the "inference about Θ" to be made if ξ is performed and yields outcome X=x. Specification of I requires a prior specification of Ξ. An i.p. may have a simple description, for example "Give the observed value of the minimum variance unbiased estimator of Θ in ξ", or "Quote the posterior density of Θ given X=x, for a standard normal prior distribution". Alternatively, an i.p. may be entirely ad hoc, so long as it specifies the particular inference to be made in every relevant situation.

4. PRINCIPLES OF INFERENCE

There appears to have been considerable confusion over the meaning and implications of some principles of inference, in spite of the clarity which Birnbaum (1962, 1964, 1969, 1972) brought to the subject. Birnbaum's analysis rested on an undefined concept of the "evidential meaning" provided by the pair (ξ,x) and this has come in for some criticism (e.g. Barndorff-Nielson, 1976: reply to discussion). We modify Birnbaum's structure slightly so as to escape this difficulty. We shall regard the principles as tests which a given inference pattern must undergo.

We shall only consider conformity principles which prescribe conditions under which we should require equality of two inferences $I(\xi,x)$ and $I(\xi',x')$. Such a principle will state that if (ξ,x), $(\xi',x')\varepsilon\mathcal{S}$ bear a certain relation to each other, than I should satisfy $I(\xi,x)=I(\xi',x')$. Moreover, some of the principles also demand closure properties for Ξ: that if I allows inference from an experiment ξ, it should also allow inference from certain derived experiments. If a particular principle of inference is regarded as valid, and the given i.p. I does not satisfy its requirements, then I must be discarded, or, if possible, modified so as to do so. However, if I satisfies all the relevant principles, this is not in itself to be regarded as a sufficient cause to use I.

We now turn to consider some specific principles of inference. Most of these follow Birnbaum (1962, 1972) or Basu (1975) with minor modifications. The Censoring Principle below is due to Pratt (1961, 1962). I believe the Reduction Principle to be new.

(1) Distribution Principle (DP). If $\xi,\xi'\varepsilon\Xi$, $\xi:X\varepsilon\mathcal{X}$, $X\sim P_\theta$, and $\xi':X'\varepsilon\mathcal{X}$, $X'\sim P_\theta$, then require $I(\xi,x)=I(\xi',x)$.

Informally, this says that the only aspects of an experiment which are relevant to inference are the sample space and the family of distributions over it. While implicitly accepted by most schools of inference, DP is rejected by Fraser (1968), who insists that the internal structuring of an experiment, whereby an error variable is transformed into the observation, must be taken into account. Its rejection is also implicit in Kalbfleisch's (1975) distinction between mathematical and experimental ancillaries. The following example indicates how Fraser's structural inference violates DP.

Example 1. The unknown parameter is (σ_1,τ,ϕ) $(\sigma_1,\phi>0)$, and an experiment ξ yields observation of n bivariate random variables, typically (X_1,X_2). These are produced from independent standard normal errors e,f by the operation

$$X_1 = \sigma_1 e$$
$$X_2 = \tau e+\phi f.$$

Different pairs (e,f) are independent.

This is an example of the zero-mean progression model (Fraser, 1968). The implied distribution of (X_1,X_2) is bivariate normal with mean $\underset{\sim}{0}$, $\mathrm{var}(X_1)=\sigma_1^2$, $\mathrm{var}(X_2)=\sigma_2^2$, $\mathrm{cov}(X_1,X_2)=\zeta$ say; where $\sigma_2^2=\tau^2+\phi^2$ and $\zeta=\sigma_1\tau$. Denote the probability density of all the observations by $f(\underset{\sim}{x}|\sigma_1,\tau,\phi)$. Since the right-invariant measure on the progression group has element $d\sigma_1\, d\tau\, d\phi/(\sigma_1^2\phi)$, it follows (Fraser, 1961) that the structural distribution of the parameters based on data $\underset{\sim}{x}$ has density element proportional to $f(\underset{\sim}{x}|\sigma_1,\tau,\phi)\sigma_1^{-2}\,\phi^{-1}\,d\sigma_1\,d\tau\,d\phi$.

Now consider a second experiment ξ', similar to ξ except that the typical pair (X_1',X_2') is generated by

$$X_1' = \tau'\,e'+\phi'\,f'$$
$$X_2' = \sigma_2\,e'$$

where e', f' are again independent standard normal, and $\tau' = \zeta/\sigma_2$, $\phi' = \sigma_1\phi/\sigma_2$. Given the parameters, the distribution of the data is the same in both ξ and ξ'. Now the structural analysis of ξ' yields a structural density element proportional to $f(\underset{\sim}{x}|\sigma_1,\tau,\phi)\,\sigma_2^{-2}\,\phi'^{-1}\,d\sigma_2 d\tau' d\phi'$, or equivalently $f(\underset{\sim}{x}|\sigma_1,\tau,\phi)(\tau^2+\phi^2)^{-1}\,\phi^{-1}\,d\sigma_1\,d\tau\,d\phi$. But this differs from that based on observing the identical data in ξ.

(2) Transformation Principle (TP). Let $\xi:X\varepsilon\mathcal{X}$, and let $t:\mathcal{X}\to\mathcal{Y}$ be one-to-one. Then TP requires: (i) $\xi\varepsilon\Xi \Rightarrow \xi^T\,\varepsilon\Xi$ and (ii) $I(\xi,x) = I(\xi^T,t(x))$.

This is a reasonable requirement that inference made from data should not depend on the way the data are expressed, but it is often violated in practice. A statistician may be tempted to assume normality, for example, for the data as actually presented, unless there is sufficient evidence to the contrary. If the data were to be transformed before presentation, he might well end up with a different inference.

If we insist that an i.p. I should satisfy both DP and TP, some strong consequences appear. Thus we have

Theorem 1. Let $\xi \in \Xi$, $\xi : X \in \mathcal{X}$, $X \sim P_\theta$, and let G be a group of transformations of $\mathcal{X}$ into itself which leave the model invariant, so that, if $\Theta = \theta, gX \sim P_\theta$ for all $g \in G$. Assume that I satisfies DP and TP. Then $I(\xi, x) = I(\xi, x')$ whenever x, x' are in the same orbit of G (i.e. $x = gx'$ for some $g \in G$).

Proof. Suppose $x' = gx$ ($g \in G$). Then $I(\xi, x') = I(\xi^{gX}, x')$ (DP), and $I(\xi^{gX}, x') = I(\xi^{gX}, gx)$ $= I(\xi, x)$ (TP).

Corollary. Suppose $\mathcal{X}$ is a discrete space, and $x, x' \in \mathcal{X}$ are such that $P_\theta(X=x) \equiv P_\theta(X=x')$. Then if I satisfies DP and TP, $I(\xi, x) = I(\xi, x')$. (This result is the principle of mathematical equivalence (MP): Birnbaum, 1972).

Proof. Consider the transformation $g: \mathcal{X} \to \mathcal{X}$ for which $g(x) = x'$, $g(x') = x$, and $g(y) = y$ ($y \neq x, x'$). The group generated by g leaves the problem invariant, and x, x' are in the same orbit.

In fact the above corollary does not depend on discreteness, and would appear to imply that, for continuous models, $I(\xi, x)$ should not depend on the data x at all. This objection is akin to one raised by Joshi (1976) against the Sufficiency Principle. We shall not consider such matters further, except to mention that they do not arise in discrete models, which may be taken to approximate continuous models.

We illustrate the above results with two examples.

Example 2. Let $\Theta = (\mu, \sigma^2)$ and suppose $\xi : \underset{\sim}{X} = (X_1, \ldots, X_n)'$, a random sample from the normal distribution $N(\mu, \sigma^2)$. Let G be the group of "rotations about the unit line": a typical element is an (nxn) orthogonal matrix A having $A\underset{\sim}{1} = \underset{\sim}{1}$, and acting as $\underset{\sim}{x} \to A\underset{\sim}{x}$. The model is invariant under G. Now $\underset{\sim}{x}, \underset{\sim}{y}$ belong to the same orbit under G if and only if $\Sigma x_i = \Sigma y_i$ and $\Sigma x_i^2 = \Sigma y_i^2$. If this holds, DP and TP imply that the same inference should be made whether the data in ξ be $\underset{\sim}{x}$ or $\underset{\sim}{y}$. Thus inference from $\underset{\sim}{x}$ within ξ can only depend on $(\Sigma x_i, \Sigma x_i^2)$.

Notice that the concept of sufficiency is not required in the above argument, although statistics derived in this way may be shown to be sufficient (Basu, 1970).

Example 3. Let $\Theta \in (0,1)$, and let ξ produce data (Z,X) with distribution (when $\Theta = \theta$) given by: $P(Z=1) = \pi$ $(0<\pi<1)$, $P(Z=2) = 1-\pi$; given Z=1, X has the binomial distribution $B(n;\theta)$, while given Z=2, X has the negative binomial distribution $NB(k;1-\theta)$ (k,n integers, $1 \leq k \leq n$). Then $P_\theta(X=n-k, Z=1) = \pi\binom{n}{k}\theta^{n-k}(1-\theta)^k \equiv \frac{n}{k}\left(\frac{\pi}{1-\pi}\right)P_\theta(X=n-k, Z=2)$. If now $\pi = k/(n+k)$, $P_\theta(X=n-k, Z=1) \equiv P_\theta(X=n-k, Z=2)$, so any i.p. I satisfying DP and TP must make the same inference for X=n-k whether Z=1 or Z=2.

In particular, suppose I quotes the (unique) unbiased estimator of Θ based on the conditional distributions given Z. Thus $I(\xi,(x,1))=x/n$, $I(\xi,(x,2))=x/(x+k-1)$, and $I(\xi,(n-k,1)) \neq I(\xi,(n-k,2))$. Hence I must violate either DP or TP. (Compare Birnbaum, 1972).

(3) Reduction Principle (RP). Let $\xi \in \Xi$, $\xi: X \varepsilon \mathfrak{X}$. Consider an i.p. I, and let $T=t(X)$ be a statistic satisfying the following definition.

Definition. T is reductive for I in ξ if $I(\xi,x_1)=I(\xi,x_2)$ whenever $t(x_1)=t(x_2)$. (Thus I depends on the data only through the value of T).
Then RP requires: (i) $\xi^T \in \Xi$ and (ii) $I(\xi,x)=I(\xi^T,t(x))$.

Informally RP says that, if the inference made in a given experiment depends only on a certain function of the raw data, then the same inference should be made if only that function is made available. RP differs somewhat from the other principles, in that the reductive statistics are themselves defined in terms of the particular i.p. under consideration. Clearly, RP$\Rightarrow$TP.

Example 4. Let $\Theta=(\mu,\sigma^2)$, and suppose I is "Quote $\hat{\sigma}^2$, where $(\hat{\mu},\hat{\sigma}^2)$ is the maximum likelihood estimate of (μ,σ^2)". Consider the experiment ξ of Example 2. Then $I(\xi,\underset{\sim}{x})=s/n$, with $s=s(\underset{\sim}{x})=\Sigma(x_i-\bar{x})^2$. Hence $S=s(\underset{\sim}{X})$ is reductive for I in ξ. Now ξ^S produces a model $S \sim \sigma^2\chi^2_{n-1}$, so that $I(\xi^S,s)=s/(n-1)$, and hence I violates RP. (I am grateful to John Pratt for this Example).

Example 5. (Stone and Dawid, 1972; Dawid, Stone and Zidek, 1973). Let $\xi:(X,Y)$, with density $f(x,y|\theta,\phi)=\theta\phi^2 \exp - \phi(\theta x+y)$ $(x,y>0;\theta\ ,\phi>0)$. Let $I(\xi,(x,y))$ yield the marginal posterior density of Θ, given (x,y), based on the uniform (improper) prior density $\pi(\theta,\phi) \propto 1(\theta,\phi>0)$.

Then $I(\xi,(x,y))=2z\theta/(\theta+z)^3$, where $z=y/x$, and so $Z=Y/X$ is reductive for I in ξ. Moreover, ξ^Z:Z with density $f(z|\theta,\phi)=\theta/(\theta+z)^2$ (depending on θ alone).

Although we have not specified $I(\xi^Z,z)$, a reasonable choice might be the posterior density for Θ based on data $Z=z$ and using some fixed prior density $\pi(\theta)$. This would yield $I(\xi^Z,z)=\pi(\theta)f(z|\theta)/g(z)=\pi(\theta)\theta/\{(\theta+z)^2g(z)\}$, with $g(z)$ chosen to normalize the distribution. We find that it is impossible to have $I(\xi,(x,y)) \equiv I(\xi^Z,y/x)$, no matter what choice of $\pi(\theta)$ is used. So, in an intrinsic way, the improper Bayes inference $I(\xi,(x,y))$ violates RP.

(4) Weak Sufficiency Principle (WSP). For $\xi \in \Xi$, WSP requires $I(\xi,x_1)=I(\xi,x_2)$ if there exists a statistic $T=t(X)$ which is sufficient for Θ in ξ, and having $t(x_1)=t(x_2)$. Thus WSP requires that no distinction be made between observations which yield the same value for a sufficient statistic.

The existence, in suitably regular models, of a minimal sufficient statistic $T^*=t^*(X)$ in ξ leads to a simpler statement of WSP: require $I(\xi,x_1)=I(\xi,x_2)$ if $t^*(x_1)=t^*(x_2)$. An alternative description is: require $I(\xi,x_1)=I(\xi,x_2)$ if (ξ,x_1) and (ξ,x_2) yield proportional likelihood functions for Θ (Lehmann and Scheffé, 1950).

In Example 3, WSP implies $I(\xi,(n-k,1))=I(\xi,(n-k,2))$, whatever be the value of π, so that the i.p. I of that example violates WSP for any π.

(5) Strong Sufficiency Principle (SSP). For $\xi \in \Xi$, $T=t(X)$ sufficient for Θ in ξ, SSP requires: (i) $\xi^T \in \Xi$ and (ii) $I(\xi,x)=I(\xi^T,t(x))$. This says that inference should be unchanged if only the value of a sufficient statistic is made available.

For example, suppose Ξ satisfies (i) of SSP and the minimal sufficient statistic for any $\xi \in \Xi$ is complete. Let $I(\xi,x)$ give the value of the (unique) minimum variance unbiased estimator of Θ. Then, by the Rao-Blackwell theorem, I satisfies SSP.

We have SSP $\Rightarrow$ WSP, TP, while (WSP$\wedge$RP) $\Rightarrow$ SSP.

Reduction by sufficiency. Take Ξ to satisfy (i) of SSP. Suppose an i.p. J is found to violate SSP. A simple modification leads to a new i.p. I, defined by $I(\xi,x)=J(\xi^{T^*},t^*(x))$, where $T^*=t^*(X)$ is minimal sufficient in ξ. (This is unambiguous so long as J satisfies TP.) Thus I is "Apply J to the model based on the minimal sufficient statistic", and it is easily shown that I will satisfy SSP. Note that this method operates on the underlying experiment, reduction (not to be confused with RP) meaning the replacement of ξ by a new frame of reference, here ξ^{T^*}. In particular, no restriction is put on the nature of the inference to be made in ξ^{T^*}, which may involve, for example, sample-space concepts such as significance level and unbiasedness.

(6) Ancillarity Principle (AP). Let $S=s(X)$ be ancillary for Θ in ξ. Then AP requires: (i) $\xi_{S=s} \in \Xi$, and (ii) $I(\xi,x)=I(\xi_{S=s(x)},x)$.

Like the sufficiency principles, AP requires that no account be taken, for inference about Θ, of variation that bears no relationship to the value of Θ.

In certain very special circumstances, it may happen that, for every $\xi \in \Xi$ there exists a maximal ancillary statistic $S^*=s^*(X)$, such that any other ancillary is a function of S^*. If so, a reduction by conditioning, exactly parallel to reduction by sufficiency, may be effected. Suppose Ξ satisfies (i) of AP. Given an i.p. J, define $I(\xi,x)=J(\xi_{S^*=s^*(x)},x)$, that is, apply J to the model conditioned on the maximal ancillary. Then I satisfies AP, and any form of inference in $\xi_{S^*=s^*}$ is admissible.

Normally there will be no maximal ancillary (Basu, 1959), and this approach fails. But another modification is sometimes available. Suppose that, for each $\xi \in \Xi$, the minimal sufficient statistic T^* is boundedly complete. Then Basu (1955) shows that, given Θ, T^* is independent of any ancillary S, so that $\xi^{T^*}_{S=s}$ and ξ^{T^*} both yield the same distribution for T^*. So, (assuming J satisfies DP), the reduction by sufficiency to T^* will unambiguously produce an i.p. that automatically satisfies AP.

(7) Conditionality Principle (CP). Let $\xi_1, \xi_2 \in \Xi$ and let ξ be the mixture $p_1 \xi_1 + p_2 \xi_2$ obtained by randomly choosing λ with known probability $p_\lambda (p_1+p_2=1)$, and then performing ξ_λ. Then CP requires: (i) $\xi \in \Xi$ and (ii) $I(\xi, (\xi_\lambda, x)) = I(\xi_\lambda, x)$.

Note that CP is implied by AP if and only if Ξ is closed under mixtures.

(8) Censoring Principle (CeP). Let $\xi : X \in \mathcal{X}$, let $A \subseteq \mathcal{X}$, $\phi \notin \mathcal{X}$. Define $t : \mathcal{X} \to A \cup \{\phi\}$ by $t(x) = x (x \in A)$, $t(x) = \phi$ $(x \notin A)$, and let $T = t(X)$. Then CeP requires: (i) $\xi \in \Xi \Rightarrow \xi^T \in \Xi$, and (ii) for $x \in A$, $I(\xi, x) = I(\xi^T, x)$.

(9) Likelihood Principle (LP). Let $f(x|\theta)$, $f'(x'|\theta)$ be the likelihood functions for Θ based respectively on (ξ, x), (ξ', x'). Then LP requires: if $\xi, \xi' \in \Xi$ and $f(x|\theta) \equiv cf'(x'|\theta)$ $(c>0)$, then $I(\xi, x) = I(\xi', x')$.

Although LP does not, perhaps, have much immediate appeal, it derives its importance from Birnbaum's (1962) result that, if I satisfies both WSP and CP, then I must satisfy LP. Equivalently, if Ξ is closed under mixtures, then $WSP \wedge AP \Rightarrow LP$. Under the same condition, Birnbaum (1972) shows that $MP \wedge AP \Rightarrow LP$. An alternative route is via CeP: if Ξ is closed under post-randomisation, and I satisfies SSP and CeP, then I satisfies LP (Birnbaum, 1964).

These results are unpalatable to many statisticians, who would like to be able to apply the method of reduction so as to satisfy SSP and AP, say. This would then be applicable to any i.p., and in particular to one that proceeds by integration over the sample space. We have seen this to be possible in some special cases, but it cannot be so in general. For if Ξ is closed under mixtures, then SSP and AP together imply LP, and the links between entirely different experiments in Ξ provided by LP are at odds with the wish to use an arbitrary i.p. within each reduced experiment.

All the other principles, except RP, follow from LP. Bayesian inference satisfies LP if and only if the prior distribution for Θ does not depend on ξ. Note that LP does not require that inference proceed by "examination of the likelihood function". In particular, if Θ and Φ are two different quantities for which we have proportional likelihood functions, there need be no connection between the inferences made for Θ and Φ. Furthermore, LP, in this form, does not say that data yielding equal likelihood values for $\Theta=\theta_1$ and $\Theta=\theta_2$ provide equal "support" for θ_1 and θ_2. Thus Fraser's (1976) interpretation of an example of Stone (1976), as a "counter-example to the likelihood principle", is not relevant here.

5. EVADING THE LIKELIHOOD PRINCIPLE

We concentrate on Birnbaum's original result that, if Ξ is closed under mixtures then WSP$\wedge$AP$\Rightarrow$LP. (Similar considerations apply for the results MP$\wedge$AP$\Rightarrow$LP, SSP$\wedge$CeP$\Rightarrow$LP). There are two possible escape routes from the implications of this result. The first is to discard either WSP or AP, perhaps replacing these by something weaker.

This was the approach of Durbin (1970), who weakened AP so as to apply only for ancillary functions of the minimal sufficient statistic. A different attack of this kind is due to Kalbfleisch (1975). Kalbfleisch distinguished between experimental ancillaries, which are ancillary by virtue of the physical set-up of the experiment, and mathematical ancillaries which are so by virtue of assumptions about the distributions $\{P_\theta\}$. He suggested that WSP apply only in minimal experiments, for which there are no experimental ancillaries.

An entirely different approach is to discard the closure properties of Ξ necessary for the argument, so that, for example, we do not admit mixtures of experiments in Ξ as candidates for the application of our i.p. This raises the question "What is Ξ ?" If we perform ξ_0 and observe x_0, what is the appropriate basis for comparison? If we are tempted to make a particular inference $I(\xi_0,x_0)=I_0$, the principles tell us that, for certain hypothetical situations (ξ,x) $(\xi \in \Xi)$, we should also infer I_0, and the test of $I(\xi_0,x_0)=I_0$ is whether, in good conscience, we would agree to do so. The smaller Ξ, the weaker the implications of the various principles.

It might be reasonable to restrict Ξ to experiments derived from (ξ_0,x_0), for example of the form $\xi_{0\ ,S=s(x_0)}^{T}$. In a sense, every such experiment has actually been performed, so the comparison of such situations is not too hypothetical. Suppose we consider AP and SSP, and further require T to be sufficient and S ancillary in ξ_0. Although $\Xi' = \{\xi_{0\ ,S=s}^{T}\}$ is not closed under mixtures, it is easily seen that SSP$\wedge$AP$\Rightarrow$LP in Ξ'. However, LP may not be very restrictive if Ξ' is small enough. For example, if there exists a boundedly complete sufficient statistic T^* in ξ_0, then any inference using the observed value of T^* and its sampling distributions satisfies LP (within Ξ').

Unfortunately this last condition is very special, and fails if, for example ξ_0 is itself a mixture experiment. In that case, a weakening of SSP as suggested by

Kalbfleisch (1975) may be brought to bear. Suppose that, after conditioning on the experimental ancillaries, there exists a boundedly complete sufficient statistic T^*. Then any inference from the conditional model for T^* is acceptable. This is equivalent to redefining ξ_0 by conditioning on experimental ancillaries.

It might be thought that this idea could lead to a mathematical formulation of the intuitive notion of experimental ancillary - one for which the above condition holds - but this is easily seen to be non-unique. Indeed, any purely mathematical definition would be in accord with DP, and so allow the reasoning of Example 3 (with $\pi = k/(n+k)$) to rule out the conditionally unbiased estimator of Θ. But the structure of ξ might well be such that Z can be regarded (intuitively) as an experimental ancillary, which would permit this i.p.

It is clear that Kalbfleisch's method cannot, in general, yield a satisfactory reduction applicable to any i.p., since a "minimal experiment" can itself have an arbitrary probability structure. It does not appear possible to find any method of reduction that will work for any Ξ', although we have seen that special methods may work for special cases.

6. BEYOND REDUCTIONISM

The idea that the principles of inference can always be satisfied by the method of reduction is widespread. If this were true, then we could start with any well-favoured i.p., and, by applying it to a suitably reduced experiment, derive an acceptable modification. Unfortunately, this is not generally possible.

The following example shows how an apparently reasonable method of reduction, motivated by AP and SSP, need not necessarily produce an i.p. that will satisfy these principles. Suppose that, for each $\xi \in \Xi$, the minimal sufficient statistic T^* has the property that there exists a maximal ancillary S^* within ξ^{T^*}. (This is trivially so if there are no ancillaries based on T^*). Then it appears reasonable to reduce ξ to $\xi^{T^*}_{S^*=s^*}$, so that any i.p. J is replaced by $I(\xi,x)=J(\xi^{T^*}_{S^*=s^*},t^*)$, where $X=x \Rightarrow T^*=t^*,\ S^*=s^*$.

Example 6. Consider $\xi_0=\xi$ of Example 3 with $\pi = \frac{1}{2}, k=2, n=4$, and suppose Ξ contains $\{\xi_0, \xi_0^T, \xi_{0,Z=1}, \xi_{0,Z=2}\}$. Here $T=t(X,Z)$ is minimal sufficient in ξ, and may be taken as : $t(x,z) = (x,z)\ (x \neq 2)$; $t(2,1) = t(2,2) = 0$. Within $\xi_{0,Z=z}$, T and X are equivalent. It is shown in Dawid (1975) that there are no non-trivial ancillaries in ξ_0^T, and this also hold for $\xi_{0,Z=1}$, $\xi_{0,Z=2}$, for which no further reduction by sufficiency is needed. Thus the proposed reduction is available, as $I(\xi,x) = J(\xi^{T^*},t^*)\ (\xi \in \Xi)$.

Let J be given by the ordinate test of $\Theta = \frac{1}{2}$ (Barndorff-Nielson, 1976). Thus

$J(\xi,x) = P\{p(X) \leqslant p(x)\}$, where p(.) is the probability mass function and P the distribution for X in ξ when $\Theta=\frac{1}{2}$. In this unreduced form, J satisfies neither WSP nor AP. We find $J(\xi_0,(4,1))=15/128$, $J(\xi_{0,Z=1},4)=16/128$.

The proposed reduction may be seen to leave both these inferences unchanged, and so, in particular, I continues to violate AP. The moral is that it may not be possible to satisfy the conformity principles by unambiguously specifying a method for experimental reduction and using an arbitrary i.p. in the reduced experiment. If the principles are to be taken seriously, they severely delimit the kind of i.p. which may be used.

7. DISCUSSION

It is not only the "deep" principles of inference, such as WSP and AP, that lead to conclusions unpalatable to many statisticians, as Example 3 makes clear. Some standard inference patterns can be ruled out by DP and TP alone. It appears to me that any attempt to modify the principles, so as to allow such excluded inferences, must start with DP. Most statisticians would feel, when they have given the set of possible distributions for the data for different values of Θ, they have described an experiment in sufficient detail. However, this attitude is hard to justify, and it may well be that additional information is needed to produce valid inferences. Fraser introduces such information into his structural models, while Kalbfleisch depends on outside considerations to distinguish experimental from mathematical ancillaries. Once such a stand is taken, many of the principles presented here may be regarded as irrelevant, and satisfactory weaker forms introduced. Unfortunately, Fraser's methods are applicable only in very special kinds of experiments, while Kalbfleisch's suggestions are vague and do not always lead to unambiguous results. Perhaps a satisfactory general method cannot be found. Nevertheless, it is surely valuable to think of the practical (rather than merely probabilistic) structure of an experiment as relevant for inference.

REFERENCES

Barndorff-Nielson, O. (1976). Plausibility inference (with Discussion). J. Roy. Statist. Soc. B, 38, 103-31.

Basu, D. (1955). On statistics independent of a complete sufficient statistic. Sankhyā, 15, 377-80.

Basu, D. (1959). The family of ancillary statistics. Sankhyā, 21, 247-56.

Basu, D. (1970). On sufficiency and invariance. Chapter 4 of Bose et al. (1970).

Basu, D. (1975). Statistical information and likelihood (with Discussion). Sankhyā A, 37, 1-71.

Birnbaum, A. (1962). On the foundations of statistical inference (with Discussion). J.Amer. Statist. Ass., 57, 269-326.

Birnbaum, A. (1964). The anomalous concept of statistical evidence. Courant Inst. of Math. Sci., technical report IMM-NYU 332.

Birnbaum, A. (1969). Concepts of statistical evidence. In Morgenbesser et al., (1969), 112.

Birnbaum, A. (1972). More on concepts of statistical evidence. J. Amer. Statist. Assoc. 67, 858-61.

Bose, R.C., Chakravarti, I.M., Mahalanobis, P.C., Rao, C.R., and Smith, K.J.C. (1970). Essays in Probability and Statistics. University of North Carolina Press, Chapel Hill.

Dawid, A.P. (1975). On the concepts of sufficiency and ancillarity in the presence of nuisance parameters. J.R. Statist. Soc. B, 37, 248-58.

Dawid, A.P., Stone, M., and Zidek, J.V. (1973). Marginalization paradoxes in Bayesian and structural inference (with Discussion). J. Roy. Statist. Soc. B, 35, 189-233.

Durbin, J. (1970). On Birnbaum's theorem on the relation between sufficiency, conditionality and likelihood. J. Amer. Statist. Ass., 65, 395-8.

Fraser, D.A.S. (1961). The fiducial method and invariance. Biometrika, 48, 261-80.

Fraser, D.A.S. (1968). The Structure of Inference. J. Wiley & Sons, New York.

Fraser, D.A.S. (1976). Comment on Stone (1976). J. Amer. Statist. Ass., 71, 122-3.

Joshi, V.M. (1976). A note on Birnbaum's theory of the likelihood principle. J. Amer. Statist. Ass., 71, 345-6.

Kalbfleisch, J.D. (1975). Sufficiency and conditionality. Biometrika, 62, 251-9.

Lehmann, E.L., and Scheffé, H. (1950). Completeness, similar regions and unbiased tests, Part I. Sankhyā, 10, 305-40.

Morgenbesser, S., Suppes, P., and White, M. (1969). Essays in Honor of Ernest Nagel. St. Martin's Press, New York.

Pratt, J.W. (1961). Review of Testing Statistical Hypotheses by E.L. Lehmann (Wiley, 1959). J. Amer. Statist. Ass., 56, 163-7.

Pratt, J.W. (1962). Comments on A. Birnbaum's "On the foundations of statistical inference". J. Amer. Statist. Ass., 57, 314-5.

Stone, M. (1976). Strong inconsistency from uniform priors. J. Amer. Statist Ass., 71, 114-116.

Stone, M., and Dawid, A.P. (1972). Un-Bayesian implications of improper Bayes inference in routine statistical problems. Biometrika, 59, 369-75.

Recent Developments in Statistics
J.R.Barra et al., editors

Invited Paper

A BAYESIAN ANALYSIS OF SOME TIME-VARYING MODELS

A.F.M. Smith

Department of Statistics and Computer Science
University College London

A Bayesian approach to the stability of regression relationships over time is considered for the case in which at most one change in relationship occurs. The method of analysis, based upon posterior probabilities of alternative linear models, is shown to extend straightforwardly to cases of serially correlated errors, autoregressive and moving average processes. Numerical illustrations are provided.

1. INTRODUCTION

We shall consider a Bayesian approach to the problem of investigating the stability over time of the regression model

$$y_t = \underline{x}'_t \underline{\beta}^{(t)} + \varepsilon_t , \qquad t = 1,\ldots,T, \tag{1}$$

where, at time t, y_t is the observation on the dependent variable, $\underline{x}_t$ is the column vector of observations on k regressor variables (including, possibly, a constant), $\underline{\beta}^{(t)}$ is the column vector of unknown regression coefficients and ε_t is the error term, assumed normally distributed with mean zero and unknown variance σ^2.

Initially, we shall assume the regressor variables to be non-stochastic and the errors to be independent. In fact, we shall see later that the approach we shall adopt turns out to be easily extended to more general situations. In particular, it will enable us to deal straightforwardly with serially correlated errors, autoregressive and integrated moving average processes.

A recent, important contribution to the study of model (1) is that of Brown, Durbin and Evans (1975), who derive significance tests for the null hypothesis of equality of the $\underline{\beta}^{(t)}$, $t = 1,\ldots,T$. The test procedures given by these authors are based on cusums of recursive residuals, and provide an omnibus screening for a wide range of departures from the null hypothesis.

In this present paper, we shall be concerned with the special case of situations in which it may be assumed that the regression relationship has either remained stable, or has undergone a single abrupt change at some unknown time point $t = r$, $1<r<T$. In other words, either

$$\underline{\beta}^{(1)} = \ldots = \underline{\beta}^{(T)} = \underline{\beta}_1 \, , \quad \text{or}$$

$$\underline{\beta}^{(1)} = \ldots = \underline{\beta}^{(r)} = \underline{\beta}_1 \quad \text{and} \quad \underline{\beta}^{(r+1)} = \ldots = \underline{\beta}^{(T)} = \underline{\beta}_2 \, ,$$

with r, $\underline{\beta}_1$ and $\underline{\beta}_2$ unknown. For this special case, we shall derive a Bayesian alternative to the Brown, Durbin and Evans approach: we shall then show that the Bayesian analysis enables us to go beyond the restrictive assumptions of independent errors and non-stochastic regressor variables (assumptions which cannot easily be dropped in the non-Bayesian analysis).

2. A BAYESIAN APPROACH

2.1 Mathematical Formulation

If we adopt the notation

$$\underline{y}'_r = (y_1, \ldots, y_r) \, , \; \underline{\tilde{y}}'_{T-r} = (y_{r+1}, \ldots, y_T) \, ,$$

$$\underline{X}'_r = (\underline{x}_1, \ldots, \underline{x}_r) \, , \; \underline{\tilde{X}}'_{T-r} = (\underline{x}_{r+1}, \ldots, \underline{x}_T) \, ,$$

we see that, if an abrupt change occurs at time $t = r$, the regression model defined by (1) may be rewritten in the form

$$\underline{y} \sim N(\underline{A}_r \underline{\beta} \, , \, \sigma^2 \underline{I}_T) \, , \tag{2}$$

where

$$\underline{A}_r = \begin{bmatrix} \underline{X}_r & 0 \\ 0 & \underline{\tilde{X}}_{T-r} \end{bmatrix} , \quad \underline{\beta} = \begin{bmatrix} \underline{\beta}_1 \\ \underline{\beta}_2 \end{bmatrix} , \tag{3}$$

$\underline{y} = \underline{y}_T$ and $\underline{I}_T$ is the $T \times T$ identity matrix. If no change occurs, that is if $r = T$, then (2) still holds, but with $\underline{A}_T = \underline{X}_T$ and $\underline{\beta} = \underline{\beta}_1$.

The linear model defined by (2) will be denoted by M_r, and we shall investigate the set of alternative linear models corresponding to $k<r<T-k$ and $r = T$. Inference about the position of the change (if any) is then equivalent to comparing the relative probabilities of these models. This is achieved by means of Bayes theorem, which states that

$$p(M_r | \underline{y}) = \frac{p(\underline{y}|M_r)p(M_r)}{\sum_r p(\underline{y}|M_r)p(M_r)} \, , \tag{4}$$

where

$$p(\underline{y}|M_r) = \int\cdots\int p(\underline{y}|M_r, \underline{\beta}, \sigma^2)p(\underline{\beta}, \sigma^2|M_r)\, d\underline{\beta}d\sigma^2 , \tag{5}$$

and $p(\underline{\beta}, \sigma^2|M_r)$ is the joint prior density for $\underline{\beta}$ and σ^2 under the assumption of M_r, $p(M_r)$ being the prior probability mass on Model M_r. In fact, we shall use the joint prior form

$$p(\underline{\beta}, \sigma^2|M_r)\, d\underline{\beta}d\sigma^2 = (2\pi)^{-p/2}\sigma^{-2}d\underline{\beta}d\sigma^2 , \tag{6}$$

where $p = 2k$ if $k<r<T-k$, $p = k$ if $r = T$. This is simply the usual improper prior element for the normal linear model, modified by a weight factor, $(2\pi)^{-p/2}$, reflecting the number of parameters in model M_r. The problem of choosing an appropriate weight factor in a situation like this is one which is largely unresolved, and we have nothing to contribute at the present time that goes beyond the ad hoc rationale advanced by Halpern (1972).

If $\hat{\underline{\beta}}_{(r)} = (\underline{A}'_r\underline{A}_r)^{-1}\underline{A}'_r\underline{y}$ denotes the least squares estimate of $\underline{\beta}$ in M_r, and $S_r = (\underline{y} - \underline{A}_r\hat{\underline{\beta}}_{(r)})'(\underline{y} - \underline{A}_r\hat{\underline{\beta}}_{(r)})$ denotes the corresponding residual sum of squares, then, omitting irrelevant constants,

$$p(\underline{y}|M_r, \underline{\beta}, \sigma^2) \propto (\sigma^2)^{-T/2}\exp\left\{-\frac{1}{2\sigma^2}\left[S_r + (\underline{\beta} - \hat{\underline{\beta}}_{(r)})'\underline{A}'_r\underline{A}_r(\underline{\beta} - \hat{\underline{\beta}}_{(r)})\right]\right\}. \tag{7}$$

Multiplying this by (6), and integrating with respect to $\underline{\beta}$ and σ^2, we can see that (4) can be summarized in the form

$$p(M_r|\underline{y}) \propto \Gamma\left(\frac{T-p}{2}\right)|\underline{A}'_r\underline{A}_r|^{-\frac{1}{2}}\left(\frac{S_r}{2}\right)^{-\left(\frac{T-p}{2}\right)} p(M_r) , \tag{8}$$

where $\Gamma(.)$ denotes the standard gamma function. If $\tilde{S}_r$, $\tilde{\tilde{S}}_{T-r}$ denote the residual sums of squares from a least squares fit to the first r and last $T-r$ observations, respectively, then, from (8), the ratio $p(M_r|\underline{y})/p(M_r)$ is easily seen to be proportional to

$$\begin{aligned} &\Gamma\left(\frac{T-2k}{2}\right) |\underline{X}'_r\underline{X}_r|^{-\frac{1}{2}}|\tilde{\underline{X}}'_{T-r}\tilde{\underline{X}}_{T-r}|^{-\frac{1}{2}}\left(\frac{\tilde{S}_r + \tilde{\tilde{S}}_{T-r}}{2}\right)^{-\left(\frac{T-2k}{2}\right)}, \quad k<r<T-k ,\\ &\Gamma\left(\frac{T-k}{2}\right) |\underline{X}'_T\underline{X}_T|^{-\frac{1}{2}}\left(\frac{S_T}{2}\right)^{-\left(\frac{T-k}{2}\right)}, \quad r = T , \end{aligned} \tag{9}$$

the first line of which generalizes an expression given by Ferreira (1975, equation 3.4) for the special case of a switching straight line regression.

We note that the computation of the quantities required in (9) is greatly facilitated by exploiting well known recursive relationships such as

$$|\underline{X}'_r\underline{X}_r| = |\underline{X}'_{r-1}\underline{X}_{r-1}| \cdot (1 + \underline{x}'_r(\underline{X}'_{r-1}\underline{X}_{r-1})^{-1}\underline{x}_r) . \tag{10}$$

A number of other useful results of this type are summarized in the paper by Brown, Durbin and Evans (1975; equations (2), (3), (4) and (5)).

Inferences about the parameters $\underline{\beta}_1$, $\underline{\beta}_2$ and σ^2, or predictive distributions for future observations, follow straightforwardly from standard Bayesian manipulations and the details will not be given here. Basically, we obtain weighted averages, with weights given by the $p(M_r|\underline{y})$, of the well-known forms derived from normal theory conditional on the assumption of a particular model M_r.

So far as related Bayesian work is concerned, we note that applications to changing sequences of random variables in non-regression contexts have been considered by, for example, Broemeling (1974) and Smith (1975), and that the general problem of posterior probabilities for alternative linear models has been studied by Lempers (1971). A number of useful, general references to the change point problem are given in the paper by Brown, Durbin and Evans, and in the published discussion of that paper.

2.2 An example

Cameron and Nash (1974) investigated a series of 43 quarterly observations relating to the staff requirements of an organization expressed as a function of workload. The model eventually considered took the form of a multiple regression equation with a constant term and three variables: in our notation, $k = 4$, $T = 43$. The data was kindly made available by the Civil Service Department in London, and an application of (9), with $p(M_T) = \frac{1}{2}$ and the remaining prior probability divided equally over the range $4<r<39$, resulted in the posterior probabilities as shown in Table 1 (those less than 10^{-5} being omitted).

TABLE 1

r	26	27	28	29	30	31
$P(M_r\|\underline{y})$	.00002	.00178	.89494	.10285	.00037	.00003

A change in the regression relationship after the 28th quarter is strongly indicated, and this conclusion is robust against a wide range of specifications of $p(M_r)$. In this case, the inference based upon posterior probabilities agrees with that obtained by Brown, Durbin and Evans (1975; Section 3, Example 3) using a plot of the log-likelihood ratio of Quandt (1958, 1960). In fact, it is now known that, in this case, a major administrative change was introduced during the 27th quarter.

In addition to the retrospective analysis of a series, we can use the Bayesian approach to monitor a series "on-line". We shall illustrate this using the Cameron-Nash data by considering, after 18, 22, 26, 30 observations, and so on, the current ratio of posterior to prior probability on "no change". For the case in which we assign equal prior probabilities to each of the actual possible change points, we obtain the approximate results shown in Table 2.

TABLE 2

T	18	22	26	30	> 30
$\approx p(M_T\|\underline{y})/p(M_T)$	17	17	10	3×10^{-4}	$< 10^{-5}$

2.3 Serially correlated errors

Suppose we now weaken the assumptions of model (1) and consider errors ε_t generated by a first order autoregressive process with parameter ρ, $|\rho|<|$. Suppose further that we denote the observations by z_t , and assume the process to begin at $t = 0$ with known z_o (for a general discussion of possible initial conditions, see Zellner and Tiao, 1964).

It follows that if we define $y_t = z_t - \rho z_{t-1}$, $t = 1,\ldots,T$, and write $\underline{y}'(\rho) = (y_1,\ldots,y_T)$, then, for given ρ ,

$$\underline{y}(\rho) \sim N(\underline{A}_r(\rho)\underline{\beta} \; , \; \sigma^2\underline{I}_T) \; , \tag{11}$$

where, for $k<r<T-k$, we have $\underline{\beta}' = (\underline{\beta}_1, \underline{\beta}_2)$ and

$$\underline{A}_r(\rho) = \begin{bmatrix} \underline{X}_r - \rho\underline{X}^*_r & 0 \\ -\rho\underline{x}'_r & \underline{x}'_{r+1} \\ 0 & \underline{\tilde{X}}_{T-r+1} - \rho\underline{\tilde{X}}^*_{T-r+1} \end{bmatrix} \tag{12}$$

with $(\underline{X}^*_r)' = (\underline{x}_o,\ldots,\underline{x}_{r-1})$, $(\underline{\tilde{X}}^*_{T-r+1})' = (\underline{x}_r,\ldots,\underline{x}_{T-1})$; for $r = T$, we have $\underline{\beta} = \underline{\beta}_1$ and $\underline{A}_T(\rho) = \underline{X}_T - \rho\underline{X}^*_T$.

Noting that the Jacobian of the transformation from $\underline{z}$ to $\underline{y}$ is unity (with z_o constant), and denoting by $p(\rho)$ the prior density for ρ , a straightforward extension of the analysis of the previous section shows that (8) is now replaced by

$$p(M_r, \rho|\underline{z}) \propto \Gamma\left(\frac{T-p}{2}\right)|\underline{A}'_r(\rho)\underline{A}_r(\rho)|^{-\frac{1}{2}} \left[\frac{S_r(\rho)}{2}\right]^{-\frac{(T-p)}{2}} p(M_r)p(\rho) \; , \tag{13}$$

where $S_r(\rho)$ is the residual sum of squares resulting from a least squares analysis of (11). The required marginal densities are easily obtained from (13): integration with respect to ρ can be well approximated by taking a suitable discretization of the range -1 to $+1$.

3. SHIFT IN A FIRST ORDER AUTOREGRESSIVE PROCESS

3.1 The model

Let us suppose that

$$\begin{aligned} z_1 &= L + \varepsilon_1 \\ z_t &= L + \rho \sum_{s=1}^{t-1} \rho^{s-1} \varepsilon_{t-s} + \varepsilon_t , \qquad t = 2,\ldots,r, \\ z_t &= L + \delta + \rho \sum_{s=1}^{t-1} \rho^{s-1} \varepsilon_{t-s} + \varepsilon_t , \qquad t = r+1,\ldots,T, \end{aligned} \tag{14}$$

with L, δ, ρ and r unknown $(|\rho| < 1)$, and $\varepsilon_t \sim N(o, \sigma^2)$, independent, with σ^2 unknown. In other words, we assume a stationary first order autoregressive process with mean L and an unknown shift δ occurring immediately after an unknown time point $t = r$. The same model was considered by Box and Tiao (1965) in order to investigate a possible change at a specified time point.

Following Box and Tiao, if we make the transformations $y_1 = z_1$, $y_t = z_t - \rho z_{t-1}$, $t = 2,\ldots,T$, then $\underline{y}'(\rho) = (y_1,\ldots,y_T)$ again satisfies (11), but with, for $1<r<T-1$, $\underline{\beta}' = (L, \delta)$ and

$$\underline{A}'_r(\rho) = \left[\begin{array}{cccc:cccc} 1 & (1-\rho) & \ldots & (1-\rho) & (1-\rho) & \ldots & \ldots & (1-\rho) \\ 0 & 0 & \ldots & 0 & 1 & (1-\rho) & \ldots & (1-\rho) \end{array}\right] , \tag{15}$$

$$\longleftarrow r \longrightarrow \quad \longleftarrow T-r \longrightarrow$$

whereas, for $r = T$, $\underline{\beta} = L$ and $\underline{A}'_T(\rho)$ consists only of the first row of (15). It follows that inference concerning r and ρ can again be based upon (13), taking $p = 1$ if $r = T$, $p = 2$ otherwise.

3.2 An example

Table 3 displays twenty simulated observations, independent and normally distributed with unit variance, the first ten having mean zero, the remainder mean one.

TABLE 3

$(T = 20\ ,\ r = 10\ ;\ L = 0\ ,\ \delta = 1\ ,\ \rho = 0\ ,\ \sigma^2 = 1)$

t	1	2	3	4	5	6	7	8	9	10
z_t	0.44	0.91	-0.02	-1.42	1.26	-1.02	-0.81	1.66	1.05	0.97
t	11	12	13	14	15	16	17	18	19	20
z_t	2.14	1.22	-0.24	1.60	0.72	-0.12	0.44	0.03	0.66	0.56

Figure 1 shows the resulting posterior distribution for ρ , calculated using a uniform prior distribution over 39 equally spaced points, together with a uniform prior over the set of possible change points.

FIGURE 1

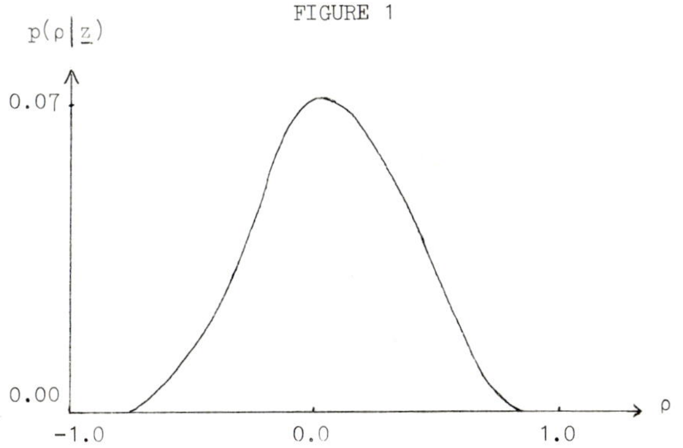

Table 4 summarizes part of the posterior distribution over the possible change points, both for the marginal distribution and for that conditional on the assumption $\rho = 0$

TABLE 4

r	4	5	6	7	8	9	10	11	12	13	14
$p(M_r\|\underline{z})$	.059	.037	.079	.330	.088	.051	.040	.026	.027	.030	.028
$p(M_r\|\underline{z}, \rho = 0)$	.058	.034	.084	.322	.086	.057	.044	.025	.025	.029	.028

We note that the marked closeness between the marginal distribution and that conditional on the mode for ρ appears, from this and other examples we have studied, to be not untypical in this situation.

4. SHIFT IN AN INTEGRATED MOVING AVERAGE PROCESS

4.1 The model

Let us suppose that

$$\begin{aligned} z_1 &= L + \varepsilon_1 , \\ z_t &= L + \gamma \sum_{s=1}^{t-1} \varepsilon_{t-s} + \varepsilon_t , \qquad t = 2,\ldots,r, \\ z_t &= L + \delta + \gamma \sum_{s=1}^{t-1} \varepsilon_{t-s} + \varepsilon_t , \qquad t = r+1,\ldots,T, \end{aligned} \tag{16}$$

where γ is assumed known, L, δ and r are unknown, and $\varepsilon_t \sim N(0, \sigma^2)$, independent, with σ^2 unknown. Again, we are considering a model used by Box and Tiao (1965) to test for a shift at a <u>specified</u> time point.

Following Box and Tiao, if we make the transformation

$$\begin{aligned} y_1 &= z_1 \\ y_t &= z_t - \gamma \sum_{s=o}^{t-2} (1-\gamma)^s z_{t-s-1} , \qquad t = 2,\ldots,T, \end{aligned} \tag{17}$$

then, for $1<r<T-1$, $\underline{y}$ satisfies (2) with $\underline{\beta}' = (L, \delta)$ and

$$\underline{A}'_r = \begin{bmatrix} 1 & (1-\gamma) & \ldots & (1-\gamma)^{r-1} & (1-\gamma)^r \ldots\ldots & (1-\gamma)^{T-1} \\ 0 & 0 & \ldots & 0 & 1 \quad (1-\gamma) \ldots & (1-\gamma)^{T-r-1} \end{bmatrix}, \tag{18}$$

whereas, for $r = T$, $\underline{\beta} = L$ and $\underline{A}'_T$ consists solely of the first row of (18). The results of section 2 can now be applied directly, and the methods of sections 2.3 and 3 can be adapted to deal with the case of unknown γ.

4.2 An example

Box and Tiao (1965) present 50 observations simulated from a process with $\gamma = 0.3$, $L = 5$, $\delta = 1.5$, $\sigma^2 = 1$ and $T = 50$, $r = 25$. Figure 2 displays these observations, together with the posterior probabilities for the M_r. In Figure 3, we present the results obtained from the same basic data set, but with a shift of $\delta = 2.0$. In each case, a uniform prior was taken over the set of possible change points.

FIGURE 2

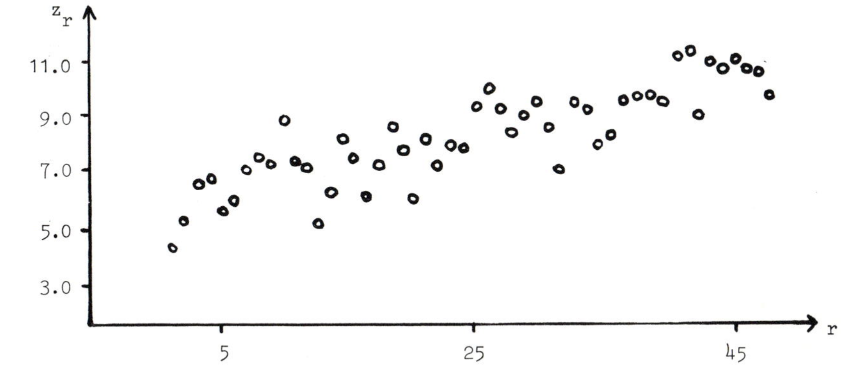

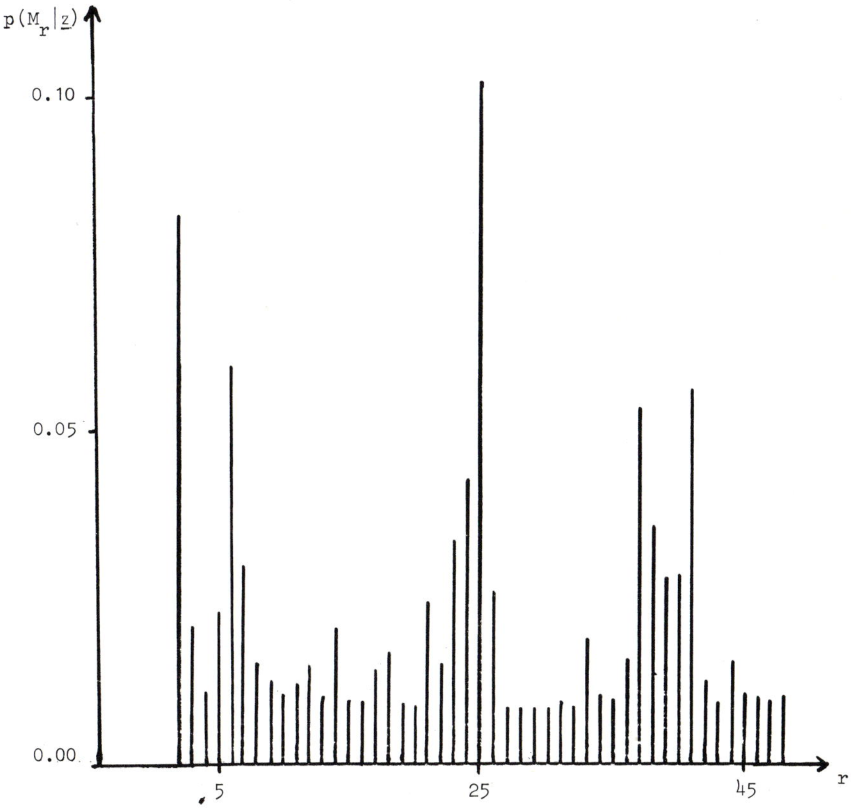

FIGURE 3

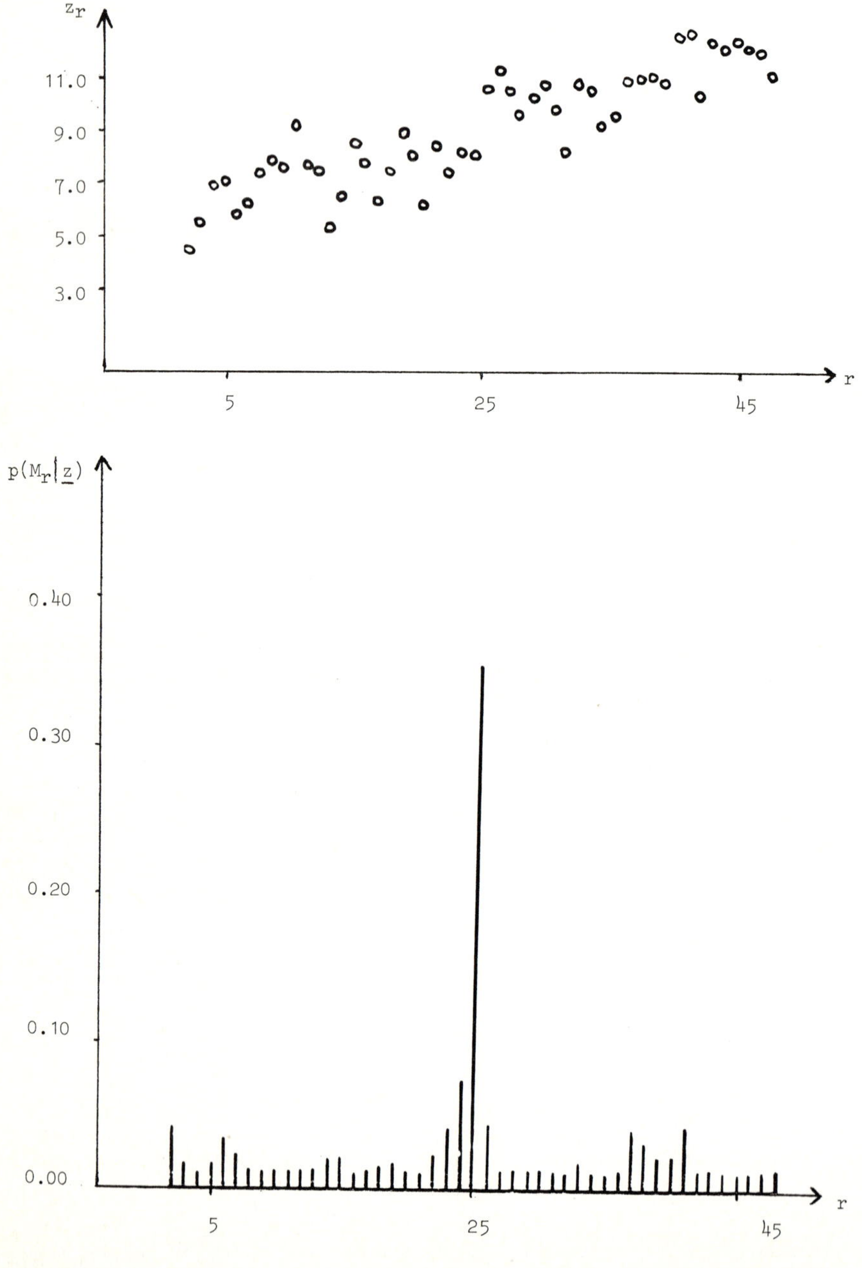

REFERENCES

Box, G.E.P. and Tiao, G.C. (1965). A change in level of a non-stationary time series. Biometrika, 52, 181-92.

Broemeling, L.D. (1974). Bayesian inferences about a changing sequence of random variables. Commun.Statist., 3, 243-55.

Brown, R.L., Durbin, J. and Evans, J.M. (1975). Techniques for testing the constancy of regression relationships over time (with Discussion). J.R.Statist.Soc. B, 37, 149-92.

Cameron, M.H. and Nash, J.E. (1974). On forecasting the manpower requirements of an organization with homogeneous workloads. J.R.Statist.Soc. A, 137, 200-18.

Halpern, E. (1972). A Bayesian approach to polynomial regression. J.Amer. Statist.Ass., 68, 137-43.

Ferreira, P.E. (1975). A Bayesian analysis of a switching regression model. J.Amer.Statist.Ass., 70, 370-74.

Lempers, F.B. (1971). Posterior probabilities of alternative linear models. Rotterdam University Press.

Quandt, R.E. (1958). The estimation of the parameters of a linear regression system obeying two separate regimes. J.Amer.Statist.Ass., 53, 873-80.

Quandt, R.E. (1960). Tests of the hypothesis that a linear regression system obeys two separate regimes. J.Amer.Statist.Ass., 55, 324-30.

Smith, A.F.M. (1975). A Bayesian approach to inference about a change point in a sequence of random variables. Biometrika, 62, 407-16.

Zellner, A. and Tiao, G.C. (1964). Bayesian analysis of the regression model with autocorrelated errors. J.Amer.Statist.Ass., 59, 763-78.

Recent Developments in Statistics
J.R.Barra et al., editors

Invited Paper

INFINITE DIMENSIONAL EXPONENTIAL TYPE STATISTICAL SPACES (GENERALIZED EXPONENTIAL FAMILIES)

J.-L. SOLER
Université Scientifique et Médicale de Grenoble
I.R.M.A., B.P. 53
38041 - Grenoble Cédex - France

SUMMARY .- An extension of the notion of Exponential Families is proposed for a precise description of some statistical spaces arising especially in statistical inference from stochastic processes and for which the observation may be regarded as an infinite dimensional vector. Essential tools for the purpose i.e. the Dual of a Measurable Vector Space and the Dual of a Probability Vector Space are introduced. Several examples of applications are given. A characterization, principal properties, and perspectives of utilization of this generalization are given.

1 - INTRODUCTION

Several statistical problems are based on observations of stochastic processes (random functions, point processes, etc.). The recent development of the theory of probability allows us to consider these observations as ones of random variables with values in general vector spaces (of functions, of measures, etc.) the laws of which are well-defined and may depend on a parameter which also runs through a subset of an infinite dimensional vector space. So, we are concerned with statistical spaces which often appear as infinite dimensional extensions of those of multivariate analysis. Their study shows that most of them appear as exponential statistical spaces (or exponential families) in a sense which generalizes the classical finite-dimensional notion.

Usually, a statistical space is said to be of exponential type, if it corresponds to the observation of a random vector in $(R^k, \mathcal{B}(R^k))$ whose distribution belongs to a family $\{P_\theta ; \theta \in \Theta\}$ of equivalent probability laws such that :

$$x \in R^k,\ \theta_0 \in \Theta,\ \left[dP_\theta / dP_{\theta_0}\right](x) = k(\theta) \exp\{\sum_{j=1}^{s} T_j(x)\, Q_j(\theta)\}.\ h(x),\ \text{for } \theta \in \Theta ;\ s \geq 1$$

By means of a reparametrization, the law of the sufficient statistic : $T(x) = (T_1(x),\dots,T_s(x))$ satisfies :

$$(1) \quad t \in R^s,\ \left[dP_{\theta,T}/dm\right](t) = c(\tilde\theta).\exp\{<t,\tilde\theta>\},\ \tilde\theta \in \tilde\Theta \subset R^s,$$

where m is some dominating measure on $(R^s, \mathcal{B}(R^s))$. (see [1], [2]).

The generalization to the infinite dimensional case is based on the fact that in (1) the parameter $\tilde\theta$ may be identified with a linear functional of the observation t.
So, it is easy to construct, in an artificial way, such generalized exponential families ; briefly, it suffices to consider a family $\{P_\theta ; \theta \in \Theta\}$ of probability measures, absolutely continuous w.r. to P_0 on some vector space E, the parameter θ

running through some adequate subset Θ containing the origin of a vector space F which is put in separated duality with E. The probabilities are defined on the σ-field $\mathcal{C}(E,F)$ generated by the cylindrical subsets of E w.r. to this duality, their density w.r. to P_0 taking the form (1), where the scalar product is replaced by the bilinear form on E x F. In a previous stage of this work, this has been proposed in 1973 by the author in [5]. Other authors, S. Johansen and more recently H.D. Brunk and co-authors (see the reference in [3]) have independently exploited the same idea.

But, although this model is nice to study, its limits appear very rapidly when looking for its applications. For instance, as we shall see, unfortunately an infinite dimensional Gaussian statistical space with unknown mean is not of this type. In this paper, this problem is solved by considering the dual space of a probability vector space whose elements are called linear random variables and constitute a widened notion of that of measurable linear functionals. A new model is proposed which appears as the true infinite-dimensional generalization of the model of exponential families, and covers the "natural" cases of applications.

Section 2 deals with the study of the dual of a probability vector space. The axiomatic definitions of the infinite dimensional exponential type statistical spaces are given in Section 3. In Section 4 several examples of applications are given and the general properties of the canonical model such as characterization, completeness, etc. are investigated in Section 5. Concluding remarks concerning these models and the perspectives of their use provided by their analytical properties are made in Section 6. For the sake of shortness many proofs and extended developments are omitted and may be found in the paper [6] or in [8].

2 - THE DUAL OF A MEASURABLE VECTOR SPACE, OF A PROBABILITY VECTOR SPACE

a) Définitions : Given a vector space E over R and a σ-field $\mathcal{A}$ of subsets of E, the pair $(E,\mathcal{A})$ will be called "Measurable Vector Space" (M.V.S.) if the mappings : $(\lambda,x) \to \lambda.x$ of R x E onto E and $(x,y) \to x+y$ of E x E onto E are measurable when R is endowed with its Borel σ-field. The vector space $(E,\mathcal{A})^m$ of all the measurable linear forms on $(E,\mathcal{A})$ will be called the "Dual of the M.V.S. $(E,\mathcal{A})$". If the canonical duality between E and $(E,\mathcal{A})^m$ is separated (i.e. $\langle x,f\rangle = 0,\ \forall f \in (E,\mathcal{A})^m \Rightarrow x = 0$ in E) then $(E,\mathcal{A})$ will be called "Separated" (S.M.V.S.). In this case, $\mathcal{C}(E,(E,\mathcal{A})^m)$ will denote the sub-σ-field of $\mathcal{A}$ generated by the elements of $(E,\mathcal{A})^m$ which is called the "Weakened σ-field of $\mathcal{A}$" and coincides with the σ-field generated by the cylindrical subsets of E w.r. to the above duality. We have $(E,\mathcal{C}(E,(E,\mathcal{A})^m)^m = (E,\mathcal{A})^m$. See [5] and [8] for a detailed study of this theory of "measurable duality", especially for a characterization of the dual of certain types of M.V.S.

Now, if P is any probability on $(E,\mathcal{A})$ and if $\mathcal{A}$ contains all the P-negligible subsets of E, the triplet $(E,\mathcal{A},P)$ is called a "Probability Vector Space" (P.V.S.).

Definition 2.1. - The Dual of the P.V.S. $(E,\mathcal{A},P)$ is the vector space $L(E,\mathcal{A},P)$ of the equivalence classes of the elements of the vector space $\mathcal{L}(E,\mathcal{A},P)$ of all the $\mathcal{A}$-measurable real functionals which are defined and are linear on a vector subspace of probability 1 of E. These elements will be called "linear real random variables" (l.r.r.v.).

Such functionals have been already considered by several authors (see [4]).

Evidently, the quotient space $(E,\mathcal{A})^m/P$ is contained in $\mathcal{L}(E,\mathcal{A},P)/P = L(E,\mathcal{A},P)$. In spite of the terminology, the spaces E and $\mathcal{L}(E,\mathcal{A},P)$ or E and $L(E,\mathcal{A},P)$ generally cannot be put in duality by a canonical bilinear form, since $\mathcal{L}(E,\mathcal{A},P)$ is not a subspace of the algebraic dual E^* of E (except of course if E is finite dimensional) ; but it can be shown that if a subspace $\mathcal{F}$ of $\mathcal{L}(E,\mathcal{A},P)$ has a countable basis, there exists a measurable subspace $E_{\mathcal{F}}$ of probability 1 of E which is in duality with $\mathcal{F}$ and is independent of this basis (see [6]).

In the following, we shall call "<u>σ-field generated by an equivalence class of a random element on $(E,\mathcal{A},P)$</u>", the smallest sub-σ-field of $\mathcal{A}$ containing all the P-negligible subsets of E, which yields all its versions measurable. In this way, $\mathcal{A}_L$ will denote the σ-field generated by the elements of $L(E,\mathcal{A},P)$, and $(E,\mathcal{A}_L,P_{\mathcal{A}_L})$ will be called the "<u>Weakened P.V.S. of $(E,\mathcal{A},P)$</u>". Obviously, $L(E,\mathcal{A}_L,P_{\mathcal{A}_L}) = L(E,\mathcal{A},P)$. Let $L_0(E,\mathcal{A},P)$ denote the vector space of all the equivalence classes of r.r.v. on $(E,\mathcal{A},P)$ endowed with the metrizable topology of the convergence in probability.

<u>Proposition 2.1.</u>- $L(E,\mathcal{A},P)$ <u>is a closed subspace of</u> $L_0(E,\mathcal{A},P)$ <u>and of any</u> $L_p(E,\mathcal{A},P)$, $1 \leq p < \infty$, <u>containing</u> $L(E,\mathcal{A},P)$. <u>If</u> $(E,\mathcal{A})^m/P$ <u>is dense in</u> $L(E,\mathcal{A},P)$ <u>for any of these topologies,</u> $L(E,\mathcal{A},P)$ <u>is exactly the set of all the (equivalence classes of) r.r.v. which are weak limits a.e. of sequences of elements of the dual</u> $(E,\mathcal{A})^m$ <u>of the M.V.S.</u> $(E,\mathcal{A})$.

b) <u>Functionals associated with a P.V.S.</u> : The complex function Φ_P defined on $L(E,\mathcal{A},P)$ by : $\xi \in L(E,\mathcal{A},P)$, $\Phi_P(\xi) = \int_E e^{i\xi}\, dP$, will be called the "<u>Fourier Transform of the P.V.S.</u> and it can be interpreted as an extension of the Fourier transform φ_P of P which is defined for all $y \in (E,\mathcal{A})^m$ by : $\varphi_P(y) = \int_E e^{i\langle x,y\rangle}\, dP(x)$, since this value depends only on the equivalence class of y. In the same way, the convex functional L_P defined by : $\xi \in L(E,\mathcal{A},P)$, $L_P(\xi) = \int_E e^{\xi}\, dP$ on the non-empty convex : $D(E,\mathcal{A},P)$ of $L(E,\mathcal{A},P)$ where this integral is finite will be called the "<u>Real Laplace Transform of the P.V.S</u>". As for the preceding functional, this latter one can be interpreted as an extension of the real Laplace transform ℓ_P of P. Finally the "<u>Cumulant Transform of the P.V.S. $(E,\mathcal{A},P)$</u>" will be the convex functional C_P defined on the same domain by $C_P = \operatorname{Log} L_P$. These functionals are continuous for the L_0 (or eventually the L_p) topology.

c) <u>The conjugate map of a linear random vector</u> : By extension of Definition 2.1., (an equivalence class of) a map U taking its values in a M.V.S. $(E_1,\mathcal{A}_1)$ which is measurable, defined and linear on a subspace of probability 1 of E will be called a "<u>linear random vector</u>" on the P.V.S. $(E,\mathcal{A},P)$.

<u>Proposition 2.2.</u>- <u>There exists a unique linear mapping</u> U^* <u>of</u> $L(E_1,\overline{\mathcal{A}}_1,PoU^{-1})$ <u>into</u> $L(E,\mathcal{A},P)$, <u>where</u> $\overline{\mathcal{A}}_1$ <u>denotes the σ-field obtained by completing</u> $\mathcal{A}_1$ <u>with the</u> PoU^{-1}-<u>negligible subsets of</u> E_1. U^* <u>is continuous for the topology of the convergence in probability on each of these spaces and verifies</u> : $\xi_1 \in L(E_1,\overline{\mathcal{A}}_1,PoU^{-1})$, $U^*(\xi_1) = \xi_1 o U$.

d) Projective properties : Let $\mathfrak{F}(L)$ denote the set of the finite dimensional subspaces of $L(E,\mathcal{A},P)$. If $N\in\mathfrak{F}(L)$, then $\xi_N = (\xi_1,\dots,\xi_n)$ defines a linear random vector on $(E,\mathcal{A},P)$, if $\{\xi_j\ ;\ j=1,\dots,n\}$ is a basis of N. Identifying any element of $L(R_n,\mathcal{B}(R_n),\ P\circ\xi_N^{-1})$ with an element $\alpha = (\alpha_1,\dots,\alpha_n)$ of R^n by the relation : $\alpha(u) = \langle\alpha,u\rangle$, $u\in R^n$ then, $\xi_N^*(\alpha) = \alpha\circ\xi_N = \sum_{j=1}^{n}\alpha_j\,\xi_j$. Thus, ξ_N^* is an isomorphism of $L(R^n,\mathcal{B}(R^n),\ P\circ\xi_N^{-1})$ onto N. Now, let $\mathcal{A}_N$ be the σ-field generated by ξ_N on E ; then $\mathcal{A}_L = \bigvee_{N\in\mathfrak{F}(L)}\mathcal{A}_N$ since $L(E,\mathcal{A},P) = \bigcup_{N\in\mathfrak{F}(L)} N$. So :

Proposition 2.3.- There exists a projective system of finite dimensional P.V.S. $(E_N,\mathcal{B}(E_N),\ P\circ\xi_N)_{N\in\mathfrak{F}(L)}$ which determines entirely the weakened P.V.S. $(E,\mathcal{A}_L,P_{\mathcal{A}_L})$.

e) Examples of duals of P.V.S. : -2.1.- Let $(\Omega,\mathfrak{F},P)$ be an arbitrary complete probability space, and let T be the mapping : $\omega\to\delta_\omega$ from Ω to the vector space E of bounded measures with finite support on Ω, endowed with the greatest σ-field $\mathcal{A}$ which yields T measurable (δ_ω being the Dirac measure on $\omega\in\Omega$).

Proposition 2.4.- There exists an isomorphism j of the dual $L(E,\mathcal{A},P\circ T^{-1})$ onto the vector space $L_0(\Omega,\mathfrak{F},P)$ of the equivalence classes of r.r.v. on $(\Omega,\mathfrak{F},P)$ such that : $j(\xi) = \xi\circ T$, $\xi\in L(E,\mathcal{A},P\circ T^{-1})$.

By the identification, the Fourier transform of the P.V.S. $(E,\mathcal{A},P\circ T^{-1})$ is defined by : $X\in L_0(\Omega,\mathfrak{F},P)$, $\Phi_{P\circ T^{-1}}(X) = \int_E e^{iX}\,dP$. The law of T appears as a generalization of the Bernoulli distribution. Indeed for $F\in\mathfrak{F}$ the r.r.v. 1_F is identified with a l.r.r.v. on $(E,\mathcal{A},P\circ T^{-1})$ of a Bernoulli distribution of parameter $P(F)$ since its characteristic function is given by : $\varphi(t) = \Phi_{P\circ T^{-1}}(t.1_F) = (P(F^c) + P(F)e^{it})$, $t\in R$.
A generalized multinomial distribution : $\mathcal{M}_{(\Omega,\mathfrak{F})}(P\ ;\ n)$ with parameters P and n will be the probability distribution of the random vector $T^{(n)}$ from the product probability space $(\Omega,\mathfrak{F},P)^n$ into $(E,\mathcal{A})$ defined by $T^{(n)}(\omega_1,\dots,\omega_n) = \sum_{j=1}^{n} T(\omega_j)$, $\omega_j\in\Omega$, $j = 1,\dots,n$.
The P.V.S. $(E,\mathcal{A},\mathcal{M}_{(\Omega,\mathfrak{F})}(P\ ;\ n))$ has the same dual so that its Fourier transfom can be defined by : $X\in L_0(\Omega,\mathfrak{F},P)$, $\Phi_{\mathcal{M}}(X) = (\int_E e^{iX}dP)^n$ which shows that each measurable partition $(1_{F_1},\dots,1_{F_k})$ of Ω can be identified with a linear random vector on this P.V.S. whose distribution is multinomial with parameters $P(F_1)$, ..., $P(F_k)$ and n (see [8]).

-2.2.- Let $(E,\mathcal{A})$ be a S.M.V.S. and P_0 be a centered Gaussian probability on $(E,\mathcal{A})$ whose Fourier transform is : $\varphi_{P_0}(y) = \exp\{-\frac{1}{2}Q_0(y)\}$, $y\in(E,\mathcal{A})^m$ where Q_0 is a positive quadratic form on $(E,\mathcal{A})^m$. The quotient space $(E,\mathcal{A})^m/P_0$ is contained in $L_2(E,\bar{\mathcal{A}},P_0)$; the Hilbert space H constituted by its closure in this space is isometric to a Hilbert space $\mathcal{H}(Q_0)$ of functionals on $(E,\mathcal{A})^m/P_0$ by means of the transformation :

$Z \in H \to E_{P_0}(ZY) = h_Z(Y)$, $Y \in (E,\mathcal{A})^m/P_0$. We have :

Proposition 2.5.- If $\mathcal{H}(Q_0) \subset E$, then $L(E,\bar{\mathcal{A}},P_0) \equiv H$.

The inclusion $H \subset L(E,\bar{\mathcal{A}},P_0)$ results from Proposition 2.1, the converse, results from a theorem of Rozanov ([4], p. 146) which insures that every l.r.r.v. on $(E,\bar{\mathcal{A}},P_0)$ is a limit in quadratic mean of a sequence of measurable linear functionals on this P.V.S.

This characterization has interesting consequences, when considering $(E,\bar{\mathcal{A}},P_0)$ as the P.V.S. induced by a centered Gaussian random function. For example, the dual of the standard Wiener P.V.S. $(C_0[0,1], \mathcal{B}(C_0[0,1], W)$ can be shown to be constituted by the stochastic integrals : $\{\int_0^1 f(t)\, d\xi_t,\ f \in L_2([0,1], \mathcal{B}([0,1]), dt)\}$, where $(\xi_t)_{t\in[0,1]}$ is the canonical standard Brownian motion on this probability space. (See [8] for other examples).

f) Quadratic real random variables : Let $(E,\mathcal{A},P)$ be a P.V.S. Every measurable functional q which is defined P-a.e. on E and verifying : $q(\alpha x) = \alpha^2 q(x)$ for all x belonging to its domain of definition, and all real α, will be called a "quadratic real random variable (q.r.r.v.) on the P.V.S.". Using a similar argument to that of Proposition 2.1, it can be proved that the vector space $Q(E,\mathcal{A},P)$ of the equivalence classes of q.r.r.v. on $(E,\mathcal{A},P)$ is a closed subspace of $L_0(E,\mathcal{A},P)$ or of any $L_p(E,\mathcal{A},P)$ $1 \leq p < \infty$ containing it.

Our purpose is to show that $Q(E,\mathcal{A},P)$ can be identified with the dual of a P.V.S. :

Let χ be the mapping defined on E by $\chi(x) = x \otimes x$, $x \in E$ and denote by E_χ the vector space spanned by $\chi(E)$ in the symmetric tensor product space $E \odot E$, endowed with the greatest σ-field which yields χ measurable, then can be proved (see [6], [8]) :

Proposition 2.6.- The mapping $u \to u \circ \chi$ is an isomorphism of $L(E_\chi,\mathcal{A}_\chi,P\circ\chi^{-1})$ onto $Q(E,\mathcal{A},P)$.

Corollary 1.- The symmetric tensor product of $L(E,\mathcal{A},P)$ by itself can be identified with a vector subspace of $L(E_\chi,\mathcal{A}_\chi,P\circ\chi^{-1})$.

Corollary 2.- With the assumptions of Proposition 2.5 in example 2 above, the Hilbert space $H \tilde{\odot} H$ (which is the completion of the prehilbertian space $H \odot H$) is isomorphic to a vector subspace of the dual $L(E_\chi,\bar{\mathcal{A}}_\chi,P\circ\chi^{-1})$, and every element $\zeta \in H \tilde{\odot} H$ can be written $\zeta = \xi \circ \chi$ where ξ is a unique l.r.r.v. on the P.V.S. $(E_\chi,\bar{\mathcal{A}}_\chi,P\circ\chi^{-1})$.

This fact will be decisive in an application of the following theory. The Fourier transform of $P \circ \chi^{-1}$ has been calculated by the author in [7] for a Gaussian probability P in a quite general case, providing a infinite dimensional extension of the Wishart distribution.

3 - EXPONENTIAL TYPE STATISTICAL SPACES

a) The canonical case : Let $(E,\mathcal{A},\mathcal{P})$ be a Vector Statistical Space (V.S.S.), $\mathcal{P}$ is a family of equivalent probabilities on a S.M.V.S. $(E,\mathcal{A})$. It will be always supposed that $\mathcal{A}$ contains all the P-negligible subsets of E, $P\in\mathcal{P}$. The $\mathcal{P}$-equivalence reduces to the P-equivalence for any $P\in\mathcal{P}$, then $(E,\mathcal{A})^m/\mathcal{P}$, $L(E,\mathcal{A},\mathcal{P})$, $L_0(E,\mathcal{A},\mathcal{P})$, $L_\infty(E,\mathcal{A},\mathcal{P})$ will denote respectively spaces of equivalence classes of "linear real statistics" (l.r.s.), of real statistics, of essentially-bounded real statistics on the V.S.S. which are just the corresponding r.r.v. on each P.V.S. $(E,\mathcal{A},P)$, $P\in\mathcal{P}$. Similarly, a Linear vector statistic will be a linear random vector on each of these P.V.S. Finally, we can suppose that $\mathcal{A}$ reduces to the weak σ-field $\mathcal{A}_L$ since the dual $L(E,\mathcal{A},\mathcal{P})$ is not affected, as we have seen above.

Definition 3.1.- A V.S.S. $(E,\mathcal{A},\mathcal{P})$ whose family $\mathcal{P}$ of hypotheses are equivalent probabilities will be said of "Canonical Exponential Type" (C.E.T.) iff for some fixed probability $P^*\in\mathcal{P}$ we have : $\{\mathrm{Log}(dP/dP^*)\ ;\ P\in\mathcal{P}\}\subset L(E,\mathcal{A},\mathcal{P})\oplus R$.

It is easy to see that this definition does not depend on P^*. It implies that for each $P\in\mathcal{P}$, there exist a unique element $\xi_P\in L(E,\mathcal{A},\mathcal{P})$ and a real k_P such that :

$$dP/dP^* = \exp\{\xi_P + k_P\}\ ;\ \text{where } k_P = -\mathrm{Log}\, E_{P^*}(e^{\xi_P}) = -\mathrm{Log}\, L_{P^*}(\xi_P) = -C_{P^*}(\xi_P).$$

Let u be the injective mapping : $P\to\xi_P$ of $\mathcal{P}$ into $L(E,\mathcal{A},\mathcal{P})$, setting $\Theta = u(\mathcal{P})$, u is a one-to-one correspondance between $\mathcal{P}$ and Θ, and $u(P^*) = 0$. For $\theta\in\Theta$, set $P_\theta = u^{-1}(\theta)$, then we have equivalently :

Definition 3.2.- A V.S.S. $(E,\mathcal{A},\mathcal{P}=\{P_\theta\ ;\ \theta\in\Theta\})$ is said to be of C.E.T. iff :

(i) - Θ is a subset containing the origin of the convex $D(E,\mathcal{A},P_0)\subset L(E,\mathcal{A},P_0)$, which is the domain of definition of the real Laplace transform of the P.V.S. $(E,\mathcal{A},P_0)$.

(ii) - The P_θ ; $\theta\in\Theta$ are equivalent and $\mathcal{A}$ contains all the P_θ-negligible subsets of E.

(iii) - For every $\theta\in\Theta$, $dP_\theta/dP_0 = \exp(\theta)/L_{P_0}(\theta)$.

Θ will be called the "natural parameter" of the C.E.T. V.S.S. $(E,\mathcal{A},\mathcal{P})$ and u the "coding transformation w.r. to P^*." The C.E.T. V.S.S. $(E,\mathcal{A},\mathcal{P}=\{P_\theta\ ;\ \theta\in\Theta\})$ will be said to be of "convex type" [resp. "saturated"] if Θ is convex [resp. if, moreover, $\Theta = D(E,\mathcal{A},P_0)$] ", to be of "finite dimensional parameter" if $sp(\Theta)$ is finite dimensional in $L(E,\mathcal{A},\mathcal{P})$, and to be "reduced" if $\mathcal{A}$ reduces to the minimal sufficient σ-field of the statistical space, which is generated by the family $\{\theta\ ;\ \theta\in\Theta\}$ of l.r.s.

This is what should be called an "abstract" or an "implicit" generalized canonical exponential family as announced in Section 1, and the previous generalization ([5]) may be recognized in the following natural restrictive definition :

Definition 3.3.- A C.E.T. V.S.S. $(E,\mathcal{A},\mathcal{P})$ will be said "Explicit" if moreover : $\{\mathrm{Log}(dP/dP^*)\ ;\ P\in\mathcal{P}\}\subset(E,\mathcal{A})^m/\mathcal{P}\oplus R$, for a fixed $P^*\in\mathcal{P}$.

Using the coding transformation u w.r. to P^*, it can be written $(E,\mathfrak{a},\mathcal{P} = \{P_\theta ; \theta \in \Theta\})$ where $\Theta \subset (E,\mathfrak{a})^m/\mathcal{P} \subset L(E,\mathfrak{a},\mathcal{P})$. But then, there exists a version of the l.r.s. $\theta \in \Theta$ of the form : $x \to <x,\theta>$ for an element, which is still denoted by θ of the dual $(E,\mathfrak{a})^m$ of the S.M.V.S. $(E,\mathfrak{a})$, where $<.,.>$ denotes the canonical bilinear form on $E \times (E,\mathfrak{a})^m$. Thus, we have equivalently :

Definition 3.4.- With the assumptions of Definition 3.2, the C.E.T. V.S.S. $(E,\mathfrak{a},\mathcal{P} = \{P_\theta ; \theta \in \Theta\})$ will be said "Explicit", if for every $\theta \in \Theta$, there exists a version of the density of P_θ w.r. to P_0 of the form :

$$(dP_\theta/dP_0)(x) = \exp(<x,\theta>)/\ell_{P_0}(\theta), \quad x \in E, \quad \theta \in (E,\mathfrak{a})^m.$$

b) The general case :

Definition 3.5.- A statistical space $(\mathfrak{X},\mathfrak{F},\mathcal{P})$ will be said to be of "Exponential Type" (E.T.) [resp. of "Explicit Exponential Type"] if there exists a sufficient statistic T taking on its values in a S.M.V.S. $(E,\mathfrak{a})$ and whose image statistical space $(E,\mathfrak{a},\mathcal{P}_T = \{PoT^{-1} ; P \in \mathcal{P}\})$ is of C.E.T. [resp. of Explicit C.E.T.].

It is easy to see what this definition implies, and using the coding transformation of the image C.E.T. V.S.S. associated with T, we can deduce a parametrized version of this definition which will be useful to recognize in practice these generalized exponential families :

Definition 3.6.- A Statistical space $(\mathfrak{X},\mathfrak{F},\mathcal{P} = \{P_\theta ; \theta \in \Theta\})$ is said to be of E.T. [resp. of Explicit E.T.] if there exists a statistic T with values in a S.M.V.S. $(E,\mathfrak{a})$ such that :

(i) - Θ is a subset, containing the origin of the convex $D(E,\mathfrak{a},P_0 oT^{-1}) \subset L(E,\mathfrak{a},\mathcal{P}_T)$ which is the domain of definition of the real Laplace transform $L_{P_0 oT^{-1}}$ of the P.V.S. $(E,\mathfrak{a},P_0 oT^{-1})$ [resp. of the convex :

$$\{y \in (E,\mathfrak{a})^m : \ell_{P_0 oT^{-1}}(y) = \int_E e^{<t,y>} dP_0 oT^{-1}(y) < \infty\}].$$

(ii) - The P_θ ; $\theta \in \Theta$ are equivalent on the sub-σ-field of $\mathfrak{F}$ generated by T (which contains the negligible subsets of $\mathfrak{X}$ w.r. to the restrictions of the P_θ on it).

(iii) - For every $\theta \in \Theta$, $dP_\theta/dP_0 = \exp(\theta oT)/L_{P_0 oT^{-1}}(\theta)$. [resp. $(dP_\theta/dP_0)(x) = \exp(<T(x),\theta>)/\ell_{P_0 oT^{-1}}(\theta)$; $x \in \mathfrak{X}$].

Remarks 3.1.- By the characterization of the dual of a finite dimensional P.V.S., every classical exponential family determines an E.T. statistical space in the above sense.

3.2.- If $(\mathfrak{X},\mathfrak{F},\mathcal{P})$ is of E.T. w.r. to the sufficient statistic T, the same is true for the product statistical space $(\mathfrak{X},\mathfrak{F},\mathcal{P})^n$, $n \geq 1$, w.r. to the sufficient statistic $T^{(n)} : \mathfrak{X}^n \to E$ defined by : $T^{(n)}(x_1,\ldots,x_n) = \sum_{i=1}^{n} T(x_i)$, $x_i \in \mathfrak{X}$, $i = 1,\ldots,n$.

3.3.- Every statistical space, corresponding to an n-sample whose probability laws are equivalent can be shown to be of E.T. w.r. to the sufficient statistic constituted by the empirical probability of the sample : Indeed, let us suppose first that $n = 1$, and let $(\Omega,\mathfrak{F},\mathcal{P})$ be an arbitrary statistical space such that $\mathfrak{F}$ contains all the negligible subsets of Ω w.r. to the equivalent probabilities in $\mathcal{P}$. Taking $T : \omega \to \delta_\omega$ with values in the S.M.V.S. $(E,\mathfrak{a})$ described in Example 2.1, Proposition-2.4 implies that for every fixed $P^* \in \mathcal{P}$ and every $P \in \mathcal{P}$, $dP/dP^* = \exp(\xi_P o T)$ where ξ_P is the element of $L(E,\mathfrak{a},\mathcal{P}_T)$ corresponding to the r.r.v. $\mathrm{Log}(dP/dP^*) \in L_0(\Omega,\mathfrak{F},\mathcal{P})$, which shows that $(\Omega,\mathfrak{F},\mathcal{P})$ is of E.T. w.r. to T. Now, by Remark 3.2, the statistical space $(\Omega,\mathfrak{F},\mathcal{P})^n$ corresponding to the n-sample $(\omega_1,\ldots,\omega_n)$, is of E.T. w.r. to the sufficient statistic $T^n(\omega_1,\ldots,\omega_n) = \sum_{i=1}^{n} \delta_{\omega_i}$ or equivalently, w.r. to the empirical probability : $\mathbb{P}_n = (\sum_{i=1}^{n} \delta_{\omega_i})/n$. The C.E.T. V.S.S. induced by $T^{(n)}$ being : $(E,\mathfrak{a}, \{\mathcal{M}_{(\Omega,\mathfrak{F})}(P;n) \; ; \; P \in \mathcal{P}\})$, which will be called the "Empirical multinomial canonical statistical space" of the n-sample $\omega_1, \ldots, \omega_n$. This result is not surprising, since we admit that a statistical space may be parametrized by an infinite dimensional parameter. It may be regarded as the one of the sufficiency theory which ensures that the order statistic is sufficient. Analogously, the generalized E.T. statistical spaces are of some interest when they are non trivial as we shall see in the examples below.

4 - EXAMPLES OF EXPONENTIAL TYPE STATISTICAL SPACES

a) Gaussian statistical spaces with unknown mean : In the context of Example 2.2, and with the assumption of Proposition 2.5, for every $m \in E$ let P_m be the image of the Gaussian probability P_0 by the measurable mapping : $x \to x+m$, $x \in E$, which determines a Gaussian probability on $(E,\mathfrak{a})$ with mean m and variance Q_0. It is well-known that P_m and P_0 are equivalent iff $m \in \mathcal{H}(Q_0)$ and, $dP_m/dP_0 = \exp\{\psi(m) - E[\psi^2(m)]/2\}$, where ψ is an isomorphism of $\mathcal{H}(Q_0)$ onto $H \equiv L(E,\bar{\mathfrak{a}},P_0)$. Then,

Proposition 4.1.- The Gaussian statistical space with unknown mean : $(E,\bar{\mathfrak{a}}, \{P_m \; ; \; m \in \mathcal{H}(Q_0) \subset E\})$ is of C.E.T., the coding transformation u w.r. to P_0 being defined by : $u(P_m) = \psi(m)$.

It is saturated since $\Theta = u\{P_m \; ; \; m \in \mathcal{H}(Q_0)\} = L(E,\bar{\mathfrak{a}},P_0) = D(E,\bar{\mathfrak{a}},P_0)$.

Application to signal detection : Let $x = (x_t \; ; \; t \in [0,1])$ be an observed sample of a stochastic process of the form : $X_t = m(t) + W_t$, during the time interval $[0,1]$ where $(W_t, \; t \in [0,1])$ is a standard Brownian noise and $m = (m(t) \; ; \; t \in [0,1])$ an unknown deterministic and continuous signal vanishing at 0. The statistical space associated with x is easily seen to be $(C_0[0,1], \mathcal{B}(C_0[0,1]), \{P_m \; ; \; m \in C_0[0,1]\})$ where $C_0[0,1]$ is the Banach space of all the continuous real functions vanishing at 0, on $[0,1]$, and $\mathcal{B}(C_0[0,1])$ its Borel-σ-field, P_0 being the Wiener probability. It can be shown (see [8]) that $(C_0[0,1], \mathcal{B}(C_0[0,1]))^m$ reduces to the topological dual of $C_0[0,1]$ that is, the space $M[0,1]$ of all the bounded Radon measures on $[0,1]$, and it is known that in this case, $\mathcal{B}(C_0[0,1]) = \mathcal{C}(C_0[0,1], M[0,1])$. Here $\mathcal{H}(Q_0) = \{f \in C_0[0,1] : \dot{f}$ exists, $f(o) = 0$ and $\int_0^1 \dot{f}^2(t)\,dt < \infty\}$. Thus, if we assume that $m \in \mathcal{H}(Q_0)$ the statistical space becomes the C.E.T. one :

$(C_0[0,1], \overline{\mathcal{B}}(C_0[0,1]), \mathcal{P} = \{P_m ; m \in \mathcal{H}(Q_0)\})$ with $dP_m/dP_\theta = \exp(\int_0^1 \dot{m}(t)\,d\xi_t - \frac{1}{2}\int_0^1 \dot{m}^2(t)\,dt)$; by the characterization of H, we deduce its natural parameter : $\theta = u(P_m) = \int_0^1 \dot{m}(t)\,d\xi_t$ and $L_{P_0}(\theta)$.

This C.E.T V.S.S. becomes Explicit by supplementary assumptions on m (that is on θ). Indeed, if moreover $\dot{m}$ is of bounded variation on $[0,1]$ the stochastic integral admits a version (see [9]) :

$$\int_0^1 \dot{m}(t)\,dt = x(1)\,m(1) - \int_0^1 x(t)\,d\dot{m}(t), \quad x \in C_0[0,1]$$

So, denoting by $\mathcal{P}'$ the sub-family of $\mathcal{P}$ obtained by this restriction, the C.E.T. V.S.S. $(C_0[0,1], \overline{\mathcal{B}}(C_0[0,1]), \mathcal{P}')$ is Explicit : setting for $\theta = u(P_m)$ the measure of $M[0,1]$ defined by : $\theta = m(1)\,\delta_1 - \int_. d\dot{m}$, the family $\mathcal{P}'$ can be parametrized by $\theta \in u(\mathcal{P}')$ and we have :

$$x \in C_0[0,1], \ (dP_\theta/dP_0)(x) = \exp(<x,\theta>)/\ell_{P_0}(\theta)$$

where $<x,\theta> = \int_0^1 x(t)\,d\theta(t)$ is the canonical bilinear form on $C_0[0,1] \times M[0,1]$. It is easy to verify that $\ell_{P_0}(\theta) = \exp\{Q_0(\theta)/2\}$ with $Q_0(\theta) = \int_0^1\int_0^1 \inf(s,t)\,d\theta(s)\,d\theta(t)$ (see [8]).

b) Gaussian statistical spaces with unknown variance : In the same conditions as in a), let $\mathcal{P}_{Q_0}$ denote the family of all the centered Gaussian probabilities on $(E,\mathcal{A})$ which are equivalent to P_0. Then,

Proposition 4.2.- The Gaussian statistical space with unknown variance :

$$(E,\overline{\mathcal{A}},\mathcal{P}_{Q_0})$$

is of E.T. w.r. to the sufficient statistic : $\chi : x \to x \otimes x$ from E into $E \otimes E$.

Proof. It is well known that if $P_Q \in \mathcal{P}_{Q_0}$, $dP_Q/dP_0 = \exp(\zeta_Q)/E_{P_0}(\exp \zeta_Q)$ where ζ_Q is a unique element of the Hilbert space $H \tilde{\otimes} H$. Applying Corollary 2 of Proposition 2.6, we have : $dP_Q/dP_0 = \exp(\xi_Q \circ \chi)/L_{P_0 \circ \chi^{-1}}(\xi_Q)$ where ξ_Q is a unique element of the dual of the P.V.S. $(E_\chi, \overline{\mathcal{A}}_\chi, P_0 \circ \chi^{-1})$ which completes the proof.

Application : Let $x = (x_t ; t \in [0,1])$ be an observed sample, during the time interval $[0,1]$ of a stochastic process $(X_t ; t \in [0,1])$ satisfying the stochastic differential equation : $dX_t = \alpha(t)\,X_t\,dt + dW_t$, $X_0 = 0$, where $(W_t ; t \in [0,1])$ is defined as above and $\alpha(t)$ is an unknown time varying coefficient. Provided that the unknown real function $\alpha = (\alpha(t)\ ;\ t \in [0,1])$ belongs to $L_2([0,1], \mathcal{B}([0,1]), dt)$, the theory of stochastic differential equations allows us to consider the observation x as the one of a random variable X with values in the S.M.V.S. $(C_0[0,1], \mathcal{B}(C_0[0,1]))$, whose probability law P_α is equivalent to P_0 (which reduces to the Wiener probability) and satisfies :

(1) $dP_\alpha/dP_0 = \exp(\int_0^1 \alpha(t)\ \xi_t\ d\,\xi_t - \frac{1}{2}\int_0^1 \alpha^2(t)\ \xi_t^2\ dt)$.

We shall see, by assuming for α to belong to a subspace $\mathcal{A}$ of $L_2([0,1], \mathcal{B}([0,1]), dt)$ that the statistical space associated with the observation x :

(2) $(C_0[0,1], \overline{\mathcal{B}}(C_0[0,1]), \mathcal{P} = \{P_\alpha\ ;\ \alpha \in \mathcal{A}\})$

is of Explicit E.T. w.r. to the sufficient statistic : $T : x \to x^2(.)$, $x \in C_0[0,1]$ with values in the positive cone of $C_0[0,1]$.

Using a theorem of Cameron and Martin (see [9]), it can be shown that if α is a continuous real function of bounded variation on $[0,1]$ (that is $\alpha \in \mathcal{A}$), then,

$$(\int_0^1 \alpha(t)\ \xi_t\ d\,\xi_t)(x) = \frac{1}{2}(\int_0^1 \alpha(t)\ dx^2(t) - \int_0^1 \alpha(t)\ dt),\quad x \in C_0[0,1],\ P_0\text{-a.e.}$$

is a version of the stochastic integral in (1). $\int_0^1 \alpha(t)\ dx^2(t)$ being considered as a Riemann-Stieljes integral : $\lim_{n\to\infty} \sum_{i=1}^{n} \alpha(\frac{i}{n})\left[x(\frac{i}{n}) + x(\frac{i-1}{n})\right]\left[x(\frac{i}{n}) - x(\frac{i-1}{n})\right]$.

Then, the likelihood function of the statistical space (2) is :

$$\alpha \in \mathcal{A},\ x \in C_0[0,1],\ (dP_\alpha/dP_0)(x) = \exp \tfrac{1}{2}\{\alpha(1)x^2(1) - \int_0^1 x^2(t)\,d\alpha(t) - \int_0^1 \alpha^2(t)x^2(t)\,dt - \int_0^1 \alpha^2(t)\,dt\}$$

The factorization theorem shows that the statistic $T(x) = x^2(.)$ is sufficient. Now it remains to point out that this likelihood function can be expressed as :

$$\alpha \in \mathcal{A},\ x \in C_0[0,1],\ (dP_\alpha/dP_0)(x) = \exp(\langle T(x), \theta(\alpha)\rangle / c(\alpha)$$

taking as natural parameter the bounded Radon measure on $[0,1]$ defined by :

$$\theta = \theta(\alpha) = \frac{1}{2}\left[\alpha(1)\ \delta_1 - \int_{\cdot} d\alpha - \int_{\cdot} \alpha^2(t)\ dt\right].$$

Hence, it follows that the family $\mathcal{P}$ can be parametrized by $\theta \in \Theta \subset M[0,1]$ and (2) becomes : $(C_0[0,1], \overline{\mathcal{B}}(C_0[0,1]), \{P_\theta\ ;\ \theta \in \Theta\})$ with for $\theta \in \Theta$, $x \in C_0[0,1]$,

$$(dP_\theta/dP_0)(x) = \exp(\langle T(\alpha), \theta\rangle)/L_{P_0 o T^{-1}}(\theta)$$

where $L_{P_0 o T^{-1}}(\theta) = c(\alpha) = \exp(\frac{1}{2}\int_0^1 \alpha(t)\ dt)$. The C.E.T. V.S.S. induced by T :

$(C_0[0,1], \overline{\mathcal{B}}(C_0[0,1]), \mathcal{P}_T = \{P_\theta o T^{-1}\ ;\ \theta \in \Theta\})$ will be described by deriving the probability law of T, when $\theta = 0$. As a particular case of the results of [7] about the generalized Wishart distribution, we have its characteristic functional :

$\varphi_T(\nu) = E[\exp(i\langle T, \nu\rangle)] = \det(\mathbb{1} - 2iB_\nu)^{-1/2}$, $\nu \in M[0,1]$ where $\det(\mathbb{1} - 2iB_\nu)$ denotes the

Fredholm determinant at 2i of the self-adjoint nuclear operator B_ν in $\mathcal{H}(Q_0)$ which is defined by :

$$f \in \mathcal{H}(Q_0),\ B_\nu(f) = \int_0^1 f(t)\ \inf(.,t)\ d\nu(t).$$

Thus, $L_{P_0 o T^{-1}}(\theta) = \varphi_T(-i\theta) = \det(\mathbb{1} + 2B_\theta)^{-1/2}$ from which we deduce that $\det(\mathbb{1} + 2B_{\theta(\alpha)}) = \exp(-\int_0^1 \alpha(t)\ dt)$.

The results of [7] allow us to envisage the construction of more general C.E.T. V.S.S. from the infinite dimensional Wishart distribution.

c) Other examples : It can be shown that the statistical space associated with an observation of several point processes, especially Poisson processes are E.T. statistical spaces corresponding to the observation of a random variable in a space of measures (see [8]).

5 - THE CANONICAL EXPONENTIAL TYPE STATISTICAL SPACES PROPERTIES

Let us consider hereafter, a C.E.T. V.S.S. endowed with its natural parametrization, that is a V.S.S. : $(E,\mathcal{A},\mathcal{P} = \{P_\theta\ ;\ \theta \in \Theta\})$ satisfying Definition 3.2.

a) Integrability :

Proposition 5.1.- $L(E,\mathcal{A},\mathcal{P}) \subset \bigcap_{\theta \in \Theta \cap \overset{\circ}{D}(E,\mathcal{A},P_0)} L_1(E,\mathcal{A},P_\theta)$

where $\overset{\circ}{D}(E,\mathcal{A},P_0)$ denotes the algebraic interior of the convex $D(E,\mathcal{A},P_0)$.

This proposition asserts that every l.r.s. is P_θ-integrable provided that θ is an internal point of the convex $D(E,\mathcal{A},P_0)$. This results from the fact that $\theta \in \overset{\circ}{D}(E,\mathcal{A},P_0) \Longleftrightarrow 0 \in \overset{\circ}{D}(E,\mathcal{A},P_\theta)$, from $L(E,\mathcal{A},P_\theta) = L(E,\mathcal{A},\mathcal{P})$, $\forall \theta \in \Theta$ and from :

Lemma 5.1.- If $(E,\mathcal{A},P)$ is any P.V.S, then $L(E,\mathcal{A},P) \subset L_1(E,\mathcal{A},P)$ provided that 0 is an internal point of the convex $D(E,\mathcal{A},P)$. (See [6], [8]).

Moreover, if $\xi \in L(E,\mathcal{A},\mathcal{P})$ and $\theta \in \overset{\circ}{D}(E,\mathcal{A},P)$, $E_{P_\theta}(\xi) = C'_{P_0}(\theta,\xi)$ which is the value at θ of the derivative in the direction ξ of the Cumulant transform of the P.V.S. $(E,\mathcal{A},P_0)$.

b) The image of a C.E.T. V.S.S. by a linear vector statistic and characterization : Let $(E_1,\mathcal{A}_1)$ be a S.M.V.S. and U be a linear vector statistic on the C.E.T. V.S.S. $(E,\mathcal{A},\mathcal{P})$ with values in $(E_1,\mathcal{A}_1)$, the image statistical space being $(E_1,\mathcal{A}_1,\mathcal{P}_U = \{P_\theta o U^{-1}\ ;\ \theta \in \Theta\})$. We can always suppose that $\mathcal{A}_1$ contains all the $P_0 o U^{-1}$-negligible subsets of E_1. By Proposition 2.2 which defines the conjugate mapping U^* of U, an element θ_1 of $L(E_1,\mathcal{A}_1,P_0 o U^{-1})$ belongs to $D(E_1,\mathcal{A}_1, P_0 o U^{-1})$ iff $\theta_1 o U = U^*(\theta_1)$ belongs to $D(E,\mathcal{A},P_0)$. Thus, $D(E_1,\mathcal{A}_1,P_0 o U^{-1}) = U^{*-1}[D(E,\mathcal{A},P_0)]$ and $L_{P_0 o U^{-1}}(\theta_1) = L_{P_0}(U^*(\theta_1))$.

Set $\Theta_1 = U^{*-1}(\Theta)$ and $\mathcal{P}_1 = \{P_{\theta_1} : dP_{\theta_1}/dP_o oU^{-1} = \exp(\theta_1)/L_{P_o oU^{-1}}(\theta_1) \ ; \ \theta_1 \in \Theta_1\}$.

It can be proved that : $\mathcal{P}_1 \subset \mathcal{P}_U$ (see [6]), which asserts that the image statistical space $(E_1, \mathcal{A}_1, \mathcal{P}_U)$ "contains" the C.E.T. one $(E_1, \mathcal{A}_1, \mathcal{P}_1)$. (Several counter-examples show that equality generally does not hold).

Consequences :

1 - Sufficiency - It is easy to see that U is a sufficient statistic on the "sub-statistical space" : $(E, \mathcal{A}, \mathcal{P}' = \{P_{U^*(\theta_1)} \ ; \ \theta_1 \in \Theta_1\})$ of $(E, \mathcal{A}, \mathcal{P})$, and its image by U is the C.E.T. V.S.S. $(E_1, \mathcal{A}_1, \mathcal{P}_1)$. We can summarize this fact in the following diagram :

$$\begin{array}{ccc} (E,\mathcal{A},\mathcal{P}) & \supset & (E,\mathcal{A},\mathcal{P}') \\ U \downarrow & & \downarrow \ U \text{ sufficient} \\ (E_1,\mathcal{A}_1,\mathcal{P}_U) & \supset & (E_1,\mathcal{A}_1,\mathcal{P}_1) \end{array}$$

As a corollary we obtain :

Proposition 5.2.- If U^* is surjective of Θ_1 onto Θ, then U is sufficient and the image statistical space is of C.E.T.

2 - Similarity - If U is a similar statistic, then $\mathcal{P}_U = \{P_o oU^{-1}\}$ which implies that $\mathcal{P}_1 = \{P_o oU^{-1}\}$; so, $\Theta_1 = U^{*-1}(\Theta) = \{0\}$ which is equivalent to : $\theta_1 \neq 0 \Rightarrow U^*(\theta_1) \notin \Theta$, or also : $U^*(L(E_1, \mathcal{A}_1, P_o oU^{-1})) \cap \Theta = \{0\}$. This implies that if 0 is an internal point of Θ (provided the algebraic interior of Θ is non-empty) $U^*(L(E_1, \mathcal{A}_1, P_o oU^{-1})) = 0$, thus $U^* \equiv 0$ and $U \equiv 0$ is trivial :

Proposition 5.3.- If the origin of $L(E, \mathcal{A}, \mathcal{P})$ is an internal point of Θ, every similar linear vector statistic reduces to 0.

3 - Characterization of the C.E.T. V.S.S. - Let N be a finite dimensional subspace of $L(E, \mathcal{A}, \mathcal{P})$. Taking for U the linear vector statistic : $\xi_N = (\xi_1, \ldots, \xi_n)$ in $(R^n, \mathcal{B}(R^n))$ where $\{\xi_i, \ i=1,\ldots,n\}$ is a basis of N the image statistical space is $(R^n, \mathcal{B}(R^n), \{P_\theta o\xi_N^{-1} \ ; \ \theta \in \Theta\})$ and as we have pointed out in section 2, d), ξ_N^* is an isomorphism of $L(R_n, \mathcal{B}(R_n), P_o o\xi_N^{-1})$ onto N. Then, $\Theta_N = \xi_N^{*-1}(\Theta)$ is such that : $\xi_N^*(\Theta_N) = \Theta \cap N$. Thus, if $\Theta \subset N$, ξ_N^* is surjective of Θ_N onto Θ and applying Proposition 5.2. ξ_N is sufficient, so :

Proposition 5.4.- If the C.E.T. statistitical space $(E, \mathcal{A}, \mathcal{P})$ is of finite dimensional parameter, there exists a sufficient linear vector statistic with values in a finite dimensional S.M.V.S. whose image is of C.E.T.

This proposition reduces the study of an E.T. statistical problem whose parameter is of finite dimension to a classical finite dimensional one, even if it is based on an infinite dimensional observation.

Proposition 5.5.- (Characterization) - With the notations of Proposition 2.3, a V.S.S. $(E,\mathcal{A}, \mathcal{P} = \{P_\theta \ ; \ \theta \in \Theta\})$ verifying conditions i) and ii) of Definition 3.2, is of C.E.T. iff the projective system :

$$(1) \quad (E_N, \mathcal{B}(E_N), \{P_\theta \circ \xi_N^{-1} \ ; \ \theta \in \Theta \cap N\})_{N \in \mathcal{F}(L)}$$

is composed of classical finite dimensional C.E.T. statistical spaces.

The generic element of (1) is the image statistical space of the sub-C.E.T. V.S.S. $(E,\mathcal{A},\mathcal{P}_N = \{P_\theta \ ; \ \theta \in \Theta \cap N\})$ by the sufficient statistic ξ_N and the necessity follows from Proposition 5.4. Conversely, since $\Theta = \bigcup_{N \in \mathcal{F}(L)} \Theta \cap N$, the Laplace transform L_{P_0} is well defined by its values on the subsets of Θ which are intersections of Θ with the finite dimensional subspaces of $L(E,\mathcal{A},\mathcal{P})$ and then,

$$\forall \theta \in \Theta \cap N, \ dP_\theta/dP_0 = (dP_\theta \circ \xi_N^{-1} \ / \ dP_0 \circ \xi_N^{-1}) \circ \xi_N = \exp(\theta) \ / \ L_{P_0}(\theta)$$

since, by assumption ξ_N is sufficient on $(E,\mathcal{A},\mathcal{P}_N)$. This proposition has been given in the explicite case in [5].

c) Spaces of real statistics and completion : Let $V_1(E,\mathcal{A},\mathcal{P})$ be the vector subspace of $L_1(E,\mathcal{A},P_0)$ spanned by the family $\{dP_\theta/dP_0 \ ; \ \theta \in \Theta\}$ of r.r.v.. Set $U_1(E,\mathcal{A},\mathcal{P}) = \bigcap_{\theta \in \Theta} L_1(E,\mathcal{A},P)$.

Obviously, $L_\infty(E,\mathcal{A},\mathcal{P}) \subset U_1(E,\mathcal{A},\mathcal{P})$; we also know, by Proposition 5.1 that if $\Theta \subset \overset{\circ}{D}(E,\mathcal{A},P_0)$, $L(E,\mathcal{A},\mathcal{P}) \subset U_1(E,\mathcal{A},\mathcal{P})$.

For every $\xi \in U_1(E,\mathcal{A},\mathcal{P})$ [resp. $\xi \in L_\infty(E,\mathcal{A},\mathcal{P})$] and $\eta \in V_1(E,\mathcal{A},\mathcal{P})$ the mapping : $(\xi,\eta) \to E_{P_0}(\xi\cdot\eta)$ is a bilinear form (denoted by $\langle .,. \rangle_{\mathcal{P}}$) on $U_1(E,\mathcal{A},\mathcal{P}) \times V_1(E,\mathcal{A},\mathcal{P})$ [resp. $L_\infty(E,\mathcal{A},\mathcal{P}) \times V_1(E,\mathcal{A},\mathcal{P})$] which put respectively these two spaces in duality. These dualities are always separated in $V_1(E,\mathcal{A},\mathcal{P})$ that is : $\langle \xi,\eta \rangle_{\mathcal{P}} = 0$ for all $\xi \in U_1(E,\mathcal{A},\mathcal{P})$ [resp. for all $\xi \in L_\infty(E,\mathcal{A},\mathcal{P})$] $\Rightarrow \eta = 0$ in $V_1(E,\mathcal{A},\mathcal{P})$. Indeed, setting $\eta = \sum_{i=1}^n \lambda_i \frac{dP_{\theta_i}}{dP_0}$, for every $A \in \mathcal{A}$, we have $\sum_{i=1}^n \lambda_i P_{\theta_i}(A) = 0$ which implies that $\lambda_i = 0$, $i = 1,\ldots,n$ if the statistical space is not degenerate). Nevertheless the assumption $\langle \xi,\eta \rangle_{\mathcal{P}} = 0$, for all $\eta \in V_1(E,\mathcal{A},\mathcal{P})$ does not imply necessarily that $\xi = 0$, which expresses that the considered duality may be non separated in $U_1(E,\mathcal{A},\mathcal{P})$ [resp. $L_\infty(E,\mathcal{A},\mathcal{P})$] ; besides, this fact can be expressed in terms of "completion" of the statistical space :

Proposition 5.6.- The statistical space $(E,\mathcal{A},\mathcal{P})$ is complete [resp. boundedly complete] iff the duality between the spaces $U_1(E,\mathcal{A},\mathcal{P})$ [resp. $L_\infty(E,\mathcal{A},\mathcal{P})$] and $V_1(E,\mathcal{A},\mathcal{P})$ is separated.

The proof is easy.

Lemma 5.2.- Assuming that the C.E.T. V.S.S. $(E,\mathcal{A},\mathcal{P})$ is reduced, then it is complete [resp. boundedly complete] iff, for every finite dimensional subspace N of $L(E,\mathcal{A},\mathcal{P})$, the finite dimensional C.E.T. V.S.S. $(E_N, \mathcal{B}(E_N), \{P_\theta \circ \xi_N^{-1} ; \theta \in \Theta \cap N\})$ is complete [resp. boundedly complete].

Proof. With the developments preceding Proposition 5.4, the property is equivalent to the fact that for every $N \in \mathcal{F}(L)$, ξ_N is a complete [resp. boundedly complete] sufficient statistic on the "sub-statistical space" $(E,\mathcal{A},\mathcal{P}_N = \{P_\theta ; \theta \in \Theta \cap N\})$, with values in $E_N \equiv R^n$ if the dimension of N is n, or also, that the restricted statistical space $(E,\mathcal{A}_N,\mathcal{P}_N)$ is complete [resp. boundedly complete], where $\mathcal{A}_N$ denotes the σ-field generated by ξ_N and $\mathcal{P}_N$ now denotes the family of the restrictions to this σ-field of the $\{P_\theta ; \theta \in \Theta \cap N\}$. Let $V_1(E,\mathcal{A}_N,\mathcal{P}_N)$ be the vector space of r.r.v. spanned by the family $\{e^\theta ; \theta \in \Theta \cap N\}$. It is a subspace of $L_1(E,\mathcal{A}_N,P_0)$ since if $\theta \in N$, θ is $\mathcal{A}_N$-mesurable. And since $\Theta = \bigcup_{N \in \mathcal{F}(L)} \Theta \cap N$, it is easy to show that :

$$V_1(E,\mathcal{A},\mathcal{P}) = \bigcup_{N \in \mathcal{F}(L)} V_1(E,\mathcal{A}_N,\mathcal{P}_N). \quad (1)$$

Now, set $U_1(E,\mathcal{A}_N,\mathcal{P}_N) = \bigcap_{\theta \in \Theta \cap N} L_1(E,\mathcal{A}_N,P_\theta)$ and assume that for every $N \in \mathcal{F}(L)$ the statistical space $(E,\mathcal{A}_N,\mathcal{P}_N)$ is complete [resp. boundedly complete], that is, by Proposition 5.6 :

$$\{\forall\, \xi \in U_1(E,\mathcal{A}_N,\mathcal{P}_N) \,[\text{resp. } \forall\, \xi \in L_\infty(E,\mathcal{A}_N,\mathcal{P}_N)],\ \int_E \xi\eta\, dP_0 = 0,\ \forall\, \eta \in V_1(E,\mathcal{A}_N,\mathcal{P}_N)\} \Rightarrow \xi = 0.$$

Let $\xi \in U_1(E,\mathcal{A},\mathcal{P})$ [resp. $\xi \in L_\infty(E,\mathcal{A},\mathcal{P})$] and assume that $\int_E \xi\eta\, dP_0 = 0$, $\forall \eta \in V_1(E,\mathcal{A},\mathcal{P})$. Then, by (1), for every $N \in \mathcal{F}(L)$ we have $\int_E \xi\eta\, dP_0 = 0$ $\forall \eta \in V_1(E,\mathcal{A}_N,\mathcal{P}_N)$, thus $\int_E E_{P_0}^{\mathcal{A}_N}(\xi).\eta\, dP_0 = 0$, $\forall \eta \in V_1(E,\mathcal{A}_N,\mathcal{P}_N)$. Since the conditional expectation $E_{P_0}^{\mathcal{A}_N}(\xi) \in U_1(E,\mathcal{A}_N,\mathcal{P}_N)$ [resp. $L_\infty(E,\mathcal{A}_N,\mathcal{P}_N)$], we have : $E_{P_0}^{\mathcal{A}_N}(\xi) = 0$, $\forall N \in \mathcal{F}(L)$, thus, $\xi = 0$ since $\mathcal{A} = \mathcal{A}_L = \bigvee_{N \in \mathcal{F}(L)} \mathcal{A}_N$.

The necessity is obvious, and completes the proof of the lemma.

Proposition 5.7.- If the origin of $L(E,\mathcal{A},\mathcal{P})$ is an internal point of Θ, then the reduced C.E.T. V.S.S. $(E,\mathcal{A},\mathcal{P} = \{P_\theta ; \theta \in \Theta\})$ is complete.

Proof. 0 is an internal point of Θ iff for every $N \in \mathcal{F}(L)$, $\Theta \cap N$ has a non-empty interior in N. Then, the proposition derives directly from Lemma 5.2, and the well-known sufficient condition for a classical finite dimensional C.E.T. statistical space to be complete (see [2] theorem 1, p. 167).

We point out that the sufficient condition for completeness is of algebraic nature and does not require any topology on $L(E,\mathcal{A},\mathcal{P})$.

d) Conditioning properties : We summarize here some other properties which should be useful. Let N_1 and N_2 be two supplementary subspaces of $L(E,\mathcal{A},\mathcal{P})$, so that every element $\xi \in L(E,\mathcal{A},\mathcal{P})$ can be expressed as $\xi = \xi_1(\xi) + \xi_2(\xi)$ with $\xi_i(\xi) \in N_i$, $i = 1,2$. Set $\Theta_1 = pr_1(\Theta) = \{\theta_1(\theta) \ ; \ \theta \in \Theta\}$ and let $\mathcal{A}_i$ be the σ-field generated by the elements of N_i, $i = 1,2$. $P_{\mathcal{A}_1}$ will denote the restriction of any probability P on $(E,\mathcal{A})$ to the sub-σ-field $\mathcal{A}_1$, and $P^{\mathcal{A}_2}_{\mathcal{A}_1}$ will denote the conditional probability w.r. to $\mathcal{A}_2$ of P on $(E,\mathcal{A}_1)$.

Proposition 5.8.- If $\Theta_1 \subset D(E,\mathcal{A},P_0)$, there exists a unique probality P_2 on $(E,\mathcal{A}_2)$ such that for every $\theta \in \Theta$, setting $\theta = \theta_1 + \theta_2$ $(\theta_i \in N_i,\ i = 1,2)$ we have :
$dP_{\theta,\mathcal{A}_2} / dP_2 = \exp(\theta_2) / L_{P_0}(\theta)$, $\theta \in \Theta$.

Proposition 5.9.- If $\Theta_1 \subset D(E,\mathcal{A},P_0)$ and if $P^{\mathcal{A}_2}_{0,\mathcal{A}_1}$ is regular, setting $\theta = \theta_1 + \theta_2$ $(\theta_i \in N_i,\ i = 1,2)$ we have :
$dP^{\mathcal{A}_2}_{\theta,\mathcal{A}_1} / dP^{\mathcal{A}_2}_{0,\mathcal{A}_1} = \exp(\theta_1) / E^{\mathcal{A}_2}_{P_0}(e^{\theta_1})$, $\theta \in \Theta$.

Corollary.- If N_1 is finite dimensional and if $\Theta_1 \subset D(E,\mathcal{A},P_0)$, the σ-field $\mathcal{A}_1$ is conditionally similar w.r. to $\mathcal{A}_2$ for the parameter θ_2, and for every $x \in E$ the restricted (to $\mathcal{A}_1$) conditional statistical space

$$(E,\mathcal{A}_1,\ \{P^{\mathcal{A}_2}_{\theta,\mathcal{A}_1}\ ;\ \theta \in \Theta\})$$

is of C.E.T. with natural parameter θ_1.

This results from Jirina's theorem which implies that the $P^{\mathcal{A}_2}_{\theta,\mathcal{A}_1}$ are regular since $\mathcal{A}_1$ is generated by a finite dimensional random vector. Moreover they do not depend on θ_2 and they define a generalized exponential family with natural parameter θ_1 by Proposition 5.9.

Consequently, if θ_2 represents a nuisance parameter and if there exists a statistic which generates $\mathcal{A}_2$, then, conditionally to it, the statistical space restricted to $\mathcal{A}_1$ becomes a classical finite dimensional C.E.T. one which does not depend on θ_2. Especially if N_1 is one-dimensional, we are able to expect a generalization of Lehmann's conditional test method which may give optimal tests for a scalar parameter in presence of an infinite dimensional nuisance one.

6 - CONCLUDING REMARKS

As the several examples of applications prove, this generalization to the infinite dimensional case of the exponential families are useful in the construction of precise mathematical models for statistical problems where infinite dimensional vector observations are considered, being generally those of sample paths of random processes.

Moreover, this formalism allows us to use the powerful tools of functional analysis for a theoritical approach of these problems.

Particularly, the analytical properties of the C.E.T. statistical spaces provide a proper treatment of estimation theory : briefly, for the unbiased estimation of funtionals of the natural parameter $\theta \in \Theta$, we can use the fact that if

$\Theta + \Theta \subset D(E,\mathcal{A},P_0)$ -which implies that $\{\frac{dP_\theta}{dP_0}\}_{\theta\in\Theta} \subset L_2(E,\mathcal{A},P_0)$- then $K(\theta,\theta') = L_{P_0}(\theta+\theta')$ defines a symmetric positive definite kernel on $\Theta \times \Theta$, and thus, a reproducing kernel Hilbert space of functionals on Θ which appear to be just the admissible ones ; they are estimated by real statistics belonging to a well specified Hilbert space. For the maximum likelihood estimation in the Explicit C.E.T. statistical spaces, a theory can be developed by means of the convex analysis tools, as has been established by Barndorf-Nielsen O., in [1] in the finite dimensional case. This theory particularly enlightens the fact that an infinite dimensional parameter generally does not admit a maximum likelihood estimator. These parts of the work will be published in [8] ; they are not detailed here, since we restrict ourselves to the investigation of the theoretical aspects and the foundations of this generalization.

In this study, the systematic use of M.V.S. and their duals permits to avoid the consideration of any topology on the vector space of the possible observations. Indeed, the σ-field generated by the observable events in the experiment is what is first considered by the statistician.

We point out that, as in probability theory, the generalization of the finite-dimensional concepts and methods of mathematical statistics may be not immediate, requiring an investigation of new tools.

Finally, this theory should constitute a useful tool for the investigation in a parametric way of some classical non parametric problems (such as the estimation of densities e.g.). Since it admits an infinite dimensional parametrization, it contributes to unify the theory of statistical spaces (see Remarks 3.3). Proposition 5.4 which gives a coherence property of these generalized exponential families is in agreement with this point of view.

BIBLIOGRAPHY

[1] Barndof-Nielsen, O. (1970). "Exponential families, Exact theory", various Publ. Series, n° 19.
[2] Barra, J.R. (1971). "Notions fondamentales de Statistique Mathématique", Dunod, Paris.
[3] Johansen, S. (1976). "Homomorphisms and general exponential families", Congrès Européen des Statisticiens, Grenoble.
[4] Rozanov, J.A. (1971). "Infinite dimensional Gaussian distributions". Amer. Math. Soc. Providence.
[5] Soler, J.-L. (1973). "m-dual d'un espace vectoriel mesurable et structures statistiques de type exponentiel", CR. Acad. Sc., Paris, t. 277, sér. A, 49-52.
[6] Soler, J.-L. (1976). "Infinite Dimensional Exponential Type Statistical Spaces (Generalized Exponential Families)", Univ. Sc. et Méd. de Grenoble.
[7] Soler, J.-L. (1975). "La loi de probabilité de Wishart associée à un élément aléatoire gaussien décentré à valeurs dans un espace de Fréchet séparable. Application aux fonctions aléatoires gaussiennes". C.R. Acad. Sc., Paris, t. 281, Sér. 1, 471-474.
[8] Soler, J.-L. (to appear). "Introduction aux structures statistiques infinidimensionnelles", Univ. Sc. et Méd. de Grenoble.
[9] YEH, J. (1973). "Stochastic processes and Wiener integrals", M. Dekker wic. N.Y.

Recent Developments in Statistics
J.R.Barra et al., editors

Invited Paper

AN ALGEBRAIC THEORY FOR NORMAL STATISTICAL MODELS WITH UNKNOWN COVARIANCE

Søren Tolver Jensen
Institute of Mathematical Statistics
University of Copenhagen

We shall define a general hypothesis for the covariance of a normally distributed observation and give an explicit solution of the problem of likelihood inference. The hypotheses are restrictive, but, as far as we know, they include all examples of patterned covariance matrices for which the distribution problems have been solved explicitly. Most of the results described here have been obtained in collaboration with my colleagues Steen Andersson and Hans Brøns.

It is important to describe the hypotheses in an invariant way. The sample space is an N-dimensional real vector space V and the observation X is normally distributed on V. For sake of simplicity we shall assume throughout that the mean is zero. Let V^* be the dual of V and $V \odot V$ the vector space of symmetric tensors of order two. The covariance of X is by definition $E(X \otimes X)$ which is an element of $V \odot V$. The space $V \odot V$ can be identified with the space of symmetric linear mappings of V^* into V and the space $(V \odot V)^*$ with the space of symmetric linear mappings of V into V^*. In this way the inverse covariance becomes an element in $(V \odot V)^*$, and the normal distribution with mean zero and covariance Σ gets the density

(1) $$\frac{1}{n(\Sigma)} e^{-\frac{1}{2}\Sigma^{-1}(x \otimes x)}$$

with respect to a Lebesgue measure on V. $n(\Sigma)$ is a norming constant which for any choice of a basis in V is proportional to $(\det \Sigma)^{\frac{1}{2}}$.

The hypotheses we shall consider are those which are linear in both the covariance and the inverse covariance. They are given by two subspaces $L \subseteq V \odot V$ and $M \subseteq (V \odot V)^*$ with the property that $\Sigma \in L$ if and only if $\Sigma^{-1} \in M$ for all regular $\Sigma \in V \odot V$. The parameter space for the covariance is the set L_+ of positive definite elements of L. These normal distributions constitute an exponential family because M is a linear space, and the likelihood equation becomes

(2) $$\forall \delta \in M : \quad \delta(\Sigma) = \delta(x \otimes x).$$

Since L too is a linear space, the equation is a linear equation which can be solved easily, and the unique solution $y = t(x)$ is a quadratic function of x (i.e. a linear function of $x \otimes x$).

The hypotheses can also be characterized by means of Jordan algebras. For definition and properties of Jordan algebras see [3].

Let Σ_0 be any fixed element in L_+. In $V \odot V$ we define the composition $\Sigma_1 * \Sigma_2 = \frac{1}{2}(\Sigma_1\Sigma_0^{-1}\Sigma_2 + \Sigma_2\Sigma_0^{-1}\Sigma_1)$. With this composition the vector space $V \odot V$ is a formally real Jordan algebra. That is, $*$ is commutative, $*$ is distributive with respect to $+$, $(\Sigma_1^2 * \Sigma_2) * \Sigma_1 = \Sigma_1^2 * (\Sigma_2 * \Sigma_1)$, and $\Sigma_1^2 + \Sigma_2^2 = 0$ implies $\Sigma_1 = \Sigma_2 = 0$.

Since $(\partial\Sigma^{-1}/\partial\Sigma)(\Sigma_0) = -\Sigma^{-1}\Sigma_0\Sigma^{-1} \in M$ for $\Sigma \in L_+$ it follows that $\Sigma\Sigma_0^{-1}\Sigma \in L$ for $\Sigma \in L_+$, which shows that L is a Jordan subalgebra of $V \odot V$. A theorem of Jordan, von Neumann and Wigner now gives the structure of L. The Jordan algebra L is a product

(3) $$L = J_1 \times \ldots \times J_k$$

of Jordan algebras where each of the J_i's is isomorphic to one of the following simple Jordan algebras: (i) the set $\mathbb{R}$ of the real numbers, (ii) $\mathbb{R} \times W$, where W is a real vector space of dimension $m \geqq 2$ with a positive definit form Φ and composition $(\lambda_1,w_1) * (\lambda_2,w_2) = (\lambda_1\lambda_2 + \Phi(w_1,w_2), \lambda_2 w_1 + \lambda_1 w_2)$. or (iii) the Hermitian $p \times p$-matrices $H_p(D)$, $p \geqq 2$, where D is either the set $\mathbb{R}$ of the real numbers, the set $\mathbb{C}$ of the complex numbers, or the set $\mathbb{H}$ of the quaternions, and composition $A * B = \frac{1}{2}(AB + BA)$.

A canonical form of the hypotheses is given in [4]. The sample space V decomposes uniquely

(4) $$V = V_1 + \ldots + V_k$$

into a direct sum of subspaces such that, for the corresponding decomposition $X = (X_1,\ldots,X_k)$ of the observation, the variables $X_1,\ldots,X_k$ are independently distributed, and their covariances vary functionally independently. The parametrization of the covariance of X_i is given by an injective Jordan algebra homomorphism τ_i of J_i into $V_i \odot V_i$. Depending on the type of the simple Jordan algebra J_i, the parameter space is either (i) the set of the positive real numbers, (ii) the set $\{(\lambda,w) \in \mathbb{R} \times W \mid 0 < \lambda \wedge \Phi(w,w) < \lambda^2\}$, or (iii) the set of positive definite $p \times p$-matrices.

It is possible to describe the structure in more detail, see [4], but to derive the distribution of the m.l.estimator we only need the Jordan structure.

The parameter space L_+ acts linearly on V by

$$g(a)(x) = a\,\Sigma_0^{-1}(x)\ , \quad x \in V,\ a \in L_+,$$

and linearly on L by

$$P(a)(\Sigma) = a\,\Sigma_0^{-1}\,\Sigma\,\Sigma_0^{-1}\,a\,, \quad \Sigma \in L,\ a \in L_+.$$

Moreover, the family $\{P(a) \mid a \in L_+\}$ acts transitively on L_+, and the following relations hold

(5)
$$(P(a)(\Sigma))^{-1}(x \otimes x) = \Sigma^{-1}(g(a)^{-1}(x) \otimes g(a)^{-1}(x)),$$
$$n(P(a)(\Sigma)) = n(\Sigma)\,|\det g(a)|,\quad t(g(a)(x)) = P(a)(t(x)).$$

In the following it is assumed that the m.l. estimator for Σ exists (i.e. $Y = t(X) \in L_+$) with probability one (the alternative is that it never exists). It then follows from (2) and (5) that the distribution of $Y = t(X)$ has density

(6)
$$\frac{n(y)}{n(\Sigma)}\, e^{-\frac{1}{2}\Sigma^{-1}(y)}$$

with respect to an invariant measure on L_+. Since L_+ acts transitively on itself, the distribution is uniquely determined. If L is a simple Jordan algebra an invariant measure has density

$$n(y)^{-2\dim L/\dim V}$$

with respect to a Lebesgue measure on L_+. In case (i) $L = \mathbb{R}$ we get the χ^2/f - distribution, in case (ii) $L = \mathbb{R} \times W$ the density

$$\text{const}\ \frac{(z^2 - \Phi(z_1,z_1))^{N/4-m/2-1/2}}{(\lambda^2 - \Phi(w,w))^{N/4}}\ \exp\left(-\frac{N}{2(\lambda^2 - \Phi(w,w))}\ (\lambda z - \Phi(w,z_1))\right),$$

$0 < z$ and $\Phi(z_1,z_1) < z^2$, and in case (iii) the Wishart distributions.

Let $L_2 \subseteq L_1$ be two Jordan subalgebras of $V \odot V$. It is then possible to test the hypothesis H_2: $\Sigma \in L_2$ against the hypothesis H_1: $\Sigma \in L_1$. Let Y_1 and Y_2 be the two m.l.estimators. It follows from (2) that the LRT-statistic is $Q = n(Y_1)/n(Y_2)$. Since $n(Y_1)/n(\Sigma) = Q\,n(Y_2)/n(\Sigma)$ and Q and Y_2 are independent, we obtain

$$E(Q^\alpha) = E((n(Y_1)/n(\Sigma))^\alpha)/E((n(Y_2)/n(\Sigma))^\alpha),\ \alpha > 0.$$

If Y has the distribution (6) the moments $E((n(Y)/n(\Sigma))^\alpha) = \Pi_{i=1}^k\ c(\alpha,J_i,V_i)$ where $c(\alpha,J_i,V_i)$ depends only on $N_i = \dim V_i$ and the type of the Jordan algebra J_i:

(i)
$$(2/N_i)^{\alpha N_i/2}\ \Gamma(N_i/2 + \alpha N_i/2)\ /\ \Gamma(N_i/2),$$

(ii) $$(4/N_i)^{\alpha N_i/2} \frac{\Gamma(N_i/4 + \alpha N_i/4)\ \Gamma(N_i/4 - m/2 + 1/2 + \alpha N_i/4)}{\Gamma(N_i/4)\ \Gamma(N_i/4 - m/2 + 1/2)},$$

(iii) $$(2p/N_i)^{\alpha N_i/2} \prod_{j=1}^{p} (\Gamma(N_i/2p - q(j-1)/2 + \alpha N_i/2p) / \Gamma(N_i/2p - q(j-1)/2)),$$

$q = \dim D$. Thus the distribution of Q depends only on the decompositions (3) and (4) for the two hypotheses H_1 and H_2.

REFERENCES.

[1] Anderson, T.W. (1969). Statistical inference for covariances with linear structure. In Proceedings of the Second International Symposium on Multivariate Analysis, ed. P.R. Krishanaiah (Academic Press, New York).

[2] Andersson, S. (1975). Invariant normal models. Ann.Statist. 3, 132-154.

[3] Jacobson, N. (1968). Structure and Representation of Jordan Algebras. Amer.Math.Soc. (Providence, Rhode Island).

[4] Jensen, S.T. (1975). Covariance hypotheses which are linear in both the covariance and the inverse covariance. Preprint No. 1, Institute of Mathematical Statistic, University of Copenhagen.

Recent Developments in Statistics
J.R.Barra et al., editors

Invited Paper

STRONG INVARIANCE PRINCIPLES

G. Tusnády
Mathematical Institute of the
Hungarian Academy of Sciences

1. HISTORICAL BACKGROUND

Let the i.i.d.r.v.-s $x_1, x_2, \ldots$ be uniformly distributed on $[0,1]$, and let

$$a_n(t) = \sum_{i=1}^{n} J(x_i < t)$$

where $J(s < t)=1$ if $s < t$, and $=0$ otherwise. In his famous "heuristic approach" Doob proposed to approximate the process a_n by

$$b_n(t) = nt + \sum_{i=1}^{n} (w_i(t)-tw_i(1)),$$

where the w_is are i.i.d. Wiener processes, i.e. they are Gaussian processes with 0 expectation and with covariance function

$$E\, w_i(t)w_i(s) = \min(t,s).$$

It is easy to see that one can construct the processes w_i as an a.s. uniform limit of continuous functions (cf. Breiman (1968)). Let

$$\{\varepsilon_{ij},\ 1 \le j \le 2^i, \qquad i = 0,1,\ldots\}$$

be independent standard normal variables, $k(t)=1-|t-1|$ if $0 \le t \le 2$, and $=0$ otherwise, and let

$$v_i(t) = \sum_{j=1}^{2^i} \varepsilon_{ij}\, k(2^i t-2j+2)$$

Then the series $\sum_{i=1}^{\infty} 2^{-\frac{i+1}{2}} v_i(t)$ converges uniformly with probability 1 on $[0,1]$, and the sum

$$w(t) = v_0(t) + \sum_{i=1}^{\infty} 2^{-\frac{i+1}{2}} v_i(t)$$

has the desired covariance. (Note that $w(t)-v_o(t)=w(t)-tw(1)$, a so-called Brownian-Bridge.)

Let $C(0,1)$ be the metric space of continuous functions on $[0,1]$ with metric

$$d(f,g) = \sup_{0\le t\le 1} |f(t)-g(t)|,$$

and let $\tilde{a}_n$ be the piecewise linear function fitting a_n at the x_i s. Both the processes $n^{-1/2}(\tilde{a}_n(t)-nt)$, $n^{-1/2}(b_n(t)-nt)$ define a probability measure on $C(0,1)$, let us denote them by A_n, B resp. (B does not depend on n.) The Prohorov distance of probability measures given on a metric space (S,d) is

$$\rho(P,Q) = \inf\{\varepsilon:\ P(F) < Q(F^{\varepsilon})+\varepsilon,\ \text{for any closed } F\subset S\},$$

where $F^{\varepsilon}=\{u\in S: \exists\, v\in F,\ d(u,v) < \varepsilon\}$. It is an equivalent form of the weak invariance principle due to Kolmogorov, Erdös, Kac, Doob, Donsker and Prohorov (cf. Billingsley (1968)) that $\rho(A_n,B)$ tends to 0.

As far as this point the processes a_n,b_n had not been defined on the same probability space, only their distributions, the measures generated by them were taken into consideration. It is a natural idea to define them on the same probability space in such a way that they get as close to each other as possible. Let us say that we couple the processes a_n,b_n by defining them on the same probability space. Any coupling determines a joint distribution of the processes a_n,b_n, which is quite arbitrary, only its marginal distributions are prescribed: they are the given distributions of the processes a_n,b_n. In the proof of his new version of the theorem of iterated logarithm Strassen (1965b) used an embedding of the partial sums of i.i.d.r.v.-s into the Wiener process. (Embedding is a special coupling: one of the coupled processes is produced by a deterministic transformation from the other process.) The basic idea of this embedding is to use the Skorohod representation of partial sums of i.i.d.r.v.-s as the values of a Wiener process at appropriate random time points. On applying Strassen's result Brillinger (1969) and Breiman (1968) have constructed an embedding of order $n^{1/4}$ for a_n,b_n. In the meantime Strassen (1965a) and Dudley (1968) have shown that once the distance $\rho(P,Q)$ of measures on (S,d) is small, then there is a measure ν on $S\times S$ such that the marginals of ν is P and Q respectively, and the random variable $d(s_1,s_2)$ is small with respect to ν. In our case this theorem means that for any $\delta > 0$ there is a coupling of a_n,b_n such that

$$P(\sup_{0\le t\le 1} |a_n(t)-b_n(t)| > n^{1/2}(\rho(A_n,B)+\delta)) < \rho(A_n,B)+\delta.$$

Hence the existence of reasonable couplings of a_n, b_n is a corollary of the weak invariance principle. On the other hand, any coupling of a_n, b_n gives an upper estimation for $\rho(A_n, B)$.

There was much work done on determining the exact magnitude of $\rho(A_n, B)$. It turned out, the best result that can be achieved by using the Skorohod-Strassen embedding, is of order $n^{-1/4}$. Csörgö and Révész were the first who constructed a better embedding. Their method was the quantile transformation (also called the probability integral transformation). For the sake of a better view let us speak about embedding of partial sums into a Wiener process. The distribution of one partial sum is near to the distribution of the corresponding increment of the Wiener process because of the central limit theorem. If we are given two variables, ξ and η such that their distribution is near to each other, it is reasonable to embed ξ into η by the transformation $\xi = T(\eta)$, where the function T is defined by

$$P(\xi < T(y)) = P(\eta < y).$$

Thus we can form blocks from the partial sums, and embed each of them into the corresponding increment of the Wiener process.

Let us formulate here the result of Csörgö and Révész in the multidimensional case. Let K be the k-dimensional unit cube

$$K = \{t : 0 \le t \le 1\},$$

where the inequality $s < t$ means the validity of all the corresponding inequalities between the coordinates of s and t. Let the i.i.d.r.v.-s $x_1, x_2, \ldots$ be uniformly distributed now on K, and set

$$a_n(t) = \sum_{i=1}^{n} \mathrm{J}(x_i < t),$$

$$b_n(t) = n\,\lambda(t) + \sum_{i=1}^{n} (w_i(t) - \lambda(t) w_i(1)),$$

where $\lambda(t)$ is the product of the coordinates of t and the w_i s are i.i.d. Gaussian processes on K with expectation 0 and with covariance function

$$E\, w_i(t) w_i(s) = \lambda(\min(t,s)),$$

where again the minimum is taken coordinatewise.

THEOREM (Csörgö, Rêvész (1975)). For any k there is a version of k-dimensional processes a_n, b_n such that

$$\sup_{t \in K} | a_n(t) - b_n(t) | \underset{a.s.}{=} O(n^{1/2-1/2k+4} (\log n)^2).$$

2. THE CONDITIONAL QUANTILE TRANSFORMATION

The quantile transformation is better than the Skorohod embedding only in such cases when the first few moments of the x_i s coincide with the corresponding moments of the normal distribution. Motivated by the above mentioned proof for the existence of a continuous version of the Wiener process, I proposed to couple the processes a_n, b_n (in the one dimensional case)in the following way. Suppose, the process b_n is given in the form

$$b_n(t) = nt + n^{1/2} \sum_{i=1}^{\infty} 2^{-(i+1)/2} \sum_{j=1}^{2^i} \varepsilon_{ij} k(2^i t - 2j + 2)$$

where the ε_{ij} s and k are the same as in the introduction. Let us determine first $a_n(1/2)$ as a function of ε_{11}. The distribution of $a_n(1/2)$ is binomial with parameters $n, 1/2$. The quantile transformation in this case is defined for any real z by

$$T_n(z) = \{\inf j: \sum_{i=0}^{j} \binom{n}{i} 2^{-n} \geq \emptyset(z)\},$$

where $\emptyset$ is the standard normal distribution.

LEMMA. For any z, n we have

$$|T_n(z) - \frac{1}{2}(n + z\sqrt{n})| < z^2 + 1.$$

The proof of this lemma is elementary and is based on Euler's approximation (see also Feller (1945))

$$\log n! = n \log n - n + \frac{1}{2} \log 2\pi n + \vartheta_n / 12n,$$

where $0 < \vartheta_n < 1$. The lemma implies that choosing

$$a_n(1/2) = T_n(\varepsilon_{11})$$

we have

$$|a_n(1/2) - b_n(1/2)| < \varepsilon_{11}^2 + 1.$$

The next step in the construction is the determination of $a_n(1/4)$ and $a_n(3/4)$. In this step we cannot neglect the fact that $a_n(1/2)$ is already determined. Our task is to construct such a process a_n that the joint distribution of the variables

$$a_n(t_1),\ a_n(t_2), \ldots, a_n(t_j)$$

for any $0 < t_1 < t_2 < \ldots < t_j < 1$ coincide with a prescribed distribution, namely that the variables.

$$a_n(t_1),\ a_n(t_2)-a_n(t_1), \ldots, n-a_n(t_j)$$

should have a polinomial distribution with parameters n and

$$t_1,\ t_2-t_1\ , \ldots,\ 1-t_j.$$

Thus we have to fit the conditional distribution of $a_n(1/4)$, $a_n(3/4)$ on the condition that $a_n(1/2)$ equals to the above determined $T_n(\varepsilon_{11})$, say $a_n(1/2)=m$. One can easily see, that we are lucky: the variables $a_n(1/4)$, $a_n(3/4)-a_n(1/2)$ are conditionally independent, and their conditional distribution is again binomial with parameters $m,\ 1/2$ and $n-m,\ 1/2,$ respectively. On choosing

$$a_n(1/4) = T_m(\varepsilon_{11}),$$

$$a_n(3/4) = m + T_{n-m}(\varepsilon_{22}),$$

we have

$$\left|\left[a_n(1/4) - \frac{1}{2}a_n(1/2)\right] - \frac{2m}{n}\left[b_n(1/4) - \frac{1}{2}b_n(1/2)\right]\right| \le \varepsilon_{21}^2+1,$$

$$\left|\left[a_n(3/4)- \frac{1}{2}(a_n(1/2)+a_n(1))\right] - \frac{2n-2m}{n}\left[b_n(3/4)- \frac{1}{2}(b_n(1/2)+b_n(1))\right]\right| \le \varepsilon_{22}^2 + 1,$$

i.e. the "curtosis" (the differences of type $f(x)- \frac{1}{2}[f(x+\Delta)+f(x-\Delta)]$ for some Δ) of the processes a_n, b_n are close to each other.

The next step is to determine

$$a_n(1/8),\ a_n(3/8),\ a_n(5/8),\ a_n(7/8)$$

in the same way, and so on, up to

$$\{a_n((2j-1)2^{-N}),\ j=1,2,\dots,2^{N-1}\},\ \text{where}\ N=[\log n].$$

All the variables $a_n(j2^{-N})$ have been determined in at most $\log n$ steps, so we can expect that

$$\sup_{0<j<2^N} |a_n(j2^{-N})-b_n(j2^{-N})| \sim \log n.$$

The only annoying nuisance is the presence of the square root term on the left hand sides in the above inequalities. This is the consequence of the difference between the conditional variances of the processes a_n, b_n. Fortunately, it does not make much disturbance, and we can prove the following

THEOREM (Komlós, Major, Tusnády (1975)). For $k=1$ there is a version of the processes a_n, b_n such that

$$P(\sup_{0<t<1} |a_n(t)-b_n(t)| > C\log n+z) < De^{-\lambda z}$$

holds true for any z, where C, D, λ are positive constants.

COROLLARY (Komlós, Major, Tusnády (1974)). There are positive constants c_1, c_2 such that

$$c_1 \frac{\log n}{\sqrt{n}} < \rho(A_n, B) < c_2 \frac{\log n}{\sqrt{n}}$$

holds true for any n.

As a matter of fact only the upper part of the estimation is a corollary of the above theorem. The lower estimation is a consequence of Bártai's "stochastic geyser theorem": Suppose that some measurements produced some process $\xi(t)$, and we have to decide wether

$$\xi(t) = n^{-1/2}(a_n(t)-nt)+\delta(t),$$

or

$$\xi(t) = n^{-1/2}(b_n(t)-nt)+\delta(t),$$

where $\delta(t)$ is the error of our measuring device , and all we know about the error is the bound $\sup_{0\le t\le 1} |\delta(t)| \le \Delta$. Then this can be done with a large probability if Δ is small compared to $\log n$.

3. THE MULTIDIMENSIONAL CASE

Let $k > 1$, and let us partition the k-dimensional vectors X_i, t to $(k-1)$ and 1 - dimensional parts: $X_i = (Y_i, Z_i)$, $t=(u,v)$, and let the corresponding partition of K be $K=(L,I)$, where $I=(0,1)$. Suppose that we have already coupled the $(k-1)$-dimensional processes

$$a_n(u,1),\ b_n(u,1),$$

and should like to extend their coupling into k-dimension. We shall now use dyadic scheme on the k-th coordinate, i.e. first we couple the $(k-1)$-dimensional processes

$$c_n(u) = a_n(u,1/2) - \frac{1}{2} a_n(u,1),$$

$$d_n(u) = b_n(u,1/2) - \frac{1}{2} b_n(u,1)$$

under the condition that the processes $a_n(u,1)$, $b_n(u,1)$ are given, then we couple step by step the appropriate differences of a_n, b_n in the same way. The process d_n is a $(k-1)$-dimensional Wiener process on L, and it is independent of $b_n(u,1)$. The conditional model for c_n is

$$c_n(u) = \sum_{i=1}^{n} \xi_i J(Y_i < u),$$

where the ξ_i s are i.i.d.r.v.-s with

$$P(\xi_i=1) = P(\xi_i = -1) = \frac{1}{2},$$

since $\xi_i = \text{sign}\,(Z_i - \frac{1}{2})$.

If $k=2$, the coupling of c_n, d_n is just the same as the coupling of partial sums of i.i.d.r.v.-s and the Wiener process. In this way I proved the following

THEOREM (Tusnády, 1976a). For $k=2$ there is a version of a_n, b_n such that

$$P(\sup_{t\in K} |a_n(t) - b_n(t)| > (C \log n + z)\log n) < De^{-\lambda z}$$

holds true for any z, where C,D,λ are positive constants.

If $k > 2$, the structure of partial sums involved in c_n is much more complicated. That is why a straightforward extension of the one-dimensional case gives an order of magnitude n^α only for $k > 2$. Still I have some hope for an embedding of order $(\log n)^\alpha$. Although I do not have a final result, let me sketch a vague idea: Let the process e_n be defined by

$$e_n(u) = \sum_{i=1}^{n} \varepsilon_i J(Y_i < u),$$

where the ε_i s are independent standard normal variables. I want to find the coupling of the conditional distributions of d_n and e_n given $Y_1,\dots,Y_n$, which is minimizing the

$$E\left(\int_L (d_n - e_n)^2 du \mid Y_1,\dots,Y_n\right).$$

The finite dimensional version of this problem is the following.

We are given two m-dimensional covariance matrices Σ_1,Σ_2. Let W_1,W_2 be m-dimensional random normal vectors with 0 expectation, and covariance Σ_1,Σ_2, respectively. What is the embedding of W_1,W_2 for which

$$E\|W_1 - W_2\|^2 \to \min\ ?$$

<u>LEMMA.</u> Given Σ_1 and Σ_2, $W_i=A_i V$, $(i=1,2)$, where the A_i's are such that $A_iA_i^* = \Sigma_i$, $i=1,2$, and V is an m-dimensional random normal vector with 0 expectation and covariance I. Then $E\|W_1 - W_2\|^2$ is minimal among all A_1,A_2 with $A_iA_i^* = \Sigma_i$ if and only if $A_1^*A_2 = A_2^*A_1$.

Professor Olkin kindly called my attention to the fact that this lemma is due to Green and Schönemann (cf. Golub (1968)). Let H be L_2 space on L, and E the m-dimensional Euclidean space. The operator

$$(Af)(u) = \int_L f(s)J(s < u)ds, \qquad u\in L$$

maps H into H, and the operators

$$(Bf)(i) = n^{-1/2}\int_L f(s)J(s < Y_i)ds, \quad i=1,2,\dots,n,$$

$$(Uf)(i) = \int_L f(s)u_i(s)ds, \qquad i=1,2,\dots,n$$

map H into E where $u_1,u_2,\dots,u_n$ are orthonormal functions on L. The above lemma leads to the equation

$$A^*BU = U^*B^*A.$$

Let us denote A^*BU by C, then

$$C^2 = A^*BB^*A,$$

which does not depend on U. My aim is to calculate C and U, and show that U determines such a coupling that with some $\alpha > 0$

$$E((d_n(u) - e_n(u))^2 | Y_1, Y_2, \ldots, Y_n) \sim (\log n)^\alpha$$

holds true for all $u \in L$ and for the majority of the possible values of $Y_1, Y_2, \ldots$ $\ldots, Y_n$. Both the processes d_n, e_n are Gaussian, so the covariance of their difference is near to the real magnitude of $\max_{u \in L} |d_n(u) - e_n(u)|$. (Let us mention here the results of Driscoll (1973), which are connected with this problem.) The embedding of c_n into e_n seems to be a much harder problem. Both the operators BB^* and B^*B have n eigenvalues, and the magnitude of these eigenvalues is strongly connected with our problem. A rough calculation shows that they are about

$$\lambda_i = \frac{1}{i^2}, \qquad i = 1, 2, \ldots, n.$$

If it is really the case, then the difference $c_n - e_n$ may be again about $(\log n)^\alpha$ with some $\alpha > 0$.

4. POSSIBLE EXTENSIONS

Let H be now the L_2 space on K and, for any $f \in H$, let us define $a_n f, b_n f$ by

$$a_n f = \sum_{i=1}^{n} f(X_i), \quad b_n f = \int_K f(s) b_n(ds).$$

For any subset $\mathrm{F} \subset H$ the embedding of a_n, b_n on F means the determination of the variables $a_n f, b_n f$ for any $f \in \mathrm{F}$. Especially, when F consists of indicator functions of subsets of K:

$$\mathrm{F} = \{f\colon\ F \subset K,\ f(s) = 1 \ \text{ iff } \ s \in F\},$$

then the embedding on F means the embedding of the empirical frequencies of the sample elements $X_1, X_2, \ldots, X_n$ in $F \in \mathrm{F}$. Following Dudley (1973) we shall say that the family F is GB (Gaussian bounded) if there is a bounded version of the process $b_n f$, $f \in F$. The boundedness of F is connected with the metric entropy: let us say that $f_1, f_2, \ldots, f_N$ is an ε-covering family of F, if for any $f \in F$ there is an i such that $\|f - f_i\| < \varepsilon$. Let $N(\varepsilon)$ be the minimal number of elements in an ε-covering family of F. The maximum of N independent standard normal variables is of the order $\sqrt{2 \log N}$, thus one can expect that F is GB, if there is a sequence of numbers ε_j such that the series $\sum_{j=1}^{\infty} \varepsilon_{j-1}(2 \log N(\varepsilon_j))^{1/2}$ is convergent. Indeed, if $\limsup_{\varepsilon \to 0} \log \log N(\varepsilon)/\log(1/\varepsilon) < 2$, then F is GB, and if $\limsup_{\varepsilon \to 0} \log \log N(\varepsilon)/\log(1/\varepsilon) > 2$, then F is surely not GB. The metric entropy of indicator functions of certain subsets of K is connected with the smoothness of the boundary of these subsets. Roughly speaking the boundary of k-dimensional GB sets has to be at least k-times differentiable.

THEOREM (Révész (1976a)). Let F_k be the family of those subsets of K which have bounded k-times differentiable boundary. There is a version of sequences $a_n f$, $b_n f$ on F_k such that

$$\sup_{f \in F_k} |a_n f - b_n f| \underset{a.s.}{=} O(n^{1/2 - 1/12k+14}).$$

It should be noted however that there are GB sets such that there is no reasonable embedding on them. The following example is due to Dudley (personal communication). Let $F = \{F_i,\ i=1,2,\ldots\}$ be a sequence of subsets of K such that the events $\{X \in F_i,\ i=1,2,\ldots\}$ are totally independent, where X is uniformly distributed on K, and $\lambda(F_i) \sim 1/(\log i)^2$. Then F is GB and

$$P(\limsup_{n \to \infty} \sup_i n^{-1/2}(a_n F_i - n\lambda(F_i)) = \infty) = 1.$$

The investigation of the structural properties of embedding-ensuring families is only one possibility to extend our results for abstract spaces. A more effective method would be the extension of the characteristic functions. Our proofs are based on calculating moment generating functions, but it is in a certain sense indirect. It would be much easier to conclude to closeness of processes directly from the closeness of their moment generating functional (see Abramov (1976), Kuelbs (1975)).

As we have seen, the first embedding was used for proving the theorem of iterated logarithm. Major investigated the weakest assumption on moments needed for useful embeddings (cf. Halász, Major (1966) and Major (1976a), (1976b)). Révész (1976b) investigated the iterated logarithm theorem in the multidimensional case.

The investigation of the processes a_n, b_n were originally motivated by testing hypotheses. The weak invariance principle gives us an effective tool to calculate limit distributions of different functionals. Nowadays however not only the limit distributions are needed. For frequently used statistics we have the whole asymptotic expansion of their distribution. They were produced by *ad hoc* calculations, and no general method is known for solving such problems. A better view on asymptotic properties of the empirical measure could help the investigation of the power of tests for goodness of fit. Csörgö and Rêvész (1974 and 1976) investigated the quantile process rather than the empirical measure. Perhaps it would be better to handle the empirical measure as a real probability measure and to investigate its closeness to the real measure in some comprehensive way, not only through a branch of frequencies. One possibility of such type is, the Prohorov distance between the empirical and the real measures for testing hypotheses. No result is known about the distribution of this statistic , only some estimations were given for its expectation (cf. Dudley (1968)).

REFERENCES

Abramov, V.A. (1972) Estimates for the Lévy-Prohorov distance (Russian). Teor. Verojatnost. i Primenen. 21, 406-410.

Bârtfai, P. (1966) Die Bestimmung der zu einem wiederkehrenden Prozess gehörenden Verteilungsfunktion aus den mit Fehlern behafteten Daten einer einzigen Realisation. Studia Sci. Math. Hungar. 1, 161-168.

Billingsley, L. (1968) Convergence of probability measures. Wiley.

Breiman, L. (1968) Probability. Addison-Wesley.

Brillinger, D.R. (1969) An asymptotic representation of the sample df. Bull. Amer. Math. Soc. 75, 545-547.

Csörgö, M., Rêvész, P. (1974) Some notes on the empirical distribution function and the quantile process. Coll. Math. Soc. J. Bolyai 11. Limit theorems of Probability theory 59-71.

Csörgö, M., Rêvész, P. (1975) A new method to prove Strassen type laws of invariance principle I-II. Z. Wahrscheinlichkeitstheorie verw. Geb. 31, 255-269.

Csörgö, M., Rêvész, P. (1976) On strong approximation of the quantile process. Ann. Statistics.

Driscoll, M.F. (1973) The reproducing kernel Hilbert space structure of the sample paths of a Gaussian process. Z. Wahrscheinlichkeitstheorie verw. Geb. 26, 309-316.

Dudley, R.M. (1968) Distances of probability measures and random variables. Ann. Math. Statist. 39, 1563-1572.

Dudley, R.M. (1972) Speeds of metric probability convergence, Z. Wahrscheinlichkeitstheorie verw. Geb. 22, 323-332

Dudley, R.M. (1973) Sample functions of Gaussian processes, Ann. Probability 1, 66-103.

Golub, G.H. (1968) Least squares, singular values and matrix approximations. Applikace Matematiky 13, 44-51.

Halász, G., Major, P. (1976) Reconstructing the distribution from partial sums of samples. Ann. Probability

Komlós J., Major P., Tusnády, G. (1974) Weak convergence and embedding. Coll. Math. Soc. J. Bolyai 11. Limit theorems of probability theory, 149-165.

Komlós, J., Major, P., Tusnády, G. (1975) An approximation of partial sums of independent RV's and the sample DF. I-II. Z. Wahrscheinlichkeitstheorie verw. Geb. 32, 111-131 and 34, 33-58.

Kuelbs, J. (1975) Sample path behavior for Brownian motion in Banach spaces. Ann. Probability 3, 247-261.

Major, P. (1976a) The approximation of partial sums of independent RV's. Z. Wahrscheinlichkeitstheorie verw. Geb. 35, 213-220.

Major, P. (1976b) Approximation of partial sums of i.i.d.r.v.-s when the summands have only two moments. Z. Wahrscheinlichkeitstheorie verw. Geb. 35, 221-229.

Révész, P. (1976a) On strong approximation of the multidimensional empirical process. Ann. Probability

Révész, P. (1976b) Three theorems on multivariate empirical process.

Strassen, V. (1965a) The existence of probability measures with given marginals. Ann. Math. Statist. 36, 433-439.

Strassen, V. (1965b) Almost sure behaviour of sums of independent random variables and martingales. Proc. 5th Berkeley Sympos. Math. Statist. Probab. Vol. II. (part 1) 315-343.

Tusnády, G. (1976a) A remark on the approximation of the sample DF in the multidimensional case. Periodica Math. Hung.

Tusnády, G. (1976b) On asymptotical optimal tests. Ann. Statistics.

Feller, W. (1945) On the normal approximation to the binomial distribution. Annals of Math. Statistics 16, 319-329.

PART TWO: Selected Contributed Papers

A

Mathematical Statistics and Statistical Inference

Recent Developments in Statistics
J.R.Barra et al., editors

ON THE STRUCTURE AND APPLICATIONS OF INFINITE DIVISIBILITY, STABILITY AND SYMMETRY IN STOCHASTIC INFEPENCE

R. Ahmad and A.M. Abouammoh,
University of Strathclyde, Glasgow.

1. INTRODUCTION

The concepts of infinite divisibility, stability, symmetry and unimodality besides being useful and well-known in probability and statistics, are basic to many inference problems. Some of these concepts are also related to stationary independent stochastic processes, infinitesimal random systems, exchangeable and spherical exchangeable processes, the random sample idea, and nonparametric methods, which are fundamental in theory and applications and occur frequently in many real life situations. Among a wide range of applications, one finds these ideas arising in economic time series, industry, meterology, quantum physics and atomic nuclei, fluid and liquid analyses, signal detection, biology, biomedical and other scientific fields.

It is most common in many spheres that random phenomena take place due to the effect of a large number of causes, where each single cause individually has only an insignificantly small effect on the phenomena, but the total aggregate effect is worth probabilistic-statistical study. Here, one can recall the basic derivations of the Gaussian processes or normality. This brings to mind the concepts of infinitesimal random systems, infinite divisibility, and sums of independent or dependent random variables. There is a large amount of work in this area, for example Lévy (1937), Gnedenko-Kolmogorov (1954), Loève (1963) Lukacs (1970), Feller (1971), Petrov (1975), and many other references contained therein. In a recent paper Steutel (1973) has surveyed some of the work on infinite divisibility during the last fifteen years.

In the problems of filtering, forecasting, prediction, and control, the underlying processes are assumed to be stable in some sense. Also, the concept of symmetry, besides being useful in the above situations, is basic to some nonparametric problems. For estimating unknown probability density functions, one often assumes the underlying family to be 'smooth' in some sense, that is unimodal, symmetric or stable. Finally, in many 'smoothing' operations in mathematical and sometimes even 'nonmathematical' sciences, one employs the above concepts via convolutions, mixing, averaging, or some other suitable transformations. Similarly, these concepts and the ideas of exchangeability and spherical exchangeability, occur in psychological and learning experiments, quantum physics, solar wind and nozal gas flow etc., to the spread of many bacteriological colonies, biological populations, some other growth phenomena, polutants, ambisonic music system and Bayesian statistics, see Ahmad (1974, 1975a). Further, see the role played by unimodality in Barndorff-Nielsen (1976) as applied to the plausibility inference which is in a sense complimentary to likelihood inference and its many variations. Finally, note that the largest family of infinite divisible distributions which includes symmetric stable, stable, some unimodal (strongly unimodal, universal), and many variants of the so called $\mathcal{L}$-class distributions, is itself contained in the totality of limit laws of the Central Limit Problems, see Loève (1963, page 297). Thus, one is in the realm of asymptotic statistics which is playing an ever greater role in the theory and applications of probability and statistics.

The object of this paper is to unify, extend wherever possible, and give interrelationships of the above concepts with their many and varied applications. The effort in this direction is not claimed to be exhaustive and definitive, rather we have tried to deal with those aspects of the above ideas which will be of some interest to applied mathematical statisticians in a broad sense.

2. THE FUNDAMENTAL MEASURE-THEORETIC AND FUNCTIONAL-ANALYTICAL RESULTS

In the sequel, we shall briefly discuss some of the fundamental functional-measure-theoretic results which form the basis for many important concepts in probability theory. These results are well known, though scattered throughout the literature. For brevity and completeness, we restrict our choice to the ideas of basic import, which are used in this paper.

Definition 2.1. A function $h(x)$ $(-a < x < a)$ belongs to the class Σ_a, if it is hermitian positive (that is, $\sum_{\alpha=1}^{n} \sum_{\beta=1}^{n} h(x_\alpha - x_\beta)\xi_\alpha \bar{\xi}_\beta \geq 0$, x's real and ξ's complex) in $(-a,a)$ and is continuous at the point $x = 0$.

The following result was obtained by Bochner in 1932 for $a = \infty$ and by Krein (1940) for $a < \infty$.

Lemma 2.1. In order that a function $\phi(x)$ $(-a < x < a)$ may have a representation of the form $\phi(x) = \int e^{ixt}\, dF(t)$, where $F(t)$ is a nondecreasing function of bounded variation, it is necessary and sufficient that $\phi(x) \in \Sigma_a$.

Notice the importance of the above result in the form of characteristic functions in the probability theory and also in Laplace-Fourier-Stieltjes transforms. In fact, we have the following corollary of the above lemma.

Corollary 2.1. The family $\Sigma_\infty = \{\phi\}$, with the normalization $\phi(0) = 1$, coincides with the aggregate of all characteristic functions for all probability distributions.

Definition 2.2. A function $h^*(x)$ $(-a < x < a)$ belongs to the class Σ_a^* $(a \leq \infty)$, if for every positive integer it is the nth root of some function $h(x)$ of the class Σ_a.

Clearly, $\Sigma_a^* \subset \Sigma_a$, and as we shall see later the family Σ_∞^* with the normalization $h^*(0) = 1$ coincides with the class of all infinitely divisible characteristic functions.

Now, let a function $h(\underset{\sim}{x}) = h(x_1, \ldots, x_m) = h(x)$, $x \in E^m(-\infty < x_1, \ldots, x_m < \infty)$ say belongs to the class $\Sigma(E^m)$, if it is continuous at 0 and is hermitian positive, that is, for any positive integer n and any real vectors x_k $(k = 1, \ldots, n)$ $x_k \in E^m$, $\sum_{\alpha=1}^{n} \sum_{\beta=1}^{n} h(x_\alpha - x_\beta)\xi_\alpha \bar{\xi}_\beta \geq 0$ for complex ξ's. The following result was originally proved by Bochner in 1933 but a proof can be found in Bochner (1955).

Lemma 2.2. Any function $\phi(x) \in \Sigma(E^m)$ permits the representation $\phi(x) = \int_{E^m} e^{ixt'}\, dF_\mu(t)$, where $F(t)$ is a monotonic point function generated by a certain measure $\mu(A)$, $A \subset E^m$. The measure is uniquely determined by the function $\phi(x)$. Conversely, any function $\phi(x)$ which has the above representation belongs to the class $\Sigma(E^m)$.

Let $G_a = \{g(x): -a < x < a, a \leq \infty\}$ be a class of functions such that they are continuous at zero, $g(0) = 0$, $g(-x) = \overline{g(x)}$, and generate a hermitian positive kernel $K(x,y)$ $(0 \leq x, y < a)$ by the formula $K(x,y) = g(x) + g(-y) - g(x-y)$.

Theorem 2.1. (i) If $g(x) \in G_a$, then $\exp(-g(x)) \in \Sigma_a$. More precisely $g(x) \in G_a$ implies $\exp(-g(x)) \in \Sigma^*_\infty$. Note $\Sigma^*_a \subset \Sigma_a$ $(a \leq \infty)$. (ii) For any $h^*(x) \in \Sigma^*_a$, there exists $g(x) \in G_\infty$ such that $h^*(x) = h^*(0)\exp(-g(x))$ $(-a < x < a)$. (iii) Any function $h^*(x) \in \Sigma^*_a$ $(a < \infty)$ can be continued into a function of the class Σ^*_∞. Similarly, any function $g(x) \in G_a$ $(a < \infty)$ can be continued into a function of the class G_∞. (iv) Any function $g(x) \in G_a$ $(a \leq \infty)$ permits the representation

$$g(x) = i\gamma x + \int_{-\infty}^{\infty}\left[1 + \frac{iux}{1+u^2} - e^{iux}\right] u^{-2}\, dF(u), \tag{2.1}$$

$(-a < x < a)$ where γ is real and $F(u)$ is a nondecreasing point function such that $\int_{-\infty}^{\infty}(1+u^2)^{-1}\, dF(u) < \infty$. Here $g(x)$ determines γ and for $a = \infty$, $F(u)$ is determined essentially uniquely. Conversely, any function $g(x)$ which permits the form (2.1) belongs to G_a.

Proof. In cases (i), (iii) and (iv) the proofs can be found in Krein (1944) and Akhiezer-Glazman (1957) with slight modifications. In particular, the first part of (iv) is based on the well known Lévy-Khintchine canonical representation, which we shall discuss in the next section. Since part (ii) of the theorem plays an important role in the later sections, we shall prove it for completeness.

Without loss of generality assume $h^*(0) = 1$. By definition one can find a function $h_n(x) \in \Sigma_a$ for any positive integer n such that $h^*(x) = [h_n(x)]^n$ $(-a < x < a)$. By (iii) the function $h_n(x)$ can be extended into a function $\tilde{h}_n(x) \in \Sigma_\infty$, so we can construct the function $g_n(x) = n[1-\tilde{h}_n(x)]$ which belongs to family G_∞.

Choose $\varepsilon < a/2$ so that for $-\varepsilon \leq x \leq \varepsilon$ the value of $h^*(x)$ is sufficiently close to unity. Let $q(x) = \mathrm{Im}\,\ell n\, h^*(x)$, then for $-\varepsilon \leq x \leq \varepsilon$, we have:

$$\mathrm{Re}[g_n(x)] = n\mathrm{Re}[1-(h^*(x))^{1/n}] = n[1-|h^*(x)|^{1/n}\cos q(x)n^{-1}],$$

$$\mathrm{Im}[g_n(x)] = -n|h^*(x)|^{1/n}\sin q(x)n^{-1}.$$

Therefore, $|\mathrm{Re}[g_n(x)]| \leq \ell n|h^*(x)|^{-1} + (2n)^{-1}q^2(x)$, $|\mathrm{Im}[g_n(x)]| \leq |q(x)|$.

The above inequalities imply that $\max\limits_{0\leq x\leq \varepsilon} |g_n(x)| \leq \delta(\varepsilon)$, and also that for $-\varepsilon^* \leq x^* \leq \varepsilon^*$ $(\varepsilon^* \leq \varepsilon)$ $|\mathrm{Re}[g_n(x^*)]| + |\mathrm{Im}[g_n(x^*)]| \leq \delta(\varepsilon^*)$,

where the quantity, $\delta(\varepsilon^*) = \underset{-\varepsilon^*\leq x\leq \varepsilon^*}{\mathrm{supremum}} [\ell n|h^*(x)|^{-1} + |q(x)| + q^2(x)/2]$,

is independent of n and goes to zero with ε^*.

Next, one can easily establish the sequence $\{g_n(x)\}$ $(-\infty < x < \infty$, $n = 1, 2, \ldots)$ is equicontinuous and uniformly bounded in every finite interval. Hence, there exists a subsequence $\{g_{n_k}(x)\}_{k=1}^{\infty}$ and such a continuous function $g(x)$ $(-\infty < x < \infty)$ that $\lim\limits_{k\to\infty} g_{n_k}(x) = g(x)$ is uniform in every finite interval. The function $g(x) \in G_\infty$, and also for $-a < x < a$: $g(x) = \lim\limits_{k\to\infty} n_k[1 - (h^*(x))^{1/n_k}] = \ell n\, h^*(x)$, which proves the result.

At this stage, one immediately notices the central role played by the classes of functions Σ, Σ^* and G_a, G_∞ in probability and statistics. Furthermore, these ideas motivate semigroups in probability and functional analysis.

3. THE BASIC CANONICAL REPRESENTATIONS:-

In this section we shall unify the most basic and important canonical representation results. In particular we investigate the role of infinite divisibility including the solution of limit problems for sums of independent random variables.

Definition 3.1. (i) The distribution $F(x)$ is said to be infinitely divisible if for each natural number n there exists $F_n(x)$ such that $F = F_n^{n*}$, the nth fold convolution. (ii) The distribution function F is called stable if, for any $a_1, a_2 > 0$ and any b_1, b_2, there exist $a > 0$ and b such that $F(a_1x+b_1) * F(a_2x+b_2) = F(ax+b)$. (iii) The distribution F is said to be symmetric around c, if for every $d \geq 0$, $F(c-d) = 1 - F(c+d)$, that is, $P(X-c \leq -d) = P(X-c \geq d)$, for every $d \geq 0$. (iv) The distribution F is called unimodal if there exists at least one a such that $F(x)$ is convex in $x < a$ and concave in $x > a$.

Let $X_1, \ldots, X_j, \ldots$ be independent identically distributed random variables with distribution F, and set $S_n = \sum_{j=1}^{n} X_j$. In general, it is not possible to find normalizing constants A_n, B_n such that $B_n^{-1}(S_n-A_n)$ converges to any nondegenerate distribution. In fact, there is no subsequence for some distributions such that $B_{n_k}^{-1}(S_{n_k}-A_{n_k})$ converges in distribution. For example, the (infinitely divisible) distribution with characteristic function:

$$\log \phi(t) = \int_{-\infty}^{-\varepsilon} (\cos tx-1)d(4 \log|x|)^{-1} + \int_{\varepsilon}^{\infty} (\cos tx-1)d(4 \log x)^{-1} \tag{3.1}$$

see Ibragimov-Linnik (1971, p. 267). It is well known that the general stochastic process X_t in R^m with stationary independent increments has the following canonical representation:

$$\log E \exp[i\langle u, X(t)\rangle] = t[i\langle u, a\rangle - u'Au/2 + \oint (\exp(i\langle u, x\rangle)-1-i\langle u, g(x)\rangle)d\mu(x)], \tag{3.2}$$

where $\langle .,.\rangle$, denotes the inner product, $\oint$ denotes integration over the domain $R^m-\{0\}$, $[g(x)]_j$ is 1, -1, x_j according as $x_j \geq 1$, $x_j \leq -1$, $|x_j| < 1$, x_j is the jth coordinate of x, a is a real vector, A is a symmetric nonnegative definite matrix, and the Lévy measure $\mu(\cdot)$ is a Borel measure on $R^m - \{0\}$ such that $\oint |g(x)|^2 d\mu(\cdot) < \infty$. Thus, given such a triple (a,A,μ), there exists a corresponding process with stationary independent increments having the representation (3.2). From this representation, if $\int_{|x|>1} |x| d\mu < \infty$, then it is easy to see that $E(X(t)) = t[a+\int(x-g(x))d\mu(x)]$; and also, if $\int|x|^2 d\mu(x) < \infty$, then $E|X(t) - EX(t)|^2 = t[\text{trace }(A) + \int|x|^2 d\mu(x)]$.

The representation (3.2) has a strong resemblance with the Lévy-Khintchine canonical representation of an infinitely divisible characteristic function. The equivalence of such representations we shall rigorously prove in the next section via the stochastic semigroup theory. But now we give various interrelationships of such representations. The most useful and often quoted is the Lévy-Khintchine representation given below.

Lemma 3.1. A function $\phi(t)$ is an infinitely divisible characteristic function, if and only if, it admits the representation:

$$\log \phi(t) = iat + \int_{-\infty}^{\infty} \left(e^{itx} - 1 - \frac{itx}{1+x^2}\right) \frac{1+x^2}{x^2}\, dG(x), \tag{3.3}$$

where a is a real constant, $G(x)$ is nondecreasing bounded left continuous function, and the function under the integral sign is equal to $-t^2/2$ for $x = 0$. This representation is unique if $G(-\infty) = 0$.

Proof. See Gnedenko-Kolmogorov (1954) or Petrov (1975).

Examples 3.1. Clearly given (a,G) one can construct the infinitely divisible distribution via the representation (3.3), and conversely. We now present an expression for the pair (a,G) for some important infinitely divisible distributions. For the normal distribution $N(\mu,\sigma^2)$, $a = \mu$, and $G(x) = 0$, if $x \leq 0$, and $G(x) = \sigma^2$ if $x \geq 0$. For the extended Poisson distribution, $P(\alpha,\beta,\lambda)$ with density function $p(t) = P(X = \alpha + \beta k) = (k!)^{-1} \lambda^k e^{-\lambda}$, $k = 0, 1, \ldots$, one notices that $a = \alpha + (\beta\lambda)(1+\beta^2)^{-1}$, $G(x) = 0$, if $x \leq \beta$, and $G(x) = (\beta^2\lambda)(1+\beta^2)^{-1}$, if $x > \beta$. For the degenerate distribution with characteristic function e^{ict}, $a = c$, the degeneracy point, and $G(x) \equiv 0$. We remark that a distribution F carried by a finite interval is not infinitely divisible except if it is concentrated at one point.

An alternative canonical representation for an infinitely divisible distribution is the so called Lévy representation. This states that a function $\phi(t)$ is an infinitely divisible characteristic function, if and only if, it admits the representation

$$(3.4)\qquad \log \phi(t) = iat - \sigma^2 t^2/2 + \int_{-\infty}^{\infty} (e^{itx} - 1 - \frac{itx}{1+x^2}) dL(x),$$

where a is a real constant, $\sigma^2 \in R_+^1$, the Lévy function L(x) is nondecreasing in $(-\infty,0)$ and $(0,+\infty)$, and satisfies the condition $L(-\infty) = 0 = L(+\infty)$, also $\int_{-\varepsilon}^{+\varepsilon} x^2 dL(x) < \infty$ for every $0 < \varepsilon < \infty$. This representation like (3.3) is unique.

We recall the expressions EX(t) and Var X(t) and the fact that if $E(X^2) < \infty$, $\phi''(0)$ exists. In the later case the converse assertion is also true, that is, a function $\phi(t)$ is an infinitely divisible characteristic function with a finite variance, if and only if, it admits the canonical representation:

$$(3.5)\qquad \log \phi(t) = iat + \int_{-\infty}^{\infty} x^{-2}(e^{itx} - 1 - itx) dK(x),$$

where a is a real constant, the Kolmogorov function K(x) is nondecreasing, and the integrand is defined to be $-t^2/2$ at $x = 0$. In the sequal, the functions $\mu(x)$, $G(x)$, $L(x)$, $K(x)$ appearing in (3.2), (3.3), (3.4), (3.5) are respectively called the Lévy spectral Borel measure, Lévy-khintchine, Lévy, and Kolmogorov functions.

In fact, (3.3) can be identified uniquely by the triple (a,σ^2,L) in (3.4) as follows. Let

$$(3.6)\qquad C(x) = \begin{cases} \int_{-\infty}^{x} y^2(1+y^2)^{-1} dL^-(y), & x < 0 \\ \int_{x}^{\infty} y^2(1+y^2)^{-1} dL^+(y), & x > 0 \end{cases}$$

where L^- and L^+ are increasing continuous functions such that $\int_{-\varepsilon}^{0} y^2 dL^-(y) + \int_{0}^{\varepsilon} y^2 dL^+(y) < \infty$ and $L^-(-\infty) = L^+(+\infty) = 0$. Then by replacing G(x) in (3.3) by that in (3.6), we get

$$(3.7)\qquad \log \phi(t) = iat - \sigma^2 t^2/2 + \int_{-\infty}^{0} (e^{itx} - 1 - \frac{itx}{1+x^2}) dL^-(x) + \\ + \int_{0}^{\infty} (e^{itx} - 1 - \frac{itx}{1+x^2}) dL^+(x)$$

which is an alternative Lévy canonical representation equivalent to the form (3.4).

Let the double sequence of r.v.'s $\{X_{n_j}, j = 1,2, \ldots\}$ be such that random variables in each row are independent and let F_{n_j}, ϕ_{n_j} be the distribution and characteristic function of X_{n_j}.

Definition 3.2. The double sequence above is said to be <u>uniformly infinitesimal random system</u>, infinitesimal system for short (or holospoudic sequence, see

Chung (1974)), if and only if, $\lim_{n\to\infty} \max_{1\leq j\leq n_j} P(|X_{n_j}| > \varepsilon) = 0$.

A necessary and sufficient condition for this to hold is that for every $t \in R^1$: $\lim_{n\to\infty} \max_{1\leq j\leq n_j} |\phi_{n_j}(t) - 1| = 0$, which is equivalent to the condition that

$$\lim_{n\to\infty} \max_{1\leq j\leq n_j} \int_{-\infty}^{\infty} x^2(1+x^2)^{-1} dF_{n_j} = 0.$$

We denote by the class $\mathcal{L}$ or the $\mathcal{L}$- functions the family of all distribution functions which are limits of distributions of sums $Y_n = B_n^{-1} \sum_{i=1}^{n} X_i - A_n$ where $\{X_n\}$ is a sequence of independent random variables such that $\{X_{n_j} = B_n^{-1} X_j\}$ $(1 \leq j \leq n)$ form an infinitesimal system, and $A_n \in R$, $B_n \in R^+$, the positive real line. The infinite smallness assumption in above is the necessary condition for $F(x)$ defined by above to be infinitely divisible. Now let

Ω_0 = the class of all infinitely divisible distributions;

$\Omega_1 \equiv \mathcal{L}$ = the family of the Lévy-Khintchine and Gnedenko-Kolmogorov $\mathcal{L}$-functions as defined above;

Ω_2 = the family of all stable distributions;

Ω_3 = the class of all symmetric stable distributions.

To give a unified structure and interrelationships of the Ω-families, we need the following result.

Theorem 3.1. Let F be a distribution and ϕ its characteristic function. Then

(a) $F\varepsilon\Omega_2$, if and only if, (σ^2,L) in the representations (3.4) or (3.7) satisfy one of the following conditions; (i) $L(x) \equiv 0$, (ii) $\sigma^2=0$, $L(x) = c|x|^{-\alpha}$ for $x < 0$, $L(x) = -dx^{-\alpha}$ for $x > 0$ $(0 < \alpha < 2$, $c \geq 0$, $d \geq 0$, $c + d > 0)$.

(b) $F \in \Omega_2$, if and only if, ϕ has the representation:

$$\log \phi(t) = iat - c|t|^{\alpha}(1 + i\beta t|t|^{-1} g(\alpha,t)), \qquad (3.8)$$

where a is real, $c \geq 0$, $0 < \alpha \leq 2$, $|\beta| \leq 1$, and $g(\alpha,t)$ equals $\tan \pi\alpha/2$ for $\alpha \neq 1$ and $2/\pi \tan|t|$, if $\alpha = 1$.

The proofs of the above results can be found in Gnedenko-Kolmogorov (1954) and Ibragimov-Linnik (1971). On the other hand, an infinitely divisible distribution F belongs to the class $\mathcal{L}$, if and only if, in (3.7) the Lévy spectral function L^+ and L^- (or L in (3.4)) are continuous at every point $x \neq 0$ and have left and right derivatives and $xL^{+\prime}$, $xL^{-\prime}$ are nonincreasing, where L' denotes either the left or right derivative. With these considerations, we see that $\Omega_2 \subset \Omega_1 \subset \Omega_0$. Furthermore, if we assume that the X's have the same distribution, then the class $\mathcal{L}^*(\equiv\Omega_1^*)$, the $\mathcal{L}$-class obtained by this extra restriction, is such that $\Omega_1^* \equiv \Omega_2$. In fact, the stable distributions are the simplest version of $\mathcal{L}$-distributions. We remark that the degenerate and normal distributions belong to the class $\mathcal{L}$, but $P(\alpha,\beta,\lambda)$ does not since L is discontinuous at $\beta = 0$. Now, we give a method via the canonical representations by which one can construct an example of a distribution function belonging to Ω_i but not to Ω_{i+1} for i = 0, 1, 2, respectively.

Example 3.2. In this example we shall show how one can construct a distribution of class Ω_0 but not an $\mathcal{L}$-function. From (3.7) we choose Lévy spectral functions such that L^+ and L^- are increasing continuous functions and at least one of the $xL^{+\prime}(x)$, $xL^{-\prime}(x)$ is increasing. Choose $xL^{-\prime}(x) = \lambda x$, or zero for $x \leq 1$, or $x > 1$, $L^+(x) = 0$ for $x < 0$ and $a = \lambda \log\sqrt{2}$. Then the corresponding characteristic function can be written as

$$\log \phi(t) = \lambda/it(e^{it}-it-1).$$

This is the simplest version of distribution in Ω_o but not in Ω_1. To obtain an $\mathcal{L}$-distribution function but not stable we choose $xL^{-\prime}(x)$ and $xL^{+\prime}(x)$ as non-increasing, nondecreasing functions of x. For simplicity take $L^-(x) = 0$ for $x < 0$, $L^+(x) = \lambda \log x$ for $x \leq 1$ and 0 if $x > 0$ and $a = \lambda\, \pi/4$. Then

$$\phi(t) = \exp\left[\lambda \int_0^t (e^{ix}-1) \frac{dx}{x}\right]$$

is an $\mathcal{L}$-characteristic function but not stable. To get a characteristic function which belongs to Ω_2 but not to Ω_3, one can proceed similar to the above two examples. It is rather complicated to express stable densities in closed form. However, series expansions are available.

For many other interesting examples of Ω- classes and domains of attraction for the class Ω_2 we refer the reader to Lévy (1937) and Feller (1971). To summarize we have seen that the following is valid.

Theorem 3.2. (i) $\Omega_3 \sqsubset \Omega_2 \sqsubset \Omega_1 \sqsubset \Omega_o$. (ii) If the defining sequence for the class Ω_1 is that of independent identically distributed random variables, then $\Omega_1 \equiv \Omega_2$.

After these preliminaries and some basic canonical representations of the above classes, we shall start on a unified algebraic structure approach, but first a few remarks about the family $\mathcal{L}$ and its subfamilies are in order. Some authors for example Loève (1963, page 322) and Urbanik (1973) have called $\mathcal{L}$-functions as self-decomposable distributions. Here, a distribution F is said to be self-decomposable if for every $\alpha \in (0,1)$, there exists a characteristic function ϕ_α such that $\phi_F(t) = \phi_F(\alpha t)\, \phi_\alpha(t)$. This definition could be thought of as a necessary and sufficient condition for an $\mathcal{L}$-function. Also, compare this with the definition of a stable distribution of the family $\Omega_2 \sqsubset \mathcal{L}$. If $\mathfrak{X}$ is a metric space and $\mathcal{M}(\mathfrak{X})$ the space of all probability measures on the Borel field generated by $\mathfrak{X}$, $\mathbb{B}_{\mathfrak{X}}$, then the class Ω_o forms a closed semigroup on $\mathcal{M}(\mathfrak{X})$. In fact, with the topology of weak convergence and the multiplication defined by the convolution, $\mathcal{M}^{\omega,*}$ becomes a topological group. The infinitely divisible distribution structure of Banach and Hilbert space valued random variables is treated in Gikhman-Skorohod (1974). In the next section we shall briefly see some connection between infinitesimal random systems defined on finite and infinite dimensional function spaces. However, with slight modifications in terminology, the main results of that section such as theorems 4.2, 4.3, and 4.4. remain valid in Banach and Hilbert spaces, or indeed any other suitable extended space.

Now, let $X_1, X_2, \ldots$ be independent random variables such that for a suitable choice of $B_n > 0$ and A_n the sequence of distributions of, $Y_n = B_n^{-1} \sum_{i=1}^{n} X_i - A_n$, has a limit distribution F(x). This problem of limit sums is very old, since the beginning of the eighteenth century, but a considerable progress has been made during the second quarter of this century. Now, define S_m $(m = 0, 1, 2, \ldots, \;)$ inductively as follows. Let S_o be the class of sequences $\{X_n\}$ of independent random variables generating convergent triangular arrays. Then S_o is the class of sequence $\{X_n\}$ of independent random variables such that for a suitable chosen random constants $A^*_n, n^{-1} \sum_{i=1}^{n} X_i - A^*_n$ has a limit distribution. Define S_m $(m = 1, 2, \ldots, \infty)$ to be the class of all sequences $\{X_n\}$ such that $\{X_n\} \in S_o$ and for every positive real constant γ the triangular array $X_{nj} = X_{[\gamma n]+j}$ is equivalent to an array generated by a sequence from S_{m-1}, here $[\gamma n]$ denotes the integral part of the number γn. Let $S_\infty = \bigcap_{i=0}^{\infty} S_i$ and $L_\infty = \bigcap_{i=0}^{\infty} L_i$, where L_i is the set of all possible limit distributions of normal sums $n^{-1} \sum_{i=1}^{n} X_i - A^*_n$ where

$\{X_n\} \in S_i$ and $a_n \in \mathcal{R}$. We define L_{-1} to be the set of all probability measures on $\mathcal{R}$. Obviously, S_m form a contracting sequence of sets and hence L_m form a decreasing sequence of limiting distributions.

The case L_o was solved by Lévy, see Loève (1963, page 326), which is our class Ω_1 containing all stable laws. The structure of L_2 has been discussed by Koroljuk-Zolotarev in 1961, where they established that in this case F(x) must be a convolution of two stable distributions. For $r > 2$ Zinger in 1965 showed that F(x) is a convolution of r stable distributions. Notice that $\mathcal{L}_\infty$ is the smallest class containing all stable distributions, that is the family Ω_2. Furthermore, S_∞ as expected form the class of all slowly varying functions. Thus to summarise, we have the following.

Theorem 3.3. (i) $\Omega_3 \sqsubset \Omega_2 \sqsubseteq L_\infty \sqsubset \cdots \sqsubset L_1 \sqsubset L_o = \Omega_1 \sqsubset \Omega_o \sqsubset \Omega_{CLP} \sqsubset \tilde{\Sigma}_\infty = \{F_\phi: \phi(0) = 1\}$, see corollary 2.1. (ii) A probability measure μ belongs to L_m ($m = 0, 1, \ldots, \infty$), if and only if, for every $\alpha \in (0,1)$, there exists a probability measure $\mu_\alpha \in L_{m-1}$ such that $\mu = \tilde{\mu}_\alpha * \mu_\alpha$, where $\tilde{\mu}_\alpha$ is defined by $\tilde{\mu}_\alpha(A) = \mu(\alpha^{-1}A) = \mu\{\alpha^{-1}x: x \in A\}$.

Notice that the restriction proposed by Urbanik (1973) is essentially a stability condition. Under these circumstances Urbanik was able to prove that a function ϕ is the characteristic function of a probability measure from L_∞, if and only if,

$$(3.9)\quad \log \phi(t) = iat - \int_{-2}^{2} [|t|^{|x|}(\cos \pi x/2 - it|t|^{-1} \sin \pi x/2) + itx] \frac{\mu(dx)}{1-|x|}$$

where a is a real constant, μ is a finite Borel measure on $(-2,0) \sqcup (0,2)$, and the integral is defined as its limiting value $(\pi/2)|t| + it \log |t| - it$ for $x = 1$, and $(\pi/2)|t| - it \log |t| + it$ when $x = -1$.

4. A UNIFIED EXTENDED ALGEBRAIC STRUCTURE APPROACH.

Let us recall the well known and very useful functional equation:

$$(4.1)\quad f(s)f(t) = f(s+t),\ f(0) = 1;\quad s,t \geqslant 0,$$

which is satisfied by $f(t) = e^{\lambda t}$, λ any constant. This equation brings in mind the Poissonian and Compound Poissonian distributions on one hand; and the study of exponential functionals on infinite-dimensional function spaces, that is on the other hand, the semigroup theory of operators. In fact, we shall see that this ties up very neatly with the so called compound Poissonian semigroups, stochastic semigroups, Markovian semigroups, and the famous Hille-Yosida-Phillips theorem in functional analysis. Before we proceed further, we need some terminology. Let

C_B = the class of all bounded continuous functions,

C_∞ = the class of all continuous functions with finite limits $f(-\infty)$ and $f(+\infty)$,

C_o = the class of all continuous functions vanishing at $\pm\infty$,

C = the class of all continuous functions,

C^∞ = $\{f: f \in C_\infty$ and f has derivatives of all order in $C_\infty\}$
= the family of infinitely differentiable functions.

Clearly, $C \sqsupset C_B \sqsupset C_\infty \sqsupset C^\infty$, C_o is the closure of C_K, the class of continuous functions $\{f\}$ vanishing outside a compact set K_f, with respect to uniform convergence.

A <u>convolution semigroup</u> $\{\mathcal{G}(t),\ t > 0\}$ is a class of operators associated with probability distributions $\{G_t\}$ on R and satisfying: $\mathcal{G}(s+t) = \mathcal{G}(s)\mathcal{G}(t)$. In fact, if

$$(4.2)\quad G_s * G_t = G_{s+t} \text{ and } \mathcal{G}(t)u(x) = \int_{-\infty}^{\infty} u(x-y)dG_t(y),$$

then $\mathcal{G}(s+t) = \mathcal{G}(s)\mathcal{G}(t)$ is equivalent to $G_{s+t} = G_s * G_t$. Note that for $\{\mathcal{G}(t)\}$ the domain is C_o, and these operators are transition operators, that is, $0 \leq u \leq 1$ implies $0 \leq \mathcal{G}(t)u \leq 1$ and $\mathcal{G}(t)1 = I$, where I is the identity operator. The convolution semigroup $\mathcal{G}(t)$ is said to be continuous if $\mathcal{G}(t) \to I$ as $t \to 0+$. In this case we write $\mathcal{G}(0) = I$. Finally, an operator $T: C^\infty \to C^\infty$ is said to generate the convolution semigroup $\{\mathcal{G}(t)\}$ if as $t \to 0+$, $t^{-1}(\mathcal{G}(t) - I) \to T$. Equivalently, one can say T is the generator.

Theorem 4.1. If $\{\mathcal{G}(t)\}$ and $\{\mathcal{G}^*(t)\}$ are two continuous semigroups having the same generator T, then $\mathcal{G}(t) = \mathcal{G}^*(t)$ for all $t \geq 0$.

Proof. By the Hille-Yosida-Phillips theorem there exists a real number λ^* such that $L(\lambda,T) = (\lambda I-T)^{-1}$ exists and is bounded for every $\lambda > \lambda^*$. So, we find that

$$L(\lambda,T) = \int_0^\infty e^{-i\lambda t}\, T_G(t)u\, dt = \int_0^\infty e^{-i\lambda t}\, T_{G*}(t)u\, dt.$$

Hence, for any f in the set of continuous linear functionals in the space of definition, $f[T_G(t)u]$ and $f[T_{G*}(t)u]$ have the same Laplace-Stieltjes transforms, and since both are $O(e^{at})$ for some finite a coincide. Since f is arbitrary the result is proved.

Now, we are in a position to state and prove the Basic Structure Theorem which unifies and inter-relates the various concepts involved. In what follows

$$Q_t = e^{-\lambda t} \sum_{k=0}^{\infty} \frac{(\lambda t)^k}{k!} G^{k*},$$ with an arbitrary probability distribution G, and $\lambda > 0$,

defines a compound Poisson process. Clearly, the family of such distributions forms a semigroup since it satisfies $Q_{s+t} = Q_s * Q_t$, s, t > 0. An alternative version of the above is the following. Let $Y_1, Y_2, \ldots$ be independent random variables with distribution G, and suppose N(t) is a Poisson process with parameter t and independent of Y_j. Then Q_t is the distribution of $\sum_{j=1}^{N(t)} Y_j$. The compound Poisson processes have many applications such as traffic jams, fishing, risk theory and insurance, water or oil reservoir contents, customers arriving at a server, signal detection, and so on. Notice that in the definitions of infinite divisibility and compound Poisson processes the assumption of the same distribution is no restriction, since there exists an appropriate group which generates the whole family of distributions involved.

Theorem (Basic Structure) 4.2. The following families of probability distributions or corresponding measures are identical.

(a) M_1 = the class of all infinitely divisible distributions.

(b) M_2 = limits of sequences of infinitely divisible distributions.

(c) M_3 = limit distributions of row sums in infinite random systems $\{X_{jn_j}\}$ where the variables X_{jn_j} of the nth row have a common distribution.

(d) M_4 = the distributions of increments in processes with stationary independent increments, that is, distributions associated with continuous convolution semigroups.

(e) M_5 = limit of sequences of compound Poisson distributions, that is, distributions generated by compound-Poissonian semigroups.

Let $M_1 \equiv M_2 \equiv M_3 \equiv M_4 \equiv M_5 \equiv \Omega$. Thus, for a given infinitesimal random system there exists a unique (upto an equivalence class) generator T which generates the family Ω via the appropriate convolution semigroup, and conversely.

Proof. For each n let G_n be an infinitely divisible distribution. By definition G_n is the distribution of the sum of n independent identically distributed random variables, and so the sequence $\{G_n\}$ generates an infinitesimal system in M_3. Thus, $M_2 \subseteq M_3$.

Next, we show $M_3 \subseteq M_4$. Let T_n be the operator induced by the distribution $G_{X_{k,n}}(\cdot)-b_n = F_n(x+b_n)$. To show that the distribution of $\sum_{j=1}^{n_k} X_{jn_j} - n_k b_{n_k}$ tends to the distribution G_1 associated with $\mathcal{G}(I)$ it is sufficient to show [see Feller (1971) pages 311-312] that as n runs through $\{n_k\}$, $n(\mathcal{G}_n - I) \to T$, where $Tu(x) = \int_{-\infty}^{\infty} y^{-2}[u(x-y) - u(x) + T_s^*(y)\, u'(x)]d\mu(y)$ with μ an appropriate canonical measure. Here, the truncation function T_s^* is the continuous monotone function such that $T_s^*(x) = x$ when $|x| \leqslant s$ and $T_s^*(x) = \pm s$ when $|x| \geqslant s$, where $s > 0$ is arbitrary but fixed. The uniqueness of the semigroup containing G_1 shows that the limit relations are valid for an arbitrary approach $n \to \infty$. This proves that $M_3 \subseteq M_4$.

Now, we show that $M_4 \subseteq M_5$. Since the operator $T_h = \frac{\mathcal{G}(h)-I}{h}$ for fixed $h > 0$ generates a compound-Poissonian semigroup of operators $\mathcal{G}_h(t)$, as $h \to 0+, T_h \to T$, and hence $\mathcal{G}_h(t) \to \mathcal{G}(t)$. Thus, G_t is the limit of compound-Poisson distribution, and consequently $M_4 \subseteq M_5$. On the other hand, $M_5 \subseteq M_2$ trivially. Finally, $M_1 \subseteq M_2$ and $M_1 \supseteq M_4$. This completes the proof of the theorem.

If $\{\phi_n, n \geqslant 1\}$ is a sequence of infinitely divisible characteristic functions converging everywhere to characteristic function ϕ, then ϕ is infinitely divisible. By employing this fact we can construct a wide class of characteristic functions in Ω as follows. For each $\lambda > 0$ and real u, consider the function $h(t;\lambda,u) = \exp(\lambda \exp(iut - 1))$, which is an infinitely divisible characteristic function. Clearly, $\exp[\sum_{j=1}^{k} \lambda_j(\exp(iu_j t-1))]$ is also infinitely divisible. Furthermore, if H is a bounded increasing function, $\phi^*(t) = \exp[\int_{-\infty}^{\infty}(\exp(itu)-1)dH(u)]$ is an infinitely divisible characteristic function. Thus, for every $\phi \in \Omega$, there exists a double array of real numbers $\{\lambda_{n_j}, u_{n_j}\}$, $1 \leqslant j \leqslant k_n$, $1 \leqslant n$, where $\lambda_j > 0$, such that

$$\phi^*(t) = \lim_{n\to\infty} \prod_{j=1}^{k_n} h(t;\lambda_{n_j}, u_{n_j}).$$

The converse is also true. Notice that the above discussion leads naturally to the well-known result due to De Finetti which states that ϕ is in Ω, if and only if, $\phi(t) = \lim_{n\to\infty} \exp(\lambda_n(\phi_n^*-1))$ where λ_n is a real number, ϕ_n^* a characteristic function.

A measure μ on (R^1,B^1) with $\mu(R^1) \leqslant 1$ is called a <u>defective probability measure</u> (dpm). The sequence $\{\mu_n, n \geqslant 1\}$ of dpm's is said to <u>converge vaguely to a dmp μ</u> (written $\mu_n \xrightarrow{v} \mu$), if and only if, there exists a dense subset D of R^1 such that for every a in D, b in D, a < b: $\mu_n((a,b]) \to \mu((a,b])$. The general criterion for vague convergence is as follows. If $\{\mu_n\}$ and μ are pm's then $\mu_n \xrightarrow{v} \mu$, if and only if, one of the two following conditions is satisfied:

(a) for every f in $\tilde{L}$: $\underline{\lim}_n \int f(x)d\mu_n(x) \geqslant \int f(x)d\mu(x)$;

(b) for every g in $\tilde{U}$: $\overline{\lim}_n \int g(x)d\mu_n(x) \leqslant \int g(x)d\mu(x)$;

where $\tilde{L}$ = {bounded lower semicontinuous functions, that is, for every

$$x \text{ in } R^1:\ f(x) \leqslant \underline{\lim}_{\substack{y\to x\\ y\neq x}} f(y)\},$$

and $\tilde{U}$ = {bounded upper semi-continuous functions}.

Thus, in view of the above considerations, the class Ω coincides with the closure, with respect to the vague convergence of convolutions of a finite number of Poisson distribution functions of the form $h(t;\lambda,u)$ above.

Recall that under Weak Law of Large Numbers (WLLN) and Central Limit Theorems (CLT) sums of independent random variables converge in distribution to a degenerate distribution with characteristic function $\exp(\lambda it)$ and normal distribution with characteristic function $\exp(ait-bt^2)$. Note that both of these characteristic functions are exponentials of polynomials of the first and second degree in (it). Other interesting cases of limiting distributions are: Poisson with characteristic function $\exp(\lambda e^{it}-1)$, $\lambda > 0$; and symmetric stable distributions of exponent α with characteristic function $\exp(-\lambda|t|^{\alpha})$, $0 < \alpha < 2$, $\lambda > 0$. For $\alpha = 1$ (2) we have the Cauchy (normal) distribution. The above considerations lead us to the following unified structure of such distributions. But first we need the theorem below, which can also be deduced from the theorem 2.1.

Theorem 4.3. Let a complex-valued function ϕ of real variable t be given. Assume $\phi(0) = 1$ and that ϕ is continuous in $(-\infty,\infty)$ and does not vanish in the interval. Then there exists a unique single-valued function ψ of t in $(-\infty,\infty)$ with $\psi(0) = 0$ that is continuous and satisfies $\phi(t) = e^{\psi(t)}$. The function ψ is called the distinguished logarithm, and $e^{\psi(t)/n}$ the distinguished nth root of ϕ.

Proof. It follows by employing slightly modified arguments as in Chung (1974; Theorem 7.6.2, pages 241-242), or can be obtained by using part (ii) of the theorem 2.1.

Thus, there is a one-to-one correspondence among an infinitely divisible distribution F, ϕ_F, and ψ_ϕ. Furthermore, if one is considering the relationship between infinitely divisible probability distributions and helical varieties in Hilbert spaces, see Masani (1973), then ψ is uniquely determined by a suitable kernel $\mathcal{K}_\psi$. There is a slight departure from uniqueness when ψ is complex-valued. For a helical variety (or helix), h, the corresponding chordal covariance kernel $\mathcal{K}^*_h$ is translation-invariance. This fact, under appropriate spaces, connects helical varieties with random processes or random fields with wide-sense stationary increments. Consequently, there is correspondence, although not one-one, between ϕ_ψ and h. Notice that the very idea of helix comes from Brownian motion, and is examplified by all random processes with stationary increments. Of course, only in cases dominated by normality do strict-sense and wide-sense concepts merge. In general, the strict-sense concepts such as stationarity, Markovian, martingale property etc., belong to the theory of probability, and the wide-sense concepts are in the domain of the theory of Hilbert and Banach spaces, as pointed out by Masani (1973).

Now, let $\tilde{\mathbf{H}}$ be a maximal group of measure-preserving transformations, which generates classes $\{h\}$ or $\{\psi\}$ discussed above. For a distribution F, whether in Ω or not, if there exists at least one h in $\tilde{\mathbf{H}}$ such that $\tilde{F} = F_{\tilde{h}}$ is in Ω, then we call such a distribution F an $\tilde{\mathbf{H}}$-potential generator. Clearly, the group $\tilde{\mathbf{H}}$ excludes some distributions. For example, for no $\tilde{h}$, $\tilde{F} \equiv U[-a,a]$ where $U_{[-a,a]}$ is a uniform distribution on $[-a,a]$ which is not infinitely divisible since its characteristic function is $\frac{\sin at}{at}$ which vanishes for some real t, although it has many divisors.

Theorem (Extended Equivalent Class Structure) 4.4. Let Ω, $\mathcal{G}$, and $\tilde{\mathbf{H}}$ be as defined above. Then $\tilde{\mathbf{H}}$ generates Ω, that is, for every G in Ω, there exists at least one $\tilde{F} = F_{\tilde{h}}$ such that $\mathcal{G}_G \equiv \mathcal{G}_{\tilde{F}}$, $h \in \tilde{\mathbf{H}}$.

Proof. By the very structure of the group $\tilde{\mathbf{H}}$, there exists a continuous homomorphism; $\tilde{\phi}: \tilde{\mathbf{H}} \to \mathcal{G}$. Therefore, by the Basic Structure Theorem for any distribution G generated by $\mathcal{G}$, there exists $F_{\tilde{\phi}}$ which is an $\tilde{\mathbf{H}}$-potential generator. Hence, to the family $\{Q(t)\}$ there corresponds a class $\{\tilde{Q}(t;\tilde{\phi},h)\}$ which generates an appropriate semigroup. Consequently, from $\mathcal{G}_{\tilde{Q}}$ one can construct the whole class of infinitely divisible distributions. This proves $\Omega(\tilde{\mathbf{H}};\ \{\tilde{F}\}) \equiv \Omega \equiv \Omega(\mathcal{G})$, where $\Omega(\mathcal{G})$ is the family of all infinitely divisible distributions generated by $\mathcal{G}$. Notice that the family $\{\tilde{Q}\}$ has some resemblance to the class $h(t;\ .,.)$ discussed after the Basic Structure Theorem.

Corollary 4.1. (i) If $F \in \Omega$ and $F * G \underset{n}{=} H \in \Omega$, then $G \in \Omega$. (ii) If F_j ($j = 1, 2, \ldots, n < \infty$) is in Ω, then $\sum_{j=1}^{n} w_j F_j \in \Omega$, where $0 \leqslant w_j \leqslant 1$ and $\sum_{j=1}^{n} w_j = 1$. (iii) If ϕ_n converges weakly to ϕ, ϕ_n an infinitely divisible characteristic function, then by virtue of the theorem 4.3, ϕ is an infinitely divisible.

It was pointed out earlier that the set of all probability measures $\mathcal{M}$ with convolution as multiplication and weak (or vague) convergence forms a topological group $\mathcal{M}^{*,w}$. Let Σ be a group and denote by $\mathbb{B}_\Sigma$ its Borel field and $\mathcal{M}(\Sigma)$ as the space of all measures defined on $\mathbb{B}_\Sigma$. Some notable Indian and Russian probabilists have introduced an extended definition of infinite divisibility. That is, μ is infinitely divisible if there exists a single element $\sigma \in \Sigma$ such that $\mu = \mu_n^{n*} * \sigma$, $\mu_n \in \mathcal{M}(\Sigma)$. A measure $\mu \in \mathcal{M}(\Sigma)$ is called idempotent if $\mu * \mu = \mu$. Let Σ be a complete separable metric space and μ an idempotent measure on Σ. Then it is well known that there exists a compact subgroup $\Sigma^* \subseteq \Sigma$ such that μ is the normalized Haar measure of Σ^*. A Haar measure is a left (or right) invariant Borel measure which is not identically zero. Furthermore, a Haar measure is not unique, that is, for each $a \in R_+^1$, $a\mu$ is a Haar measure if μ is. Clearly, every locally compact topological group has a Haar measure. Let ϕ_μ be the characteristic function of an infinitely divisible measure μ. If $\phi(\bar{u}) = 0$ for some character $\bar{u}$ then μ has an idempotent factor. So, the normalized Haar measure of a compact subgroup Σ^* is an example of an infinitely divisible distribution. This fact is a consequence of the result that such a measure is idempotent. Thus, if we have a locally compact (vague) topological group, then it has a Haar measure, and this measure should generate other infinitely divisible distributions. Notice the relationship between this observation and our extended equivalent class algebraic structure theorem.

In considering the structure of stochastic independence, Bell (1958) has given interrelationship among various concepts of independence: (a) sigma-independence; (b) almost-sigma-independence; (c) stochastic-independence; and (d) quasi-sigma-independence. The solution proposed by Bell is as follows: no two of the conditions are equivalent, and $a \to b \to c \to d$. The result above motivates one to seek classes larger than the class Ω which, in some appropriate sense, exhibit essentially a similar structure as the class of infinitely divisible distributions. To the best of the authors' knowledge no results on these lines exist in the literature. By definition, F is infinitely divisible, if and only if, for each n it can be represented as the distribution of the sum $S_n = \sum_{j=1}^{n} X_{jn}$ of n independent random variables with a common distribution F_n. In view of this fact, we propose that 'independence' should be replaced by some 'weaker form of independence' or some 'tractable form of dependence', and then some 'weaker form of infinite divisibility' behaviour should be studied. We hope to take up these aspects in a further paper. We may recall that the class Ω belongs to that of limit laws of the central limit problem. If we denote the later family by Ω_{CLP}, then in the above we are talking about a family Ω' or $\bar{\Omega}$ such that $\Omega \subset \Omega' \subset \Omega_{CLP} \subset \bar{\Omega}$. We may also notice that similar results as theorems 4.2 and 4.4 hold even in nonstationary processes with independent increments, provided a suitable continuity condition is imposed to assure the existence of an infinitesimal random system.

5. ON THE UNIMODALITY OF CLASSES Ω_1, Ω_2 AND Ω_3:

The unimodality of a process is one of the nice and useful behaviour properties which help obtain better statistical inferences. From the definition we have noticed that $F'(x)$ is continuous everywhere with a unique finite maximum either a single point or a connected interval. It is known, Gnedenko-Kolmogorov (1954, p. 160), that a distribution $F(x)$ is unimodal with vertex at $x = 0$, if and only if, its characteristic function $\phi(t)$ can be represented in the form: $\phi(t) = t^{-1}\int \phi^*(s)ds$, $-\infty < t < \infty$, where $\phi^*(t)$ is some characteristic function. For example, $\phi(t) = (1+|t|^{\alpha})^{-1}$, $0 < \alpha \leqslant 2$ is the characteristic function of a unimodal distribution. The structure of unimodality in the literature can be summed up as follows.

Lemma 5.1. (i) If F and G are (symmetric unimodal) distributions, then so is $F * G$. (ii) All stable distributions with characteristic functions given by $\exp(-|t|^{\alpha})$ are unimodal. Consequently all members of the class Ω_3 are unimodal. (iii) All distributions of the class Ω_2, that is, all stable distributions are unimodal.

Definition 5.1. A distribution $\Gamma(x)$ is said to be strongly unimodal, if and only if, the convolution of F with any unimodal distribution is unimodal. [For example, the normal and Wishart distributions are strongly unimodal].

Lemma 5.2. A distribution from the class Ω_1 is unimodal if at least one of the following conditions is valid: (i) F is symmetric; (ii) F is stable; (iii) F is symmetric stable; (iv) either L^- or L^+ is zero; (v) $F = N * G$, where N is a normal distribution and G any unimodal distribution, or equivalently $F = G_s * G^*$, where G_s is a strongly unimodal distribution and G^* an arbitrary unimodal distribution.

In the Russian edition of Gnedenko-Kolmogorov book, which appeared in 1949, there appeared a theorem due to Gnedenko, stating that every L-function is unimodal. This result depended on an incorrect theorem due to Lapin in his 1947 thesis, which states that the convolution of two unimodal distributions with vertex at zero is unimodal with vertex at zero. Chung (1953) gave two counter examples and presented a correct version of Lapin's result. Ibragimov (1957) gave some examples of distributions in Ω_1 which were not unimodal, but Sun (1967) proved the unimodality of Ibragimov's examples. With these historical remarks, in the sequel we give some new results.

Theorem 5.1. Let $F(x) = F_1(x) * F_2(x)$ be an $\mathcal{L}$-function with the Lévy spectral functions L^- and L^+ such that the corresponding distribution functions F_1 and F_2, respectively, satisfy

$$(5.1) \qquad V(x) = \tau_1 * F_2 + \tau_2 * F_1 - F_1 * F_2 + s(F_1' * F_2),$$

V is a distribution, where $\tau_i(x) = F_i(x) - F_i'(x)$ $(i = 1, 2)$ and s is the vertex of $F_1 * F_2$, $V(x)$ may be either left or right continuous, and $F_i'(x)$ is the left derivative. Then F_1 and F_2 are unimodal and consequently F is unimodal. Notice that in fact $F = N(a,\sigma^2) * F_1 * F_2$ where N has trivial Lévy spectral measures.

Proof. Clearly $L^-(x)$ and $L^+(x)$ with respect to $F(x)$ are absolutely continuous. If we let $xL^{-\prime}(x) = \delta(x)$ and $xL^{+\prime}(x) = \delta^*(x)$, then the Lévy canonical representation can be written in terms of $\delta(x)$ and $\delta^*(x)$. The Lévy spectral functions $L^-(x)$ and $L^+(x)$ have right and left derivatives for every x and the functions $\delta(x)$ and $\delta^*(x)$ are nonincreasing, here $L^{-\prime}(x)$ and $L^{+\prime}$ denote either right or left derivative, possibly different ones at different points.

Now, $\phi(t)$ can be written as $\phi(t) = \phi_o(t)\,\phi_1(t)\,\phi_2(t)$, where ϕ_1 and ϕ_2 are characteristic functions of F_1 and F_2, and ϕ_o is the characteristic function of $N(a,\sigma^2)$. Since

$$\phi_2(t) = \exp\Big[\int_0^\infty (e^{itx} - 1 - \frac{itx}{1+x^2}\,\frac{\delta^*(x)}{x}\,dx\Big],$$

it is possible to construct a sequence of nonincreasing step functions $\{\delta_j^*(x)\}$ such that $0 \leqslant \delta_1^*(x) \leqslant \dots \leqslant \delta_j^*(x) \leqslant \dots$, and $\delta_n^*(x) \to \delta^*(x)$ as $n \to \infty$ for $x > 0$. For each n let $F_{2n}(x)$ denote the $\mathcal{L}$-function with characteristic function

$$\phi_{2n}(t) = \exp\Big[\int_0^\infty (e^{itx} - 1 - \frac{itx}{1+x^2}\,\frac{\delta_n^*(x)}{x}\,dx\Big].$$

Let $G_{2n}(x)$ and $G_2(x)$ be the Lévy-Khintchine spectral functions of $F_{2n}(x)$ and $F_2(x)$, respectively, where $G(x)$ is as defined by (3.6). It is easy to see that $G_{2n}(x) \to G_2(x)$ as $n \to \infty$ for all values of x. Then from the fact that almost sure convergence in probability, we get that $F_{2n}(x) \to F_2(x)$ as $n \to \infty$, and that F_2 is unimodal. By a similar argument one can prove that $F_1(x)$ is unimodal. If any of F_1 or F_2 is with vertex $s_i \neq 0$ $(i = 1, 2)$, we can easily shift the vertex to the origin. Because of the validity of (5.1), $F_1 * F_2$ is unimodal at s and $\phi^*_{1,2}(t) = \phi_1(t)\,\phi_2(t)$. Since $\phi_o(t) = \exp(iat - t^2\sigma^2/2)$ is the characteristic function of $N(a,\sigma^2)$ which is strongly unimodal, by the definition of strong unimodality and the fact that a continuous distribution is strongly unimodal, if and only if, the logarithm of its density is concave, we can see that $\phi(t) = \phi_o(t) \times \phi^*_{1,2}(t) = \phi_o(t)\,\phi_1(t)\,\phi_2(t)$ is a characteristic function of a unimodal distribution. This completes the proof of the theorem.

Now, one can compare the following corollaries with the main results of Wolfe (1971).

Corollary 5.1. (i) If $L^-(x) = 0$ or $L^+(x) = 0$ in the theorem above, then the $\mathcal{L}$-function with the characteristic function of the Lévy canonical representation is unimodal. (ii) Any $\mathcal{L}$-function which is unimodal is a convolution of at the most three distinct unimodal distributions. The convolution of two unimodal infinitely divisible distributions need not be unimodal.

6. SOME COMMENTS:

Barndorff-Nielsen (1976) has defined the Universality concept in connection with plausibility inference. It turns out that, under some mild regularity conditions, for the regular exponential family the concepts of strong unimodality and universality are essentially the same. LeCam (1960, 1974) in constructing asymptotically sufficient estimates, uses the following approximating functions: $L(t,x;\alpha,\beta) = -\alpha^2\beta^{-2}[\exp(\beta(x-t))-1-\beta(x-t)]$ for $\alpha > 0$, $\beta \neq 0$, β in $(-\infty,\infty)$, and $L(.) = -\alpha^2(x-t)^2/2$ for $\alpha > 0$, $\beta = 0$. Comparing these functions with the integrand kernels of the basic representation in section 3, one realizes an important application of the class Ω_o and its subfamilies. For more details see Ahmad-Al-Mutair (1976). Let the summand S_n from an i.i.d. sequence with any distribution F, have distribution F_n. Then, Ibragimov-Linnik (1971, p. 267) $\exists$ a sequence $\{F_n{}^o\}$ from Ω s.t. $\sup_x |F_n{}^o(x)-F_n(x)| \leq cn^{-1/3}$, where c is an absolute constant. Notice the implication of this global approximation to the situation above.

In connection with multivariate unimodality there are several definitions such as monotone unimodal (MU), central convex (CCU), and convex unimodal (CU), see Dharmadhikari-Jogdeo (1976) and the references cited there. Here we prove a conjecture of Sherman and give an affirmative answer to two of the Dharmadhikari-Jogdeo open questions. This depends on the lemma below. Due to lack of space our discussion is brief.

Lemma 6.1. Let E denote a closed convex set. Then for any C in E and every non-negative t, the set of all vectors X such that C + tX is in E is closed convex cone independent of C.

Proof. Without loss of generality we may take an element $y \in E$ instead of the set C and a point x instead of the vector X. Call ray, for a cone which is a proper subset of a line. Now, one can easily see that the sets of elements x such that $y + tx \in E$ is a convex cone. For non-negative t the set of all x such that $y + tx \in E$ is a closed set, but the above cone is the intersection of these closed sets. Therefore, it is closed. To prove the independent part, take any element z of E. Suppose $y + tx \in E$ for $t > 0$. The set E is closed convex set, and it contains the point z and the ray with origin y and direction x. The E contains the closed convex hull of the set formed by z and this ray. Whereas this ray may be said contains the ray with origin z and direction x. Hence the considered cone is independent of y. This proves the lemma.

Let $\tilde{L}$ be a linear space, $\tilde{T}$ be a topological space, $\tilde{T}L$ be a topological linear space which can be obtained by defining the topology on subsets of the linear space $\tilde{L}$. A convex set of $\tilde{T}L$ which has an interior point is called a convex body.

Now, MU and CCU are symmetric, and for simplicity we shall take the symmertization to be around the origin. In every MU defined above we can take f(C+tX) to be bounded. Now, recall Dharmadhikari and Jogdeo (1976) corollary 3.1, which states that a distribution F in R^k is MU, if and only if, the defining property holds for all symmetric compact convex bodies C in $\tilde{R}^k$. From the above lemma, previous discussion, the corollary, and the space $\tilde{TL}$ structure, one concludes that MU implies CCU and both the families MU and CCU are closed under convolutions.

Finally, to conclude we mention some open problems. Are all univariate and multivariate $\mathcal{L}$-distributions unimodal, in some suitable sense? What is the essential structure (other than what is already known) of strong unimodality, stability, symmetry and universality in the multivariate case? Are multivariate $\mathcal{L}$-distributions absolutely continuous?

REFERENCES

1. Ahmad, R. (1974). Ann. Inst. Statist. Math., 26, 233-245.
2. Ahmad, R. (1974). Math. Operationsforsch Statist., 5, 643-656.
3. Ahmad, R. (1975). Statistical Distributions in Scientific Work (eds. G.P. Patil et al), Vol. 3, 237-248. D. Reidel, Dordrecht.
4. Ahmad, R. (1975). Proc. Inter. Symp. on Interval Math. (ed. K. Nickel). Lecture Notes in Computer Science, Vol. 29, 127-134. Springer-Verlag.
5. Ahmad, R.-Abouammoh, A.M. (1976). Arabic Scientific Res. Soc. J., 2, to appear.
6. Ahmad, R.-Al-Mutair, M.A. (1976). 1976 European Cong. of Statisticians.
7. Akhiezer, N.I.-Glazman, I.M. (1957). Commun. Soc. Math. Kharkov, 25 (in Russian).
8. Barndorff-Nielsen, O. (1976). JRSS Ser. B, 38, 103-131.
9. Bell, C.B. (1958). Illinois J. Math., 2, 415-424.
10. Bochner, S. (1955). Harmonic Analysis & the Theory of Probability. Calif. UP.
11. Chung, K.L. (1953). C.R. Acad. Paris, 236, 583-584.
12. Chung, K.L. (1974). A Course in Probability Theory. Academic Press.
13. Dharmadhikari, S.W.-Jogdeo, K. (1976). Ann. Statist., 4, 607-613.
14. Doob, J.L. (1953). Stochastic Processes. Wiley.
15. Feller, W. (1971). An Introd. to Prob. Theory and its Appls., Vol. 2. Wiley.
16. Fisz, M. (1962). Ann. Math. Statist., 33, 68-84.
17. Fisz, M. (1963). Trans. Amer. Math. Soc., 106, 185-192.
18. Gikhman, I.-Skorohod, A. (1974). The Theory of Stochastic Processes. Springer.
19. Gnedenko, B.V.-Kolmogorov, A.M. (1954). Limit Distns. for Sums of I.R.Vs. Add.-Wesley.
20. Ibragimov, I.A. (1956;1957). Theor. Prob. Appl., 1, 255-260; 2, 117-119.
21. Ibragimov, I.A.-Linnik, Yu.V. (1971). Indep. & Stat. Seq. R. Vs. Wool.-Noordhoff.
22. Krein, M.G. (1940;1944). Dokl. Akad. Nauk. SSSR, 26, 17-22; 46, 91-94.
23. LeCam, L. (1960). Univ. Calif. Publ. Statist., 3, 37-98.
24. LeCam, L. (1974). Proc. Prague Symp. Asymp. Statist. (ed. J. Hájek), 1, 179-200. Charles UP.
25. Lévy, P. (1937). Theor. De L'Add. Des. Var. Aleat. [2nd ed. 1954] Paris.
26. Loève, M. (1963). Probability Theory. D. Van Nostrand.
27. Lukacs, E. (1970). Characteristic Functions. Griffin.
28. Masani, P. (1973). Mult. Analy. - III (ed. P.R. Krishnaiah), 209-223. AP.
29. Petrov, V. (1975). Sums of Independent Random Variables. Springer.
30. Steutel, F.W. (1973). Stoch. Processes Appl., 1, 125-143.
31. Sun, T.C. (1967). Ann. Math. Statist., 37, 1296-1299.
32. Urbanik, K. (1973). Mult. Analy.- III, 225-237. Academic Press.
33. Wintner, A. (1936). Amer. J. Math., 54, 45-90.
34. Wolfe, S.J. (1971). Ann. Math. Statist., 42, 212-218.
35. Yosida, K. (1966). Functional Analysis. Springer.
36. Zinger, A. (1965). Theor. Prob. Appl., 10, 431-435.
37. Zolotarev, V.M. (1957). Theor. Prob. Appl., 2, 433-460.
38. Zolotarev, V.M.-Korolyuk, V.S. (1961). Theor. Prob. Appl., 6, 431-435.

Recent Developments in Statistics
J.R.Barra et al., editors

ASYMPTOTICALLY OPTIMAL NONPARAMETRIC TESTS FOR LINEAR MODELS WITH INTERACTION

R. Ahmad and M.A. Al-Mutair
University of Strathclyde, Glasgow

1. INTRODUCTION: There are many and varied real life situations in a wide range of scientific areas, where the linear stochastic models occur most naturally. For such problems, in nonparametric statistical hypotheses testing and estimation, (with notable exceptions) it is usually postulated that there is no interaction effect possibility for various linear models. This is so because, unlike parametric or classical inference, it is not quite clear what an 'interaction' is when one considers nonparametric testing in general linear models. Since interaction hypotheses are common in practice, in this paper we consider nonparametric testing and estimation for such problems by assuming a large class of distributions Ω, including continuous distributions, underlying the linear models. The class Ω is such that it includes all continuous distributions on the real line, and the distributions which can reasonably be approximated by an infinitely divisible distribution, see Ahmad-Abouammoh (1976). Notice this family includes 'differentially asymptotically normal' and more general exponential families of LeCam (1960,1974), which are basic to contiguity ideas and their extensions. Note that under Ω if the ordering of given data is not unique, then the level of significance is not exact but approximate.

2. THE BASIC DISTRIBUTION-FREE TESTING STRUCTURE: These techniques can be used whenever the underlying distributions are unknown, and this is due to lack of physical knowledge, or where apriori knowledge of physical considerations, indicate uncertainties in the observed data. In what follows, a statistic T is a measurable real-valued function of the sample data point Z, and it has its own induced probability distribution. In nonparametric testing or decision theory the null and alternative hypotheses classes are denoted by $\Omega(H)$ and $\Omega(K)$ respectively. A decision function or test function, ψ, is a measurable real-valued function on R_Z satisfying $0 \leq \psi(Z) \leq 1$ and $\psi(Z) = P\{\text{Reject } H|Z\}$. Now, consider the model:

$$(2.1)\quad f(\alpha,\beta,\gamma,\varepsilon) = \mu + \alpha_i + \beta_j + \gamma_{ij} + \varepsilon_{ijk} \quad (1\leq i\leq c,\ 1\leq j\leq r,\ 1\leq k\leq n)$$

or levels could be continuous where ε_{ijk} are independent identically distributed as a stochastic process with a distribution F in Ω. Here, one can think of μ, α's, β's, and γ's as a general constant, α-treatment, β-treatment, and interaction levels, with ε's as random error. With a broad interpretation, the above model is quite general and embraces many testing problems, for example one-sample, 2-sample, k-sample, randomness and independence. As usual, assume

$$(2.2)\qquad \int \alpha(t)dt = \int \beta(s)ds = \int \gamma(t,s)dt = \int \gamma(t,s)ds = 0$$

where the integrals have to be replaced by sums if the levels are discrete and throughout this paper each integral or sum is extended over the domain of definition of the integrand or summand unless otherwise stated.

The study of distribution-free tests is essentially a study of the appropriate similar sets and similar test functions with respect to the pertinent null hypothesis class of distributions. The idea of similar tests, that is, tests of Neyman structure, and a rigorous formulation of the basic concepts, is due to Scheffé (1943), Lehmann and Stein (1949), and Bell and Doksum (1967). These tests are conditional tests, given the orbit, which is a complete sufficient statistic for the problem. The family of these permutation tests is very large, therefore one desires to have a smaller subfamily of rank-order tests which is more tractable. These tests are Chernoff-Savage, and Hájek-Šidák type general scores rank-order tests. It was shown in Ahmad (1974b,d) that all DF test statistics for various hypotheses are functions of the Bell-Pitmann (BP) or permutation statistics.

Now if we denote by $R_{(.)}$ and $Z_{(.)}$, respectively, the rank vector or matrix and order statistics of the sample data point Z, then one notes that there is one-to-one correspondence between the rank vectors and BP functions. For 'no row effect' and 'no column effect' hypotheses one can characterize all DF tests as in Ahmad (1974a,d) by using the following fundamental lemma in connection with orbits.

Lemma 2.1. (i) If Z is the generic data point, then $R_{(.)}$ and $Z_{(.)}$ are stochastically independent and form a sufficient statistic for the problem. (ii) If G is the joint null hypothesis distribution for Z, then $P(Z^*|S(Z),G) = (C(S))^{-1}$, for Z^* in S(Z), and zero elsewhere. Furthermore, the orbit is a complete sufficient statistic for $\Omega(H)$, the null hypothesis family.

Proof. First one notes that S(Z) is complete w.r.t. $\Omega(H)$ for 'no row' and 'no column' effect hypotheses. This can be shown by a similar argument as in Bell, Blackwell and Breiman (1960). Then (i) and (ii) are slight modifications of the Lemma 3.1 and the Theorem 2.2 of Ahmad (1974a,d) respectively Q.E.D.

Let S_N denote the symmetric permutation group of N! permutations of the integers {1,2,.., N}. The relevant groups for 'no row effect' and 'no column effect', respectively, are $\times_1^c S_r$ and $\times_1^r S_c$ with cardinalities $(r!)^c$ and $(c!)^r$. For any group S we define the orbit of Z as $S(Z) = \{\lambda Z : \lambda \varepsilon S\}$. Note that in the univariate case the orbit is equivalent to the order statistics. Constructing similar sets and tests consists of choosing a fixed proportion of the points of almost every (a.e.) orbit, and is accomplished by BP statistics.

Definition 2.1. (i) A statistic, T, is DF with respect to (w.r.t.) the null hypothesis class $\Omega(H)$ if there exists a single distribution Q such that $P(T \leqslant t|F) = Q_T(t)$ for all real t and all F in $\Omega(H)$. (ii) A set, A, is non-trivial similar w.r.t. $\Omega(H)$ if there exists an $\alpha (o < \alpha < 1)$ such that $P(A|F) = \alpha$ for all F in $\Omega(H)$; and an α-Scheffé set if A contains exactly $\alpha C(S)$ points of almost all orbits, C(S) is the cardinality of the group S. (iii) A test function, $\psi(Z)$, is similar if there exists an $\alpha(o < \alpha < 1)$ such that $\int \psi(Z)dF(Z) = \alpha$ for all F in $\Omega(H)$; and an α-Lehmann-Stein function if $\sum_{\lambda\varepsilon S} \psi(\lambda Z) = \alpha C(S)$ for almost all $\Omega(H)$.

Definition 2.2. (i) A real valued measurable function, V, is called a BP function w.r.t. S if $P_F(V(Z) = V(\lambda Z)) = o$ for all F in $\Omega(H)$, and all non-identity elements λ of S. (ii) If V is a BP function, then $R(V(Z)) = \sum_{\lambda\varepsilon S} \varepsilon(V(Z) - V(\lambda Z))$ is called a BP statistic induced by the BP function V, where $\varepsilon(y) = 1$ if $y \geqslant o$, and zero otherwise.

Theorem 2.1. (i) A statistical test function $\psi(Z)$ is α-similar w.r.t. $\Omega(H)$, if and only if, $\psi(Z)$ is an α-Lehmann-Stein function. (ii) A statistic T is DF w.r.t. $\Omega(H)$, if and only if, there exists a measurable function q such that $T \equiv q(R(V(Z)))$.

Proof. (i) The orbit is a complete sufficient statistic for $\Omega(H)$, and a necessary and sufficient condition for the test function $\psi(Z)$ to be α-similar is that $\psi(Z)$ possesses Neyman structure. Therefore $E(\psi(Z) \mid S(Z), G) = \alpha$, for all G in $\Omega(H)$. But from the basic property of the orbit S(Z), $E(\psi(Z) \mid S(Z), G) = (C(S))^{-1}. \sum_{\lambda\varepsilon S} \psi(\lambda Z)$, for all G in $\Omega(H)$, and this proves (i). (ii) This follows from (i). Q.E.D.

From this theorem one derives immediately the fundamental results concerning similar sets by allowing ψ to be the indicator function of a set, and choosing a BP function to select the points on each orbit. To this end one has the following corollary.

Corollary 2.1. (i) If a set, A, is similar w.r.t. $\Omega(H)$, then P(A) is one of the values $t(C(S))^{-1}$, t = 1,2,..., C(S). (ii) The following three conditions are equivalent: (a) A is a set similar w.r.t. $\Omega(H)$ and $P(A) = \alpha$; (b) $\psi = I_A$ is a classical test function of size α; (c) A is an α-Scheffé set.

From the above results which characterize the families of similar sets and DF statistics, it is clear that the similar tests are conditional tests, given the orbit which is a complete sufficient statistic for the problem.

Definition 2.3. (i) A statistic $T(Z)$ is called strongly distribution-free (SDF) w.r.t. $\Omega(H \cup K)$ and $G^*(Z)$ if it is invariant over each equivalent class of $\Omega(H \cup K)$ under $G^*(Z)$. (ii) A set, A, is called SDF if its indicator function, I_A, is an SDF statistic.

Theorem 2.2. Under the randomness hypothesis: (i) each rank statistic is SDF w.r.t. $\Omega(H \cup K)$ and G^*; (ii) each SDF statistic is DF, and (iii) similar results hold for sets.

Proof. Clearly under G^*, $R_{(.)}$, $Z_{(.)}$, and V are all invariant. Consequently, under the hypothesis a test statistic T is SDF, if and only if, there exists a Borel measurable function g such that $T \equiv g(R_{(.)})$. Now, by the complete sufficiency of the orbit and the theorem 2.1, (i) and (ii) follow. The part (iii) is proved, if we let I_A to be the indicator function of the set A.

Corollary 2.2. If T is a rank statistic, then any test based on T has constant power over every equivalent class of $\Omega(H \cup K)$ under G^*. Thus a SDF statistic divides $\Omega \times \Omega$ into $\bigcup_{j \in J} \Omega_j$, J an index set, such that the power function $\pi_{\Omega_j} = \pi_j$ for all distributions F in Ω_j, $j \in J$.

Let Z_N be the generic data point from a measurable space $(\mathcal{X}_N, \mathcal{B}_N)$. Then similar to LeCam (1974) and Pfanzagl (1974), though with slight modifications to allow for nonparametric setting, we define asymptotic similarity and sufficiency, and consequently give some asymptotically optimal results.

Definition 2.4. (i) A test function $\psi_N = \psi_N(Z_N, F)$ is said to be asymptotically α-similar w.r.t. $\Omega_N(H)$, if $E\psi_N = \alpha + \varepsilon_N$, where $\varepsilon_N \to 0$ as $N \to \infty$, for every F in $\Omega_N(H)$. (ii) Let $\mathcal{B}_N^*$ be a sigma-algebra such that $\mathcal{B}_N^* \subset \mathcal{B}_N$. The sequence $\{\mathcal{B}_N^*\}$ is called ε_N-asymptotically sufficient w.r.t. $\Omega_N(H)$, if there exists $\bar{\Omega}_N(H)$ such that: (a) for $\bar{\Omega}_N(H)$ the sigma-algebra $\mathcal{B}_N^*$ is sufficient, and (b) for every $\theta \varepsilon \Theta \subset R^k$ and every $c \varepsilon (1,\infty)$, $\lim_{N\to\infty} \sup_{|\theta^*-\theta| < c\varepsilon_N} ||\bar{F}_{\theta^*,N}(z) - F_{\theta^*,N}(z)|| = 0$, $F \varepsilon \Omega_N$, $\bar{F} \varepsilon \bar{\Omega}_N$.

Now, consider an alternative family of distributions whose densities w.r.t. the Lebesgue measure (or counting measure) are proportional to $\exp[L(t,x,\alpha,\beta)]$, where according to LeCam (1974),

$$L(t,x,\alpha,\beta) = \begin{cases} -\alpha^2\beta^{-2}\left[e^{\beta(x-t)} - 1 - \beta(x-t)\right], & \alpha > 0,\ \beta \varepsilon (-\infty,\infty),\ \beta \neq 0, \\ -\alpha^2/2\ (x-t)^2, & \beta = 0. \end{cases}$$

For further details on this matter see Ahmad-Abouammoh (1976), and Pfanzagl (1972). By using these ideas, the concepts of asymptotic similarity and asymptotic sufficiency as defined above, one can establish the following by slightly modifying arguments in the proof of the theorem 4.2 in Ahmad (1975).

Theorem 2.3. Under $\Omega(H)$: (i) In the family of all DF tests, the MPDF level α test against a specific alternative likelihood function K_1 is of the form $\psi(Z) = 1, \delta$, or 0 according as $R(V_1(Z))$ is >, =, or < $c(\alpha, S(Z), C(Z))$, where V_1 is a BP function w.r.t. S, whose ordering on a.e. orbit is consistent with that of K_1. This test is asymptotically (A) MPDF, that is, AMPDF, if the alternative density K_1' is proportional to $\exp[L(.)]$ as defined above.

(ii) If $K_1(z) = \prod_{i=1}^{N} K_i'(x_i)$, where the K_i' are not all equal, then the MPDF (AMPDF) test has power $>\alpha(>\alpha+\varepsilon_N)$. A necessary and sufficient condition for the power to be greater than the level of significance is that K_1 is not invariant under the appropriate group S.

(iii) For the alternative family $\{K_1 : K_1(z) \propto \exp[a(\theta,L)V_1(z)+b(z,\theta,L)]$ with $a(\theta,L) > 0$ and V_1 a BP function$\}$, $\psi(Z)$ is; (a) AMPDF if b is invariant under S for all θ, (b) ALMPDF if $b = b_1+b_2$ and b_1 is invariant under S and $b_2(z,\theta,L) = o(a(\theta))$, and $a(\theta)$ is continuous with $a(\theta_o) = 0$. (L = locally).

(iv) If $\{K_1(z) = \exp[a(\theta)V_1(z) + b(z,\theta)]\}$ with similar conditions, then in cases (a), (b) the test is MPDF, LMPDF, respectively.

Note that by letting $\alpha^* = \alpha+\varepsilon_N$, an α^*-similar DF test becomes α-similar ADF test. The terms AMPDF and ALMPDF may be somewhat obscure and unfamiliar, but we have used them since $\{\exp[q(L,\theta,Z)]\}$ provide 'super-smooth' approximations to a wide class of distributions. As alternative terminology one could think asymptotically S-optimal (ASO) DF and ALSO DF tests, respectively.

3. TREATMENT OF INTERACTIONS IN TESTING THEORY: Here we will propose a rather general treatment of interactions in the two-way layouts with one observation per cell, and consequently prove certain related results. The extension to higher-way layouts and with more than one observation per cell is similar and can be carried out likewise.

3.1. Treatment Of Interactions In The Linear Model. Let us first consider the following problem in the linear set-up. We have the null hypothesis $H : \alpha_i = \beta_j = \gamma_{ij} = 0$, for all i,j, which is to be tested against the alternative hypothesis $K : \alpha_i = \beta_j = 0$ for all i,j, and $\gamma_{ij} \neq 0$ for some i,j. In other words, $H : F_{ij} = F$ for some $F \in \Omega$ and for all i,j; $K : F_{ij} = F(x-\gamma_{ij})$ for some $F \in \Omega$ and for all i,j where not all γ_{ij} are equal to zero. The class of all distribution-free tests for this problem would be the distribution-free tests for randomness of these rc random observations, and is given by the following theorem.

Theorem 3.1. (i) T is distribution-free with respect to $\Omega(H)$, if and only if, T is equivalent to a function of a basic Bell-Pitman statistic, that is, if and only if there exists a Bell-Pitman function V and a measurable function W such that $T \equiv W(R(V))$. (ii) Each distribution-free statistic T has a discrete null hypothesis distribution with probabilities which are integral multiples of $(rc!)^{-1}$. (iii) For any preassigned discrete distribution G with probabilities which are integral multiples of $(rc!)^{-1}$, there exists a distribution-free statistic with null hypothesis distribution G.

Proof. (i) If V is a Bell-Pitman function, then the set $\{Z : R(V(Z)) = j\}$ is similar and so is the set $\{Z : W(R(V(Z))) = w\}$, so that $W(R(V(Z)))$ is distribution-free. Conversely, if T is distribution-free with respect to $\Omega(H)$, then T has a discrete distribution, that is, there exist real numbers $a_1 < a_2 < \ldots < a_q$, $1 \leqslant q \leqslant (rc!)$, and integers $o = b_o < b_1 < \ldots < b_q = (rc!)$ such that $P\{T \leqslant a_1 | H\} = o$, and $P\{T \leqslant a_j | H\} = b_j\ (rc!)^{-1}$. Therefore, one can order the points of almost all orbits $S(Z)$, so that $T(Z_i) = a_j$ for $b_{j-1} < i \leqslant b_j$. Next, define $V(Z_i) = i$ and $W(i) = a_j$ for $b_{j-1} < i \leqslant b_j$. Then V is a Bell-Pitman function and $T(Z) = W(R(V(Z)))$ almost everywhere $\Omega(H)$, and this proves (i).

(ii) Let T be a distribution-free statistic. Then $A_t = \{Z : T(Z) \leqslant t\}$ is a similar set, and by the complete sufficient property of the orbit, $P(A_t) = j\ (rc!)^{-1}$, where $j = 0,1,\ldots, (rc!)$, and this proves (ii).

(iii) If G assigns probabilities $p_1,\ldots,p_q$ to the points $a_1 < a_2 < \ldots < a_q$, with $\sum_{j=1}^{q} p_j = 1$, then there are integers $o = b_o < b_1 < \ldots < b_q = (rc!)$ such that $\sum_{j=1}^{i} p_j = b_i\ (rc!)^{-1}$, $i = 0,1,\ldots, q$. Let V be an arbitrary Bell-Pitman function and let $W(j) = a_i$ for $b_{i-1} < j \leqslant b_i$. Then $W(R(V(Z)))$ is the required distribution-free statistic. Q.E.D.

Next, we consider another problem, namely, $H : \beta_j = \gamma_{ij} = 0$, for all i,j, against $K : \beta_j = 0$, for all j and at least one $\gamma_{ij} \neq 0$. We again recognize that this is a problem of testing randomness within each row. Now, we apply the analogue of theorem 3.1 with the group of transformations in this case being the permutation group $S = \overset{r}{\underset{1}{\times}} S_c$ with cardinality $(c!)^r$. Similarly, for the case $H : \alpha_i = \gamma_{ij} = 0$, for all i,j, versus $K : \alpha_i = o$ for all i and at least one $\gamma_{ij} \neq 0$, we get the class of distribution-free tests by the above theorem, the group will now be $S = \overset{c}{\underset{1}{\times}} S_r$ with cardinality $(r!)^c$. Note that all the results of theorem 3.1 can be restated for the above two cases, by merely replacing (rc!) by their respective group cardinalities.

3.2. A General Treatment Of Interactions. In the sequel we consider tests based on the interaction contrasts $\{C_{ijkm}\}$ defined by $C_{ijkm} = X_{ij} - X_{im} - X_{kj} + X_{km}$ $(i \neq k$ and $j \neq m)$. We treat four types of hypotheses concerning these contrasts. The first two of these are as follows: (i) $H^1 : X^*_{ij} = \mu + \alpha_i + \beta_j + \varepsilon_{ij}$, where $\forall$ i,j the ε_{ij}'s are identically distributed with an arbitrary distribution in Ω.
(ii) The contrasts $\{C_{ijkm}\}$ are identically distributed with an arbitrary distribution in Ω. For the third hypothesis first one needs the following definition.

Definition 3.1. Let $A_t = \{(x_1,\ldots,x_p) : x_t < 0\}$ and set $A_t^1 = A_t$ and $A_t^o = R^p - A_t$. Then the set $B(j_1,\ldots,j_p) = \bigcap_{t=1}^{p} A_t^{j_t}$ is called a p-tant, where $j_t = 0$ or 1.

(iii) H^3 : For each contrast C_q and each p-tant B_j, $P(C_q \in B_j) = 2^{-p}$. (iv) H^4 : Each contrast has median zero.

The main difficulty in constructing distribution-free tests here is that the contrasts are not mutually independent unless they have no X's in common. The first is then to consider families of non-overlapping contrasts, that is, contrasts based on disjoint subjects of four variables $\{X_{ij}, X_{im}, X_{kj}$, and $X_{km}\}$ each with $i \neq k$ and $j \neq m$. The contrasts formed from these sets are then mutually independent. Of course, to ensure independence we have sacrificed $(rc!(4!(rc-4)!)^{-1}) - [\frac{r}{2}][\frac{c}{2}]$ of the total contrasts, where $[\frac{j}{2}]$ denotes the largest integer $\leq \frac{j}{2}$. However, with these $q = [\frac{r}{2}][\frac{c}{2}]$ mutually independent contrasts one can easily prove the following.

Theorem 3.2. Let $C_1,\ldots,C_q$ be q mutually independent contrasts found from a partition of the r × c table, as described above. Then, under H^1 and H^2 each test of randomness of the C's is a distribution-free test. In particular, each rank statistic in the C's is distribution-free. Furthermore, under H^3 and H^4, the minimal sigma-algebra containing sets of the form $\{C_j > 0\}$ is similar. All statistics measurable with respect to this sigma-algebra are distribution-free.

The partition employed above has one major disadvantage. It makes use of only a fraction of the total possible contrasts, and, hence the test will have power equal to the level of significance against alternatives with exactly those interactions not 'covered' by the contrasts of the partition. Another approach is to treat dependence directly. In the case of r x c ($r \geq 2$ and even) divide $X_{ij}(i = 1,2,\ldots,2r^*=r)$ randomly, and define $Y_{kj} = X_{\alpha_k j} - X_{\beta_k j}$ $(k = 1,2,\ldots,r^*$; and $\alpha_k \neq \beta_k)$, and $(\alpha_1,\ldots,\alpha_{r^*}; \beta_1,\ldots,\beta_{r^*})$ simply denote randomly chosen partition observations as $\{X_{ij}; i(j) = 1,2,\ldots,2r^*(c)\}$. Thus, when r is even, in the manner above, one can prove the following for r x c table.

Theorem 3.3. Under H^1 and H^2, (i) $Y_{kj}(k(j) = 1,2,\ldots,r^*(c))$ are independently and identically distributed; and hence (ii) each Bell-Pitman statistic, and consequently each rank statistic in the Y's is distribution-free.

For H^3 one can use any multinomial test. One such test of wide applicability is the chi-square test. Suppose $C_1,\ldots,C_q$ are the total number of independent contrasts.

Theorem 3.4. The test statistic, $T = \sum_{j=1}^{M} (n_j - q2^{-p})^2 q^{-1}2^p$, is distribution-free, and has, under H^3, an asymptotic chi-square distribution with (M-1) degrees of freedom, where n_j is the number of contrasts falling in the jth p-tant, B_j, and of course, there are total $M = 2^p$ such p-tants.

One should note that the n_j may be too small for reasonable approximation by the chi-square distribution. Hence, under such circumstances, one may wish to use unions of p-tants as the partition sets.

4. MOST POWERFUL SIMILAR REGIONS: As before let Z be the generic data point in the problems of goodness-of-fit, 2-sample, k-sample, randomness, and 2-way layouts. Consider regions B in the Euclidean space R_Z and two distributions $F_1 \varepsilon \Omega(H_o, R_Z)$ and $F_2 \varepsilon \Omega(H_1, R_Z)$. We have already called $\int_B dF_1 = P_1(B)$, the size of B, and $\int_B dF_2 = P_2(B)$, the power of B. Let $\mathcal{B}_Z$ be an additive family of regions in R_Z, that is, this family is closed under unions, intersections and differences. In the sequel, all the regions considered belong to the family $\mathcal{B}_Z$. Notice that R_Z is R_N for randomness hypothesis with $Z = (X_1,\ldots,X_N)$, and $R_m \times R_n$ for 2-sample problem with $Z = (X_1,\ldots,X_m; Y_1,\ldots,Y_n)$, and similarly for other hypotheses.

Definition 4.1. A region B is called a most powerful region (MPR) of the family $\mathcal{B}_Z$ if $P_2(B) \geqslant P_2(\tilde{B})$ for all $B \varepsilon \mathcal{B}_Z$ for which $P_1(B) = P_1(\tilde{B})$.

In a similar manner as Mann (1960) in the classical case, we shall assume that our regions satisfy the following two conditions: (i) If B is any region of size α and if $\tilde{\alpha} < \alpha$, then B has a subregion of size $\tilde{\alpha}$. This condition implies: (i*) If B is any region of size α and if $\alpha = \sum_{i=1}^{m} \alpha_i$, then $B = \bigcup_{i=1}^{m} B_i$, where B_i has size α_i.

Now, it is easy to establish the following result.

Lemma 4.1. (i) Let B be a MPR of size α and B_1 any subregion of B of size $\alpha_1 \leqslant \alpha$. Let B_2 be any region of size α_1 and $B \cap B_2 = \phi$, the null set. Then $P_2(B_2) \leqslant P_2(B_1)$. Otherwise, the region $B - B_1 + B_2$ would have size α and higher power than B.

(ii) Let B be a MPR of size α and B^* any subregion of B of size $\alpha^* \leqslant \alpha$. Let $\tilde{B}$ be any region of size $k\alpha^*$ such that $\tilde{B} \cap B$ is empty, then $kP_2(B^*) \geqslant P_2(\tilde{B})$.

By using the above lemma, and modified arguments of Mann (1960) to allow for nonparametric problems, we have the following basic result.

Theorem 4.1. (i) Let B be a MPR of size α. Suppose $B_1,\ldots,B_m$ are m regions with $P_1(B_i) = \alpha_i$ and let $p_1,\ldots,p_m$ be m nonnegative numbers such that

$$\sum_{i=1}^{m} p_i\alpha_i = \alpha, \quad \sum_{i=1}^{m} p_i = 1. \quad \text{Then,} \quad \sum_{i=1}^{m} p_i P_2(B_i) \leqslant P_2(B).$$

(ii) Let $\mathfrak{X}$ be a space and μ a probability measure defined over this space. For every $Z \varepsilon \mathfrak{X}$ let B_Z be a region in R_Z. Let $\alpha = \int P_1(B_Z)d\mu$ and assume that $\int P_2(B_Z)d\mu < \infty$. If B is a MPR of size α, then $\int P_2(B_Z)d\mu \leqslant P_2(B)$.

An immediate consequence of the first part of the theorem above is the following.

Corollary 4.1. Let $\{B_i\}$ be an infinite sequence of regions in $\mathcal{B}_Z$, and $\{p_i\}$ a sequence of nonnegative numbers such that $\sum_{i=1}^{\infty} p_i P_1(B_i) = P_1(B)$, and $\sum_{i=1}^{\infty} p_i = 1$, where B is a MPR. Then $\sum_{i=1}^{\infty} p_i P_2(B_i) \leq P_2(B)$.

To prove the last result, choose q so that $\sum_{i=q+1}^{\infty} p_i < \varepsilon$, $\varepsilon > 0$. Then

$$\sum_{i=1}^{\infty} p_i P_2(B_i) \leq \sum_{i=1}^{q} p_i P_2(B_i) + \varepsilon \leq P_2(B) + \varepsilon.$$

The second part of the theorem follows by applying the corollary to a MP subregion of B of size $\alpha-\delta$, $\delta > 0$; and if $\delta < 0$ one chooses a MPR B* of size $\alpha+\delta$ containing B, and then applying the corollary.

The application of the above ideas in our problems is as follows. Let T be a distribution-free test statistic for testing $\Omega(H_o)$ against $\Omega(H_1)$, and suppose T_o is the critical value. If $P(T \geq T_o|F) = \alpha$ for all $F \in \Omega(H_o)$, then the region $\{T \geq T_o\}$ is a most powerful region of size α (or asymptotically MPR) of the Class $\mathcal{B}_Z$.

5. SOME OPEN PROBLEMS AND COMMENTS: The characterization of all similar sets or all distribution-free tests, needs precise statements of the completeness properties as in Bell, Blackwell and Breiman (1960) of the null hypotheses families defined above. Since these families are defined in terms of interaction contrasts, the usual product space completeness results do not directly apply. Thus, one open research problem is to establish these completeness results for interaction contrasts hypotheses, H^1, H^2 and H^3. Alternatively, perhaps one could establish asymptotic completeness and then construct asymptotically DF tests on the lines of Neyman (1954), LeCam (1960,1974), and Pfanzagl (1974).

In the late 1950's and early 1960's Lehmann and Hodges, in a series of papers, obtained many interesting and efficient nonparametric test and estimation procedures, which were essentially competitors to the Student's t-test and the classical tests applied to analysis of variance situations. In considering tests based on median contrasts Lehmann (1963) shows that the estimate $\hat{\theta}$ of any contrast θ is symmetrically distributed about the true value of the contrast, and hence in particular is unbiased, if either block distributions are symmetric or all sample sizes are all equal. In fact, the symmetry assumption can be replaced by either montone unimodality or central convex unimodality, see Ahmad-Abouammoh (1976). The modified assumption implies symmetry and an extra 'smoothness' property of unimodality. This helps enlarge the underlying parameter space by convoluting the error process by a strongly unimodal or symmetric measure. For these larger families the above procedures are valid exactly or in a modified form. Here essentially one gets asymptotic chi-squared distributions.

The construction of various DF tests for interaction hypotheses can be carried out similar to Ahmad (1974c,d). Now, let $\Omega(C)$ be a subfamily of the class Ω satisfying a version of the Cramér condition:

$$\lim_{t\to\infty} \sup \sup_{\theta} \int \exp(i \sum_{j=1}^{k} t_j x_j) dF_\theta(x_1,\ldots,x_k) < 1.$$

By using Pfanzagl (1974) ideas, it is possible to develop a general method for the construction of asymptotic distribution-free asymptotically optimal median-unbiased critical regions and confidence intervals. An estimator-sequence for $g(\theta) \in \Theta_g \subset R^k$, $\{T_j(Z|F_\theta), j \in J\}$ is called median-unbiased of order $o(n^{-r})$ if uniformly on compact subsets of the parameter space

$$\tfrac{1}{2} - P\{T_n(Z) \geq g(\theta) \mid F_\theta\} \leq o(n^{-r}), \text{ and } \tfrac{1}{2} - P\{T_n(Z) \leq g(\theta) \mid F_\theta\} \leq o(n^{-r}),$$

for every F_θ in $\Omega_c(H)$. Notice that such tests are not generally completely distribution-free since the variance of the asymptotic distribution depends on the population distribution.

It is clear from the previous discussion and development that to test for 'no interaction effect' hypotheses one could use any rank statistic for randomness or independence of the non-overlapping contrasts C_j's. A large class of nonparametric tests in practice is constructed as follows. First for given independent random variables $Z = (X_1,\ldots,X_N)$ with X_j having a distribution in Ω one considers the random rank vector or matrix $(R_1^*,\ldots,R_N^*)$. Then one chooses some scores $a(1),\ldots,a(N)$ and regression coefficients $d_1,\ldots,d_N$, and finally one lets $W_N = \sum_{j=1}^{N} d_j a(R_j^*)$. Now one seeks the conditions under which the linear rank statistics W_N are asymptotically normal. Once this aspect is settled, then one takes a suitable function g of W_N and constructs a test statistic $T \equiv g(W_N)$ which has usually a well-known asymptotic distribution, for example, chi-square distribution. Since interaction hypothesis reduces to randomness hypothesis which is a special case of the k-sample problem, to construct various tests one can employ the techniques of Ahmad (1974c,d) either in univariate or multivariate setting.

Finally, we may remark that by using group-invariance principle and the ideas of contiguity as developed by LeCam (1960), one can also construct asymptotically similar tests as proposed by Neyman (1954). In this context our theorem 2.3 is relevant, and has many potentially efficient and robust applications.

REFERENCES:

1. Ahmad, R. (1971). J. Indian Statist. Assoc., 9, 1-7.
2. Ahmad, R. (1972). Trabajos de Estadistica, 23, 51-60.
3. Ahmad, R. (1973). Bull. Inter. Statist. Inst., 45(1), 87-93.
4. Ahmad, R. (1974a). Ann. Inst. Statist. Math., 26, 233-245.
5. Ahmad, R. (1974b). Math. Operationsforsch. Statist., 5, 643-656.
6. Ahmad, R. (1974c). Colloq. Math. Soc. J. Bolyai, 9, 51-60.
7. Ahmad, R. (1974d). Proc. Prague Symp. on Asymptotic Statistics (ed. J. Hájek), 1, 371-402. Charles Univ. Press.
8. Ahmad, R. (1975). Statistical Distributions in Scientific Work (eds. G.P.Patil et al.), 3, 237-248. Reidel Pub. Co., Holland.
9. Ahmad, R. - Abouammoh, A.M. (1976). 1976 European Congress of Statisticians.
10. Ahmad, R. - Al-Mutair, M.A. (1976). Arabic Scientific Research Society J., 2, in press.
11. Bell, C.B. - Blackwell, D. - Breiman, L. (1960). Ann. Math. Statist., 31, 795-797.
12. Bell, C.B. - Doksum, K. (1967). Ann. Math. Statist., 38, 429-446.
13. LeCam, L. (1960). Univ. Calif. Publ. Statist., 3, 37-98.
14. LeCam, L. (1974). Proc. Prague Symp. on Asymptotic Statistics (ed. J. Hájek), 1, 179-200.
15. Lehmann, E.L. - Stein, C. (1949). Ann. Math. Statist., 20, 28-45.
16. Lehmann, E.L. (1959). Testing Statistical Hypotheses. Wiley.
17. Lehmann, E.L. (1963). Ann. Math. Statist., 34, 957-966.
18. Neyman, J. (1954). Trabajos de Estadistica, 5, 161-168.
19. Pfanzagl, J. (1972). Ann. Math. Statist., 43, 553-568.
20. Pfanzagl, J. (1974). Proc. Prague Symp. on Asymptotic Statistics (ed. J. Hájek), 1, 201-272.
21. Scheffé, H. (1943). Ann. Math. Statist., 14, 305-322.

Recent Developments in Statistics
J.R.Barra et al., editors

LEAST SQUARES AS MARKOV ESTIMATORS FOR REGRESSION COEFFICIENTS

H.T. Amundsen,
University of Oslo, Norway.

When the distribution of each regressand in a regression problem belongs to a regular exponential (Darmois-Koopmans) family of distributions, and the joint distribution satisfies the regularity conditions for the multiparameter Cramér-Rao inequality, then the weighted least squares estimators are minimum variance unbiased provided that the solution does not involve unknown parameters.

A number of results are known about the conditions for least squares (LS) and weighted least squares (WLS) estimators to be minimum variance linear unbiased (MVLUE) and minimum variance unbiased (MVUE). In this note we give a proposition about MVUE-properties when the distribution of the regressand belongs to a regular exponential (Darmois-Koopmans) family of distributions.
Consider the nx1 vector $Y=(Y_1,\dots,Y_i,\dots,Y_n)'$ of random variables, the nxk matrix x of observations on k regressor variables, with rank k<n and the kx1 vector β of regression coefficients. We assume that the regression of Y on x is linear,

$$E(Y|x) = x\beta \ , \tag{1}$$

with covariance matrix $E(Y-x\beta)(Y-x\beta)' = \Sigma$.
By (1) the LS estimators of β and any estimable function of β are unbiased. For $\Sigma=\sigma^2 I_n$ they are also MVLUE, according to the Gauss Markov theorem. However, T.W. Anderson (1962) showed that when the distribution of Y_i is an arbitrary member of the class of all distributions, then the LS estimators are not MVUE in general, but may be so in some very special cases.
On the other hand, C.R. Rao showed in 1959, see Ch. 5a.2, p. 258 in his book (1965), that if the LS estimator is MVUE, then $Y_i-x_i'\beta$ is normally distributed, provided that $Y_i-x_i'\beta$ for i=1,2,...,n are identically distributed with finite moments of all orders.
Now there are examples of non-normal distributions for which LS estimators are MVUE, e.g. in Poisson and binomial cases.
For an arbitrary Σ, conditions on x for LS estimators to be MVLU have been established, see Watson (1967) and Haberman (1975) with references.
As to the generalized estimators

$$\bar{\beta} = (x'\Sigma^{-1}x)^{-1}(x'\Sigma^{-1}Y), \tag{2}$$

Aitken's theorem establishes that they are MVLUE when Σ is known or known up to a multiplicative constant. This does not apply when Σ is unknown and estimated.
In several special cases it has been shown that LS or WLS estimators are identical to maximum likelihood (ML) estimators. Bradley (1973) showed this to be true when Y_i for i=1,2,...,n are independent and distributed according to a member of a regular exponential family of distributions. It follows that the asymptotic properties of the WLS estimators can be deduced for this case. It seems worthwhile also to look into their small sample properties, utilizing the Cramér-Rao lower bound, having in mind the existence of estimators satisfying this for distributions admitting sufficient statistics.

We assume then, that each Y_i has a distribution satisfying the regularity conditions for the exponential family, see Lehman (1959), and that the density function can be written in the form used by Bradley,

$$f(y_i;\theta_i,\tau_i) = h(y_i)\ \exp\left[P_i(\theta_i)y_i - Q_i(\theta_i)\right] ,$$

where

$$\theta_i = EY_i = x_i'\beta .$$

τ_i are possible nuisance parameters, appearing in the functional forms in the exponent, while h is independent of the parameters.

In the family $\mathcal{F}$ satisfying these conditions, derivation of $\int f(y,\theta_i,\tau_i)dy = 1$ under the integral sign with respect to θ is permitted. Setting Q' and P' for the derivatives w.r. to θ, we obtain

$$EY_i = Q_i'(\theta_i)/P_i'(\theta_i) \quad \text{which gives} \quad Q_i'(\theta_i) = \theta_i P_i'(\theta_i), \tag{3}$$

and

$$\operatorname{var} Y_i = EY_i^2 - \theta_i^2 = \left(P_i'(\theta_i)\right)^{-1} . \tag{4}$$

Further, we assume that the simultaneous density function f_n is in the family $\mathcal{C}$ that satisfies the regularity conditions for the multiparameter Cramér-Rao inequality to hold. Then, if $\xi^*(Y)$ is a statistic, and if

$$\frac{\partial \log f_n}{\partial \xi} = E\left(\frac{\partial \log f_n}{\partial \xi_j}\ \frac{\partial \log f_n}{\partial \xi_q}\right)(\xi^* - \xi) \tag{5}$$

holds, ξ^* is the minimum variance bound unbiased (MVBU) estimator of ξ.
Setting now,

$$f_n(y_1,y_2,\ldots,y_n) = \prod_{i=1}^{n} f(y_i,\theta_i,\tau_i) , \tag{6}$$

we obtain the kx1 vector

$$V = \frac{\partial \log f_n}{\partial \beta} = \left(\sum_i P_i'(\theta_i)x_{ij}(y_i - \theta_i)\right) ,$$

and the kxk matrix

$$M = \left(E\ \frac{\partial \log f_n}{\partial \beta_j}\ \frac{\partial \log f_n}{\partial \beta_q}\right) = \left(\sum_i P'(\theta_i)x_{ij}x_{iq}\right) .$$

Defining the kx1 vectors

$$V_1 = \left(\sum_i P_i'(\theta_i)x_{ij}\theta_i\right) = M\beta ,$$

and

$$\beta^* = M^{-1}(V+V_1) \tag{7}$$

we have

$$M(\beta^* - \beta) = M(M^{-1}(V+V_1) - \beta) = V,$$

so that (5) is satisfied. The following proposition then holds

PROPOSITION: If $f\in\mathcal{F}$, $f_n\in\mathcal{C}$, and β^* defined by (7) is an estimator, then it is an WLS- and an MVBU-estimator.

Now

$$V = x'\ \frac{\partial P}{\partial \theta}\ (Y - \theta) \quad \text{and} \quad M = x'\ \frac{\partial P}{\partial \theta}\ x$$

where $\frac{\partial P}{\partial \theta}$ is the diagonal matrix of $P_i'(\theta_i)$-values. In view of (4) and (6) we have $\frac{\partial P}{\partial \theta} = \Sigma^{-1}$, so that

$$\beta^* \equiv (x'\Sigma^{-1}x)^{-1}x'\Sigma^{-1}Y = \bar{\beta} .$$

We see that if the solution of (2) does not involve unknown parameters, then $\bar{\beta}$ is MVBU for $f\in\mathcal{F}$, $f_n\in\mathcal{C}$ when the Y_i are independent. Examples of this situation, where unknown factors of Σ cancel out, are

1. Σ is known, i.e. $P_i'(\theta_i) = P_i'$ not depending on θ and known.
2. Σ not known, but $P_i'(\theta_i)$ independent of i and θ, as e.g. for the normal distribution. Seen in conjunction with Rao's result quoted above, this implies

   Corollary: The only $f \in \mathcal{F} \cap \mathcal{C}$ with all moments finite and P' not dependent on θ is the normal distribution.
3. $P_i'(\theta_i) = ck_i$ with k_i known and c an unknown constant.
4. $P_i' = \phi(\beta)k_i$, where ϕ does not depend on i. Examples are the Poisson with EY_i = var $Y_i = \beta t_i$, say, and the exponential distribution with $EY_i = \beta t_i$, var $Y_i = \beta^2 t_i^2$.
5. Examples where (4) give solutions that are identical to LS and are MVBU are cases where
   i) the θ_i are not different for all i, so that the vector θ of different expectations can be estimated directly by LS, and
   ii) where $P_i'(\theta_i) = P'(\theta_i)$ depend on i only through θ_i,
   iii) the number of coefficients in β is equal to the number in θ. The normal equations for $\hat{\theta}$ will partition into subsets

   $$2\hat{P}_j'\hat{\theta}_j - 2\hat{P}_j'\bar{Y}_j = 0, \quad \text{so that } \hat{\theta}_j = \bar{Y}_j ,$$

   where $\bar{Y}_j$ is the mean of those Y_i-variables having expectation θ_j. Likewise, the P_j' will cancel out in the system for θ^* corresponding to (7), so that $\hat{\theta}=\theta^*$. When there is an 1-1-correspondence between the members of θ and the members of β, it follows that $\hat{\beta}=\beta^*$ and fulfills the conditions in the proposition. An example of this is the binomial situation in Amundsen (1974), section 3.

An extension to multinomial situations is possible, see Amundsen (1976).

REFERENCES.

Amundsen, H.T. (1974). Binary variable multiple regressions. Scand. J.Stat.1,59-70.

Amundsen, H.T. (1976). Binary regressions for a polytomeous regressand. Scand. J. Stat. 3, 39-41.

Anderson, T.W. (1962). Least squares and best unbiased estimates. Ann. Math. Stat., 33, 266-272.

Bradley, E.L. (1973). The equivalence of maximum likelihood and weighted least squares estimates in the exponential family. J. Am. Stat. Ass., 68, 199-200.

Habermann, S.J. (1975). How much do Gauss-Markov and least squares differ? A coordinate free approach. Annals of Statistics, 3, 982-990.

Lehmann, E.L. (1959). Testing statistical hypotheses. John Wiley and Sons.

Rao, C.R. (1965). Linear statistical inference and its applications. John Wiley and Sons.

Watson, G.S. (1967). Linear least squares regression. Ann. Math. Stat. 38, 1679-1699.

Recent Developments in Statistics
J.R.Barra et al., editors

SPECTRAL ANALYSIS OF DEMOGRAPHIC TIME SERIES: A COMPARISON OF TWO SPECTRAL MODELS

A. BASILEVSKY
University of Kent
Department of Quantitative
Social Science
Canterbury, England

D. HUM
University of Manitoba
Department of Economics
Winnipeg, Canada

AND

THE MANITOBA BASIC ANNUAL INCOME EXPERIMENT

The paper presents an eigenvector method to decompose a given social time series into its spectral components such as trend, cycle, and seasonal variation and compares its performance to the classic Fourier transform spectral model normally used in time series analysis. It is found that the eigenvector results are more straightforward to interpret since the decomposition is given directly in the time domain. Also the original series need not be detrended prior to the analysis since the trend term is itself calculated simultaneously with other terms, and the procedure can be employed with both long as well as relatively short time series. Finally, an illustration of the procedure is presented by means of a demographic time series.

INTRODUCTION

Social scientists confronted with problems of analysing behavioural patterns in time series data often carry out a linear regression analysis, whereby a given series $X(t)$ is viewed as dependent upon a set of k explanatory variables. The predicted value $\hat{X}(t)$ can then be considered as that portion of $X(t)$ which is accounted for by the explanatory variables. This approach creates difficulties concerning the choice of regressor variables which, for time series data are frequently highly intercorrelated. The high correlation between social time series is usually due to the presence of underlying time functions such as trend, cyclical or periodic variation, and other movements which tend to influence many time series simultaneously. Clearly in such a situation it is inefficient to use regression analysis to explain the behaviour of $X(t)$.

There also has been much interest in applying the alternative procedures of spectral analysis to decompose X(t) directly into a finite number of additive time components or functions (see, for example, Nerlove (1964), Granger et al. (1970)). For social science applications it is thought that this method will enable a simpler and perhaps more easily understood characterization of the dynamic process which gave rise to the original series. For example, we can postulate a linear (or log linear) decomposition of the type

$$X(t) = T(t) + C(t) + S(t) + N(t) \tag{1}$$

where X(t) is a given quarterly time series and T(t), C(t) and S(t) are respectively a trend, a cycle consisting of irregular periodic movements, and a seasonal term representing periodic within-year variations. N(t) denotes a residual vector composed of unidentified terms (including error) commonly found in social data. The decomposition can be of interest for several reasons. Firstly, we may be more interested in a particular aspect of a time series, say the cyclic term, if we wish to forecast the turning points of economic activity. Secondly, it may be necessary to de-seasonalize a time series and to estimate the proportion of observed variance due to seasonal factors. Thirdly, it is frequently of interest to determine empirically whether a given time series contains a particular time component, for example whether stock prices contain cyclic or seasonal movements (see Granger et al., 1970). Also, more refined applications of spectral analysis enable investigators to estimate lags between components of two series, measure association between series, or formulate causal models and test hypotheses concerning social change (Coen et al., 1969). Finally, the components of X(t) will usually be intercorrelated and it is often of some interest to be able to observe the extent of such intercorrelation. For example, if employers implement decisions to lay off labour due to the cyclical slackening of economic activity during peak seasonal unemployment periods, as is frequently the case in agriculture, the seasonal term of unemployment time series will not be independent of the business cycle. When the series is de-seasonalized bias is introduced unless the dependence of the seasonal term on the cycle is taken into account.

In what follows we describe a new method of spectral analysis based on eigenvector decomposition of a symmetric matrix and compare its usefulness to the usual Fourier transform procedure.

THE KARHUNEN-LOÈVE SPECTRAL DECOMPOSITION

Let X(t) be a continuous Gaussian (normal) random process with zero mean and autocovariance function $K(t,s) = E\{X(t)\,X(s)\}$. Given an actual realization of X(t) suppose we wish to decompose the process into two portions as $X(t) = L(t) + N(t)$ where the "signal" portion $L(t)$ can be viewed as some linear combination of p time functions $\phi_i(t)$, i.e.

$$L(t) = \sum_{i=1}^{p} \alpha_i \Phi_i (t) \tag{2}$$

The residual portion $N(t)$ is assumed to be independent of $L(t)$ and thus of the α_i and $\Phi_i(t)$, and it is convenient to select the $\Phi_i(t)$ as ortho-normal functions. Also the α_i are specified to be orthogonal, $E(\alpha_i\alpha_j) = \sigma_i^2$ $(i=j)$, and $E(\alpha_i) = E(\alpha_j) = 0$. It is easy to show (Basilevsky, 1973) that the autocovariance function can be expressed as

$$K(t,s) = \sum_{i=1}^{\infty} \lambda_i \Phi_i(t)\Phi_i(s) \tag{3}$$

where $a \leqslant t, s \leqslant b$ and $\lambda_i = E(\alpha_i^2) = \sigma_i^2$. Multiplying (3) by $\Phi_j(s)$ and integrating over $[a,b]$ yields the stochastic integral equation

$$\int_a^b K(t,s)\Phi_j(s)ds = \sum_{i=1}^{\infty} \int_a^b \lambda_i \Phi_i(t)\Phi_i(s)\Phi_j(s)ds = \lambda_j \Phi_j(t) \tag{4}$$

Within the context of the theory of homogeneous integral equations and Hilbert spaces the sets λ_i and $\Phi_i(t)$ can be considered as non-zero eigenvalues and orthonormal eigenfunctions, respectively. Also the α_i are uncorrelated time dependent random variables defined by

$$\alpha_i = \int_a^b X(t)\Phi_i(t)dt \tag{5}$$

$$i = 1, 2, \ldots$$

Since $\lambda_i = E(\alpha_i^2)$ the variances of the α_i can be obtained conveniently as the eigenvalues of (4).

It is instructive to indicate the Fourier transform equivalent of (4) since it is common to represent the spectrum of a process in the frequency domain. The spectral density function can be written as

$$\int_{-\infty}^{\infty} K(t,s)\, e^{-iws} ds = F(w,t) \tag{6}$$

where $i^2 = -1$ and w is frequency measured in radians per unit time. In this form the Fourier transform is seen as a special case of the Karhunen-Loève integral equation (4), where $\Phi_j(s) = e^{-iws}$ a periodic function. However in (4) the functions $\Phi_j(s)$ need not be restricted to be periodic. In practice since a social process rarely yields more than a single realization K(t,s) is estimated from a single finite time series. Although equation (4) does not require that the time series be detrended (Basilevsky, 1973) it still requires the assumption of stationarity since K(t,s) must be estimated from a stationary series.

THE DISCRETE CASE

Since social time series are usually given in discrete form equation (4) must be modified in order to accommodate sample estimates of α_i, $\Phi_i(t)$, λ_i and the autocovariance function K(t,s). Consider a realization of a Gaussian random process, X(t), observed at (m+1) equidistant time points. The values $X(t_0)$, $X(t_1)$, ..., $X(t_m)$ are then distributed as a multivariate normal distribution with covariance matrix V_m containing $\text{cov}\{X(t), X(s)\} = \hat{\sigma}(t,s)$. When second moments are time invariant (weak stationarity) the population matrix V_m is the Toeplitz matrix

$$V^*_m = \begin{bmatrix} \sigma(0) & & & \\ \sigma(1) & \sigma(0) & & \\ \sigma(2) & \sigma(1) & \sigma(0) & \\ \vdots & \vdots & \ddots & \\ \sigma(m) & \sigma(m-1) & \dots & \sigma(0) \end{bmatrix}$$

where upper triangular elements are omitted. The structure of V^*_m can be simplified by calculating its eigenvalues $\lambda^*_1, \lambda^*_2, \dots, \lambda^*_{m+1}$ and eigenvectors $P^*_1(t), P^*_2(t), \dots, P^*_{m+1}(t)$. Wise (1955) has shown that the eigenvalues trace out the spectral curve of X(t) as $m \to \infty$, within a finite interval $[a,b]$, so that the eigenvalues form a discrete approximation to the continuous spectral density function of X(t). Wise also showed that when X(t) is periodic (circular) with period m (in which case $\sigma(k) = \sigma(m-k)$) we have $\lambda^*_k = \lambda_{m-k}$ for some k, and the eigenroots can be expanded into a finite Fourier series.

In practice the usefulness of the Toeplitz matrix is limited since social time series rarely exhibit such simple structure. Rather, we observe the more general sample covariance structure $\hat{V}_m$ with elements $\sigma(i,j) = \text{cov}\{X(t-j)X(t-i)\}$, $(i,j = 0,1,2, \ldots, m)$. The problem then becomes to estimate, from a single time series, the unobserved time functions which have given rise to the structure of V_m^* and to identify these functions in terms of empirical behaviour. It therefore is natural to consider a principal components analysis of matrix V_m^*. As an illustration consider the simple case where a finite time series X(t) is known to consist of a stochastic trend T(t) and a random term N(t) such that $\text{cov}\{T(t),N(t)\} = 0$. Then (1) can be written as

$$X(t) = b\Upsilon(t) + N(t) \tag{7}$$

where $b\Upsilon(t) = T(t)$ and $\sigma(i,j) = \text{cov}\{X(t-j), X(t-i)\} = b^2\sigma_y^{\ 2} + \sigma_n^{\ 2} = \text{var}(T) + \text{var}(N)$, and b is some coefficient to be estimated. The off-diagonal elements of the sample covariance matrix $\hat{V}_m$ depend on two independent (but not necessarily uncorrelated in small samples) terms T(t) and N(t), and a principal components analysis will yield a dominant eigenroot λ, followed by m low-order (and insignificantly different) eigenroots, which reflect the degree of error in the data. The first time function (principal component) $Z_1(t)$, which can be considered as a discrete sample estimate of the random variable α_1, will then recover the structure of the trend term T(t). For example if T(t) is linear $Z_1(t)$ will be linear, and so on. More generally when X(t) includes T additional independent time functions we obtain (r+1) principal components $Z_1(t)$, $Z_2(t)$, ..., $Z_{r+1}(t)$.

THE ESTIMATION PROCEDURE

Given a weakly stationary time series X(t) which consists of N = n + m observations, consider the generated set of m+1 lagged column vectors x(t), x(t-1), ..., x(t-m) each consisting of n observations. This is illustrated in figure 1 below where "+" denotes the presence of an observation and blanks denote deletions. The resultant data matrix is then (n x (m+1)) where n = 10, m = 5. Note that the initial value for t is set equal to m+1, where m is predetermined arbitrarily. In practice the value of m will depend on the number of observations available and the desired length, number and smoothness of the time functions. For instance, if it is suspected that X(t) contains five time functions then the number of lagged vectors should exceed five.

N	x(t-5)	x(t-4)	x(t-3)	x(t-2)	x(t-1)	x(t)
1	+					
2	+	+				
3	+	+	+			
4	+	+	+	+		
5	+	+	+	+	+	
6	+	+	+	+	+	+
7	+	+	+	+	+	+
8	+	+	+	+	+	+
9	+	+	+	+	+	+
10	+	+	+	+	+	+
11		+	+	+	+	+
12			+	+	+	+
13				+	+	+
14					+	+
15						+

Figure 1: Lagged vector; n=10, m=5

A typical lagged vector x(t-i) can be expressed in terms of a sub-set of elements of X(t) as $x(t-i) = (x_{1-i}, x_{2-i}, \ldots, x_{n-i})$, $i = 0,1,2,\ldots,m$, and the $(u,v)^{th}$ element of $\hat{V}_m$ is then

$$\hat{\sigma}(u,v) = \frac{1}{n-1} \sum_{t=1}^{n} x(t-u)x(t-v) \qquad u,v = 0,1,\ldots,m \tag{8}$$

where the x(t-i) are expressed as deviations about their sample means

$$\bar{x}(t-i) = \frac{1}{n} \sum_{t=1}^{n} x(t-i) \qquad i = 0,1,\ldots,m \tag{9}$$

For the special case where the autocovariance matrix is Toeplitz, (8) reduces to $\hat{\sigma}(u) = \frac{1}{n-1} \sum_{t=1}^{n} x(t-u)x(t)$.

Sample estimates of the $\{\phi_i\}$ $\{\lambda_i\}$, and $\{\alpha_i\}$, denoted by $\{Z_i\}$, $\{l_i\}$ and $\{P_i\}$ respectively are obtained from the principal components normal equations

$$P^T \hat{V}_m P = \Lambda \tag{10}$$

where Λ = diag $(1, l_2, \ldots, l_{m+1})$, P and Z =XP consist of column eigenvectors and time functions, respectively, and X is the (n x(m+1)) lagged data matrix with columns vectors x(t), x(t-1),...,x(t-m). If superscript T denotes matrix transposition the covariance matrix is simply estimated as $\hat{V}_m = \frac{1}{n-1}(X^T X)$. It is also convenient to standardize the Z_i to unit variance by the transformation $\Lambda^{-\frac{1}{2}}Z$ in which case $\Lambda^{-\frac{1}{2}}P = A = (a_{i,j})$. Equation (10) then becomes the discrete analogue of (4), where $\{\phi_i\}$ are replaced by the eigenvectors $\{P_i\}$, the $\{\lambda_i\}$ correspond to eigenvalues $\{l_i\}$ and the random variables $\{\alpha_i\}$ are estimated by the principal components $\{Z_i\}$. The elements a_{ij} of matrix A can be further converted to correlation coefficients by dividing the rows of A by the standard deviations of corresponding lagged variables (equation (8)). Alternatively, the matrix $\hat{V}_m$ can be converted directly to a correlation matrix $\hat{R}_m$ and correlation coefficients obtained directly from the eigenvector matrix, as is the usual practice in principal components analysis. The weakness of this procedure is that the resulting correlations cannot be compared to those obtained from matrix $\hat{V}_m$, and statistical testing of the eigenroots (see below) is more difficult. Also since the lagged vectors x(t), x(t-1), ..., x(t-m) are expressed in the same natural units the correlation matrix loses much of its rationale. Consequently in what follows the discussion is based on the sample covariance matrix.

AGGREGATION OF THE DECOMPOSED TIME FUNCTIONS

The eigenvector spectral decomposition will frequently yield functions (principal components) with very similar time structures. This is especially true of periodic behaviour where a particular cycle may be decomposed into a number of equal-frequency time functions. However, for purposes of substantive identification and interpretation it is preferable to represent similar time behaviour by a single time function. Consequently we are faced with the additional problem of aggregating time functions. We indicate briefly how the results of Girshick (1939) and Anderson (1963) may be adopted for this purpose.

Although the functions $\{Z_i\}$ are uncorrelated normal variates it does not follow that each function measures distinct time behaviour. For example, seasonal peaks of births may be captured by more than one distinct principal component in the sense that all are needed to reconstruct the seasonal behaviour of the data. This is because of the statistical fact that orthogonal eigenvectors (and hence the corresponding time functions) may contain highly correlated elements, either within a particular eigenvector or between eigenvectors (Girshick, 1939; Jackson et al., 1973).

Let the ith element of the jth population eigenvector Π_j be π_{ij} $(i, j = 1,2,\ldots,m)$. Population covariances are then given by $\alpha_j = \lambda_j^{\frac{1}{2}}\Pi_j$ where λ_j is the jth eigenroot. Let α_{ij} denote a sample value of a_{ij}. Girshick (1939) has shown that for large samples the covariance between any two elements a_{ik} and a_{jh} is given by

$$\sigma^2_{ik,jh} = E(a_{ik}-\alpha_{ik})(a_{jh}-\alpha_{jh}) = \begin{cases} \dfrac{\lambda_i}{n} \displaystyle\sum_{\substack{s=1 \\ i\neq s}}^{r} \dfrac{\lambda_s}{(\lambda_s-\lambda_i)^2} \alpha_{sk}\alpha_{sh} & i = j \\ \dfrac{\lambda_j}{n} \dfrac{\lambda_i}{(\lambda_j-\lambda_i)^2} \alpha_{ih}\alpha_{jk} & i \neq j \end{cases} \qquad (11)$$

The first equation yields variances/covariances in the same eigenvector while the second gives variances/covariances between elements of different eigenvectors. Also if the original variables are normally distributed $(a_{ik}-\alpha_{ik})$ will also be normally distributed with zero mean and variance σ^2_{ik}. On the null hypothesis that $\alpha_{ik} = 0$ the ratio a_{ik}/σ_{ik} is a standard normal variate. The important point of equation (22) is that the covariances of a_{ik} and a_{jh} are largely dependent on the eigenvalue differences. It follows that any two principal components $Z_i(t)$ and $Z_j(t)$ which correspond to insignificantly different eigenroots l_i and l_j can be aggregated into a single time component since their elements will be highly correlated. It is consequently sufficient to test the equality of the eigenroots alone and this may be done most conveniently by Anderson's (1963) likelihood ratio χ^2 statistic given by

$$\chi^2 = -n \sum_{j=1}^{r} \log_e l_j + nr \log_e \left(\sum_{j=1}^{r} \frac{l_j}{r} \right) \qquad (12)$$

with $\frac{1}{2}r(r+1)-1$ degrees of freedom (large n) where r is the number of eigenroots to be tested.

COMPARISON OF THE EIGENVECTOR AND FOURIER ANALYSES

For purposes of comparison we present an example drawn from demographic quarterly time records of the number of registered births (birth rates are not available) for two administrative parishes in Jamaica, 1880-1938 (see Hum et al., 1975; Spencer et.al., 1976a, 1976b). The data as well as their estimated spectral properties are shown in figures 2-5. Our underlying hypothesis is that fluctuations in the number of births depend to a large extent on the uneven distribution of the work-load due to seasonality, cyclic movements and a long-term trend caused by changes in the mode of production. Variations in births can then be understood and partially explained in terms of economic criteria as well as more traditional socio-cultural determinants. Thus parishes given over to the cultivation of highly seasonal crops for staple exports would have a relatively marked seasonal pattern of workload as times of intense activity (involving also women) would alternate with periods of unemployment. On the other hand, parishes in which agricultural activities are organized for small domestic production often employ crop rotation practices to ensure continuous supplies of domestic foodstuffs. This leads to a more evenly distributed workload rhythm. Our example, being parishes primarily devoted to the cultivation of seasonal (plantation) crops would be expected to have a significant seasonal component (Hum et al., 1975).

Figures 2-5 reveal fairly regular movements in the number of births, which can be seen to consist of seasonal and cyclic behaviour as well as a more drawn out tendency for the birth level to rise towards a peak just before World War I and then to decline. Although there is a relatively short-lived increase around 1920, due to a wave of Jamaican migrant labour returning from Cuban sugar plantations, the long-term tendency following World War I is for births to decline. The seasonal component (figure 4) clearly represents fluctuations in births due to the intense seasonal nature of plantation work, which normally involved both men and women in one way or another. Since the amplitude is slowly decreasing over time we may assume that seasonality has tended to decline in importance over the years. Also peak births (see figure 2) lead periods of low work activity by 3 quarters, so that we may suppose that conception has occurred during slack periods during which migrant labour has largely withdrawn from the plantations. Thus seasonality can be assumed to be largely related to birth rates rather than to the level of births. The exact role of the cycle (figure 3) is more problematic since it most likely reflects the influence of economic factors on both the birth rate as well as the population level. Thus during tough periods of the business cycle (due to a slackening of foreign trade) there will be a rise in unemployment, which might in turn cause people to postpone marriages and to favour emigration of young men in search of work.

We now consider the method of calculating the three time components or functions. The seasonal component of the data is captured by the first two principal components which correspond to insignificantly different eigenroots (95% confidence interval) as determined by Anderson's criteria (12). The non-linear trend is given by a single principal component, and a cycle term with an approximate $2\frac{1}{2}$ year period is represented by the next three principal components with the same equal frequency and insignificantly different latent roots (table 1).

EIGENROOTS	Corresponding Principal Components	χ^2	Time Functions
$\lambda(1) = 286872.0$	Z(1)	.10	S(t): Seasonality
$\lambda(2) = 284684.1$	Z(2)		S(t): Seasonality
$\lambda(3) = 127786.6$	Z(3)	-	T(t): Trend
$\lambda(4) = 58758.6$	Z(4)		T(t): Trend
$\lambda(5) = 55943.6$	Z(5)	4.0	C(t): Cycle
$\lambda(6) = 45054.2$	Z(6)		C(t): Cycle

Table 1: Eigenroots, Principal Components and Time Functions of Matrix $\hat{V}_m$.

Let Z(1), Z(2), ..., Z(m+1) denote the principal components of matrix $\hat{V}_m$ and let x(t) represent the zero-lag or the most recent segment of the births series, of length n-m+1. In our example we take m+1 = 32 quarters so that x(t) represents the years 1888-1938. x(t) can then be represented in terms of the principal components as

$$x(t) = a_{11}Z(1) + a_{12}Z(2) + \ldots + a_{1,m+1}Z(m+1) \tag{13}$$

where the a_{ij} are the appropriate elements of matrix A and are specific to each lagged vector x(t-i). The seasonal, the trend and the cycle terms are then constructed by aggregating the relevant principal components as

$$\begin{aligned} S(t) &= .322\ Z(1) - .557\ Z(2) \\ T(t) &= .263\ Z(3) \\ C(t) &= -.108\ Z(4) + .335\ Z(5) + .247\ Z(6), \end{aligned} \tag{14}$$

yielding $\hat{x}(t) = S(t) + T(t) + C(t)$ as the explained portion of x(t). Since principal components are standardized independent normal variates the contribution to explained variance of each time function can be obtained from the coefficients of (14). The functions S(t), T(t), and C(t) then account for 41.6%, 6.8%, and 18.5% respectively of the variance of x(t). Note that the eigenroots only indicate the average explanatory power of a principal component over the entire set x(t), x(t-1), ..., x(t-m) and not for any particular lag.

It is instructive to compare the eigenvector decomposition to the usual Fourier transform model (6). Figure 5 depicts the two spectra where frequencies are denoted by a solid line and the first nine latent roots appear as triangles. The first important difference between the two procedures is the discontinuous nature of the eigenroot spectrum. The main effect of the discontinuity is to concentrate more of the variance of the cyclic component ($2\frac{1}{2}$ year period). Since this cycle corresponds to .17 cycles per quarter this makes it easier to identify important components of a time series. Thus the "leakage" of power frequently encountered with Fourier transforms is reduced. The second important difference is that the time functions of a series which correspond to the latent roots can themselves be observed in the time domain, which enables a more complete description of a time series. Thirdly, the latent roots are ranked in descending order of proportion of total variance explained while the Fourier spectrum is given in terms of continuous frequency (cycles/time).

The overall structure of the two spectra, however, is fairly consistent. The two latent roots which measure seasonal variance (and therefore correspond to .25 cycles/quarter) are situated at the seasonal peak of the Fourier spectrum, and the existence of a low frequency (0 cycles/quarter) trend is indicated equally well by both spectra. As already noted above the representation of low frequency cycles is less consistent due to power leakage and a tendency of the eigenroots to concentrate similar terms (whose frequencies are not very different) into a smaller number of components. Also, both methods agree fairly closely as to the level of "noise" or error variance which is present in the time series and which is picked up by low-order latent roots (and high frequency periodic Fourier functions). Finally, the principal components procedure does not "average out" irregular features peculiar to an individual time series, such as variations in amplitude or cycle length, as in the Fourier transform model. The irregular periodic features are restricted to cyclic time functions and this can provide a methodology for social analysis to investigate time series thought to reflect patterns of social change.

REFERENCES

Anderson, T.W. (1973). Assymptotic Theory for Principal Components Analysis; Annals of Math. Statist., 34, 122-148.

Basilevsky, A. (1973). Spectral Decomposition in Time Series by the Method of Principal Components: Prices of Industrial Shares and the Level of Unemployment, University of Kent (Canterbury) Discussion Series.

Coen, P.G., Gomme, E.D., and Kendall, M.G. (1969). Lagged Relationships in Economic Forecasting; J. Roy. Statist. Soc. (A), 132, 133-163.

Girshick, M.A. (1939). On the Sampling Theory of Roots of Determinantal Equations, Annals of Math. Statist., 10, 203-224.

Granger, C.W.J. and Morgenstem, O. (1970). Predictability of Stock Market Prices, Heath-Lexington.

Hum, D., Lobdell, R., and Spencer, B. (1975). Plantations, Staple Exports and the Seasonality of Births in Jamaica: 1880-1938. Presented at the Western Economic Association Meeting, San Diego.

Jackson, J.E. and Hearne, F.T. (1973). Relationships among coefficients of Vectors used in Principal Components; Technometrics, 15, 601-610.

Spencer, B. and Hum, D. (1976a). Workload and Seasonal Variation in Birthrates: Some International Comparisons; Intern. J. of Comparative Sociol., 17, No. 1.

Spencer, B., Hum, D. and Deprez, P. (1976b). Spectral Analysis and the Study of Seasonal Fluctuations in Historical Demography; J. of European Econ. History (in press).

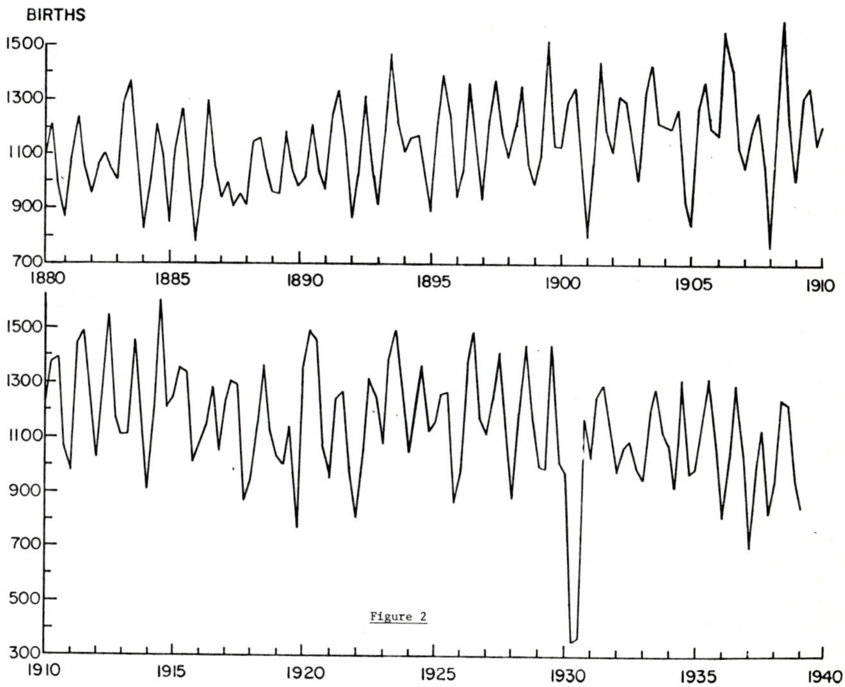

Figure 2

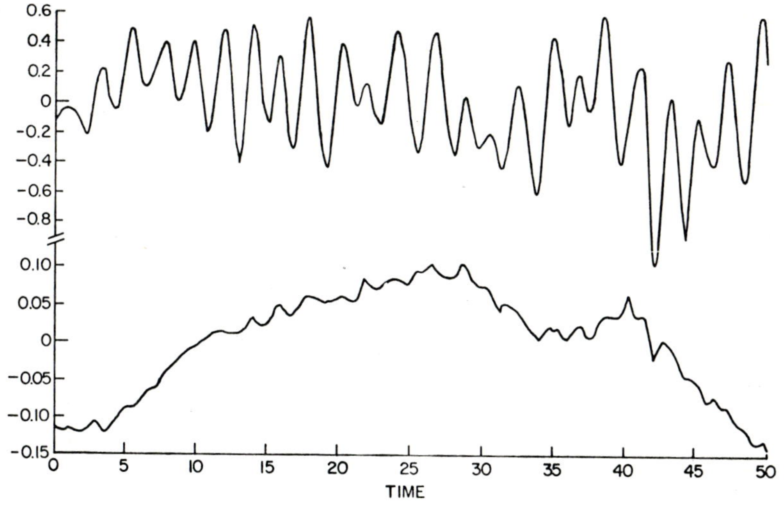

Figure 3

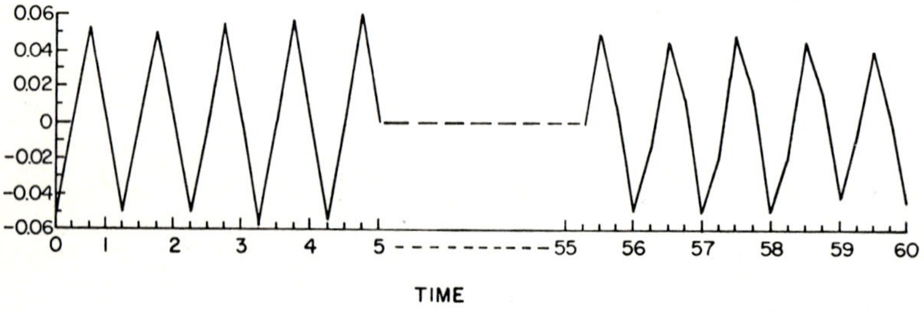

Figure 4

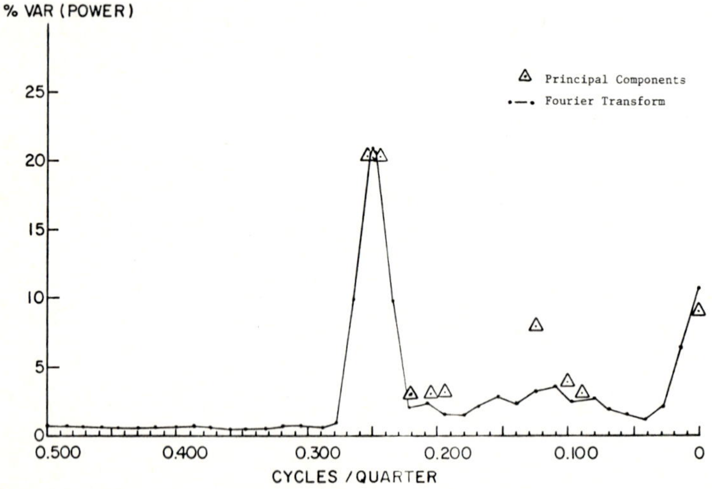

Figure 5

Recent Developments in Statistics
J.R.Barra et al., editors

INFERENCES ABOUT THE RATIO OF NORMAL MEANS: A BAYESIAN APPROACH TO THE FIELLER-CREASY PROBLEM

José M. Bernardo
Departamento de Estadística
Facultad de Ciencias
Universidad de Valencia
Valencia, Spain

In biological assay work the problem arises of estimating the ratio of the means of two normal populations. The problem gave rise to a controversy among fiducialists and confidence-intervalists and it is usually known as the Fieller-Creasy problem.

A Bayesian solution to such a problem is presented here making use of a new method, developed by the author, which provides appropriate reference 'non-informative' prior distributions.

THE PROBLEM

In biological assay work one is often interested in the relative power of two treatments or drugs, and the following problem suggests itself. Suppose that two samples $x=\{x_1,x_2,\ldots,x_n\}$ and $y=\{y_1,y_2,\ldots,y_m\}$ are available from two independent normal populations with unknown means μ,η and common unknown variance σ^2. The problem is to make inferences about the value of $\psi=\mu/\eta$, the ratio of the means.

This problem was discussed in a Symposium on Interval Estimation held by the Royal Statistical Society back in 1954. E.C. Fieller and M.A. Creasy presented there two different solutions that both claimed to be fiducial. Fieller's solution is difficult to accept for it can lead, for instance, to a 'confidence' interval consisting in the *whole* real line. Kappenman, Geisser and Antle (1970) showed that Creasy's solution may be reproduced from a Bayesian point of view by using the usual 'non-informative' prior $p(\mu,\eta,\sigma)\propto\sigma^{-1}$.

However, the uncritical use of standard so-called 'non-informative' priors has been increasingly questioned. Although Jeffreys' (1939/67) prior is often accepted in the one-dimensional continuous case, no similarly acceptable results seem to exist in the case of several parameters. A key reference is Dawid, Stone and Zidek (1973) and ensuing discussion.

Our view, developed in Bernardo (1976) is that in *each* particular inference problem a *reference* posterior distribution may be produced which only uses information provided by the data and the model specification. Such a reference posterior may be used as an *origin* with respect to which posteriors obtained from informative priors may be compared, in order to assess the relative importance of the initial opinions in the final inference. In this paper, we obtain the reference posterior distribution which corresponds to the Fieller-Creasy problem.

DEFINITION OF A REFERENCE POSTERIOR

Consider a vector θ of parameters and an experiment E providing some data z which give information about θ. Suppose that we are interested in the value of some *real* function $\psi=\psi(\theta)$ of the parameters. Without loss of generality, one may assume that ψ is the first component of θ, for otherwise an appropriate transformation could be made to achieve such a situation. Then, $\theta=\{\psi,\omega\}$, $\omega=\{\omega_1,\omega_2,\ldots,\omega_k\}$.

The basic idea underlying our construction of a reference posterior distribution for ψ is as follows. First, generalizing earlier work by Lindley (1956), the expected information about ψ to be provided by E when the prior density of θ is $p(\theta)$ is defined to be

$$I^{\psi}\{E,p(\theta)\} = \iint p(z,\theta)\log\frac{p(\psi|z)}{p(\psi)}d\theta\,dz$$

whenever the integral exists. Similarly, the *residual* expected information about each of the ω_i's when ψ is known may be defined as

$$I^{\omega_i|\psi}\{E,p(\theta)\} = \iint p(z,\theta)\log\frac{p(\omega_i|\psi,z)}{p(\omega_i|\psi)}d\theta\,dz$$

Now, let $E(n)$ be the experiment which consists of n independent replications of E and consider $I^{\psi}\{E(n),p(\theta)\}$, the expected information about ψ to be provided by $E(n)$ when the prior density of $\theta=\{\psi,\omega\}$ is $p(\theta)=p(\psi)p(\omega|\psi)$. By performing infinite replications of E one could expect to know θ, and therefore ψ. Thus, the number $I^{\psi}\{E(\infty),p(\theta)\}$, if it exists, measures the amount of *missing information* about ψ that one could obtain by repeating E when the prior is $p(\theta)$. It seems natural to define *diffuse* opinions about ψ relative to E as those described by that density $\pi(\psi)$ maximizing the missing information about ψ, $I^{\psi}\{E(\infty),p(\theta)\}$, for any fixed $p(\omega|\psi)$, providing such a density exists.

All prior densities of the form $\pi(\psi)p(\omega|\psi)$ will be called *ψ-diffuse relative to E*; they differ in the opinions they describe about the value of ω given ψ. We propose to select as reference prior the more 'ω-diffuse' among them that we define to be the one maximizing *each* of the missing residual informations about the ω_i's given ψ, $I^{\omega_i|\psi}\{E(\infty),\pi(\psi)p(\omega|\psi)\}$, when the marginal prior of ψ is $\pi(\psi)$, providing such a density $\pi(\omega|\psi)$ exists. The reference posterior distribution of ψ will be that obtained by the formal use of Bayes theorem with the reference prior $\pi(\theta)=\pi(\psi)\pi(\omega|\psi)$ just described.

Only a slight generalization of earlier work by Stone (1958) is necessary to establish that if, as is usual, the posterior distribution of ψ is asymptotically normal with precision $nh_0(\hat\theta)$ where $\hat\theta$ is the maximum likelihood estimator of θ then for any positive prior $p(\theta)$, i.e. such that $p(\theta)>0$ for any θ, and $n\to\infty$,

$$I^{\psi}\{E(n),p(\theta)\} = \frac{1}{2}\log\frac{n}{2\pi e} + \int p(\theta)\log\frac{h_0(\theta)^{1/2}}{p(\psi)}d\theta + o(1)$$

It follows (Bernardo, 1976) that if $h_0(\theta)$ may be decomposed such that

$$h(\theta)^{1/2} = \pi(\psi)f(\omega)$$

then the missing information about ψ is maximized, for any fixed $p(\omega|\psi)$, when $p(\psi)=\pi(\psi)$.

Similarly, if the posterior distribution of ω_i given ψ is asymptotically normal with precision $nh_i(\hat\theta)$ and

$$h_i(\theta)^{1/2} = \pi(\omega_i)g(\theta_i^*)$$

where θ_i^* contains all the components of θ except ω_i, then the missing information about ω_i given ψ is maximized when $p(\omega_i|\psi)=\pi(\omega_i)$ for any fixed $p(\omega_j|\psi)$, $j\neq i$, and for any fixed $\pi(\theta)$. Therefore, in such a case the ψ-reference prior is

$$\pi(\psi)\prod_i\{\pi(\omega_i)\}$$

and the reference posterior for ψ is that obtained from such a prior.

THE REFERENCE POSTERIOR OF THE RATIO OF TWO MEANS

Consider again the Fieller-Creasy problem. Here, $\theta=\{\mu,\eta,\sigma\}$ and $\psi=\mu/\eta$. The problem may be reparametrized in terms of $\zeta=\{\psi,\eta,\sigma\}$. It is known (Walker, 1968) that the posterior distribution of ψ is asymptotically normal with precision matrix $nF(\hat{\zeta})$ where $\hat{\zeta}$ is the maximum likelihood estimator of ζ and $F(\zeta)$ is Fisher's information matrix of typical element

$$-\int p(z|\zeta)\frac{\partial^2}{\partial\zeta_i\partial\zeta_j}\log p(z|\zeta)\,dz$$

In this case, the matrix is easily found to be

$$F\begin{pmatrix}\psi\\ \eta\\ \sigma\end{pmatrix} = \frac{1}{\sigma^2}\begin{pmatrix}\eta^2 & \psi\eta & 0\\ \psi\eta & 1+\psi^2 & 0\\ 0 & 0 & 4\end{pmatrix}$$

It follows that the asymptotic posterior distribution of ψ and those of η and σ given ψ are normal with respective precisions (see e.g. Graybill, 1961, ch. 3) $nh_0(\hat{\zeta})$, $nh_1(\hat{\zeta})$, $nh_2(\hat{\zeta})$, where

$$h_0(\zeta) = \eta^2(1+\psi^2)^{-1}\sigma^{-2}$$
$$h_1(\zeta) = (1+\psi^2)\sigma^{-2}$$
$$h_2(\zeta) = 4\sigma^{-2}$$

Thus, the ψ-diffuse densities are those of the form

$$p(\psi,\eta,\sigma) \propto (1+\psi^2)^{-1/2}p(\eta,\sigma|\psi)$$

and the ψ-reference prior is $\pi(\psi,\eta,\sigma)=(1+\psi^2)^{-1/2}\sigma^{-1}$ or, in terms of the original parametrization

$$\pi(\mu,\eta,\sigma) = (\mu^2+\eta^2)^{-1/2}\sigma^{-1} \tag{1}$$

The reference posterior distribution of $\psi=\mu/\eta$ after the samples $x=\{x_1,x_2,\ldots,x_n\}$ and $y=\{y_1,y_2,\ldots,y_m\}$ have been observed may now be produced via Bayes Theorem. Our reference prior combines nicely with the likelihood function so that, unlike the posterior density for ψ obtained from the usual prior $p(\mu,\eta,\sigma)=\sigma^{-1}$, the reference posterior distribution for ψ may be obtained in closed form.

Indeed, the joint density of the two samples is

$$p(x,y|\mu,\eta,\sigma) = (2\pi)^{-(n+m)/2}\sigma^{-(n+m)}\exp\left[\frac{1}{2\sigma^2}\{S^2+n(\bar{x}-\mu)^2+m(\bar{y}-\eta)^2\}\right] \tag{2}$$

where $\bar{x}=\Sigma x_i/n$, $\bar{y}=\Sigma y_i/m$, and $S^2=\Sigma(x_i-\bar{x})^2+\Sigma(y_i-\bar{y})^2$. Combining (1) and (2), the joint posterior density of $\theta=\{\mu,\eta,\sigma\}$ is

$$p(\mu,\eta,\sigma|x,y) \propto (\mu^2+\eta^2)^{-1/2}\sigma^{-(n+m+1)}\exp\left[\frac{1}{2\sigma^2}\{S^2+n(\bar{x}-\mu)^2+m(\bar{y}-\eta)^2\}\right]$$

so that, since the Jacobian of the one-to-one transformation from $\theta=\{\mu,\eta,\sigma\}$ to $\zeta=\{\psi,\eta,\sigma\}$ is $1/|\eta|$, the joint posterior density of $\zeta=\{\psi,\eta,\sigma\}$ is

$$p(\psi,\eta,\sigma|x,y) \propto (1+\psi^2)^{-1/2}\sigma^{-(n+m+1)}\exp\left[\frac{1}{2\sigma^2}\{S^2+n(\bar{x}-\psi\eta)^2+m(\bar{y}-\eta)^2\}\right] \tag{3}$$

Using gamma-related integral results, σ is easily integrated out from (3) to get

$$p(\psi,\eta|x,y) \propto (1+\psi^2)^{-1/2}\left[S^2+\frac{nm(\bar{x}-\psi\bar{y})^2}{m+\psi^2 n}\right]^{-(n+m)/2}\cdot$$
$$\cdot\left[1+\left\{\frac{m+\psi^2 n}{S^2+\dfrac{nm(\bar{x}-\psi\bar{y})^2}{m+\psi^2 n}}\right\}\left\{\eta-\frac{m\bar{y}+\psi n\bar{x}}{m+\psi^2 n}\right\}^2\right]^{-(n+m)/2} \tag{4}$$

Finally, using Student-t related integral results, the parameter η may be integrated out from (4) to obtain

$$p(\psi|x,y) = C\ (1+\psi^2)^{-1/2}(m+\psi^2 n)^{-1/2}\{S^2 + \frac{nm(\bar{x}-\psi\bar{y})^2}{m+\psi^2 n}\}^{-(n+m-1)/2} \tag{5}$$

where C is a normalizing constant. The density (5) is our reference posterior density for the ratio of the means.

This could be compared with the posterior density of ψ that one obtains from the usual so-called 'non-informative' prior

$$p(\mu,\eta,\sigma) = \sigma^{-1} \tag{6}$$

Using a similar procedure to that followed above, (6) and (2) may be combined to obtain the posterior density of $\theta=\{\mu,\eta,\sigma\}$ which is

$$p(\mu,\eta,\sigma|x,y) \propto \sigma^{-(m+n+1)}\exp\left[\frac{1}{2\sigma^2}\{S^2+n(\bar{x}-\mu)^2+m(\bar{y}-\eta)^2\}\right]$$

so that the corresponding posterior density of $\zeta=\{\psi,\eta,\sigma\}$ is

$$p(\psi,\eta,\sigma|x,y) \propto |\eta|\ \sigma^{-(m+n+1)}\exp\left[\frac{1}{2\sigma^2}\{S^2+n(\bar{x}-\psi\eta)^2+m(\bar{y}-\eta)^2\}\right] \tag{7}$$

As before, σ is easily integrated out of (7) to obtain an expression for the posterior distribution of ψ and η which is the same as (4) except that the factor $(1+\psi^2)^{-1/2}$ is substituted by $|\eta|$. To obtain the posterior distribution of the parameter of interest ψ one would now integrate out η. However, because of the factor $|\eta|$, this integral can no longer be evaluated in terms of Student-t intergral results and the posterior density of ψ cannot be expressed in closed form as above.

Our reference posterior density for the ratio of the means, given by (5), is always proper provided $m\geq 1$ and $n\geq 1$. It can have one or two modes which, as the constant C may be determined numerically. Clearly, it is symmetric about the origin when either $\bar{x}=0$ or $\bar{y}=0$. This was to be expected since, in either case, there is no information to decide on the sign of ψ. This feature is not obtained with the usual prior (6). When $n=m=1$ and $x=y=0$, the reference density (5) reduces to a Cauchy density function.

DISCUSSION

The reference density (5) has been studied using Montecarlo methods. For instance, random samples of size 5 were independently generated from Normal distributions with means $\mu=3$ and $\eta=1$ and variance $\sigma^2=1$. The results of three of those and the corresponding values for the statistics $\bar{x}$, $\bar{y}$ and S^2 are described below.

1.	x={3.31, 3.90, 3.22, 2.00, 2.88}	$\bar{x}$ = 3.062	S^2 = 3.1608
	y={0.49, 0.64, 1.58, 1.53, 0.57}	$\bar{y}$ = 0.962	
2.	x={3.01, 3.16, 4.31, 2.62, 3.38}	$\bar{x}$ = 3.296	S^2 = 3.7728
	y={1.37, 0.17, 0.18, 0.74, 1.42}	$\bar{y}$ = 0.776	
3.	x={1.55, 3.33, 3.87, 1.10, 3.69}	$\bar{x}$ = 2.708	S^2 = 7.7297
	y={1.18, 1.30, 1.04, 1.67, 0.68}	$\bar{y}$ = 1.174	

In Figure 1, the corresponding posterior distributions of ψ, calculated from (5) estimating numerically the corresponding proportionality constants, are plotted together. All of them seem to assess a reasonable high posterior density to the true value $\psi=3$.

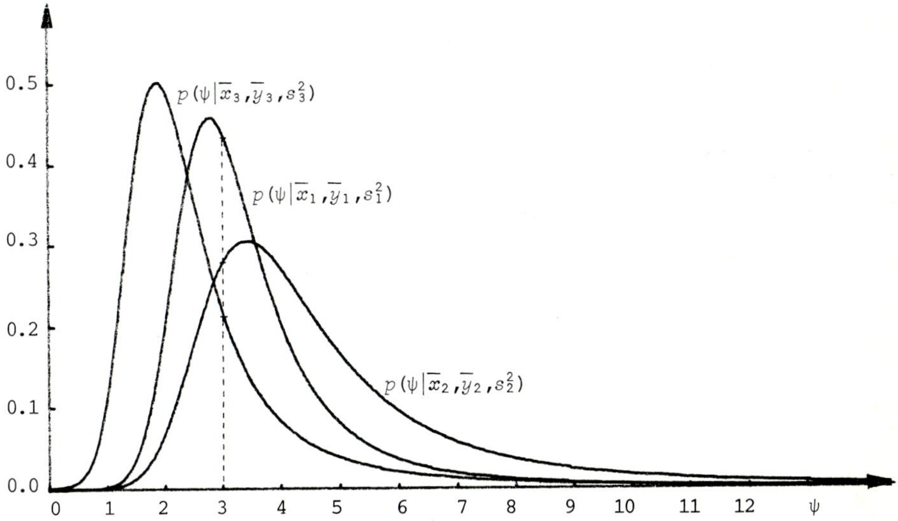

Figure 1. Example of posterior distributions of ψ.

No claim of 'objectivity' (whatever that may mean) is made for the set of inferences about ψ that could be produced from (5). It is only argued that (5) is a useful *reference* density to look at for it gives a feeling of the sort of values of ψ that the data $z=\{x,y\}$ are supporting. One should compare (5) with the posterior density of ψ obtained from an informative prior which describes one's initial information on the particular problem dealt with. From this comparison, one would obtain information about the relative importance of data and prior information in the final conclusions.

ACKNOWLEDGEMENTS

I am grateful to Professor D.V. Lindley for his useful comments and to Dr. J. Basulto and Mr. I. Zabala for their help with the computations.

REFERENCES

Bernardo, J.M. (1976). *The use of information in the design and analysis of scientific experimentation*. Ph.D. Thesis, University of London.

Creasy, M.A. (1954). Limits for the ratio of the means. *J. Roy. Statist. Soc. Ser. B 16*, 186-194 (with discussion).

Dawid, A.P., Stone, M. and Zidek, J.V. (1973). Marginalization Paradoxes in Bayesian and Structural Inference. *J. Roy. Statis. Soc. Ser. B 35*, 189-233 (with discussion).

Fieller, E.C. (1954). Some problems in interval estimation. *J. Roy. Statist. Soc. Ser. B 16*, 175-185 (with discussion).

Graybill, F.A. (1961). *An Introduction to Linear Statistical Models*. (McGraw-Hill, New York)

Jeffreys, H. (1939/67). *Theory of Probability.* (Clarendon Press, Oxford).

Kappenman, R.F., Geisser, S. and Antle, C.E. (1970). Bayesian and fiducial solutions to the Fieller-Creasy problem. *Sankhya B 32*, 331-340.

Lindley, D.V. (1956). On a measure of the Information provided by an Experiment. *Ann. Math. Statist. 27*, 986-1005.

Stone, M. (1958). *Studies with a measure of information.* Ph.D. Thesis, University of Cambridge.

Walker, A.M. (1969). On the asymptotic behaviour of posterior distributions. *J. Roy. Statist. Soc. Ser. B 31*, 80-88.

Recent Developments in Statistics
J.R.Barra et al., editors

AN APPLICATION OF PRIESTLEY'S P(λ) TEST TO THE SOUTHEND TIDE HEIGHTS DATA

R. J. Bhansali
Department of Computational and Statistical Science
University of Liverpool, England

The P(λ) test suggested by Priestley (1962a, b) is applied to detect unknown periodic terms in the non-tidal component of the Southend tide heights. This was obtained by estimating and removing the tidal component from the original observations using the Harmonic method of Doodson (1921). Five harmonic terms of period ranging from 12 to 146 hours are shown to be significant. The estimated spectrum of the residuals obtained after removing these harmonic terms suggests that an autoregressive model may be used for non-parametrically predicting their future values. The univariate method used here thus provides an alternative to a model-building, cross-spectral approach described by Cartwright (1967) for predicting tide heights.

1. INTRODUCTION

To many research workers, a primary motivation for contemplating the use of statistical spectral analysis is to be able to detect "hidden" periodic terms whose presence is not suspected before carrying out the analysis. If strictly periodic terms of frequencies $\theta_1, \theta_2, \ldots, \theta_k$ say are present in the data, then the estimated spectrum will show peaks at these frequencies. Thus a visual examination of the estimated spectrum can help in identifying unknown periodic terms. However one difficulty with this purely visual approach is that not all the peaks in the estimated spectrum are necessarily caused by strictly periodic terms. It is, for example, well-known that the spectrum of an autoregressive process can have peaks if some of the roots of its characteristic polynomial are complex-valued. It will therefore be necessary in practice to decide whether a peak in the estimated spectrum implies the presence of a strictly periodic term in the observed time series. By its very nature the question of detecting unknown periodic terms is easily tackled in the context of statistical hypothesis testing. A first known test for hidden periodicities was proposed by Schuster in 1891. Ever since his work a number of different tests have been proposed, notably those by Hannan (1961) and Priestley (1962), for testing the hypothesis of a continuous spectrum against the alternative that the spectrum has continuous as well as discrete components. In this paper we apply Priestley's P(λ) test to Southend tide heights data. Although this test has been known in the literature for the last 14 years, not many accounts of its application to practical time series have been published so far. Priestley himself applied his test procedure only to simulated time series which contained just one periodic term. The P(λ) test detects 5 harmonic terms and conveniently estimates the continuous part of the spectrum of the Southend tide heights. However, for this particular series, $P(\lambda)$ had a number of peaks. Hence the twin requirements of testing each peak of $P(\lambda)$ in order of frequency and re-estimating $P(\lambda)$, at each stage, by removing the effect of a significant peak, were found to be computationally arduous.

2. THE P(λ) TEST

Let x_t; $t = 1,\dots,T$ denote the observed time series. Priestley's test is designed to test the hypothesis that $A_i = 0$ ($i = 1,\dots,K$) in the model

$$x_t = \sum_{i=1}^{K} A_i \cos (t\lambda_i + \phi_i) + y_t \qquad (2.1)$$

where $\{A_i\}$ and $\{\lambda_i\}$ are unknown constants, the $\{\phi_i\}$ are independent and rectangularly distributed on $(-\pi,\pi)$ and y_t is a linear process with a continuous spectral density function. If the A_i's are all zero then the correlogram of x_t will eventually damp out to zero. On the other hand, if one or more of the A_i's are non-zero, then the correlogram would not damp out and in its tail behave like a linear combination of cosine terms of the same frequency as those of the harmonic terms. Priestley's test exploits this property of the correlogram of x_t. A double weighting function technique is adopted to carry out the Fourier analysis of the tail of the correlogram of x_t.

Denote by $f_m(\lambda)$ and $f_n(\lambda)$ two 'windowed' estimates of the spectral density function of x_t obtained using the truncation points m and n respectively where $n > 2m$. The windows used for forming these estimates may be the same or may be different. In the simulation experiments described by Priestley (1962b), the Bartlett window was used for forming $\hat{f}_n(\lambda)$ while the truncated window was used for forming $\hat{f}_m(\lambda)$. One will then compute

$$P(\lambda) = \hat{f}_n(\lambda) - \hat{f}_m(\lambda) \qquad (\lambda = \tfrac{2\pi j}{T};\ j = 0, 1,\dots,[\tfrac{T}{2}])$$

and plot $P(\lambda)$ against λ. If the A_i's are non-zero then $P(\lambda)$ will, in general have several well-defined peaks. Priestley recommended testing each peak in order of frequency. Suppose that the first significant peak is detected at $\omega_0 = \frac{2\pi p}{T}$ for some $p \neq 0, [\frac{T}{2}]$. The amplitude of the corresponding harmonic term is estimated using

$$\hat{A}_0^{\,2} = \frac{8\pi P(\omega_0)}{n-2m}.$$

The corresponding frequency is of course estimated to be ω_0. The effect of this periodic term is then removed by correcting the sample covariances (C_s, say) by computing

$$C_s^{(1)} = C_s - \tfrac{1}{2} \hat{A}_0^{\,2} \cos (s\,\omega_0).$$

For testing whether any more harmonic terms are present, one will recompute P(λ) using $C_s^{(1)}$ and examine its peaks again in order of frequency. If the significance level normally used is α then the significance level used for testing the kth peak in order of frequency should be α/k. This procedure is repeated until no further peaks in P(λ) are shown to be significant at the chosen level of significance.

3. ANALYSIS OF THE SOUTHEND TIDE HEIGHTS DATA

The P(λ) test described in the last section was applied by the author to hourly observations on tide heights at Southend-on-Sea, England for September 1969 - March 1971. T, the number of data points in the observed time series, equalled 11424. The author's interest in this time series arose from a statistical enquiry he received from Mr. M. Amin of the Institute of Oceanographic Science, Bidston, Merseyside concerning improvement of the prediction of shallow water tide heights. The basic method used at this Institute for predicting shallow water tides is due to Doodson (1921). This consists of projecting the tidal part of the series into the future by first estimating amplitudes of known harmonic terms of different frequencies using least squares regression. The frequency of the harmonic terms comprising the tidal part are known because of the constancy of the motions of the moon and the sun (Cartwright, 1967). However in shallow waters, especially in estuaries, non-tidal effects are often well-developed. These usually take the form of meteorological perturbations and their interaction with thermal and other slowly varying conditions of the sea. These non-tidal effects distort the profile of shallow water tides and thus make their prediction complicated and difficult. Southend was chsen because it is sited at the mouth of the Thames estuary and the prediction of the tide heights there were of interest to the Port of London Authority.

Let x_t t = 1,..., 11424 denote the non-tidal component of the observed tide heights at Southend. This was obtained by subtracting out the tidal component from the observed tide heights. The estimated spectrum of x_t is shown in Figure 1. The Bartlett window with truncation point n = 1428 was used to estimate the spectrum. With a sample of 11424 observations we felt justified in smoothing the spectrum quite heavily by choosing the truncation point to be about $12\frac{1}{2}$% of the sample size. A much larger value of n(= 2856) was also tried but the resulting spectrum was found to be highly unstable especially in the low frequencies. We only show the spectrum when the frequency $\lambda \epsilon (0, 0.35\pi)$ because for $\lambda \epsilon (0.35\pi, \pi)$ the estimated spectrum was found to be very close to the λ-axis indicating that these frequencies were relatively unimportant in explaining the non-tidal component.

The estimated spectrum has several peaks including a major peak around $\lambda = 0.164\pi$. For reasons already given, it is necessary to test whether these peaks imply the presence of un-detected harmonic terms before predicting this series by a model with a continuous spectrum. A plot of $P^*(\lambda)$ $(= P(\lambda)/C_0)$ is shown in Figure 2.

The truncated window with m = 357 was used to obtain $\hat{f}_m(\lambda)$. For convenience in computing J_q, we decided to choose m so that T/m was an integer. We decided to work with an overall significance level of 12%. Hence the significance level used for testing the kth peak was 0.12/k.

$P^*(\lambda)$ has several peaks. Testing the peaks in order of frequency, the peak at $\lambda = 0.004\pi$ and the next two peaks were found to be not significant. However the peak at $\lambda = 0.014\pi$ was found to be significant at the 0.1% level. This peak was removed and $P^*(\lambda)$ was re-estimated. $P^*(\lambda)$ again had a number of peaks which were tested in order of frequency. The peak at $\lambda = 0.0193\pi$ was the first peak found to be significant at the 2% level. This procedure was continued until all the significant peaks shown in Table 1 were detected and their effect removed.

Table 1

Harmonic terms detected by the $P(\lambda)$ test

Harmonic Term Number	Estimated Frequency (Multiple of $2\pi/11424$)	Estimated Amplitude	Level of Significance
1	78	0.154	0.0008
2	110	0.119	0.0192
3	242	0.098	0.0102
4	340	0.092	0.0082
5	938	0.293	0.0009

The Bartlett spectrum obtained after removing these 5 harmonic terms is shown in Figure 3. Compared to the original spectrum, the spectrum of the residuals after removing these five harmonic terms does not have any major peaks, though their behaviour near the origin is quite similar. The estimated spectrum decreases as the frequencies increase, suggesting that the residuals obtained after removing the five harmonic terms may be predicted from their past. The spectral factorisation procedure discussed by Bhansali (1974) was applied for non-parametrically fitting an autoregressive model to the residuals. However, in the interest of space, an account of the application of this technique is not given here.

ACKNOWLEDGEMENTS

The author is grateful to Mr. M. Amin for supplying him with the data on the Southend tide heights. Thanks are also due to Mr. N.M. Shah and Mr. I. Sheikh for computational assistance.

REFERENCES

Bhansali, R.J. (1974). Asymptotic properties of the Wiener-Kolmogorov predictor I. J. R. Statist. Soc., B, 36, 61-73.

Cartwright, D.E. (1967). Time series analysis of tides and similar motions of sea surface. J. Appl. Prob., 4, 103-112.

Doodson, A.T. (1921). The harmonic development of the tide-generating potential. Proc. Royal Soc., A, 100, 305-329.

Hannan, E.J. (1961). Testing for a jump in the spectral function. J.R. Statist. Soc., B, 23, 394-404.

Priestley, M.B. (1962a). Analysis of stationary processes with mixed spectra - I. J. R. Statist. Soc., B, 24, 215-233.

Priestley, M.B. (1962b). Analysis of stationary processes with mixed spectra - II. J. R. Statist. Soc., B, 24, 511-529.

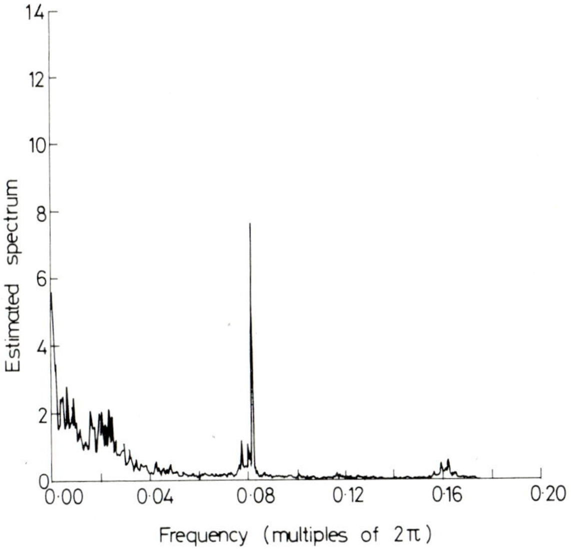

Figure 1. Bartlett spectrum for Southend tide heights (before removing harmonic terms)

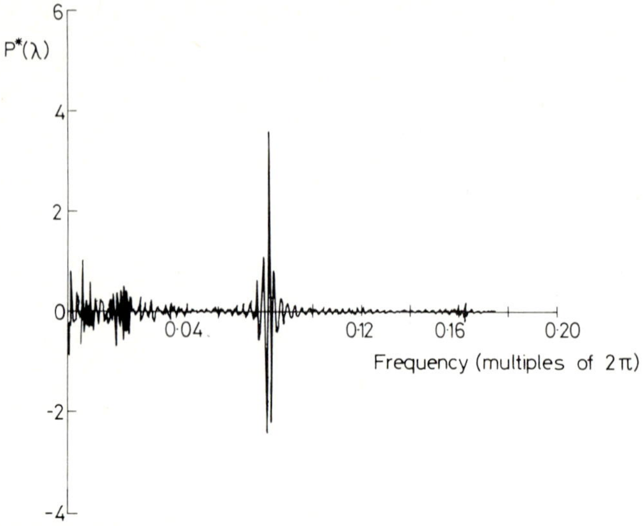

Figure 2. $P^*(\lambda)$ for Southend tide heights (before removing harmonic terms.)

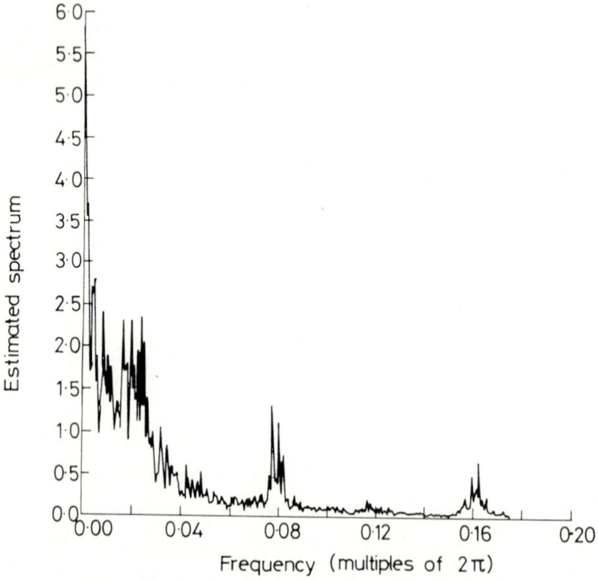

Figure 3. Bartlett spectrum for Southend tide heights (after removing harmonic terms)

Recent Developments in Statistics
J.R.Barra et al., editors

ON THE APPROXIMATION OF ARBITRARY SECOND ORDERED STOCHASTIC PROCESSES BY CONTINUOUS STOCHASTIC PROCESSES

J. Bulatović
Mathematical Institute, Belgrade, Yugoslavia

PRELIMINARY NOTIONS AND DEFINITIONS

Let $X=\{X(t),\ t\in R\}$ be an arbitrary stochastic process of the second order, i.e. $\|X(t)\|<\infty$ for each $t\in R$, where R is the set of all real numbers, and the scalar product and the norm are defined as in [4]. The Hilbert space spanned by variables $X(s)$, $s\leq t$ ($s<t$) we shall denote by $H(X;t)$ ($H(X;t-0)$); the Hilbert space spanned by variables $X(s)$ for all $s\in R$ we shall denote by $H(X)$. We shall suppose that (for all processes in this paper) the following conditions are satisfied: a) $H(X;-\infty)=0$, and b) $X(t-0)$ exists for all t; it can be shown that the space $H(X)$ is separable, [1].

Denote by $E_X(t)$ the projection operator from $H(X)$ onto $H(X;t)$; it is easy to see that $E_X=\{E_X(t),\ t\in R\}$ is a resolution of the identity of the space $H(X)$. It is well known, [4], that E_X uniquely determines the so-called Hida-Cramér representation of X:

$$X(t)=\sum_{n=1}^{N}\int_{-\infty}^{t}g_n(t,u)dZ_n(u),\quad t\in R. \tag{1}$$

We say that $\mathbf{Z}=(Z_1,\dots,Z_N)$ represents an *innovation process of* X; the number N, which may be a positive integer or equal to infinity, we call the *multiplicity of* X.

Any element $x\in H(X)$ determines the measure μ_x, which is induced by the function $F_x(t)=\|E_X(t)x\|^2$, $t\in R$. Let us put $M_X=\{\mu_x,\ x\in H(X)\}$. We introduce a partial ordering in M_X by saying that μ_x is *subordinated* to μ_y, and writing $\mu_x<\mu_y$, whenever μ_x is absolutely continuous with respect to μ_y. If μ_x and μ_y are mutually absolutely continuous, we say that they are *equivalent*, and we write $\mu_x\sim\mu_y$. The *spectral type* ρ_x is equal to the set of all elements $\mu\in M_X$ which are equivalent to μ_x; in the set $M_X/\sim$ a partial ordering is introduced in the obvious way: we say that ρ_x is *subordinated* (*equal*) to ρ_y, and we write $\rho_x<\rho_y$ ($\rho_x=\rho_y$), if the corresponding relation holds for each $\mu\in\rho_x$ and each $\nu\in\rho_y$.

Each process Z_n, $n=\overline{1,N}$, of the innovation process of X defines the measure μ_n and the spectral type ρ_n, which are induced by the function $F_n(t)=\|Z_n(t)\|^2$, $t\in R$. We say that $\boldsymbol{\rho}=(\rho_1,\ldots,\rho_N)$ is the *spectral type of* X; note that the inequalities $\rho_1>\ldots>\rho_N$ hold.

The information, which the process X gives at the moment t, is equal to the space $H(X;t)$. We say that the processes X and Y *give the same information* if the equality $H(X;t)=H(Y;t)$ holds for each t. The information given by Y *is separated from* X if there exists a family of spaces $\{H_t, t\in R\}$, such that $H_t\subset H(X;t)$ and $H_t=H(Y;t)$ for each t.

All the other notions, concerning measures, spectral types, and multiplicities of spectral types, are defined as in [6].

RESULTS

Let Y be an arbitrary process and let us suppose that the process X is defined by $X(t)=BY(t)$, $t\in R$, where B is some linear operator on $H(Y)$. We can consider the operator B as the filter which transforms the input signal Y into the output signal X. The problems which we shall discuss are: 1. Is there a process Y_0, continuous in quadratic mean, such that its spectral type $\boldsymbol{\rho_0}=(\rho_1^0,\ldots,\rho_M^0)$ is equal to the given spectral type $\boldsymbol{\rho_Y}=(\rho_1^Y,\ldots,\rho_M^Y)$ (in the sense that $\rho_i^0=\rho_i^Y$ for $i=\overline{1,M}$); 2. Under which conditions the information, given by Y, can be separated from X if the maximal spectral type of Y is known in advance.

The first problem is solved by the following theorem.

THEOREM 1. For any given spectral type $\boldsymbol{\rho_Y}=(\rho_1^Y,\ldots,\rho_M^Y)$ there exists a stochastic process Y_0, continuous in quadratic mean, such that $\boldsymbol{\rho_Y}$ is its spectral type, i.e. $\boldsymbol{\rho_0}=\boldsymbol{\rho_Y}$.

P r o o f. Let $A_1,\ldots,A_M$ be a family of disjoint subsets of R, such that for each n ($n=\overline{1,M}$) and any a, b ($a<b$), the ordinary Lebesgue measure of $A_n\cap[a;b]$ is positive (about a construction of such a family, see, for example, [2], [3]). There exist mutually orthogonal processes $Z_1,\ldots,Z_M$ with orthogonal increments, such that the function $F_n(t)=\|Z_n(t)\|^2$, $t\in R$, induces the spectral type ρ_n^Y, $n=\overline{1,M}$ (see [2]). Let $\chi_n(t)$, $t\in R$, be the indicator-function of the set A_n. It is easy to see that the process Y_n, defined by

$$Y_n(t)=\int_{-\infty}^{t}\chi_n(u)Z_n(u)du,\ t\in R,$$

has the spectral type ρ_n^Y; the processes $Y_1,\ldots,Y_M$ are, obviously, mutually orthogonal. Define now the process Y_0 by

$$Y_0(t)=\sum_{n=1}^{M}c_nY_n(t),\ t\in R,$$

(factors c_n, $n=\overline{1,M}$, insure the finiteness of the norm when $M=\infty$). It is clear that we have $H(Y_0;t)\subset\sum_{n=1}^{M}\oplus H(Z_n;t)$, $t\in R$. From the fact that any fixed t belongs to one and only one set A_n $(n=\overline{1,M})$, we have

$$Y_0'(t)=c_n Z_n(t),\quad t\in A_n,$$

where Y_0 denotes the first derivative in quadratic mean of Y_0. Since $H(Y_0';t)\subset H(Y_0;t)$ for each t, and as A_n is everywhere dense in R for each $n=\overline{1,M}$, from the last equality we get $H(Z_n;t)\subset H(Y_0;t)$ for each t and each $n=\overline{1,M}$. Hence $H(Y_0;t)=\sum_{n=1}^{M}\oplus H(Z_n;t)$, $t\in R$, which means that the spectral type of Y_0 is equal to ρ_Y. It is easy to see that the correlation function $r(s,t)$ of Y_0 is continuous; from that it follows that Y_0 is continuous in quadratic mean, [5]. QED

We say that the process Y is *submitted* to the process X if $H(Y;t)\subset \subset H(X;t)$ for each t; Y is *fully submitted* to X if it is submitted and $H(Y)\ominus H(Y;t)\subset H(X)\ominus H(X;t)$ for each t. If Y is fully submitted to X (and if $\boldsymbol{\rho}_Y=(\rho_1^Y,\ldots,\rho_M^Y)$ and $\boldsymbol{\rho}_X=(\rho_1^X,\ldots,\rho_N^X)$ are spectral types of Y and X, respectively), then $\boldsymbol{\rho}_Y<\boldsymbol{\rho}_X$, in the sense that $M\leq N$ and $\rho_i^Y<\rho_i^X$ for $i=\overline{1,M}$ (see, for example, [7]).

In the following we shall always suppose that the process Y is fully submitted to the process X.

LEMMA 1. If for each t the statement

$$\text{if } \operatorname{mult}\rho_{X(t)}>M, \text{ then } \inf\{\rho_1^Y,\ \rho_{X(t)}\}=0 \tag{2}$$

is true, then for each $x\in H(X)$ the following statement is also true:

$$\text{if } \operatorname{mult}\rho_x>M, \text{ then } \inf\{\rho_1^Y,\ \rho_x\}=0. \tag{3}$$

P r o o f. Suppose that (3) is not true, i.e. that there exists $x\in H(X)$ such that $\operatorname{mult}\rho_x\geq M+1$ and $\inf\{\rho_1^Y,\ \rho_x\}=\mu\neq 0$; from the last equality follows $\mu<\rho_x$, which means (by reason of $\operatorname{mult}\rho_x\geq M+1$) that $\operatorname{mult}\mu\geq M+1$. Let us show that there exists t, such that $0\neq\rho_{X(t)}<\mu$ -that will imply $\operatorname{mult}\rho_{X(t)}\geq\operatorname{mult}\mu\geq M+1$, which is in contradiction to (2). Let $0\neq z\in H(X)$ be an arbitrary element with the property $\rho_z=\mu$ (such an element exists because $\mu\in M_X/\sim$). It cannot be $(z,X(t))=0$ for all t - from this assumption it follows that z is orthogonal to $H(X)$, which is impossible. Hence, there exists $t=\bar{t}$, such that the equality $(z,X(\bar{t}))=0$ does not hold, which means that z can be written as $z=P_L z+z_1$, where z_1 is orthogonal to L and P_L is the projection operator from $H(X)$ onto the space L spanned by $X(\bar{t})$. From $P_L z=aX(\bar{t})$, $a\neq 0$, it follows that $z=aX(\bar{t})+z_1$, $(z_1,X(\bar{t}))=0$ and $\rho_z=\sup\{\rho_{aX(\bar{t})},\ \rho_{z_1}\}$. But, by reason of $\rho_{aX(\bar{t})}=\rho_{X(\bar{t})}$, we have $\rho_{X(\bar{t})}<\rho_z=\mu$. Hence we infer that (3) holds for each $x\in H(X)$.

LEMMA 2. If for each t the statement

$$\left.\begin{array}{l}\text{if } \inf\{\rho_{X(t)},\ \rho_1^Y\}\neq 0, \text{ then } \inf\{\rho_{X(t)},\ \rho_1^Y\}<\rho_i^Y \text{ for } i=\overline{1,k},\\ \qquad\text{where } k=\operatorname{mult}\ \inf\{\rho_{X(t)},\ \rho_1^Y\},\end{array}\right\} \tag{4}$$

is true, then for each $x\in H(X)$ the following statement is also true:

$$\left.\begin{array}{l}\text{if } \inf\{\rho_x, \rho_1^Y\}\neq 0, \text{ then } \inf\{\rho_x, \rho_1^Y\}<\rho_i^Y \text{ for } i=\overline{1,k}, \\ \text{where } k=\text{mult } \inf\{\rho_x, \rho_1^Y\}.\end{array}\right\} \quad (5)$$

P r o o f. Suppose that (5) is not true, or, equivalently, that there exists $x\in H(X)$ such that $\inf\{\rho_x, \rho_1^Y\}=\sigma\neq 0$ and $\inf\{\sigma, \rho_k^Y\}=0$, where k has the meaning from (5). However, it can be shown (as in LEMMA 1) that there exists t such that $0\neq\rho_{X(t)}<\sigma$, which means that $\inf\{\rho_{X(t)}, \rho_1^Y\}=\rho_{X(t)}\neq 0$; but, from $\rho_{X(t)}<\sigma$ and $\inf\{\sigma, \rho_k^Y\}=0$, it follows $\inf\{\inf\{\rho_{X(t)}, \rho_1^Y\}, \rho_k^Y\}=0$, which is in contradiction to (4) (indeed, from $\inf\{\rho_{X(t)}, \rho_1^Y\}=\rho_{X(t)}<\sigma$ it follows mult $\inf\{\rho_{X(t)}, \rho_1^Y\}\geq\text{mult}\,\sigma=k$).

QED

Suppose that the output signal X is known, i.e. that the Hida-Cramér representation (1) of X is known. The spectral type of the input signal Y is $\boldsymbol{\rho}_Y=(\rho_1^Y,\ldots,\rho_M^Y)$.

THEOREM 2. The following statements are equivalent:

(I) The statements (2) and (4) are true for each t.

(II) For each t, the space $H(Y;t)$ is equal to the set of all elements of $H(X;t)$ whose spectral types are subordinated to ρ_1^Y.

(III) The process X can be represented in the form

$$X(t)=Y(t)+Z(t), \quad t\in R,$$

where Y and Z are mutually orthogonal processes with mutually orthogonal spectral types.

P r o o f. Let us prove that (I) is a sufficient condition for (II). Let us put

$$\rho_i=\inf\{\rho_1^Y, \rho_i^X\}, \quad i=\overline{2,N}. \quad (6)$$

It is clear that we have $\rho_1^Y>\rho_2>\ldots>\rho_N$. Let us show that $\rho_i=0$ for $i=\overline{M+1,N}$. Suppose that it is not true, i.e. that it is, for example, $\rho_{M+1}\neq 0$; from that it follows that mult $\rho_{M+1}\geq M+1$. Hence, by reason of LEMMA 1, it follows that the following equality is true:

$$\inf\{\rho_1^Y, \rho_{M+1}\}=0. \quad (7)$$

However, from (6) it follows that $\rho_{M+1}<\rho_1^Y$, i.e. $\inf\{\rho_1^Y, \rho_{M+1}\}=\rho_{M+1}\neq 0$, which is in contradiction to (7). So, $\rho_i=0$ for $i=\overline{M+1,N}$, and we have the sequence $\rho_1^Y>\rho_2>\ldots>\rho_M$, each member of which is defined by (6); it is easy to see that the inequality $\rho_M\neq 0$ is satisfied. Now, we must show that $\rho_i=\rho_i^Y$ for $i=\overline{2,M}$. It is clear that $\rho_i^Y<\rho_i$, $i=\overline{2,M}$. If $\rho_{i'}^Y\neq\rho_{i'}$ for some $i=i'$, then there exists a spectral type $\sigma_{i'}\neq 0$ such that

$$\inf\{\sigma_{i'}, \rho_{i'}^Y\}=0, \quad \rho_{i'}=\sigma_{i'}+\rho_{i'}^Y \;; \quad (8)$$

hence $\sigma_{i'}<\rho_{i'}$, which, by reason of (6), implies $\sigma_{i'}<\rho_1^Y$. From the last inequality, in accordance with LEMMA 2, it follows that $\sigma_{i'}<\rho_i^Y$, $i=\overline{1,k}$,

$k=\mathrm{mult}\,\sigma_{i'}$, which is in contradiction to (8) (indeed, from (8) follows $\mathrm{mult}\,\sigma_{i'}>i'$, which means that it must be $\inf\{\sigma_{i'}, \rho^Y_{i'}\}=0$ and $\sigma_{i'}<\rho^Y_{i'}$, at the same time, which is impossible). Consequently, the equality $\rho_i=\rho^Y_i$ holds for each $i=\overline{2,M}$. From the fact that the cyclic spaces $M(z_i)=H(Z_i)$, $i=\overline{1,N}$ (such that $\rho_{z_i}=\rho^X_i$), which determine the Hida-Cramér decomposition of X, are uniquely determined, it follows that the cyclic subspaces, which determine the Hida-Cramér decomposition of Y and whose spectral types are $\rho^Y_1,\ldots,\rho^Y_M$, are also uniquely determined ([4], Ch. X).

Let us prove that (I) is a necessary condition for (II). Let us put

$$H(Y)=\{x\in H(X);\ \rho_x<\rho^Y_1\}. \tag{9}$$

The assumption that (I) does not hold is equivalent to the assumption that at least one of the statements (2) and (4) does not hold. From the assumption that (2) does not hold, it follows that there exists $t=t^*$, such that $\mathrm{mult}\,\rho_{X(t^*)}>M$ and $\inf\{\rho^Y_1, \rho_{X(t^*)}\}=\mu\neq 0$; hence $\mu<\rho^Y_1$ and $\mathrm{mult}\,\mu>M$, which is impossible (indeed, in that case $H(Y)$ contains at least $M+1$ mutually orthogonal cyclic subspaces, whose spectral types are equal to μ, which is in contradiction to the assumption that the multiplicity of Y is equal to M). From the assumption that (4) does not hold, it follows that there exists $t=t^*$, such that $\inf\{\rho_{X(t^*)}, \rho^Y_1\}\neq 0$, and that, for example, the inequality $\inf\{\rho_{X(t^*)}, \rho^Y_1\}<\rho^Y_k$ ($k=\mathrm{mult}\,\inf\{\rho_{X(t^*)}, \rho^Y_1\}$) does not hold, i.e. that there exists $\sigma>0$ with the properties $\sigma<\inf\{\rho_{X(t^*)}, \rho^Y_1\}$ and $\inf\{\sigma, \rho^Y_k\}=0$; but, from that it follows that there exists $x\in H(X)$, $x\neq 0$, such that $\rho_x=\sigma$, $\inf\{\rho_x, \rho^Y_1\}=\rho_x\neq 0$, $\mathrm{mult}\,\rho_x\geq k$, and $\inf\{\rho_x, \rho^Y_k\}=0$, which is in contradiction to (9) (namely, the last relations mean that the right side in (9) contains a non-zero element, orthogonal to $H(Y)$, which is impossible). The equivalence of the statements (I) and (II) is proved.

Let us prove that (II) is a sufficient condition for (III). Let us put

$$Z(t)=P_{H(X)\ominus H(Y)}X(t),\quad t\in R,$$

where the space $H(Y)$ is defined by (9). It is clear that the processes Y and Z are orthogonal. From the fact that the process Y is fully submitted to the process X, it follows that $H(Y)$ (and also $H(X)\ominus H(Y)$) reduces the resolution of the identity E_X (see [7]), which implies $\boldsymbol{\rho}_Z<\boldsymbol{\rho}_X$. It is easy to see that $H(Z)=H(X)\ominus H(Y)$, which means (by reason of (9)) that the spectral types $\boldsymbol{\rho}_Z$ and $\boldsymbol{\rho}_Y$ are orthogonal.

Let us prove that (II) is a necessary condition for (III). Let us put $H_0=\{y\in H(X);\ \rho_y<\rho_1^Y\}$; it is clear that $H(Y)\subset H_0$. Suppose that there exists an element $0\neq y\in H_0$, which is orthogonal to $H(Y)$; by reason of $H(X)=H(Y)\oplus H(Z)$, it follows that $y\in H(Z)$, which implies $\inf\{\rho_y,\ \rho_1^Y\}=0$; but, the last relation is in contradiction to the condition $y\in H_0$. Thus, the equality $H(Y)=H_0$ holds.

THEOREM 3. Let any of the statements (I), (II), (III) be satisfied. If $\mathbf{Z_X}=(Z_1^X,\dots,Z_N^X)$ is an innovation process of $\mathbf{X}$, then an innovation process of Y is defined by

$$\mathbf{Z}_n^0(t)=E_X(t)z_n^0,\quad t\in R,\quad n=\overline{1,M},$$

where

$$z_n^0=\int_{-\infty}^{\infty}\chi_{S_1}(t)dZ_n^X(t),\quad n=\overline{1,M}, \tag{10}$$

$S_1=\left\{t;\ \dfrac{d\mu_1^Y(t)}{d\mu_1^X(t)}>0\right\}$, χ_{S_1} is the indicator-function of the set S_1, μ_1^Y and μ_1^X are arbitrary measures which belong to ρ_1^Y and ρ_1^X, respectively, and $\dfrac{d\mu_1^Y}{d\mu_1^X}$ is the Radon-Nikodym derivative of the measure μ_1^Y with respect to the measure μ_1^X.

P r o o f. The function χ_{S_1} is, according to the Radon-Nikodym theorem, μ_1^X-measurable, which means that the integral (10) exists for all $n=\overline{1,M}$. It is clear that $Z_n^0(t)\in H(Z_n^X;t)$, $t\in R$, $n=\overline{1,M}$. Let us show that Z_n^0, for each $n=\overline{1,M}$, induces the spectral type ρ_n^Y. Let us put: $F_n^0(t)=\|Z_n^0(t)\|^2$, $F_n^X(t)=\|Z_n^X(t)\|^2$, $t\in R$, $n=\overline{1,M}$. These functions induce the measures μ_n^0 and μ_n^X $(n=\overline{1,M})$, defined by

$$\mu_n^0(A)=\int_A dF_n^0(t),\quad A\in B,$$
$$\mu_n^X(A)=\int_A dF_n^X(t),\quad A\in B,$$

where B is the Borel σ-field on R. According to (10) we have

$$\mu_n^0(A)=\int_A \chi_{S_1}(u)d\mu_n^X(u),\quad A\in B,\quad n=\overline{1,M},$$

which, by reason of the Radon-Nikodym theorem, means that $\mu_n^0<\mu_n^X$, i.e. $\rho_n^0<\rho_n^X$, $n=\overline{1,M}$. However, according to the definition of S_1, we have $\rho_n^0<\rho_1^Y$, which, together with the previous relation, gives $\rho_n^0<\inf\{\rho_1^Y,\ \rho_n^X\}$, i.e., according to THEOREM 2,

$$\rho_n^0<\rho_n^Y,\quad n=\overline{1,M}. \tag{11}$$

Let us show that, for any $A\in B$, from $\mu_n^0(A)=0$ it follows that $\mu_n^Y(A)=0$ (μ_n^Y is arbitrary measure of ρ_n^Y) - from that it follows that $\rho_n^Y<\rho_n^0$, which, together with (11), gives $\rho_n^0=\rho_n^Y$. Since $\rho_n^Y<\rho_n^X$, we have

$$\mu_n^Y \sim \int \chi_{S_n} d\mu_n^X, \quad n=\overline{1,M},$$

where $S_n=\left\{t; \frac{d\mu_n^Y(t)}{d\mu_n^X(t)}>0\right\}$. From $\rho_n^X<\rho_1^X$ and $\rho_n^Y<\rho_1^Y$ $(n=\overline{1,M})$ it follows that $S_n \subset S_1$ i.e. $\chi_{S_n} \leq \chi_{S_1}$ for all $n=\overline{1,M}$. Let $A \in \mathcal{B}$ be an arbitrary set for which the equality $\mu_n^0(A)=0$ holds; we have

$$0 \leq \int_A \chi_{S_n}(t)\, d\mu_n^X(t) \leq \int_A \chi_{S_1}(t)\, d\mu_n^X(t)=0,$$

i.e. $\mu_n^Y(A)=0$. From that it follows that the equality $\rho_n^0=\rho_n^Y$ holds. Hence, the processes $Z_1^0,\ldots,Z_M^0$ determine the spectral types $\rho_1^Y,\ldots,\rho_M^Y$, i.e. the cyclic subspaces $M_1^0,\ldots,M_M^0$, determined by $z_1^0,\ldots,z_M^0$, have the spectral types $\rho_1^Y,\ldots,\rho_M^Y$; for each $n=\overline{1,M}$, the cyclic space M_n^0 is the subspace of the cyclic space M_n^X, whose spectral type is ρ_n^X. That means that the cyclic spaces $M_1^0,\ldots,M_M^0$ are uniquely determined, [6].

COROLLARY. Let us suppose that all assumptions of THEOREM 3 are satisfied. Then the process Y_0, defined by

$$Y_0(t)=\sum_{n=1}^{M} c_n \int_{-\infty}^{t} \chi_n(u) Z_n^0(u)\, du, \quad t \in R,$$

is continuous in quadratic mean and gives the same information as Y; the functions $\chi_n(t)$ and the processes $Z_1^0,\ldots,Z_M^0$ have the same meaning as in THEOREM 1 and THEOREM 3, respectively.

REMARK

If the stochastic process X satisfies the conditions $H(X;-\infty)=0$ and $H(X;t-0)=H(X;t)$, $t \in R$, then it can be shown that the space $H(X)$ is, in general, non-separable. (Note: we accept the continuum hypothesis.) But, in this case, also, the family E_X represents the resolution of the identity of $H(X)$, and there exists a decomposition of $H(X)$ into the sum of mutually orthogonal cyclic subspaces; but, the number of these subspaces is, in general, equal to $\aleph_1$. It can be shown without dificulties that the above theorems, about the separation of the information, given by Y, from X, hold (with necessary changes concerning X) when M is a positive integer or equal to infinity.

REFERENCES

1. Bulatović, J. and Ašić, M. (in press). *The separability of the Hilbert space generated by a stochastic process*, J. Multivar. Anal.

2. Cramer, H. (1964). *Stochastic processes as curves in Hilbert space*, Theor. Probability Appl., 9, 169-179.

3. Cramer, H. (1967). *A contribution to the multiplicity theory of stochastic processes*, Proc. Fifth Berkeley Symp. Math. Stat. and Prob., Vol. II, 215-221, Berkeley.

4. Cramer, H. (1971). *Structural and statistical problems for a class of stochastic processes*, Princeton University Press, Princeton, New Jersey.

5. Cramer, H. and Leadbetter, M. R. (1967). *Stationary and related stochastic processes*, John Wiley, New York.

6. Plesner, A. I. (1965). *Spectral theory of linear operators*, (in Russian), Nauka, Moskow.

7. Rozanov, Yu. A. (1974). *Theory of innovation processes*, (in Russian), Nauka, Moskow.

Recent Developments in Statistics
J.R.Barra et al., editors
© North-Holland Publishing Company,(1977)

ON THE NOTION OF BALANCE IN BLOCK DESIGNS

Tadeusz Caliński
Department of Mathematical and Statistical Methods
Academy of Agriculture
Poznań, Poland

A general definition of balance for block designs is given. The generalization lies in the arbitrariness of the inner product relation used in normalizing the contrasts that are to be estimated with the same variance. It appears that the two most common definitions of balance, concerning the so-called "variance balance" and "efficiency balance", are particular cases of the introduced general definition of balance.

0. INTRODUCTION

There have been several suggestions about how to define balance in block designs. Most of the definitions proposed are equivalent for equi-replicate designs but are not necessarily such, if the treatments are replicated in unequal numbers. A particular distinction has recently been made between two notions of balance, the "variance balance" and the "efficiency balance". It appears, however, that any variance-balanced design is in a sense an efficiency-balanced design and, v.v., any efficiency-balanced design is in a sense a variance-balanced design. The confusion is caused by the different ways of defining contrasts that are to be estimated with the same variance, as well as by the different ways of defining an efficiency factor that is to be equal for all estimable contrasts.

In the present paper a generalized definition of balance is proposed and it is shown that the two rival definitions are its particular cases. A special role played by some system of eigenvalues and eigenvectors chosen in correspondence with the definition of balance is indicated. Also, it is explained in which sense a block design may be termed variance-balanced and in which it may be called efficiency-balanced.

1. SOME ASSUMPTIONS AND NOTATION

Let $\underline{y}$ be an $n\times 1$ vector of observations from an experiment in which v treatments are applied to n plots arranged in b blocks.

Assume the model

(1.1) $$\underline{y} = \underline{1}\alpha + \underline{D}'\underline{\beta} + \underline{\Delta}'\underline{\gamma} + \underline{\eta}\ ,$$

where $\underline{1}$ is the $n\times 1$ unit vector, $\underline{D}$ is an $b\times n$ matrix with a row for each block and a column for each plot, such that an element is 1 if the plot is in the block and is 0 otherwise, $\underline{\Delta}$ is an $v\times n$ matrix with a row for each treatment and a column for each plot, such that an element is 1 if the plot receives the treatment and 0 otherwise; furthermore, α is a general parameter, $\underline{\beta}$ is a vector of block parameters, $\underline{\gamma}$ is a vector of treatment parameters and $\underline{\eta}$ is a vector of errors. Also, assume that

(1.2) $$E(\underline{\eta}) = 0 \quad \text{and} \quad E(\underline{\eta}\underline{\eta}') = \sigma^2\underline{I}$$

($\underline{I}$ being an identity matrix).

The following relations hold between the matrices $\underline{D}$, $\underline{\Delta}$, the $v\times v$ incidence matrix $\underline{N}$ of the block design, the vector $\underline{k}$ of block sizes and the vector $\underline{r}$ of treatment replications: $\underline{D}'\underline{1} = \underline{1}$, $\underline{D}\underline{1} = \underline{k} = \underline{N}'\underline{1}$, $\underline{D}\underline{D}' = \underline{k}^{\delta}$, $\underline{\Delta}'\underline{1} = \underline{1}$, $\underline{\Delta}\underline{1} = \underline{r} = \underline{N}\underline{1}$, $\underline{\Delta}\underline{\Delta}' = \underline{r}^{\delta}$ and $\underline{\Delta}\underline{D}' = \underline{N}$, where $\underline{x}^{t\delta}$ represents a diagonal matrix with diagonal elements $x_1^t, x_2^t, \ldots,$ if $x_1, x_2, \ldots$ are the elements of a vector $\underline{x}$.

Under these assumptions and with this notation, the least squares normal equation for $\underline{\gamma}$ (also called the adjusted normal equation) is

(1.3) $$(\underline{r}^{\delta} - \underline{N}\underline{k}^{-\delta}\underline{N}')\underline{\gamma}^{o} = \underline{T} - \underline{N}\underline{k}^{-\delta}\underline{B}\ ,$$

where $\underline{T} = \underline{\Delta}\underline{y}$ and $\underline{B} = \underline{D}\underline{y}$. The equation can be presented in a more convenient way by introducing the symmetric matrix $\underline{\phi} = \underline{I} - \underline{D}'\underline{k}^{-\delta}\underline{D}$, for which

(1.4) $$\underline{\phi}\underline{\phi} = \underline{\phi} \quad \text{and} \quad \underline{\phi}\underline{1} = \underline{0} = \underline{\phi}\underline{D}'\ .$$

In fact, with $\underline{\phi}$, the equation (1.3) may be written

(1.5) $$\underline{\Delta}\underline{\phi}\underline{\Delta}'\underline{\gamma}^{o} = \underline{\Delta}\underline{\phi}\underline{y} \quad \text{or} \quad \underline{A}\underline{\gamma}^{o} = \underline{Q}\ ,$$

where

(1.6) $$\underline{A} = \underline{\Delta}\underline{\phi}\underline{\Delta}' = \underline{r}^{\delta} - \underline{N}\underline{k}^{-\delta}\underline{N}'$$

and

$$Q = \Delta \phi y = T - N k^{-\delta} B \ . \tag{1.7}$$

Also note that, from (1.4), $A1 = 0$ and so A is singular. Let its rank be denoted by $h(\leq v-1)$. (See also Pearce, Caliński and Marshall, 1974.)

2. SOME BASIC RESULTS

The following results (taken from Caliński and Pearce, 1976) will be utilized in defining the notion of balance in block designs. Some of the results are well known.

Theorem 1. A necessary and sufficient condition for a linear function of the treatment parameters, $c'\gamma$, to be estimable is the equality

$$c'A^{-}A = c' \ , \tag{2.1}$$

where A^{-} is any generalized inverse (g-inverse) of A.

Proof. Equality (2.1) is a necessary and sufficient condition for the function $c'\gamma^{o}$ to have a unique value for all solutions γ^{o} of (1.5), as follows from Theorem 2.3.1 in Rao and Mitra (1971). But it is known (see e.g. Rao, 1973, section 4a.2) that the uniqueness holds if and only if there exists a linear function of y with expectation $c'\gamma$, and this means that the function $c'\gamma$ is estimable. (qed)

Remark 1. It can easily be seen that if (2.1) is satisfied by one g-inverse of A it is also satisfied by any other.

Corollary 1. The only linear functions of the treatment parameters that are estimable are treatment contrasts, though not necessarily all (i.e., if $c'\gamma$ is estimable, then $c'1 = 0$, but not the opposite).

Proof. This result follows immediately from the fact that (2.1) implies $c'1 = c'A^{-}A1 = c'A^{-}\Delta\phi 1 = 0$, on account of (1.4). (qed)

The following system of eigenvalues and eigenvectors will be utilized subsequently.

Let A be as in (1.6) and let X be an arbitrary (real) $v \times v$ positive definite symmetric matrix. Then, if

$$(2.2) \qquad \underline{A}\underline{w}_i = \lambda_i \underline{X}\underline{w}_i \ ,$$

the quantity λ_i is called an eigenvalue and $\underline{w}_i$ an eigenvector of $\underline{A}$ with respect to $\underline{X}$. It can be proved, by Theorems 6.3.1 and 6.3.2 in Rao and Mitra (1971), that there are precisely h non-zero eigenvalues of $\underline{A}$ with respect to $\underline{X}$, i.e. as many as there are non-zero eigenvalues of $\underline{A}$ with respect to $\underline{I}$, and, furthermore, that corresponding eigenvectors can be chosen to be X-orthonormal in pairs, i.e., to satisfy

$$(2.3) \qquad \underline{w}_i'\underline{X}\underline{w}_i = 1 \quad \text{and} \quad \underline{w}_i'\underline{X}\underline{w}_j = 0$$

for $i \neq j$ and $i,j = 1, 2, \ldots, v$. Also, since $\underline{A}\underline{1} = \underline{0}$, the last (say) eigenvector $\underline{w}_v$ may be taken equal to $\underline{1}/(\underline{1}'\underline{X}\underline{1})^{1/2}$. Hence $\underline{1}'\underline{X}\underline{w}_i = 0$ for $i = 1, 2, \ldots, v-1$.

<u>Lemma 1</u>. Let $\underline{w}_1, \underline{w}_2, \ldots, \underline{w}_h, \ldots, \underline{w}_v$ be a set of X-orthonormal eigenvectors of $\underline{A}$ with respect to $\underline{X}$, the first h of them corresponding to the non-zero eigenvalues of $\underline{A}$ (with respect to $\underline{X}$) and $\underline{w}_v = \underline{1}/(\underline{1}'\underline{X}\underline{1})^{1/2}$. Then a linear function $\underline{c}'\underline{\gamma}$ is estimable if and only if

$$(2.4) \qquad \underline{c} = \sum_{i=1}^{h} l_i \underline{X}\underline{w}_i \ ,$$

where $l_1, l_2, \ldots, l_h$ are some coefficients ($\underline{c}'\underline{\gamma}$ thus being an estimable treatment contrast).

<u>Proof</u>. Since the vectors $\underline{X}\underline{w}_1, \underline{X}\underline{w}_2, \ldots, \underline{X}\underline{w}_{v-1}$ are linearly independent and all X-orthogonal to $\underline{1}$, they span the subspace of all vectors that define treatment contrasts. So, for any contrast $\underline{c}'\underline{\gamma}$ there exists a set of coefficients $l_1, l_2, \ldots, l_{v-1}$ such that $\underline{c} = \Sigma_{i=1}^{v-1} l_i \underline{X}\underline{w}_i$. But, on account of (2.2) and (2.3), it can be verified that

$$(2.5) \qquad \underline{A} = \sum_{i=1}^{h} \lambda_i \underline{X}\underline{w}_i\underline{w}_i'\underline{X} \ ,$$

and hence

$$(2.6) \qquad \underline{A}\Big(\sum_{i=1}^{h} \underline{w}_i\underline{w}_i'/\lambda_i\Big)\underline{c} = \sum_{i=1}^{h} l_i \underline{X}\underline{w}_i \ .$$

Also, it can easily be checked that

$$(2.7) \qquad \sum_{i=1}^{h} \underline{w}_i\underline{w}_i'/\lambda_i$$

is a g-inverse of $\underline{A}$, and thus, on account of Remark 1, the condition (2.1) is equivalent to (2.6), which is satisfied if and only if $\underline{c}$ is of the form (2.4). (That $\underline{c}'\underline{\gamma}$ is then a contrast follows immediately from the fact that $\underline{1}'\underline{A} = \underline{0}$ implies $\underline{1}'\underline{X}\underline{w}_i = 0$ for any $\underline{w}_i$ corresponding to a non zero λ_i .) (qed)

Lemma 1 may also be expressed equivalently as

Lemma 2. The vectors $\underline{X}\underline{w}_1$, $\underline{X}\underline{w}_2$,..., $\underline{X}\underline{w}_h$, corresponding by (2.2) to the non-zero eigenvalues λ_1 , λ_2 ,..., λ_h , span the subspace of all vectors defining estimable treatment contrasts. (It is an X^{-1}-orthonormal basis of the subspace.)

This result is essential to prove the following

Theorem 2. Let $\underline{W} = [\underline{w}_1 , \underline{w}_2 ,..., \underline{w}_h]$ be a matrix such that its columns are those X-orthonormal eigenvectors of $\underline{A}$, with respect to $\underline{X}$, which correspond to the non-zero eigenvalues of $\underline{A}$, with respect to $\underline{X}$, λ_1 , λ_2 ,..., λ_h . Then any set of estimable treatment contrasts may be written as a vector $\underline{C}'\underline{\gamma} = \underline{L}'\underline{W}'\underline{X}\underline{\gamma}$, where $\underline{L}' = [\underline{l}_1 , \underline{l}_2 ,..., \underline{l}_h]$ is a matrix of h columns, its Best Linear Unbiased Estimate (BLUE) may be written as

$$\widehat{\underline{C}'\underline{\gamma}} = \underline{L}'\widehat{\underline{W}'\underline{X}\underline{\gamma}} = \sum_{i=1}^{h} \underline{l}_i\underline{w}_i'\underline{Q}/\lambda_i \ , \tag{2.8}$$

and its covariance matrix as

$$\mathrm{var}(\widehat{\underline{C}'\underline{\gamma}}) = \sigma^2\underline{L}'\underline{\lambda}^{-\delta}\underline{L} = \sigma^2 \sum_{i=1}^{h} \underline{l}_i\underline{l}_i'/\lambda_i \ , \tag{2.9}$$

where $\underline{\lambda} = [\lambda_1 , \lambda_2 ,..., \lambda_h]$.

Proof. That $\underline{C}'\underline{\gamma}$ is $\underline{L}'\underline{W}'\underline{X}\underline{\gamma}$ follows from Lemma 2. By the theory of least squares (see e.g. Theorem 7.2.4 in Rao and Mitra, 1971), and on account of (1.5), the BLUE of $\underline{W}'\underline{X}\underline{\gamma}$ is $\underline{W}'\underline{X}\underline{\gamma}^o = \underline{W}\underline{X}\underline{A}^-\underline{Q}$.

But, applying (2.7) as $\underline{A}^-$, the BLUE may be written

$$\widehat{\underline{W}'\underline{X}\underline{\gamma}} = [\underline{w}_1'\underline{Q}/\lambda_1 \ , \ \underline{w}_2'\underline{Q}/\lambda_2 \ ,..., \ \underline{w}_h'\underline{Q}/\lambda_h] \ . \tag{2.10}$$

Furthermore, since from (1.7),(1.2) and (1.4), $\mathrm{var}(\underline{Q}) = \sigma^2\underline{A}$, the covariance matrix of (2.10) is

$$\mathrm{var}(\underline{W}'\underline{X}\underline{\gamma}) = \sigma^2\underline{W}'\underline{X}\underline{A}^-\underline{A}(\underline{A}^-)'\underline{X}\underline{W} = \sigma^2\underline{\lambda}^{-\delta}, \tag{2.11}$$

on account of (2.5), (2.7) and (2.3). (Note that, because of Theorem 1 and Remark 1, the covariance matrix is invariant for any choice of $\underline{A}^-$.) Finally, from (2.10) and (2.11), the formulae (2.8) and (2.9) follow, respectively. (qed)

Now note that if $\underline{c}$ defines an estimable treatment contrast, then, from (2.4),

$$\underline{c}'\underline{X}^{-1}\underline{c} = \sum_{i=1}^{h} l_i^2 \ . \tag{2.12}$$

Hence the following

Corollary 2. An estimable treatment contrast $\underline{c}'\underline{\gamma}$ is X^{-1}-normalized (i.e. such that $\underline{c}'\underline{X}^{-1}\underline{c} = 1$) if and only if $\underline{c} = \underline{XWl}$, where $\underline{l} = [l_1, l_2, \ldots, l_h]'$ is a normalized ($\underline{I}$-normalized) vector.

3. A GENERAL DEFINITION OF BALANCE

The results given in the previous section give rise to the following general definition of balance in block designs.

Definition 1. A block design is said to be X^{-1}-balanced if every X^{-1}-normalized estimable linear function of treatment parameters can be estimated with the same variance.

A condition for this type of balance is given by the following

Theorem 3. A block design is X^{-1}-balanced if and only if the non-zero eigenvalues of its matrix $\underline{A}$ with respect to $\underline{X}$ are all equal.

Proof. Let $\underline{c}'\underline{\gamma}$ be estimable. Then (2.4) holds and, from (2.9),

$$\text{var}(\underline{c}'\hat{\underline{\gamma}}) = \sigma^2 \sum_{i=1}^{h} l_i^2/\lambda_i \ . \tag{3.1}$$

Now, on account of Corollary 2, it is evident that the variance (3.1) is constant for every X^{-1}-normalized estimable function $\underline{c}'\underline{\gamma}$ if and only if $\lambda_1 = \lambda_2 = \ldots = \lambda_h$. (qed)

In connection with Theorem 3, the following result is applicable.

Corollary 3. If a block design is X^{-1}-balanced, then the BLUE of any vector (set) of estimable treatment contrasts $\underline{C}'\underline{\gamma}$ has the covariance matrix of the form

$$\text{var}(\widehat{\underline{C}'\underline{\gamma}}) = (\sigma^2/\lambda)\underline{C}'\underline{X}^{-1}\underline{C} \ , \tag{3.2}$$

where λ is the unique non-zero eigenvalue of $\underline{A}$ with respect to X , with multiplicity h .

Proof. By Theorems 2 and 3, $\text{var}(\widehat{\underline{C}'\underline{\gamma}}) = (\sigma^2/\lambda)\underline{L}'\underline{L}$, where $\lambda = \lambda_1 = \lambda_2 = \ldots = \lambda_h$. Also, Theorem 2 implies $\underline{C}'\underline{X}^{-1}\underline{C} = \underline{L}'\underline{W}'\underline{XWL} = \underline{L}'\underline{L}$, on account of (2.3). Hence the result (3.2). (qed)

Remark 2. If $\underline{C}'\underline{X}^{-1}\underline{C} = \underline{I}$, i.e., the set of contrasts is X^{-1}-orthonormal, the covariance matrix (3.2) takes the form $\text{var}(\widehat{\underline{C}'\underline{\gamma}}) = (\sigma^2/\lambda)\underline{I}$. This indicates that the precision of an X^{-1}-balanced block design can by measured by σ^2/λ .

A useful result for connected designs (i.e. designs with $h = v - 1$, as shown by Bose, 1950) is the following

Corollary 4. A connected block design is X^{-1}-balanced if and only if its matrix $\underline{A}$ is of the form

$$\underline{A} = \lambda[\underline{X} - \underline{X}\underline{1}\underline{1}'\underline{X}/(\underline{1}'\underline{X}\underline{1})] , \tag{3.3}$$

where $$\lambda = \frac{n - tr(\underline{N}\underline{k}^{-\delta}\underline{N}')}{tr(\underline{X}) - \underline{1}'\underline{X}\underline{X}\underline{1}/(\underline{1}'\underline{X}\underline{1})} .$$

Proof. If a design is X^{-1}-balanced, then, from (2.5) and by Theorem 3, $\underline{A} = \lambda\underline{X}(\Sigma_{i=1}^{h}\underline{w}_i\underline{w}_i')\underline{X}$. But, from (2.2) and (2.3), $\Sigma_{i=1}^{v}\underline{w}_i\underline{w}_i' = \underline{X}^{-1}$ and one can take $\underline{w}_v = \underline{1}/(\underline{1}'\underline{X}\underline{1})^{1/2}$. Hence, $h = v - 1$ implies (3.3) and this implies (3.4). To go the other way, suppose that (3.3) holds. This implies that any non-zero eigenvalue of $\underline{A}$ with respect to $\underline{X}$ must be λ , as $\underline{1}'\underline{X}\underline{W}_i = 0$ for $i = 1 , 2 ,..., v-1$. Hence, by Theorem 3, the design is X^{-1}-balanced. (qed)

4. TWO RIVAL DEFINITIONS

As particular cases of the above general notion of balance, two definitions often applied are those for which $\underline{X} = \underline{I}$ and $\underline{X} = \underline{r}^{\delta}$. In fact, if $\underline{X} = \underline{I}$, Definition 1 is reduced to that of Vartak (1963):

Definition 2. A block design is said to be balanced (i.e. I-balanced) if every normalized (I-normalized) estimable linear function of treatment parameters can be estimated with the same variance.

This definition is an extension of that originally used by Tocher (1952) and Rao (1958) for connected designs. It has been applied by many authors, e.g. by Pearce (1964) and Raghavarao (1971) .

In correspondence with Definition 2, Theorem 3 implies

Corollary 5. A block design is balanced (I-balanced) if and only if the non-zero eigenvalues of its matrix $\underline{A}$ (with respect to $\underline{I}$) are all equal.

This has originally been proved by Vartak (1963), again as an extension of a similar result given by Rao (1958) for connected designs.

Furthermore, Corollary 4 gives the following

Corollary 6. A connected block design is balanced (I-balanced) if and only if its matrix $\underline{A}$ is of the form

(4.1) $$\underline{A} = \lambda(\underline{I} - v^{-1}\underline{1}\underline{1}'),$$

where $\lambda = [n - \mathrm{tr}(\underline{N}\underline{k}^{-\delta}\underline{N}')]/(v-1)$.

Also, this result has originally been given by Rao (1958) .

Now, if $\underline{r}^{\delta}$ is substituted for the matrix $\underline{X}$, another useful definition emerges.

Definition 3. A block design is said to be $r^{-\delta}$-balanced if every $r^{-\delta}$-normalized estimable linear function of treatment parameters can be estimated with the same variance.

This definition has implicitly been used by Nair and Rao (1948), Jones (1959) and Caliński (1971). A necessary and sufficient condition for this type of balance folows from Theorem 3 as

Corollary 7. A block design is $r^{-\delta}$-balanced if and only if the nonzero eigenvalues of its matrix $\underline{A}$ with respect to $\underline{r}^{\delta}$ are all equal.

Again, Corollary 4 yields the following result for connected desings.

Corollary 8. A connected block design is $r^{-\delta}$-balanced if and only if its matrix $\underline{A}$ is of the form

(4.2) $$\underline{A} = \lambda[\,\underline{r}^{\delta} - \underline{r}\underline{r}'/n\,],$$

where $\lambda = [n - \mathrm{tr}(\underline{N}\underline{k}^{-\delta}\underline{N}')]/(n - n^{-1}\underline{r}'\underline{r})$.

This result may be traced back to Jones (1959), whose matrix $\underline{M} = \underline{r}^{-\delta}\underline{N}\underline{k}^{-\delta}\underline{N}'$ is related to $\underline{A}$ by the equation $\underline{r}^{-\delta}\underline{A} = \underline{I} - \underline{M}$. (See also Caliński, 1971, p. 281 .)

5. SOME COMMENTS ON TERMINOLOGY

Block designs obeying Definition 2 have been termed "variance-balanced" by Hedayat and Federer (1974), and those obeying Definition 3 have been termed "efficiency balanced" by Puri and Nigam (1975).

These terms, although very convenient, are not completely satisfactory, since each of the two types of designs can be considered as variance-balanced in view of Definition 1. Moreover, each of them can be regarded as efficiency-balanced.

To see their efficiency balance, note that for an orthogonal block design, i.e. a design of $\underline{N} = \underline{r}\underline{k}'/n$ (as defined by Pearce, 1970), $\underline{A} = \underline{r}^{\delta} - \underline{r}\underline{r}'/n$ and its unique non-zero eigenvalue with respect to $\underline{r}^{\delta}$ is 1, with multiplicity $(v - 1)$. Hence, on account of Remark 2, its precision is given by σ^2. Thus, for any efficiency-balanced ($\underline{r}^{-\delta}$-balanced) block design, the unique non-zero eigenvalue of $\underline{A}$ with respect to $\underline{r}^{\delta}$ is a common efficiency factor of the design for all estimable treatment contrasts, relative to an orthogonal block design with the same $\underline{r}$. (This way of defining efficiency has been adopted by Pearce, 1970.) On the other hand, for an equireplicate orthogonal block design, i.e. a design of $\underline{N} = r\underline{1}\underline{k}'/n = \underline{1}\underline{k}'/v$, where r is the common number of replications, $\underline{A} = r\underline{I} - r^2\underline{1}\underline{1}'/n$ and its unique non-zero eigenvalue (with respect to $\underline{I}$) is r, with multiplicity $(v - 1)$. Hence, on account of Remark 2, its precision is given by σ^2/r. Thus, for any variance balanced (I-balanced) block design the unique non-zero eigenvalue of $\underline{A}$ (with respect to $\underline{I}$) divided by n/v is the common efficiency factor of the design for all estimable treatment contrasts, relative to an equi-replicate orthogonal block design with the same n. (This way of defining efficiency has been adopted by Atiqullah, 1961.)

So it is now evident that the so-called "efficiency-balanced" designs are also variance-balanced for $\underline{r}^{-\delta}$-normalized estimable treatment contrasts and that they are not efficiency-balanced relative to an orthogonal design of a different replication vector. Similarly, it is clear that the so-called "variance-balanced" designs are also efficiency-balanced relative to an equi-replicate orthogonal design with the same number of plots, and that they are not variance-balanced for estimable treatment contrasts unless they are normalized in the sense of the usual Euclidean inner product relation, or proportionally to it.

REFERENCES

Atiqullah, M. (1961). On a property of balanced designs. Biometrika 48, 215.

Bose, R.C. (1950). Least Squares Aspects of Analysis of Variance. Institute of Statistics, University of North Carolina.

Caliński, T. (1971). On some desirable patterns in block designs (with discussion). Biometrics 27, 275.

Caliński, T. and Pearce, S.C. (1976). A general theory of the intrablock analysis. (In preparation .)

Hedayat, A. and Federer, W.T. (1974). Pairwise and variance balanced incomplete block designs. Ann. Inst. Statist. Math. 26, 331.

Jones, R.M. (1959). On a property of incomplete blocks. J. Roy. Statist. Soc. Ser. B 21, 172.

Nair, K.R. and Rao, C.R. (1948). Confounding in asymmetrical factorial experiments. J. Roy. Statist. Soc. Ser. B 10, 109.

Pearce, S.C. (1964). Experimenting with blocks of natural sizes. Biometrics 20, 699.

Pearce, S.C. (1970). The efficiency of block designs in general. Biometrika 57, 339.

Pearce, S.C., Caliński, T. and Marshall, T.F. de C. (1974). The basic contrasts of an experimental design with special reference to the analysis of data. Biometrika 61, 449.

Puri, P.D. and Nigam, A.K. (1975). On patterns of efficiency-balanced designs. J. Roy. Statist. Soc. Ser. B 37, 457.

Raghavarao, D. (1971). Constructions and Combinatorial Problems in Design of Experiments. Wiley, New York.

Rao, C.R.and Mitra, S.K. (1971). Generalized Inverse of Matrices and Its Applications. Wiley, New York.

Rao, C.R.(1973). Linear Statistical Inference and Its Applications. 2-nd ed. Wiley, New York.

Rao, V.R. (1958). A note on balanced designs. Ann. Math. Statist. 29, 290.

Tocher, K.D. (1952). The design and analysis of block experiments. J. Roy. Statist. Soc. Ser. B 14, 45.

Vartak, M.N. (1963). Disconnected balanced designs. J. Indian Statist. Assoc. 29, 290.

Recent Developments in Statistics
J.R.Barra et al., editors

ON A DISTRIBUTION-FREE TEST OF INDEPENDENCE BASED ON GINI'S RANK ASSOCIATION COEFFICIENT

Donato Michele Cifarelli*
University of Pavia - University "L.Bocconi"
Milano, Italy

and

Eugenio Regazzini*
University of Torino - University "L.Bocconi"
Milano, Italy

INTRODUCTION

Let $(X_1,Y_1),\dots,(X_n,Y_n)$ be independent and identically distributed random vectors over $\mathcal{R}^2$. If R_i is the rank of X_i among the X's and S_i is the rank of the corresponding Y_i, then Gini's rank association coefficient [4] is defined to be

$$G = \frac{2}{D} \sum_{i=1}^{n} \{|n+1-R_i-S_i| - |R_i-S_i|\} \tag{1}$$

if ties do not occur, and where

$$D = \begin{cases} n^2 & n \text{ even} \\ n^2-1 & n \text{ odd.} \end{cases}$$

G satisfies the following inequalities

$$-1 \le G \le 1$$

and

$$G = -1 \quad \text{iff} \quad S_i = n+1-R_i \quad \forall i$$
$$G = 1 \quad \text{"} \quad S_i = R_i \quad \forall i$$

It is known that Gini's rank association coefficient is a statistic, like Kendall's and Spearman's rank correlation coefficients for testing independence between two random variables (r.v.).
According to Gini's definition, the statistic G belongs to the class of measures of concordance together with Kendall's and Spearman's rank correlation coefficients. G has been investigated by many Authors: F.Savorgnan [11] , V.Amato [1] , T.Salve-

*The sections 4,6,7,8 are due to the first author: 2,3,5 are due to the second author.

mini [10], O.Cucconi [3], A.Rizzi [9], A.Herzel [5].

In this paper we will give some rules (section 2) to obtain the exact distribution of G as well as a new hint concerning the normal approximation to the distribution in the case of independence (section 3). Furthermore, the asymptotic distribution is derived, for an interesting class of cumulative distribution functions (c.d.f.) of (X_i, Y_i) (section 7). In the case of independence, tables of the distribution of G are given in [9], by simulation, when $n \leq 30$, while the exact distribution is given in [11] for $n \leq 6$ and in [10] for n=7. We present tables for n=8,9,10.

In section 4 we propose a theoretical measure of the strength of the association of X and Y.

In section 5 we shall give a U-statistic, which is its consistent and M.V. U. estimator.

We shall study (section 5) the connection between G and the theoretical measure of association defined in section 4 and demonstrate that G is asymptotically unbiased for it. In section 8 we shall study some more asymptotic properties of G in order to test independence. For many details in proofs we shall quote [2].

SOME PROPERTIES OF G UNDER INDEPENDENCE

Let $(X_1,Y_1), \ldots,(X_n,Y_n)$ be a random sample from a population with the c.d.f. F(x,y). In the following the c.d.f. F(x,y) of (X,Y) is assumed to be continuous. We will denote by $\mathcal{F}$ the class of continuous c.d.f.'s F(x,y) having continuous joint and marginal density distribution functions (p.d.f.):

$$f(x,y) = \frac{\partial^2 F(x,y)}{\partial x\, \partial y}, \quad f_1(x) = \int f(x,y)dy, \quad f_2(y) = \int f(x,y)dx.$$

Let (X'_j, Y_{ji}), i=1,...,n, be an ordering of (X_i,Y_i) with $X'_1 < X'_2 < \ldots < X'_n$, where $(j_1,\ldots,j_n)$ is a permutation of $(1,\ldots,n)$.

We denote by $\Pi_{n\upsilon}$,(υ=1,2,...,n!), the n! possible rankings of samples of size n from a bivariate population.

Definition 1. We say that between X and Y exists no association when:

$$P(\Pi_{n\upsilon}) = \frac{1}{n!} \qquad \upsilon = 1,\ldots,n! \tag{2}$$

It is known that if $F(x,y) = F_1(x).F_2(y)$, where F_1,F_2 are the marginal c.d.f. of F, then (2) holds.

We have also the following

Theorem 1. (Hoeffding [7]). If F(x,y) is in $\mathcal{F}$ and (2) holds for some $n \geq 5$, then

$$F(x,y) = F_1(x) \,.\, F_2(y).$$

Therefore, under the stated conditions, the properties of no association and independence are equivalent.

Now we state some simple properties of the distribution of G when (2) holds, $n \geq 5$,

$F\varepsilon\mathcal{F}$. The proofs of these properties are given in [2].

Theorem 2. If $n \geq 5$, $F\varepsilon\mathcal{F}$, (2) holds, then the probability frequency function (p.f.f.) of

$$D.G = Z = \sum_{i=1}^{n} \{|n+1-R_i-S_i| - |R_i-S_i|\}$$

is symmetric with $E(Z^{2k+1})=0$, $k=0,1,\ldots$. The probability generating function of Z is

$$\Psi(t) = \frac{1}{n!}\sum a^{t(q_{1j_1}+\ldots+q_{nj_n})}$$

where the summation extends over all n! permutations of $(1,2,\ldots,n)$ and

$$(q_{ij}) = (|n+1-i-j|-|i-j|)\qquad i,j=1,\ldots,n.$$

The above theorem has been used to give tables of p.f.f. for n=8,9,10.

Theorem 3. (Cucconi [3]). If $n \geq 5$, $F\varepsilon\mathcal{F}$, (2) holds, then the variance of G is

$$\mathrm{Var}(G) = \begin{cases} \dfrac{2}{3(n-1)}\left(1+\dfrac{2}{n^2}\right) & n \text{ even} \\ \dfrac{2}{3(n-1)}\left(1+\dfrac{4}{n^2-1}\right) & n \text{ odd.} \end{cases} \tag{3}$$

ASYMPTOTIC DISTRIBUTION OF G UNDER INDEPENDENCE

Considerations about the asymptotic distribution of G have been made in [1],[3] and [5]. All these Authors have stated asymptotic normality of G without providing a rigorous proof. To prove the next result concerning the asymptotic normality of G it is sufficient to verify that in our case the conditions of the following theorem are fulfilled.

Theorem 4. (Hoeffding [8]). Let $b_n(i,j)$, $i,j=1,2,\ldots,n$, be real numbers defined for every positive integer n and let $(M_1,\ldots,M_n)$ be a r.v. which takes each permutation of $(1,\ldots,n)$ with the same probability 1/n!. Let us consider:

$$L_n = \sum_{i=1}^{n} b_n(i,M_i);$$

$$d_n(i,j) = b_n(i,j) - \frac{1}{n}\sum_{t=1}^{n} b_n(t,j) - \frac{1}{n}\sum_{S=1}^{n} b_n(i,S) + \frac{1}{n^2}\cdot\sum_{t,S} b_n(t,S).$$

If

$$\lim_{n\to+\infty} \frac{\max\limits_{1\leq i,\, j\leq n} d_n^2(i,j)}{\frac{1}{n}\cdot\sum_{i,j} d_n^2(i,j)} = 0 \tag{4}$$

then the r.v. L_n is asymptotically normally distributed with mean value

$$E(L_n) = \frac{1}{n}\sum_{i,j} b_n(i,j) \quad\text{and variance}\quad \mathrm{Var}(L_n) = \frac{1}{n-1}\sum_{i,j} d_n^2(i,j).$$

From the preceding theorem we obtain

Theorem 5. If $n \geq 5$, $F \varepsilon \mathcal{F}$ and (2) holds, then $S = (\frac{3n}{2})^{\frac{1}{2}}.G$ is asymptotically normally distributed with mean value 0 and variance 1.

Proof of Theorem 5. Let us define

$$b_n(i,j) = \frac{2}{D}\sqrt{\frac{3n}{2}}\{|n+1-i-j|-|i-j|\} \qquad i,j=1,2,\ldots,n$$

$$(M_1,\ldots,M_n) = (j_1,\ldots,j_n)$$

$$L = \sum_{i=1}^{n} b_n(i,j_i) = \frac{2}{D}\sqrt{\frac{3n}{2}}\sum_{i=1}^{n}\{|n+1-i-j_i|-|i-j_i|\} = \sqrt{\frac{3n}{2}}\,G.$$

Since

$$d_n(i,j) = b_n(i,j), \qquad \max_{1 \leq i,j \leq n} d_n^2(i,j) = \frac{6n(n-1)^2}{D^2}$$

and

$$\frac{1}{n}\sum_{i,j=1}^{n} d_n^2(i,j) \sim \frac{n^4}{D^2} \; (n \to +\infty)$$

the condition (4) is fulfilled. Furthermore, according to theorems 2 and 3 we have $E(S)=0$, $\lim_{n \to +\infty} Var(S)=1$.

A THEORETICAL MEASURE OF ASSOCIATION

Consider the problem of testing the hypothesis H_0, that is $F(x,y) \varepsilon \mathcal{F}_1 \subset \mathcal{F}$, where $\mathcal{F}_1$ is the class of c.d.f. in $\mathcal{F}$ for which (2) holds with $n \geq 5$.

Generally speaking, testing H_0 is made easy if:

i) a functional $I(F)$, $F \varepsilon \mathcal{F}$, exists which satisfies $I(F)=0$ iff $F \varepsilon \mathcal{F}_1$;

ii) a consistent estimator of $I(F)$ exists for each $F \varepsilon \mathcal{F}$.

In the following we propose a functional $I(F)$ which has property i) partially as well as property ii). The functional is:

$$I(F) = 4\int_0^1 \{ F(F_1^{-1}(x), F_2^{-1}(x)) + F(F_1^{-1}(1-x), F_2^{-1}(x)) - x \}\, dx \qquad (5)$$

where F_1^{-1}, F_2^{-1} are the inverse functions of the marginal c.d.f. F_1, F_2.

The functional (5) is a theoretical measure of the association between the two random variables X and Y, as the following theorem shows.

Theorem 6. $-1 < I(F) < 1$. Moreover,

1) $I(F) = 0$ if $F \varepsilon \mathcal{F}_1$
2) $I(H_{+1}) = 1$ if $H_{+1}(x,y) = Min(F_1(x), F_2(y))$
3) $I(H_{-1}) = -1$ if $H_{-1}(x,y) = Max(F_1(x)+F_2(y)-1,0)$
4) $I(F) = -I(\psi)$ if $\psi(x,y) = F_1(x)-F(x,-y)$ or $\psi(x,y) = F_2(y)-F(-x,y)$
5) $I(F) = I(\xi)$ if $\xi(x,y) = P\{\phi_1(X) \leq x;\ \phi_2(Y) \leq y\}$

for each couple of monotonic increasing (or decreasing) functions ϕ_1 and ϕ_2 such that $\xi \varepsilon \mathcal{F}$.

Proof of Theorem 6. See [2].

Note that I(F) does not satisfy property i) completely as shown in the following example.

Let F(x,y) and G(x,y) be two c.d.f. in $\mathcal{F}$ and suppose $I(F)=-I(G)=-\eta$, $\eta \neq 0$. If we define $H=1/2(F+G)$, we have:

$$I(H) = 0, \; H \varepsilon \mathcal{F},$$

but, H may not be in $\mathcal{F}_1$.

THE MINIMUM VARIANCE UNBIASED ESTIMATOR (M.V.U.E.) OF I(F)

The aim of this section is to find a M.V.U.E. of I(F) when the marginal c.d.f.are known, resorting to the Hoeffding method [6]. We are to find a symmetric kernel ψ_0, that is a symmetric function, in the sense specified below, which depends on m (degree) variables and which satisfies the condition of regularity of the functional I(F), that is:

$$E(\psi_0) = I(F) \qquad \forall F \varepsilon \mathcal{F}$$

Given a sample of size $n \geq m$, let us consider the statistic

$$U = \{\binom{n}{m}^{-1}\} \Sigma \psi_0(X_{i_1}, Y_{i_1}; \ldots ; X_{i_m}, Y_{i_m})$$

where the summation extends over all $\binom{n}{m}$ combinations $(i_1, \ldots, i_m)$. If the size of the sample is 2, let us put:

$$\psi(X_1,Y_1; X_2,Y_2) = S(1-F_1(X_1)-F_2(Y_1)) [S(X_2-X_1)-S(Y_1-Y_2)] - $$
$$- S(F_1(X_1)-F_2(Y_1)) [S(X_1-X_2)-S(Y_1-Y_2)],$$

where

$$S(u) = \begin{cases} 0 & u < 0 \\ 1 & u \geq 0; \end{cases}$$

furthermore let

$$\psi_0(X_1,Y_1; X_2,Y_2) = \frac{1}{2}\{\psi(X_1,Y_1; X_2;Y_2) + \psi(X_2,Y_2; X_1,Y_1)\}.$$

The function ψ_0 so defined is symmetric in the sense that

$$\psi_0(X_1,Y_1; X_2,Y_2) = \psi_0(X_2,Y_2; X_1,Y_1)$$

The following theorem shows that an unbiased estimator of I(F) exists, that is I(F) is a regular functional.

Theorem 7. I(F) is a regular functional $\forall F \varepsilon \mathcal{F}$ of degree 2 and its kernel is given by

$$T_1 = 4\psi_0.$$

Proof of Theorem 7. We have to show that

$$E(T_1) = 4 E(\psi_0) = I(F) \qquad \forall F \varepsilon \mathcal{F}.$$

Note that:

$$E(\psi) = P\{1-F_1(X_1)>F_2(Y_1);\ X_2>X_1\} - P\{1-F_1(X_1)>F_2(Y_1);\ Y_1>Y_2\} -$$
$$-\Big[P\{F_1(X_1)>F_2(Y_1);\ X_1>X_2\}- P\{F_1(X_1)>F_2(Y_1);\ Y_1>Y_2\}\Big];$$
$$P\{F_1(X_1)>F_2(Y_1);\ X_1>X_2\} = \int_C F_1(x_1)f(x_1;y_1)dx_1dy_1;$$
$$C =\{(x_1,y_1)\in\mathcal{R}^2 | F_1(x_1)>F_2(y_2)\}$$
$$P\{F_1(X_1)>F_2(Y_1);\ Y_1>Y_2\} = \int_C f(x_1;y_1)F_2(y_1)dx_1dy_1;$$

hence:

$$P\{F_1(X_1)>F_2(Y_1);\ X_1>X_2\} - P\{F_1(X_1)>F_2(Y_1);\ Y_1>Y_2\}=$$
$$= \frac{1}{2} - \int_0^1 F(F_1^{-1}(z),\ F_2^{-1}(z))dz\cdot$$

analogously we obtain:

$$P\{1-F_1(X_1)>F_2(Y_1);\ X_2>X_1\}- P\{1-F_1(X_1)> F_2(Y_1);\ Y_1>Y_2\}=$$
$$=\int_0^1 F(F_1^{-1}(1-z),\ F_2^{-1}(z))dz.$$

The function:

$$T = \frac{4}{n(n-1)} \sum_{i\neq j} \psi(X_i,Y_i;\ X_j,Y_j)$$

is a U-statistic and according to a well-known result, we have:

$$E(T) = I(F),\ Var(T)\sim \frac{64}{n}\cdot\xi= \frac{64}{n}\cdot \underset{(X_1,Y_1)}{Var}\left\{\underset{(X_2,Y_2)}{E}(\psi_0)\right\}, \quad n\to+\infty .$$

In the case of independence, we have, after some simple calculations,

$$Var(T)\sim \frac{2}{3n}, \qquad n\to+\infty$$

hence:

$$Var(T)\sim Var(G), \qquad n\to+\infty .$$

If $F\in\mathcal{F}$ and $\xi\neq 0$, T is a consistent estimator of I(F).

As to the asymptotic distribution of the r.v., $\sqrt{n}(T-I(F))$, from a property relating to U-statistics [6], we obtain immediately:

Theorem 8. If $\xi\neq 0$, then the r.v.

$$\sqrt{n}(T-I(F))$$

is asymptotically normally distributed with mean value given by 0 and variance given by 64ξ.

To conclude this section, we note that T is a symmetric function of the n couples (X_i,Y_i) and therefore it is a function of the order statistics, which are sufficient statistics. Completeness of the latter in the class of absolutely continuous distributions yields completeness of T. From the Lehmann-Scheffé theorem, we then deduce that the statistic T is the M.V.U.E. of I(F).

CONNEXIONS BETWEEN G AND T

In this section we shall show that the statistics G and T defined in the above section are connected quite simply. Let us consider a new expression of G equivalent to (1):

$$G = \frac{4}{D} \sum_{i \neq j} \{ S(n+1-R_i-S_i) \left[S(X_j-X_i)-S(Y_i-Y_j) \right] -S(R_i-S_i) \cdot \left[S(X_i-X_j)-S(Y_i-Y_j) \right] \}. \tag{6}$$

(1), (6) and the definition of T imply, see [2]

$$G-T \simeq \frac{2}{n} \sum_{i=1}^{n} \{ |1- \frac{R_i}{n+1} - \frac{S_i}{n+1} | - | \frac{R_i}{n+1} - \frac{S_i}{n+1} | - |1-F_1(X_i)-F_2(Y_i)| + |F_1(X_i)-F_2(Y_i)| \}.$$

Let us put:

$$A_n = \frac{2}{n} \cdot \sum_{i=1}^{n} \{ |F_1(X_i)-F_2(Y_i)| - | \frac{R_i}{n+1} - \frac{S_i}{n+1} | \}$$

$$B_n = \frac{2}{n} \cdot \sum_{i=1}^{n} \{ |1 - \frac{R_i}{n+1} - \frac{S_i}{n+1} | - |1-F_1(X_i)-F_2(Y_i)| \}.$$

We can prove

<u>Theorem 9.</u> The r.v. (G-T) converges to zero in probability.

<u>Proof of Theorem 9.</u> It is not difficult to obtain the inequality

$$P\{|A_n| \leq \varepsilon\} \geq P\{ \frac{2}{n} \sum_{i=1}^{n} |F_1(X_i) - \frac{R_i}{n+1}| < \frac{\varepsilon}{2} \} + P\{ \frac{2}{n} \sum_{i=1}^{n} |F_2(Y_i) - \frac{S_i}{n+1}| < \frac{\varepsilon}{2} \} -1.$$

Hence:

$$P\{|A_n| \leqq \varepsilon\} \geqq 1 - \frac{4}{n\varepsilon} \sum_{i=1}^{n} E(|F_1(X_i) - \frac{R_i}{n+1}|) + E(|F_2(Y_i) - \frac{S_i}{n+1}|).$$

It is easy to see that $\exists K > 0$ such that

$$E(|F_1(X_i) - \frac{R_i}{n+1}|) < \frac{K}{n} \qquad i=1,2,\dots,n$$

$$E(|F_2(Y_i) - \frac{S_i}{n+1}|) < \frac{K}{n}$$

Then:

$$\lim_{n \to +\infty} P\{|A_n| \leq \varepsilon\} = 1.$$

Similar arguments imply:

$$\lim_{n \to +\infty} P\{|B_n| \leq \varepsilon\} = 1.$$

The result follows from the fact that $(G-T) \simeq A_n + B_n$.

It is easy to obtain:

Theorem 10. The rank association coefficient G is an asymptotically unbiased estimator of $I(F) \in (-1,1)$.

Proof of Theorem 10. $G-I(F) \simeq T-I(F)+A_n+B_n$.

From the fact that $T-I(F)$ converges to zero in probability and from Theorem 9, $G-I(F)$ converges to zero in probability,

$$|E(G-I(F)| \leq E(|G-I(F)|) \quad \text{and}$$

$$|G-I(F)| \leq 2, \forall n,$$

hence by virtue of a well-known necessary and sufficient condition for the convergence in probability of a sequence of uniformly bounded r.v.'s, we have

$$\lim_{n \to +\infty} E(|G-I(F)|) = 0,$$ what implies the result.

THE ASYMPTOTIC DISTRIBUTION OF G IN THE GENERAL CASE

In a preceding section we established that $\sqrt{n}G$ is asymptotically normally distributed as n tends to infinity, when X and Y are independent r.v.'s. We shall now discuss the asymptotic normality of $\sqrt{n}(G-I(F))$ for an interesting class of c.d.f.'s.

We have:

Theorem 11. Let (X,Y) be a r.v. with c.d.f. F.

If

i) $F \in \mathcal{F}$;

ii) $\xi > 0$;

then the r. v.

$$\sqrt{n}\ (\ G - I(F)\ )$$

is asymptotically normal with mean given by zero and variance given by $64\,\xi$.

Proof of Theorem 11. See [2].

FURTHER COMMENTS ABOUT THE USE OF G TO TEST THE HYPOTHESIS OF INDEPENDENCE

In the preceding sections we have shown that G is an asymptotically unbiased and consistent estimator of the functional (5). For Spearman's B_S and Kendall's B_K rank correlation coefficients we have:

$$\lim_{n \to +\infty} E(B_S) = \rho(F) = 3(4\iint F(x,y)\cdot f_1(x)\cdot f_2(y)\ dx\ dy-1)$$

$$E(B_K) = \tau(F) = 4\iint F(x,y)\ dF(x,y) - 1.$$

These functionals are proposed as a measure of the strength of the association between the two characteristics and they vanish when the two characteristics are independent.

Unluckily, these functionals vanish in situations different from that of independence, hence B_S, B_K are not suitable tests to omnibus alternatives, as they are consistent only when the alternative to the hypothesis of independence satisfies the inequalities

$$I(F) \neq 0, \qquad \rho(F) \neq 0, \qquad \tau(F) \neq 0$$

respectively.

In this section we shall consider the Pitman efficiency of the tests based on Gini's coefficient G relative (ARE) to the tests based on the correlation coefficient when F is the bivariate normal distribution.

Suppose F is the bivariate normal distribution $(\mu_1 \mu_2 \sigma_1 \sigma_2 \rho)$ with correlation coefficient $\rho \neq 0$. Let r be the sample correlation coefficient.

We have

$$ARE(G,r) = \frac{8}{\pi^2} \cong 0.81$$

For calculations see [2].

It is well-known that the asympotic efficiency of the test based on B_k and B_S relative to the test based on r is

$$ARE(B_k,r) = ARE(B_S,r) = \frac{9}{\pi^2} ;$$

this implies, in the normal case:

$$ARE(G,r) < ARE(B_k,r) = ARE(B_S,r).$$

However, in our opinion, this is not a sufficient reason for preferring B_k or B_S to Gini's G under any circumstance, as Pitman's efficiency is a measure of the power of a test only when the sample size tends to infinity.

TABLE 1. - The p.f.f. and the c.d.f. of $D \cdot G = Z = \sum_{i=1}^{n} \{ |n+1-R_i-S_i| - |R_i-S_i| \}$ in the case of independence.

Z=	n = 8		n = 9		n = 10	
	p.f.f.	c.d.f.	p.f.f.	c.d.f.	p.f.f.	c.d.f.
0	0.07217258	0.53608614	0.06471008	0.53235465	0.05419312	0.52709627
2	0.07921624	0.61530238	0.06342590	0.59578055	0.05608906	0.58318532
4	0.06408727	0.67938966	0.06133708	0.65711761	0.05228836	0.63547367
6	0.06790674	0.74729639	0.05652557	0.71364313	0.05227733	0.68775100
8	0.05796131	0.80525768	0.05340884	0.76705194	0.04715939	0.73491037
10	0.05079365	0.85605133	0.04763007	0.81468195	0.04519731	0.78010768
12	0.03938492	0.89543623	0.04151234	0.85619426	0.03953703	0.81964469
14	0.03373016	0.92916638	0.03547178	0.89166600	0.03610091	0.85574555
16	0.02500000	0.95416635	0.02958278	0.92124873	0.03074074	0.88648629
18	0.01904762	0.97321397	0.02335758	0.94460630	0.02661679	0.91310304
20	0.01259930	0.98581314	0.01816027	0.96276653	0.02171186	0.93481487
22	0.00773809	0.99355119	0.01334876	0.97611529	0.01776152	0.95257634
24	0.00394345	0.99749464	0.00955136	0.98566663	0.01372134	0.96629763
26	0.00173611	0.99923074	0.00648148	0.99214810	0.01074653	0.97704411
28	0.00059524	0.99982595	0.00410328	0.99625134	0.00797839	0.98502249
30	0.00014881	0.99997473	0.00221561	0.99846691	0.00581900	0.99084145
32	0.00002527	1.00000000	0.00102513	0.99949199	0.00394180	0.99478322
34			0.00037478	0.99986672	0.00250551	0.99728870
36			0.00010747	0.99997419	0.00143959	0.99872828
38			0.00002206	0.99999619	0.00074598	0.99947423
40			0.00000381	1.00000000	0.00033620	0.99981040
42					0.00013200	0.99994236
44					0.00004299	0.99998534
46					0.00001130	0.99999660
48					0.00000220	0.99999875
50					0.00000125	1.00000000

REFERENCES

[1] Amato, V., Sulla distribuzione dell'indice di cograduzione del Gini. Statistica (1954), 505-519.

[2] Cifarelli, D.M. and Regazzini, E., Alcune osservazioni sull'indice di cograduazione del Gini. Unpublished manuscript circulated through personal communication.

[3] Cucconi, O, La distribuzione campionaria dell'indice di cograduazione del Gini. Statistica (1964), 143-151.

[4] Gini, G., L'ammontare e la composizione della ricchezza delle nazioni. Torino, F.lli Bocca, (1914).

[5] Herzel, A., Sulla distribuzione campionaria dell'indice di cograduzione del Gini. Metron 30 (1972), 137-153.

[6] Hoeffding, W., A class of statistics with asymptotically normal distribution. The Annals of Mathematical Statistics. 19 (1948), 293-325.

[7] Hoeffding, W., A non parametric test of independence. The Annals of Mathematical Statistics. (1948), 546-557.

[8] Hoeffding, A combinatorial central limit theorem. The Annals of Mathematical Statistics. 22 (1951), 558-566.

[9] Rizzi, A, Distribuzione dell'indice di cograduazione del Gini. Metron. 29 (1971), 63-73.

[10] Salvemini, T., Sui vari indici di cograduzione. Statistica (1951), 133-154.

[11] Savorgnan, F., Sulla formazione dei valori dell'indice di cograduzione. Studi Economico-Giuridici dell'Università di Cagliari (1915).

Recent Developments in Statistics
J.R.Barra et al., editors

SOME CONVERGENCE PROBLEMS IN FACTOR ANALYSES

J. DAUXOIS and A. POUSSE (*)

ABSTRACT :

Principal component analysis of a measurable random function is defined through the notion of principal component analysis of a probability on a separable Hilbert space. Such analysis being necessarily undertaken through a sample, we are led to investigate the effect of sampling.

We introduce the concept of canonical analysis of two closed subspaces of a real separable Hilbert space and consider as a particular case the non linear canonical analysis of two random variables. We study some problems of convergence arising in practical applications.

* Laboratoire de Statistique - E.R.A. 591 C.N.R.S.
Université Paul Sabatier - 31077 TOULOUSE Cédex FRANCE.

INTRODUCTION

Consider a random phenomenon underlying a random function. A descriptive study by means of factor analysis requires a sample either "statistical" or of another type (e.g. for a random process). Although the obtained analysis is in fact a mere description of the data, it is nevertheless usually interpreted as an approximating description of the phenomenon. In doing so, this analysis is considered as approximating the studied random function. Is such a practice justified ? To answer this question, one must first define the concept of principal component analysis (P.C.A.). Secondly, legitimize the substitution of this P.C.A. $\mathfrak{F}$ by a P.C.A. $\mathfrak{F}_n$ obtained through sampling. In other words one must show that, as the size of the sample indefinitely increases, the C.A. $\mathfrak{F}_n$ sequence converges, in a sense to be defined, to the P.C.A. $\mathfrak{F}$.

In § 1.1., we define the notion of P.C.A. of a probability π on a separable real Hilbert space and its Borel σ-field. It is based on the P.C.A. of an operator which in turn leads to the spectral analysis of the covariance operator (assumed to exist) of the probability π. It is associated with a "duality scheme" which is convenient for applications. It also enables the definition of the P.C.A. of a measurable random function. In this particular case, we find both the P.C.A. of r.v. in the sense of Hotelling and the P.C.A. used in data analysis. The problem of the convergency of the P.C.A. obtained through sampling is treated in § 2.1. .

The linearity constraint which is usually imposed in factor analysis generally causes a great loss of information. Therefore the concept of non linear canonical analysis (N.L.C.A.) is to be investigated. We outline here some aspects of N.L.C.A.. Introducing in § 1.2. the concept of canonical analysis of two closed subspaces of a separable real Hilbert space, we study a unique model including in particular the linear canonical analysis of r.v. and the non linear canonical analysis of r.v..

A numerical treatment can be achieved through the use of a "discretisation" procedure. By discretisation we mean that qualitative variables are substituted for the considered r.v. . N.L.C.A. is thus replaced by a correspondence analysis. Finer and finer discretisationslead to satisfactory approximations of the studied non linear analysis if, for appropriate splitting, the associated sequence of correspondence analyses converges. However, again in this case, the investigated statistical variables are only known through the data set (of their observed values on a sample). This leads to another approximation. Therefore a new problem of convergency arises and must be studied. Both problems of convergency (resulting from discretisation and from sampling) are exposed in a simple case in § 2.3. .

Some answers to the problems of sensitivity of factor analysis to additional observations and/or sequential sampling may also be derived.

§ 1 . FACTOR ANALYSIS

1.1. PRINCIPAL COMPONENT ANALYSIS OF A PROBABILITY - APPLICATIONS

1.1.1. PRINCIPAL COMPONENT ANALYSIS OF AN OPERATOR

Let H and H' be two real separable Hilbert spaces . H (resp. H') is identified with its topological dual, as for any Hilbert space. Let U be a non-zero bounded linear operator on H to H', and U^* its adjoint.

Definition 1 : *We call principal component analysis (P.C.A.) of U any couple*
$(\{\lambda_i\}_{i \in I}, \{u_i\}_{i \in I})$, *where :*

* *I is $\mathbb{N}^*$ or a set $\{1,\ldots,n\}$, $n \in \mathbb{N}^*$*
* *$\{\lambda_i\}_{i \in I}$ is a decreasing sequence of non negative real numbers*
* *$\{u_i\}_{i \in I}$ is a sequence of elements of H',*

which satisfies :

(i) $\forall (i,j) \in I^2$ $\langle u_i, u_j \rangle_{H'} = \delta_{ij}$ (the Kronecker delta)

(ii) $\forall i \in I$ $\lambda_i = \sup \{ \|U^* u\|_H / \|u\|_{H'} , u \in H'$ and $\forall j < i \langle u, u_j \rangle_{H'} = 0 \}$

If $I = \{1,\ldots,n\}$, the P.C.A. is called a P.C.A. of order n.
If the P.C.A. satisfies
   (iii) $\{u_i\}_{i \in I}$ is a basis of H'
it is called a "total" P.C.A.
Hence, the first step of the P.C.A. of U is the research of a normed element u of H' if it exists, whose image by U^* has maximal norm in H, and so on ... under constraint of orthonormality.
The following duality scheme can be associated with this P.C.A.

$$\begin{array}{ccc} H' & \xleftarrow{U} & H \\ I \Updownarrow V & & I \Updownarrow W \\ H' & \xrightarrow{U^*} & H \end{array}$$

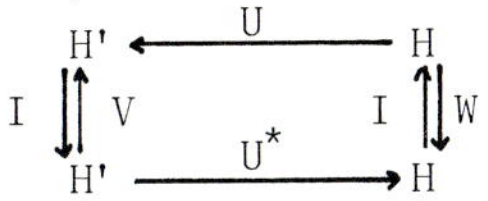

$V = U \circ U^*$
$W = U^* \circ U$

I : identity operator

The P.C.A. of U is obtained by spectral analysis of the bounded self adjoint operator V. In particular if U is compact, then V is and, by classical properties of its spectral analysis, $\{\lambda_i\}_{i \in I}$ is the full decreasing sequence (i.e. where each λ_i is repeated according to its multiplicity) of its eigenvalues and $\{u_i\}_{i \in I}$ is an associated orthonormal sequence of eigenvectors. Adding to it an orthonormal basis of the kernel of V, if required, we obtain a "total P.C.A.".
$\{\lambda_i\}_{i \in I}$ is then called the sequence of the "principal values" and $\{u_i\}_{i \in I}$ the sequence of the "principal factors" of U. Putting : $\forall i \in I$, $f_i = U^* u_i / \|U^* u_i\|$, we obtain an orthonormal sequence $\{f_i\}_{i \in I}$, called sequence of the "principal components" of U.
U can be written : $U = \sum_{i \in I} \sqrt{\lambda_i} \, f_i \otimes u_i$ in $\mathcal{L}(H,H')$.

Remark : We can generalize the previous definition of P.C.A. using the spectral function of V : then any bounded operator admits a P.C.A. which is said to be a "global P.C.A.", although the preceding P.C.A. (called "iterative" P.C.A.) may not exist (see [2.2.]).

1.1.2. PRINCIPAL COMPONENT ANALYSIS OF A PROBABILITY

Let E be a real separable Hilbert space, and $\mathcal{B}_E$ its Borel σ-field. Let $\langle .,. \rangle$ and

$\nu = \|.\|$ be the inner product and the associated norm of E. π being a probability on $(E, \mathcal{B}_E)$, let $L^2(\pi)$ (i.e. $L^2(E,\mathcal{B}_E,\pi)$) be the real Hilbert space of (equivalence classes of) random variables f defined on $(E,\mathcal{B}_E,\pi)$ for which f^2 is π-integrable. We denote by $(.,.)$ and $|||.|||$ its usual inner product and associated norm. All the L^2 spaces considered here are separable.
From now on, unless otherwise stated, we suppose that π has a second moment, i.e. satisfies the condition

$$\text{(i)} \quad \int_E \|e\|^2 \, d\pi(e) < +\infty$$

π is said to be centered if it also verifies :

$$\text{(ii)} \quad \int_E e \, d\pi(e) = 0$$

1.1.2.1. DEFINITION AND EXISTENCE

a) Case of centered probability

Let U be the linear operator on $L^2(\pi)$ into E defined by :

$$f \in L^2(\pi) \to Uf = \int_E e\, f(e)\, d\pi(e)$$

Since $|f|$ and ν belong to $L^2(\pi)$, it follows that :

$$\|Uf\| = \|\int_E e\, f(e)\, d\pi(e)\| \leq \int_E |f(e)|\, \|e\|\, d\pi(e) = (|f|,\nu)$$

Hence $\|Uf\| \leq |||f||| \cdot |||\nu|||$, and U is a well defined and bounded operator. We easily see that its adjoint U^* is the operator on E into $L^2(\pi)$ defined by :

$$e' \in E \to U^*e' = \langle .,e' \rangle$$

Definition 2 : *We call principal component analysis of the probability π (which admits a second moment and is centered) the principal component analysis (as defined in § 1.1.1.) of the operator U.*

The following duality scheme can be associated with this principal component analysis of π :

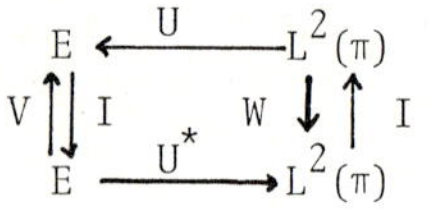

$$V = U \circ U^*$$

$$W = U^* \circ U$$

The operators V and W are easily made explicit by

$$e' \in E \to Ve' = \int_E e(U^*e')\, e\, d\pi(e) = \int_E e\langle e,e'\rangle\, d\pi(e)$$

$$f \in L^2(\pi) \to Wf = \langle .,\int_E ef(e)\, d\pi(e)\rangle = \int_E \langle e,.\rangle\, f(e)\, d\pi(e).$$

Then V is the covariance operator of π, therefore it is nuclear. It follows that U and U^* are Hilbert-Schmidt operators, and that W is also nuclear.
According to § 1.1.1., the P.C.A. of π is total and given by spectral analysis of the covariance operator of π.
We known (see [12]) that any covariance operator V can be associated with a unique gaussian centered probability whose covariance operator is V. The P.C.A. of π can also be considered as the study of the gaussian measure μ which has the same covariance operator as π. So we can say that μ and π are "equivalent" in the sense of Fisher's information in inferential statistics, or in the sense of information in descriptive statistics. It is in this way that the statisticians have tried, originally, to transform contingency tables of data by means of factor analysis (see [9] for example).

b) Case of a non centered probability :

We call P.C.A. of the probability π admitting a second moment and a mean m ($\neq 0$) the P.C.A. of the centered measure $\pi^{(m)}$, which is the translated probability of π (i.e. the P.C.A. of $\pi^{(m)} \doteq \delta_{-m} * \pi$).

1.1.2.2. PARTICULAR CASES

a) Principal component analysis of a random variable valued in a Hilbert space :

Let X be a r.v. defined on $(\Omega,\mathcal{A},P)$ and valued in $(E,\mathcal{B}_E)$, whose squared norm is P-integrable, and centered. We denote by $\mathcal{B}$ the complete sub σ-field of $\mathcal{A}$ induced by X, and P_X the distribution of X on $(E,\mathcal{B}_E)$. By the hypotheses made on X, it follows that P_X satisfies (i) and (ii).

Definition 3 : *We call linear P.C.A. of X the principal component analysis of the probability* P_X .

We can define a linear operator $\mathfrak{J}$ from $L^2(\Omega,\mathcal{B},P)$ into $L^2(E,\mathcal{B}_E,P_X)$ by :

$$\mathfrak{J} : f \in L^2(\Omega,\mathcal{B},P) \longrightarrow g \in L^2(E,\mathcal{B}_E,P_X) \;/\; f = g \circ X$$

$\mathfrak{J}$ is an unitary operator. We can now complete the duality scheme, associated with the P.C.A. of P_X as follows :

$$\begin{array}{ccccccc} E & \xleftarrow{U} & L^2(E,\mathcal{B}_E,P_X) & \xleftarrow{\mathfrak{J}} & L^2(\Omega,\mathcal{B},P) & \xleftarrow{E^{\mathcal{B}}} & L^2(\Omega,\mathcal{A},P) \\ V\uparrow\downarrow I & & I\uparrow\downarrow W & & I\uparrow\downarrow W' & & I\uparrow\downarrow W'' \\ E & \xrightarrow[U^*]{} & L^2(E,\mathcal{B}_E,P_X) & \xrightarrow[\mathfrak{J}^{-1}]{} & L^2(\Omega,\mathcal{B},P) & \xrightarrow[i]{} & L^2(\Omega,\mathcal{A},P) \end{array}$$

where i is the canonical imbedding from $L^2(\Omega,\mathcal{B},P)$ into $L^2(\Omega,\mathcal{A},P)$. We denote by U' (resp. U") the operator $U \circ \mathfrak{J}$ (resp. $U \circ \mathfrak{J} \circ E^{\mathcal{B}}$), and :

$$W' = (U')^* \circ U' \quad , \quad W'' = (U'')^* \circ U''$$

We have : $V = U \circ U^* = U' \circ (U')^* = U'' \circ (U'')^*$

Hence :

Proposition 1 : *The P.C.A. of the r.v. X is the P.C.A. of any one of the operators* U, U', U''.

U' is the operator :

$$f \in L^2(\Omega,\mathcal{B},P) \rightarrow U'f = \int_E e g(e)\, dP_X(e) = \int_\Omega X\, g \circ X\, dP = E[Xf]$$

and its adjoint $(U')^* = \mathfrak{J}^{-1} \circ U^*$:

$$e \in E \longrightarrow (U')^* e = \langle e,X(.)\rangle \in L^2(\Omega,\mathcal{B},P).$$

Since, for any f of $L^2(\Omega,\mathcal{A},P)$: $E(Xf) = E[E^{\mathcal{B}}(Xf)] = E[XE^{\mathcal{B}}(f)]$, U" is the operator :

$$f \in L^2(\Omega,\mathcal{A},P) \longrightarrow U''f = E[Xf].$$

This operator U" is generally used to obtain the P.C.A. of X.

- If $E = \mathbb{R}^p$, let $X = (X_1,\ldots,X_p)$.

$$\forall f \in L^2(P) \qquad U''f = E(Xf) = (E(X_1 f),\ldots,E(X_p f))$$

$$\forall x = (x_1,\ldots,x_p) \in \mathbb{R}^p \qquad (U'')^* x = \langle x,X(.)\rangle = \sum_{i=1}^{p} x_i X_i$$

We retrieve the P.C.A. of $(X_1,\ldots,X_p)$, as defined by Hotelling, when the metric of E is the usual euclidian one with matrix I.
This definition of the P.C.A. of X is thus an extension of Hotelling's.

b) Principal component analysis of a measurable random function :

$(\Omega,\mathcal{A},P)$ (resp. $(T,\mathcal{E},\mu)$) is a probability space (resp. a measure space and μ is bounded). We denote by H a separable Hilbert space, and $(X_t)_{t\in T}$ (or X) a random function valued in $(H,\mathcal{B}_H)$, belonging to $L^2_H(P\otimes\mu)$ and centered. X' is the r.v. defined on $(\Omega,\mathcal{A},P)$ (valued in $L^2_H(\mu)$) by :

$$\forall\omega\in\Omega \qquad X'(\omega) = g \quad , \text{ with } g = X(\omega,.)$$

Thus :

$$U''f = E(X'f) = \int_\Omega X(\omega,.)f(\omega)dP(\omega) \in E = L^2_H(\mu)$$

$$(U'')^*g = \langle g,X'(.)\rangle = \int_T \langle g,X_t\rangle_H \, d\mu(t)$$

Definition 4 : *We call P.C.A. of $(X_t)_{t\in T}$ the P.C.A. of X' or, equivalently, the P.C.A. of U".*

$V = U'' \circ U''^*$ is the covariance operator of $P_{X'}$. $W = U''^* \circ U''$ is the "Escoufier's operator" of $(X_t)_{t\in T}$. they are nuclear operators, and the P.C.A. of $(X_t)_{t\in T}$ is equivalent to the spectral analysis of V (which gives the principal values and principal factors) or of W (which gives the principal values and principal components). The kernel $X(\omega,t)$ of U" has the orthogonal decomposition:

$$X(\omega,t) = \sum_{i\in I} \sqrt{\lambda_i}\, f_i(\omega)\, u_i(t) \ ,$$

where λ_i are the principal values, f_i the normed principal components and u_i the normed principal factors.

A particular case :
If $H = \mathbb{R}$ and $T = \{1,\dots,p\}$, $(X_t)_{t\in T}$ becomes $X = (X_1,\dots,X_p)$ and we obtain, as in a) the Hotelling's P.C.A.

c) P.C.A. of statistical variables :

Such a P.C.A. appears as a particular case of b) : Ω is a finite set (moreover, T is usually $\{1,\dots,p\}$).
Let $\Omega = \{\omega_1,\dots,\omega_n\}$ $(n\in\mathbb{N}^*)$, $\mathcal{A}$ be the σ-field $\mathcal{P}(\Omega)$ of all subsets of Ω, and P be defined by : $\forall\, i=1,\dots,n \quad P[\{\omega_i\}] = \frac{1}{n}$. We denote by 1_i the indicator of $\{\omega_i\}$. The space $L^2(\Omega,\mathcal{A},P)$ is the space of linear combinations of r.v. 1_i, and can be identified with $\mathbb{R}^n$, whose metric defining matrix is $\frac{1}{n}I_n$.
Therefore, we can write the duality scheme :

$$\begin{array}{ccc} L^2_H(T,\mathcal{E},\mu) & \xleftarrow{\quad U \quad} & L^2(\Omega,\mathcal{A},P) \simeq \mathbb{R}^n \\ I \big\downarrow \quad \big\uparrow V & & W \big\downarrow \quad \big\uparrow I \\ L^2_H(T,\mathcal{E},\mu) & \xrightarrow[\quad U^* \quad]{} & L^2(\Omega,\mathcal{A},P) \end{array} \qquad \begin{array}{l} V = U \circ U^* \\ \\ W = U^* \circ U \end{array}$$

U and its adjoint U^* being defined (as in b)) by :

$$f = \sum_{i=1}^n a_i 1_i \longrightarrow g = Uf \ / \ g(t) = \frac{1}{n}\sum_{i=1}^n a_i X(\omega_i,t)$$

$$g\in L^2_H(T,\mathcal{E},\mu) \longrightarrow U^*g = \sum_{i=1}^n \langle X(\omega_i,.) \ , g\rangle_T \, 1_i$$

It then follows that :

$$g\in L^2_H(T,\mathcal{E},\mu) \longrightarrow Vg = \frac{1}{n}\sum_{i=1}^n \langle X(\omega_i,.),g\rangle_T \ X(\omega_i,.) = \frac{1}{n}\sum_{i=1}^n X(\omega_i,.)\otimes X(\omega_i,.)$$

If $H = \mathbb{R}$ and $T = \{1,\ldots,p\}$, with $\mathcal{E} = \mathcal{P}(T)$ and : $\forall j \in T$ $\mu[\{j\}] = 1$, we obtain the classical P.C.A. of p real statistical variables $(X_1,\ldots,X_p)$. $L^2_H(T, \mathcal{E}, \mu)$ can be identified with $\mathbb{R}^p$, and U^* is made explicit by :

$$u = (u_1,\ldots,u_p) \in \mathbb{R}^p \longrightarrow U^*u = \sum_{i=1}^{p} u_i X_i$$

1.2. CANONICAL ANALYSIS OF TWO CLOSED SUBSPACES OF A HILBERT SPACE

1.2.1. GENERAL DEFINITION

Let H be a separable real Hilbert space, and let H_i (i=1,2) be a closed subspace .

Theorem : *H is isomorphic to a Hilbertian integral* $\int^{\otimes}_{[0,1]} H(\lambda)\, d\mu(\lambda)$ *so that*

1. $H_i \simeq \int^{\otimes}_{[0,1]} H_i(\lambda) d\mu(\lambda)$ *for fields* $H_i(.)$ *of* $H(.)$ $(i=1,2)$.
2. *For* $i = 1,2$, *and* $\lambda \in [0,1]$, *if* $P_i(\lambda)$ *denotes the hermitian projection of* $H(\lambda)$ *onto* $H_i(\lambda)$, *the operator* $\frac{1}{\sqrt{\lambda}} P_1(\lambda)$ *is, for any* $\lambda > 0$, *an isomorphism of* $H_2(\lambda)$ *onto* $H_1(\lambda)$ *and its inverse operator is* $\frac{1}{\sqrt{\lambda}} P_2(\lambda)$.

If P_i (i = 1,2) is the hermitian projection of H onto H_i , and V_i the restriction of the linear operator $P_{3-i} \circ P_i$ to H_i , it can be shown that the spectra $\sigma(.)$ of V_1 and V_2 may possibly differ by zero only. We denote by Λ the set $\sigma(V_1) \cup \{0\}$ $= \sigma(V_2) \cup \{0\}$, and ν the restriction of μ on Λ , then :

$$H_i \simeq \int^{\otimes}_{\Lambda} H_i(\lambda) d\nu(\lambda)$$

Definition 5 : *We call canonical analysis (C.A.) of* H_1 *and* H_2 , *the quadruple* $(\Lambda, \nu, \{H_1(\lambda)\}_{\lambda\in\Lambda}, \{H_2(\lambda)\}_{\lambda\in\Lambda})$.

1.2.2. COUNTABLE CANONICAL ANALYSIS

If the spectrum of V_1 or V_2 is countable, the C.A. of H_1 and H_2 is said to be countable. In that case, the measure μ is the counting measure and the C.A. gives a privileged decomposition (and not uniquely an isometric representation) of H_1 and H_2. In that case , the C.A. of H_1 and H_2 may be considered as a triplet

$$(\{\rho_i\}_{i\in I}, \{f_i\}_{i\in I} \cup B_1^o, \{g_i\}_{i\in I} \cup B_2^o)$$

where B_i^o is a basis of the kernel of V_i , and

(i) $\{f_i\}_{i\in I}$ (resp. $\{g_i\}_{i\in I}$) is an orthonormal basis of $H_1 \ominus \mathrm{Ker}\, V_1$ (resp. $H_2 \ominus \mathrm{Ker}\, V_2$),

(ii) for any i of I, $P_2 f_i = \rho_i g_i$ and $P_1 g_i = \rho_i f_i$,

(iii) for any (i,j) of I^2, $\langle f_i, g_j\rangle = \rho_i\, \delta_{ij}$

An important particular case of countable C.A. is the "compact C.A." : that is when one of the operators V_i is compact. It can be proved that, if the C.A. is compact, H_1+H_2 is a closed subspace of H.

Proposition 2 : *The compact C.A. of* H_1 *and* H_2 *is given by spectral analysis of one of the following compact operators* :
$V_1 = P_1 \circ P_{2|H_1}$, $V_2 = P_2 \circ P_{1|H_2}$, $P_1 \circ P_2$, $(P_1 + P_2 - I)_{|H_1+H_2}$
(where I is the identity operator).

The compact C.A. of H_1 and H_2 is said to be a Hilbert-Schmidt C.A., when $(P_1+P_2-I)_{|H_1+H_2}$ is a Hilbert-Schmidt operator . Consequently V_i is a nuclear operator. This type of C.A. is frequently used.

Proposition 3 : *If, for j=1,2, the sequence $\{H_j^n\}_{n\in\mathbb{N}}$ of closed subspaces of H converges uniformly to H_j, and if, for any n, H_1^n and H_2^n have a compact C.A., then H_1 and H_2 have a compact C.A.*

1.2.3. APPLICATIONS :

Let $(\Omega,\mathcal{A},P)$ be a probability space . The Hilbert space (for the usual inner product) $L^2 = L^2(\Omega,\mathcal{A},P)$ is assumed separable.

1.2.3.1. LINEAR CANONICAL ANALYSIS OF TWO RANDOM VARIABLES :

For i=1,2, let $X_i = (X_i^j)_{j\in\{1,2\ldots p_i\}}$ be a random variable such that, for any j, X_i^j belongs to L^2 . Let F_i be the subspace spanned by 1_Ω and $\{X_i^j\}_{j\in\{1,2\ldots p_i\}}$. The dimensions of these subspaces are finite . F_i is a closed subspace of L^2 and the hermitian projection of L^2 onto F_i is compact. Then we call linear canonical analysis (L.C.A.) of the r.v. X_1 and X_2 the C.A. (necessarily compact) of the two subspaces F_1 and F_2. L.C.A. includes as a particular case the C.A. defined by Hotelling and, when considering a finite Ω , the standard C.A. used in data analysis.

1.2.3.2. NON LINEAR CANONICAL ANALYSIS OF TWO RANDOM VARIABLES :

Let X_i (i=1,2) be a r.v. defined on $(\Omega,\mathcal{A},P)$ with values in a measurable space $(\Omega_i,\mathcal{A}_i)$. Let $\mathcal{B}_i$ be the complete sub σ-field generated by X_i (i.e. the sub σ-field of $\mathcal{A}$ generated by $X^{-1}(\mathcal{A}_i)$ and the P-negligible sets of Ω) , and $L^2(\mathcal{B}_i)$ be the closed subspace of L^2 of the $\mathcal{B}_i$-measurable r.v.

Definition 6 : *We call non linear canonical analysis (N.L.C.A.) of the r.v. X_1 and X_2 (or of (X_1,X_2)), the C.A. of the subspace $L^2(\mathcal{B}_1)$ and $L^2(\mathcal{B}_2)$ of L^2.*

The σ-fields $\mathcal{B}_1$ and $\mathcal{B}_2$ playing a fundamental part, we will also call it C.A. of the σ-fields $\mathcal{B}_1$ and $\mathcal{B}_2$; this being motivated by the fact that an iterative method like Hotelling's leads in a quite natural way to the notion of generated σ- field.

Let $E^{\mathcal{B}_i}$ (i = 1,2) be the "conditional expectation given $\mathcal{B}_i$" operator . By 1.2.2., the N.L.C.A. (if it is compact) of X_1 and X_2 is obtained , for example, by the spectral analysis of the linear operator $E^{\mathcal{B}_1} \circ E^{\mathcal{B}_2}{/}_{L^2(\mathcal{B}_1)}$.

Let P_{X_i} (i = 1,2) (respectively $P_{(X_1,X_2)}$) be the probability induced by X_i (resp. (X_1,X_2)) on $(\Omega_1 \times \Omega_2 , \mathcal{A}_1 \otimes \mathcal{A}_2)$.

Proposition 4 : *(X_1,X_2) have a Hilbert-Schmidt C.A. if and only if $P_{(X_1,X_2)}$ is absolutely continuous with respect to $P_{X_1} \otimes P_{X_2}$, and if the Radon-Nikodym derivative $f = dP(X_1,X_2)/d(P_{X_1} \otimes P_{X_2})$ belongs to $L^2(\Omega_1 \times \Omega_2, \mathcal{A}_1 \otimes \mathcal{A}_2 , P_{X_1} \otimes P_{X_2})$.*

1.2.3.3. CORRESPONDENCE ANALYSIS

Correspondence analysis may be presented as a particular case of C.A. of two r.v. X_1 and X_2 , where the σ-field $\mathcal{B}_\ell$ (ℓ =1,2) generated by X_ℓ is finite. Thus, the σ-field $\mathcal{B}_1$ (resp. $\mathcal{B}_2$) is generated by a partition $\{B_1^i\}_{i=1\dots p_1}$ (respectively $\{B_2^j\}_{j=1\dots p_2}$) of the set Ω, and the space $L^2(\mathcal{B}_1)$ (resp. $L^2(\mathcal{B}_2)$) is also the subspace F_1 (resp. F_2) spanned by the family of indicators $\{1_{B_1^i}\}_{i=1\dots p_1}$ (respectively $\{1_{B_2^j}\}_{j=1\dots p_2}$).

Without loss of generality, it may be assumed that, for any i and j, B_1^i and B_2^j are not P-negligible (see [2.2.]).

Let us put $p_{ij}=P(B_1^i \cap B_2^j)$, $p_{i.} = \sum_{j=1}^{p_2} p_{ij}$ and $p_{.j} = \sum_{i=1}^{p_1} p_{ij}$. For any point f (resp. g) of $L^2(\mathcal{B}_1)$ (resp. $L^2(\mathcal{B}_2)$), let f(i) (resp. g(j)) be the value taken by f (resp. g) on B_1^i (resp. B_2^j). If $(\Omega_\ell, \mathcal{P}(\Omega_\ell))$ is the measurable space where Ω_ℓ is the finite set of the values of X_ℓ and $\mathcal{P}(\Omega_\ell)$ the σ-field of the whole subsets of Ω_ℓ, we denote by π the probability induced by (X_1,X_2) on the product space $(\Omega_1\times\Omega_2, \mathcal{P}(\Omega_1) \otimes \mathcal{P}(\Omega_2))$ and we call π_1 and π_2 the marginal probabilities of π. It is easily seen that proposition (4) applies. In particular, $d\pi/d(\pi_1 \otimes \pi_2)$ takes the value $p_{ij}/p_{i.}p_{.j}$ at the point of $\Omega_1 \times \Omega_2$ which is the image of $B_1^i \cap B_2^j$ by (X_1,X_2). The C.A. of (X_1,X_2) - in other words the correspondence analysis of the contingency table of the p_{ij} - is obtained by the spectral analysis of the operator $V_1 = E^{\mathcal{B}_1} \circ E^{\mathcal{B}_2}{}_{|L^2(\mathcal{B}_1)}$. The matrix of V_1 relative to the bases $\{1_{B_1^i}\}$ and $\{1_{B_2^j}\}$ of $L^2(\mathcal{B}_1)$ and $L^2(\mathcal{B}_2)$ respectively is : $D_1^{-1}\,{}^tT\,D_2^{-1}\,T$, where $T = (p_{ij})_{i=1\dots p_1;j=1\dots p_2}$, $D_1 = \mathrm{diag}(p_{i.})_{i=1\dots p_1}$ and $D_2 = \mathrm{diag}(p_{.j})_{j=1\dots p_2}$.

For further details, the reader is referred to [2.2.].

§ 2 . CONVERGENCE WHEN SAMPLING AND DISCRETISATING

2.1. CONVERGENCES OF FACTOR ANALYSES

2.1.1. CONVERGENCE OF SELF ADJOINT OPERATORS AND SPECTRAL ANALYSIS

Let E be a real separable Hilbert space, T and $\{T_n\}_{n\in\mathbb{N}}$ a family of bounded self adjoint operators on E to E. We denote by the same symbol $\|.\|$ the norm of E and that of the Banach space $\mathcal{L}(E)$. π and π_n are the spectral functions of T and T_n respectively. σ is the spectrum of T.

2.1.1.1. UNIFORM CONVERGENCE

We assume that T_n converges uniformly to T, i.e. :

$$\lim_{n\to\infty} \|T_n - T\| = 0$$

We know that, for any interval Δ of $\mathbb{R}$ whose bounds do not belong to σ, the projection $\pi_n(\Delta)$ converges uniformly to $\pi(\Delta)$ (see [13]). If λ is an isolated eigenvalue of T, with finite order of multiplicity m, it exists a sequence $\{\lambda_n\}_{n\in\mathbb{N}}$ of eigenvalues of T_n which converges to λ. The eigenspace of dimension m associated with λ is the uniform limit of some eigenspaces of the T_n, the sum of the dimensions of these spaces being m for n sufficiently large.

In the sequel we shall be particularly concerned with cases where, for any n, T_n is a compact operator. T, as uniform limit of T_n, is also compact. For any n of $\mathbb{N}$, the spectrum of T_n is finite or countable, it admits zero as a unique possible accumulation point and its eigenvalues have a finite order of multiplicity. These results can be completed (see [4]), as follows.

LEMMA 1 : *Let T_n, T be compact operators, and let $T_n \to T$ in the uniform operator topology. Let $\lambda_m(T)$ be an enumeration of the non-zero eigenvalues of T, each repeated according to its multiplicity. Then there exist enumeration $\lambda_m(T_n)$ of the non-zero eigenvalues of T_n, with repetitions according to multiplicity, such that :*

$$\lim_{n\to\infty} \lambda_m(T_n) = \lambda_m(T) \qquad m \geq 1$$

the limit being uniform in m.

2.1.1.2. STRONG CONVERGENCE

We assume that T_n converges in the strong operator topology to T, i.e.

$$\forall f \in E \quad \lim_{n\to\infty} ||T_n f - Tf|| = 0$$

Then we know that any open interval containing a point of the spectrum of T contains a point of the spectrum of T_n, for n sufficiently large (see [8]).

If T and T_n, for any n of $\mathbb{N}$, are compact and if λ is a non-zero eigenvalue of T, it exists therefore a sequence $\{\lambda_n\}_{n\in\mathbb{N}}$ converging to λ such that, for any n, λ_n is an eigenvalue of T_n. The sum of the orthogonal projection operators on the eigenspaces associated with the values λ_n (which exist for n sufficiently large) converges in the strong operator topology to the orthogonal projection operator on the eigenspace of T associated with λ.

2.1.2. CONVERGENCES OF FACTOR ANALYSES : DEFINITIONS

All the factor analyses we study here can be obtained by spectral analysis of a self adjoint operator defined on a real separable Hilbert space E. These operators being bounded, they belong to the Banach space $\mathcal{L}(E)$.

Let $\mathcal{F}$ (resp. $\mathcal{F}_n$; $n \in \mathbb{N}$) be a factor analysis, obtained by the spectral analysis of a self adjoint operator T (resp. T_n) of $\mathcal{L}(E)$. We assume that, for each type of analysis, T and T_n are similar operators.

Definition 7 : *The sequence $\{\mathcal{F}_n\}_{n\in\mathbb{N}}$ converges uniformly (resp. strongly) to $\mathcal{F}$ if a sequence $\{T_n\}_{n\in\mathbb{N}}$ converges, in the uniform topology (resp. strong topology) of $\mathcal{L}(E)$, to T.*

2.2. CONVERGENCE OF LINEAR PRINCIPAL COMPONENT ANALYSES

We remain in the general case of the P.C.A. of a random function, which admits as particular case the more classical P.C.A. of random variables or statistical variables.

The hypotheses will be the same as in § 1.1.2.2. b).

2.2.1. STATISTICAL SAMPLING

X can be considered as a r.v. defined on $(\Omega, \mathcal{A}, P)$ and valued in $(E = L^2_H(T, \mathcal{E}, \mu), \mathcal{B}_E)$, where $\mathcal{B}_E$ is the borel σ-field of E. Let $\{X_i\}_{i=1\ldots n}$ be a sequence of n independent r.v., each being distributed as X. It is interesting to note that the sequence $\{X_i(\omega)\}_{i=1\ldots n}$ of the "trajectories" obtained in E for any ω of Ω is used, in common

utilisation of P.C.A., in a very special way : the P.C.A. resulting from the sample is the one obtained when Ω is a finite set $\{\omega_1 \dots \omega_n\}$ and P such that : $P[\{\omega_i\}] = \frac{1}{n}$
This P.C.A. is given by the spectral analysis of an operator $V_n(\omega)$ on $\mathcal{L}(E)$ while the linear P.C.A. of X is given by that of V.
Although we might possibly look for a "best" estimator of V, we shall focus on the currently used operator V_n to study its converging properties.
The case where Ω is finite has been studied in 1.1.2.2. c).
Since $L^2(\Omega, \mathcal{A}, P)$ ($\simeq \mathbb{R}^n$) is varying with n, W cannot be used in a study of convergence, we shall therefore work with V.

2.2.1.1. SAMPLING AND RANDOM OPERATORS . STUDY OF THE CONVERGENCE

For any ω of Ω, we consider the selfadjoint operator defined on E from the sample $\{X_i\}_{i=1\dots n}$, according to § 1.2.2.2. c), by :

$$V_n(\omega) = \frac{1}{n} \sum_{i=1}^{n} X_i(\omega) \otimes X_i(\omega)$$

V_n is a mapping on $(\Omega, \mathcal{A}, P)$ into $\sigma_2 = \sigma_2(E,E)$, the Hilbert space of the Hilbert-Schmidt operators on E into itself . Let $\mathcal{B}_{\sigma_2}$ be the Borel σ-field of σ_2. We can easily prove the following result :

Lemma 2 : *If E and E' are two real Hilbert spaces, the mapping ψ : $(x,y) \longrightarrow y \otimes x$ on the hilbertian sum $E \oplus E'$ into $\sigma_2(E',E)$ is continuous.*

Since the mapping on E into $E \oplus E$: $x \longrightarrow (x,x)$ is continuous, the continuity of the mapping on E into σ_2 : $x \to x \otimes x$ follows.

Proposition 5 : *V_n is an integrable r.v. on $(\Omega, \mathcal{A}, P)$ into $(\sigma_2, \mathcal{B}_{\sigma_2})$.*

From the previous lemma, it follows that, for any i of $\{1,\dots,n\}$, $X_i \otimes X_i$ is measurable. Therefore V_n is a r.v. on $(\Omega, \mathcal{A}, P)$ into $(\sigma_2, \mathcal{B}_{\sigma_2})$. The space σ_2 is here separable (see [6]).

The operator V, from which we obtain the P.C.A. of X, is defined by :

$$g \in E \to h = Vg \; / \; h(t) = \int_\Omega \langle X(\omega,.),g\rangle_T \; X(\omega,t) \; dP(\omega)$$

Let $E(X \otimes X)$ be this operator of σ_2 , hence defined by :

$$\forall g \in E \qquad E(X \otimes X)(g) = \int_\Omega (X(\omega) \otimes X(\omega))(g) \; dP(\omega)$$

$E(X \otimes X)$ is proved to be the mean of the r.v. $X \otimes X$ on $(\Omega, \mathcal{A}, P)$ into $(\sigma_2, \mathcal{B}_{\sigma_2})$ (see [10]). Similarly the $X_i \otimes X_i$ are integrable, accordingly V_n is.

Proposition 6 : *The sequence $\{V_n\}_{n \in \mathbb{N}^*}$ converges almost surely to V in σ_2.*

The r.v. $\{X_i\}_{i=1\dots n}$ are independent, and so are $\{X_i \otimes X_i\}_{i=1\dots n}$. Since σ_2 is separable and the $X_i \otimes X_i$ such that : $\forall i=1\dots n \quad E(X_i \otimes X_i) = E(X \otimes X) = V$, the proposition follows immediately from the strong law with strong convergence for r.v. on a real separable Banach space (see [7] here used in the case of a Hilbert space).

Corollary 1 : *The sequence $\{V_n\}_{n \in \mathbb{N}^*}$ converges a.s. to V in $\mathcal{L}(E)$ (for the uniform topology of operators).*

According to the definition given in § 2.1.2., we have also

Corollary 2 : *The linear principal component analysis obtained through the sample $\{X_i\}_{i=1\dots n}$ converges uniformly, a.s., when n increases, to the linear principal component analysis of $(X_t)_{t \in T}$.*

Note 1 . A priori other modes of convergence might have been considered. However, either convergence in quadratic mean or stochastic convergence, which are standard in most estimation problems, seem to be less satisfying : considering independant observations - therefore for a given ω in the preceding model - we look for $V_n(\omega)$ and, more precisely, for its spectral decomposition.
Our main concern is then to see whether the $V_n(\omega)$ converge to V a.e. in $\mathcal{L}(E)$ or not. This is exactly a problem of a.s. convergence.
Convergence in the mean, order α, in σ_2 may also be of some interest.
Since $\| X \otimes X \|_2 = \|X\|^2$, the hypothesis of theorem 6.4.3. in [7] are satisfied.
Therefore convergence in mean, order 1, of $\{V_n\}_{n\in\mathbb{N}^*}$ to V is proved in σ_2 and thus in $\mathcal{L}(E)$.
Moreover if $\|X\|$ admits a 4th moment, $E[\| X \otimes X \|_2^2] = E[\| X \|^4] < +\infty$ and quadratic mean convergence of $\{V_n\}_{n\in\mathbb{N}}$ to V is also proved.
This last result was pointed out in [3] (but a.s. convergence was not considered).

Note 2 . Since the operators V, $X(\omega) \otimes X(\omega)$ and $V_n(\omega)$, for any ω of Ω, are nuclear, $X \otimes X$ and V_n are also r.v. defined on $(\Omega, \mathcal{A}, P)$ and valued in the Banach space $\sigma_1 = \sigma_1(E,E)$ of the nuclear operators on E into itself. Since E is separable, σ_1 also is (see [6]).
It can also be proved that $X \otimes X$ and V_n are integrable in σ_1 and that $E(V_n)=E(X\otimes X)$ for any n of $\mathbb{N}$. The strong law with strong convergence in the separable Banach space σ_1 then gives the following proposition :

the sequence $\{V_n\}_{n\in\mathbb{N}}$ converges a.s. to V in σ_1.

This result is stronger than proposition 6.

2.2.2. SAMPLING ON T

Let us now assume that are given n values of X at independent times $t_1 \dots t_n$. Is the linear P.C.A. of the r.v. $X_{t_1} \dots X_{t_n}$, for n sufficiently large, a "good approximation" of that of $(X_t)_{t\in T}$? This problem is in some sense symmetric to the preceding one and can be analogously solved by exchange of probability spaces $(\Omega, \mathcal{A}, P)$ and $(T, \mathcal{E}, \mu)$ in the duality scheme.

2.2.2.1. STUDY OF THE CONVERGENCE

X is regarded here as a r.v. defined on $(T, \mathcal{E}, \mu)$ and valued into $L^2_H(\Omega, \mathcal{A}, P) = E'$, a real separable Hilbert space with its Borel σ-field $\mathcal{B}_{E'}$. Let $\{X^i\}_{i=1\dots n}$ be a sequence of independent r.v. defined on $(T, \mathcal{E}, \mu)$, valued into $(E', \mathcal{B}_{E'})$ and distributed like X . X has an integrable squared norm, but is not necessarily centered. The linear non centered P.C.A. of $X = \{X_\omega\}_{\omega\in\Omega}$ leads to the following duality scheme :

$$\begin{array}{ccc} L^2_H(\Omega, \mathcal{A}, P) & \xleftarrow{\tilde{U}} & L^2(T, \mathcal{E}, \mu) \\ I \downarrow \quad \uparrow \tilde{V} & & \tilde{W} \downarrow \quad \uparrow I \\ L^2_H(\Omega, \mathcal{A}, P) & \xrightarrow{\tilde{U}^*} & L^2(T, \mathcal{E}, \mu) \end{array}$$

where : $\tilde{V} = E_\mu[X \otimes X]$ (E_μ is the integral with respect to μ).
A study of the case where T is finite shows that the P.C.A. obtained from $\{X^i\}_{i=1\dots n}$ is given by spectral analysis of $\tilde{V}_n = \frac{1}{n} \sum_{i=1}^{n} X^i \otimes X^i$. Therefore all the results of § 2.2.1.2. hold.

Proposition 7 : *The linear principal component analysis obtained from the sample $\{X^i\}_{i=1\dots n}$ converges uniformly a.s., when n increases, to the principal component analysis of X.*

Analogously convergence in the mean, order 1, and convergence in quadratic mean (if $\|X\|_{E'}$ admits a 4th moment with respect to μ) of $\{\tilde{V}_n\}_{n\in\mathbb{N}^*}$ to $\tilde{V}$ may also be proved.

2.2.2.2. CASE OF A REAL RANDOM FUNCTION

If the Hilbert space H is $\mathbb{R}$, the duality scheme is symmetric, and we see that the operator $\tilde{U}$ is the operator U^* of § 2.2.1. . Hence :

$$\tilde{U}^* = U \quad , \quad \tilde{V} = W \; , \; \tilde{W} = V \; , \; \tilde{V}_n = W_n \; .$$

The previous study of convergence provides an answer to the problem of stability of P.C.A. when sampling in a population of real r.v. . This problem, originally raised by Hotelling, was studied in some aspects in [5].

2.2.3. SPEED OF CONVERGENCE

We can briefly give some indications concerning the speed of convergence. If $\|X\|$ admits a 4th moment, we can apply to the sequence $\{X_n \otimes X_n\}_{n\in\mathbb{N}^*}$ of r.v. into the real separable Hilbert space σ_2 the central limit theorem (see [7]). If we denote by

$$Y_n = \frac{1}{\sqrt{n}} \sum_{i=1}^{n} (X_i \otimes X_i - V) = \sqrt{n}\,(V_n - V) \; ,$$

the distribution of Y_n converges weakly to a normal distribution in σ_2 . The speed of convergence of V_n to V is therefore in the order of $1/\sqrt{n}$.

2.3. NON LINEAR CANONICAL ANALYSIS . DISCRETISATION AND SAMPLING

We will restrict our study to the N.L.C.A. of two real r.v. X and Y.

2.3.1. DISCRETISATION

Let $P_{(X,Y)}$ be the probability induced on $(\mathbb{R}^2, \mathcal{B}_{\mathbb{R}^2})$ by (X,Y), and let $\mathcal{B}$ (resp. $\mathcal{C}$) be the complete sub σ-field of $\mathcal{A}$ generated by X (resp. Y). The N.L.C.A. of (X,Y), in other words the C.A. of the σ-fields $\mathcal{B}$ and $\mathcal{C}$, is given by the spectral analysis of $V = E^{\mathcal{B}} \circ E^{\mathcal{C}}{}_{|L^2(\mathcal{B})}$ for instance. The explicit description of this operator V is often difficult, and moreover often does not allow a direct obtainment of the spectral analysis of V - although it may be obtained in some cases (see [2.1.]). Even the simplifying assumption that the studied C.A. is Hilbert-Schmidt, which implies that V is an integral operator, does not enable to explicit the spectral analysis of V. Furthermore direct processing by computer is difficult. Nevertheless if the dimensions of $L^2(\mathcal{B})$ and $L^2(\mathcal{C})$ are finite and not too large , the problem can be solved by matricial calculus and thus processed by computer. It is especially the case when the σ-fields $X^{-1}(\mathcal{B}_{\mathbb{R}})$ and $Y^{-1}(\mathcal{B}_{\mathbb{R}})$ are composed of a finite number of unnegligible events, the C.A. is then a correspondence analysis.
Such an approximation of the N.L.C.A. may be obtained by the following procedure, termed "discretisation" :

Let $A = \{A_i\}_{i\in I=\{1,2,\dots,k\}}$ and $B = \{B_j\}_{j\in J=\{1,2,\dots,\ell\}}$ be two partitions of $\mathbb{R}$ composed of borelian sets. For every (i,j) of I × J, we denote by p_{ij} the probability $P_{(X,Y)}(A_i \times B_j)$. The N.L.C.A. will be "approximated" by correspondence analysis of the contingency table of the p_{ij}. $\tau(A)$ and $\tau(B)$ are substituted for $\mathcal{B}$ and $\mathcal{C}$. If the sequence of the C.A. obtained for finer and finer partitions converges (as defined in 2.1. for instance) then it will be correct to speak of an approximation.

2.3.2. CONVERGENCY BY DISCRETISATION

2.3.2.1. SOME USEFUL LEMMAS

The first lemma is suggested by classical results of perturbation theory. The proofs are omitted [see [2.2.].]

Lemma 3 : *Let $\{\pi_m\}_{m\in\mathbb{N}}$ be a sequence of hermitian projections on a separable real Hilbert space H, which converges strongly to an hermitian projection π. If $T : H \to H$ is a compact linear operator, then the sequence $\{\pi_m \circ T\}_{m\in\mathbb{N}}$ converges uniformly to $\pi \circ T$.*

Lemma 4 : *Moreover if $\{T_m\}$ is an uniformly convergent sequence of compact operators belonging to $\mathcal{L}(H)$, $\{\pi_m \circ T_m\}_{m\in\mathbb{N}}$ and $\{T_m \circ \pi_m\}_{m\in\mathbb{N}}$ converges uniformly to $\pi \circ T$ and $T \circ \pi$ respectively.*

Corollary : *Let $\{\pi_m\}_{m\in\mathbb{N}}$ and $\{\pi'_m\}_{m\in\mathbb{N}}$ be two sequences of hermitian projections on H, which converge strongly to the hermitian projections π and π' respectively. Let $\{T_m\}_{m\in\mathbb{N}}$ be a sequence of compact operators which converges to T in the Banach space $\mathcal{L}(H)$. Then, $\{\pi_m \circ T_m \circ \pi'_m\}_{m\in\mathbb{N}}$ converges uniformly to $\pi \circ T \circ \pi'$.*

2.3.2.2. CORRESPONDENCE ANALYSIS CONVERGENCY

Let $\{\sigma_m\}_{m\in\mathbb{N}}$ and $\{\Sigma_m\}_{m\in\mathbb{N}}$ be two increasing sequences of partitions of $\mathbb{R}$ composed by elements of the borel σ-field $\mathcal{B}_{\mathbb{R}}$ (this means that for every integer m, σ_{m+1} (resp. Σ_{m+1}) is a subpartition of σ_m (resp. Σ_m)). For any m of $\mathbb{N}$, we denote by $\mathcal{D}_m$ and $\mathcal{E}_m$ the finite σ-fields of $\mathcal{B}_{\mathbb{R}}$ generated by σ_m and Σ_m respectively. Let $\mathcal{B}_m$ (resp. $\mathcal{C}_m$) be the complete sub σ-field of $\mathcal{A}$ generated by $X^{-1}(\mathcal{D}_m)$ (resp. $Y^{-1}(\mathcal{E}_m)$) and the P-negligible events of $\mathcal{A}$. We assume that the sequence of partitions is such that the σ-field generated by $\bigcup_m \mathcal{D}_m$ or $\bigcup_m \mathcal{E}_m$ is $\mathcal{B}_{\mathbb{R}}$. Such partitions can easily be built ([2.2.])

Lemma 5 : *$\{\mathcal{B}_m\}_{m\in\mathbb{N}}$ (resp. $\{\mathcal{C}_m\}_{m\in\mathbb{N}}$) is strongly convergent to the complete sub σ-field of $\mathcal{A}$ generated by X (resp. Y).*

Let $\mathcal{B}$ (resp. $\mathcal{C}$) be the complete sub-σ-field of $\mathcal{A}$ generated by X (resp. Y), and let $\mathcal{B}'$ (resp. $\mathcal{C}'$) be the σ-field generated by $\bigcup_{m\in\mathbb{N}} \mathcal{B}_m$ (resp. $\bigcup_{m\in\mathbb{N}} \mathcal{C}_m$). It is easily seen that : $\mathcal{B} = \mathcal{B}'$ and $\mathcal{C} = \mathcal{C}'$. And the result then evolves from a classical proposition (see [1] or [10]).

Proposition 8 : *The sequence of canonical analyses $\mathcal{F}_m$ of the σ-fields $\mathcal{B}_m$ and $\mathcal{C}_m$ is uniformly convergent to the N.L.C.A. of (X,Y) when it is compact.*

In some way, this result is the best we can hope for : $\mathcal{F}_m$ is a compact C.A., for any m . Thus, if $\{\mathcal{F}_m\}$ is supposed to be uniformly convergent to a C.A. $\mathcal{F}$ (it is here the best convergency concept), according to § 1.2. prop. 3, $\mathcal{F}$ is necessarily a compact C.A. .
According to § 1.2. prop. 2, $E^{\mathcal{B}} \circ E^{\mathcal{C}}$ and $E^{\mathcal{C}} \circ E^{\mathcal{B}}$ are compact operators. From lemmas 3 and 4, it follows that $E^{\mathcal{B}_m} \circ E^{\mathcal{C}_m} = E^{\mathcal{B}_m} \circ E^{\mathcal{B}} \circ E^{\mathcal{C}} \circ E^{\mathcal{C}_m}$ converges uniformly to $E^{\mathcal{B}} \circ E^{\mathcal{C}}$. Using again lemma 3 for $T_m = E^{\mathcal{B}_m} \circ E^{\mathcal{C}_m}$ and $\pi_m = E^{\mathcal{B}_m}$, we prove the uniform convergency to $E^{\mathcal{B}} \circ E^{\mathcal{C}} \circ E^{\mathcal{B}}$ of the sequence of the compact self-adjoint operators $E^{\mathcal{B}_m} \circ E^{\mathcal{C}_m} \circ E^{\mathcal{B}_m}$. Since the spectrum of $E^{\mathcal{B}_m} \circ E^{\mathcal{C}_m} \circ E^{\mathcal{B}_m}$ (resp. $E^{\mathcal{B}} \circ E^{\mathcal{C}} \circ E^{\mathcal{B}}$)

is also the spectrum (possibly without 0) of $E^{\mathcal{B}_m} \circ E^{\mathcal{C}_m}{}_{|L^2(\mathcal{B}_m)}$ (resp. $E^{\mathcal{B}} \circ E^{\mathcal{C}}{}_{|L^2(\mathcal{B})}$), the C.A. of $\mathcal{B}_m$ and $\mathcal{C}_m$ converge uniformly to the N.L.C.A. of (X,Y).
It follows that, if we use a partition of $\mathbb{R}$ composed of sufficiently small classes, we can find a correspondence analysis that will be a satisfying approximation of the investigated compact N.L.C.A. . Nevertheless this approximation remains mainly theoritical since it requires the precise knowledge of $P_{(X,Y)}$: only estimates of the p_{ij} (obtained through a sample of size n) are at hand.
Hence possible effects of sampling are to be investigated.

2.3.3. SAMPLING

We denote by Z the couple (X,Y) . Let $\{Z_p\}_{p \in M}$ $(M = \{1,2,\ldots,n\}$, $n \in \mathbb{N}^*)$ be a sequence of independent r.v. such that $P_Z = P_{Z_p}$ (for any p) (i.e. a sample). Let $\{A_i\}_{i \in I}$ and $\{B_j\}_{j \in J}$ be two partitions of $\mathbb{R}$ (composed of borel sets) which generate the σ-fields $\mathcal{D}$ and $\mathcal{E}$ respectively. For any ω of Ω , and any (i,j), we denote by :

$$p_{ij}^{(n)}(\omega) = \frac{1}{n} \sum_{p=1}^{n} 1_{[A_i \times B_j]}(Z_p(\omega)),$$

(So $p_{ij}^{(n)}(\omega)$ is the ratio of the values of $Z_p(\omega)$ on $A_i \times B_j$) and :

$$p_{i.}^{(n)}(\omega) = \sum_{j \in J} p_{ij}^{(n)}(\omega) \ , \ p_{.j}^{(n)}(\omega) = \sum_{i \in I} p_{ij}^{(n)}(\omega).$$

For any ω of Ω , let μ_n^ω be the discrete probability defined on $(\mathbb{R}^2, \mathcal{B}_{\mathbb{R}^2})$ by :

$$\forall p \in M \ , \ \mu_n^\omega[Z_p(\omega)] = \frac{1}{n} \operatorname{card}\{ t \in \{1,\ldots,n\} \ / \ Z_t(\omega) = Z_p(\omega)\}.$$

We know (see [12]) that, for P-almost every ω of Ω, the probability μ_n^ω converges weakly to the distribution P_Z of Z, and thus :

$$\forall (i,j) \in I \times J, \ p_{ij} = P_Z(A_i \times B_j) = \lim_{n \to \infty} \mu_n^\omega(A_i \times B_j) = \lim_{n\to\infty} p_{ij}^{(n)}(\omega) \quad \text{a.e. (1)}$$

It is no loss of generality to assume that, for every i of I and every j of J, $P_Z(A_i \times \mathbb{R})$ and $P_Z(\mathbb{R} \times B_j)$ are not zero. Then, for any $i \in I$ (resp. every $j \in J$), $p_{i.}$ (resp. $p_{.j}$) is not zero.
Since $p_{i.} = \lim_{n\to\infty} p_{i.}^{(n)}(\omega)$ a.e. (resp. $p_{.j} = \lim_{n\to\infty} p_{.j}^{(n)}(\omega)$ a.e.), $p_{i.}^{(n)}(\omega)$ (respectively $p_{.j}^{(n)}(\omega)$) is different from zero for n sufficiently large and for P-almost every ω of Ω.
The correspondence analysis given by the sample may be considered as the linear principal component analysis of the density of the restriction to $\mathcal{D} \otimes \mathcal{E}$ of μ_n^ω with respect to its marginal probabilities :

$$\frac{p_{ij}^{(n)}(\omega)}{p_{i.}^{(n)}(\omega) p_{.j}^{(n)}(\omega)}$$

$(i \in I, \ j \in J)$. It may be also considered as a linear C.A. of indicators, and then is obtained by the spectral analysis of a selfadjoint operator $R_n(\omega)$ (cf. [2.2.]) on $\mathbb{R}^J$, whose matrix is :

$$\left(\frac{1}{\sqrt{p_{.i}^{(n)}(\omega) \ p_{.j}^{(n)}(\omega)}} \sum_{\ell \in I} \frac{p_{\ell i}^{(n)}(\omega) p_{\ell j}^{(n)}(\omega)}{p_{\ell .}^{(n)}(\omega)} \right)_{(i,j) \in J^2}$$

From (1), $\{R_n(\omega)\}_{n \in \mathbb{N}}$ converges uniformly (for almost every ω) to the selfadjoint operator R which has for matrix : $\left(\frac{1}{\sqrt{p_{.i}\, p_{.j}}} \sum_{\ell \in I} \frac{p_{\ell i}\, p_{\ell j}}{p_{\ell .}} \right)_{(i,j) \in J^2}$. Its spectral analysis gives the correspondence analysis of $\mathcal{D}$ and $\mathcal{E}$ associated with P_Z.

Hence :

Proposition 9 : *The sequence of (correspondence) analyses of $\mathcal{D}$ and $\mathcal{E}$ given by sampling converges uniformly, a.e., to the (correspondence analysis) of $\mathcal{D}$ and $\mathcal{E}$ associated with P_Z .*

REFERENCES

[1] N.I. AKHIEZER and J.M. GLAZMAN, *Theory of linear operators in Hilbert space* t.1 et t.2 . Frederick Ungar Publishing Co, New-York, 1961.

[2] J. DAUXOIS and A. POUSSE

[2.1.] *Une extension de l'analyse canonique . Quelques applications.* Ann. de l'Institut Henri Poincaré, vol. XI, N° 4, p. 355-379, 1975

[2.2.] *Analyses factorielles en Calcul des Probabilités et en Statistique: essai d'étude synthétique.* Thèse de Doctorat . Université Paul Sabatier . Toulouse . 1976.

[3] J.C. DEVILLE, *Méthodes Statistiques et numériques de l'analyse harmonique.* Annales de l'INSEE, N° 15, Paris, 1974.

[4] N. DUNFORD and J.T. SCHWARTZ, *Linear operators,* tomes 1 et 2 Interscience publishers, New York, 1964.

[5] Y. ESCOUFIER, *Echantillonnage dans une population de variables aléatoires réelles.* Thèse de Doctorat . Publ. de l'ISUP, vol. XIX, fasc. 4, 1970.

[6] I.C. GOHBERG and M.G. KREIN, *Introduction à la théorie des opérateurs linéaires non autoadjoints dans un espace hilbertien.* Dunod . Paris . 1971.

[7] U. GRENANDER, *Probabilities on algebraic structures.* Almquist et Wiksell, Stockholm, 1963.

[8] T. KATO, *Perturbation theory for linear operators.* Springer Verlag, New-York, 1966.

[9] M.G. KENDALL and A. STUART, *The advanced theory of Statistics,* tomes 1 et 2, Griffin, London, 2° édition, 1967.

[10] J. NEVEU, *Martingales à temps discret.* Masson et Cie, Paris , 1972.

[11] J.P. PAGES, *A propos des opérateurs d'Y. Escoufier.* Note du C.E.A., Fontenay, 1975.

[12] K.R. PARTHASARATHY, *Probability measures on metric spaces.* Academic Press, New-York, 1967.

[13] F. RIESZ and B. NAGY, *Leçons d'analyse fonctionnelle.* Gauthier-Villars, Paris, 1965.

[14] H.H. SCHAEFER, *Banach lattices and positive operators.* Springer-Verlag, Berlin, 1974.

Recent Developments in Statistics
J.R.Barra et al., editors

TESTS ON THE DISPERSION PARAMETER IN VON MISES DISTRIBUTIONS

S. DEGERINE
Université Scientifique et Médicale de Grenoble
I.R.M.A., B.P. 53
38041 - Grenoble Cédex - France

The S-dimensional von Mises distribution is presented as the canonical exponential distribution with respect to the uniform distribution on the unit sphere of R^S. When the modal direction is known, tests on the dispersion parameter are given by tests in scalar exponential structure. Otherwise we use one-sided and two-sided tests for Polya type distributions jointly with the invariance principle.

NOTATION

If E is a topological space, $\mathcal{B}_E$ is the σ-field of its Borel sets. For any positive integer k, k' stands for (k-2)/2, 1_k for the identity matrix of R^k and λ_k for the Lebesgue measure on R^k. λ_+ denotes the Lebesgue measure on R^+.

Σ_s is the unit sphere of R^s :

$$\Sigma_s = \{x \in R^s \; ; \; ||x|| = 1\}.$$

If H is a linear subspace of R^s, $H^{\perp}$ is the orthogonal of H in R^s and x_H the orthogonal projection of x on H.

The indicator function $\mathbf{1}_A$ of a set A is defined by :

$$\mathbf{1}_A(x) = 1 \quad \text{if} \quad x \in A \text{ else } 0$$

A function f(x) is said to be radial if it depends only on $||x||$, and the corresponding function on R^+ is also denoted by f.

1 - UNIFORM DISTRIBUTION ON THE UNIT SPHERE OF R^S

Let $A_s = [0,\pi]^{s-2} \times [0,2\pi[$ and let T be the transformation from A_s into R^s defined by :

$$a \in A_s \longrightarrow x = T(a) \in R^s$$

with :

$$x_j = \cos a_j \prod_{i=1}^{j-1} \sin a_i, \; j = 1,\dots,s-1$$

$$x_s = \prod_{i=1}^{s-1} \sin a_i .$$

Let ν be the probability measure on $(A_s, \mathcal{B}_{A_s})$ with density :

$$\frac{d\nu}{da} = \frac{\Gamma(s'+1)}{2\Pi^{s'+1}} \prod_{i=1}^{s-2} (\sin a_i)^{s-i-1}$$

The uniform distribution U_s on Σ_s is the singular probability measure on $(R^s, \mathcal{B}_{R^s})$ defined by :

$$\forall B \in \mathcal{B}_{R^s},\ U_s(B) = \nu(T^{-1}(B)).$$

The Laplace transform L_{U_s} of U_s is given by :

$$\forall \xi \in C^s,\ L_{U_s}(\xi) = \int_{R^s} e^{-<\xi,x>}\, dU_s(x) = {}_0F_1(s'+1\ ;\ \frac{1}{4} <\xi,\bar{\xi}>),$$

where ${}_0F_1\ (\nu,\xi)$ is the generalized hypergeometrical series :

$${}_0F_1(\nu,\xi) = \Gamma(\nu) \sum_{n=0}^{\infty} \frac{\xi^n}{\Gamma(\nu+n)n!},\ (\nu,\xi) \in C^2.$$

The characteristic function $\Omega_s(t)$ of U_s is $L_{U_s}(-it)$, so we have :

$$\forall t \in R^s,\ \Omega_s(t) = 2^{s'}\ \Gamma(s'+1)\ \frac{J_{s'}(||t||)}{||t||^{s'}}$$

where $J_\nu(\xi)$ is the Bessel function of order ν :

$$J_\nu(\xi) = \sum_{n=0}^{\infty} \frac{(-1)^n (\xi/2)^{\nu+2n}}{n!\ \Gamma(\nu+n+1)},\ (\nu,\xi) \in C^2$$

We say that a probability distribution P on $(R^s, \mathcal{B}_{R^s})$ is spherical if and only if it is invariant under rotations. Its characteristic function φ is radial and an inversion formula given by Van Der Vaart ([9]) aids us to prove :

1.1 - Lemma : Let φ be the characteristic function of a spherical distribution P on $(R^s, \mathcal{B}_{R^s})$. If the following limit exists λ_s almost everywhere then

$$\lim_{T\to+\infty} \frac{1}{(2\Pi)^{s'+1}} \int_0^T \varphi(\rho)\ \frac{J_{s'}(\rho||x||)}{(\rho||x||)^{s'}}\ \rho^{s-1}\ d\rho$$

is the density of the absolutely continuous component of P with respect to λ_s.

Let $X_1, \ldots, X_n$ be n independent random vectors with the same distribution U_s on $(R^s, \mathcal{B}_{R^s})$ and let us denote their resultant by X.

The probability density functions of X and $||X||$ are given by Kluyver for s = 2 ([4]), by Lord Rayleigh for s = 3 ([6]) and by G.N. Watson for all s ([10]). Here lemma 1.1 yields the results.

1.2 - Proposition : For $n \geq 2$ the distribution Q_n of X is absolutely continuous with respect to λ_s with density :

$$q_n(x) = \frac{[2^{s'}\,\Gamma(s'+1)]^n}{(2\Pi)^{s'+1}} \int_0^\infty \left[\frac{J_{s'}(\rho)}{\rho^{s'}}\right]^n \frac{J_{s'}(\rho||x||)}{(\rho||x||)^{s'}} \rho^{s-1}\, d\rho$$

For $n \geq 1$ the distribution ${}_hQ_n$ of the orthogonal projection Y of X on a h-dimensional linear subspace of R^s is absolutely continuous with respect to λ_h with density :

$${}_hq_n(y) = \frac{[2^{s'}\,\Gamma(s'+1)]^n}{(2\Pi)^{h'+1}} \int_0^\infty \left[\frac{J_{s'}(\rho)}{\rho^{s'}}\right]^n \frac{J_{h'}(\rho||y||)}{(\rho||y||)^{h'}} \rho^{h-1}\, d\rho$$

The value of the Weber-Schafheitlin integrals ([10], ch. XIII) implies the absolute continuity of Q_2 and ${}_hQ_1$; the other cases follow by convolution.

The distribution $\tilde{Q}_n$ (resp. ${}_h\tilde{Q}_n$) of $R = ||X||$ (resp. $V = ||Y||$) is absolutely continuous with respect to λ_+ with density :

$$\tilde{q}_n(r) = \frac{2\Pi^{s'+1}}{\Gamma(s'+1)}\, r^{s-1}\, q_n(r) \quad \left(\text{resp. } {}_h\tilde{q}_n(v) = \frac{2\Pi^{h'+1}}{\Gamma(h'+1)}\, v^{h-1}\, {}_hq_n(v)\right)$$

2 - S-DIMENSIONAL VON MISES DISTRIBUTION

2.1 - Definition : A probability distribution P_θ on $(R^s, \mathcal{B}_{R^s})$ is an s-dimensional Von Mises distribution if it is absolutely continuous with respect to U_s with density :

$$\frac{dP_\theta}{dU_s} = \frac{e^{<\theta,x>}}{L_{U_s}(-\theta)}, \quad \theta \in R^s$$

The norm k of θ is a dispersion parameter and the unit vector $M = \frac{\theta}{k}$ is the modal direction.

The Laplace transform L_{U_s} is a radial function on R^s :

$$L_{U_s}(-\theta) = L_{U_s}(k) = 2^{s'}\,\Gamma(s'+1)\,\frac{I_{s'}(k)}{k^{s'}}$$

where $I_\nu(\xi)$ is the modified Bessel function of the first kind and the νth order :

$$I_\nu(\xi) = \sum_{n=0}^{\infty} \frac{(\xi/2)^{\nu+2n}}{n!\Gamma(\nu+n+1)}, \quad (\nu,\xi) \in C^2.$$

The s-dimensional Von Mises distribution is written s-D.M.D. (θ) or s-D.M.D. (k,M). Let X_0 be a random vector on $(R^s, \mathcal{B}_{R^s})$ with s-D.M.D. (k,M) distribution, then we have ([1], Ch. X) :

(i) The characteristic function of X_0 is :

$$\forall t \in R^s, \varphi(t) = \frac{k^{s'}\,{}_0F_1[s'+1\ ;\ 1/4(k^2+2ik \langle M,t\rangle - ||t||^2)]}{2^{s'}\Gamma(s'+1)I_{s'}(k)}$$

(ii) All the moments of X_0 exist and :

$$E(X_0) = \frac{I_{s'+1}(k)}{I_{s'}(k)}\ M$$

$$\Lambda_{X_0} = \frac{I_{s'+2}(k)}{I_{s'}(k)}\ M\,{}^tM + \frac{1}{k}\ \frac{I_{s'+1}(k)}{I_{s'}(k)}\ 1_s - E(X_0)\,{}^tE(X_0)$$

where Λ_{X_0} is the covariance matrix of X_0.

(iii) Let X (resp. R) be the resultant (resp. the norm of the resultant) of a sample $X_1, \ldots, X_n$ from s-D.M.D. (k,M). For $n \geq 2$, the distribution P_θ^{*n} of X (resp. $\tilde{P}_k^n$ of R) is absolutely continuous with respect to λ_s (resp. λ_+) with density :

$$f_\theta^{*n}(x) = \frac{e^{\langle\theta,x\rangle}}{[L_{U_s}(k)]^n}\ q_n(x) \quad \left(\text{resp. } \tilde{f}_k^n(r) = \frac{2^{s'}\Gamma(s'+1)}{[L_{U_s}(k)]^n}\ \frac{I_{s'}(kr)}{(kr)^{s'}}\ \tilde{q}_n(r)\right)$$

3 - TESTS ON THE DISPERSION PARAMETER

Stephens computed tests of $k = k_0$ against $k > k_0$, or $k = k_0$ against $k \neq k_0$ in R^2 ([7]) and R^3 ([8]) ; Mardia studied their properties ([5]). We present one-sided and two-sided tests in R^s.

The resultant of an n-sample from s.D.M.D. (k,M) is a P-minimal complete sufficient statistic that induces the following canonical exponential structure :

$$\left\{R^s, \mathcal{B}_{R^s}\ ;\ \frac{dP_{(k,M)}}{dQ_n} = \frac{e^{k\langle M,x\rangle}}{L_{Q_n}(k)}, \ k \in R^+, \ M \in \Sigma_s\right\}$$

with :

$$L_{Q_n}(k) = [L_{U_s}(k)]^n.$$

When the modal direction M is known, the statistic $Y(x) = \langle M,x \rangle$ is P-minimal sufficient and induces a scalar exponential structure :

$$\{R, \mathcal{B}_R \ ; \ \frac{dP_k}{d_1Q_n} = \frac{e^{ky}}{L_1Q_n(k)} \ , \ k \in R^+\}$$

where :

$$L_1Q_n(k) = L_{Q_n}(k)$$

One-sided and two-sided tests are then given by J.R. Barra ([1], Ch. XI).

Otherwise, we assume that the modal direction M is in an h-dimensional linear subspace of R^s, $1 \leq h \leq s$. The orthogonal projection of x on this subspace is a P-minimal complete sufficient statistic inducing :

$$\{R^h, \mathcal{B}_{R^h} \ ; \ \frac{dP_{(k,M)}}{d_hQ_n} = \frac{e^{k \langle M,x \rangle}}{L_{Q_n}(k)}, \ k \in R^+, \ M \in \Sigma_h\} \qquad (1)$$

Inference problems on k remain invariant under the group of rotations of R^h. $R(x) = ||x||$ is a maximal invariant statistic inducing the structure :

$$\{R^+, \mathcal{B}_{R^+} \ ; \ \frac{dP_k}{d_h\tilde{Q}_n} = \frac{L_{U_h}(kr)}{L_{Q_n}(k)}, \ k \in R^+\} \qquad (2)$$

There is a natural one-to-one correspondance between invariant tests and tests defined on the structure (2).

As this latter one is strictly Polya 3 ([2]), one-sided and two-sided tests are given by S. Karlin in [3]. In fact we have here a natural extension of unilateral and bilateral tests on exponential structure as described in detail in [2]. These tests lead to optimum tests among the invariant ones.

3.1 - Proposition : Let K_0 and K_1 be two separate hypotheses in R^+ and let Φ_2 be a test on (2) being admissible among the invariant tests. The test Φ_1 on (1) corresponding to Φ_2 is admissible for testing K_0 against K_1.

If Ψ is a test on (1) better than Φ_1, then the test $\tilde{\Psi}$ given by :

$$\tilde{\Psi}(r) = \int_{R^h} \Psi(rx) \ dU_h(x)$$

is better than Φ_2, since its power function is :

$$\beta_{\tilde{\Psi}}(k) = \int_{R^h} \beta_\Psi(ku) \ dU_h(u).$$

This is incompatible with admissibility of Φ_2, so Φ_1 is admissible.

3.2 - Proposition : Let K_0 and K_1 be two separate hypotheses in R^+. If there exists a U.M.P. (resp. U.M.P. unbiased) α level test Φ of K_0 against K_1 on (1), then the invariant test $\tilde{\Phi}$ defined on (1) by :

$$\tilde{\Phi}(x) = \int_{R^h} \Phi(||x||u)\ dU_h(u)$$

is a U.M.P. (resp. U.M.P. unbiased) $\tilde{\alpha}$ level test with $\tilde{\alpha} \leq \alpha$.

It is easy to see the inequality $\tilde{\alpha} \leq \alpha$. If Φ is U.M.P., then we have :

$$\beta_{\tilde{\Phi}}(\theta) \leq \beta_{\Phi}(\theta),\ \theta \in K_1 \times \Sigma_h.$$

Therefore $\beta_{\tilde{\Phi}}$ satisfies :

$$\beta_{\tilde{\Phi}}(\theta) = \int_{R^h} \beta_{\tilde{\Phi}}(||\theta||u)\ dU_h(u) \leq \int_{R^h} \beta_{\Phi}(||\theta||u)\ dU_h(u) = \beta_{\tilde{\Phi}}(\theta), \theta \in K_1 \times \Sigma_h.$$

The power functions β_{Φ} and $\beta_{\tilde{\Phi}}$ are continuous so they coincide on $K_1 \times \Sigma_h$. Let Ψ be another test at level $\alpha' \leq \tilde{\alpha}$. α' is less than or equal to α, so we have $\beta_{\Psi}(\theta) \leq \beta_{\Phi}(\theta) = \beta_{\tilde{\Phi}}(\theta)$ on $K_1 \times \Sigma_h$ and $\tilde{\Phi}$ is U.M.P. The proof is similar for a U.M.P. unbiased test.

We now consider the invariant tests on (1) associated with unilateral and bilateral tests on (2). These tests are admissible (prop. 3.1). In fact, these are the only ones which could be U.M.P. or U.M.P. unbiased, except, perhaps, when the alternative is equal to the adherent points of the null hypothesis. This result is a consequence of proposition 3.2 jointly with the special form of the tests (indicator functions) and the corresponding properties of the power functions ([2]).

REFERENCES

[1] Barra, J.R. (1971). "Notions fondamentales de statistiques mathématiques", Dunod éditeur, Paris.

[2] Degerine, S. (1975). "Etude des structures statistiques associées aux lois de Von Mises", Thèse, Université Scientifique et Médicale de Grenoble.

[3] Karlin, S. (1957). "Polya type distribution II", Ann. Math. Statist., 28, 281-308.

[4] Kluyver, J.C. (1906). "A local probability theorem", Ned. Akad. Wet. Proc., A8, 341-350.

[5] Mardia, K.V. (1972). "Statistics of directional data", Academic Press, London and New York.

[6] Rayleigh Lord (1919). "On the problem of random vibrations, and of random flights in one, two or three dimensions", Phil. Mag., (6), 37, 321-347.

[7] Stephens, M.A. (1964). "The testing of unit vectors for randomness", J. Amer. Statist. Ass., 50, 160-167.

[8] Stephens, M.A. (1967). "Tests for the dispersion and for the modal vector of a distribution on a sphere", Biometrika, 56, 149-160.

[9] Van Der Vaart, H.R. (1967). "Determining the absolutely continuous component of a probability distribution from its Fourier-Stieltjes transform", Arkiv för Mathematik, Band 7, n° 24, 331-342.

[10] Watson, G.N. (1944). "A treatise on the theory of Bessel functions", (2nd ed.), Cambridge University Press.

Recent Developments in Statistics
J.R.Barra et al., editors

CENTRAL LIMIT THEOREMS FOR DENSITY ESTIMATORS OF A L^2-MIXING PROCESS

Michel DELECROIX
Université des Sciences et techniques de Lille I
U.E.R. de Mathématiques Pures et Appliquées
B.P. 36 - 59650 - VILLENEUVE D'ASCQ
FRANCE

I - PRELIMINARIES.

Let (ξ_i), $i \in \mathbb{N}$, be a sequence of random variables defined on a probability space (Ω,A,P) and taking values in a measure space $(E,\mathcal{B},\mu)$, where μ is a σ-finite measure. Define M_o^t and M_{t+n}^∞ as the σ-fields generated by $\xi_o,\xi_1,\ldots,\xi_t$ and $\xi_{t+n},\xi_{t+n+1},\ldots$. Throughout the paper, we shall assume that :

1) The sequence (ξ_i), $i \in \mathbb{N}$, is stationary : for any choice of natural integers t_1, $t_2,\ldots,t_k$, ℓ, with $k \in \mathbb{N}^*$ and $t_1 < t_2 < \ldots < t_k$, the vector $(\xi_{t_1},\ldots,\xi_{t_k})$ has the same distribution as $(\xi_{t_1+\ell},\ldots,\xi_{t_k+\ell})$.

2) The sequence (ξ_i), $i \in \mathbb{N}$, is L^2-mixing : for any pair (X,Y) of real-valued random variables of finite variances, measurable over M_o^t and M_{t+k}^∞ respectively, we have $|\text{corr}(X,Y)| \leqslant \alpha(k)$, with $\alpha(k) \downarrow 0$ as $k \to \infty$.

We remark that the L^2-mixing condition, introduced by Stein (1), is weaker than the "Φ-mixing condition" used among others by Doob (2) and Billingsley (3), and stronger than the "mixing condition" of Rosenblatt (4).

We call P' the common law of the ξ_i, and assume moreover that P' is absolutely continuous with respect to μ $(P' << \mu)$. We can then estimate the Radon-Nikodym derivative $\frac{dP'}{d\mu}$. Using the method defined independently by Bleuez and Bosq ((5)(6)), and Földes and Revesz ((7),(8)), we shall consider estimators of the form :

$$\hat{f}_n(t) = \frac{1}{n} \sum_{i=0}^{n-1} K_{r(n)}(\xi_i,t) \ , \quad t \in E, \quad n \in \mathbb{N}^*$$

where . (K_r), $r \in I$, is a family of real-valued, $\mathcal{B}^2$-measurable functions defined on E^2

. I is an unbounded subset of $\mathbb{R}^+$

. $r(n) \to \infty$ as $n \to \infty$.

If we choose a family (K_r), $r \in I$, fix any t, $t \in E$, and define $f_n(t)$ and S_n (*) as : $f_n(t) = E[\bar{K}_{r(n)}(\xi_o,t)] = \int_E K_{r(n)}(x,t) \frac{dP'}{d\mu}(x)d\mu(x)$.

$$S_n = [\hat{f}_n(t) - f_n(t)]\{\sigma(\hat{f}_n(t))\}^{-1}$$

we can study the limit law of S_n, as $n \to \infty$, which is the first purpose of this paper.

II - A LIMIT LAW FOR S_n.

We introduce first three hypotheses about the family (K_r), $r \in I$, which are usually satisfied, as seen in ((5),(6)) :

H1) $\forall x,\ x \in E,\ |K_r(x,t)| \leq A_1 r^{\alpha}$

H2) $\int_E K_r(x,t) \frac{dP'}{d\mu}(x)d\mu(x) \to \ell$ as $r \to \infty$

H3) $\int_E K_r^2(x,t).\frac{dP'}{d\mu}(x)d\mu(x) \geq A_2 r^{\alpha}$ $(A_2 > 0)$.

[In H1), H3), we suppose r to be large enough $(r \geq r_o)$. On the other hand α denotes a strictly positive number attached to the family (K_r)].

Under the three assumptions, S_n is well defined for any large enough n, since we have $\sigma(\hat{f}_n(t)) > 0$... Moreover, we can write immediately that, for any large enough n, if H2) and H3) hold, we have :

$$E(X'^2_{j,n}) \geq A'_2 \frac{[r(n)]^{\alpha}}{n^2}$$

and if H1) holds, we have :

$$|X'_{j,n}| \leq A'_1 \frac{[r(n)]^{\alpha}}{n}$$

where : $X'_{j,n} = \frac{1}{n}\{K_{r(n)}(\xi_j,t) - E[\bar{K}_{r(n)}(\xi_j,t)]\}$.

We define then $X_{j,n}$ as : $X_{j,n} = X'_{j,n}.\{E[(\sum_{j=0}^{n-1} X'_{j,n})^2]\}^{-1}$, it is clear that $S_n = \sum_{j=0}^{n-1} X_{j,n}$, and $(X_{j,n})$, $n \in \mathbb{N}$, $j = 0,\ldots,n-1$, is a double sequence of real-valued random variables, centered as expectations.

Moreover $(X_{j,n})$ is stationary and L^2-mixing, with the same $\alpha(k)$ as the sequence (ξ_i), $i \in \mathbb{N}$, in an easily generalized definition of I. An easy

(*) As t is fixed, we shall omit t in $S_n(t)$.

computation gives :

LEMMA 1.- Under the assumptions H1), H2), H3) and if $\sum_{i=1}^{\infty} \alpha(i) < \frac{1}{2}$, there exist K_1, K_2, K_3, $K_1 > 0$, $K_2 > 0$, $K_3 > 0$, such that, for any large enough n :

$$K_1 \leq n.\sigma^2(X_{j,n}) \leq K_2 \quad \text{and} \quad |X_{j,n}| \leq K_3\left(\frac{[r(n)]^{\alpha}}{n}\right)^{1/2}.$$

To obtain the limit law of S_n, it is enough to use Philipp's central limit theorem [(9) p. 161 - Theorem 1] : Philipp proved it for a Φ-mixing double sequence of random variables $(X_{j,n})$, but it is not very difficult to see that the result is still true with our L^2-mixing one, provided $\sum_{i=1}^{\infty} \alpha(i) < \frac{1}{2}$, $i^2\alpha(i) \to 0$ $(i \to \infty)$ and, obviously, lemma 1 hold... Finally, it yields :

THEOREM 1.- Under the six assumptions : H1, H2, H3, $\sum_{i=1}^{\infty} \alpha(i) < \frac{1}{2}$, $i^2\alpha(i) \to 0$ $(i \to \infty)$, $\frac{[r(n)]^{4\alpha}}{n} \to 0$ $(n \to \infty)$, we have :

$$S_n \xrightarrow{\mathcal{L}} N(0,1).$$

III - A SECOND CENTRAL LIMIT THEOREM.

Troughout this paragraph, we always suppose that (ξ_i) is a stationary and L^2-mixing process, with $P' \ll \mu$, but Q, the common law of the vectors (ξ_i, ξ_{i+1}) is absolutely continuous with respect to $\mu \otimes \mu$: $Q \ll \mu \otimes \mu$.

Using the same family (K_r), $r \in I$, as in I, we can define :

. $\hat{g}_n$, a new estimator (*) of $\frac{dP'}{d\mu}$, as :

$$\hat{g}_n(t) = \frac{1}{\alpha(n)} \sum_{j=0}^{\alpha(n)} [K_{r(n)}(\xi_{2j}, t)], \quad t \in E,$$

where $\alpha(n)$ is a natural integer equal to $\frac{n-2}{2}$ when n is even, to $\frac{n-3}{2}$ when n is odd.

. $\hat{h}_n$, an estimator of $\frac{dQ}{d(\mu\otimes\mu)}$, as :

$$\hat{h}_n(t,t') = \frac{1}{\alpha(n)} \sum_{j=0}^{\alpha(n)} [K_r(n)(\xi_{2j}, t) K_{r(n)}(\xi_{2j+1}, t')], (t,t') \in E^2.$$

$\hat{h}_n/\hat{g}_n$ can be used as an estimator of the transition density of the process (ξ_i).

Our second purpose is to find the limit law of T_n, with

(*) As seen in the definition, $\hat{g}_n$ is very little different from $\hat{f}_n$.

$$T_n = U_n . (V_n)^{-1}$$
$$U_n = \{\hat{h}_n(t,t') - E[\hat{h}_n(t,t')]\}.\{\sigma[\hat{h}_n(t,t')]\}^{-1}$$
$$V_n = \{\hat{g}_n(t) - E[\hat{g}_n(t)]\}.\{\sigma[\hat{g}_n(t)]\}^{-1}$$

t,t' being fixed in E.

To obtain the limit law of T_n, we need first two more assumptions about the family (K_r), $r \in I$:

$$H_4)\ \iint_{E^2} K_r(x,t).K_r(x',t').\ \frac{dQ}{d\mu\otimes\mu}(x,x')d\mu\otimes\mu(x,x') \to \ell \qquad \text{as} \quad r \to \infty.$$

$$H5)\ \iint_{E^2} [K_r(x,t).K_r(x',t')]^2\ \frac{dQ}{d\mu\otimes\mu}(x,x')d\mu\otimes\mu(x,x') \gtrsim A_3\ r^{2\alpha},$$

r being supposed to be large enough.

We introduce then the following notations :

$$A^{t_1t_2} = t_1\ U_n + t_2\ V_n \quad \text{(where } (t_1,t_2) \text{ is any pair of real numbers)}$$

$$W_{h,n} = t_1\ U'_{h,n}.\{E[(\sum_{h=0}^{\alpha(n)} U'_{2h,n})^2]\}^{-1/2}$$
$$+\ t_2\ V'_{h,n}.\{E[(\sum_{h=0}^{\alpha(n)} V'_{2h,n})^2]\}^{-1/2}$$

$$U'_{h,n} = \frac{1}{\alpha(n)}\ [K_{r(n)}(\xi_h,t).K_{r(n)}(\xi_{h+1},t')]$$
$$-\ \frac{1}{\alpha(n)}\ E\{[K_{r(n)}(\xi_h,t)][K_{r(n)}(\xi_{h+1},t')]\}$$

$$V'_{h,n} = \frac{1}{\alpha(n)}\ \{K_{r(n)}(\xi_h,t) - E[K_{r(n)}(\xi_h,t)]\}.$$

It is clear that $A_n^{t_1t_2} = \sum_{j=0}^{\alpha(n)} W_{2j,n}$, and $(W_{2j,n})$, $n \in \mathbb{N}$, $j = 0,\dots,\alpha(n)$, is a double sequence of random variables, stationary and L^2-mixing. It follows :

LEMMA 2.- Under the six assumptions : H1), H2), H3), H4), H5) and $\sum_{i=1}^{\infty} \alpha(i) < \frac{1}{2}$, we have $\text{cov}(U_n,V_n) \to 0$ as $n \to \infty$, provided $\frac{r(n)^{3\alpha}}{n} \to 0$, as $n \to \infty$.

LEMMA 3.- Under the six assumptions of lemma 2, there exist K_4, K_5, K_6, $K_4 > 0$,

$K_5 > 0$, $K_6 > 0$ so that, for any large enough n :

$$K_5 \leqslant n.E(W_{2j,n}^2) \leqslant K_6 \quad \text{and} \quad |W_{2j,n}| \leqslant K_4 [r(n)]^{\alpha}.n^{-1/2}.$$

Using again Philipp's theorem, we obviously obtain :

LEMMA 4.- Under the six assumptions of Lemma 2, if we suppose that $i^2\alpha(i) \to 0$ as $i \to \infty$, and $\frac{r(n)^{8\alpha}}{n} \to 0$ as $n \to \infty$, we have :

$$A_n^{t_1 t_2} \xrightarrow{\mathcal{L}} N(0, \sqrt{t_1^2 + t_2^2})$$

To complete our proof, we remark that, by Cramer-Wold's theorem we have : $(U_n, V_n) \xrightarrow{\mathcal{L}} (X,Y)$ where (X,Y) is a pair of independent gaussian random variables, centered as expectations, and whose variances are equal to 1.
It follows finally that $U_n(V_n^{-1}) \xrightarrow{\mathcal{L}} C$, where C is a Cauchy random variable (see Billingsley (3), theorem 5-1 p. 30), and we can sum up the whole discussion as :

THEOREM 2.- Under the eight hypotheses : H1), H2), H3), H4), H5), $\sum_{i=1}^{\infty} \alpha(i) < \frac{1}{2}$, $i^2\alpha(i) \to 0$ as $i \to \infty$, $\frac{r(n)^{8\alpha}}{n} \to 0$ as $n \to \infty$, we have : $U_n(V_n)^{-1} \xrightarrow{\mathcal{L}} C$, where C is a Cauchy random variable.

IV - TWO REMARKS.

1) Proofs of the theorems are obviously too long to be more detailed. For a complete proof, one can see (10), but we had supposed there that (ξ_i) is a Φ-mixing process, and used a special kind of family K_r (case of orthogonal function method).

2) In the same way, we cannot detail here examples of families (K_r).
We shall only remark that the class of estimators obtained with such families contains both the "convolution kernel method" estimators and the "orthogonal functions method" estimators (see (5) and (6)).

V - REFERENCES.

[3] BILLINGSLEY P. - "Convergence of probability measures" Wiley New-York.

[5] BLEUEZ J. and BOSQ D. - Comptes rendus de l'Académie des Sciences de Paris, Série A, 1976, p. 63.

[6] BLEUEZ J. and BOSQ D. - Comptes rendus de l'Académie des Sciences de Paris, Série A, 1976, p. 1023.

[10] DELECROIX M. - "Sur l'estimation des densités marginales et de transition d'un processus stationnaire et mélangeant", (Thèse de 3ème cycle, Université des Sciences et Techniques de Lille, 15/1/1975).

[2] DOOB J.L. - "Stochastic processes" - Wiley New-York.

[7] FOLDES A. - "Density estimation for dependent sample", Studia Scientarium Mathematicarum Hungarica 9 (1974), p. 443-452.

[8] FOLDES A. and REVESZ P. - "A general method for density estimation", Studia Scientarium Mathematicarum Hungarica 9 (1974), p. 81-92.

[9] PHILIPP W. - "The central limit problem for mixing sequences of random variables", Z. Wahrscheinlichkeitstheorie verw. Gebiet. 12 (1969), p. 155-171.

[4] ROSENBLATT M. - "A central limit theorem and a mixing condition", Proc. Nat. Acad. Sciences U.S.A. 42, (1956) p. 43-47.

[1] STEIN C. - "A bound for the error in the normal approximation to the distribution of a sum of dependent random variables", Sixth Berkeley Symposium (vol. 2, p. 583-602).

Recent Developments in Statistics
J.R.Barra et al., editors

ASYMPTOTIC NORMALITY OF M-ESTIMATORS ON WEAKLY DEPENDENT DATA

C. Deniau, G. Oppenheim, M.C. Viano
U.E.R. de Mathématiques
Université René Descartes
12 rue Cujas, 75005 Paris

INTRODUCTION

Studies on the robustness of location parameters have led to the construction of now classical estimators : M-estimators (estimators of modified maximum likelihood), L-estimators (linear combinations of order statistics) and R-estimators (estimators constructed from rank-tests). Their asymptotic properties, established for independent random variables, have been extended to the case of weak dependence by Gastwirth-Rubin (1975) for L-estimators and the Hodges-Lehmann's R-estimator. Here, we are mainly interested in the asymptotic properties of M-estimators under weak dependence conditions. The results are deduced from central limit theorems of sums of weakly dependent random variables asymtotically identically distributed.

1. DEFINITIONS AND PRELIMINARY RESULTS

A. Coefficients of dependence

Given a real stationary process $X = (X_1, \ldots, X_n, \ldots)$ on $(\Omega, \mathcal{A}, \mathbb{P})$ we define two coefficients of dependence as follows (Iosifescu-Theodorescu (1969)) :

$$\alpha(n) = \sup \{ |\mathbb{P}(A \cap B) - \mathbb{P}(A)\mathbb{P}(B)| \; ; \; B \in \mathcal{A}_o^k , \; A \in \mathcal{A}_{k+n}^{\infty} \}$$

$$\varphi(n) = \sup \{ \text{ess sup } |\mathbb{P}(B|\mathcal{A}_o^k) - \mathbb{P}(B)| \; ; \; B \in \mathcal{A}_{k+n}^{\infty} \}$$

where $\mathcal{A}_i$ is the σ-algebra generated by X_i, $\mathcal{A}_\ell^m = \bigvee_{i=\ell}^{m} \mathcal{A}_i$.

The convergence to zero of $\alpha(n)$ and $\varphi(n)$ as $n \to \infty$ expresses the weak dependence of any pair of events sufficiently distant in time. A number of studies on weakly dependent processes have shown that if the coefficients of dependence decrease fast enough towards zero, most of the classical asymptotic results remain valid (Iosifescu-Theodorescu (1969), Ibragimov (1962)).

B. Preliminary results

First, we recall the result which will later be needed, we omit its proof.

Theorem 1 (Ibragimov (1962))

Given a stationary process X such that :

1) $\sum_{n=1}^{\infty} \alpha(n) < \infty$ with, for all $n \geq 1$, $\alpha(n) < \dfrac{k}{n \text{ Log } n}$ where k is a constant and $\alpha(n)$ decreases.

2) X_i is ($\mathbb{P}$. a.s.) bounded.

Then, $\sigma^2 = \text{Var } X_1 + 2 \sum_{2}^{\infty} \text{Cov}(X_1, X_j)$ and $n^{-1/2} \sum_{i=1}^{n} [X_i - E(X_i)] \xrightarrow{\mathcal{D}} N(0, \sigma^2)$.

This result is easily extended to asymptotically stationary random variables $f_n(X_i)$ under the following conditions :

Lemma 1

Given a stationary process X with marginal distribution P and a sequence of measurable functions $\{f_n\}$ from $(\mathbb{R}, \mathcal{B}(\mathbb{R}))$ into $(\mathbb{R}, \mathcal{B}(\mathbb{R}))$ such that :

1) $\sum_{n=1}^{\infty} \alpha(n) < \infty$ with for all $n \geq 1$, $\alpha(n) < \frac{k}{n \text{ Log } n}$, where k is a constant and $\alpha(n)$ decreases.

2) f_n converges towards f uniformly in the following sense :
If $A_n(\varepsilon) = \{x \, ; \, |f_n(x) - f(x)| < \varepsilon\}$,
$(\forall \varepsilon > 0)\ (\exists n_o \in N)\ (\forall n \geq n_o) : P(A_n(\varepsilon)) = 1$

3) f is (P.a.s.) bounded.

Then, $\sigma^2 = \text{Var } f(X_i) + 2 \sum_{i=2}^{\infty} \text{Cov } (f(X_1), f(X_i)) < \infty$

and $n^{-1/2} \sum_{i=1}^{n} [f_n(X_i) - Ef_n(X_i)] \xrightarrow{\mathcal{D}} N(0,\sigma^2)$.

C. Definition of the M-estimators

Given a function Ψ from R into R, if Ψ is monotone non decreasing with positive and negative values, we set :

$$T_n^{**}(X_1,\ldots,X_n) = \inf\{t \, ; \, \sum_{i=1}^{n} \Psi(X_i - t) < 0\}$$

$$T_n^{*}(X_1,\ldots,X_n) = \sup\{t \, ; \, \sum_{i=1}^{n} \Psi(X_i - t) > 0\} .$$

The M-estimator T_n associated with Ψ is defined by : $T_n = \frac{T_n^* + T_n^{**}}{2}$.

The family of the M-estimators includes most of the usual estimators of translation parameters : Mean $(\Psi(x) = x)$, Median $(\Psi(x) = 2\, I_{]-\infty,0[}(x) - 1$, $\Psi(0) = 0)$ maximum likelihood estimator ($\Psi = -\frac{f'}{f}$, where f is the absolutely continuous density of the independent variables X_i).

The a.s. convergence of T_n and the asymptotic normality of $n^{1/2}T_n$ have been proved (Hubert (1964)), when the X_i are independent and stationary, under fairly general conditions relating Ψ to the distribution of X_i.

2. MAIN RESULTS

Lemma 1 allow us to examine the asymptotic behaviour of T_n when the X_i are weakly dependent.

Theorem 2

If X is a stationary process and Ψ a monotone non decreasing function with positive and negative values and if :

1) $\sum_{n=1}^{\infty} \alpha(n) < \infty$ with for all $n \geq 1$, $\alpha(n) < \frac{k}{n \text{ Log } n}$, where k is a constant and $\alpha(n)$ decreases.

2) Ψ is (P. a. s.) bounded and $\lambda(t) = E(\Psi(X_i - t))$ vanishes in a single point t_o where its derivative $\lambda'(t_o)$ exists and is different from zero.

3) Ψ satisfies the following continuity condition :

$$\text{if } A_k(\varepsilon) = \{x \; ; \; |\Psi(x+h) - \Psi(x)| < \varepsilon\},$$
$$(\forall \varepsilon > 0)\ (\exists h_o > 0)\ (\forall h,\ 0 \leq h \leq h_o) : P\{A_h(\varepsilon)\} = 1$$

Then : i) T_n converges a.s. towards t_o as $n \to \infty$

ii) $\sigma^2 = E\,\Psi^2(X_1 - t_o) + 2 \sum_{i=2}^{\infty} E[\Psi(X_1 - t_o).\Psi(X_i - t_o)] < \infty$

iii) $n^{1/2}(T_n - t_o) \xrightarrow{\mathcal{D}} N\left(0,\ \left(-\frac{\sigma}{\lambda'(t_o)}\right)^2\right)$

<u>Note 1</u> : Other conditions than condition 1, for instance Withers'ones (1975), would to the asymptotic normality of $\sqrt{n}\, T_n$.

<u>Note 2</u> : In lemma 1 we assume that f is bounded. This assumption has led us to a bounded Ψ. In the forthcoming lemma 1' more restrictive conditions on X allow us to relax this assumption.

<u>Lemma 1'</u>

Given X a stationary process and $\{f_n\}$ a sequence of mesurable functions defined on $(\mathbb{R}, \mathcal{B}(\mathbb{R}))$ with values in $(\mathbb{R}, \mathcal{B}(\mathbb{R}))$ such that :

1) $\sum_{n=1}^{\infty} \sqrt{\varphi(n)} < \infty$ and $\varphi(n)$ decreases.

2) $E(f(X_i))^2 < \infty$.

3) If $n \to \infty$, $f_n(X_i)$ converges in quadratic mean towards $f(X_i)$.

Then : i) $\sigma^2 = \mathrm{Var}\, f(X_1) + 2 \sum_{i=2}^{\infty} \mathrm{Cov}(f(X_1), f(X_i)) < \infty$.

ii) $n^{-1/2} \sum_{i=1}^{n} [f_n(X_i) - E(f_n(X_i))] \xrightarrow{\mathcal{D}} N(0, \sigma^2)$.

<u>Theorem 2'</u>

Given X a stationary process and Ψ a function from $(\mathbb{R}, \mathcal{B}(R))$ to $(\mathbb{R}, \mathcal{B}(\mathbb{R}))$ monotone, non decreasing with positive and negative values.

If : 1) $\sum_{n=1}^{\infty} \sqrt{\varphi(n)} < \infty$ and $\varphi(n)$ decreases.

2) $\lambda(t) = E\,\Psi(X_i - t)$ exists for any t, vanishing in a single point t_o, where the derivative $\lambda'(t_o)$ exists and is not equal to zero ; $E\,\Psi^2(X_i - t_o) < \infty$

3) $\Psi(X_i - t_o)$ is continuous at t_o in quadratic mean.

Then the results of theorem 1 are valid.

<u>Application</u>

As an immediate consequence we compare the asymptotic variances of T_n tor the following two models :

Model 1. X is a weakly dependent stationary process satisfying condition 1 of theorem 2 (or 2').

Model 2. X is an independent stationary process with same marginal as in model 1. If Ψ satisfies conditions 2 and 3 of theorem 2, for both models $\sqrt{n}\, T_n$ is asymptotically normally distributed and the variances ratio is :

$$\frac{V_1}{V_2} = 1 + 2 \sum_{2}^{\infty} \operatorname{corr} \left(\Psi(X_1 - t_o) , \Psi(X_j - t_o) \right) .$$

Numerical computations of this ratio would, as in Gastwirth-Rubin (1975) and Oppenheim-Viano (1971-1975), lead to an evaluation of the robustness of M-estimators for weakly dependent data.

3. PROOF OF THE RESULTS

Proof of lemma 1

We proceed in two steps :

a) Theorem 1 applies directly to the process $(f(X_i))_{i\in N}$ since f is mesurable and f is (P. a. s.) bounded. It follows that :

$$\sigma^2 < \infty \quad \text{and} \quad n^{-1/2} \sum_{i=1}^{n} [f(X_i) - Ef(X_i)] \xrightarrow{\mathcal{D}} N(0,\sigma^2)$$

b) It is sufficient, then, to prove the convergence in quadratic mean towards zero of :

$$n^{-1/2} \sum_{i=1}^{n} [f_n(X_i) - Ef_n(X_i)] - n^{-1/2} \sum_{i=1}^{n} [f(X_i) - Ef(X_i)] \quad ;$$

That is : $\lim\limits_{n\to\infty} \operatorname{Var} \left\{ n^{-1/2} \sum\limits_{i=1}^{n} [f_n(X_i) - f(X_i)] \right\} = \lim\limits_{n\to\infty} \left[n^{-1} \sum\limits_{i,j=1}^{n} C_n^{i,j} \right] = 0$

where $C_n^{i,j} = \operatorname{Cov} [f_n(X_i) - f(X_i) , f_n(X_j) - f(X_j)]$.

From condition 2 of convergence :

$(\forall\varepsilon > 0)(\exists\, n_o(\varepsilon) \in N)(\forall n \geq n_o(\varepsilon)) : \mathbb{P}(\{|f_n(X_i) - f(X_i)| < \varepsilon\} \cap \{|f_n(X_j) - f(X_j)|\} < \varepsilon) = 1$

then, for $n > n_o(\varepsilon)$, $|C_n^{i,j}| \leq 4\, \varepsilon^2\, \alpha(|i-j|)$ is obtained by the application of Ibragimov's lemma 1.2 (1962) to the couple of variables :

$$Z_n = f_n(X_i) - f(X_i) \quad \text{and} \quad T_n = f_n(X_j) - f(X_j) .$$

It follows that :

$$(\forall\varepsilon > 0)(\forall n > n_o(\varepsilon)) : \operatorname{Var} \left\{ n^{-1/2} \sum_{i=1}^{n} [f_n(X_i) - f(X_i)] \right\} < 4\, \varepsilon^2 \sum_{i,j=1}^{n} \frac{\alpha(|i-j|)}{n}$$

where $\sum\limits_{i,j=1}^{n} \frac{\alpha(|i-j|)}{n}$ converges (condition 1). The variance, therefore, converges towards zero.

Proof of theorem 2

i) Almost sure convergence of T_n to t_o

Ergodicity of the process X, as a result of condition 1 assures the a.s. convergence of $n^{-1} \sum_{i=1}^{n} \Psi(X_i - \theta)$ to $\lambda(\theta)$ for any θ. It suffices, hence, to remark that if we note :

$$A = \bigcap_k \left(\left\{ \lim_{n\to\infty} n^{-1} \sum_i \Psi(X_i - t_o - \frac{1}{k}) = \lambda(t_o + \frac{1}{k}) \right\} \cap \left\{ \lim_{n\to\infty} n^{-1} \sum_i (X_i - t_o + \frac{1}{k}) = \lambda(t_o - \frac{1}{k}) \right\} \right)$$

$$B = \left\{ (\forall k)\ (\exists n_o \in N)\ (\forall n > n_o) : t_o - \frac{1}{k} < T_n < t_o + \frac{1}{k} \right\}$$

$$C = \left\{ \lim_{n\to\infty} T_n = t_o \right\}$$

then : $A \subset B \subset C$, and $\mathbb{P}(C) = 1$ since $\mathbb{P}(A) = 1$.

ii) Asymptotic normality

Let $f_n(X_i) = \Psi\left(X_i - t_o - \frac{a}{\sqrt{n}} \right) \quad 1 \leq i \leq n$;

$f(X_i) = \Psi(X_i - t_o)$.

f_n , f and the process X satisfies the conditions of lemma 1, hence :

$$\sigma^2 < \infty \quad \text{and} \quad n^{-1/2} \sum_{i=1}^{n} \left(\Psi\left(X_i - t_o - \frac{a}{\sqrt{n}} \right) - \lambda\left(t_o + \frac{a}{\sqrt{n}} \right) \right) \xrightarrow{\mathcal{D}} N(0,\sigma^2).$$

In order to study $\mathbb{P}\{n^{1/2}(T_n - t_o) < a\}$ it is sufficient to remark that for all $a \in R$ and $k > 0$, if :

$$H_n = \mathbb{P}\left\{ n^{-1/2} \sum \left[\Psi\left(X_i - t_o - \frac{a + \frac{1}{k}}{n^{1/2}} \right) - \lambda\left(t_o + \frac{a + \frac{1}{k}}{n^{1/2}} \right) \right] \leq -n^{1/2} \lambda\left(t_o + \frac{a + \frac{1}{k}}{n^{1/2}} \right) \right\}$$

$$K_n = \mathbb{P}\left\{ T_n \leq \frac{a}{n^{1/2}} + t_o \right\}$$

$$L_n = \mathbb{P}\left\{ n^{-1/2} \sum \left[\Psi\left(X_i - t_o - \frac{a}{n^{1/2}} \right) - \lambda\left(t_o + \frac{a}{n^{1/2}} \right) \right] \leq -n^{1/2} \lambda\left(t_o + \frac{a}{n^{1/2}} \right) \right\}$$

then : $H_n \leq K_n \leq L_n$.

From the existence of $\lambda'(t_o)$ it follows that :

$$\lim_{n\to\infty} \left[n^{1/2} \lambda\left(t_o + \frac{a}{n^{1/2}} \right) \right] = a\,\lambda'(t_o).$$

It, then, becomes easy to prove that :

$$\lim_{n\to\infty} H_n = \Phi\left(-\left(a + \frac{1}{k} \right) \lambda'(t_o) \right)$$

$$\lim_{n\to\infty} L_n = \Phi\ (-a\,\lambda'(t_o))$$

$$\lim_{n\to\infty} K_n = \Phi\ (-a\,\lambda'(t_o))$$

Hence $n^{1/2}(T_n - t_o) \xrightarrow{\mathcal{D}} N\left(0, \left(-\frac{\sigma}{\lambda'(t_o)} \right)^2 \right)$

Proof of lemma 1'

Theorem 1 is not sufficient to prove lemma 1', it should be replaced by theorem (1.5) of Ibragimov (1962). The proof is then the same as the previous one for lemma 1, with an inequality about the covariance $C_n^{i,j}$ resulting from condition 2 and lemma 1.7 of Ibragimov (1962).

Proof of theorem 2'

This proof, similar to that of theorem 2, is omitted.

REFERENCES

Huber, P. (1964), Robust estimation of a location parameter, Ann. Math. Stat. 35, n°1, 73-101.
Gastwirth, J.L., Rubin, H. (1975), The behaviour of robust estimators on dependent data, Ann. Stat. 3, n°5, 1070-1100.
Ibragimov, I.A. (1962), Some limit theorems for stationary processes. Theory of probability and its applications, 11, n°4, 349-382.
Iosifescu, M., Theodorescu, R. (1969), Random processes and learning, Springer-Verlag.
Oppenheim, G., Viano, M.C. (1974), Robustesse de tests de comparaisonsde fréquences vis à vis de modèles markoviens de dépendance, Publications de l'Institut de Statistique des Universités de Paris, 1971 - 1.2, 95-119.
Oppenheim, G., Viano, M.C. (1975), Robustesse de tests de comparaisons de moyennes vis à vis de la non-indépendance des observations, Math. Operationsforsch.und Statistik, 6 , 197-211.
Withers, C.S. (1975), Convergence of empirical processes of mixing random variables on [0,1], Ann. Stat. 1975, 3-5, 1101-1108.

Recent Developments in Statistics
J.R.Barra et al., editors

A NOTE ON SOME STATIONARY , DISCRETE-TIME , SCALAR LINEAR PROCESSES

S. DOSSOU-GBETE , P. ETTINGER , A. de FALGUEROLLES (*)

A class of linear models useful in representing many types of time series are the autoregressive integrated moving average (ARIMA) models ([3]). The purpose of this paper is to study the approximation of a class of second order, stationary, discrete-time, scalar processes - with known autocovariance function - by means of finite moving averages. A theoretical justification of overfitting MA(q) models is provided.

Let $(x_t)_{t\in Z}$ be a second order stationary, discrete-time, scalar process with mean μ_x, autocovariance function σ_x and autocorrelation ρ_x. It is further assumed that there exists a white noise process $(\varepsilon_t)_{t\in Z}$ with zero mean and variance $\sigma_\varepsilon(o)$ such that :

$$(1) \qquad x_t - \mu_x \overset{q.m.}{=} \sum_{j=o}^{+\infty} \theta_j \varepsilon_{t-j} \qquad (\theta_o = 1) .$$

Then $(x_t)_{t\in Z}$ has an absolutely continuous spectral distribution function F_x (see [5] p. 499). Let f_x be its spectral density.

As a matter of terminology, processes satisfying (1) and such that :

i) $\sum_{j=o}^{+\infty} |\theta_j| < +\infty$

ii) $\Theta(z) = \sum_{j=o}^{+\infty} \theta_j z^j \neq 0$ for any $|z| \leq 1$

are to be called hereafter $\overline{\text{ARMA}}$ processes.

Let $\Psi(z) = [\Theta(z)]^{-1} = \sum_{j=o}^{+\infty} \psi_j z^j \quad (|z| \leq 1)$.

Then for any $\overline{\text{ARMA}}$ process

a) $\varepsilon_t \overset{q.m.}{=} \sum_{j=o}^{+\infty} \psi_j (x_{t-j} - \mu_x) \iff x_t - \mu_x \overset{q.m.}{=} \sum_{j=o}^{+\infty} \theta_j \varepsilon_{t-j}$,

b) $\sum_{k=o}^{+\infty} |\sigma_x(k)| < +\infty$,

c) $f_x(\phi) = \frac{1}{2\pi} (\sigma_x(o) + 2 \sum_{k=1}^{+\infty} \sigma_x(k) \cos k\phi) > 0$.

(*) Laboratoire de Statistique , E.R.A.-C.N.R.S. N° 591
Université Paul Sabatier 118, route de Narbonne
31077 TOULOUSE Cédex FRANCE

Remark :

Any ARMA(p,q) process is an $\overline{\text{ARMA}}$.

I - Pure moving-average processes of finite order q-(MA(q))

Let $(x_t)_{t\in Z}$ be a finite, invertible, pure moving average process of order q :

$$x_t - \mu_x = \sum_{j=o}^{q} \theta_j \varepsilon_{t-j}$$

$$\varepsilon_t \overset{q.m.}{=} \sum_{j=o}^{+\infty} \psi_j (x_{t-j}-\mu_x) \quad .$$

Since its autocovariance function σ_x and its autocorrelation function ρ_x cut off after lag q,

$$f_x(\phi) = \frac{1}{2\pi} (\sigma_x(o) + 2 \sum_{k=1}^{q} \sigma_x(k) \cos k\phi) > 0$$

and

$$\frac{2\pi}{\sigma_x(o)} f_x(\phi) = 1 + 2 \sum_{k=1}^{q} \rho_x(k) \cos k\phi > 0 \quad .$$

I.1. Determination of $\sigma_\varepsilon(o)$ and the θ_j given $\{\sigma_x(k) \mid k = 0,1,...,q\}$

Let $P_q(z) = \sigma_x(q) z^{2q} + \ldots + \sigma_x(1) z^{q+1} + \sigma_x(o) z^q + \sigma_x(1) z^{q-1} + \ldots + \sigma_x(q)$ and $\Theta(z) = \sum_{j=o}^{q} \theta_j z^j$.

It is proved (see [2] p. 224-226) that

$$z^q \Theta(\frac{1}{z}) = \prod_{\ell=1}^{q} (z-z_\ell)$$

where the z_ℓ are the q roots of $P_q(z)$ strictly less than one in absolute value. A simple algorithm for this factorization is given in Wilson [7].

Furthermore :

$$\sigma_\varepsilon(o) = \frac{\sigma_x(o)}{\sum_{j=1}^{q} \theta_j^2} = \frac{\sigma_x(q)}{\theta_q} \quad ;$$

I.2. Restrictions on the autocovariances of a pure MA(q) process

Let $\underline{\alpha} = (\alpha_o,\alpha_1,\ldots,\alpha_q) \in R^{q+1}$. $A_q(\underline{\alpha})$ denotes the following real symmetric matrix of order q+1 :

$$A_q(\underline{\alpha}) = \begin{bmatrix} \alpha_o & \alpha_1 & \cdots & \alpha_q \\ \alpha_1 & \alpha_o & \cdots & \alpha_{q-1} \\ \vdots & \vdots & & \vdots \\ \alpha_q & \alpha_{q-1} & \cdots & \alpha_o \end{bmatrix}$$

Let $\ell'_q(\underline{\alpha})$ and $\ell''_q(\underline{\alpha})$ be respectively the smallest and the largest eigenvalue of $A_q(\underline{\alpha})$.

Theorem :

$$\sigma_x(o)\ \ell'_q(\underline{\alpha}) \leq \alpha_o\ \sigma_x(o) + 2 \sum_{k=1}^{q} \alpha_k\ \sigma_x(k) \cos k\phi \leq \sigma_x(o)\ \ell''_q(\underline{\alpha}) \qquad \forall \phi \in [-\pi,+\pi].$$

If $\underline{\alpha} \in (R^+)^{q+1}$

$$\alpha_o\ \sigma_x(o) + 2 \sum_{k=1}^{q} \alpha_k\ |\sigma_x(k)| \leq \ell''_q(\underline{\alpha}).$$

Particular choices of $\underline{\alpha}$ (for example $\underline{\alpha} = (0,\ldots,0,1,0,\ldots,0)$, $\underline{\alpha} = (1,\ldots,1,0,\ldots,0)$ or $\underline{\alpha} = (1,\ldots,1)$) and of ϕ (for example $\phi \in \{0,\pi\}$) provide necessary conditions which are to be satisfied by $\{\sigma_x(k) \mid k \leq q\}$ to ensure that σ_x is associated with a pure MA(q).

Remark : In any case $\ell'_q(\underline{\alpha})$ and $\ell''_q(\underline{\alpha})$ can be easily computed by standard numerical procedures (see [6]). However for some $\underline{\alpha}$ (see [1], [4] and [6]) a general formula can be derived : for example with $\underline{\alpha} = (1,\ldots,1)$,

$$0 \leq \sigma_x(o) + 2 \sum_{k=1}^{q} \sigma_x(k) \cos k\phi \leq (q+1)\ \sigma_x(o)$$

and

$$\sigma_x(o) + 2 \sum_{k=1}^{q} |\sigma_x(k)| \leq (q+1)\ \sigma_x(o).$$

II - The approximation of $\overline{\text{ARMA}}$ processes by finite moving average

Let $(x_t)_{t\in Z}$ be an $\overline{\text{ARMA}}$ process.

Lemma 1 :

There exists an integer number q_o such that for any integer p

$$\sigma_x(o) + 2 \sum_{k=1}^{q_o+p} \sigma_x(k) \cos k\phi > 0 \qquad \forall \phi \in [-\pi,+\pi].$$

From lemma 1, $P_{q_o+p}(z)$ (as defined in I.1.) has no zero on the unit circle.

Let $z_\ell(q_o+p), \ell \in \{1,\ldots,q_o+p\}$, be the q_o+p roots of P_{q_o+p} which are strictly less than one in module and the $\theta_j(q_o+p)$, $j \in \{0,\ldots,q_o+p\}$, be defined as follows :

$$z^{q_o+p} \sum_{j=o}^{q_o+p} \theta_j(q_o+p)\, z^{-j} = \prod_{\ell=1}^{q_o+p} (z-z_\ell(q_o+p))$$

Lemma 2 :

There exists an $\overline{\text{ARMA}}$ process $(\varepsilon_t^*(q_o+p))_{t\in Z}$ - not necessarily a white noise process - such that :

$$x_t-\mu_x \doteq \sum_{j=o}^{q} \theta_j(q_o+p)\, \varepsilon_{t-j}^*(q_o+p) \quad (\theta_o = 1)$$

and

$$\varepsilon_t^*(q_o+p) \overset{q.m.}{=} \sum_{j=o}^{+\infty} \nu_j(q_o+p)\, \varepsilon_{t-j} \qquad (\nu_o(q_o+p) = 1)$$

Let $f^*_{q_o+p}(\phi)$ be the spectral density of $(\varepsilon_t^*(q_o+p))_{t\in Z}$

$$f^*_{q_o+p}(\phi) = \frac{\sigma_\varepsilon(o)}{2\pi} \quad \lambda(q_o+p) \frac{\sigma_x(o) + 2 \sum_{k=1}^{+\infty} \sigma_x(k) \cos k\phi}{\sigma_x(o) + 2 \sum_{k=1}^{q_o+p} \sigma_x(k) \cos k\phi}$$

where

$$\lambda(q_o+p) = \frac{\sigma_x(o)}{\sigma_\varepsilon(o)\{\sum_{j=o}^{q_o+p} \theta_j^2(q_o+p)\}} .$$

Theorem :

i) $\lim_{p\to+\infty} \lambda(q_o+p) = 1$

ii) $\varepsilon_t^*(q_o+p) \xrightarrow{q.m.} \varepsilon_t$ with $p \to +\infty$

iii) $\lim_{p\to+\infty} \theta_j(q_o+p) = \theta_j$

(with $\theta_j(q_o+p) = 0$ for $j > q_o+p$) .

A detailed proof of the theorem can be found in [6].

Corollary :

Let $y_t(q_o+p) - \mu_x = \sum_{j=o}^{q_o+p} \theta_j(q_o+p)\, \varepsilon_{t-j}$. Then

$$y_t(q_o+p) \overset{q.m.}{\to} x_t \quad \text{with } p\to+\infty .$$

References :

[1] ANDERSON, O.D. (1975) . Moving Average Processes. The Statistician, vol 24, No. 4, 283-297.

[2] ANDERSON, T.W. (1971) . The Statistical Analysis of Time Series . New-York : Wiley.

[3] BOX, G.E.P. and JENKINS, G.M. (1970). Time Series Analysis, Forecasting and Control . San Francisco : Holden-Day.

[4] DAVIES, N. , PATE, M.B. and FROST, M.G. (1974) . Maximum autocorrelations for moving average processes , Biometrika , 61 , 199-200.

[5] DOOB, J.L. , (1953,1959) . Stochastic Processes . New-York : Wiley.

[6] DOSSOU-GBETE, S. , ETTINGER, P. et de FALGUEROLLES, A. (1976) . Contribution à l'étude des processus ARMA. Publications du Laboratoire de Statistique No. 05-76, Université Paul Sabatier . Toulouse.

[7] WILSON, G. (1969) . Factorization of a covariance generating function . SIAM J. Numerical Analysis , 6 , 1-7 .

Recent Developments in Statistics
J.R.Barra et al., editors

A NON PARAMETRIC TEST IN COVARIANCE ANALYSIS

D. Dugué (Paris)

I. THEOREM I. X and Y being two normal variables $N(0,1)$ and (Z_1,Z_2) a random vector, if X, Yz, (Z_1, Z_2) are independent, the variables $W = \dfrac{X Z_1 \pm Y Z_2}{\sqrt{Z_1^2 + Z_2^2}}$ are normal variables $N(0,1)$ whatever may be the law of the vector (Z_1,Z_2).

PROOF.

$$E(\exp iuW) = \int dF(z_1,z_2) \int \exp\left[\frac{iu\, z_1}{\sqrt{z_1^2+z_2^2}}\, x - \frac{1}{2} x^2\right] dx \int \exp\left[\pm \frac{iu\, z_2}{\sqrt{z_1^2+z_2^2}}\, y - \frac{1}{2} y^2\right] dy$$

$$= \int \exp - \frac{1}{2} u^2 \frac{z_1^2}{z_1^2+z_2^2} \quad \exp - \frac{1}{2} u^2 \frac{z_2^2}{z_1^2+z_2^2} \quad dF(z_1,z_2)$$

$$= \exp - \frac{1}{2} u^2 .$$

II. Let us take X a bidimensional normal vector with a covariance matrix $\Sigma = \begin{pmatrix} \sigma_{11} & \sigma_{12} \\ \sigma_{21} & \sigma_{22} \end{pmatrix}$, (the distribution matrix Σ^{-1} being $\begin{pmatrix} \varphi_{11} & \varphi_{12} \\ \varphi_{21} & \varphi_{22} \end{pmatrix}$ and a null mean vector; let us put $R = \begin{pmatrix} R_{11} & R_{12} \\ R_{21} & R_{22} \end{pmatrix}$ the Wilks matrix $\sum_{i=1}^{n} X_i^t X_i$ $(n \geq 2)$.

After Wilks the law of R is :

$$C \exp - \frac{1}{2} (\varphi_{11} R_{11} + 2\varphi_{12} R_{12} + \varphi_{22} R_{22}) [\det R]^{\frac{n-3}{2}} \, dR_{11}\, dR_{12}\, dR_{22}$$

which may be written :

$$C \exp - \frac{1}{2} \left\{ \frac{R_{11}}{\sigma_{11}} + \varphi_{22} \left[\sqrt{R_{11}} \left(\frac{R_{12}}{R_{11}} - \frac{\sigma_{12}}{\sigma_{11}}\right)\right]^2 + \varphi_{22} \left[R_{22} - \frac{R_{12}^2}{R_{11}}\right]\right\} R_{11}^{\frac{h-3}{2}} \left[R_{22} - \frac{R_{12}^2}{R_{11}}\right]^{\frac{h-3}{2}} dR_{11} \ldots$$
$$\ldots dR_{12}\, dR_{22}$$

Putting

$$X = \sqrt{\frac{R_{11}}{\sigma_{11}}} \qquad Y = \sqrt{\varphi_{22}} \sqrt{R_{11}} \left[\frac{R_{12}}{R_{11}} - \frac{\sigma_{12}}{\sigma_{11}}\right] \qquad Z = \sqrt{\varphi_{22}} \left[R_{22} - \frac{R_{12}^2}{R_{11}}\right]^{\frac{1}{2}}$$

the law of X, Y, Z is :

$$C (\exp - \tfrac{1}{2} X^2) [X^2]^{\frac{n}{2}-1} dX^2 (\exp - \tfrac{1}{2} Y^2) dY (\exp - \tfrac{1}{2} Z^2) (Z^2)^{\frac{n-1}{2} - 1} dZ^2$$

So a) X, Y, Z are independent

b) X^2, Z^2 are χ^2 with n (resp $n-1$) degrees of freedom

c) Y is a normal variable $N(0,1)$

So $\frac{Y}{Z}$ is a Student t with median value $\frac{\sigma_{12}}{\sigma_{11}}$; this parametric result was obtained by Sir R.A. Fisher .

With the help of theorem I we can give a non parametric result useful for comparison of covariance between two sets of n' and n'' independent results .

X', Y', Z' and X'', Y'', Z'' being the variables defined for each group after theorem I

$\frac{Y'X'' - Y''X'}{\sqrt{X'^2 + X''^2}}$ is a variable $N(0,1)$

With the hypothesis of identity of the two covariance matrices

$$\frac{Y'X'' - Y''X'}{X'^2 + X''^2} = \sqrt{\varphi_{22}}\ \frac{\sqrt{R'_{11}}\sqrt{R''_{11}}}{\sqrt{R'_{11} + R''_{11}}} \left[\frac{R'_{12}}{R'_{11}} - \frac{R''_{12}}{R''_{11}}\right]$$

Z' and Z'' are independent of this ratio. $Z'^2 + Z''^2$ is a χ^2 with $n' + n'' - 2$ degrees of freedom .

Let us take :

$$\frac{1}{\sqrt{Z'^2 + Z''^2}}\ \frac{Y'X'' - Y''X'}{\sqrt{X'^2 + X''^2}} = \frac{\sqrt{R'_{11}}\sqrt{R''_{11}}}{\sqrt{R'_{22} - \frac{R'^2_{12}}{R'_{11}} + R''_{22} - \frac{R''^2_{12}}{R''_{11}}}}\ \frac{1}{\sqrt{R'_{11} + R''_{11}}} \left[\frac{R'_{12}}{R'_{11}} - \frac{R''_{12}}{R''_{11}}\right] = Q$$

Q is a centred Student t with $n' + n'' - 2$ degrees of freedom. So :

$$P(Q < x) \equiv C \int_{-\infty}^{x} \frac{dt}{(1+t^2)^{\frac{n'+n''-1}{2}}}$$

Recent Developments in Statistics
J.R.Barra et al., editors

SUFFICIENT CONVERGENCE CONDITIONS FOR SOME TESTS IN THE CASE OF NOT NECESSARILY INDEPENDENT OR EQUIDISTRIBUTED DATA

Jean GEFFROY
University Pierre et Marie Curie
Paris (France)

1. Let θ be an unknown parameter for which we assume the general hypothesis $H(\theta \in \Theta)$, and consider the null hypothesis $H_0(\theta \in \Theta_0)$, where Θ and Θ_0 are given sets.

We shall define, for every $n = 1,2,\ldots$, a test T_n of H_0 against $H_1(\theta \in \Theta_1 = \Theta - \Theta_0)$ on the basis of a random sequence $\{X_n\}$ $(n \in \mathbb{N}^*)$. Each random variable X_n takes its values in an arbitrary measurable space $(\mathcal{X}_n, \mathcal{B}_n)$ and we suppose that the unknown joint distribution of $(X_1, X_2, \ldots, X_n, \ldots)$ belongs to a given family $\{P_\theta\}$ $(\theta \in \Theta)$ of probability laws on $(\mathcal{X}^{(\infty)}, \mathcal{B}^{(\infty)}) = \bigotimes_{n=1}^{\infty} (\mathcal{X}_n, \mathcal{B}_n)$.

Clearly, it is possible to choose a set $A_1 \in \mathcal{B}_1$ and, for every $j \geq 2$, an application $A_j : \mathcal{X}^{(j-1)} = \bigotimes_{i=1}^{j-1} \mathcal{X}_i \to \mathcal{B}_j$ satisfying the following conditions :

(a) the functions $\xi_j : \mathcal{X}^{(j)} \to \{0,1\}$ defined by putting :

$$\text{i)}\ (\forall x_1 \in \mathcal{X}_1),\quad \xi_1(x_1) = 1 \Longleftrightarrow x_1 \in A_1$$

$$\text{ii)}\ (\forall j \geq 2), (\forall (x_1,\ldots,x_j) \in \mathcal{X}^{(j)}),\ \xi_j(x_1,\ldots,x_j) = 1 \Leftrightarrow x_j \in A_j(x_1,\ldots,x_{j-1})$$

are $\mathcal{B}^{(j)}$-measurable ;

(b) there exist two monoton non-decreasing sequences $\Theta_{0,j} \subset \Theta_0$ and $\Theta_{1,j} \subset \Theta_1$ $(j=1,2,\ldots)$, converging respectively towards Θ_0 and Θ_1, and such that :

$$(1) \qquad (\forall j \in \mathbb{N}^*)\ ,\ \bar{\omega}_j \leq \bar{\omega}'_j$$

where $\bar{\omega}_j$ and $\bar{\omega}'_j$ are defined by :

$$\bar{\omega}_1 = \sup\{E_\theta(\xi_1) / \theta \in \Theta_{0,1}\}\ ;\ \bar{\omega}'_1 = \inf\{E_\theta(\xi_1) / \theta \in \Theta_{1,1}\}$$

$$\bar{\omega}_j = \sup\{E_\theta(\xi_j / \xi_1,\ldots,\xi_{j-1}) / \theta \in \Theta_{0,j}\ ,\ (\xi_1,\ldots,\xi_{j-1}) \in \mathcal{X}^{(j-1)}\}$$

$$\bar{\omega}'_j = \inf\{E_\theta(\xi_j / \xi_1,\ldots,\xi_{j-1}) / \theta \in \Theta_{1,j}\ ,\ (\xi_1,\ldots,\xi_{j-1}) \in \mathcal{X}^{(j-1)}\}$$

Definition Let $\{A_j\}$ $(j=1,2,\ldots)$ be a family of sets as above. We shall say that a sequence of tests $T_n (n=1,2,\ldots)$ is associated with $\{A_j\}$ if, for every n, the critical set of T_n is :

$$C_n = \{(x_1,\ldots,x_n) / \sum_{j=1}^{n} (\bar{\omega}'_j - \bar{\omega}_j)\xi_j(x_1,\ldots,x_j) > \frac{1}{2} \sum_{j=1}^{n} (\bar{\omega}'^2_j - \bar{\omega}^2_j)\}$$

Concerning such a sequence of tests, our fundamental result is the following :

THEOREM 1 Let $\alpha(T_n,\theta)$ $(\theta \in \Theta_0)$ and $\beta(T_n,\theta)$ $(\theta \in \Theta_1)$ be respectively the significance level and the power of T_n.

Define the functions N_o and N_1 by putting

$$(\forall \theta \in \Theta_o)\ ,\ N_o(\theta) = \max\{i / (i=0) \cup (\theta \notin \Theta_{o,i})\}$$

$$(\forall \theta \in \Theta_1)\ ,\ N_1(\theta) = \max\{i / (i=0) \cup (\theta \notin \Theta_{1,i})\}$$

Then, we have

(2) $$(\forall \theta \in \Theta_o, \forall n > N_o(\theta)),\ \alpha(T_n,\theta) \leq \exp\left[\sum_1^{N_o(\theta)} (\bar{\omega}'_j - \bar{\omega}_j) - \frac{1}{4}\sum_1^n (\bar{\omega}'_j - \bar{\omega}_j)^2\right]$$

(3) $$(\forall \theta \in \Theta_1, \forall n > N_1(\theta)),\ \beta(T_n,\theta) \geq 1-\exp\left[\sum_1^{N_1(\theta)} (\bar{\omega}'_j - \bar{\omega}_j) - \frac{1}{4}\sum_1^n (\bar{\omega}'_j - \bar{\omega}_j)^2\right]$$

(with the obvious convention $\sum_1^o = 0$)

Proof

For each $n \in \mathbb{N}^*$, we introduce the random variable

$$Y_n = \sum_{j=1}^n (\bar{\omega}'_j - \bar{\omega}_j)\xi_j(X_1,\ldots,X_j)$$

The definition of C_n implies

$$P_\theta\left[Y_n > \frac{1}{2}\sum_{j=1}^n (\bar{\omega}'^2_j - \bar{\omega}^2_j)\right] = \alpha(T_n,\theta) \text{ (resp. } \beta(T_n,\theta))$$

if $\theta \in \Theta_o$ (resp. $\theta \in \Theta_1$).

Putting $\gamma_j = \bar{\omega}'_j - \bar{\omega}_j$, it is useful to note that

$$\frac{1}{2}\sum_{j=1}^n (\bar{\omega}'^2_j - \bar{\omega}^2_j) = \sum_{j=1}^n \gamma_j(\bar{\omega}_j + \frac{1}{2}\gamma_j) = \sum_{j=1}^n \gamma_j(\bar{\omega}'_j - \frac{1}{2}\gamma_j)$$

So, we can write

- if $\theta \in \Theta_o$, $\alpha(T_n,\theta) = P_\theta\left[Y_n > \sum_{j=1}^n \gamma_j(\bar{\omega}_j + \frac{1}{2}\gamma_j)\right]$

- if $\theta \in \Theta_1$, $1-\beta(T_n,\theta) = P_\theta\left[Y_n \leq \sum_{j=1}^n \gamma_j(\bar{\omega}'_j - \frac{1}{2}\gamma_j)\right]$

The inequalities (2) and (3) are then a clear consequence of the following lemma.

<u>LEMMA 1</u> $\xi_n (n=1,2,\ldots)$ being a sequence of simultaneous Bernoulli variates and γ_n a sequence of numbers in $[0,1]$,consider the sequence $Y_n = \sum_1^n \gamma_j \xi_j$.

i) $\bar{\omega}_1, \bar{\omega}_2, \ldots, \bar{\omega}_n, \ldots$ being numbers in $[0,1]$ and n_o a non-negative integer, suppose that the conditions

$$E(\xi_1) \leq \bar{\omega}_1\ ;\ E(\xi_2/\xi_1) \leq \bar{\omega}_2\ ;\ldots;\ E(\xi_n/\xi_1,\ldots,\xi_{n-1}) \leq \bar{\omega}_n\ ;\ldots$$

are satisfied whenever $n > n_o$. For these values of n, we have

$$\Pr\{Y_n > \sum_1^n \gamma_j(\bar{\omega}_j + \frac{1}{2}\gamma_j)\} \leq \exp(\sum_1^{n_o} \gamma_j - \frac{1}{4}\sum_1^n \gamma_j^2)$$

ii) $\bar{\omega}'_1, \bar{\omega}'_2, \ldots, \bar{\omega}'_n, \ldots$ being numbers in $[0,1]$ and n_1 a non-negative integer, suppose that the conditions

$$E(\xi_1) \geq \bar{\omega}'_1 \; ; \; E(\xi_2 / \xi_1) \geq \bar{\omega}'_2 \; ; \ldots ; \; E(\xi_n / \xi_1, \ldots, \xi_{n-1}) \geq \bar{\omega}'_n \; ; \ldots$$

hold whenever $n > n_1$. For these values of n, we have

$$\Pr\{Y_n \leq \sum_1^n \gamma_j(\bar{\omega}'_j - \frac{1}{2}\gamma_j)\} \leq \exp\left(\sum_1^{n_1} \gamma_j - \frac{1}{4}\sum_1^n \gamma_j^2\right)$$

Proof A) Assuming the hypothesis of (i), let c be an arbitrary number. The Markov inequality applied to $\exp(Y_n)$ gives us

$$(4) \quad \Pr\{Y_n > c\} \leq e^{-c} E[\exp(Y_n)] = e^{-c} E\left[\exp\left(\sum_1^{n_0} \gamma_j\xi_j\right) \exp\left(\sum_{n_0+1}^{n} \gamma_j\xi_j\right)\right]$$

$$\leq e^{-c} \exp\left(\sum_1^{n_0} \gamma_j\right) E\left[\exp\left(\sum_{n_0+1}^{n} \gamma_j\xi_j\right)\right]$$

It is easy to see that, for every $n > n_0+1$

$$(5) \qquad E\left[\exp\left(\sum_{n_0+1}^{n} \gamma_j\xi_j\right)\right] \leq \prod_{n_0+1}^{n} [\bar{\omega}_j(e^{\gamma_j}-1)+1]$$

This obvious is $n = n_0+1$. So, it suffices to show that, if (5) is true for some n, then it is true for (n+1). In this aim, we can write

$$(6) \qquad E\left[\exp\left(\sum_{n_0+1}^{n+1} \gamma_j\xi_j\right)\right] = E\left\{E\left[\exp\left(\sum_{n_0+1}^{n+1} \gamma_j\xi_j\right) / \xi_1, \ldots, \xi_n\right]\right\}$$

$$= E\left\{\exp\left(\sum_{n_0+1}^{n} \gamma_j\xi_j\right) E\left[\exp(\gamma_{n+1}\xi_{n+1}) / \xi_1, \ldots \xi_n\right]\right\}$$

But we have

$$E[\exp(\gamma_{n+1}\xi_{n+1}) / \xi_1, \ldots, \xi_n] = e^{\gamma_{n+1}} \Pr(\xi_{n+1}=1 / \xi_1, \ldots, \xi_n) + \Pr(\xi_{n+1}=0 / \xi_1, \ldots, \xi_n)$$

$$= (e^{\gamma_{n+1}}-1)\, E(\xi_{n+1} / \xi_1, \ldots, \xi_n) + 1$$

$$\leq \prod_{n_0+1}^{n+1} [\bar{\omega}_j(e^{\gamma_j}-1)+1]$$

which achieves to establish (5). This relation and (4) imply

$$\Pr\{Y_n > c\} \leq \exp\left(-c + \sum_1^{n_0} \gamma_j\right) \prod_{n_0+1}^{n} [\bar{\omega}_j(e^{\gamma_j}-1)+1]$$

Putting $c = \sum_1^n \gamma_j(\bar{\omega}_j + \frac{1}{2}\gamma_j)$ in this relation, it becomes, after a slight transformation

(7) $$\Pr\{Y_n > \sum_1^n \gamma_j (\bar{\omega}_j + \frac{1}{2}\gamma_j)\} \leq \exp\left[\sum_1^{n_o} \gamma_j (1-\bar{\omega}_j) - \frac{1}{2}\sum_1^n \gamma_j^2\right] \times \prod_{n_o+1}^{n} \left[\bar{\omega}_j\, e^{\gamma_j(1-\bar{\omega}_j)} + (1-\bar{\omega}_j)\, e^{-\gamma_j\bar{\omega}_j}\right]$$

From the elementary inequalities

$$\exp\left[\gamma_j(1-\bar{\omega}_j)\right] \leq 1+\gamma_j(1-\bar{\omega}_j)+\gamma_j^2(1-\bar{\omega}_j)^2$$

$$\exp(-\gamma_j\bar{\omega}_j) \leq 1-\gamma_j\,\bar{\omega}_j + \gamma_j^2\,\bar{\omega}_j^2$$

it follows that

$$\bar{\omega}_j \exp\left[\gamma_j(1-\bar{\omega}_j)\right] + (1-\bar{\omega}_j)\exp(-\gamma_j\bar{\omega}_j) \leq 1 + \bar{\omega}_j(1-\bar{\omega}_j)\,\gamma_j^2$$

$$\leq 1 + \frac{1}{4}\gamma_j^2 \leq \exp(\frac{1}{4}\gamma_j^2)$$

Then, we deduce from (7)

$$\Pr\{Y_n > \sum_1^n \gamma_j(\bar{\omega}_j + \frac{1}{2}\gamma_j)\} \leq \exp(\sum_1^{n_o}\gamma_j - \frac{1}{2}\sum_1^n \gamma_j^2 + \frac{1}{4}\sum_{n_o+1}^{n}\gamma_j^2) \leq \exp(\sum_1^{n_o}\gamma_j - \frac{1}{4}\sum_1^n \gamma_j^2)$$

and so, the first part of our lemma is proved.

B) To prove (ii), we can introduce the variables : $\bar{\xi}_j = 1-\xi_j$ and $\bar{Y}_n = \sum_1^n \gamma_j\bar{\xi}_j$, and put $\bar{\omega}_j = 1-\bar{\omega}'_j$.

The variables $\bar{\xi}_j$ satisfy the same relations as the ξ_j in (i), and consequently we have

(8) $$\Pr\{\bar{Y}_n > \sum_1^n \gamma_j(\bar{\omega}_j + \frac{1}{2}\gamma_j)\} \leq \exp(\sum_1^{n_o}\gamma_j - \frac{1}{4}\sum_1^n \gamma_j^2)$$

But it is readily seen that

$$\{\bar{Y}_n > \sum_1^n \gamma_j(\bar{\omega}_j + \frac{1}{2}\gamma_j)\} = \{Y_n \leq \sum_1^n \gamma_j(\bar{\omega}'_j - \frac{1}{2}\gamma_j)\}$$

and in fact (8) is the desired relation.

<u>THEOREM 2</u> If it is possible to choose the family of sets $\{A_j\}$ in such a way that

$$\sum_1^\infty (\bar{\omega}'_j - \bar{\omega}_j)^2 = \infty$$

then there exists a convergent sequence of tests of H_0 against H_1.

<u>Proof</u> By the preceding theorem, the sequence of tests T_n associated with the family $\{A_j\}$ is convergent.

2. The case where the X_n are independent data deserves special attention, because

it is frequently encountered in practical applications. It is characterized by

$$P_\theta = \bigotimes_{n=1}^{\infty} P_{\theta,n}$$

where $P_{\theta,n}$ is the projection of P_θ on $(\mathcal{X}_n, \mathcal{B}_n)$.

Then a particular choice of the $\{A_j\}$ leads to some interesting results. This choice consists in taking, for every $j>2$, the set $A_j(x_1,..,x_{j-1})$ independently from $(x_1,...x_{j-1})$, which is clearly possible (e.g., $A_j = \mathcal{X}_j$ is a trivial solution).

ξ_j is now the indicator function of the event $\{X_j \in A_j\}$, and hereby the variables $\xi_1, \xi_2, \ldots \xi_n, \ldots$ are independent. As to the definition of $\bar{\omega}_j$ and $\bar{\omega}'_j$, it reduces to $\bar{\omega}_j = \sup\{P_{\theta,j}(A_j)/\theta \in \Theta_{o,j}\}$; $\bar{\omega}'_j = \inf\{P_{\theta,j}(A_j)/\theta \in \Theta_{1,j}\}$.

More restrictively, consider the situation in which H_0 and H_1 are simple hypotheses. We can put $\Theta_0 = \{0\}$ and $\Theta_1 = \{1\}$. It is well known that the "variational distance" d_V between two measures is always reached, and so there exists, for every $j \in \mathbb{N}^*$, a set $A_j \in \mathcal{B}_j$ such that

$$\begin{aligned} P_{1,j}(A_j) - P_{0,j}(A_j) &= \max\{|P_{1,j}(B) - P_{0,j}(B)| / B \in \mathcal{B}_j\} \\ &= d_V(P_{1,j}, P_{0,j}) \end{aligned}$$

We have $(\forall j \in \mathbb{N}^*)$, $\bar{\omega}_j = P_{0,j}(A_j)$ and $\bar{\omega}'_j = P_{1,j}(A_j)$ and $N_o = N_1 = 0$.

The next result is an immediate consequence of theorem 1.

THEOREM 3 Under the conditions just mentioned, the significance level $\alpha(T_n)$ and the power $\beta(T_n)$ of the tests T_n associated with the sequence $\{A_j\}$ satisfy

$$\alpha(T_n) \leq \exp\left[-\frac{1}{4}\sum_1^n d_V^2(P_{0,j}, P_{1,j})\right]$$

$$\beta(T_n) \geq 1-\exp\left[-\frac{1}{4}\sum_1^n d_V^2(P_{0,j}, P_{1,j})\right]$$

COROLLARY a) (T_n) is a convergent sequence of tests if

$$\sum_1^\infty d_V^2(P_{0,n}, P_{1,n}) = \infty$$

b) $(\mathcal{X}_n, \mathcal{B}_n, P_{0,n})$ and $(\mathcal{X}_n, \mathcal{B}_n, P_{1,n})$ $(n=1,2,\ldots)$ being arbitrary probability spaces, we have

$$d_V\left(\bigotimes_1^n P_{0,j}, \bigotimes_1^n P_{1,j}\right) \geq 1 - 2\exp\left[-\frac{1}{4}\sum_1^n d_V^2(P_{0,j}, P_{1,j})\right].$$

3. Our final step is to investigate the "sample case", i.e. where the X_n are independent and equidistributed. Then, each statistical model $(\mathcal{X}_n, \mathcal{B}_n, \{P_{\theta,n}\}_{\theta \in \Theta})$ is isomorphic to a given model $(\mathcal{X}, \mathcal{B}, \{P_\theta\}_{\theta \in \Theta})$.

THEOREM 4 Suppose that the family $\{P_\tau\}_{\theta \in \Theta}$ and the two hypothes s H_0 and H_1 satisfy the condition

$$(\exists B \in \mathcal{B}) , (\forall \theta \in \Theta_0) , (\forall \theta' \in \Theta_1) , P_\theta(B) < P_{\theta'}(B)$$

Then, there exists a convergent sequence of tests of H_0 against H_1.

Proof Putting $\bar{\omega} = \sup \{P_\theta(B)/\theta \in \Theta_0$, we shall distinguish two cases

a) Let $(\forall \theta \in \Theta_0) , P_\theta(B) < \bar{\omega}$.

Then, we can choose a sequence $\gamma_n \in]0,\bar{\omega}[$ converging towards 0 and such that $\sum_1^\infty \gamma_n^2 = \infty$. The sequence

$$\Theta_{o,n} = \{\theta \in \Theta_o / P_\theta(B) < \bar{\omega} - \gamma_n\}$$

converges towards Θ_o and

$$(\forall n) , (\forall \theta \in \Theta_{o,n}) , (\forall \theta' \in \Theta_1) , P_\theta(B) < \bar{\omega} - \gamma_n < \bar{\omega} \leq P_{\theta'}(B)$$

The conclusion follows directly from Theorem 2.

b) Let $(\exists \theta_0 \in \Theta_0) , P_{\theta_0}(B) = \bar{\omega}$.

γ_n being chosen as above, we see that the sequence

$$\Theta_{1,n} = \{\theta' \in \Theta_1 / P_{\theta'}(B) \geq \bar{\omega} + \gamma_n\}$$

converges towards Θ_1 and that

$$(\forall n) , (\forall \theta \in \Theta_o) , (\forall \theta' \in \Theta_{1,n}) , P_\theta(B) \leq \bar{\omega} < \bar{\omega} + \gamma_n \leq P_{\theta'}(B)$$

Once again, the Theorem 2 gives us the desired result.

THEOREM 5 H_0 being the simple hypothesis $(\theta = \theta_0)$, suppose that a finite or countable family $\{A_j\}_{j \in J}$ in $\mathcal{B}$ is such that

(9) $$(\forall \theta \in \Theta_1) , (\exists j(\theta)) , P_{\theta_0}[A_{j(\theta)}] \neq P_\theta[A_j(\theta)]$$

Then, there exists a convergent sequence T_n of tests of H_0 against H_1.

Proof Without any loss in generality, we can admit that $J = \mathbb{N}^*$ and that

$$(\forall j) , (\exists j') , A_{j'} = A_j^c \; (= \mathfrak{X} - A_j) .$$

This latter property permits to precise the relation (9) as follows

$$(\forall \theta \in \Theta_1) , (\exists j(\theta)) , P_{\theta_0}[A_{j(\theta)}] < P_\theta[A_{j(\theta)}]$$

So, if we put, for each $r \in \mathbb{N}^*$, $\Theta_{1,r} = \{\theta \in \Theta_1 / P_{\theta_0}(A_r) < P_\theta(A_r)\}$, we have

(10) $$\Theta_1 = \bigcup_{r=1}^{\infty} \Theta_{1,r}$$

By Theorem 4, there exists convergent sequence of tests of the hypothesis H_0 against the hypothesis $(\theta \in \Theta_{1,r})$. The conclusion is easily deduced therefrom and from (10).

It may be interesting to formulate the following straightforward consequence of the preceding theorem.

THEOREM 6 Suppose that the σ-algebra $\mathcal{B}$ has a countable basis, and that

$P_\theta \neq P_{\theta'}$ if $\theta \neq \theta'$

Then, for every $\theta_0 \in \Theta$, there exists a convergent sequence of tests T_n of the hypothesis $(\theta = \theta_0)$ against $(\theta \neq \theta_0)$.

REFERENCES

Geffroy, J. (1976). Inégalités pour le niveau de signification et la puissance de certains tests reposant sur des données quelconques; C.R. Acad. Sc. Paris, 282, série A, 1299.

Hillion, A. (1976). Sur l'intégrale de Hellinger et la séparation asymptotique; C.R. Acad. Sc. Paris, 283, série A, 61.

Moché, R. (1977). Thèse de Doctorat de l'Univ. de Lille I.

Recent Developments in Statistics
J.R.Barra et al., editors

CONSISTENCY OF MAXIMUM LIKELIHOOD ESTIMATORS OF THE FACTOR ANALYSIS MODEL, WHEN THE OBSERVATIONS ARE NOT MULTIVARIATE NORMALLY DISTRIBUTED

R.D. Gill
Department of Mathematical Statistics
Mathematisch Centrum
Amsterdam

A new proof of the consistency of maximum likelihood factor analysis estimation (i.e. maximum likelihood using the assumption of multivariate normality) is given which uses only the existence of 2nd moments and the uniqueness of the model.

Special attention is given to the problem of "Heywood Cases". The generality of the proof is such as to enable it to be adapted to many other situations.

The proof demonstrates the unsuitability of other much used methods of factor analysis, for which the property of consistency does not hold.

Introduction

Anderson & Rubin (1956) gave a proof of the asymptotic normality of maximum likelihood estimators, based on the assumption of asymptotic normality of the sample covariance matrix and various regularity conditions, of which the most important is identification of the model. Here a proof is given of a weaker property based on weaker assumptions though retaining identification; one which is perhaps clearer than theirs in that we consider directly the maximizing problem instead of transforming to a simultaneous equation problem obtained by differentiating a log likelihood function. An advantage of the procedure applied here is that consistency (both weak and strong) is also proved when the true parameters lie on the boundary of their permissable region (obviously no nondegenerate normal distribution about the true value is possible when the estimates too are constrained to be in this region). The procedure of Lawley & Maxwell (1971) for dealing with so called Heywood cases (i.e. maximum likelihood attained on the boundary) coincides with the maximization problem considered here (at least in so far as their method correctly recognizes a boundary case, or in general, succeeds in finding a global maximum and not just a local one). Their method is motivated by making new assumptions about the model if a Heywood case is indicated; that this is the same as true maximum likelihood is not proved here; but consists in observing that as parameters approach the border, so too does the likelihood function converge to their new likelihood function defined on the border.

Our method here is to show that if the model is identified, then the function, which when given a sample covariance matrix supplies maximum likelihood estimates, is continuous at the true covariance matrix (where it supplies the true parameter values). Then convergence in probability or almost surely of the sample covariance matrix to the true one implies the same kind of convergence of the estimates.

Assumptions, basic and simplifying

Independent observations are made of a p-component random vector $\underline{x}$ (random variables are underlined) possessing the p×p non-singular covariance matrix Σ_0. Then the correlation matrix of $\underline{x}$, say Γ_0, also exists, and the sample correlation matrix $\underline{C}_n$ based on n observations of $\underline{x}$ converges in probability (and a.s.) to Γ_0.

Now under the factor analysis model $\Gamma_0 = \Lambda_0\Lambda_0' + \Psi_0$, where Λ_0 is a $p\times m$ real matrix, and Ψ_0 is diagonal with non-negative diagonal elements. It is assumed that $m < p$ is known; and what is very important, that given this m, Ψ_0 is unique: i.e. if it is also so that $\Gamma_0 = \Lambda\Lambda' + \Psi$ with Λ $p \times m$ etc., then $\Psi = \Psi_0$. Λ_0 can now also be made unique, for instance by requiring the "above diagonal" elements of Λ (of which there are $\frac{1}{2}m(m-1)$) to be zero. (For a discussion of identification problems, see Anderson & Rubin (1956)).

We consider the maximum likelihood estimation process applied to $\underline{C}_n$ as if it were a sample covariance matrix. Estimates, say $(\underline{\Psi}_n,\underline{\Lambda}_n)$, obtained in this way would have to be scaled to make them correspond to (Ψ_0,Λ_0), the parameters in the model for the correlation matrix Γ_0. However it turns out that $\mathrm{diag}(\underline{\Lambda}_n\underline{\Lambda}_n'+\underline{\Psi}_n) = \mathrm{diag}(\underline{C}_n) = I$, so the scaling never has to be made. We can accordingly write $\underline{\Gamma}_n = \underline{\Lambda}_n\underline{\Lambda}_n' + \underline{\Psi}_n$ for an M.L. (maximum likelihood) estimate of Γ_0.

A suitable function to be maximized is $f(\underline{C}_n;\Psi,\Lambda)$ (by choice of Ψ,Λ), defined by

$$f(C;\Psi,\Lambda) = \mathrm{logdet}\,(C\Gamma^{-1}) - \mathrm{trace}\,(C\Gamma^{-1}) + p, \text{ and } \Gamma = \Lambda\Lambda' + \Psi,$$

where Λ,Ψ are real matrices; Λ is $p \times m$, Ψ is diagonal with $\Psi_{ii} \geq 0\ \forall i$; and Λ and Ψ are further restricted so that Γ is positive definite. C is symmetric and also positive definite. It is easy to prove that $C\Gamma^{-1}$ has all eigenvalues positive; i.e. $\det(C\Gamma^{-1}) > 0$. Suppose that these eigenvalues are ϕ_i, $1 \leq i \leq p$; then

$$f(C;\Psi,\Lambda) = \textstyle\sum_{i=1}^{p} (\log \phi_i - \phi_i + 1).$$

Now

$$\begin{aligned} &(\log \phi-\phi+1) \leq 0 && 0 < \phi < \infty \\ &(\log \phi-\phi+1) \to -\infty && \phi \to 0 \quad \text{or} \quad \phi \to \infty \\ &\log \phi-\phi+1 = 0 \iff \phi = 1. \end{aligned}$$

So

$$\begin{aligned} f(C;\Psi,\Lambda) &= 0 \iff \phi_i = 1\ \forall i \\ &\iff C\Gamma^{-1} = I \\ &\iff C = \Gamma = \Lambda\Lambda' + \Psi. \end{aligned}$$

We can remove the restriction of Γ to being positive definite (for instance, Γ is singular if more than m Ψ_{ii}'s are equal to zero), by noting that as $\det(\Gamma) \to 0$, smallest eigenvalue $(C\Gamma^{-1}) \to 0$, so $f \to -\infty$. f is therefore considered as an extended real valued function taking values in $[-\infty,0]$ (this does not affect the result of the maximization of f). As such it is continuous, when $[-\infty,0]$ has the natural topology generated by the usual open intervals together with the intervals $[-\infty,a)$ and $(a,0]$.

As remarked before it can be proved that $\mathrm{diag}(\underline{\Gamma}_n) = \mathrm{diag}(\underline{C}_n) = I$. If $\Gamma = \Lambda\Lambda' + \Psi$ then $\Gamma_{ii} = \sum_j \Lambda_{ij}^2 + \Psi_{ii}$; but $\Psi_{ii} \geq 0$; hence $|(\underline{\Lambda}_n)_{ij}| \leq 1$ and $|(\underline{\Psi}_n)_{ij}| \leq 1\ \forall i$ and j. (Lawley & Maxwell (1971) prove this result essentially for the non-Heywood case; but the argument can also be repeated for their method when some of the diagonal values of an M.L. Ψ_n are zero.) So we may restrict the maximization to taking place over (Ψ,Λ) with $\Psi_{ii} \leq 1$ and $|\Lambda_{ij}| \leq 1\ \forall i,j$, and our solution is the same as theirs.

Ψ and Λ can now be considered as lying in closed, bounded subspaces of their respective Euclidean spaces. It is convenient to do the same for C, the remaining argument of f,to make possible the analytic lemma which we shall shortly prove to be directly applied. We already know that C is a correlation matrix, so we can assume $C_{ii} = 1$, $|C_{ij}| \leq 1$, $C_{ij} = C_{ji}\ \forall i,j$; we also require C to be positive definite, i.e. the smallest eigenvalue of $C > 0$. ($\underline{C}_n$ may be nonsingular, in which case the f specified here was not defined). Now Γ_0 is nonsingular, so suppose smallest eigenvalue $(\Gamma_0) > c > 0$. Then by continuity of the smallest eigenvalue as a function of C, and by convergence (in probability and a.s.) of $\underline{C}_n$ to Γ_0, we only need look at $\underline{C}_n$ with smallest eigenvalue $(\underline{C}_n) \geq c$.

We collect the ingredients as follows: Let

$$\mathcal{C} = \{C \mid C \text{ a } p\times p \text{ real matrix, } C_{ij} = C_{ji},\ C_{ii} = 1,$$
$$|C_{ij}| \leq 1, \text{ eigenvalues } (C) \geq c\}$$

and

$$\mathcal{P} = \{(\Psi,\Lambda) \mid \Psi\ p\times p \text{ real, } \Psi_{ij} = 0\ i\neq j,\ 0 \leq \Psi_{ii} \leq 1,$$
$$\Lambda\ p\times m \text{ real, } \Lambda_{ij} = 0\ j > 1,\ |\Lambda_{ij}| \leq 1\}$$

$\mathcal{C}$ and $\mathcal{P}$ are compact.

Then $f\colon \mathcal{C}\times\mathcal{P}\to[-\infty,0]$ is a continuous function on a compact space. Let $\mathcal{Y}$ be the set function on $\mathcal{C}$ defined by

$$\mathcal{Y}(C) = \{(\Psi,\Lambda) \mid f(C;\Psi,\Lambda) = \sup_{(\Psi,\Lambda)\in\mathcal{P}} f(C;\Psi,\Lambda)\}.$$

By compactness of $\mathcal{P}$ and continuity of f, $\mathcal{Y}(C)$ is closed and nonempty as a subset of $\mathcal{P}$ for each $C \in \mathcal{C}$.

$\mathcal{Y}(\underline{C}_n)$ contains the maximum likelihood estimates of (Ψ_0,Λ_0) and $\mathcal{Y}(\Gamma_0) = \{(\Psi_0,\Lambda_0)\}$ (uniqueness of Ψ_0).

We shall show that $\mathcal{Y}\colon \mathcal{C} \to \{\text{closed nonempty subsets of } \mathcal{P}\} = \mathcal{F}$ say, is continuous, where the suitable metric on $\mathcal{F}$ is the Hausdorff distance $\rho(A,) = \sup_{a\in A, b\in B}\{d(a,b)\}$ and d is the ordinary Euclidean distance on $\mathcal{P}$.

Hence, as $P(\underline{C}_n\in\mathcal{C}) \to 1$ and $\underline{C}_n \overset{P}{\to} \Gamma_0$,

$$\mathcal{Y}(\underline{C}_n) \overset{P}{\to} \mathcal{Y}(\Gamma_0);$$

i.e. the distance of the furthest maximum likelihood estimate of (Ψ_0,Λ_0) to (Ψ_0,Λ_0) itself converges in probability to zero (and by analogous arguments, almost surely).

<u>Some analysis</u>

c, c_0, c' will be points in $\mathcal{C}$; c_0 being fixed and playing the role of Γ_0. Similarly p, p_0, $p' \in \mathcal{P}$; z, z_0, z' will be the corresponding points in $\mathcal{C}\times\mathcal{P}$. $\mathcal{C}$ and $\mathcal{P}$ are closed, bounded Euclidean subspaces; instead of f we shall equivalently consider

$$g = \frac{f}{1-f} \qquad 0 \geq f > -\infty$$

$$g = -1 \qquad f = -\infty;$$

$g\colon \mathcal{C}\times\mathcal{P}\to[-1,0]$ is continuous and hence uniformly continuous. $\mathcal{F}$ is the set of closed non-empty subsets of $\mathcal{P}$ endowed with the Hausdorff metric. $\mathcal{Y}\colon \mathcal{C}\to\mathcal{F}$ is defined by

$$\mathcal{Y}(c) = \{p\colon g(c,p) = \sup_{p'\in\mathcal{P}} g(c,p')\}.$$

d is the Euclidean metric on $\mathcal{C}$, $\mathcal{P}$ or $\mathcal{C}\times\mathcal{P}$ as indicated by its arguments. In particular note that

$$d^2(z,z') = d^2(c,c') + d^2(p,p').$$

c_0 is such that

$$\mathcal{Y}(c_0) = \{p_0\} \quad\text{and}\quad g(c_0,p_0) = 0.$$

<u>LEMMA</u>. *With the above notations and definitions, $\mathcal{Y}$ is continuous at c_0, i.e. given $\varepsilon > 0\ \exists\delta$:*

$$d(c,c_0) < \delta \quad\text{and}\quad p \in \mathcal{Y}(c) \Rightarrow d(p,p_0) < \varepsilon.$$

<u>PROOF</u>. Choose $\varepsilon > 0$.

$$\exists\eta(\varepsilon) > 0\colon d(p,p_0) \geq \varepsilon \Rightarrow g(c_0,p) \leq g(c_0,p_0) - \eta$$

because the supremum of $g(c_0,\cdot)$ outside of a neighbourhood of p_0 must be strictly less than $g(c_0,p_0)$, p_0 being the unique maximizing value in $\mathcal{P}$, and $g(c_0,\cdot)$ being uniformly continuous.

$$\exists\delta(\eta) > 0: \delta < \varepsilon \text{ and } d^2(z,z') \le 2\delta \Rightarrow |g(z)-g(z')| \le \frac{\eta}{3},$$

because g is uniformly continuous.

Suppose now c is such that $d(c,c_0) < \delta$. If p satisfies $d(p,p_0) < \delta$, then $d((c,p),(c_0,p_0)) < \sqrt{2\delta} \Rightarrow g(c,p) \ge g(c_0,p_0) - \eta/3$.
If p satisfies $d(p,p_0) \ge \varepsilon$, then $d((c,p),(c_0,p)) < \delta \Rightarrow g(c,p) \le g(c_0,p) + \eta/3 \le$ $\le (g(c_0,p_0) - \eta) + \eta/3$ $(d(p,p_0) \ge \varepsilon)$, i.e. $g(c,p) \le g(c_0,p_0) - 2\eta/3$. So $g(c,\cdot)$ attains values $\ge g(c_0,p_0) - \eta/3$ when $d(p,p_0) < \delta$, but is bounded above by $g(c_0,p_0) - 2\eta/3$ when $d(p,p_0) \ge \varepsilon > \delta$.

So $p \in \mathcal{Y}(c) \Rightarrow d(p,p_0) < \varepsilon$. □

Conclusion and some heuristic comments

The idea of this paper is that the result will be comforting to those applying factor analysis with the only technique presently available for it which has some statistical justification, i.e. maximum likelihood methods applied under the assumption of multivariate normality. Of course, how quickly convergence occurs can presumably be as slow as anything one may suggest; the computational method used need not give an absolute maximum*); also, somewhere along the line, the number of factors must be specified.

We can see from the above proof that if the model holds with the given number of factors, then the maximum likelihood criterion used here will converge to zero; otherwise it will presumably converge to some value less than zero representing the "closest" m-factor covariance matrix to the true covariance matrix. If this closest distance can be given an empirical relevance - i.e. closer than ε is to all intents and purposes the same as exactly zero, and if the closest factor model is essentially unique, then the problem of the number of factors is also asymptotically solved. Note too that the normal theory likelihood ratio test statistic is basically the estimated smallest distance blown up by the number of observations; under the null hypothesis asymptotically χ^2 distributed.

REFERENCES

Anderson, T.W., and Rubin, H. (1956). Statistical Inference in Factor Analysis. Proc. Third Berkeley Symp. Math. Statist. Probab., 5, 111-150.

Lawley, D.N., and Maxwell A.E. (1971). Factor Analysis as a Statistical Method. (Butterworths, London).

Prins, H.J., and Van Driel, O.P. (1974). Estimating the Parameters of the Factor Analysis Model without the Usual Constraints of Positive Definitness. Proc. Symp. Computational Stat., Vienna.

*) For an approach giving computational and interpretational advantages based on a broader specification of the model, see Prins and Van Driel (1974).

Recent Developments in Statistics
J.R.Barra et al., editors

GENERALISING THE YAGLOM LIMIT THEOREMS

P.J. Green
School of Mathematics
University of Bath
Bath, United Kingdom.

In this paper we present generalisations of the so-called Yaglom conditional limit theorems to the general branching process counted by the values of a random characteristic, as proposed by Jagers (1974). Even when restricted to the special case of the usual population-size process, our results are stronger than those previously available.

INTRODUCTION.

Any branching process may be classified as subcritical, critical or supercritical according to whether the mean offspring number is less than, equal to, or greater than unity. In the critical and subcritical cases extinction is certain, so the first-order limit behaviour of the process is uninteresting. But consideration of second-order behaviour - the state of the process at time t conditional on extinction not occurring before t - yields the so-called Yaglom limit theorems for the population size process. It is appealing to look for similar results for other features of the branching process obtained by counting the individuals in a more general fashion: this paper presents two such theorems.

The branching model we use is the general branching process in which individuals can give birth throughout their lifetimes, an arbitrary degree of age-dependence of fertility being permitted. The "branching property" is that individuals behave from their births in an independent and identically distributed manner. The model includes as very special cases the Galton-Watson process, the Bellman-Harris process, and Sevast'yanov's age-dependent process. Each individual's effect will be measured in a generalised way by an arbitrary non-negative random function of the time since its birth, the total effect of the population being the sum of the individual effects. Special cases include the number of living individuals, the number in a given age-range, the integral of the population size process, and the accumulated number of births or deaths - all features that have received attention in the literature. But it is clear that we may thus model in a very general way the impact of the branching process on its environment.

We will use the formulation of the general branching process given by Jagers (1969). The idea of counting the population as above, by the values of a "random characteristic" is also due to Jagers (1974), who obtained limit theorems for the resulting process in the supercritical case only.

Conditional limit theorems for the population size in this general process were obtained by Ryan (1968) for the subcritical case, and by Durham (1971) and Holte (1974) in the critical case. The theorems we present are however stronger than these, even when restricted to the special case of the population size.

Similar theorems to those given here appear in the author's thesis (1976a). These apply to the Bellman-Harris model only, and to a restricted type of random characteristic.

RESULTS.

We use the notation of Jagers (1974). In summary, each individual x of the process is given a triple $(\lambda_x, \xi_x, \chi_x)$; λ_x is a non-negative random variable, the lifespan of x : ξ_x is a point process on $[0,\infty)$, $\xi_x(t)$ being the number of births to x by age t : χ_x is a non-negative stochastic process (the random characteristic), $\chi_x(t)$ being the effect of x at age t, zero for $t<0$. All triples are independently and identically distributed; we omit the subscript x when referring to an arbitrary triple.

The process we study is

$$X(t) = \sum_x \chi_x(t-\sigma_x)$$

where σ_x is the birth time of x. A special case is obtained by taking χ_x to be the indicator of $[0,\lambda_x)$, yielding the population size process, $Z(t)$. The extinction time is $T = \inf\{t : Z(t) = 0\}$. We will consider limit theorems for $\{X(t) \mid T>t\}$ as $t\to\infty$.

Throughout, we make the assumption that

$$P\{\chi(t) = 0 \mid \lambda \leq t\} = 1,$$

so that only live individuals contribute to $X(t)$. We believe that the theorems remain true with minor changes under slightly weaker assumptions; if, however, there is a substantial contribution to $X(t)$ from "dead" individuals, as in the case with the birth count or integral of the process, qualitatively different behaviour is found (see Green, 1976a).

Some further notations and regularity conditions are required. We write $\mu(t) = E(\xi(t))$ for the reproduction function, assumed non-lattice, and $a = \mu(\infty)$ for the expected family size. The case $1<a<\infty$ is covered by Jagers: we consider only $a \leq 1$, and require $\mu(0)<1$. We write $L(t) = P\{\lambda \leq t\}$. Using ^ to denote Laplace-Stieltjes transform, the Malthusian parameter α, the root of $\hat{\mu}(\alpha) = 1$, is assumed to exist - this is no restriction if $a = 1$, when $\alpha = 0$. We let $\beta = \int t e^{-\alpha t} \mu(dt)$, assumed finite, denote the expectation of the resulting associated distribution, or "average age at child-bearing".

Finally there is a weak smoothness assumption on χ: we suppose that $\chi(t)$ is almost surely almost everywhere continuous in t, and require $e^{-\alpha t}E(\chi(t))$ to be directly Riemann integrable, writing $\bar{\chi} = \beta^{-1}\int e^{-\alpha t} (\chi(t))\, dt$.

We now state our theorems: all the assumptions above remain in force.

Theorem 1.

If $\quad a<1,$

$$E(\hat{\xi}(\alpha) \log \xi(\infty)) < \infty$$

and

$$\int t e^{-\alpha t} L(dt) < \infty ,$$

then

$$P\{X(t) \leq y \mid T > t\} \overset{D}{\to} F(y)$$

as $t \to \infty$, where F is a proper distribution with finite expectation $C^{-1}\bar{\chi}$; here $C = \lim_{t\to\infty} e^{-\alpha t} P\{T > t\}$, which is a finite positive constant (see Ryan, 1968).

Theorem 2.

If $a = 1$,

$\sigma^2 = \mathrm{Var}(\xi(\infty)) < \infty$,

$t^2(1 - L(t)) \to 0$ and $t^2(1 - \mu(t)) \to 0$ as $t \to \infty$,

and

$E(\chi(t)^2)$ is bounded and $\to 0$ as $t \to \infty$,

then

$$P\left\{\frac{X(t)}{t} \leq y \mid T > t\right\} \overset{D}{\to} 1 - \exp\left\{\frac{-2\beta}{\sigma^2\bar{\chi}} y\right\}$$

as $t \to \infty$.

It is interesting to compare these results, when specialised to the population size process, with those previously available.

In the subcritical case, the expectation of the limit law was not previously known. In the critical case, Holte's theorem requires an unnatural assumption on the rate of convergence of $E(Z(t))$ to $\beta^{-1}\int t\, L(dt)$, while Durham assumes that all moments of $\xi(\infty)$ are finite.

There appears to be no previous work on conditional limit theorems for the random characteristic model.

Proofs of these theorems proceed from the observation that under the condition $P\{\chi(t) = 0 \mid \lambda \leq t\} = 1$, we have

$$E(e^{-\theta X(t)} \mid T > t) = 1 - H(\theta,t)/P\{T > t\} ,$$

where

$$H(\theta,t) = E(1 - e^{-\theta X(t)}) .$$

It is easy to obtain a renewal type equation in the form

$$H(\theta,t) = \theta\, E(\chi(t)) + Q(\theta,t) + \int_{[0,t]} H(\theta,t-u)\mu(du)$$

where the term $Q(\theta,t)$ involves H to second and higher order. But bounds on H, and hence on its effect on Q may be derived, so that standard renewal theory may be utilised to give the required asymptotic behaviour of H.

Full details of the proofs will appear elsewhere (Green, 1976b).

REFERENCES.

Durham, S.D. (1971). Limit theorems for a general critical branching process. J.Appl.Prob. 8, 1-16.

Green, P.J. (1976a). Random variables on branching process paths. Ph.D. Thesis, University of Sheffield.

Green, P.J. (1976b). Conditional limit theorems for general branching processes. To appear.

Holte, J.M. (1974). Limit theorems for critical general branching processes. Ph.D. Thesis, University of Wisconsin.

Jagers, P. (1969). A general stochastic model for population development. Skand.Aktuar.Tidskr. 1, 84-103.

Jagers, P. (1974). Convergence of general branching processes and functionals thereof. J.Appl.Prob. 11, 471-478.

Ryan, T.A. (1968) On age-dependent branching processes. Ph.D. Thesis, Cornell University.

Recent Developments in Statistics
J.R.Barra et al., editors

ON A CLASS OF BRANCHING PROCESSES

G. GREGOIRE
Université des Sciences Sociales - U.E.R. des Sciences Economiques
Domaine Universitaire
38400 - St Martin d'Hères - France

SUMMARY : Laplace functionals are used to obtain sufficient conditions for the existence of a class of measure-valued (possibly infinite) branching processes ; these branching processes are constructed by an extension of some cluster processes.

1 - INTRODUCTION

In some space (the real line or R^d) particles are distributed at random and constitute what we call the "underlying" (or "primary") point distribution. Let us suppose that every point of the underlying point distribution gives rise to a new point distribution (a cluster), independently of all other particles and of the probability law of the primary distribution. The superposition of all the point distributions coming from the particles of the underlying point distribution is the "secondary" point distribution, and we have constructed the so-called cluster process.

Various problems on the probability law of the "offspring" of each particle and those of the primary and secondary point distributions are studied by many authors. Prekopa, Szasz [5], Matthes [6] for instance are concerned with the Poisson primary distribution. Conditions are given so that the secondary distribution be also of Poisson type or compound Poisson type. For some distributions of offspring, invariant primary distributions are studied. But generally, the number of offspring of each particle is a-s bounded, and the bound is the same for all particles, Goldman [7] gives a theorem for the Poisson primary distribution with a homogeneous distribution of offspring which is only a-s finite.

Here, we consider σ-finite measures instead of point distributions, and each point of the primary random measure gives rise to a σ-finite measure. In this setting, we shall say that a cluster process exists when it can be extended in a measure-valued branching process ; we study this problem of existence by way of Laplace functionals. This is done through the generalization of a Jirina formula ([3], [4]), which concerns only the finite case.

2 - RANDOM MEASURES

For a general theory of random measures, the reader is referred to Jagers [8]. However, to make the present paper self-contained, we recall some of the basic definitions and properties.

Let X be R^d (but X may be a second countable locally compact space), M the set of Radon (i.e. σ-finite) positive measures on X, and M_p the subset of M constituted by the measures of the form $\sum_{i\in I} \alpha_i \delta_{x_i}$ (the α_i ($i\in I$) are integers, δ_x is the Dirac measure on x, and (x_i, $i\in I$) a locally finite sequence on X).

$\mathcal{M}$ will be the σ-field on M generated by the mappings : $\mu \to (\mu(A_1),\ldots,\mu(A_n))$, for all n and $A_1,\ldots,A_n$ bounded (i.e. relatively compact) Borel sets. It is known that $\mathcal{M}$ is the Borel σ-field for the vague topology on M.

A random measure on X is a measurable function from a probability space $(\Omega,\mathcal{S},P)$ into $(M,\mathcal{M})$. A point process is a random measure with values in M_p.

Let us recall that there is a one-to-one correspondance between the probability distributions P on $(M,\mathcal{M})$, and the systems of finite dimensional distributions $P_{\{A_1,\ldots,A_n\}}(\cdot)$ on R_+^n, indexed on all the finite systems of bounded Borel sets of X, and satisfying :

(C_1) A_1, A_2 bounded Borel sets of X, $A_1 \cap A_2 = \emptyset$, then :

$$P_{\{A_1,A_2,A_1 \cup A_2\}}\{(x,y,z)\ \ R_+^3 \ /\ x+y = z\} = 1.$$

(C_2) A, A_n $(n \in N)$ bounded Borel sets of X, $A_n \subset A_{n+1}$ ($\forall n$ N), $A = \bigcup_n A_n$, then :

$$P_{\{A \setminus A_n\}} \to \delta_0 \qquad \text{(vaguely or narrowly).}$$

(C_3) A bounded Borel set, then :

$$P_{\{A\}}(R_+) = 1.$$

We shall say that a random measure with probability law P is integrable if the measure on X defined by :

$$EP(A) = \int_M \nu(A)\ P(d\nu)$$

for all A Borel sets of X, is in M.

Finally, we define the Laplace transform of a probability distribution P on $(M,\mathcal{M})$ by :

$$\varphi_P(f) = \int_M e^{-\langle f,\nu\rangle}\ P(d\nu), \text{ with } \langle f,\nu\rangle = \int_X f(x)\ \nu(dx).$$

The functions f are taken in the space $\mathcal{G}^+$ of all bounded positive measurable functions on X, but often, we may restrict ourselves to the space $\mathcal{C}_K^+$ of positive continuous functions with compact support.

Notice that, if we take $f = \sum_{i=1}^{n} \lambda_i 1_{A_i}$, with A_i disjoint bounded Borel sets and $\lambda_i \geq 0$, then we may define ;

$$\varphi(A_1,\ldots,A_n\ ;\ \lambda_1,\ldots,\lambda_n) = \varphi_P\Big(\sum_{i=1}^{n} \lambda_i 1_{A_i}\Big)$$

with $1_A(x) = 1$ if $x \in A$, and $1_A(x) = 0$ if $x \notin A$.

The function $\varphi(A_1,\ldots,A_n\ ;\ \cdot)$ is just but the Laplace transform $\varphi_{\{A_1,\ldots,A_n\}}(\cdot)$ of the probability distribution $P_{\{A_1,\ldots,A_n\}}$ on R_+^n.

We call Logarithmic functional, the function : $\Psi(f) = -\mathrm{Log}\,\varphi(f)$.

An interesting result that will be used is the following one :

If the sequence $(\varphi_n, n \in N)$ of the Laplace functionals of the probabilities P_n on $(M,\mathcal{M})$ converges pointwise on $\mathcal{C}_K^+$ towards a function $\varphi : \mathcal{C}_K^+ \to [0,1]$ and if $\lim_{t\downarrow 0} \varphi(tf) = 1$ for all $f \in \mathcal{C}_K^+$, then the sequence of probabilities P_n converges narrowly on $(M,\mathcal{M})$ towards a probability P, whose Laplace transform is φ.

3 - CLUSTER, BRANCHING PROCESSES WITH MEASURE-VALUED STATES

Here, "branching process" designates a homogeneous discrete time-parameter branching process with states in M. Such a branching process is completely defined by :

* P^1, the initial distribution on $(M,\mathcal{M})$.
* $P(\mu,U)$ a transition function on $M \times \mathcal{M}$ satisfying :

$$P(\mu_1+\mu_2,.) = P(\mu_1,.) * P(\mu_2,.) \qquad (* \text{ denotes convolution}).$$

$\psi^1(f)$, $\Psi(\mu,f)$ will denote respectively the logarithmic functionals of P^1, $P(\mu,.)$. $\Psi(\delta_x,f)$ as a function of $x \in X$, will also be denoted by $\Psi(x,f)$ (and $\Psi(.,f)$ will denote the function $x \to \Psi(x,f)$.).

Following Jirina, we call "regular" the branching processes we are concerned with, i.e. those which satisfy :

$$\forall f \in \mathcal{G}^+,\ \Psi(\mu,f) = \int_X \Psi(x,f)\ \mu(dx) \qquad P^1(d\mu) \text{ a.s.} \tag{3.1}$$

A straightforward consequence of (3.1) is that, in terms of finite dimensional distributions, we have :

$\forall n$, $\forall A_1, \ldots, A_n$ bounded Borel sets of X,

$$\forall \lambda \in R_+^n,\ \Psi_{\{A_1,\ldots,A_n\}}(\mu,\lambda) = \int_X \Psi_{\{A_1,\ldots,A_n\}}(x,\lambda)\ \mu(dx) \quad P^1(d\mu) \text{ a.s.} \tag{3.2}$$

The most interesting consequence of (3.1) or (3.2) is that we can derive a formula for the Logarithmic functional $\psi^2(f)$ of the distribution of the second generation :

$$\psi^2(f) = \psi^1(\Psi(.,f)) \qquad \forall f \in \mathcal{G}^+ \tag{3.3}$$

and therefore

$$\psi^2_{\{A_1,\ldots,A_n\}}(\lambda) = \psi^1(\Psi_{\{A_1,\ldots,A_n\}}(.,\lambda)) \qquad \forall \lambda \in R_+^n \tag{3.4}$$

for all $n \geq 0$ and $A_1, \ldots, A_n$ bounded Borel sets of X.

We give examples, in section 5, for the use of formulas (3.3) and (3.4).

We complete this section by the definition of the existence of a cluster process.

A cluster process is given by :

* P^1, the primary distribution on $(M,\mathcal{M})$.
* $P(x,U)$ a transition function on $X \times \mathcal{M}$.

It is said non-degenerated (or it exists), if the transition function $P(x,U)$ can be extended, P^1 almost surely, in a transition function $P(\mu,U)$ on $M \times \mathcal{M}$ by way of the functional defined by :

$$\forall f \in G^+, \Psi(\mu,f) = \int_X \Psi(x,f)\ \mu(dx).$$

If this is the case, then (3.3) and (3.4) yield the secondary probability distribution. In the next section we discuss sufficient conditions under which the extension is possible.

4 - SUFFICIENT CONDITIONS FOR EXISTENCE

Theorem : If, P^1 almost surely, for all $f \in C_K^+$, $\int_X \Psi(x,f)\ \mu(dx) < +\infty$, and if there exists a sequence $(\mu_n,\ n \in N)$ of bounded positive measures with finite supports such that for all $f \in C_K^+$, $\int_X \Psi(x,f)\ \mu_n(dx)$ tends to $\int_X \Psi(x,f)\ \mu(dx)$, then the cluster process exists and relations (3.3) and (3.4) are satisfied.

The proof results from the fact that, if $\mu_n = \sum_{i=1}^{m} \alpha_i\ \delta_{x_i}$ $(\alpha_i \in R_+)$, then the function $f \to \int_X \Psi(x,f)\ \mu_n(dx)$ is a Logarithmic functional ; thus the function $f \to \int_X \Psi(x,f)\ \mu(dx)$ is a limit of Logarithmic functionals, and for completing the proof we have only to show that $\int_X \Psi(x,tf)\ \mu(dx)$ tends towards 0 when t tends towards 0 ; but this results from the monotone convergence theorem.

Remarks

1) If the function $\mu \to \int \Psi(x,f)\ \mu(dx)$ for all f in C_K^+, is vaguely continuous everywhere on M, then the hypotheses of the above theorem are satisfied. But this means that $\Psi(.,f)$ belongs to C_K^+, and this is not very interesting. Indeed, let us suppose that the primary random measure is a point distribution, the hypotheses on Ψ mean that only particles located in a fixed compact set give rise to offspring and the problem is reduced to a problem of finite measures.

2) If P^1 is concentrated on M_p, it is clear that we have only to verify that, almost surely, the integral $\int_X \Psi(x,f)\ \mu(dx)$ is convergent. But this condition is in stochastic form and therefore not easy to verify. So we give the following corollaries which can be deduced from the theorem.

Corollary 1 : If the primary random measure is concentrated on M_p, (i.e. if it is a point process) and integrable with expectation-measure EP^1, then the condition $\forall f \in C_K^+$, $\int_X \Psi(x,f)\ EP^1(dx) < +\infty$ is a sufficient condition for the existence of the cluster process.

Let S' stand for the space of positive tempered measures on R^d, i.e. the space of positive measures μ which satisfy :

$$\exists k \in N \quad \int_X \frac{\mu(dx)}{(1+|x|^2)^k} < +\infty \quad \text{with } |x|^2 = \sum_{i=1}^{d} x_i^2,\ x = (x_1,\dots,x_d).$$

$\mathcal{S}$ is the space of all infinitely differentiable functions f on R^d such that $|x|^k \partial^p f(x)$ vanishes at infinity for all $k \in N$ and $p \in N^d$.

Then we have :

Corollary 2 : If the primary random measure is integrable, concentrated on $\mathcal{S}' \cap M_p$, and if, for all $f \in C_K^+$, the function $x \to \Psi(x,f)$ belongs to $\mathcal{S}$, then the cluster process exists.

As an application of this latter corollary, note that the samples of a stationary Poisson process on the real line X = R are almost surely in $\mathcal{S}' \cap M_p$:

$$EP^1(dx) = adx \Rightarrow \int_R \frac{1}{1+x^2} EP^1(dx) = \int_R \frac{1}{1+x^2} adx < +\infty.$$

Thus $\int_R \frac{1}{1+x^2} \mu(dx) < +\infty \quad P^1(d\mu)$ a.s.

5 - EXAMPLES

In this section, for simplicity's sake, X will be the real line R.

1) Random motion of the particles of a stationary Poisson process.
Let P_0 be a probability distribution on X, we associate with P_0 the discrete random measure defined by :

$$P_{\{A_1,\ldots,A_n\}}(e_i) = P_0(A_i) \; ; \; P_{\{A_1,\ldots,A_n\}}(0) = 1 - P_0(\bigcup_{i=1}^{n} A_i)$$

with $\{e_i\}_{i=1,\ldots,n}$ the canonical basis of R^n, and $0 = (0,0,\ldots,0)$.

The corresponding Logarithmic functional is :

$$\Psi(f) = -\mathrm{Log} \int e^{-f(y)} P_0(dy), \; f \in C_K^+$$

Then let us consider a homogeneous cluster process with P^1 a stationary Poisson process with parameter a, and P(x,.) defined by :

$$P_{\{A_1,\ldots,A_n\}}(x,e_i) = P_0(\bigcup_{i=1}^{n} A_i - x) \; ; \; P_{\{A_1,\ldots,A_n\}}(x,0) = 1 - P_0(\bigcup_{i=1}^{n} A_i - x)$$

$$\Psi(x,f) = -\mathrm{Log} \int e^{-f(x+y)} P_0(dy)$$

We easily see : $\int \Psi(x,f) \mu(dx) < \infty \quad P^1(d\mu)$ - a.s.

This results from the convergence of : $\int \Psi(x,f) EP^1(dx) = -\int \mathrm{Log}[\int e^{-f(x+y)} P_0(dy)] adx$

Observe that, for $f \in C_K^+$, we have :

$$\int e^{-f(x+y)} P_0(dy) = 1 - P_0(K-x) + \int_{K-x} e^{-f(x+y)} P_0(dy), \quad \text{with K, support of f}$$

$$\geq 1 - P_0(K-x) \, |1-e^{-S}| \qquad S = \sup f(x) > 0$$

and the convergence of $\int \Psi(x,f) \mu(dx)$ is made clear.

The secondary probability distribution is given by :

$$\Psi^1(f) = \int (1 - e^{-f(x)})\ adx$$

$$\Psi^2(f) = \Psi^1(\Psi(.,f))$$

$$= \int (1 - e^{\mathrm{Log}\int e^{-f(y+x)}P_0(dy)})\ adx$$

$$= \int (1 - \int e^{-f(y+x)}\ P_0(dy))\ adx$$

$$= \int (1 - e^{-f(x)})\ adx$$

$$= \Psi^1(f)$$

2) Stationary Poisson process and geometric clustering.
Let us define a geometric random measure by :

$$\varphi_{\{A\}}(\lambda) = [1+\nu(A)(1-e^{-\lambda})]^{-1}\ ;\ \varphi_{\{A_1,A_2\}}(\lambda_1,\lambda_2) = \Big[1+\nu(A_1)(1-e^{-\lambda_1})+\nu(A_2)(1-e^{-\lambda_2})$$

$$-\nu(A_1\cap A_2)(1-e^{-\lambda_1})(1-e^{-\lambda_2})\Big]^{-1}$$

and so on ... (with ν, a σ-finite measure).

$$\Psi(f) = -\mathrm{Log}\Big[1 + \int (1 - e^{-f(x)})\ \nu(dx)\Big]\ , \qquad f\in C_K^+$$

Then suppose P^1 to be like in 1), and $P(x,.)$ of the above geometric type with measure parameter defined by $\nu(x,A) = e^{-x}\ \nu_1(A)\ 1_{R_+}(x)$ (ν_1 a σ-finite measure on X).

We have

$$\Psi_{\{A\}}(x,\lambda) = -\mathrm{Log}[1 + e^{-x}\ 1_{R_+}(x)\ \nu_1(A)(1 - e^{-\lambda})]^{-1}$$

$$\Psi(x,f) = -\mathrm{Log}[1 + \int (1 - e^{-f(y)})\ e^{-x}\ \nu_1(dy)]\ , \qquad f\in C_K^+$$

$$= -\mathrm{Log}[1 + e^{-x} \int (1 - e^{-f(y)})\ \nu_1(dy)]$$

The function $x \to \Psi(x,f)$ is in S and we may use the corollary 2 for stating the existence of the cluster process.

We may then use (3.4) for computing the secondary probability distribution :

$$\Psi^2_{\{A\}}(\lambda) = \Psi^1(\Psi_{\{A\}}(.,\lambda)) = \int (1 - e^{-\mathrm{Log}[1-e^{-x}1_{R_+}(x)\nu_1(A)(1-e^{-\lambda})]})\ adx$$

$$= \int_0^{+\infty} \frac{e^{-x}(1-e^{-\lambda})\nu_1(A)}{1+e^{-x}(1-e^{-\lambda})\nu_1(A)}\ adx$$

$$= -\mathrm{Log}[1 + \nu_1(A)(1 - e^{-\lambda})]^{-a}$$

and $P^2_{\{A\}}$ is a negative binomial distribution.

In the same way we compute :

$$\Psi^2_{\{A_1,A_2\}}(\lambda_1,\lambda_2) = -\mathrm{Log}\left[1+\nu(A_1)(1-e^{-\lambda_1})+\nu(A_2)(1-e^{-\lambda_2})-\nu(A_1\cap A_2)(1-e^{-\lambda_1})(1-e^{-\lambda_2})\right]^{-a},$$

and all the $\Psi^2_{\{A_1,\ldots,A_n\}}(\lambda_1,\ldots,\lambda_n)$.

P^2 is a negative binomial random measure.

REFERENCES

[1] Harris, T.E. (1963). The theory of branching processes, Springer-Verlag.

[2] Bauer, H. (1972). Probability Theory and Elements of Measure Theory, International Series in Decision Processes, H.R.W.

[3] Jirina, M. (1964). Branching processes with measure-valued states, 3rd Prague Conference on Information Theory,... 333-357.

[4] Jirina, M. (1966). Asymptotic behaviour of measure-valued branching processes, Rozpravy Ceskoslovenske Akad. Ved. Rada. Mat. Prirod. Ved. 76, n° 3, 1-52.

[5] Szasz, D. (1967). On the general branching process with continuous time-parameter Studia Scient. Math. Hungarica 2, 227-247.

[6] Matthes, K. (1972). Infinitely divisible point processes. Stochastic Point Processes : Statistical Analysis, Theory and Applications, Lewis, Wiley-Interscience, 384-404.

[7] Goldman, J.R. (1967). Infinitely divisible point processes in R^r, Journal of Mathematical Analysis and Applications 17, 133-146.

[8] Jagers, P. (1974). Aspects of random measures and point processes. Advances in probability and related topics, 3, Peter Ney, Sidney Port, (Marcel Dekker New-York), 179-239.

Recent Developments in Statistics
J.R.Barra et al., editors

MARKOV DECISION PROCESSES AND QUASI-MARTINGALES

Luuk P.J. Groenewegen and Kees M. Van Hee
Eindhoven University of Technology
Department of Mathematics
Eindhoven, The Netherlands

0. Abstract. It is shown in this paper that (super-)martingales play an important role in the theory of Markov decision processes.
For excessive functions (with respect to a charge) it is proved that the value of the state at time t converges almost surely under each Markov strategy, which implies that the value function in the state at time t converges to zero (a.s), if an optimal strategy is used. Finally a well-known characterization of optimality is proved using martingales.

1. Quasi-martingales.

Quasi-martingales have been introduced by Fisk (1965) as continuous time stochastic processes having a decomposition into the sum of a martingale and a process having almost all sample functions of bounded variation. In this paper we give essentially the same definition for the discrete time case.

Let $\mathbb{N}$ be the set $\{0,1,2,...\}$ and let $(\Omega,A,\mathbb{P})$ be a probability space and $\{A_t, t \in \mathbb{N}\}$ an increasing sequence of σ-fields contained in A. All stochastic processes in this section are defined on $(\Omega,A,\mathbb{P})$ and have values in the set of real numbers with the Borel-σ-field on it. Moreover they are *adapted* to $\{A_t, t \in \mathbb{N}\}$.

Definition 1.1. A stochastic process $\{V_t, t \in \mathbb{N}\}$ is called a *quasi-(super)martingale* (QSPM) if there exist two stochastic processes $\{B_t, t \in \mathbb{N}\}$ and $\{S_t, t \in \mathbb{N}\}$ such that

i) $$\sum_{t\in\mathbb{N}} |B_t| < \infty \quad \mathbb{P}\text{-a.s.},$$

ii) $\{S_t, t \in \mathbb{N}\}$ is a (super)martingale ,

iii) $$V_t = S_t + \sum_{k=0}^{t-1} B_k \quad \mathbb{P}\text{-a.s.}$$

($\{V_t, t \in \mathbb{N}\}$ is said to be a QSPM w.r.t. $\{B_t, t \in \mathbb{N}\}$) .

Lemma 1.2. Let $\{B_t, t \in \mathbb{N}\}$ and $\{V_t, t \in \mathbb{N}\}$ be stochastic processes with

$\sum_{t \in \mathbb{N}} |B_t| < \infty$ $\mathbb{P}$-a.s. Then:

$$\{V_t,\ t \in \mathbb{N}\} \text{ is a QSPM w.r.t. } \{B_t,\ t \in \mathbb{N}\} \text{ iff } \mathbb{E}^{A_t} V_{t+1} \le B_t + V_t \quad \mathbb{P}\text{-a.s.}$$

Proof. Define

$$S_t := V_t - \sum_{k=0}^{t-1} B_k, \qquad t \in \mathbb{N} .$$

Then

$$\mathbb{E}^{A_t} S_{t+1} \le S_t \quad \text{iff} \quad \mathbb{E}^{A_t} V_{t+1} - \sum_{k=0}^{t} B_k \le V_t - \sum_{k=0}^{t-1} B_k .$$ □

Lemma 1.3. Let $\{V_t\}$ be a QSPM w.r.t. $\{B_t\}$. Assume furthermore

i) $\limsup_{t \in \mathbb{N}} \mathbb{E} V_t^- < \infty$

ii) $\mathbb{E} \sum_{k=0}^{\infty} B_k^+ < \infty$

then V_t converges $\mathbb{P}$-a.s. for $t \to \infty$.

Proof. Defining S_t as in the proof of lemma 1.2, the convergence of S_t and therefore V_t follows from

$$\limsup_{t \in \mathbb{N}} \mathbb{E}\ S_t^- \le \limsup_{t \in \mathbb{N}} \mathbb{E}\ V_t^- + \limsup_{t \in \mathbb{N}} \mathbb{E} \sum_{k=0}^{t-1} B_k^+ < \infty$$

and theorem IV-1-2 of Neveu (1972). □

The results of lemma 1.2 and 1.3 can partially be found in MacQueen (1965).

2. Markov decision processes. In this section we first sketch the framework of convergent dynamic programming, see Hordijk (1974[a], 1974[b]).

Let S be a countable set, called the *state space*.

Let E be the set of *Markov transition functions* on S, i.e. $P \in E$ implies $P : S \times S \to [0,1]$, $\sum_{j \in S} P(i,j) = 1$ for all $i \in S$.

It is assumed that E is endowed with a metric such that E is a Polish space. Let P be a Borel subset of E with the *product property:* if $P_1, P_2, P_3 \ldots \in P$ and if $A_1, A_2, A_3 \ldots$ is a partition of S, then there is a $P \in P$ such that for all $i \in S$:

$$P(i,\cdot) = P_k(i,\cdot) \quad \text{if } i \in A_k$$

Finally let r be a real measurable function on $S \times P$, called the *reward function*. It is assumed that for all $i \in S$:

$$r_{P_1}(i) = r_{P_2}(i) \quad \text{if} \quad P_1(i,\cdot) = P_2(i,\cdot) \quad \text{for } P_1,P_2 \in \mathcal{P}$$

(notation: $r_P(i) = r(i,P)$).
The triplet $(S,\mathcal{P},r)$ is called the *Markov decision process*.
A *Markov strategy* R is a sequence $(P_0,P_1,P_2,\ldots)$ with $P_k \in \mathcal{P}$ $(k = 0,1,2,\ldots)$; the set of all Markov strategies is denoted by M.
Each pair $i \in S$, $R \in M$ determines an (inhomogeneous) Markov chain $\{X_n, n=0,1,2,..\}$ and a probability $\mathbb{P}_{i,R}$ on the sample space. The expectation w.r.t. $\mathbb{P}_{i,R}$ will be denoted by $\mathbb{E}_{i,R}$. Let $\mathcal{F}_n$ be the usual σ-field on the sample space generated by $X_0,\ldots X_n$, $(n = 0,1,2,\ldots)$.

Definition 2.1.

i) A function $g : S \times \mathcal{P} \to \mathbb{R}$ is called a *charge* iff

$$\mathbb{E}_{i,R}[\sum_{n=0}^{\infty} |g_{P_n}(X_n)|] \qquad \text{for all } i \in S,\ R \in M.$$

Let g be such a charge.

ii) A function $f : S \to \mathbb{R}$ is called *superharmonic w.r.t. (a charge) g* iff $f(X_n)$ is integrable w.r.t. $\mathbb{P}_{i,R}$ for all i,R,n and

$$f \geq g_P + Pf \qquad \text{for all } P \in \mathcal{P}.$$

iii) A function $f : S \times\to \mathbb{R}$ is called *excessive w.r.t. (a charge) g* iff f is superharmonic w.r.t. g and

$$f(i) \geq \sum_{n=0}^{\infty} \mathbb{E}_{i,R}[g_{P_n}(X_n)] \qquad \text{for all } i,R.$$

Assumption 2.2.

i) The reward function r is a charge and

ii) $$\sup_{R\in M} \mathbb{E}_{i,R} \sum_{n=0}^{\infty} \{r_{P_n}(X_n)\}^+ < \infty$$

(recall: $x^+ := \max(0,x), x^- := (-x)^+$).

Definition 2.3.

i) The *value function* v of $(S,\mathcal{P},r)$ is a real function on S:

$$v(i) := \sup_{R\in M} \mathbb{E}_{i,R}[\sum_{n=0}^{\infty} r_{P_n}(X_n)] \qquad \text{for all } i \in S$$

ii) A strategy $R \in M$ is called *optimal* if this supremum is attained for R in all $i \in S$.

iii) A strategy $R = (P_0, P_1, P_2, \ldots) \in M$ is called *conserving* if $v(X_n) = r_{P_n}(X_n) + (P_n v)(X_n)$ $\mathbb{P}_{i,R}$-as for all $i \in S$ and R is called *equalizing* if

$$\lim_{n\to\infty} \mathbb{E}_{i,R}[v(X_n)] = 0 \quad \text{for all } i \in S .$$

Property 2.4. The value function v satisfies Bellman's optimality equation:

$$v(i) = \sup_{P\in \mathcal{P}} \{r_P(i) + Pv(i)\} .$$

The proof of this is a direct consequence of a theorem of Van Hee (1975).

Property 2.5. Let g be a charge and f a superharmonic function w.r.t. g then $\lim_{n\to\infty} \mathbb{E}_{i,R} f(X_n)$ exists for all $R \in M$, $i \in S$ and the following assertions are equivalent:

i) f is excessive w.r.t. g

ii) $\lim_{n\to\infty} \mathbb{E}_{i,R} f^-(X_n) = 0$ for all $R \in M$, $i \in S$

iii) $\lim_{n\to\infty} \mathbb{E}_{i,R} f(X_n) \geq 0$ for all $R \in M$, $i \in S$.

For a proof see Hordijk (1974[a]) th. 2.17 .

Remark 2.6. It is obvious from 2.4 that the value function v is superharmonic w.r.t. r and by its definition it is clear that

$$v(i) \geq \mathbb{E}_{i,R}[\sum_{n=0}^{\infty} r_{P_n}(X_n)] \qquad \text{for all } i \in S,\ R \in M ,$$

hence v is excessive w.r.t. r.

Property 2.7. Let f be a superharmonic function w.r.t. g. Then for all $R \in M$, $t \in \mathbb{N}$ and $i \in S$ it holds that

$$f(X_t) \geq g_{P_t}(X_t) + \mathbb{E}^{F_t}_{i,R} f(X_{t+1})$$

The proof is trivial.

Theorem 2.8. Let f be an excessive function w.r.t. a charge g.
For any $i \in S$, $R \in M$ $\{f(X_t),\ t \in \mathbb{N}\}$ is a quasi-supermartingale w.r.t. $\{g_{P_k}(X_k),\ k \in \mathbb{N}\}$ and $f(X_t)$ converges $\mathbb{P}_{i,R}$-a.s. (for $t \to \infty$).

Proof. Fix $i \in S$, $R \in M$. By 2.7 we have $f(X_t) \geq g_{P_t}(X_t) + \mathbb{E}_{i,R}^{F_t} f(X_{t+1})$.
Since g is a charge we have

$$\sum_{k=0}^{\infty} |g_{P_k}(X_k)| < \infty \qquad \mathbb{P}_{i,R}\text{-a.s.}$$

So lemma 1.2 shows that $\{f(X_t),\ t \in \mathbb{N}\}$ is QSPM w.r.t. $\{g_{P_k}(X_k),\ k \in \mathbb{N}\}$.
From g being a charge and property 2.5 ii) it follows that all conditions of lemma 1.3 are fulfilled, which proves the theorem. □

Theorem 2.9. Let f be an excessive function w.r.t. a charge g.
The supermartingale

$$\{S_t := f(X_t) + \sum_{k=0}^{t-1} g_{P_k}(X_k),\ t \in \mathbb{N}\}$$

is regular, for any $i \in S$, $R \in M$.
(For a definition of regularity see Neveu (1972)).

Proof. We only have to check that S_t^- converges in L^1-sense.

$$S_t^- \leq f^-(X_t) + \sum_{k=0}^{t-1} g_{P_k}^-(X_k)$$

hence

$$S_t^- - f^-(X_t) \leq \sum_{k=0}^{t-1} |g_{P_k}(X_k)|$$

On the other hand, since $(a + b)^- \geq a^- - b^+$, we have

$$S_t^- \geq f^-(X_t) - \left[\sum_{k=0}^{t-1} g_{P_k}(X_k)\right]^+ \geq f^-(X_t) - \sum_{k=0}^{t-1} g_{P_k}^+(X_k)$$

hence

$$S_t^- - f^-(X_t) \geq - \sum_{k=0}^{\infty} |g_{P_k}(X_k)| .$$

By the dominated convergence theorem we have the L^1-convergence of $S_t^- - f^-(X_t)$ and by property 2.5 ii) we have the L^1-convergence of $f^-(X_t)$ to zero. □

Applying th. IV-5-25 in Neveu (1972) we get the following:

Corollary 2.10. Let f be an excessive function w.r.t. a charge g, and let τ and ν be stopping times w.r.t. $\{F_t,\ t \in \mathbb{N}\}$ then

$$\sum_{k=0}^{\tau-1} g_{P_k}(X_k) + f(X_\tau) \geq \mathbb{E}_{i,R}^{F_\tau}\left[\sum_{k=0}^{\nu-1} g_{P_k}(X_k) + f(X_\nu)\right]$$

$\mathbb{P}_{i,R}$-a.s. on $\{\tau \leq \nu\}$ for all $i \in S$, $R \in M$.

This result sharpens a theorem in Hordijk (1974[a]) and in Dynkin and Juschkewitsch (1969).

In Mandl (1974) a martingale is considered in connection with the average cost criterion for the optimal control of a Markov chain.

Using this construction for the total return criterion we derive in a direct way a result first proved by Hordijk (1974[a]) (th. 4.6).

Theorem 2.11. A strategy $R \in M$ is optimal iff it is conserving and equalizing.

Proof. Define a real function on $S \times P$:

$$\varphi(i,P) := r_P(i) + (Pv)(i) - v(i)$$

and a random variable

$$Y_n := r_{P_n}(X_n) + v(X_{n+1}) - v(X_n) - \varphi(X_n,P_n)$$

It is easy to see that $\mathbb{E}_{i,R}^{F_n} Y_n = 0$.

Hence

$$\mathbb{E}_{i,R}\left[\sum_{m=0}^{n} r_{P_m}(X_m) + v(X_{n+1}) - v(X_0) - \sum_{m=0}^{n} \varphi(X_m,P_m)\right] = 0$$

Therefore

$$v(i) - \mathbb{E}_{i,R}\left[\sum_{m=0}^{\infty} r_{P_m}(X_m)\right] = \lim_{n\to\infty} \mathbb{E}_{i,R}[v(X_n)] - \mathbb{E}_{i,R}\left[\sum_{m=0}^{\infty} \varphi(X_m,P_m)\right]$$

Since $\varphi(X_m,P_m) \leq 0$ $\mathbb{P}_{i,R}$-a.s. by 2.4 and since $\lim_{n\to\infty} \mathbb{E}_{i,R}[v(X_n)] \geq 0$ by 2.5 iii) we have that the left hand side equals 0 iff R is equalizing and conserving.

Remark 2.12. For each equalizing strategy R we may conclude that $v(X_n) \to 0$ $\mathbb{P}_{i,R}$-a.s. In Groenewegen (1975) this result has been proved for optimal strategies. For a conserving strategy it holds that $\{v(X_t),\ t \in \mathbb{N}\}$ is a quasi-martingale.

Remark 2.13. Let f be an excessive function w.r.t. a charge g.
Although $f(X_t)$ converges $\mathbb{P}_{i,R}$-a.s. for all $i \in S$ and $R \in M$, it is not true in general that $f(X_t)$ converges in L^1-sense.
Counterexample:
$S := \{0,1,2,...\}$. P and Q are Markov transition functions with $P(0,0) = 1$, $P(i,i+1) = i / (i+1)$ for $i \geq 1$, $P(i,0) = 1 / (i+1)$ for $i \geq 1$, $Q(i,0) = 1$.
$\mathcal{P}$ is the collection of Markov transition functions which can be generated from P and Q by using the product property. Furthermore $r_P \equiv 0$ and $r_Q(i) = i$.
It can be verified easily that the conditions 2.2 i) and ii) are fulfilled, that $v(i) = i$, that $\lim_{t\to\infty} v(X_t) = 0$ $\mathbb{P}_{i,R}$-a.s. for all R and that $\lim_{t\to\infty} \mathbb{E}_{1,R}v(X_t) = 1$ for $R = PPP...$
But it is well-known that the L^1-limit and the a.s.-limit should be equal, if both exist. So $v(X_t)$ does not converge in L^1-sense for $t \to \infty$.

Literature

Dynkin, E.B. and Juschkewtisch, A.A., (1969). Sätze und Aufgaben über Markoffsche Prozesse, Springer Verlag, Berlin.

Fisk, D.L. (1965). Quasi-Martingales, Trans. AMS 120 (1965) 369-389.

Groenewegen, L.P.J. (1975). Convergence results related to the equalizing property in Markov decision processes (1975) COSOR 1975-18, Eindhoven University of Technology.

Van Hee, K.M. (1975). Markov strategies in dynamic programming, COSOR 1975-20, Eindhoven University of Technology.

Hordijk, A. (1974[a]). Dynamic programming and Markov potential theory, Math. Centre Tract 51, Mathematisch centrum Amsterdam.

Hordijk, A. (1974[b]). Convergent dynamic programming, Techn. Report 28, Dep. Op. Res. Stanford University, Stanford.

Mandl, P. (1974). Estimation and control in Markov chains, Adv. Appl. Prob. 6, 40-60 (1974).

MacQueen, J. (1965). Some methods for classification and analysis of multivariate observations, Proc. 5th. Berkeley Symp. on Math. Stat. and Prob., vol. I, 281-297.

Neveu, J. (1972). Martingales à temps discret, Masson, Paris.

Recent Developments in Statistics
J.R.Barra et al., editors

ON THE ASYMTOTIC BEHAVIOUR OF THE GENERALIZED MULTINOMIAL DISTRIBUTIONS

B. GYIRES
Kossuth L. University
Debrecen, Hungary

In this paper we consider the asymptotic behaviour in the weak sense of the generalized multinomial and marginal multinomial distributions. As special cases we obtain from our results the limit theorems of the wellknown multinomial and marginal multinomial distributions. One of them gives a possibility to apply the Chi-square method for more general hypotheses as usual.

1. Let R_p be the $p \geqq 2$ -dimensional vector space with column vector as its elements. Let

$$A = \begin{pmatrix} a_{11} & \dots & a_{1p} \\ a_{21} & \dots & a_{2p} \\ . & \dots & . \end{pmatrix}$$

be a stochastic matrix, that is,let

$$a_{jk} \geq 0 \ , \ \sum_{k=1}^{p} a_{jk} = 1 \ (j=1,2,\dots; \ k=1,\dots,p) \ .$$

Let A_m be a finite matrix built from the first m rows of A . Let $(A_{\beta_1}^{(1)} \dots\dots A_{\beta p}^{(p)})$ be the matrix, which is built from the columns of A , namely the k -th column A_m appears β_k -times. If $\beta_k=0$, then the k -th column of A_m is missing from the matrix $(A_{\beta_1}^{(1)} \dots\dots A_{\beta p}^{(p)})$.

Definition 1.
The random vector-variable $\eta_m = \eta_m^{(k)} \varepsilon R_p$ defined on the probability space (Ω,A,P) is called a generalized multinomial distributed random vector variable generated by the matrix A_m , if

$$P(\eta_m^{(k)} = \beta_k \ (k=1,\dots,p)) =$$

$$= \frac{1}{\beta_1! \dots \beta_p!} \ \mathrm{Per}\,(A_{\beta_1}^{(1)} \dots A_{\beta_p}^{(p)}), \tag{1}$$

where $\beta_1,\dots,\beta_p$ are non-negative integers which satisfy the condition $\beta_1+\dots+\beta_p = m$.

Definition 2.
The random vector-variable $\eta_m^{(o)} = (\eta_m^{(k)}) \varepsilon R_{p-1}$ built from the first p-1 components of η_m is called generalized marginal multinomial distributed random vector-variable generated by the matrix A_m .

If all rows of A_m are equal, then η_m and $\eta_m^{(o)}$ are the well-known multinomial and marginal multinomial distribution respectively ([2], 31-32).

The aim of this paper is an investigation of the asymptotic behaviour of the sequences $\{\eta_m\}_{m=1}^{\infty}$ and $\{\eta_m^{(o)}\}_{m=1}^{\infty}$ respectively. In part 2 we give sufficient conditions for the sequence $\{\eta_m\}_{m=1}^{\infty}$ to converge weakly to a p-dimensional normal distributed random vector-variable. As an application of this theorem in part 3 a theorem is proved, which is suitable for the Chi-square method to apply for more general hypotheses as usual. In part 4 is given a necessary and sufficient condition for the sequence $\{\eta_m^{(o)}\}_{m=1}^{\infty}$ to converge weakly to a $p-1$ variate distribution with independent Poissonian components.

2. We adjoin now to the j -th row of the matrix A the random vector-variable $\xi_j=(\xi_j^{(k)})\varepsilon R_p$ defined on the probability space (Ω,A,P). Let the conditions

$$\xi_j^{(1)}+\dots+\xi_j^{(p)}=1\ ,$$

$$P(\xi_j^{(k)}=1\ ,\ \xi_j^{(1)}=0\ (\alpha=1,\dots,p\ ;\ \alpha\neq k))=a_{jk}$$

$$(k=1,\dots,p)$$

be satisfied by the components of ξ_j . It is obvious that the characteristic function of ξ_j is equal to

$$a_{j1}e^{it_1}+\dots+a_{jp}e^{it_p}\ ,\quad t=(t_k)\varepsilon R_p\ .$$

Suppose that the elements of the sequence $\{\xi_j\}_{j=1}^{\infty}$ are independent random vector-variables. Thus the characteristic function of the random vector-variable $\xi_1+\dots+\xi_m$ is equal to

$$(2)\quad \phi_m(t)=\prod_{j=1}^{m}(a_{j1}e^{it_1}+\dots+a_{jp}e^{it_p})\ .$$

On the other hand it was proved by the author ([1], Corollary 1) that (2) is also the characteristic function of the generalized multinomial random vector-variable ξ_m .
Thus

$$(3)\qquad \eta_m=\xi_1+\dots+\xi_m\ .$$

We obtain easily from (2) that([1], Corollary 1)

$$E(\eta_m^{(k)})=\sum_{j=1}^{m}a_{jk}\quad (k=1,\dots,p)$$

and

$$(4)\quad \operatorname{cov}\eta_m=\begin{pmatrix}E(\eta_m^{(1)}) & & (0)\\ & \ddots & \\ (0) & & E(\eta_m^{(p)})\end{pmatrix}-A_m^{*}A_m\ .$$

One can give the following interpretation of the result (3). Let the independent experiments with mutually exclusive and exhaustive events $E_1,\dots,E_p$ of the probability space (Ω,A,P) be given. In the j -th experiment let $P(E_k)=a_{jk}$ $(k=1,\dots,p)$. Then the probability that the event E_k $(k=1,\dots,p)$ occurs β_k -times in the first m experiments is given by (1), where $\beta_1+\dots+\beta_p=m$.

Let $r\leq p$ be the rank of the matrix

$$B_m = \begin{pmatrix} b_{11}^{(m)} \dots b_{1p}^{(m)} \\ \cdot \quad \dots \quad \cdot \\ b_{r1}^{(m)} \dots b_{rp}^{(m)} \end{pmatrix} \quad (m = 1,2,\dots)$$

with real elements. Let

$$\zeta_m=(\zeta_m^{(k)} = B_m(\eta_m - E(\eta_m))\varepsilon R_r$$

$$(m=1,2,\dots),$$

where $\eta_m=(\eta_m^{(k)})\varepsilon R_p$ is the generalized multinomial random vector-variable generated by the matrix A_m . It is obvious that

$$E(\zeta_m)= 0 \ \varepsilon R_r \ ,$$

(5) $$\Gamma_m = \text{cov}\ \zeta_m = B_m\ (\text{cov}\eta_m)B_m^* \ ,$$

$$\text{rank}\ \Gamma_m \leqq \min\ \{r,\ \text{rank cov}\ \eta_m\}\ .$$

Let $g_m(u)$, $u=(u_\beta)\varepsilon R_r$ be the characteristic function of ζ_m and let

$$S(B_m)= \sum_{j=1}^{m} \{ (\sum_{k=1}^{n} a_{jk}\ h_k(B_m))\ (\sum_{k=1}^{p} a_{jk}\ h_k^2(B_m)) + \sum_{k=1}^{p} a_{jk}\ h_k^3(B_m)\}\ ,$$

where

$$h_k(B_m)=\sqrt{\sum_{\beta=1}^{r} (b_{\beta k}^{(m)})^2} \quad (k=1,\dots,p).$$

Theorem 1.
If

(6) $$\lim_{m\to\infty} S(B_m)=0,$$

then

$$\lim_{m\to\infty} g_m(u)\exp(\frac{1}{2}\ u^*\Gamma_m u) = 1\ ,\ u\varepsilon R_r\ .$$

Proof.
By the application of the notation

$$b_k^{(m)} = \sum_{\beta=1}^{r} b_{\beta k}^{(m)}\ u_\beta \qquad (k=1,\dots,p)$$

in the characteristic function

$$g_m(u)= \exp\{ - iu^*B_m E(\eta_m)\}\ .\ E(\exp\ \{iu^*B_m\eta_m\})\ ,$$

we get

$$(7) \qquad u^* B_m E(\eta_m) = \sum_{j=1}^{m} \sum_{k=1}^{p} a_{jk} b_k^{(m)} ,$$

$$u^* B_m \eta_m = \sum_{k=1}^{p} b_k^{(m)} \eta_m^{(k)} .$$

Therefore

$$g_m(u) = \exp\{- iu^* B_m E(\eta_m)\} \cdot \phi_m (b_1^{(m)}, \dots , b_p^{(m)}) ,$$

where the function $\phi_m(t)$ is defined by (2). From the last expression

$$\log g_m(u) = - iu^* B_m(\eta_m) + \sum_{j=1}^{m} \log (\alpha_{j1} e^{ib_1^{(m)}} + \dots + a_{jp} e^{ib_p^{(m)}}) .$$

By application of the Taylor formula we get

$$\log g_m(u) = -iu^* B_m E(\eta_m) +$$

$$+ \sum_{j=1}^{m} \log[1 + i \sum_{k=1}^{p} a_{jk} b_k^{(m)} - \frac{1}{2} \sum_{k=1}^{p} a_{jk} (b_k^{(m)})^2 +$$

$$+ O(\sum_{k=1}^{p} a_{jk} (b_k^{(m)})^3) .$$

We expand the logarithm and use (7) to obtain

$$\log g_m(u) + \frac{1}{2} \sum_{j=1}^{m} \{ \sum_{k=1}^{p} a_{jk} (b_k^{(m)})^2 - (\sum_{k=1}^{p} a_{jk} b_k^{(m)})^2 \} =$$

$$= O (\sum_{j=1}^{m} S_j) ,$$

where

$$S_j = (\sum_{k=1}^{p} a_{jk} b_k^{(m)}) (\sum_{k=1}^{p} a_{jk} (b_k^{(m)})^2) +$$

$$+ \sum_{k=1}^{p} a_{jk} (b_k^{(m)})^3 \qquad (j = 1, \dots, m) .$$

Since

$$| b_k^{(m)} | \leqq h_k(B_m) (u^* u)^{1/2} ,$$

therefore

$$\sum_{j=1}^{m} |S_j| \leq S(B_m) (u^* u)^{3/2} .$$

According to the assumption (6) and the expressions (4) and (5),

$$\lim_{m\to\infty} \{\log g_m(u) + \frac{1}{2} u^* \Gamma_m u\} = 0$$

and this is the statement of our Theorem 1.

An important special case of Theorem 1 is the following Theorem:

Theorem 2.

Let $B_m = (\frac{1}{m})^{\alpha} C_m$, $\alpha > \frac{1}{3}$ and let the elements of the matrix-sequence $\{C_m\}_{m=1}$ be bounded. Then

$$\lim_{m\to\infty} g_m(u)\exp(\frac{1}{2} u^*\Gamma_m u) = 1, \ u\varepsilon R_r .$$

Proof.

Let $C = (c_{\beta k}^{(m)})$. Since $\{C_m\}_{m=1}^{\infty}$ is bounded, we see that $\frac{1}{m} S(C_m)$ is also bounded, hence

$$S(B_m) = (\frac{1}{m})^{3\alpha} S(C_m) = O((\frac{1}{m})^{3\alpha-1}) ,$$

that is, the assumption (6) is satisfied.

If $C_m = B$ (m=1,2,...), then the sequence $\{C_m\}_{m=1}^{\infty}$ is bounded obviously. We obtain therefore the following Corollary:

Corollary 1.

If

$$B_m = (\frac{1}{m})^{\alpha} B (m=1,2,\ldots) , \quad \alpha > \frac{1}{3} , \text{ rank } B = r \leqq p,$$

then

$$\lim_{m\to\infty} g_m(u)\exp(\frac{1}{2} u^*\Gamma_m u) = 1, \quad u\varepsilon R_r .$$

Theorem 3.

Let $B_m = \frac{1}{\sqrt{m}} C_m$. Assume that the limits

$$\lim_{m\to\infty} C_m = B = (b_{jk}) , \quad \text{rank } B = r,$$

(8) $$\lim_{j\to\infty} a_{jk} = a_k \ (k=1,\ldots,p) , \quad \sum_{k=1}^{p} a_k = 1$$

exist. Then

(9) $$\lim g_m(u) = \exp(-\frac{1}{2} u^*\Gamma u) , \ u \ R_r ,$$

where $\Gamma = BGB^*$, with

$$G = \begin{pmatrix} a_1 & & \\ & \ddots & (o) \\ (0) & & a_p \end{pmatrix} - a a^* , \qquad a = \begin{pmatrix} a_1 \\ \vdots \\ a_p \end{pmatrix} .$$

Proof.

Using (8) we get

$$\lim_{m\to\infty} \frac{1}{m} \sum_{j=1}^{m} a_{jk} = a_k ,$$

$$\lim_{m\to\infty} \frac{1}{m} \sum_{j=1}^{m} a_{jk} a_{jl} = a_k a_l .$$

Applying the notation

$$b_k = \sum_{\beta=1}^{r} b_{\beta k} u_\beta \qquad (k=1,\ldots,p)$$

the relations

$$\lim_{m\to\infty} \frac{1}{m} \sum_{j=1}^{m} \sum_{k=1}^{p} a_{jk} (b_k^{(m)})^2 = \sum_{k=1}^{p} a_k b_k^2 ,$$

$$\lim_{m\to\infty} \frac{1}{m} \sum_{j=1}^{m} \left(\sum_{k=1}^{p} a_{jk} b_k^{(m)} \right)^2 = \left(\sum_{k=1}^{p} a_k b_k \right)^2$$

hold and thus according to our statement

$$\lim_{m\to\infty} u^*\Gamma_m u = \sum_{k=1}^{p} a_k b_k^2 - \left(\sum_{k=1}^{p} a_k b_k \right)^2 = u^*\Gamma u .$$

As a consequence of Theorem 3, we obtain the following Corollary, well known in the literature.

Corollary 2.

If

$$B_m = \frac{1}{\sqrt{m}} B \ (m=1,2,\ldots) , \qquad \text{rank } B=r$$

and

$$a_{jk} = a_k \qquad (j=1,2,\ldots; \quad k=1,\ldots,p) ,$$

then the limit (9) holds.

3. We consider now the following important special case of Theorem 2.

Theorem 4.

Suppose that the elements of the matrix A satisfy the conditions

(10) $0 < b \leqq a_{jk} \qquad (j=1,2,\ldots; \quad k=1,\ldots,p).$

Let $B_m = \frac{1}{\sqrt{m}} C_m$, where

(11)
$$C_m = \begin{pmatrix} \frac{1}{\sqrt{\lambda_1^{(m)}}} & & (0) \\ & \ddots & \\ (0) & & \frac{1}{\sqrt{\lambda_p^{(m)}}} \end{pmatrix} \qquad (m=1,2,\ldots)$$

with

$$\lambda_k^{(m)} = \frac{1}{m} \sum_{j=1}^{m} a_{jk} \qquad (m=1,2,\ldots; \quad k=1,\ldots,p) .$$

Then

$$\lim_{m\to\infty} g_m(u) \exp(\tfrac{1}{2} u^*\Gamma_m u) = 1, \ u \in R_p$$

with

$$\Gamma_m = J - \left(\frac{\sum_{j=1}^{m} a_{jk} a_{jl}}{\sqrt{\sum_{j=1}^{m} a_{jk} \sum_{j=1}^{m} a_{jl}}} \right)_{k,l=1}^{p} ,$$

where J is the unit matrix of p -th order.

Proof.
It follows from the assumption (10) that the sequence $\{C_m\}_{m=1}^{\infty}$ is bounded, thus Theorem 2 is applicable. We can derive the expression Γ_m easily from (4) and (5) respectively.

Theorem 5.
Let the elements of the matrix A be positive numbers and let

(12) $$\lim_{j\to\infty} a_{jk} = a_k > 0 \qquad (k=1,\dots,p) .$$

If $B_m = \frac{1}{\sqrt{m}} C_m$ and the sequence $\{C_m\}_{m=1}^{\infty}$ is defined by (11), then

(13) $$\lim_{m\to\infty} g_m(u) = \exp(-\tfrac{1}{2} u^*\Gamma u) \quad u\in R_p$$

with $\Gamma = J - bb^*$, where the components of the vector $b\in R_p$ are one after another $\sqrt{a_k}$ $(k=1,\dots,p)$ and rank $\Gamma = p-1$.

Proof.
It follows from the assumption of Theorem 5 that the condition (10) is satisfied, thus Theorem 4 is applicable. If we use (11) and we take into consideration that the limit (12) holds, then in accordance with our statement

$$\lim_{m\to\infty} u^*\Gamma_m u = u^*(\delta_{kl} - \sqrt{a_k a_l})_{k,l=1}^{p} u ,$$

where δ_{kl} is the Kronecker symbol.

The following special case of Theorem 5 is a well-known statement in the literature.

Corollary 3.
Suppose that the elements of the matrix A satisfy the conditions

$$a_{jk} = a_k > 0 \qquad (j=1,2,\dots ; \quad k=1,\dots,p) .$$

Let $B_m = \frac{1}{\sqrt{m}} C_m$, where

$$C_m = \begin{pmatrix} \frac{1}{\sqrt{a_1}} & & (0) \\ & \ddots & \\ (0) & & \frac{1}{\sqrt{a_p}} \end{pmatrix} \qquad (m=1,2,\dots).$$

Then the limit (13) holds.

We can express Theorem 5 also in the following form:

Let the elements of the matrix A be positive numbers, which satisfy the condition (12). Let $B_m = \frac{1}{\sqrt{m}} C_m$ and let the elements of the sequence $\{C_m\}_{m=1}^{\infty}$ be defined by (11). Then

$$B_m(\eta_m - E(\eta_m)) \to N(0, J-bb^*), \ m \to \infty$$

and the rank of the matrix $J-bb^*$ is equal to $p-1$.

If we use Theorem 3.4.2. of the monograph [2], we obtain the following result:

Theorem 6.
Let the elements of the matrix A be positive numbers which satisfy the conditions (12). Let $B_m = \frac{1}{\sqrt{m}} C_m$ and let the elements of the sequence $\{C_m\}_{m=1}^{\infty}$ be defined by (11). Then the sequence of the random variables

$$\left\{ \sum_{k=1}^{p} \frac{\left(\eta_m^{(k)} - \sum_{j=1}^{m} a_{jk}\right)^2}{\sum_{j=1}^{m} a_{jk}} \right\}_{m=1}^{\infty}$$

converges weakly to the Chi-square distribution with degrees of freedom $p-1$.

This Theorem gives a possibility of generalizing the Chi-square test in the following way:

Let the independent experiments with mutually exclusive and exhaustive events $E_1,\ldots,E_p$ of the probability space (Ω,A,P) be given. Then the H_o hypothesis is the following:

In the j-th experiment

$$P(E_k) = a_{jk} > 0 \quad (k=1,2,\ldots,p;\ j=1,2,\ldots)$$

and

$$\lim_{j\to\infty} a_{jk} = a_k > o \qquad (k=1,\ldots,p)$$

4. Assume the matrix-sequence $\{A_m\}_{m=1}^{\infty}$ satisfies the conditions

$$A_m = (a_{jk}^{(m)}) \quad , \quad a_{jk}^{(m)} \geqq 0 \quad , \quad \sum_{k=1}^{p} a_{jk}^{(m)} = 1$$

$$(k=1,\ldots,p; \quad j=1,\ldots,m; \quad m=1,2,\ldots).$$

Let $\eta_m \varepsilon R_p$ be the generalized multinomial random vector variable generated by the matrix A_m . We see from (2) that the corresponding characteristic function is equal to

$$(14)\quad \phi_m(t) = \prod_{j=1}^{m} (a_j^{(m)} e^{it_1} + \ldots + a_{jp}^{(m)} e^{it_p}) \ , \quad t = (t_k)\varepsilon R_p \ .$$

Suppose that the random vector-variables

$$\begin{array}{cccc} \xi_{11} & & & \\ \xi_{21} & \xi_{22} & & \\ \cdot & \cdot & & \\ \xi_{m1} & \xi_{m2} & \cdots & \xi_{mm} \\ \cdot & \cdot & \vdots & \end{array}$$

satisfy the following conditions: $\xi_{mj} = (\xi_{mj}^{(k)}) \varepsilon R_p$ and the random variable ξ_{mj}^{k} has the values 0, 1, namely

$$P\,(\xi_{mj}^{(k)} = 1, \quad \xi_{mj}^{(\alpha)} = 0 \quad (\alpha=1,..,p;\ \alpha\neq k))\alpha = a_{jk}^{(m)}$$

$$(k=1,\ldots,p;\quad j=1,\ldots,m;\quad m=1,2,\ldots).$$

and the random vector-variables belong to the same row are independent. Then in accordance with (3)

$$\eta_m = \xi_{m1} + \ldots + \xi_{mm}$$

Let $\eta_m^{(o)} \varepsilon R_{p-1}$ be the generalized marginal multinomial random vector-variable generated by the matrix A_m. Substituting $t_p=0$ in (14) we get the characteristic function $\phi_m^{(o)}(t)$, $t=(t_k)\ \varepsilon R_{p-1}$ of the random vector-variable $\eta_m^{(o)}$. Thus

$$\text{(15)}\qquad \phi_m^{(o)}(t) = \prod_{j=1}^{m} [1 + a_j^{(m)}(e^{it_1} - 1) + \ldots + a_{jp-1}^{(m)}(e^{it_{p-1}} - 1)] \,.$$

Theorem 7.
The sequence of the random vector-variables $\{\eta_m\}_{m=1}^{\infty}$ converges weakly to a p-1 variate distribution with independent Poissonian components, iff the conditions

$$\text{(16)}\qquad \lim_{m\to\infty} \sum_{j=1}^{m} a_{jk}^{(m)} = \lambda_k \quad (k=1,\ldots,p-1)\,,$$

$$\text{(17)}\qquad \lim_{m\to\infty} \sum_{j=1}^{m} (1 - a_{jp}^{(m)})^2 = 0$$

are satisfied.

Proof.
It follows from (15) that

$$\text{(18)}\qquad \log \phi_m^{(o)}(t) = \sum_{k=1}^{m} (e^{it_k} - 1) \sum_{j=1}^{m} a_{jk}^{(m)} + O(S_m(t))\,,$$

where

$$S_m(t) = \sum_{j=1}^{m} [a_j^{(m)}(e^{it_1} - 1) + \ldots + a_{jp-1}^{(m)}(e^{it_{p-1}} - 1)]^2 \,.$$

In consideration of

$$\text{(19)}\qquad |S_m(t)| = \left|\sum_{\alpha=1}^{p-1}\sum_{\beta=1}^{p-1} (e^{it_\alpha} - 1)(e^{it_\beta} - 1) \sum_{j=1}^{m} a_{j\alpha}^{(m)} a_{j\beta}^{(m)}\right| \leqq 4 \sum_{j=1}^{m} (1 - a_{jp}^{(m)})^2 \,,$$

and since in (19) is an equality if

$$t_\alpha = (2k_\alpha + 1)\,\pi \quad (\alpha=1,\ldots,p-1)$$

with arbitrary integers k_α, we see from (18) that the conditions

(16) and (17) are necessary and sufficient for the existence of the relation

$$\lim_{m\to\infty} \phi_m^{(o)}(t) = \exp\{\lambda_1(e^{it_1} - 1)+\ldots+\lambda_{p-1}(e^{it_{p-1}} - 1)\} .$$

If

$$(20) \qquad a_{jk}^{m} = a_k^{(m)} , \quad m\, a_k^{(m)} = \lambda_k > 0$$

$$(k=1,\ldots,p-1;\ m=1,2,\ldots) ,$$

then simultaneously

$$a_{jk}^{(m)} = 1 - \frac{\lambda_1 +\ldots+ \lambda_{p-1}}{m} \quad (j=1,\ldots,m) ,$$

thus

$$\sum_{j=1}^{m} (1 - a_{jp}^{m})^2 = \frac{(\lambda_1 +\ldots+ \lambda_{p-1})^2}{m} \quad (m=1,2,\ldots) ,$$

that is the conditions (16) and (17) are satisfied.

We use now Theorem 7 to obtain the following well-known result:

Corollary 4.

Let $\eta_m^{(o)} \varepsilon R_{p-1}$ be the generalized marginal multinomial random vector-variable generated by the matrix A_m. If the sequence $\{A_m\}_{m=1}^{\infty}$ satisfies the condition (20) then the sequence $\{\eta_m^{(o)}\}_{m=1}^{\infty}$ converges weakly to a p-1 variate distribution with independent Poissonian components.

REFERENCES

[1] Gyires,B. (1973). Discrete distributions and permanents. PUBL.MATH., Debrecen, 20, 93-106.

[2] Lukács,E. and Laha,R.G. (1964). Applications of Characteristic Functions. (Hafner, New York).

Recent Developments in Statistics
J.R. Barra et al., editors
© North-Holland Publishing Company, (1977)

COMPARATIVE CLASSIFICATION OF BLOCK DESIGNS

M. GRAF-JACCOTTET *
Département de mathématiques.
Ecole Polytechnique Fédérale de Lausanne
Lausanne, Switzerland

The efficiency of a block design can be expressed in terms of the eigenvalues of the matrix M_0 due to Calinski. It is first shown, for type E designs of Pearce, where treatments fall into m groups of n elements, that M_0 can be reduced to Kronecker products of the form $V \otimes I + (U-V) \otimes (1/n)\underline{1}\underline{1}'$ with U and V of order m and I and $(1/n)\underline{1}\underline{1}'$ of order n. Thus the classification of type E designs is equivalent to a classification according to the number of non-zero distinct eigenvalues of M_0, i.e. those of U and V. For the more general class of type M designs, M_0 cannot be reduced in this manner. The classification of Preece is in fact a classification of type M designs which are balanced in the sense of variance or efficiency. In special cases designs can be constructed for which the matrix M_0 has specific properties.

INTRODUCTION

Two criteria can be considered for the classification of block designs: (1) precision and (2) efficiency of contrasts between the t treatments. Let $\{c_i, i=1,\ldots,t-1\}$ be a system of t-dimensional contrasts, independent with respect to the norm Ω, the treatment covariance matrix, such that $c_i' \Omega c_j = 0$, $i,j=1,\ldots,t-1$, $i \neq j$. The precision of the ith contrast is defined by:

$$\theta^{-1}(c_i) = \theta_i^{-1} = 1/(c_i' \Omega c_i) \tag{1}$$

and the efficiency

$$\varepsilon(c_i) = \varepsilon_i = c_i' r^{-\delta} c_i / (c_i' \Omega c_i) \tag{2}$$

where $r^{-\delta}$ is the diagonal covariance matrix of the treatments, if the design is orthogonal. r is the t-dimensional vector of treatment replications and $r^{-\delta}$ has as k-th diagonal element the inverse of the k-th element of r. (Pearce, 1970 and 1971, Calinski, 1971).
Generally, as Ω is supposed to be of rank t, it can be factorised into

$$\Omega = (I - M_0)^{-1} r^{-\delta} \tag{3}$$

(Calinski, 1971), where M_0 is a non-orthogonality matrix, or the zero matrix in the orthogonal case. The contrasts c_i are then related to the eigenvectors s_i of M_0

$$c_i = r^\delta s_i \tag{4}$$

---

*Present address: Institut de mathématiques, Université de Neuchâtel, Neuchâtel, Switzerland.

Thus, if $M_o s_i = \mu_i s_i$, we get $\Omega c_i = (1-\mu_i)^{-1} r^{-\delta} c_i$ and

$$c_i' \Omega c_i = (1-\mu_i)^{-1} c_i' r^{-\delta} c_i \tag{5}$$

Substitution into (1) and (2) yields

$$\Theta_i^{-1} = (1-\mu_i)/c_i' r^{-\delta} c_i \tag{6}$$

$$\varepsilon_i = 1-\mu_i \tag{7}$$

A design is said to be balanced in the Jones sense or J-balanced, if $\varepsilon(c)$ is a constant for any contrast c. This implies [Pearce 1971] that M_o has a multiple eigenvalue of order t-1 (the remaining eigenvalue being zero). A design is said to be balanced in the Vartak sense, or V-balanced, if $\Theta(c) = L\, c'c$ for any contrast c, L being a constant. For the contrast c_i defined above, relation (6) implies

$$\mu_i = 1 - L\, c_i' r^{-\delta} c_i/(c_i' c_i) \tag{8}$$

The design has both balance properties (is totally balanced) if and only if it is equireplicate ($r = r\,1$) and in this case

$$\mu_i = \mu = 1 - L\, r^{-1} \tag{9}$$

Furthermore, for the equireplicate designs, the study of the eigenvalue of M_o concerns both the precision and the efficiency .

1. RELATIONS BETWEEN THE CLASSIFICATION OF PEARCE (1963) AND THE MATRIX M_o .

The classification of Pearce (1963) can be viewed as an extension of the class of totally balanced designs (type T) to the case where M_o has several multiple eigenvalues. Thus, the preceding property of J-balance still holds for subsets of contrasts associated with the same eigenvalue of M_o .

1.1. Let us consider first the type F designs, which form the more general class of equireplicate designs. Type F is defined as follows: the treatments form m groups of n elements , such that the between- and within-group covariance matrices are of type T and Ω^{-1} is of the following structure .

Denote by I_n the n×n-identity matrix , by 1_n the n-dimensional vector of units and by $\otimes$ the Kronecker product of two matrices. We have

$$D = (a-b)I_n + b\, 1_n 1'_n \;;\quad D^* = (a^*-b^*)I_n + b^*\, 1_n 1'_n \tag{10}$$

$$\Omega^{-1} = I_m \otimes (D-D^*) + 1_m 1'_m \otimes D^* . \tag{11}$$

From (4) we have $M_o = I - r^{-\delta}\, \Omega^{-1}$ (12)

Set $\lambda = r^{-1} b^*$; $\mu = r^{-1}(b-b^*)$; $\nu = r^{-1}(a^*-b^*)$ (13)

Substitute into (11), then into (12) , and using (13) we obtain , after some transformations

$$M_o = U \otimes (I_n - J_n) + V \otimes J_n \tag{14}$$

with $J_n = I_n - n^{-1}\, 1_n 1'_n$

$$U = (mn\lambda) + m\nu)J_m \;,\; V = (mn\lambda + n\mu)I_m + m\nu\, J_m .$$

The eigenvalues and corresponding eigenvectors are

eigenvalues	eigenvectors
$mn\lambda + m\nu = \mu^u$	$\underset{\sim}{x}_h \otimes \underset{\sim}{1}_n \quad (h=1,\dots,m-1)$
$mn\lambda + n\mu = \mu_1^v$	$\underset{\sim}{1}_m \otimes \underset{\sim}{y}_k \quad (k=1,\dots,n-1)$
$mn\lambda + n\mu + m\nu = \mu_2^v$	$\underset{\sim}{x}_h \otimes \underset{\sim}{y}_k$

(15)

where x_h (or y_k) are m-1 (or n-1) independent m(or n) dimensional contrasts. We then find by (6) and (7) the corresponding precision and efficiency.

1.2. The equipartite designs (type E) are an extension of type F designs for which the condition of equality of between (or within) group covariance matrices is suspended. We will sketch the construction of $\underset{\sim}{M}_o$. Thus $\underset{\sim}{\Omega}^{-1} = (\underset{\sim}{D}_{ij})$ i,j = 1,... .., m , with $D_{ij} = (a_{ij}-b_{ij})\underset{\sim}{I}_n + b_{ij}\,\underset{\sim}{1}_n\,\underset{\sim}{1}'_n$ of type T. Set $\underset{\sim}{A} = (a_{ij})$; $\underset{\sim}{B} = (b_{ij})$.

Then $\underset{\sim}{\Omega}^{-1} = (\underset{\sim}{A} - \underset{\sim}{B}) \otimes \underset{\sim}{I}_n + \underset{\sim}{B} \otimes \underset{\sim}{1}_n\,\underset{\sim}{1}'_n$ and $\underset{\sim}{r} = \underset{\sim}{R} \otimes \underset{\sim}{1}_n$

$\underset{\sim}{R}' = (R_1\ R_2\ \dots\ R_m)$, $R_i = \Sigma(a_{ij}+(n-1)b_{ij})$ is the number of replications of the treatments from the ith group. By the above transformations we obtain for $\underset{\sim}{M}_o$ the relation (14) with

$$\underset{\sim}{U} = \underset{\sim}{I}_m - \underset{\sim}{R}^{-\delta}(\underset{\sim}{A}+(n-1)\underset{\sim}{B}) \ , \ \underset{\sim}{V} = \underset{\sim}{I}_m - \underset{\sim}{R}^{-\delta}(\underset{\sim}{A} - \underset{\sim}{B}) \tag{16}$$

The eigenvalues of $\underset{\sim}{M}_o$ are thus those of $\underset{\sim}{U}$ and $\underset{\sim}{V}$.

2. THE BALANCED DESIGNS

Another way of extending type T designs is to consider J-balance and V-balance separately. Preece (1971) bases his classification on these two conditions and on specifications of the incidence matrix $N = (n_{ij})(i=1,\dots, t\ ;\ j = 1\dots b)$ of t treatments with b blocks. The design can be binary (n_{ij} = 0 or 1), proper ($\sum_i n_{ij} = k_j = k$), equireplicate ($\sum_j n_{ij} = r_i = r$).

Set $n = \sum_{i,j} n_{ij}$.

We shall give several properties of binary balanced designs (for several constructions of non binary designs from binary designs, see John, 1964 and Preece, 1971).

2.1 Let us show first, that V- or J-balanced, binary designs cannot exist, if they are both proper and non equireplicate.

A V-balanced, non equireplicate design satisfies

$$\underset{\sim}{N}\,\underset{\sim}{k}^{-\delta}\,\underset{\sim}{N}' = \underset{\sim}{r}^{\delta} - L^{-1}(\underset{\sim}{I} - t^{-1}\underset{\sim}{1}\,\underset{\sim}{1}') \tag{17}$$

For a binary and proper design, we have on the diagonal

$$(\underset{\sim}{N}\,\underset{\sim}{k}^{-\delta}\underset{\sim}{N}')_{ii} = k^{-1}\sum_j n_{ij} = k^{-1}r_i \tag{18}$$

From (17) and (18) we find that the design is necessarily equireplicate.

For a J-balanced and non equireplicate design, we have

$$\underset{\sim}{N}\,\underset{\sim}{k}^{-\delta}\underset{\sim}{N}' = \mu\,\underset{\sim}{r}^{\delta} + n^{-1}(1-\mu)\underset{\sim}{r}\,\underset{\sim}{r}' \tag{19}$$

From (18) and (19) we again see that the design is equireplicate.

2.2 We shall now calculate the non-zero eigenvalue of $\underset{\sim}{M}_o$ for J-balanced, binary designs. For binary designs we have

$$\mathrm{tr}(\underset{\sim}{N}\,\underset{\sim}{k}^{-\delta}\,\underset{\sim}{N}') = \sum_{i,j} k_j^{-1}\, n_{ij} = \sum_j k_j^{-1}\, k_j = b \qquad (20)$$

From (19) and (20) we get

$$\mu = (n\,b - \Sigma\, r_i^{\,2})/(n^2 - \sum_i r_i^{\,2}) \qquad (21)$$

2.3 John (1964) constructs V-balanced, binary designs by adding to a BIBD (t,b,r,k,λ) one treatment, that appears in t extra blocks once and only once with each treatment of the BIBD . From (17) we then find that the parameters take the form $(t,\ t(t-1)/(2k-2)\ ,\ k(t-1)/(2k-2),\ k,\ k/2)$ and that the constant for precision is $L = 2/(t+1)$.

$\underset{\sim}{M}_o$ has two non zero eigenvalues $\mu_1 = (t+1)\ /\ (k(t+1)-2)$ of order $t-1$ and $\mu_2 = k\,\mu_1\ /\ 2$ of order 1 .

REFERENCES

Calinski, T. (1971). On some desirable patterns in block designs. Biometrics 27 , 275-292 .

Jaccottet, M. (1976). Structure et efficacité des plans d'expérience en blocs. Thèse EPF. Lausanne .

John, P.W.M. (1964). Balanced designs with unequal number of replicates. Ann. Math. Statist. 35 , 897-899 .

Jones, R.M. (1959). On a property of incomplete blocks. J. Roy. Statist. Soc. Ser. B 21 , 172-179 .

Pearce, S.C. (1963). The use and classification of non-orthogonal designs. J. Roy. Statist. Soc. Ser. A 126 , 353-377 .

Pearce, S.C. (1970). The efficiency of block designs in general. Biometrika 57 , 339-346 .

Pearce, S.C. (1971). Precision in block experiments. Biometrika 58 , 161-167 .

Preece, D.A. (1971). Combinatorial analysis and experimental design. The Statistician 21 , 77-87 .

Vartak, N.N. (1963). Disconnected balanced designs. J. Indian Statist. Assoc. 29 , 290 .

Recent Developments in Statistics
J.R.Barra et al., editors

REPRESENTATION OF FAMILIES OF PROBABILITIES BY AN ALMOST SURELY CONTINUOUS RANDOM FUNCTION

Pierre JACOB
CENTRE UNIVERSITAIRE DE VALENCIENNES
59326 VALENCIENNES (FRANCE)

I) - Let (E,d) be a separable and complete metric space, $\mathcal{B}_E$ the σ-field of Borel sets, and $\mathcal{M}(E)$ the set of probability measures on (E, $\mathcal{B}_E$), endowed with the topology of weak convergence. A.V. SKOROKHOD's useful theorem ((2) or (6)), generalized by R.M. Dudley when E is not necessarily complete, states the following result.

> Let $\{\mu_n\}$ be a sequence of probabilities weakly convergent to some probability μ. Then, there exists a probability space $(\Omega,\mathcal{A},\lambda)$ on which are defined random variables X_n with distributions μ_n, and a random variable X with distribution μ, with values in E, such that : $X_n \to X$, λ-almost surely.

This theorem has been proved independently by J. GEFFROY and O. JOUANDET. In their version, as well as in SKOROKHOD's, and in the present paper, λ is the Lebesgue measure on the Borel sets $\mathcal{A}$ of the unit interval Ω.

II)- Our first extension is the following one :

__Theorem I__

> Let $\mathcal{M}$ be a family of probability measures, dominated by a probability $\sigma \in \mathcal{M}(E)$. There exists a random function X(.,.), defined on $\mathcal{M}(E) \times \Omega$, with values in E, such that :
>
> 1) $\forall \mu \in \mathcal{M}(E)$, μ is the distribution of $X(\mu,.)$
>
> 2) The random function X(.,.) is almost surely continuous on $\mathcal{M}$: $\forall \mu \in \mathcal{M}$, $\exists \Omega\mu \in \mathcal{A}$ such that $\lambda(\Omega\mu) = 1$, and $\forall \omega \in \Omega\mu$, the sample path $X(.,\omega)$ is continuous at the point μ

__Proof__

1) Since E is separable, and $\mathcal{M}$ dominated, we can construct a sequence of partitions of E, $\{\mathcal{D}_k\}_{k \in \mathbb{N}^*}$, such that for each $k \in \mathbb{N}^*$, $\mathcal{D}_k$ is a decomposition of E into Borel sets $A_{n_1, n_2, \ldots, n_k}$ of diameters less than $1/k$, ν-continuous for each ν in $\mathcal{M}$. Moreover, each $\mathcal{D}_{k+1}$ is a refinement of $\mathcal{D}_k$ in the following meaning :

$$A_{n_1,n_2,\dots,n_k} = \bigcup_{j=1}^{\infty} A_{n_1,n_2,\dots,n_k,j}$$

Then, for each ν in $M(E)$ and each integer k, we associate to the Borel set $A_{n_1,n_2,..,n_k}$ an interval $I^{\nu}_{n_1,n_2,..,n_k}$ so defined :

$$I^{\nu}_{n_1,n_2,..,n_k} = \left[\sum_{\ell=1}^{k} \sum_{j=1}^{n_\ell -1} \nu(A_{n_1,n_2,..,n_{\ell-1},j}) ; \sum_{\ell=1}^{k} \sum_{j=1}^{n_\ell -1} \nu(A_{n_1,n_2,..,n_{\ell-1},j}) + \nu(A_{n_1,n_2,..n_k}) \right[$$

We shall denote by J^{ν}_k the partition of $[0,1[$ obtained that way.

2) k being fixed, each point ω of $[0,1[$ belongs to some interval $I^{\nu}_{n_1,n_2,..,n_k}$. Denoting by $X_k(\nu,\omega)$ an arbitrary value in $A_{n_1,n_2,..,n_k}$, we construct point by point a random variable $X_k(\nu,.)$. Since E is complete, the fundamental sequence $\{X_k(\nu,\omega)\}_{k \in \mathbb{N}^*}$ converges, for all ω, to a limit $X(\nu,\omega)$. Thus we obtain, for each $\nu \in M(E)$, a random variable $X(\nu,.)$

3) Let U be the set : $\{A_{n_1,n_2,..,n_k} : k \in \mathbb{N}^* ; (n_1,n_2,..,n_k) \in \mathbb{N}^{*k}\} \cup \emptyset$. Clearly, U is closed under the formation of finite intersections and each open set in E is a finite or countable union of elements of U. If we denote by P_k the distribution of $X_k(\nu,.)$, we obtain :

$$\forall s \geq k, \ \nu(A_{n_1,n_2,..n_k}) = \lambda(I^{\nu}_{n_1,n_2,..,n_k}) = P_s(A_{n_1,n_2,..,n_k})$$

Hence, from (1) p. 14, the sequence $\{P_k\}$ converges weakly to ν, so that ν is the distribution of $X(\nu,.)$.

4) The reader can verify this lemma :

<u>Lemma 1</u> : Let $\sigma \in M(E)$, and $\mathcal{D} = \{D_j\}_{j \in \mathbb{N}^*}$ a partition of E into σ-continuous Borel sets ; for each $\varepsilon > 0$, there exists a weak neighborhood of σ in $M(E)$, $V_{\sigma,\varepsilon}$, such that :

$$\forall \nu \in V_{\sigma,\varepsilon}, \ \sum_{j=0}^{\infty} |\nu(D_j) - \sigma(D_j)| < \varepsilon$$

5) $\mu \in M$ being fixed, we denote by $\Omega\mu$ the set of all the points of Ω which are not an end point of an interval $I^{\mu}_{n_1,n_2,..,n_k}$, $\forall k \in \mathbb{N}^*$, $\forall (n_1,n_2,..n_k) \in \mathbb{N}^{*k}$ clearly, $\lambda(\Omega\mu) = 1$.

Moreover, let us fix $\varepsilon > 0$, and $\omega \in \Omega\mu$. there exists an integer k such that $\frac{1}{k} < \varepsilon$ and an interval $I^{\mu}_{n_1,n_2,..,n_k}$ that contains ω. Let β be the distance between the point ω and the boundary of this interval. According to lemma 1, there exists a weak neighborhood of μ, $V_{\mu,\beta}$, such that : $\forall \mu \in V_{\mu,\beta}$,

$$|\mu(A_{n_1,n_2,\ldots,n_k})-\nu(A_{n_1,n_2,\ldots,n_k})|+\sum_{\ell=1}^{k}\sum_{j=1}^{n_\ell -1}|\mu(A_{n_1,n_2,\ldots,j})-\nu(A_{n_1,n_2,\ldots,j})|<\beta/2$$

in other words, if we define for each $\nu \in M(E)$: $[a_\nu, b_\nu[= I^{\nu}_{n_1,n_2,\ldots n_k}$, we have : $\forall \nu \in V_{\mu,\beta}$, $a_\nu \in]a_\mu - \beta/2 \; ; \; a_\mu + \beta/2[$ et $b_\nu \in]b_\mu - \beta/2 \; ; \; b_\mu + \beta/2[$.

Hence, $\forall \nu \in V_{\mu,\beta}$, $\omega \in I_{n_1,n_2,\ldots,n_k}$, thus $d(X(\mu\ \omega) \; ; \; X(\nu\ \omega)) \leq \frac{1}{k} < \varepsilon$

III)- The points of discontinuity of this random function proceed from the fact that the different partitions $\mathcal{D}_k (k \in \mathbb{N}^{*})$ are made of quite arbitrarily chosen Borel sets. We are going to show that, under some requirements, we can choose a numbering of these Borel sets in order to construct a random function $X(.,.)$ with continuous random paths, for all $\omega \in \Omega$. We had to strengthen the hypotheses, but we shall give an example showing that they are not really exaggerated. M is now a family of equivalent probabilities whose common spectrum is precompact. Then, for every integer k, we can assume that the partition $\mathcal{D}_k$ of the spectrum is finite, and made of Borel sets of positive ν-measure, for each probability $\nu \in M$.

For every $\nu \in M$, with each partition $\mathcal{D}_k$ we associate, following the pattern of II, a partition J^{ν}_k of $[0,1[$. Furthermore, we must assume that every couple of successive intervals of J^{ν}_k, say $I^{\nu}_{k,1}$ and $I^{\nu}_{k,2}$, can be associated with Borel sets $A_{k,1}$ and $A_{k,2}$ of $\mathcal{D}_k$, such that $\overline{A_{k,1}} \cap \overline{A_{k,2}} \neq \emptyset$. Then, we can state the following result :

Theorem II

There exists a random function $X(.,.) : M \times \Omega \to E$ such that :

1) $\forall \mu \in M$, $X(\mu,.)$ have μ as distribution.
2) $\forall \omega \in \Omega$, $X(.,\omega)$ is a random path continuous on M .

Proof :

According to the results of paragraph II, it is sufficient to show that, if ω is the end point of an interval such that $I^{\mu}_{n_1,n_2,\ldots,n_k}$, for any $k \in \mathbb{N}^{*}$ and $\mu \in M$, the sample path $X(.,\omega)$ is also continuous at μ. Let us fix $\varepsilon > 0$, $\mu \in M$, and $m \in \mathbb{N}^{*}$ such that $\frac{2}{m} < \varepsilon$. There exist two consecutive intervals of the partition J_m^μ, say $I_{m,1}^\mu$ and $I_{m,2}^\mu$, such that $I_{m,1}^\mu \cap I_{m,2} = \omega$. Let $\alpha = \lambda(I_{m,1}^\mu)$ and $\beta = \lambda(I_{m,2}^\mu)$: just as in the second paragraph, one can assert the existence of a weak neighborhood V'_μ of μ such that the lower bound of $I_{m,1}^\nu$, for every $\nu \in V'_\mu$, is less than $\alpha/2$ from the lower bound of $I_{m,2}^\nu$. Likewise, there exists a weak neighborhood V''_μ of μ such that, for every $\nu \in V''_\mu$, the upper bound of $I_{m,2}^\nu$ is less than $\beta/2$ from the upper bound of $I_{m,2}^\mu$. Consequently :

$$\forall \nu \in V'_\mu \cap V''_\mu, \quad X(\nu,\omega) \in \overline{A}_{m,1} \cup \overline{A}_{m,2}$$

while $X(\mu,\omega) \in \overline{A}_{m,2}$. But since $\overline{A}_{m,1} \cap \overline{A}_{m,2} \neq \emptyset$, we obtain :

$$d(X(\mu,\omega) \; ; \; X(\nu,\omega)) \leq 2/m < \varepsilon.$$

Example :

Let $E = [0,1]^k$ with the metric d defined by : $d(x,y) = \operatorname{Max}_{i=1}^{k} (x_i - y_i)$, $\forall x = (x_1, x_2, \ldots, x_k)$ and $y = (y_1, y_2, \ldots, y_k)$, and let $\mathcal{M}$ be the subset of $\mathcal{M}(E)$ of all the probabilities equivalent to the Lebesgue measure on $[0,1]^k$, then we are within the requirements of theorem II.Indeed, there exists a sequence of partitions $\{\mathcal{D}_m\}_{m \in \mathbb{N}^*}$ of $[0,1]^k$ such that :

1) $\forall m \in \mathbb{N}^*$, $\mathcal{D}_m = \{D_{n_1,n_2,\ldots,n_i,\ldots,n_m} \; ; \; 1 \leq n_i \leq 2^k \; ; \; i = 1, 2, \ldots, n\}$
2) $\forall x \in D_{n_1,n_2,\ldots,n_m}$ and $\forall y \in D_{n_1,n_2,\ldots,n_m+1}$, $d(x,y) < \frac{1}{2^{m-1}}$
3) $\forall x \in D_{n_1,n_2,\ldots,n_{m-1},2^k}$ and $\forall y \in D_{n_1,n_2,\ldots,n_{m-1}+1,1}$, $d(x,y) < \frac{1}{2^{m-1}}$
and so on.

IV) - In order to obtain a random function having a property of uniform continuity, we assume now that $\mathcal{M}$ is a family of probabilities dominated by a probability $\sigma \in \mathcal{M}(E)$, and that $\mathcal{M}$ is compact in $\mathcal{M}(E)$; this may happen for instance when $\mathcal{M}$ is uniformly absolutely continuous with respect to a probability $\sigma \in \mathcal{M}(E)$. (i.e. : $\forall \varepsilon > 0, \exists \delta(\varepsilon) > 0$ such that $\forall A \in \mathcal{B}_E$, $\forall \nu \in \mathcal{M}$, $\sigma(A) < \delta(\varepsilon) \Rightarrow \nu(A) < \varepsilon$) : then, $\overline{\mathcal{M}}$ satisfies the requirements. Furthermore we shall denote by ρ the metric of Prokhorov on $\mathcal{M}(E)$.

Theorem III

There exists a random function $X(.,.) : \mathcal{M}(E) \times \Omega \to E$ such that :

1) $\forall \nu \in \mathcal{M}(E)$, ν is the distribution of $X(\nu,.)$

2) $\forall \varepsilon > 0$, $\forall \varepsilon' > 0$, $\exists\ \eta(\varepsilon,\varepsilon') > 0$ such that : $\forall\ \mu \in \mathcal{M}$, $\forall \nu \in \mathcal{M}(E)$,

$\rho(\mu,\nu) < \eta(\varepsilon,\varepsilon') \Rightarrow \lambda\{d(X(\mu,.)\ ;\ X(\nu,.)) < \varepsilon\} > 1-\varepsilon'$

Proof :

1) The construction of $(X(.,.)$ is exactly the same as in theorem I. Then, we have to establish a few lemmas :

lemma 2 : Let A be a ν-continuous Borel set, for each probability ν in $\mathcal{M}$, and $A^\delta = \{x \in E : d(x,A) < \delta\}$ Then, $\lim\limits_{\delta \to 0} \sup\limits_{\nu \in \mathcal{M}} \nu(A^\delta - A) = 0$.
(This lemma 2 is a kind of Dini theorem)

lemma 3 : $\forall \varepsilon > 0$, $\forall m \in \mathbb{N}^*$, $\exists\,(k_1,k_2,..,k_m) \in \mathbb{N}^{*m}$ such that : $\forall \nu \in \mathcal{M}$,

$$\sum_{n_1 \leq k_1;..;n_m \leq k_m} \nu(A_{n_1,n_2,..,n_m}) > 1-\varepsilon$$

(The reader can refer to (5) p. 48-49 for the proof).

lemma 4 : $\forall\ \varepsilon > 0$, $\forall m \in \mathbb{N}^*$, $\exists \eta(m,e)$ such that : $\forall \nu \in \mathcal{M}(E)$; $\forall \mu \in \mathcal{M}$,

$$\rho(\nu,\mu) < \eta(m,e) \Rightarrow \sum_{(n_1,n_2,..,n_m) \in \mathbb{N}^{*m}} |\mu(A_{n_1,n_2,..,n_m}| - \nu(A_{n_1,n_2,..,n_m})| < \varepsilon$$

(This is an extension of lemma 1 ; the reader can refer to (1) p. 136 and use the lemmas 3) and 4)).

2) These preliminaries being established, let us fix $\varepsilon > 0$ and $1 > \varepsilon' > 0$. Let m be an integer such that $\frac{1}{m} < \varepsilon$; we shall denote by $J_j (j=1,2,..,k)$ some Borel sets of the partition $\mathcal{D}_m$, chosen in order to have :

$$\inf_{\nu \in \mathcal{M}} \nu(\bigcup_{j \leq k} J_j) > 1 - \varepsilon'/4$$

With each probability $\nu \in \mathcal{M}(E)$ are associated k intervals in $[0,1]$, mutually disjoint, say $L_1^\nu, L_2^\nu,.., L_k^\nu$, such that : $\forall j = 1,2,..,k$, $\lambda(L_j^\nu) = \nu(J_j)$, hence :

$$\inf_{\nu \in \mathcal{M}} \sum_{j \leq k} \lambda(L_j^\nu) > 1 - \varepsilon'/4$$

It must be noticed that, for every $\nu \in \mathcal{M}$, one can consider those intervals L_j^ν whose lengths are over $\varepsilon'/4k$: indeed, $\sum\limits_{j:\lambda(L_j^\nu) > \varepsilon'/4k} \lambda(L_j^\nu) > 1-\varepsilon'/2$.

In order to lighten an already weary notation, we shall agree that all the intervals L_j^ν have a length greater than $\varepsilon/4k$, for all $\nu \in M$.

Now, let $e = \frac{\varepsilon'}{8k}$: According to lemma 4, there exists $\eta(m,e) > o$ such that :

$$\forall \nu \in M(E) \; ; \; \forall \mu \in M, \; \rho(\mu,\nu) < \varepsilon \Rightarrow \sum_{(n_1,n_2,\ldots,n_m) \in \mathbb{N}^{xm}} |\mu(A_{n_1,n_2,\ldots,n_m}) - \nu(A_{n_1,n_2,\ldots,n_m})| < e$$

Then, if a_{2j-1}^ν (and a_{2j}^ν) are respectively the upper (and lower) bounds of L_j^ν $(j = 1,2,..,k)$, we have $a_{2j-1}^\nu \leq a_{2j-1}^\mu + e$ and $a_{2j}^\nu \geq a_{2j}^\mu - e$

Therefore, for each $\mu \in M$, we have : $\bigcap_{\nu:\rho(\mu,\nu)<\eta(m,e)} L_j^\nu \supset [a_{2j-1}^\mu + e ; a_{2j}^\nu - e]$

Let $A_{\mu,\varepsilon,\varepsilon'} = \bigcup_{j=1}^{k} \bigcap_{\nu:\rho(\mu,\nu)<\eta(m,e)} L_j^\nu$, $\forall \mu \in M$. It can be verified easily that $\lambda(A_{\mu,\varepsilon,\varepsilon'}) > 1 - \varepsilon'$, and that, if $\omega \in A_{\mu,\varepsilon,\varepsilon'}$ there exists an index j such that $\omega \in L_j^\nu$, whenever $\rho(\mu,\omega) < \eta$ (m.e). Therefore, it follows from the construction that $X_m(\nu,\omega)$ and $X_m(\mu,\omega)$ belong to the same J_j whose diameter is less than 1/m. This leads to the result : $d(X(\mu,\omega) \; ; \; X(\nu,\omega)) < \varepsilon$

BIBLIOGRAPHY

(1) P. BILLINGSLEY : Convergence of probability measures (Wiley)

(2) A.V. SKOROKHOD : Limit theorems for stochastic processes Theory. Proba. Appl. (1956) p. 261-290

(3) R.M. DUPLEY : Annals of Math. Stat. 39, n°5, 1966, p. 1563-72

(4) J. GEFFROY et O. JOUANDET : C.R.A.S.P. t. 269, 3 dec. 1969. 1077

(5) K.R. PARTHASSARATHY : Probability measures on metric spaces. (Academic Press)

(6) A.V. SKOROKHOD : Studies in the Theory of Random Processes. (Addison-Wesley)

Recent Developments in Statistics
J.R.Barra et al., editors

ABSORPTION PROBLEMS IN SEMI-MARKOV CHAINS

Jacques JANSSEN
Université Libre de Bruxelles

ABSTRACT

Let $((J_n, X_n), n \in \mathbb{N})$ be a semi-Markov chain (or a (J-X) process) with state space $\{1, \ldots, m\} \times \mathbb{R}$ $(m \geq 1)$ of kernel $\mathbb{Q} = (Q_{ij})$. We consider the following associated processes $((J_n, M_n), n \geq 0)$, $((J_n, W_n), n \geq 0)$ where :

$$M_n = \sup (S_0, S_1, \ldots, S_n), \; n \geq 0$$
$$W_0 = 0, \; W_{n+1} = \sup (0, W_n + X_{n+1}), \; n \geq 0$$
$$S_n = X_0 + X_1 + \ldots + X_n, \; n \geq 0, \; X_0 = 0 .$$

Firstly, we give explicit forms of the following probabilities :

$$\mathbb{P}(M_n \leq x, J_n = j | J_0 = i), \; \mathbb{P}(W_n \leq x, J_n = j | J_0 = i)$$

in terms of $\mathbb{Q}$.

We also show how the asymptotic behaviour is given by the resolution of two Wiener-Hopf systems of integral equations in duality for which there is an existence and unicity theorem. Finally, we treat the particular case of (J-X) processes of zero order.

RESUME

Soit $((J_n, X_n), n \in \mathbb{N})$ une chaîne semi-markovienne à espace d'états $\{1, \ldots, m\} \times \mathbb{R}$ (où $m \geq 1$) de noyau $\mathbb{Q} = (Q_{ij})$ à laquelle on associe les processus $((J_n, M_n), n \geq 0)$, $((J_n, W_n), n \geq 0)$ où :

$$M_n = \sup (S_0, S_1, \ldots, S_n), \; n \geq 0$$
$$W_0 = 0, \; W_{n+1} = \sup (0, W_n + X_{n+1}), \; n \geq 0 ,$$
$$S_n = X_0 + X_1 + \ldots + X_n, \; n \geq 0, \; X_0 = 0 .$$

Nous donnons d'abord la forme explicite des probabilités :

$$\mathbb{P}(M_n \leq x, J_n = j | J_0 = i), \; \mathbb{P}(W_n \leq x, J_n = j | J_0 = i)$$

uniquement à l'aide de $\mathbb{Q}$. Ensuite, nous montrons que le comportement asymptotique se réduit à la recherche des solutions de deux systèmes d'équations intégrales de Wiener-Hopf en dualité pour lesquelles il existe un théorème d'existence et d'unicité.
Pour terminer, nous envisageons quelques cas particuliers dont les chaînes semi-markoviennes d'ordre 0.

1. PRELIMINARIES

The aim of this communication is to extend some well-known results on random walks with independent steps to those with steps in a semi-Markov chain dependence.
Let X_n $(n \geq 1)$ denote the successive steps of the random walk, S_n $(n \geq 1)$, the position after the nth step and, for the facility define $X_o = S_o = 0$. Then if

$$M_n = \sup (S_o, S_1, \ldots, S_n), \; n \geq 0 \tag{1.1}$$

$$W_o = 0, \; W_n = (W_{n-1} + X_n)^+ , \; n \geq 1 \tag{1.2}$$

it is well-known (see for example W. Feller (1966)) that, for all n, M_n and W_n have the same distribution when the variables X_n $(n \geq 1)$ are independent and equidistributed. Moreover, if $\mathbb{E}(X_n)$ exists and is strictly negative, the distribution function $\mathcal{M}$ of the random variable lim M_n is the unique P-solution (in Spitzer's sense) of the Wiener-Hopf equation :

$$\mathcal{M}(x) = \begin{cases} 0 & , x < 0 \\ \int_{-\infty}^{x} \mathcal{M}(x-s)dQ(s) & , x \geq 0 \end{cases} \tag{1.3}$$

Q being the distribution function of the X_n.
We shall put a certain dependence of the X_n; more precisely, we accept a semi-Markov chain (or as a (J-X) process) dependence defined by the consideration of a two-dimensional stochastic process $((J_n, X_n), n \geq 0)$ with states space $I \times \mathbb{R}$, $\mathbb{R}$ being the real set and $I = \{1, \ldots, m\}$; m is a fixed positive integer. This process may be defined on the measurable space $(E, \mathcal{E})$ where $E = \prod_{i=o}^{\infty} \mathbb{R} \times I$ with the usual product σ-field $\mathcal{E}$ constructed with the Borel sets β of R and the natural σ-field of I by means of the semi-Markov matrix $\mathbb{Q}$:

$$\mathbb{Q} = (Q_{ij}) \qquad i, j = 1, \ldots, m \tag{1.4}$$

where each Q_{ij} is a mass function defined on the real line.
Let $\mathbb{P}_{i,0}$ be the unique probability measure defined on $(E, \mathcal{E})$ such that $X_o = 0$, $J_o = i$; then, from the definition of a (J-X) process, we have $\mathbb{P}_{i,0}$ - a.s. :

$$\mathbb{P}_{i,0}[J_n=j, X_n \leq x | (J_k, X_k), k = 0, \ldots, n-1] = Q_{J_{n-1}j}(x) \tag{1.5}$$

Let us recall also that the process $(J_n, n \geq 0)$ is a Markov chain of probability transition matrix P defined by

$$P = \lim_{x \to \infty} \mathbb{Q}(x) \tag{1.6}$$

2. TRANSIENT BEHAVIOUR OF THE PROCESSES $((J_n, W_n), n \geq 0)$ and $((J_n, M_n), n \geq 0)$

Let us first recall that the process $((J_n, W_n), n \geq 0)$ is a Markov process with state space $I \times \mathbb{R}^+$ such that

$$\mathbb{P}_{i,0}[J_n = k, W_n \leq y \mid (J_\ell, W_\ell), \ell = 0, \ldots, n-2, J_{n-1} = j, W_{n-1} = x]$$

$$= U_o(y)\, Q_{jk}(y-x) \tag{2.1}$$

($U_c(y)$ is the distribution function giving the mass o at $y = c$);
On the contrary, the process $((J_n, M_n), n \geq 0)$ is trivially non markovian.
For all $n \geq 0$, define :

$$\mathcal{M}^n_{ij}(x) = \mathbb{P}_{i,0}[J_n = j, M_n \leq x] \tag{2.2}$$

$$\mathcal{W}^n_{ij}(x) = \mathbb{P}_{i,0}[J_n = j, W_n \leq x] \tag{2.3}$$

The values of these probabilities are inductively given by the following relations (J. Janssen (1976)) :

$$\begin{aligned} \mathcal{M}^o_{ij}(x) &= \delta_{ij}\, U_o(x) \\ \mathcal{M}^n_{ij}(x) &= U_o(x) \sum_k \int_{-\infty}^x \mathcal{M}^{n-1}_{kj}(x-s)dQ_{ik}(s), \; n \geq 1 \end{aligned} \tag{2.4}$$

$$\begin{aligned} \mathcal{W}^o_{ij}(x) &= \delta_{ij}\, U_o(x) \\ \mathcal{W}^n_{ij}(x) &= U_o(x) \sum_k \int_{-\infty}^x \mathcal{W}^{n-1}_{ik}(x-s)dQ_{kj}(s), \; n \geq 1 \end{aligned} \tag{2.5}$$

To give condensed explicit solutions of these two recursive integral systems, we use the following notations :

$$\mathcal{M}^n(x) = (\mathcal{M}^n_{ij}(x)), \quad \mathcal{W}^n(x) = (\mathcal{W}^n_{ij}(x)) \tag{2.6}$$

$$I(x) = (\delta_{ij}\, U_o(x)) \tag{2.7}$$

If $\mathcal{K} = (K_{ij})$ and $\mathcal{L} = (L_{ij})$ are two mxm matrices of mass functions, $\mathcal{K} * \mathcal{L}$ represents the convolution of these two matrices :

$$(\mathcal{K} * \mathcal{L})_{ij}(x) = \sum_k K_{ik} * L_{kj}(x) . \tag{2.8}$$

The operator + is defined on any matrix $\mathcal{A} = (A_{ij})$ by

$$\mathcal{A}^+(x) = (A^+_{ij}(x)) \tag{2.9}$$

where, of course

$$A^+_{ij}(x) = U_o(x) \, . \, A_{ij}(x) . \tag{2.10}$$

Now, it is possible to define, inductively, for any matrix of mass functions $\mathcal{A}$, two sequences of matrices defined as follows :

$$\mathcal{A}^{[1]} = \mathcal{A}^{\{1\}} = \mathcal{A}^{+} \tag{2.11}$$

$$\mathcal{A}^{[n]} = (\mathcal{A} * \mathcal{A}^{[n-1]})^{+} , \quad n > 1 \tag{2.12}$$

$$\mathcal{A}^{\{n\}} = (\mathcal{A}^{\{n-1\}} * \mathcal{A})^{+} , \quad n > 1 \tag{2.13}$$

Of course if m = 1, then, for all n

$$\mathcal{A}^{[n]} = \mathcal{A}^{\{n\}} \tag{2.14}$$

This last relation is also true whatever m is if $\mathcal{A}(0-) = 0$ and moreover the common value is then $\mathcal{A}^{(n)}$, the n-fold convoluated of $\mathcal{A}$.
Putting now the relations (2.4) into the matrix form and using the commutativity of the convolution of mass functions, we get :

$$\mathcal{M}^{0}(x) = I(x) \tag{2.15}$$

$$\mathcal{M}^{n}(x) = (\mathbb{Q} * \mathcal{M}^{n-1}(x))^{+}, \quad n \geq 1 \tag{2.16}$$

This leads to

$$\mathcal{M}^{1}(x) = \mathbb{Q}^{+}(x) \tag{2.17}$$

$$\mathcal{M}^{2}(x) = \mathbb{Q}^{[2]}(x) \tag{2.18}$$

Now, let us suppose that

$$\mathcal{M}^{n-1}(x) = \mathbb{Q}^{[n-1]}(x) \quad . \tag{2.19}$$

Then, from (2.16), we get

$$\mathcal{M}^{n}(x) = (\mathbb{Q} * \mathbb{Q}^{[n-1]})(x)^{+} \tag{2.20}$$

and, finally

$$\mathcal{M}^{n}(x) = \mathbb{Q}^{[n]}(x) \tag{2.21}$$

Starting from (2.5), an analogous treatment gives the explicit form of $\mathcal{W}^{n}(x)$ by means of the semi-Markov kernel Q :

$$\mathcal{W}^{0}(x) = I(x) \tag{2.22}$$

$$\mathcal{W}^{n}(x) = \mathbb{Q}^{\{n\}}(x) , \quad n \geq 1 \tag{2.23}$$

Let us remark that we pass from (2.21) to (2.23) by commuting the factors of a matrix convolution product. We can then use these results to develop a formal algebraic treatment like Kingman's algebra of queues (1966) in the particular case m = 1 but here in a matrix form.

3. DUAL PROCESSES

This notion was introduced (J. Janssen (1969)) when the matrix P is aperiodic irreducible. In this case let $((\pi_1, \dots, \pi_m)$ be the unique vector of stationary probabilities and let us associate now with the kernel $\mathbb{Q}$, its dual kernel $\hat{\mathbb{Q}}$ defined by :

$$\hat{\mathbb{Q}} = D\,\mathbb{Q}^{\tau}\,D^{-1} \tag{3.1}$$

where D denotes the diagonal matrice with entries π_1^{-1} $(i = 1, \dots, m)$ (τ is the usual transposition operator).

Surrounding with the sign $\hat{}$ the quantities related to the dual kernel, we have the existence of a process $((\hat{J}_n, \hat{X}_n), n \geq 0)$ defined by $\hat{\mathbb{Q}}$ on $(E, \mathcal{E}, \hat{\mathbb{P}}_{i,o})$. For this process, we have (J. Janssen (1976)) :

$$\hat{\mathfrak{M}}^n(x) = D(\mathfrak{W}^n(x))^{\tau}\,D^{-1} \tag{3.2}$$

$$\hat{\mathfrak{W}}^n(x) = D(\mathfrak{M}^n(x))^{\tau}\,D^{-1} \tag{3.3}$$

these two relations being equivalent as $(\hat{\hat{\mathbb{Q}}}) = \mathbb{Q}$.

It follows that

$$\mathfrak{M}^n(x) = D(\hat{\mathfrak{W}}^n(x))^{\tau}\,D^{-1} \tag{3.4}$$

$$\mathfrak{W}^n(x) = D(\hat{\mathfrak{M}}^n(x))^{\tau}\,D^{-1} \tag{3.5}$$

The relations (3.2), (3.3), (2.21) and (2.23) give the explicit expressions of $\hat{\mathfrak{M}}^n$ and $\hat{\mathfrak{W}}^n$ by means of the initial kernel $\mathbb{Q}$:

$$\hat{\mathfrak{M}}^n(x) = D(\mathbb{Q}^{\{n\}}(x))^{\tau}\,D^{-1} \tag{3.6}$$

$$\hat{\mathfrak{W}}^n(x) = D(\mathbb{Q}^{[n]}(x))^{\tau}\,D^{-1} \tag{3.7}$$

from which we easily deduce the expressions of $\hat{\mathbb{Q}}^{[n]}$ and $\hat{\mathbb{Q}}^{\{n\}}$ in terms of $\mathbb{Q}^{[n]}$ and $\mathbb{Q}^{\{n\}}$:

$$\hat{\mathbb{Q}}^{[n]}(x) = D(\mathbb{Q}^{\{n\}}(x))^{\tau}\,D^{-1} \tag{3.8}$$

$$\hat{\mathbb{Q}}^{\{n\}}(x) = D(\mathbb{Q}^{[n]}(x))^{\tau}\,D^{-1} \tag{3.9}$$

4. ASYMPTOTIC BEHAVIOUR OF THE PROCESSES $((J_n, W_n), n \geq 0)$ AND $((J_n, M_n), n \geq 0)$

We recall briefly some asymptotic results valid under the assumptions of irreducibility and aperiodicity of P and moreover if the quantity

$$\varphi = \sum_i \sum_j \int_{\mathbb{R}} x\,dQ_{ij}(x) \tag{4.1}$$

exists and is strictly negative. If so, we have (J. Janssen, 1976)

$$\lim_n \mathfrak{W}^n_{ij}(x) = \pi_j\,\hat{\mathfrak{M}}_j(x)\ , \quad i, j \in I \tag{4.2}$$

$$\lim_n \mathfrak{M}^n_{ij}(x) = \pi_j\,\mathfrak{M}_i(x)\ , \quad i, j \in I \tag{4.3}$$

and consequently

$$\lim_n \hat{\mathcal{W}}^n_{ij}(x) = \pi_j \mathcal{M}_j(x) \tag{4.4}$$

$$\lim_n \hat{\mathcal{M}}^n_{ij}(x) = \pi_j \hat{\mathcal{M}}_i(x) \tag{4.5}$$

where $(\mathcal{M}_1, \ldots, \mathcal{M}_m)$, $(\hat{\mathcal{M}}_1, \ldots, \hat{\mathcal{M}}_m)$ are respectively the unique P-solutions of the Wiener-Hopf systems in duality :

$$\mathcal{M}_i(x) = U_0(x) \sum_{j=1}^{m} \int_{-\infty}^{x} \mathcal{M}_j(x-s)dQ_{ij}(s), \ i \in I \tag{4.6}$$

$$\hat{\mathcal{M}}_i(x) = U_0(x) \sum_{j=1}^{m} \int_{-\infty}^{x} \hat{\mathcal{M}}_j(x-s)d\hat{Q}_{ij}(s), \ i \in I \tag{4.7}$$

5. (J-X) PROCESSES OF ZERO ORDER

Such processes have the following property :

$$Q_{ij}(x) = Q_j(x) , \quad i, j \in I \tag{4.8}$$

or equivalently

$$Q_{ij}(x) = p_j F_j(x) , \ i, j \in I \tag{4.9}$$

where, for each $j \in I$:

$$F_j(x) = \begin{cases} \dfrac{Q_j(x)}{Q_j(+\infty)} , & \text{if } Q_j(+\infty) \neq 0 \\ U_1(x) , & \text{if } Q_j(+\infty) = 0 . \end{cases} \tag{5.0}$$

It follows that :

$$\pi_j = p_j , \ j \in I \tag{5.1}$$

and

$$\hat{Q}_{ij}(x) = p_j F_i(x) , \quad i, j \in I \tag{5.2}$$

In this case, as shown by R. Pyke (1961) in the particular case of positive (J-X) processes, the variables $(J_n, n \geq 1)$ as well as the variables $(X_n, n \geq 1)$ are i.i.d. such that

$$\mathbb{P}_{i,0}(J_n = j) = p_j \tag{5.3}$$

$$\mathbb{P}_{i,0}(X_n \leq x) = H(x) \tag{5.4}$$

where

$$H = \sum_j Q_j \tag{5.5}$$

Consequently, for all n, M_n and W_n are equidistributed so that, for all n and each i :

$$\sum_j \mathcal{M}^n_{ij}(x) = \sum_j \mathcal{W}^n_{ij}(x) \qquad (5.6)$$

The fact that the kernel $\mathbb{Q}$ has identical rows gives of course a simplified form to $\mathbb{Q}^{[n]}$ and $\mathbb{Q}^{\{n\}}$.

Define first the following functions :

$$H_j = H * Q_j \,, \quad j = 1, \ldots, m \qquad (5.7)$$

and for each $j \in I$:

$$H^1_j = H^+_j \,, \; (H * H^{n-1}_j)^+ = H^n_j \,, \quad n > 1 \qquad (5.8)$$

$$\bar{H}_j = H^+ * Q_j \,, \quad j = 1, \ldots, m \qquad (5.9)$$

$${}^{1}\bar{H}_j = \bar{H}^+_j \,, \; (\sum_{j=1}^{m} {}^{n-1}\bar{H}_j) * Q_i = {}^{n}\bar{H}_i \,, \quad n > 1 \qquad (5.10)$$

Then, a little computation shows that for each n, the matrice $\mathbb{Q}^{[n]}$ has all its rows identical to

$$(H^{n-1}_1, \ldots, H^{n-1}_m) \qquad (5.11)$$

and that all the rows of $\mathbb{Q}^{\{n\}}$ are

$$({}^{n-1}\bar{H}_1, \ldots, {}^{n-1}\bar{H}_m) \qquad (5.12)$$

It can be verified (5.6), that , for each n :

$$\sum_j H^{n-1}_j = \sum_j {}^{n-1}\bar{H}_j \qquad (5.13)$$

Now, we want to recall that all the asymptotic behaviour only depends on *one* Wiener-Hopf integral equation if of course the quantity

$$\varphi = \sum_j p_j \int x dF_j(x) \qquad (5.14)$$

is strictly negative.

In fact the systems (4.6) and (4.7) have the following solutions (see J. Janssen (1976)) :

$$\mathcal{M}_i(x) = \psi(x) \,, \quad i \in I \qquad (5.15)$$

$$\hat{\mathcal{M}}_i(x) = U_0(x) \int_{-\infty}^{x} \psi(x-s) dF_i(s) \,, \quad i \in I \qquad (5.16)$$

where ψ is the unique P-solution of the Wiener-Hopf equation :

$$\psi(x) = U_0(x) \int_{-\infty}^{x} \psi(x-s) dH(s) \qquad (.517)$$

Using results (4.2), (4.3), (4.4) and (4.5), we get :

$$\lim \mathcal{W}_{ij}^{n}(x) = p_j \; U_o(x) \int_{-\infty}^{x} \psi(x-s)dF_j(s) \quad (5.18)$$

$$\lim \mathcal{M}_{ij}^{n}(x) = p_j \; \psi(x) \quad (5.19)$$

$$\lim \hat{\mathcal{W}}_{ij}^{n}(x) = p_j \; \psi(x) \quad (5.20)$$

$$\lim \hat{\mathcal{M}}_{ij}^{n}(x) = p_j \; U_o(x) \int_{-\infty}^{x} \psi(x-s)dF_i(s) \quad (5.21)$$

6. SOME APPLICATIONS

An application of the results presented here to queueing theory and risk theory will appear shortly. For risk theory, the most interesting aspect is that it is possible to consider the heterogeneity phenomenon of the claims.
Another interesting direction is the application to sociology. This has been done by R. Ginsbergh (1971). An application to regional migration movements in Belgium has just appeared (J. Mercenier (1976).
First, we only need positive (J-X) processes to describe the situation; but if we consider costs or benefits, we obtain a structure of a general (J-X) process in which random variables as (1.1) may be useful.

REFERENCES

FELLER, W. (1966). An introduction to probability theory and its applications; Vol. 2, J. Wiley and Sons, New York.
GINSBERGH, R.B. (1971). Semi-Markov processes and mobility. Journal of Math. Sociology, 1, 233.
JANSSEN, J., (1969). Les processus (J-X). Cahiers du C.E.R.O. 11, 181.
JANSSEN, J. (1970). Sur une généralisation du concept de promenade aléatoire sur la droite réelle. Ann. de l'Inst. H. Poincaré, B, VI, 249.
JANSSEN, J. (1976). Some duality results in semi-Markov chain theory. Rev. Roumaine de Math. Pures et Appl., XXI, 429.
KINGMAN, J.F.C., On the algebra of queues. J. Appl. Probability, 3, 285.
MERCENIER, J. (1976). Modèles et processus de décision semi-markoviens : Théorie et essai d'application au phénomène de migrations en Belgique, mémoire de Licence, Université Libre de Bruxelles.
PYKE, R., (1962). Markov renewal processes or zero order and their applications to counter theory in Studies in Applied Probability and Management Science edited by K.J. Arrow, S. Karlin, H.S. Scarf, Stanford University Press, Stanford.

Recent Developments in Statistics
J.R.Barra et al., editors

HOMOMORPHISMS AND GENERAL EXPONENTIAL FAMILIES

S. Johansen
University of Copenhagen and
Oregon State University

It is shown how in any family of distributions a certain duality exists between the observation and the parameter. This duality is then used to represent the family as a generalized exponential family, and it is shown how one can extend any family to a canonical exponential family.

1. INTRODUCTION

An exponential type family is usually defined by its densities

$$e^{\alpha(\theta)'t(x) + B(x) + C(\theta)}$$

with respect to some measure μ_0. Here $\alpha(\theta) \in R^r$, $\theta \in \Theta$ and $t(x) \in R^r$, $x \in X$. Clearly the factor B(x) can be taken into μ_0 and $C(\theta)$ is just a normalizing constant. Thus the interplay between the observation and the parameter is given by the bilinear form

$$\alpha(\theta)'t(x). \tag{1.1}$$

In this formulation the parameter $\alpha(\theta)$ and the observation t(x) have a dual role, in that t(x) can be considered a linear functional on the range of $\alpha(\theta)$ and vice versa [2].

Of special interest are the models obtained by allowing $\alpha(\theta)$ to vary freely in the set

$$D = \{\alpha \mid \phi(\alpha) = \int e^{\alpha' t(x)} d\mu < \infty\}$$

and we then get a canonical (or full) model defined by its densities

$$\frac{e^{\alpha' t(x)}}{\phi(\alpha)}, \quad \alpha \in D$$

Research for this paper was supported by the National Science Foundation under grant MPS 74-07589 AOI with Oregon State University.

with respect to $d\mu = e^{B(x)}\, d\mu_0$.

The purpose of this note is to show how in any family of probability measures, a similar duality holds. The basic idea is to represent x by its likelihood function and θ by the density and show how this represents (x,θ) as a point in a pair of semigroups in duality thus giving a general formulation of (1.1). The idea of using semigroups and homomorphisms as tools in the discussion of exponential families is due to Lauritzen [3].

Once this formulation is given one has the possibility of defining a general concept of a canonical family and to show how one can extend a given family to a canonical family.

The report presents part of the work done jointly with H.D. Brunk, D. Birkes and J. Lee at Oregon State University, Corvallis, see also [1].

2. THE FORMULATION

To avoid measure theoretic difficulties the following set up will be used: X is a separable topological space and $P = \{P_\theta, \theta \in \Theta\}$ a family of probabilities on X. Let S_θ denote the support of P_θ and μ a σ-finite measure on P. We assume that P_θ has density $f(x,\theta)$ with respect to μ, where μ is equivalent to P and that $f(x,\theta) = 0,\ x \notin S_\theta$ and $f(x,\theta)$ is continuous on S_θ. This defines the density uniquely.

Let $S(X)$ denote the semigroup of finite ordered samples from X, i.e. the set of samples $(x_1,\ldots,x_n)$ with composition given by

$$(x_1,\ldots,x_n)(y_1,\ldots,y_m) = (x_1,\ldots,x_n,y_1,\ldots,y_m).$$

On the space $S(X)$ we define the equivalence relations. For $x = (x_1,\ldots,x_n) \in S(X)$ and $y = (y_1,\ldots,y_m) \in S(X)$ we define

$$x \sim y \quad \text{if} \quad \prod_{i=1}^{n} f(x_i,\theta) = \prod_{j=1}^{m} f(y_j,\theta),\ \theta \in \Theta \tag{2.1}$$

Thus two samples are equivalent if they have the same (strict) likelihood function.

Similarly we define

$$x \approx y \quad \text{if} \quad \exists\, \lambda > 0: \prod_{i=1}^{n} f(x_i,\theta) = \lambda \prod_{j=1}^{m} f(y_j,\theta),\ \theta \in \Theta. \tag{2.2}$$

Thus $x \approx y$ if x and y have the same likelihood function.

We shall also write for $\lambda > 0$

$$[x] = \lambda[y] \quad \text{if} \quad \prod_{i=1}^{n} f(x_i,\theta) = \lambda \prod_{j=1}^{m} f(y_j,\theta) \tag{2.3}$$

and we thus have

$$x \approx y \quad \text{iff} \quad \exists\, \lambda > 0\colon [x] = \lambda[y].$$

Now we have the natural mappings

$$X \xrightarrow{t_1} S(X) \xrightarrow{t_2} S(X)/\!\sim\; \xrightarrow{t_3} S(X)/\!\approx,$$

where x is mapped into the one-element sample, and t_2 and t_3 map elements into equivalence classes.

Notice that the mapping $t = t_3 \circ t_2$ is the mapping that to each sample associates its likelihood function. Thus t is minimally sufficient in the sense that it induces the minimal sufficient σ-algebra, see Loève [4].

Similarly we consider the parameter space Θ and form $S(\Theta)$ and define, for $\tau = (\tau_1,\ldots,\tau_n) \in S(\Theta)$ and $\sigma = (\sigma_1,\ldots,\sigma_m) \in S(\Theta)$ the equivalence relation

$$\tau \sim \sigma \Leftrightarrow \prod_{i=1}^{n} f(x,\tau_i) = \prod_{j=1}^{m} f(x,\theta_j),\ x \in X. \tag{2.4}$$

Thus in particular θ_1 and θ_2 (in Θ) are equivalent if they correspond to the same density. Notice that since proportional densities are equal (they integrate to 1) there is no equivalent to $\approx$ on the parameterspace.

Note also that $\sigma = (\sigma_1,\ldots,\sigma_n)$, $\sigma_i \in \Theta$, $i = 1,\ldots,n$ corresponds to a measure whose density is a product of densities in the family P.

We then have the natural mapping

$$\Theta \xrightarrow{\alpha_1} S(\Theta) \xrightarrow{\alpha_2} S(\Theta)/\!\sim$$

where the paramter $\alpha_2(\alpha_1(\theta))$ is the maximal identifiable parameter.

We can then formulate the main result as:

Proposition 1 The semigroups $S(X)/\!\sim$ and $S(\Theta)/\!\sim$ are in duality by the bihomomorphism $\circ$ defined by

$$[x] \circ [\sigma] = \prod_{i=1}^{n} \prod_{j=1}^{m} f(x_i,\sigma_j). \tag{2.5}$$

Proof Note that [] denotes the equivalence class corresponding to $\sim$, thus $[x] = t_2(x)$ and $[\sigma] = \alpha_2(\sigma)$.

Let us first remark that by the definition of $\sim$, the right hand side of (2.5) only depends on the equivalence class of x and σ.

It is easily seen that $\circ$ is a bihomomorphism, since for instance,

$$[x] \circ [\sigma\tau] = \prod_{i=1}^{n} \left[\prod_{j=1}^{m} f(x_i,\sigma_j) \prod_{k=1}^{r} f(x_i,\tau_k) \right]$$

$$= \prod_{i=1}^{n} \prod_{j=1}^{m} f(x_i,\sigma_j) \prod_{i=1}^{n} \prod_{k=1}^{r} f(x_i,\tau_k)$$

$$= ([x] \circ [\sigma])([x] \circ [\tau]).$$

If

$$[x] \circ [\tau] = [x] \circ [\sigma], \; x \in S(X)$$

then in particular for $x \in X$ we get

$$\prod_{k=1}^{r} f(x,\tau_k) = \prod_{j=1}^{m} f(x,\sigma_j), \; x \in X$$

which by definition means that $\tau \sim \sigma$ and hence $[\tau] = [\sigma]$. Thus the bihomomorphism separates points and the semigroups are in duality.

The content of (2.5) is that it generalizes (1.1) which holds for ordinary exponential families. The following examples will illustrate the formulation, but the essence of (2.5) is that the parameter can be viewed as a homomorphism on the space of (strict) likelihood functions, and the observation as a homomorphism on the space of densities.

3. SOME EXAMPLES

Example 1 Let the densities be given by

$$f(x,\theta) = \frac{e^{\alpha(\theta)' t(x)} h(x)}{\phi(\theta)} \tag{3.1}$$

with respect to Lebesgue measure μ on R. Assume further that t and h are continuous and that $h > 0$.

In order that the essential features of the approach be apparent we will also assume that the functions $\{1, \alpha(\theta), \theta \in \Theta\}$ are linearly independent, and that the functions $\{1, \ln h(x), t(x), x \in R\}$ are linearly independent.

In this case $x = (x_1,\dots,x_n)$ and $y = (y_1,\dots,y_m)$ are equivalent, $x \sim y$, if

$$\alpha(\theta)'\left(\sum_{i=1}^{n} t(x_i) - \sum_{j=1}^{m} t(y_j)\right) + \sum_{i=1}^{n} \ln h(x_i) - \sum_{j=1}^{m} \ln h(y_j) - (n-m)\ln \phi(\theta) = 0,$$

$\theta \in \Theta$.

Now since $\ln \phi(\theta)$ is strictly convex as a function of $\alpha(\theta)$, we must have $n = m$. By the independence of the vectors $\{1, \alpha(\theta), \theta \in \Theta\}$ it follows that

$$\sum_{i=1}^{n} t(x_i) = \sum_{j=1}^{m} t(y_j)$$

and

$$\sum_{i=1}^{n} \ln h(x_i) = \sum_{j=1}^{m} \ln h(y_i).$$

Hence the equivalence relation $\sim$ on $S(X)$ is induced by the mapping

$$t_2((x_1,\dots,x_n)) = (n, \sum_{i=1}^{n} t(x_i), \sum_{i=1}^{n} \ln h(x_i)),$$

and $S(X)/\sim$ can be represented by a certain subsemigroup of

$$S = (N,+) \times (R^r,+) \times (R,+).$$

Similarly $\sigma = (\sigma_1,\dots,\sigma_n)$ is equivalent to $\tau = (\tau_1,\dots,\tau_m)$ if

$$\left(\sum_{i=1}^{n} \alpha(\sigma_i) - \sum_{j=1}^{m} \alpha(\tau_j)\right)' t(x) + (n-m)\ln h(x) - \sum_{i=1}^{n} \ln \phi(\sigma_i) + \sum_{j=1}^{m} \ln \phi(\tau_j) = 0,$$

$x \in X$.

By the independence of $\{1, \ln h(x), t(x), x \in X\}$ it follows, that

$$n = m, \quad \sum_{i=1}^{n} \alpha(\sigma_i) = \sum_{j=1}^{m} \alpha(\tau_j), \quad \sum_{i=1}^{n} \ln \phi(\sigma_i) = \sum_{j=1}^{m} \ln \phi(\tau_j)$$

and hence that the equivalence $\sim$ on the parameterspace is given by the function

$$\alpha_2(\sigma_1,\dots,\sigma_n) = \left(-\sum_{i=1}^{n} \ln \phi(\sigma_i), \sum_{i=1}^{n} \alpha(\sigma_i), n\right)$$

and that the semigroup $S(\Theta)/\sim$ can be represented by a certain subsemigroup of

$$(R,+) \times (R^r,+) \times (N,+).$$

The parameter θ can be identified with homomorphism given by the coefficients

$$(-\ln\phi(\theta),\ \alpha(\theta),\ 1).$$

Notice that not all homomorphisms on S can occur, since $\phi(\theta)$ is determined from $\alpha(\theta)$ by normalization and 1 occurs as the last coeeficient. Notice also that the semigroup $S(\Theta)/\sim$ as well as the bihomomorphism between $S(\Theta)/\sim$ and $S(X)/\sim$ depend on h, the underlying measure.

Now consider (2.3). In a similar way we find that $[x] = \lambda[y]$ iff

$$(3.2)\quad n = m,\quad \sum_{i=1}^{n} t(x_i) = \sum_{j=1}^{m} t(y_j),\quad \ln\lambda = \sum_{i=1}^{n}\ln h(x_i) - \sum_{j=1}^{m}\ln h(y_j),$$

and hence $x \approx y$ iff $n = m$ and $\Sigma_{i=1}^{n} t(x_i) = \Sigma_{j=1}^{m} t(y_j)$, thus the equivalence relation is induced by the function

$$s(x_1,\ldots,x_n) = \left(n,\ \sum_{i=1}^{n} t(x_i)\right)$$

which is known to be minimally sufficient, and which is independent of the underlying measure.

Example 2 Let the densities be given by

$$(3.3)\qquad f(x,\theta) = \begin{cases}\frac{1}{\theta}, & 0 \leq x \leq \theta,\\ 0, & \theta < x < \infty,\end{cases}$$

where $0 < \theta < \infty$. Then

$$(x_1,\ldots,x_n) \sim (y_1,\ldots,y_m) \quad \text{iff}$$

$$\frac{1}{\theta^n} 1_{[0,\theta]}(x_{(m)}) = \frac{1}{\theta^m} 1_{[0,\theta]}(y_{(m)}),\ 0 < \theta < \infty,$$

which happens only if $n = m$ and $x_{(n)} = y_{(m)}$. Thus $S(X)/\sim$ can be identified with $(N,+) \times (R_+,\vee)$ and θ with the homomorphism $h(n,x) = h_1(n)\, h_2(x)$ where

$$h_1(n) = \theta^{-n},\quad h_2(x) = 1_{[0,\theta]}(x).$$

Notice that in this case $x \approx y$ iff $x \sim y$ and this corresponds to $(n,\ x_{(n)})$ being minimally sufficient.

Example 3 Let P_1 and P_2 be given by two positive continuous densities f_1 and f_2 on R, with respect to Lebesgue measure. Then $(x_1,\ldots,x_n) \sim (y_1,\ldots,y_m)$ when

$$\prod_{i=1}^{n} f_k(x_i) = \prod_{j=1}^{m} f_k(y_j), \quad k = 1,2$$

i.e. when

$$t_1(x_1,\ldots,x_n) = \left(\prod_{i=1}^{n} f_1(x_i), \prod_{i=1}^{n} f_2(x_i)\right)$$

takes on the same value at x and y. Thus $S(X)/\sim$ is a subset of $(R_+,\cdot)^2$. Further $[x] = \lambda[y]$ if

$$\prod_{i=1}^{n} f_k(x_i) = \lambda \prod_{j=1}^{m} f_k(y_j), \quad k = 1,2.$$

Thus the equivalence relation $\approx$ is induced by the likelihood ratio

$$t_2(x_1,\ldots,x_n) = \frac{\prod_{i=1}^{n} f_1(x_i)}{\prod_{i=1}^{n} f_2(x_i)}$$

which is known to be minimally sufficient.

4. THE CANONICAL FAMILY

In order to obtain the proper definition of a canonical family we shall need the following mathematical object: a semigroup S with scalar multipliers i.e. there exists an operation o an S such that

$$x \circ (y \circ z) = (x \circ y) \circ z$$

and a mapping * on $R_+ \times S \to S$ such that

$$\lambda * (x \circ y) = (\lambda * x) \circ y = x \circ (\lambda * y).$$

We let S* denote the set of homogenous homomorphisms on S, i.e. mappings $h: S \to (R_+,\cdot)$ such that

$$h(x \circ y) = h(x)\, h(y)$$

$$h(\lambda * x) = \lambda\, h(x).$$

Now let (X,μ) be a measure space and $t: X \to S$ a statistic

Definition 4.1 A canonical family determined by (μ,t,S) is given by densities with respect to μ of the form

$$(4.1) \qquad \frac{\theta(t(x))}{\int \theta(t(x))\,\mu(dx)}, \quad \theta \in D,$$

where

$$D = \{\theta \in S^* \mid \theta(t(x)) \text{ is measurable and } 0 < \int \theta(t(x))\,\mu(dx) < \infty\}.$$

The reason that we use only the homomorphisms is that we want the family (4.1) to be essentially independent of μ, in the following sense:

Proposition 4.2 Let μ_0 be equivalent to μ and $h = \frac{d\mu}{d\mu_0}$. Let $t_0(x) = h(x) * t(x)$ then the families determined by (μ,t,S) and (μ_0,t_0,S) are the same.

Proof This follows since

$$\theta(t_0(x))\,d\mu_0 = \theta(h(x) * t(x))\,d\mu_0 = h(x)\,\theta(t(x))\,d\mu_0 = \theta(t(x))\,d\mu.$$

Thus the densities are the same and the two families are identical.

We now want to show how the family $\mathcal{P}$ considered in section 2 can be extended to a canonical family and finally give some examples how this extension works.

We then start with $\mathcal{P}$ given by the densities $f(x,\theta)$ on $\mathcal{X}$ with respect to μ.

First choose S as the space of multiples of likelihood functions

$$S = \{\lambda \prod_{i=1}^{n} f(x_i,\cdot),\ \lambda > 0,\ i = 1,\ldots,n\}.$$

This is a semigroup with scalar multipliers. To each $x = (x_1,\ldots,x_n)$ we associate as before the (strict) likelihood function i.e. $t(x) = [x]$.

Thus we are generating S from $S(\mathcal{X})/\sim$ by adding all multiples of elements of $S(\mathcal{X})/\sim$.

Now consider the family $\tilde{\mathcal{P}}$ generated by (μ,t,S). The mapping $\theta \to [\theta]$ associates with each $\theta \in \Theta$ a homogeneous homomorphism on S

$$[\theta] \circ (\lambda * [x]) = \lambda[\theta] \circ [x]$$

since both sides equal $\lambda \cdot \Pi_{i=1}^{n} f(x_i,\theta)$, $x = (x_1,\ldots,x_n)$.

The density corresponding to $[\theta]$ is given by $[\theta] \circ [t(x)] = f(x,\theta)$, $x \in \mathcal{X}$ and the family P is thus imbedded into $\tilde{P}$.

Corollary 4.3 The family $\tilde{P}$ does not depend on the choice of μ.

Proof Let μ_0 be another measure equivalent to P and let $h = \frac{d\mu}{d\mu_0}$. Likelihoods from densities with respect to μ_0 are proportional to likelihoods from μ and hence the mapping

$$t_0(x) = [x]_{\mu_0}$$

i.e. the likelihood with respect to μ_0, sends X into S. Note, however, that $t_0(x) = h(x) * [x]_\mu$, where $h(x) = \Pi_{i=1}^n h(x_i)$ and hence $t_0 = h * t$ and it follows from Proposition 4.2 that the families generated by (μ,t,S) and (μ_0,t_0,S) are the same.

Let us conclude this section by considering again the examples.

Example 1 The extension of this exponential type family depends on the semigroup generated by the statistic $\{n, \Sigma_1^n t(x_i), \Sigma_1^n \ln h(x_i)\}$.

As an example consider the following special case of a normal distribution with positive mean and variance equal to 1.

$$f(x,\theta) = \frac{1}{\sqrt{2\pi}} e^{\theta x} e^{-\frac{1}{2}x^2} e^{-\frac{1}{2}\theta^2}, \quad x \in R, \ \theta > 0.$$

We want to extend this family and the space S of multiples of likelihood functions consists of functions of the form

$$c\, e^{a\theta+b\theta^2}, \quad c > 0, \quad a \in R, \quad b < 0.$$

The semigroup operations are given by

$$(a,b,c) \circ (a',b',c') = (a+a',b+b',cc')$$

and

$$\lambda * (a,b,c) = (a,b,\lambda c).$$

Homomorphisms of S have the form

$$h(a,b,c) = c^\gamma e^{a\alpha+b\beta}$$

$\alpha \in R$, $\beta \in R$, $\gamma \in R$. The homogeneous homomorphisms satisfy

$$h(a,b,\lambda c) = \lambda\, h(a,b,c)$$

and hence we have $\gamma = 1$.

For an element $h \in S^*$ we then have

$$h(t(x)) = h(x, -\tfrac{1}{2}, \sqrt{2\pi}\, e^{-\frac{1}{2}x^2}) = \frac{1}{\sqrt{2\pi}} e^{-\frac{1}{2}x^2} e^{\alpha x - \frac{1}{2}\beta}.$$

When normalized this becomes the normal density with mean $\alpha \in R$. Thus the canonical family generated by the family with positive mean is the family with arbitrary mean.

Example 2 The space of likelihoods and their multiples are of the form

$$c\, 1_{[0,\theta]}(a), \quad c > 0, \quad a > 0$$

with

$$(a,c) \circ (a',c') = (a \vee a', cc')$$

and

$$\lambda * (a,c) = (a, \lambda c).$$

A homomorphism has the form

$$h(a,c) = 1_{[0,\theta]}(a)\, c^{\alpha}$$

and the homogeneous ones have $\alpha = 1$. When normalized this gives the family we started with. Thus the uniform distributions on $[0,\theta]$, $0 < \theta < \infty$ form a canonical family.

Example 3 In this example the space S is spanned by the vectors

$$\left(\lambda \prod_{i=1}^{n} f_1(x_i),\ \lambda \prod_{i=1}^{n} f_2(x_i)\right), \quad \lambda > 0,\ x_i \in \mathcal{X},\ i = 1,\dots,n.$$

If we assume that any point in $]0,\infty[^2$ can be so represented then $S =]0,\infty[^2$ with the operations

$$(a,b) \circ (a',b') = (aa', bb')$$

$$\lambda * (a,b) = (\lambda a, \lambda b).$$

A homomorphism is of the form

$$h(a,b) = a^{\alpha} b^{\beta}$$

and the homogeneous ones satisfy $\alpha + \beta = 1$.

Thus the family generated by (f_1,f_2) has densities of the form

$$\frac{f_1^{\alpha}(x)\ f_2^{1-\alpha}(x)}{\int f_1^{\alpha}(x)\ f_2^{1-\alpha}(x)\ \mu(dx)}, \quad \alpha \in D,$$

$$D = \{\alpha \mid \int f_1^{\alpha}(x)\ f_2^{1-\alpha}(x)\ \mu(dx) < \infty\}.$$

REFERENCES

[1] Birkes,D., Brunk,H.D. and Lee,J.W. (1975) Generalized Exponential Families. Tech.Rep.45, Oregon State University,

[2] Johansen,S. (1970) Exponential Models. Notes. University of Copenhagen.

[3] Lauritzen,S.L. (1975) General Exponential Models for Discrete Observations, Scand.J.Statist.2, 23-33.

[4] Loève,M. (1960) Probability Theory. (Van Nostrand, New York).

Recent Developments in Statistics
J.R.Barra et al., editors

OPTIMAL 2^k DESIGNS OF ODD AND EVEN RESOLUTION

by
Stratis Kounias
McGill University and University of Thessaloniki
CANADA GREECE

The purpose of this paper is to study 2^k designs with respect to different optimality criteria.It is proved that orthogonal arrays generate 2^k designs which are D,G,A and E optimal.A number of theorems are proved and the case where orthogonal arrays do not exist is also studied.

I.INTRODUCTION

The theory of optimal designs in regression problems has been developed intensively for the last I5 years.A very good presentation and an extensive bibliography on the subject is in Fedorov [4].

The equivalence theorems of Kiefer [7] and Kiefer and Wolfowitz [8] initiated a series of papers on the development of numerical algorithms for the construction of D-optimal (D_s-optimal) designs.

To begin with,a regression design is called:

(i) D-optimal if it minimizes the determinant of the covariance matrix of the l.s.e.(least squares estimates) of all the parameters.

(ii) Truncated D-optimal or D_s-optimal if it minimizes the determinant of the covariance matrix of the l.s.e. of some of the parameters.

(iii) G-optimal or minimax if it minimizes the maximum variance of the l.s.e. of the response surface.

(iv) A-optimal if it minimizes the sum of the variances of the l.s.e. of all the parameters.

(v) E-optimal if it minimizes the maximum eigenvalue of the covariance matrix of the l.s.e. of all the parameters.

The analysis of a 2^k experiment is done using Yates' technique and does not concern us here.

In this paper we are interested in studying the different optimality criteria for 2^k factorial designs.Box and Hunter [2],[3] studied 2^k factorial designs of resolution III,IV and V from the point of view of aliasing and confounding.A design is of (odd) resolution 2t + 1 if all effects of order t or less are estimable whenever all effects of order higher than t are suppressed.

A design is of resolution 2t if all effects of order (t - 1) or less are estimable whenever all effects of order (t + 1) or higher are suppressed.Thus in the designs of (even) resolution 2t the effects of order t are not all estimable which means that the covariance matrix of the l.s.e. of the parameters is singular.

Plackett and Burman [9] proved that orthogonal arrays give 2^k designs minimizing simultaneously the variances of all the parameters.

In this paper we prove that:

(a) If an O.A(N,k,2,2t) exists,then it generates an N-observation 2^k design of resolution 2t + 1 which is D,G,E and A-optimal.

(b) If an O.A(N,k,2,2t-1) exists,then it generates an N-observation 2^k design of resolution 2t which is D_s-optimal.

(c) If an O.A(N,k,2,2t) does not exist then the determinant of the information matrix of the D-optimal N-observation 2^k design of resolution 2t is less than 1.

2. 2^k DESIGNS OF RESOLUTION III

We first study 2^k factorial designs in which we suppress all two or higher order interactions considering them to be negligible.

All possible 2^k treatment combinations are given by the columns of a $k \times 2^k$ array of zeros and ones.The k rows of the array correspond to the k factors F_1, $F_2,\ldots,F_k$ and the two symbols 0 and 1 to the levels of the factors.

After suppressing all interactions of order two or higher our model becomes:

$$y_{i_1,i_2,\ldots,i_k} = I + \sum_{r=1}^{k} (-1)^{i_r + 1} A_r + e_{i_1,i_2,\ldots,i_k} \qquad (1)$$

where $i_1,i_2,\ldots,i_k$ are the levels (0 or 1) of the factors $F_1,F_2,\ldots,F_k$ in the ith column (treatment) of the array.Assume the errors to be uncorrelated with the same known or unknown variance which without loss of generality is taken $\sigma^2 = 1$. The general mean I and the k effects $A_1,A_2,\ldots,A_k$ are the unknown parameters to be estimated by unweighted least squares from N observations.We examine here the continuous optimal designs which in turn give optimal N observation designs for some values of N.

Our model (1) is written:

$$y(\underset{\sim}{x}_i) = \underset{\sim}{\vartheta}'f(\underset{\sim}{x}_i) + e(\underset{\sim}{x}_i) \qquad (2)$$

where $\underset{\sim}{\vartheta}' = (I,A_1,\ldots,A_k)$,

$$f'(\underset{\sim}{x}_i) = (1,(-1)^{i_1+1},\ldots,(-1)^{i_k+1})$$ and

$x_i' = (i_1, i_2, \ldots, i_k)$, x_i is a column (treatment) of the $k \times 2^k$ array of zeros and ones.

Now let ε be a design on the observation space X assigning probability p_i to the point x_i, $\sum_{i=1}^{2^k} p_i = 1$. The probabilities are $p_i > 0$ for some points (vectors) x_i and $p_i = 0$ for the rest of the points, and let J_N be the subset of $\{1,2,\ldots 2^k\}$ containing N not necessarily different integers such that $p_i > 0$ if $i \in J_N$ and $p_i = 0$ if $i \notin J_N$.

The information matrix of the unweighted least squares estimates of $I, A_1, A_2, \ldots A_k$ is the $(k+1) \times (k+1)$ matrix $M = \{m_{ij}\}$ $i,j = 0,1,\ldots,k$:

$$M = \sum_{i \in J_N} p_i \, f(x_i) f'(x_i) \qquad (3)$$

We want to find ε such that the determinant M is maximized (D-optimality).

We give here the following:

Theorem 2.1 If an O.A(N,k,2,2) exists, then it generates an N-observation 2^k design which is D and G-optimal.

Proof. We observe that $|M| \le m_{11}, \ldots, m_{kk} = 1$ with equality only if $M = I_{k+1}$ because all diagonal elements of M are equal to 1. An O.A(N,k,2,2) generates a design with $M = I_{k+1}$, hence it is D-optimal. If on the other hand a design has $M = I_{k+1}$ then it is not difficult to prove that it is generated by an orthogonal attay. From the equivalence theorem of Kiefer and Wolfowitz [8] we conclude that these designs are G-optimal.

Q.E.D.

From the above theorem it follows, that if an O.A(N,k,2,2) does not exist, then the D-optimal N-observation 2^k design of resolution III cannot have $|M|=1$. It is known that such O.A do not exist if $N \neq 0 \pmod 4$.

Corollary 2.1 An O.A(N,k,2,2) generates an N-observation 2^k design ε^* of resolution III which is E and A-optimal.

Proof. We can show that $0 < \lambda_{min}(M) \le 1$ and the O.A gives $M = I_{k+1}$ that is $\lambda_{min}(M) = 1$

Also (Fedorov [4], theorem 2.11.1 page 139), we conclude that ε^* is A-optimal.

Q.E.D.

3. 2^k DESIGNS OF RESOLUTION 2t + 1

For a design of resolution 2t+1 we keep only the interactions of order $\le t$ and set all interactions of order $\ge t+1$ equal to zero considering them to be negligible.

The procedure is exactly the same as for resolution III designs. We are going to demonstrate it for $t = 2$ in which case we preserve the general mean, main effects and two factor interactions. The model is

$$y(\underset{\sim}{x}_i)=I + \sum_{s=1}^{k} (-1)^{i_s+1} A_s + \sum_{s=1}^{k} \sum_{s<r} (-1)^{i_s + i_r} A_{sr} + e(\underset{\sim}{x}_i),$$

where $\underset{\sim}{x}_i' = (i_1,i_2,\ldots,i_k)$ a column (treatment) of the $k \times 2^k$ array, and $i_s = 0$ if the sth factor is at low level and $i_s = 1$ if the sth factor is at high level. Also, A_{sr} denotes the interaction of the sth and rth factor $s \neq r$. Our model then is given by (2) with the $1 + k + \frac{k(k-1)}{2}$ parameters

$$\underset{\sim}{\vartheta}' = (I,A_1,\ldots,A_k,A_{12},\ldots,A_{1k},\ldots,A_{k-1,k})$$

and $f'(\underset{\sim}{x}_i) = (I,(-1)^{i_1 + 1},\ldots,(-1)^{i_k + 1},(-1)^{i_1 + i_2},\ldots,(-1)^{i_{k-1}+i_k})$.

The information matrix given by (3) is now:

$$M = \begin{bmatrix} 1 & M_{01}' & M_{02}' \\ M_{01} & M_{11} & M_{12}' \\ M_{02} & M_{12} & M_{22} \end{bmatrix}$$

where $M_{01}' = 1 \times k$, $M_{02}' = 1 \times \frac{k(k-1)}{2}$, $M_{11} = k \times k$, $M_{22} = \frac{k(k-1)}{2} \times \frac{k(k-1)}{2}$.

Let us take as before a subset of N columns (treatments) of the $k \times 2^k$ array and denote this $k \times N$ array by R with $\underset{\sim}{x}_i \in R$ if $i \in J_N$, J_N having N non-necessarily different integers from $\{1,2,\ldots,2^k\}$. Any D-optimal design ε assigns probability p_i to the point $\underset{\sim}{x}_i$ and $p_i>0$ if $i \in J_N$, $p_i = 0$ if $i \notin J_N$ for some J_N and $\sum_{i\in J_N} p_i = 1$. Then the elements of the information matrix M are given by

$$m_{rr} = 1, \qquad r = 0,1,\ldots,\frac{k(k-1)}{2}.$$

$$m_{0r} = \sum_{i\in J_N} p_i(-1)^{i_r+1}, \quad r = 1,2,\ldots,k, \qquad m_{0u} = \sum_{i\in J_N} p_i(-1)^{i_r+i_s},$$

$r < s$, $r,s = 1,2,\ldots,k$, $\quad u = r(k-2)+s+1$

$$m_{rs} = \sum_{i\in J_N} p_i(-1)^{i_r + i_s}, \qquad r,s = 1,2,\ldots,k.$$

$$m_{tu} = \sum_{i\in J_N} p_i(-1)^{i_t + i_r + i_s + 1} \quad \text{with} \quad u = r(k-2) + s + 1,\ t,r,s = 1,2,\ldots,k$$

$$m_{vu} = \sum_{i\in J_N} p_i(-1)^{i_t + i_q + i_r + i_s} \quad \text{with} \quad v = t(k-2) + q + 1,\ u = r(k-2) + s + 1 \text{ and}$$

$t,q,r,s = 1,2,\dots,k.$

Since the diagonal elements of M for any design ε of resolution $2t + 1$ are equal to unity,hence as in theorem 2.1 we conclude that:

Theorem 3.1: If an O.A(N,k,2,2t) exists then it generates an N-observation 2^k design of resolution $2t + 1$ which is D,G,E and A-optimal.

Proof: The proof is similar to the proof of theorem 2.1

Q.E.D.

4. TRUNCATED D-OPTIMAL DESIGNS

Truncated D-optimal designs have been studied mainly by Kiefer [7],Karlin and Studden [6],Atwood [1] and Silvey and Titterington [13].

Let $\underset{\sim}{\vartheta}' = (\underset{\sim}{\vartheta}_1 \vdots \underset{\sim}{\vartheta}_2)'$ be the unknown parameters and we are interested in estimating only $\underset{\sim}{\vartheta}_1$,then we want to minimize the determinant of Σ_{11} where

$$\Sigma_{11}^{-1} = M_{11} - M_{12} M_{22}^{-} M_{12} \quad \text{and}$$

$$M = \begin{bmatrix} M_{11} & M_{12} \\ M_{21} & M_{22} \end{bmatrix}$$

is the information matrix.

I design is of resolution $2t, t \geq 2$ if we can estimate all effects of order $\leq t-1$ by ignoring all effects of order $\geq t+1$. In these designs the effects of order t are not necessarily estimable but might be confounded with other effects of order t.

Theorem 4.1: A sufficient condition for a 2^k design ε of resolution 2t to be truncated D-optimal is that $M_{11} = I$ and $M_{12} = 0$.

Theorem 4.2: If an O.A(N,k,2,2t-1) exists then it generates an N-observation 2^k design of resolution 2t which is truncated D-optimal.

The proof of the above theorems if straightforward.

It is very interesting to know what happens if an O.A(N,k,2,t) does not exist as is the case for $N \neq 0 \pmod{2^t}$. This problem is still open and in that case the D-optimal might not be also G-optimal designs.

REFERENCES

[1] Atwood,C.L.,Optimal and efficient designs of experiments Ann.Math.Statist.40, 1969, 1570-1602

[2] Box,G.E.P. and Hunter,J.S., The 2^{k-p} fractional factorial designs; part I.Technometrics, Vol.3, No.3, 1961, 311-361

[3] Box,G.E.P. and Hunter,J.S., The 2^{k-p} fractional factorial designs; part II.Technometrics, Vol.3, No.4, 1961, 449-458

[4] Fedorov, V.V., Theory of Opimal Experiments. Academic Press, 1972

[5] Fisher, R.A., The theory of confounding in factorial experiments in relation to theory of groups. Ann. of Eugenics,11 , 1944, 341-353

[6] Karlin, S. and Studden, N.J., Optimal experimental designs. Ann. Math. Statist. 37, 1966, 783-815

[7] Kiefer, J., Optimum designs in regression problems, II. Ann. Math. Statist. 32, 1961 , 198-325

[8] Kiefer, J. and Wolfowitz, J., The equivalence of two extremum problems. Canadian J. Math. 12, 1960, 363-366

[9] Plackett , R.L. and Burman, J.P., Designs of optimum multifactorial experiments. Biometrika, 33, 1946, 305-325

[10] Raghavarao, D., Construction and Combinatorial Problems in Design of Experiments. J. Wiley and Sons, 1973

[11] Rao, C.R., Linear Statistical Inference and its Applications. Second Edition, J. Wiley and Sons, 1973

[12] Seiden, E. ams Zemach, R., On Orthogonal arrays. Ann. Math. Statist. 37, 1966, 355-370

[13] Silvey, S.D. and Titterington, D.M., A geometric approach to optimal design theory. Biometrika, 60, 1973, 21-32

Recent Developments in Statistics
J.R.Barra et al., editors

ON A GENERAL CHARACTERIZATION OF ALL AND CONSTRUCTION OF BEST DISTRIBUTIONFREE TESTS

Viktor Kurotschka
Freie Universität Berlin
Germany

0. Introduction

In most nonparametric testing problems one has the following situation:
Let $(X, \mathcal{B})$ be the sample space and $\mathcal{P}$ a class of probability measures on $\mathcal{B}$, $H_o: P\in\mathcal{P}_o\subset\mathcal{P}$ and $H_1: P\in\mathcal{P}_1\subset\mathcal{P}$ with $\mathcal{P}_o\cap\mathcal{P}_1 = \emptyset$, such that there exists a group S of measurable transformations of X with the property that $Pos^{-1} = P$ for all $s\in S$ and all $P\in\mathcal{P}'$ where $\mathcal{P}'$ is either $\mathcal{P}_o$ or the common boundary of $\mathcal{P}_o$ and $\mathcal{P}_1$.
In the first case when $\mathcal{P}' = \mathcal{P}_o$ the group S exhibits the hypothesis as a set consisting of some "symmetrical" members of $\mathcal{P}$.
For instance if $\mathcal{P}$ is the family of all atomfree measures on $\mathcal{B} = \mathcal{L}_N$ (=: Borel σ-algebra on R_N), $\mathcal{P}_o$ might be the subset of all stationary measures in $\mathcal{P}$ (Hypotheses of exchangeability or randomness) so that the Symmetric group S_N plays the role of S, or $\mathcal{P}_o$ might be the subset of all spherically symmetric measures in $\mathcal{P}$ (hypothesis of sphericity or spherical symmetry) so that S stands for the group O_N of orthogonal transformations of R_N. In the second case where $\mathcal{P}'$ is the common boundary of $\mathcal{P}_o$ and $\mathcal{P}_1$, the group S expresses a certain "symmetry" of members of the boundary of $\mathcal{P}_o$ and $\mathcal{P}_1$; members of $\mathcal{P}_o$ and $\mathcal{P}_1$ might be "asymmetric" in different directions. For obvious reasons we have called such a group S a symmetry group of $\mathcal{P}'$ (see [8]). In both cases the group S can be used for construction of distributionfree tests with respect to $\mathcal{P}'$, often a necessary step to obtain most powerful tests in the first case and a uniformly most powerful unbiased test in the second case.
The construction can be made considerably more transparent if one discovers another transformation - group G of measurable transformations g of X with the property that $\{Pog^{-1}; g\in G\} = \mathcal{P}'$ for every $P\in\mathcal{P}'$. Considering only $\mathcal{P}' = \mathcal{P}_o$ we called in [8] such a group G a hypothesis generating group which we obviously change here to $\mathcal{P}'$-generating group.
Such groups G exist for most concrete problems (see [8] and reference therein) and are the groups leaving the testproblem invariant. The discovery of new such groups continues (see among others [18]), providing further cases falling under the theory which was sketched by C.B. Bell and me in [8], and which unifies some

of the work of Bell and his associates (see [2] - [7], [9], [10],) related to the ideas of Lehmann and Stein [14] and to the approach of Pitman [15], [16]. In the meantime we have investigated the scope of the theory. In the course of these investigations one of my students, K. Junge (see [13]) has made valuable contributions. As soon as circumstances allow, the complete theory will be submitted for publication. This paper is intended for giving some insight into the main results and for keeping discussion on this subject alive (see for example the "open question" in [18]).

1. Characterization of distributionfree statistics and their relation to generalized rank statistics

The most general conditions on P', S and G which enabled us to prove the statements concerning a constructive characterization of all P'-distributionfree statistics (statistics having under each $P \in P'$ the same distribution) and to analyse their relation to P'-rank statistics defined as maximal invariant statistics under G because of their easily seen P'-distributionfreeness and because of the invariance property of classically defined rank statistics (see [20] and [8]) are the following assumptions:

(i) *The sample space* (X, B) *is a standard Borel-space.*
This does not restrict the scope of applications but enables us to apply Lebesgue's decomposition and isomorphism theorem from [17] to prove theorem 1.

(ii) *The group* S *and* G *are canonical,* i.e.
a) The S- and G-orbits are B-measurable,
b) for all $x, x' \in X$ the set $S(x) \cap G(x')$ contains exactly one point,
c) for all $g \in G$, $s \in G$, and $x \in X$ $g(S(x)) = S(g(x))$ and $s(G(x)) = G(s(x))$,

so that in particular for all $x \in X$
$\{S(g(x));\ g \in G\}$ and $\{G(s(x));\ s \in S\}$ are two B-measurable partitions of X cross-sectioning each other

(iii) *The symmetry group* S *is a topological group such that*
α) $L: S \times X \ni (s,x) \to s(x) \in X$ *is continuous with respect to the product topology of* $S \times X$.

β) S *is compact.*

This assumption fulfilled by all known symmetry groups in nonparametric statistics enables us

1) to define, on the Borel σ-algebra S of S a normed Haar-measure μ and to induce a uniform probability distribution on the S-orbits of X which is used to measure the amount of points exhibited from each S-orbit by the

socalled Generalized-Bell-Pitman Statistic R_h (GBP-statistic) which is defined via a socalled GBP-function h, a numerical $\mathcal{B}$-measurable function for which $P\{x \in X; h(s(x)) = h(x)\} = o$ for all $s \in S \setminus \{id\}$ and all $P \in P'$, a function which induces a linear order on P'-almost all S-orbits. Finally $R_h: X \ni x \to R_h(x) := \mu\{s \in S; h(s(x)) \leq h(x)\} \in [0,1]$ measures the amount of points on the S-orbits S(x) having an h-value not larger than the one of x.

2) to ensure that the maximal P'-order-statistic O_S defined as a maximal invariant statistic under S (for motivation see [20] and [8]) is sufficient for P', a fact which allows in particular to see the bounded completeness of O_S with respect to P' being equivalent to the Neyman structure with respect to O_S of all P'-distributionfree statistics.

(iv) *P' is S-boundedly complete*

i.e. every maximal invariant statistic under S is boundedly complete. General conditions for this property, (as well as asumption (ii) can be derived from the results in [1].

This assumption not only allows employing Neyman-structure arguments for the construction of P-distributionfree tests but also together with (ii) the application of the Berk and Bickel theorem on almost invariant statististics [11] to establish the final form of theorem 1.

Let H be the set of all GBP-functions for P' and S with the property that for P' almost all S-orbits S(x) one has:

$$R_h(x') = R_h(x'') \Longleftrightarrow x' = x'' \text{ for all } x', x'' \in S(x)$$

then up to a P'-Null set N the following map

$$g_h: X \ni x \longrightarrow g_h(x) := S(g(x)) \cap R_h^{-1}(r) \in X, \text{ where } r = R_h(x),$$

is well defined for each $x \in X' := X - N$ having the following properties: (after identifying X with X' mod P'): For each $h \in H$

(i) g_h *is a measurable transformation of* X',

(ii) $G_h := \{g_h; g \in G\}$ *is a* P'*-generating group*,

(iii) S *and* G_h *are canonical*,

(iv) R_h *is a maximal invariant statistic under* G_h,

(v) R_h *and* O_S *are stochastically independent with respect to each* $P \in P'$ *(as* R_G *and* S *are)*.

These properties show that each G_h, $h \in H$ is structurally equivalent to the original G and could have been used at the starting point to define the maximal

P'-rank statistics. This again questions the general definition of rank statistics via invariance properties, because R_h is also P'-distributionfree which is relatively easy to see. Much harder to prove is the "only if" statement of the following Theorem 1. The main idea for proving the richness of the class H, i.e. that there are sufficiently many GBP-functions h in H each inducing a different order on the S-orbits, is based on a decomposition theorem of measure spaces into canonical measure spaces and on an isomorphism theorem for the corresponding factor spaces (see [17]).

Theorem 1

Let S *be a compact symmetry group of* P' *and* P' *boundedly* S-*complete; then a statistic* T *is* P'-*distributionfree iff there exists a* GBP-*function* $h \in H$ *and a Baire function* W *so that* $T = W \circ R_h$ (P'-a.s.).

This theorem states in particular that every distributionfree statistic can be represented as a G_h-invariant statistic mod P' for some suitable $h \in H$ (which always exists !). So that from the formal point of view - by means of invariance with respect to some P'-generating group - generalized rank statistics and general P'-distributionfree statistics cannot be distinguished structurally. Positively expressed, all statements for P'-rank statistics can be derived for general P'-distributionfree statistics because according to their definition and their basic properties GBR-statistics R_h are also maximal P'-rank statistics, of course inducing different partitions of the sample space forming different cross-sections of the S-orbits for different $h \in H$.

Precisely this last fact has given rise originally to the definition of GBP-statistics R_h as an elegant instrument for selecting from each S-orbit points according to their values induced by the particular GBP-function h so that R_h can be used for the construction of best P'-similar tests, unifying the well known argument of applying the Neyman-Pearson lemma at each S-orbit separately as indicated in the next section.

2. Construction of best P'-similar tests

For the construction of a best P'-similar level α test against a simple alternative $P_1 \in P_1$, i.e. a test φ^* with $E_P\varphi^* = \alpha$ or all $P \in P'$ and E_{P_1} φ for all P'-similar level α tests, one can use (because of the assumption on the P'-order statistics O_S) the GBP-statistics R_h to select from P'-almost every S-orbit an equal amount (a set of Haar measure μ equal to α)of those points,which assign a density f_1 of P_1 to its largest values within each considered S-orbit namely by taking an $h \in H$ which preserves the order of points induced by f_1 within each of the P'-a.e. relevant S-orbits.

As result one obtains, after overcoming some technical difficulties, the following theorem:

Theorem 2

If in addition to the previous assumptions there exists a density f_1 *of* $P_1 \in P_1'$ *with respect to some S-invariant measure* ν *on* B *(in most interesting nonparametric problems the Lebesgues measure* λ_N *on* $B = \mathcal{B}_N$ *or the restriction of* λ_N *to some trace of* $_N$ *plays the role of* ν*) then*

(i) *Every most powerful P'-similar test at level* α *against* $P_1 \in P_1'$ *is equivalent to*

$$\varphi^*(X) = 1_{\{R_{h^*}(X) > 1-\alpha'\}} + (\alpha - \alpha')1_{\{R_{h^*}(X) = 1-\alpha'\}}$$

where

a) h^* *is a GBP-function preserving the order of points on P'-almost all S-orbits induced by* f_1, i.e.
for P'-almost all $x \in X$ *and P'-almost all* $x', x'' \in S(x)$
$h^*(x') < h^*(x'') \Rightarrow f_1(x') \leq f_2(x'')$
and
$f_1(x') < f_2(x'') \Rightarrow h^*(xI) < h^*(x'')$

b) $(1-\alpha')$ *is the smallest possible value of* R_{h^*} *not smaller than* $1-\alpha$.
(For infinite S there is always $\alpha' = \alpha$*).*

(ii) φ^* *is a uniformly most powerfull* G_h*-invariant level* α *test against* $P_{h^*} := \{P_1 o g^{-1};\ g \in G_{h^*}\}$.

In particular the last statement shows an analogy to rank tests. This analogy will be even clearer in the next section.

3. Generalization of Hoeffding's formula

Translating the arguments of W. Hoeffding [10] to derive the distribution of classical rank statistics under an alternative $P_1 \in P_1$ to S-orbits and G-orbits respectively G_h-orbits one obtains with some patience almost directly the following very general statement.

Theorem 3

Let f_0 *and* f_1 *be versions of densities of* $P_0 \in P'$ *respectively of* $P_1 \in P_1$ *with respect to some suitable measure* ν *on* B *with*

$$P_1\{x \in X;\ f_0(x) = 0 \text{ and } f_1(x) \neq 0\} = 0$$

then

(i) *in the case of finite symmetry groups S one has: For every GBP-function* $h \in H$ *and every* $r \in \{1, 2, \dots, \text{ord } S\}$

$$P_1[R_h(X) = r] = (\text{ord } S)^{-1} E_{P_o}\left[\frac{f_1(Q_r(X))}{f_0(Q_r(X))}\right]$$

where

Q_r *is defined pointwise by*

$Q_r(x) = x'$ *iff* $O_S(x') = O_S(x)$ *and* $R_h(x') = r$

(ii) *in the case of infinite symmetry groups* S *one has: For every* GBP-*function* $h \in \mathcal{H}$ *and every* $t \in [0,1]$

$$P_1[R_h(X) \leq t] = t \cdot E_{P_o}\left[\frac{f_1(\overline{Q}_t(X))}{f_0(\overline{Q}_t(X))}\right]$$

where

$\overline{Q}_t$ *is defined pointwise by*

$\overline{Q}_t(x) = x'$ *iff* $O_S(x') = O_S(x)$ *and* $R_h(x') = t \cdot R_h(x)$

From the definition of Q_r and $\overline{Q}_t$ one sees immediately that Hoeffding's classical formula is a special case of the statement of this theorem. The generalization given here is twofold. First it extends the formula to generalized maximal rank statistics defined as maximal invariants with respect to general P'-generating groups G providing statistics with even more than finitely many values (see as a useful example the hypothesis of spherical symmetry). Secondly it extends the formula to general distributionfree statistics enhancing the formal analogy between rank and general distributionfree statistics.

4. Conclusion

The formal analogy between rank statistics and general distributionfree statistics linked by the characterisation (Theorem 1) of all distributionfree statistics being equivalent to functions of maximal invariant statistics with respect to an appropriate one of the different but structurally equivalent groups G_h, $h \in H$ helps to understand the general structure of distributionfree statistics and their relation to general rank statistics and allows the extension of other results known for rank statistics and not so easily seen for general distributionfree statistics. As an example one may consider the extension of central limit theorems for rank statistics to corresponding invariance principles via martingale arguments which crucially depend on the invariance properties of the considered rank statistics. (see for instance [19]).

REFERENCES

[1] Bell, C.B., Blackwell, D., and Breiman, L. (1960)
On the completeness of order statistics,
Ann. Math. Statist., 31, 794-797.

[2] Bell, C.B. (1964b)
A characterization of multisample distributionfree statistics,
Ann. Math. Statist., 35, 735-738.

[3] Bell., C.B., and Doksum, K.A. (1965)
Some new distributionfree statistics,
Ann. Math. Statist., 36, 203-214.

[4] Bell, C.B., and Doksum, K.A. (1966)
"Optimal" one-sample distributionfree tests and their two-sample extensions,
Ann. Math. Statist., 37, 120-132.

[5] Bell, C.B., and Doksum, K.A. (1967)
Distributionfree tests of independence,
Ann. Math. Statist., 38, 429-446.

[6] Bell, C.B., and Donoghue, J.F. (1969)
Distributionfree tests of randomness,
Sankhya, Series A, 31, 157-176.

[7] Bell, C.B., and Haller, H. Smith (1969)
Bivariate symmetry tests: parametric and nonparametric,
Ann. Math. Statist., 40, 259-269.

[8] Bell, C.B., and Kurotschka, V.G. (1971)
Einige Prinzipien zur Behandlung Nichtparametrischer Hypothesen,
Studi di probabilita, statistica e ricerca operativa in onore di
Giuseppe Pompilj,Oderisi-Gubbio, 165-186.

[9] Bell, C.B., and Smith, Paul J. (1969)
Some nonparametric tests for the multivariate goodness-of-fit, multi-sample, independence, and symmetry problems,
Multivariate Analysis II (Edited by R. Krishnaiah) Academic Press,
New York, pp. 3-23.

[10] Bell, C.B., and Smith, Paul J. (1972)
Completeness theorems for characterizing distributionfree statistics,
Ann. Inst. Math., 24, 435-453.

[11] Berk, R.H., and Bickel, P.J. (1968)
On invariance and almost invariance,
Ann. Math. Statist., 39, 1573-1576.

[12] Hoeffding, W. (1951)
"Optimum" nonparametric tests,
Proc. Second Berkeley Symp. Math. Statist. Prob., Univ. of California
Press, Berkeley, Calif., 83-92.

[13] Junge, K.
Charakterisierungssätze verteilungsfreier Statistiken und ähnlicher
Mengen,
Diplomarbeit, Göttingen 1976.

[14] Lehmann, E.L., and Stein, C. (1949)
On the theory of some nonparametric hypotheses,
Ann. Math. Statist., 20, 28-45.

[15] Pitmann, E.J.G. (1937a)
Significance tests which may be applied to sample from any population,
Suppl. JRSS, IV, 119-130.

[16] Pitman, E.J.G. (1937b)
Significance tests which may be applied to sample from any population II, Suppl. JRSS, IV, 225-232.

[17] Rohlin, V.A. (1949)
On the fundamental ideas of measure theory, functional analysis and measure theory,
Mat. Sbornik 25 (67), 107-150,
(Engl. Überarbeitung: Amer. Math. Society 1952, Transl. Nr. 71, 1-54).

[18] Rüschendorf, L. (1976)
Hypotheses generating groups for testing multivariate symmetry,
Ann. Statist., 4, 791-795.

[19] Sen, P.H. (1975)
Rank statistics, martingales and limit theorems, (to appear).

[20] Witting, H. (1970)
On the theory of nonparametric tests. Nonparametric Techniques in Statistical Inference (Edited by M.L. Puri),
Cambridge Univ. Press, 41-49.

Recent Developments in Statistics
J.R.Barra et al., editors

IDENTIFIABILITY OF A CONTINUOUS TIME SYSTEM WITH UNKNOWN NOISE COVARIANCE

A. LE BRETON
Université Scientifique et Médicale de Grenoble
I.R.M.A., B.P. 53
38041 - Grenoble Cédex - France

In identifying parameters of a continuous-time, linear time-invariant dynamical system, with white noise affecting additively both the state and the observation, a difficulty arises when the observation noise covariance is unknown : the problem is that the likelihood functional cannot be defined. The present paper solves this difficulty : we propose a method for parameter estimation which consists in decoupling the estimation of the parameters in the observation noise covariance from that of the other paramaters in the system.

1 - INTRODUCTION

Let us consider the following continuous-time stochastic linear dynamical system

(1.1) $$X_t = \int_0^t \theta X_s \, ds + \int_0^t \Lambda \, dW_s^1$$

(1.2) $$Y_t = \int_0^t F X_s \, ds + \int_0^t G dW_s^2$$

where X_t and Y_t are n and p-dimensional "state" and "output" functions respectively; θ, Λ, F and G are n x n, n x n, p x n, p x p constant, but partially unknown matrices ; W_s^1 and W_s^2 are n x 1 and p x 1 independent standard Wiener processes. We assume $G.{}^tG > 0$ to exclude the singular case, where ^{t} denotes the transpose. One problem of system identification is to estimate the unknown parameters in the matrices θ, $\Lambda.{}^t\Lambda$, F and $G.{}^tG$, based on the observation Y_t for one realization of the experiment in $0 \leq t \leq T$.

In [1], A.V. Balakrishnan studies the problem, for a similar model containing an "input", under the essential restriction that $G.{}^tG$ be completely known in order to apply the maximum likelihood method, but he does not give any method for a previous estimation of unknown parameters in $G.{}^tG$. In [2], A. Bagchi proposes a slight modification of A.V. Balakrishnan's work for the case of unknown parameters in $G.{}^tG$: he defines a functional analogous to the Log-likelihood functional for known $G.{}^tG$, which involves an a priori guess of $G.{}^tG$ and he claims that maximizing this functional will yield consistent estimates of all the unknown parameters in the system. In both papers [1] and [2] the authors assume the "identifiability condition" of the parameters to be satisfied (i.e. the limit as T increases of the second order derivative of the concerned functional is a positive definite matrix), but they use techniques for proofs which neither provide a practical expression of the matrix in the "identifiability condition" nor give precise statistical properties of the estimates.

In the present paper, we propose to identify the model (1.1) - (1.2) with a method which consists in decoupling the estimation of the parameters in the observation noise covariance from that of the other parameters in the system. Let $\phi_{\bar{\Delta}}$ denote the vector of the unknown parameters in the covariance $G.{}^tG$ and ϕ_Δ the vector of the unknown parameters in $(\theta,\Lambda.{}^t\Lambda,F)$ and let $\phi^*_{\bar{\Delta}}$ and ϕ^*_Δ be their true values. In part 2, we describe the statistical space associated with the problem of estimation of the parameter $(\phi_{\bar{\Delta}},\phi_\Delta)$. In part 3, we show that we can dispense with the restriction in A.V. Balakrishnan's proceeding by proving that $\phi^*_{\bar{\Delta}}$ is strongly identifiable in the sense that for every finite time interval $[0,t]$ there exists a sequence of estimates $\{\hat{\phi}^N_{\bar{\Delta},t}\ ;\ N \geq 1\}$, based on the observation in $[0,t]$ needing no a priori estimate as in A. Bagchi's approach, such that :

$$\lim_{N\to+\infty} \hat{\phi}^N_{\bar{\Delta},t} = \phi^*_{\bar{\Delta}} \text{ almost surely.}$$

Therefore we can assume that $G.{}^tG$ is completely known and the problem of estimation of the remaining parameters, which we study in part 4, concerns a number of parameters smaller than that of the parameters involved in A. Bagchi's study : this is another advantage. Using proof techniques which are quite different from those used by A.V. Balakrishnan and A. Bagchi, under the usual conditions that the system is stable and completely observable, we are able to provide a practical algebraic characterization of the matrix appearing in the "identifiability condition" of the value ϕ^*_Δ in $(\theta,\Lambda.{}^t\Lambda,F)$. We show that, if the "identifiability condition" is satisfied, then ϕ^*_Δ is asymptotically identifiable in the sense that there exists a family of maximum likelihood estimates $\{\hat{\phi}_{\Delta,T}\ ;\ T > 0\}$, based on the observation in $[0,T]$, such that :

$$\lim_{T\to+\infty} \hat{\phi}_{\Delta,T} = \phi^*_\Delta \text{ in probability}$$

and $\sqrt{T}.(\hat{\phi}_{\Delta,T}-\phi^*_\Delta)$ is asymptotically normally distributed.

Further, the algebraic characterization of the matrix in the "identifiability condition" suggests a constructive approach for computing the estimates.

The results presented here are part of a study undertaken by the Author, on parameter estimation in stochastic differential models. A more complete presentation is given in Chapter 3 §2 of paper [3].

2 - THE STATISTICAL SPACE FOR PARAMETER ESTIMATION

We assume that we observe in $0 \leq t \leq T$, an "output" $y(.) = (y_t\ ;\ t \geq 0) \in C_p$ related to the unobservable "state" $x(.) = (x_t\ ;\ t \geq 0) \in C_n$ of a stochastic dynamical system, where, for every integer $k \geq 1$, C_k denotes the space of continuous functions from R_+ into R^k.

Φ denoting the subset $\Phi_{\bar{\Delta}} \times \Phi_\Delta$ of $R^r \times R^d$, $r \geq 1$, $d \geq 1$, where $\phi = (\phi_{\bar{\Delta}}, \phi_\Delta)$ takes its value, $\theta(\phi_\Delta)$, $(\Lambda.{}^t\Lambda)(\phi_\Delta)$, $F(\phi_\Delta)$, $(G.{}^tG)(\phi_{\bar{\Delta}})$ stand for the values of θ, $\Lambda.{}^t\Lambda$, F and $G.{}^tG$ respectively for the value ϕ of the parameter. We assume that there exists some $\phi^* = (\phi^*_{\bar{\Delta}}, \phi^*_\Delta)$, unknown in Φ, called "true value of ϕ", such that the probability law for $(x(.),y(.)) \in C_n \times C_p$ is that of a process $((X_t,Y_t) \; ; \; t \geq 0)$ satisfying (1.1) - (1.2) for $\theta = \theta(\phi^*_\Delta)$, $\Lambda.{}^t\Lambda = (\Lambda.{}^t\Lambda)(\phi^*_\Delta)$, $F = F(\phi^*_\Delta)$ and $G.{}^tG = (G.{}^tG)(\phi^*_{\bar{\Delta}})$.

The theory of linear filtering (see e.g. [4]) says that if $((X_t,Y_t) \; ; \; t \geq 0)$ satisfies (1.1) - (1.2) then, $\hat{X}_t$ denoting the filtered state, one has :

$$(2.1) \quad \begin{cases} d\hat{X}_t = \theta\hat{X}_t \, dt + Q(t).{}^tF(G.{}^tG)^{-1}.(dY_t - F\hat{X}_t \, dt) \\ \hat{X}_o = 0 \end{cases}$$

where $Q(t)$ is given by the Riccati equation :

$$(2.2) \quad \begin{cases} \dfrac{dQ(t)}{dt} = \theta Q(t) + Q(t).{}^t\theta + \Lambda.{}^t\Lambda - Q(t).{}^tF(G.{}^tG)^{-1}.FQ(t) \\ Q(o) = 0 \end{cases}$$

and, if $\tilde{W}_t = Y_t - \int_0^t F\hat{X}_s \, ds \; ; \; t \geq 0$,

(2.3) the process $(\tilde{W}_t \; ; \; t \geq 0)$ is a Wiener process with $E(\tilde{W}_t.{}^t\tilde{W}_s) = G.{}^tG \min(s,t)$, which is called the "innovation process".

From (2.1) one easily deduces that, if $Y = (Y_t \; ; \; t \geq 0)$,

$$\hat{X}_t = a(t,Y)$$

where, for $y(.) \in C_p$,

$$(2.4) \quad a(t,y) = Q(t).{}^tF(G.{}^tG)^{-1}.y_t - R(t).\int_0^t R^{-1}(s)[Q(s){}^t\theta + \Lambda.{}^t\Lambda].{}^tF(G.{}^tG)^{-1}.y_s \, ds$$

with

$$(2.5) \quad \begin{cases} \dfrac{dR(t)}{dt} = [\theta - Q(t).{}^tF(G.{}^tG)^{-1}.F].R(t) \\ R(o) = 1_n \text{ (identity matrix } n \times n) \end{cases}$$

Therefore, in view of (2.3), the process $(Y_t \; ; \; t \geq 0)$ admits the representation :

$$(2.6) \quad \begin{cases} dY_t = Fa(t,Y)dt + (G.{}^tG)^{1/2}dW_t \\ Y_o = 0 \end{cases}$$

where $(W_t \; ; \; t \geq 0)$ is a standard Wiener process.

So the probability law for the observation $y(.) \in C_p$ is that of a process $(Y_t \; ; \; t \geq 0)$ satisfying (2.6) for $\theta = \theta(\phi^*_\Delta)$, $\Lambda.{}^t\Lambda = (\Lambda.{}^t\Lambda)(\phi^*_\Delta)$, $F = F(\phi^*_\Delta)$ and $G.{}^tG = (G.{}^tG)(\phi^*_{\bar{\Delta}})$

in (2.2), (2.4), (2.5) and (2.6).

We are then naturally led to study the problem of estimating the parameter ϕ, in view of the "output" $y(.)$ in $[0,T]$, in the statistical space

(2.7) $(C_p, \mathcal{S}^p_T, \{\mu^{p,T}_\phi ; \phi \in \Phi\})$

where $\mathcal{S}^p_T$ (resp. $\mathcal{S}^p$) denotes the σ-algebra on C_p which is generated by the coordinate functions Π_s, $0 \leq s \leq T$ (resp. $s \geq 0$), and, for every $\phi \in \Phi$, $\mu^{p,T}_\phi$ denotes the restriction to $\mathcal{S}^p_T$ of the measure μ^p_ϕ induced on $\mathcal{S}^p$ by a process $(Y^\phi_t ; t \geq 0)$ which is a solution of (2.6) for $\theta = \theta(\phi_\Delta)$, $\Lambda.{}^t\Lambda = (\Lambda.{}^t\Lambda)(\phi_\Delta)$, $F = F(\phi_\Delta)$ and $G.{}^tG = (G.{}^tG)(\phi_{\bar\Delta})$ in (2.2), (2.4), (2.5) and (2.6).

Now we give a precise description of the family $\{\mu^{p,T}_\phi ; \phi \in \Phi\}$ of measures in the following proposition where, for $\phi_{\bar\Delta} \in \Phi_{\bar\Delta}$, $\mu^{p,T}_{\phi_{\bar\Delta}}$ denotes the measure on $\mathcal{S}^p_T$ which is the restriction of the measure $\mu^p_{\phi_{\bar\Delta}}$ induced on $\mathcal{S}^p$ by a Wiener process $(\tilde W_t ; t \geq 0)$ satisfying (2.3) with $G.{}^tG = (G.{}^tG)(\phi_{\bar\Delta})$.

2.1. Proposition : "For every $T > 0$ and $\phi = (\phi_{\bar\Delta}, \phi_\Delta) \in \Phi$, the measures $\mu^{p,T}_\phi$ and $\mu^{p,T}_{\phi_{\bar\Delta}}$ are mutually absolutely continuous, the Radon-Nikodym derivative being given by :

$$(2.8) \quad \left\{ \begin{aligned} \frac{d\mu^{p,T}_\phi}{d\mu^{p,T}_{\phi_{\bar\Delta}}} = \exp\{ & \int_0^T \langle F(\phi_\Delta)a_\phi(t,.), \left[(G.{}^tG)(\phi_{\bar\Delta})\right]^{-1} d\Pi_t\rangle \\ & - \frac{1}{2}\int_0^T \langle F(\phi_\Delta)a_\phi(t,.), \left[(G.{}^tG)(\phi_{\bar\Delta})\right]^{-1}.F(\phi_\Delta)a_{\phi_\Delta}(t,.)\rangle dt\} \end{aligned} \right.$$

where the stochastic integral is defined with respect to the measure $\mu^{p,T}_{\phi_{\bar\Delta}}$ and the random function $(a_\phi(t,.) ; t \geq 0)$ is defined on C_p by (2.4) for $\theta = \theta(\phi_\Delta)$, $\Lambda.{}^t\Lambda = (\Lambda.{}^t\Lambda)(\phi_\Delta)$, $F = F(\phi_\Delta)$ and $G.{}^tG = (G.{}^tG)(\phi_{\bar\Delta})$ in (2.2), (2.4) and (2.5)".

The proof consists simply in verifying that results on absolute continuity of measures associated with diffusion type processes (see [5], [6]) can be applied.

Proposition 2.1. is the key of our approach to parameter identification in (1.1) - (1.2). An obvious generalization of a well known result in one dimension ([7], Theorem 3.2.1) is that, if $\phi_{\bar\Delta} \neq \phi'_{\bar\Delta}$, then $\mu^p_{\phi_{\bar\Delta}}$ is singular with respect to $\mu^p_{\phi'_{\bar\Delta}}$.

Therefore, in view of Proposition 2.1., if $\phi_{\bar\Delta} \neq \phi'_{\bar\Delta}$ then μ^p_ϕ is singular with respect to $\mu^p_{\phi'}$; so, the likelihood functional cannot be defined in (2.7). Nevertheless this singularity can be used in order to identify $\phi^*_{\bar\Delta}$ as we will see in part 3. On the other hand Proposition 2.1. provides the expression for the likelihood functional in the case of known $\phi^*_{\bar\Delta}$ that we will study in part 4.

3 -PARAMETER ESTIMATION IN THE OBSERVATION NOISE COVARIANCE

Let, for every integer $N \geq 1$, $0 = t_0^N < t_1^N < \ldots < t_{K_N}^N = t$ be a subdivision of the interval $[0,t]$ such that, if $\delta_N = \max_{k=\overline{1,K_N}} |t_k^N - t_{k-1}^N|$, then $\lim_{N\to+\infty} \delta_N = 0$. We consider the sequence $\{\hat\phi^N_{\bar\Delta,t}\ ;\ N \geq 1\}$ of statistics defined on the statistical space (2.7) for $T \geq t > 0$ by :

$$
(3.1)\quad \begin{cases}
\hat\phi^N_{\bar\Delta,t} = (\hat\phi^{N,i}_{\bar\Delta,t}\ ;\ i = \overline{1,r}) \\
\hat\phi^{N,i}_{\bar\Delta,t} = \hat\sigma_t^{N,(k,\ell)(i)} \quad \text{where, for } i = \overline{1,r},\ (G.{}^tG)_{(k,\ell)(i)} \text{ is the unknown} \\
\text{coefficient in } G.{}^tG \text{ which is parametrized by } \phi_{\bar\Delta,i}, \\
\hat\sigma_t^N = \frac{1}{t}\sum_{j=1}^{K_N} (\Pi_{t_j^N} - \Pi_{t_{j-1}^N})^{2\otimes} = (\hat\sigma_t^{N,(k,\ell)}\ ;\ k,\ell = \overline{1,p})\ ;\ N \geq 1
\end{cases}
$$

Then we can prove the following result :

3.1. Proposition : "In the system (1.1) - (1.2) the (true) value $\phi^*_{\bar\Delta}$ is strongly identifiable on every finite time interval $[0,t]$ by the sequence $\{\hat\phi^N_{\bar\Delta,t}\ ;\ N \geq 1\}$ of statistics defined by (3.1), i.e. for every $\phi = (\phi_{\bar\Delta},\phi_\Delta)$ such that $\phi_{\bar\Delta} = \phi^*_{\bar\Delta}$ one has :

$$
(3.2)\qquad \lim_{N\to+\infty} \hat\phi^N_{\bar\Delta,t} = \phi^*_{\bar\Delta} \text{ in the sense of convergence in } \mu_\phi\text{-probability.}
$$

Furthermore the sequence $\{\delta_N^{-1/2}(\hat\phi^N_{\bar\Delta,t} - \phi^*_{\bar\Delta})\ ;\ N \geq 1\}$ is bounded in μ_ϕ-probability and, if δ_N approaches 0 so fast that $\Sigma\delta_N < +\infty$, then convergence occurs in (3.2) also with μ_ϕ-probability 1".

The proof is essentially based on results on the quadratic variation of a quasi-martingale (see [8] Lemma 3.1.4 and Theorem 7.2.2) and on Proposition 2.1.

3.2. Remarks :

a) By virtue of Proposition 3.1., we can assume that $\phi^*_{\bar\Delta}$ is known (i.e. $G.{}^tG$ is completely known) for the estimation of ϕ_Δ because $\phi^*_{\bar\Delta}$ can be identified previously on a finite time interval. The initial problem which concerned r+d parameters is then decoupled into a problem with r parameters and a problem with d parameters.

b) This is a theoritical result since in fact it can be argued that a true Wiener process can never be observed, that any physical measuring device must eventually cease to respond as the input frequencies increase beyond some limit. In fact, in practical situations the model for the state is not quite accurate, i.e. the noise is not exactly a white noise. So in practice, the use of the method is only limited by the accuracy limits of the model.

4 - PARAMETER ESTIMATION IN CASE OF KNOWN OBSERVATION NOISE COVARIANCE

By virtue of Proposition 2.1. and (2.8), the Log-likelihood function in case of known $G.{}^tG$ associated with the statistical space

$$(C_p,\ \mathcal{S}_T^p,\ \{\mu_{\phi_\Delta}^{p,T}\ ;\ \phi_\Delta \in \Phi_\Delta\})$$

is given by :

$$L_T(\phi_\Delta,.) = \int_0^T \langle F(\phi_\Delta).a_{\phi_\Delta}(t,.),\ (G.{}^tG)^{-1}.\ d\Pi_t\rangle$$
$$- \frac{1}{2}\int_0^T \langle F(\phi_\Delta).a_{\phi_\Delta}(t,.),\ (G.{}^tG)^{-1}.\ F(\phi_\Delta).a_{\phi_\Delta}(t,.)\rangle\ dt$$

where $(a_{\phi_\Delta}(t,.)\ ;\ t \geq 0)$ is defined on C_p by (2.4) for $\theta = \theta(\phi_\Delta)$, $\Lambda.{}^t\Lambda = (\Lambda.{}^t\Lambda)(\phi_\Delta)$, $F = F(\phi_\Delta)$ and $G.{}^tG$ in (2.2), (2.4) and (2.5). Here we do not investigate the problem of existence of a regular version of $L_T(\phi_\Delta,.)$ with successive derivatives $L_T^{(1)}(\phi_\Delta,.)$ and $L_T^{(2)}(\phi_\Delta,.)$ given by formal derivation of (4.1) ; this is solved in [3].

We make the following assumptions, $\mathcal{V}(\phi_\Delta^*)$ denoting a neighbourhood of ϕ_Δ^* :

(H1) For every $\phi_\Delta \in \mathcal{V}(\phi_\Delta^*)$, $\theta(\phi_\Delta)$ is a stable matrix.

(H2) For every $\phi_\Delta \in \mathcal{V}(\phi_\Delta^*)$, the pair $(F(\phi_\Delta), \theta(\phi_\Delta))$ is completely observable.

It is well known (see e.g. [4]) that, under (H1) and (H2), if $(Q_{\phi_\Delta}(t)\ ;\ t \geq 0)$ is given by (2.2), then :

$$\lim_{t\to+\infty} Q_{\phi_\Delta}(t) = Q_{\phi_\Delta}^\infty$$

where $Q_{\phi_\Delta}^\infty$ is the unique solution of the algebraic equation :

(4.2) $$\theta(\phi_\Delta).Q + Q.{}^t\theta(\phi_\Delta) + (\Lambda.{}^t\Lambda)(\phi_\Delta) - Q.{}^tF(\phi_\Delta).(G.{}^tG)^{-1}.\ F(\phi_\Delta)Q = 0$$

and, further :

(4.3) $$\theta(\phi_\Delta) - Q_{\phi_\Delta}^\infty.{}^tF(\phi_\Delta).(G.{}^tG)^{-1}.\ F(\phi_\Delta) \qquad \text{is a stable matrix.}$$

Now we define successively the matrices $S(\phi_\Delta)$, I, $D(\phi_\Delta)$ and J by :

(4.4) $$U(\phi_\Delta).S(\phi_\Delta) + S(\phi_\Delta).{}^tU(\phi_\Delta) = -V(\phi_\Delta).{}^tV(\phi_\Delta)$$

where

$$V_{i,j}(\phi_\Delta) = \begin{cases} (Q^\infty_{\phi_\Delta} . {}^tF(\phi_\Delta) . {}^tG^{-1})_{i,j} & \text{if } i = \overline{1,n} \;;\; j = \overline{1,p} \\ \frac{\partial}{\partial \phi_k} (Q^\infty_{\phi_\Delta} . {}^tF(\phi_\Delta) . {}^tG^{-1})_{i-kn,j} & \text{if } i = \overline{kn+1,\ (k+1)n} \;;\; k = \overline{1,d} \\ & \phantom{\text{if }} j = \overline{1,p} \end{cases}$$

$$U_{i,j}(\phi_\Delta) = \begin{cases} (\theta(\phi_\Delta))_{i,j} & \text{if } i = \overline{1,n} \;;\; j = \overline{1,n} \\ (\theta(\phi_\Delta) - Q^\infty_{\phi_\Delta} . {}^tF(\phi_\Delta) . (G . {}^tG)^{-1} . F(\phi_\Delta))_{i-kn,j-kn} & \\ & \text{if } i,j = \overline{kn+1,\ (k+1)n} \;;\; k = \overline{1,d} \\ (\frac{\partial}{\partial \phi_k} \theta(\phi_\Delta) - Q^\infty_{\phi_\Delta} . {}^tF(\phi_\Delta) . (G . {}^tG)^{-1} \frac{\partial}{\partial \phi_k} F(\phi_\Delta))_{i-kn,j-kn} & \\ & \text{if } i = \overline{kn+1,\ (k+1)n} \;;\; k = \overline{1,d} \\ & \phantom{\text{if }} j = \overline{(k-1)n+1,\ kn} \\ 0 & \text{otherwise} \end{cases}$$

(4.5) $$I_{i,j} = \begin{cases} 1 \text{ if } i = j+kn,\ k = \overline{0,d-1},\ j = \overline{1,n} \\ 1 \text{ if } i = j+(d-1)n,\ j = \overline{n+1,\ (d+1)n} \\ 0 \text{ otherwise} \end{cases}$$

(4.6) $$D_{i,j}(\phi_\Delta) = \begin{cases} (G^{-1} \frac{\partial}{\partial \phi_k} F(\phi_\Delta))_{i-kp,j-kn} & \\ \quad \text{if } kp+1 \le i \le (k+1)p,\ kn+1 \le j \le (k+1)n,\ k = \overline{0,d-1} & \\ (G^{-1} F(\phi_\Delta))_{i-kp,j-kn} & \\ \quad \text{if } kp+1 \le i \le (k+1)p,\ kn+1 \le j \le (k+1)n,\ k = \overline{d,2d-1} & \\ 0 \quad \text{otherwise} \end{cases}$$

(4.7) $$J_{i,j} = \begin{cases} 1 \text{ if } i = \overline{1,pd} \text{ and } j = i \text{ or } j = i+pd \\ 0 \text{ otherwise} \end{cases}$$

Then we are able to prove the following result :

4.1. Proposition : "Under the assumptions (H1) and (H2) we have :

(i) $\lim_{T\to+\infty} T^{-1} . L_T^{(1)}(\phi^*_\Delta,.) = 0$

(ii) $\lim_{T\to+\infty} T^{-1} . L_T^{(2)}(\phi^*_\Delta,.) = -\Gamma(\phi^*_\Delta)$

} in the sense of convergence in $\mu_{\phi^*_\Delta}$-probability

(iii) $\lim_{T\to+\infty} T^{-1/2} . L_T^{(1)}(\phi^*_\Delta,.) = N(0,\Gamma(\phi^*_\Delta))$ in the sense of convergence in distribution with respect to $\mu_{\phi^*_\Delta}$, where $N(0,\Gamma(\phi^*_\Delta))$ denotes a Gaussian vector with mean 0 and covariance $\Gamma(\phi_\Delta^*)$ given by :

$$(4.8)\quad \begin{cases} \Gamma_{i,j}(\phi_\Delta^*) = \mathrm{Tr}\, R_{ij}(\phi_\Delta^*) \ ; \ i,j = \overline{1,d} \\ (R_{i,j}(\phi_\Delta^*))_{k,\ell} = \left[(JD(\phi_\Delta^*)I)\cdot S(\phi_\Delta^*)\cdot{}^t(JD(\phi_\Delta^*)I)\right]_{(i-1)p+k,(j-1)p+\ell} \ ; \ k,\ell = \overline{1,p} \\ \text{where } S(\phi_\Delta^*),\ I,\ D(\phi_\Delta^*) \text{ and } J \text{ are given by (4.4), (4.5), (4.6) and (4.7)} \\ \text{respectively".} \end{cases}$$

The proof is based on asymptotic properties of processes which are solutions of some stochastic differential equations. It is detailed in [3].

Now we have the main proposition :

4.2. Proposition : "Under the assumptions (H1) and (H2), if the matrix $\Gamma(\phi_\Delta^*)$ defined by (4.8) is non-singular, then the (true) value ϕ_Δ^* is asymptotically identifiable by a family $\{\hat\phi_{\Delta,T} \ ; \ T > 0\}$ of estimates satisfying :

(i) for every $T > 0$, $\hat\phi_{\Delta,T}$ is $\mathcal{S}_T$-measurable

(ii) $\lim\limits_{T\to+\infty} \mu_{\phi_\Delta^*}\left[L_T^{(1)}(\hat\phi_{\Delta,T},\cdot) = 0\right] = 1$

(iii) $\lim\limits_{T\to+\infty} \hat\phi_{\Delta,T} = \phi_\Delta^*$ in the sense of convergence in $\mu_{\phi_\Delta^*}$-probability.

Further, if $\{\tilde\phi_{\Delta,T} \ ; \ T > 0\}$ is another family satisfying (i), (ii) and (iii), then one also has :

(iv) $\lim\limits_{T\to+\infty} \mu_{\phi_\Delta^*}\left[\hat\phi_{\Delta,T} = \tilde\phi_{\Delta,T}\right] = 1$

Furthermore, every family $\{\hat\phi_{\Delta,T} \ ; \ T > 0\}$ which satisfies (i), (ii) and (iii) is such that

(v) $\lim\limits_{T\to+\infty} T^{-1/2}\left[\hat\phi_{\Delta,T}-\phi_\Delta^*\right] = N(0,\Gamma^{-1}(\phi_\Delta^*))$

in the sense of convergence in distribution with respect to $\mu_{\phi_\Delta^*}$".

The proof is essentially based on the assertions in Proposition 4.1. and classical arguments in estimation by the maximum likelihood method.

4.3. Remarks :

a) In practical situations one cannot verify precisely that the "identiability condition" ($\Gamma(\phi_\Delta^*)$ being non-singular) is satisfied for it depends on the true (and unknown) value ϕ_Δ^* of ϕ_Δ. But in practice Φ_Δ is a neighbourhood of ϕ_Δ^*, possibly because an a priori estimation is available by some previous analysis of the system, so one can investigate identifiability in that neighbourhood.

b) The algebraic characterization of the matrix $\Gamma(\phi_\Delta^*)$ in the "identifiability condition" suggests the following constructive approach for computing the estimates, based on a Newton-Raphson algorithm :

$$\hat{\phi}_{\Delta,T}^{(n+1)}(y) = \hat{\phi}_{\Delta,T}^{(n)}(y) - \Gamma(\hat{\phi}_{\Delta,T}^{(n)}(y))^{-1}\left[\frac{1}{T}\nabla_T(\hat{\phi}_{\Delta,T}^{(n)}(y),y)\right]$$

where $\Gamma(\phi_\Delta)$ is defined for ϕ_Δ such as $\Gamma(\phi_\Delta^*)$ for ϕ_Δ^*, and $\nabla_T(\phi_\Delta,y)$ is obtained from $L_T^{(1)}(\phi_\Delta,y)$ by substituting $Q_{\phi_\Delta}^\infty$ to $Q_{\phi_\Delta}(t)$.

REFERENCES

[1] Balakrishnan, A.V. (1973). "Stochastic Differential Systems I", Lecture Notes in Economics and Mathematical Systems, Springer-Verlag, Vol. 84.

[2] Bagchi, A. (1975). "Continuous time systems identification with unknown noise covariance", Automatica, Vol. 11, 533-535.

[3] Le Breton, A. (1976). "Sur l'estimation de paramètres dans les modèles différentiels stochastiques multidimensionnels", Thèse Grenoble.

[4] Bucy, R.S., Joseph, P.D. (1968). "Filtering for stochastic processes with applications to guidance", Interscience Publishers, John Wiley, New-York.

[5] Girsanov, I.V. (1960). "On transforming a certain class of stochastic processes by absolutely continuous substitution of measures", Theory of Prob. and its appl., Vol. 5, 285-301.

[6] Duncan, T.E., Varaiya, P. (1971). "On the solutions of a stochastic control system", SIAM J. Control, Vol. 9, n° 3, 354-371.

[7] Yeh, J. (1973). "Stochastic Processes and the Wiener Integral", M. Dekker, New-York.

[8] Arnold, L. (1974). "Stochastic Differential Equations", Theory and applications, John Wiley, New-York.

Recent Developments in Statistics
J.R.Barra et al., editors

LINEAR MODELS AND GENERALIZED INVERSE MATRICES IN MULTIVARIATE STATISTICAL ANALYSIS

G. Lennes
Service de Statistique appliquée
Université de Liège - Belgique

In this paper, we extend a method developed by S.R. Searle in the univariate case to linear models involving p-normal variables. This method enables us to solve estimation problems for models not of full rank without introducing constraints leading to unique solutions. It also enables us to formulate and to test linear hypotheses very easily. Our generalization, as well as Saerle's developments, rests upon C.R. Rao's works in the field of generalized inverse matrices.

1- INTRODUCTION

Let A be a rectangular p x q matrix whose elements are real or complex numbers. A generalized inverse (g-inverse) of A is a q x p matrix A^- such that

$$AA^-A = A \tag{1}$$

When A is rectangular or square singular, it admits an infinity of g-inverses. Rao and Mitra (1971) have thoroughly studied their properties. Let us quote the following ones, which play a central role in this paper. We have limited ourselves to real numbers and used slightly different notations, better suited to our developments.

T1 - Let us consider the matrix equation

$$AXB = C \tag{2}$$

The necessary and sufficient condition for this equation to have a solution is that

$$AA^-CB^-B = C, \tag{3}$$

in which case the general solution is

$$X = A^-CB^- + U - A^-AUBB^- , \tag{4}$$

where U is an arbitrary matrix.

T2 - The following results hold for any choice of g-inverse involved :

$$(ZZ')(ZZ')^-Z = Z \tag{5}$$

$$Z'(ZZ')^-(ZZ') = Z' \tag{6}$$

2- GENERAL LINEAR MODEL

2.1) Notations

Suppose $\vec{x}_1, \ldots, \vec{x}_N$ are a set of N observations, $\vec{x}_\alpha$ being drawn from the p-variate non singular normal distribution $N(B\vec{z}_\alpha, \Sigma)$.

$B = (b_{ik})$ is a constant p x q matrix whose elements, together with those of $\Sigma = (\sigma_{ik})$, constitute the unknown parameters of our model. The $\vec{z}_\alpha$ are known constant vectors defining the experimental design.

Let us denote by X the p x N matrix whose columns are the $\vec{x}_\alpha$ and by Z the q x N matrix whose columns are the $\vec{z}_\alpha$. T.W. Anderson (1958) has considered the case where the rank of Z is equal to q and where $N \geqslant p + q$. In this paper, we shall study the more general situation where the rank of Z is equal to r and we shall consider more specifically the case where $r < q$.

In order to avoid introducing too many symbols, we shall denote by $\vec{x}_\alpha$ not only an observation but also the random variable associated with $N(B\vec{z}_\alpha, \Sigma)$. The context will always make clear which is the right meaning.

With this convention, our model is specified by the following expressions :

$$\mathcal{E}(X) = BZ \tag{7}$$

$$\mathrm{Cov}(\vec{x}_\alpha, \vec{x}_\beta) = \delta_{\alpha\beta}\, \Sigma \tag{8}$$

2.2) The system of maximum likelihood equations

Let us denote by L the likelihood function. Then,

$$\log_e L = -\frac{1}{2} Np \log_e 2\pi + \frac{1}{2} N \log_e |\Sigma^{-1}| - \frac{1}{2} \mathrm{tr}(\Sigma^{-1} D) \tag{9}$$

where
$$D = \sum_{\alpha=1}^{N} (\vec{x}_\alpha - B\vec{z}_\alpha)(\vec{x}_\alpha - B\vec{z}_\alpha)' = (X - BZ)(X - BZ)' \tag{10}$$

Here, B and Σ denote matrices to be chosen so as to maximize $\log_e L$. The maximum likelihood (m.l.) equations are derived by equating to zero the partial derivatives of $\log_e L$ with respect to the elements of B and Σ^{-1}. In matrix form we finally obtain

$$\begin{cases} BA = C & (11) \\ N\Sigma = (X - BZ)(X - BZ)' & (12) \end{cases}$$

where
$$A = \sum_{\alpha=1}^{N} \vec{z}_\alpha \vec{z}_\alpha{}' = ZZ'$$

$$C = \sum_{\alpha=1}^{N} \vec{x}_\alpha \vec{z}_\alpha{}' = XZ'$$

Let us now solve the m.l. system.

The first equation (11) does not admit any more a unique solution, as it was the case in the situation studied by T.W. Anderson. The order q matrix A is now singular, since rank A = rank ZZ' = rank Z = $r < q$.

Equation (11) is of the kind considered in T1. The first question to be answered is about the possibility condition (3). With our present notations, this condition may be expressed as follows

$$CA^{-}A = C \tag{15}$$

or, more explicitly,

$$XZ'(ZZ')^{-}ZZ' = XZ' \tag{16}$$

Condition (16) is nothing else than identity (6) whose two members have been left multiplied by X. It will always be satisfied and thus equation (11) will always have solutions. The general solution is given by

$$\hat{\hat{B}} = CA^{-} + U - UAA^{-} = CA^{-} + U\,(I - AA^{-}) \tag{17}$$

Setting U = 0, we obtain the particular solution $\hat{B}$

$$\hat{B} = CA^{-}$$

and $$\hat{\hat{B}} = \hat{B} + U\,(I - AA^{-}) \tag{19}$$

From (19), we deduce

$$X - \hat{\hat{B}}Z = X - \hat{B}Z + U\,(I - AA^{-})\,Z = X - \hat{B}Z, \tag{20}$$

since $(I - AA^{-})\,Z = Z - (ZZ')(ZZ')^{-}Z = 0$, as shown by (5).

Then, substituting in equation (12), we have with obvious notations

$$N\hat{\Sigma} = (X - \hat{B}Z)(X - \hat{B}Z)' = (X - \hat{\hat{B}}Z)(X - \hat{\hat{B}}Z)' = N\hat{\hat{\Sigma}} \tag{21}$$

Thus, notwithstanding the multiplicity of solutions for B, the m.l. system admits a unique solution for Σ.

2.3) Sampling distributions

From (6) and (18), we deduce

$$\mathcal{E}(\hat{B}) = \mathcal{E}(CA^{-}) = \mathcal{E}(X)\,Z'(ZZ')^{-}$$
$$= BZZ'(ZZ')^{-} = BAA^{-} \tag{22}$$

Similarly, from (6) and (19), we obtain

$$\mathcal{E}(\hat{\hat{B}}) = \mathcal{E}(\hat{B}) + U\,(I - AA^{-}) \tag{23}$$

Thus, neither $\hat{B}$ nor $\hat{\hat{B}}$ are unbiased estimates of B when A is singular. In what follows, we shall restrict our developments to $\hat{B}$, since we have

$$\hat{B} - \mathcal{E}(\hat{B}) = \hat{\hat{B}} - \mathcal{E}(\hat{\hat{B}}) \tag{24}$$

Let us denote by $\vec{\hat{b}}_i$ the row of rank i of matrix $\hat{B}$. We have shown (Lennes (1976)) that

$$\mathrm{Cov}(\vec{\hat{b}}_i, \vec{\hat{b}}_j) = \sigma_{ij}\,(A^{-})'AA^{-} \tag{25}$$

Taking into account the relations $\hat{\hat{B}}A = \hat{B}A = C$, we deduce from (21) that

$$N\hat{\Sigma} = XX' - \hat{B}A\hat{B}' = XX' - \hat{\hat{B}}A\hat{\hat{B}}' = N\hat{\hat{\Sigma}} \tag{26}$$

and $$\hat{B}A\hat{B}' = \hat{\hat{B}}A\hat{\hat{B}}' \tag{27}$$

In order to compute $\hat{B} = CA^{-}$, we need a g-inverse of A. A convenient g-inverse is afforded by

$$A^{-} = F' \begin{pmatrix} I_r & 0 \\ 0 & 0 \end{pmatrix} F, \tag{28}$$

where F is such that

$$FAF' = \begin{pmatrix} I_r & 0 \\ 0 & 0 \end{pmatrix} \tag{29}$$

Such a choice of A^{-} and formulas (26) and (27) have enabled us (Lennes (1976)) to prove the following property which extends Anderson's theorem 8.2.2 ((1958), p.183) to the case where rank $Z = r < q$.

T3- $N\hat{\Sigma}$ is distributed according to $W(\Sigma, N-r)$ and is independent of $\hat{B}$. Thus, $\frac{N\hat{\Sigma}}{N-r}$ is an unbiased estimate of Σ.

3- THE GENERAL LINEAR HYPOTHESIS

3.1) Invariant functions - Estimable functions

We have noticed that the m.l. system admits a multiplicity of solutions for B and that neither $\hat{B}$ nor $\hat{\hat{B}}$ are unbiased estimates of B. We are going now to study matrix functions such as BK, where K denotes a constant q x s matrix.

First, we intend to discover under what condition such functions are invariant. This requirement can be expressed as follows

$$\hat{\hat{B}}K = \hat{B}K, \tag{30}$$

or

$$\hat{B}K + U(I - AA^{-})K = \hat{B}K, \tag{31}$$

for every q x s matrix U.

Thus, the necessary and sufficient condition of invariance is that K be such that

$$AA^{-}K = K \tag{32}$$

Let us consider now a function BK with K satisfying (32). From (7), (19) and (32), we deduce

$$\mathcal{E}(\hat{\hat{B}}K) = \mathcal{E}(\hat{B}K) = \mathcal{E}(XZ'A^{-}K) = BAA^{-}K = BK \tag{33}$$

This formula shows that invariant functions BK are also unbiasedly estimable and that $\hat{B}K$ is such an estimate.

Notice that the class of matrix K satisfying (32) does not depend on the choice of a g-inverse of A. Similarly, $K'A^{-}K$ is invariant for any choice of A^{-} if A is symmetric. For instance, we have

$$K'A^{-}A(A^{-})'K = K'(A^{-})'AA^{-}K = K'A^{-}K \tag{34}$$

since $A^{-}A(A^{-})'$ and $(A^{-})'AA^{-}$ are also g-inverses of A when A is symmetric.

Finally, from (25) and (34), we deduce

$$\mathrm{Cov}(\vec{\hat{b}}_i K, \vec{\hat{b}}_j K) = \sigma_{ij} K'(A^{-})'AA^{-}K = \sigma_{ij} K'A^{-}K \tag{35}$$

3.2) The general linear hypothesis

Let us consider the following linear hypothesis

$$(H) : BK = M, \tag{36}$$

where M and K are given constant matrices of order p x s and q x s respectively. Moreover, K is assumed to be of rank s.

We are going now to study the model obtained by joining conditions (H) to those already specified in paragraph 2.1.

$$\text{Let } V = (v_{ih}) = BK - M \tag{37}$$

We are to maximize the following expression

$$\varphi = \log_e L + \sum_{i=1}^{p} \sum_{h=1}^{s} v_{ih}\, t_{ih}\ , \tag{38}$$

where $T = (t_{ih})$ is a p x s matrix whose elements are the Lagrange multipliers.

Equating to zero the partial derivatives of φ with respect to the elements of B, T and Σ^{-1}, we obtain

$$\begin{cases} C - BA + \Sigma TK' = 0 & (39) \\ BK = M & (40) \\ N\Sigma = (X - BZ)(X - BZ)' & (41) \end{cases}$$

3.3) The system of m.l. equations

$$\text{Let } \approx{B} = \hat{B} + W \tag{42}$$

Let us impose on $\approx{B}$ to satisfy equation (39). Since $\hat{B}A = C$, W must satisfy the following equation

$$WA = \Sigma TK' \tag{43}$$

which is of the kind considered in T1. Its possibility condition is

$$\Sigma TK'A^{-}A = \Sigma TK' \tag{44}$$

Since A is symmetric, $AA^{-}K = K$ is equivalent to $K'A^{-}A = K'$. Our condition is nothing else than this identity whose two members have been left multiplied by ΣT. It is identically satisfied and the general solution of (43) is given by

$$W = \Sigma TK'A^{-} + U\,(I - AA^{-}) \tag{45}$$

where U is an arbitrary p x q matrix. Then,

$$\approx{B} = \hat{B} + \Sigma TK'A^{-} + U\,(I - AA^{-}) \tag{46}$$

Setting U = 0, we obtain the particular solution $\tilde{B}$

$$\tilde{B} = \hat{B} + \Sigma TK'A^{-} \tag{47}$$

and $$\approx{B} = \tilde{B} + U\,(I - AA^{-}) \tag{48}$$

From this last formula, we deduce

$$X - \approx{B}Z = X - \tilde{B}Z + U\,(I - AA^{-})\,Z = X - \tilde{B}Z \tag{49}$$

since $(I - AA^{-})\,Z = 0$, as shown by (5)

Then, substituting in equation (41), we have

$$N\tilde{\Sigma} = (X - \tilde{B}Z)(X - \tilde{B}Z)' = (X - \approx{B}Z)(X - \approx{B}Z)' = N\approx{\Sigma} \tag{50}$$

Now, let us multiply the two members of (48) by K. Taking (32) into account, we have

$$\tilde{\tilde{B}}K = \tilde{B}K + U\,(I - AA^-)\,K = \tilde{B}K \tag{51}$$

Relations (50) and (51) show that it is sufficient to consider the particular solution $\tilde{B}$. Taking account of its detailed expression (47), of relation (34) and of the fact that $(X - \hat{B}Z)\,Z' = C - \hat{B}A = 0$, we obtain the following expression for $N\tilde{\Sigma}$:

$$\begin{aligned} N\tilde{\Sigma} &= (X - \hat{B}Z - \Sigma TK'A^-Z)(X - \hat{B}Z - \Sigma TK'A^-Z)' \\ &= N\hat{\Sigma} + \Sigma TK'A^-K\,(\Sigma T)' \end{aligned} \tag{52}$$

Finally, let us impose on the expression (47) of $\tilde{B}$ to satisfy equation (40). Since we have assumed that rank K = s, $K'A^-K$ is regular (Searle (1971)) and we deduce

$$\Sigma T = -\,(\hat{B}K - M)(K'A^-K)^{-1} \tag{53}$$

and

$$N\tilde{\Sigma} = N\hat{\Sigma} + (\hat{B}K - M)(K'A^-K)^{-1}(\hat{B}K - M)' \tag{54}$$

3.4) <u>Test of the general linear hypothesis</u>

In order to be able to test (H), we need the following theorem about the sampling distribution of $(\hat{B}K - M)(K'A^-K)^{-1}(\hat{B}K - M)'$ (Lennes (1976)).

T4- If (H) is true, $(\hat{B}K - M)(K'A^-K)^{-1}(\hat{B}K - M)'$ is distributed according to $W\,(\Sigma, s)$, where s is the rank of K.

To test (H) we compute the $\frac{2}{N}$ power of the likelihood ratio criterion λ

$$\lambda^{\frac{2}{N}} = \frac{|N\hat{\Sigma}|}{|N\tilde{\Sigma}|} = \frac{|N\hat{\Sigma}|}{|N\hat{\Sigma} + (\hat{B}K - M)(K'A^-K)^{-1}\,(\hat{B}K - M)'|} \tag{55}$$

The study of its sampling distribution when (H) is true rests upon the theorems T3 and T4. The results obtained by T.W. Anderson (loc.cit.,§ 8.5 and 8.6) are immediately applicable to it.

$\lambda^{\frac{2}{N}}$ is to be compared to $U_{p,\,s,\,N-r}\,(\alpha)$, the α significance point of the $U_{p,\,s,\,N-r}$ distribution.

REFERENCES

ANDERSON, T.W. (1958). An Introduction to Multivariate Statistical Analysis. Wiley, New York.

LENNES, G. (1976). Inverses généralisés, problèmes d'estimation et épreuve de l'hypothèse linéaire générale en analyse statistique multivariate. Revue de Statistique appliquée, Vol.XXIV, n°3, 5-29.

RAO, C.R. and S.K. MITRA (1971). Generalized Inverse of Matrices and its Applications. Wiley, New York.

SEARLE, S.R. (1971). Linear Models. Wiley, New York.

Recent Developments in Statistics
J.R.Barra et al., editors

NONLINEAR REGRESSION WITH NUISANCE PARAMETERS: AN EFFICIENT ALGORITHM TO ESTIMATE THE PARAMETERS

by

H.N. Linssen

Technological University
Eindhoven, the Netherlands

In some experimental situations the experimentalist simultaneously observes an unknown value of an independent variable and the value of a dependent variable related to each other by a known modelfunction. The modelfunction contains parameters, whose values are to be estimated. Given a number of observations the determination of least squares estimates is straightforward in case the values of the independent variable are known. It requires only little extra computational effort to cover also the case that they are unknown but observed. An efficient algorithm is defined.

INTRODUCTION

Suppose the unknowns η and ξ are functionally related:

$$\eta = \eta(\xi,\beta)$$

where β is a p-vector of parameters to be estimated and ξ is called the nuisance parameter. There are given n observations (X_i,Y_i) according to

$$X_i = \xi_i + \text{error}$$
$$Y_i = \eta_i + \text{error} = \eta(\xi_i,\beta) + \text{error} .$$

An estimate for β follows from

$$\text{(1)} \qquad \min_{\beta,\xi_i} \sum_i^n (\eta_i - Y_i)^2 + (\xi_i - X_i)^2 .$$

This represents an unconstrained least squares minimization problem in n + p unknowns. Commonly used nonlinear least squares algorithms (for a review see [1]) are applicable but may lead to computer storage problems and large computing times because of the large number of unknowns. A more efficient algorithm utilizes the specific structure of (1) and is described in the following section.

THE ALGORITHM

The basic algorithm employed in solving least squares problems is the Gauss-Newton algorithm. This algorithm generates, given trial values ξ_0 and β_0, the next trial values through the solution of the linear least squares problem, which comes about when (1) is linearized with respect to ξ and β around ξ_0 and β_0. We rewrite (1) as:

$$\min_{\beta,\xi} \left\| \begin{pmatrix} \eta - Y \\ \xi - X \end{pmatrix} \right\|^2$$

where $\xi = (\xi_i)$, $X = (X_i)$, $\eta = (\eta_i)$, $Y = (Y_i)$ and $\|a\|$ denotes the Euclidean norm of a.

The linearized problem can be written:

$$(2) \qquad \min_{\beta,\xi} \left\| \begin{pmatrix} \eta_0 - Y \\ \xi_0 - X \end{pmatrix} + \begin{pmatrix} \Lambda & J \\ I & 0 \end{pmatrix} \begin{pmatrix} \xi - \xi_0 \\ \beta - \beta_0 \end{pmatrix} \right\|^2 ,$$

where $J_{ij} = \partial\eta_i/\partial\beta_j$, $\Lambda_{ij} = 0$ $(i \neq j)$ and $\Lambda_{ii} = \partial\eta_i/\partial\xi_i$ $(i = 1,\ldots,n;\ j = 1,\ldots,p)$ evaluated in $\xi = \xi_0$ and $\beta = \beta_0$.

Λ is a diagonal matrix. This enables us to separate the variables β and ξ in the following way. First we remark that (2) is invariant with respect to orthogonal transformations. And then we choose the orthogonal transformation:

$$\begin{pmatrix} H\Lambda & H \\ H & -\Lambda H \end{pmatrix} \quad \text{with } H = (\Lambda^2 + I)^{-\frac{1}{2}}$$

and apply this transformation to (2) which results in:

$$(3) \qquad \min_{\xi,\beta} \left\| \begin{pmatrix} \Lambda H(\eta_0 - Y) + H(\xi_0 - X) \\ H(\eta_0 - Y) - \Lambda H(\xi_0 - X) \end{pmatrix} + \begin{pmatrix} H^{-1} & \Lambda HJ \\ 0 & HJ \end{pmatrix} \begin{pmatrix} \xi - \xi_0 \\ \beta - \beta_0 \end{pmatrix} \right\|^2 .$$

This can be written:

$$(4) \qquad \min_{\xi,\beta} \{ \| \delta_1 + H^{-1}(\xi - \xi_0) + \Lambda HJ(\beta - \beta_0) \|^2 + \| \delta_2 + HJ(\beta - \beta_0) \|^2 \}$$

where δ_1 and δ_2 represent the left side part of (3).

The first term of (4) can be made equal to zero for every value of β and the second term is only a function of β. So we can solve (4) by seperately minimizing its two terms. The two steps in the minimization are:

<u>first</u>

$$(5) \qquad \min_{\beta} \| \delta_2 + HJ(\beta - \beta_0) \|^2$$

which is a least square problem in p unknowns.

<u>second</u>: for fixed β, ξ follows from:

(6) $$\xi = \xi_0 - H(\delta_1 + \Lambda HJ(\beta - \beta_0)) .$$

The regression problem with nuisance parameters is reduced to the ordinary regression problem (5) and the simple adjustment, given by (6), of the nuisance parameters in each iteration step.

APPLICABILITY

Estimates, satisfying (1), seem appropriate in case the observational errors have zero mean, are homogeneous and independent. In case the errors are moreover normally distributed, equation (1) is the maximum loglikelihood equation. Whether the parameter estimates in the presence of nuisance parameters have the well-known asymptotic properties of maximum likelihood [2] or least squares [3] estimates is an open question. Some more general cases can be reduced to the first mentioned case. Suppose the observational model is given by

(7) $$\begin{matrix} \mathbb{E}Y_i = \eta(\xi_i,\beta);\ \mathrm{var}\ Y_i = \gamma_i^2\sigma^2 \\ \mathbb{E}X_i = \xi_i;\ \mathrm{var}\ X_i = \alpha_i^2\sigma^2 \end{matrix}\ ;\ \rho(X_i,Y_i) = \rho_i .$$

ξ_i, σ^2 and β unknown; (X_i,Y_i) independent.
α_i, γ_i and ρ_i known.
The quantities:

$$X_i^* = X_i/\alpha_i;\ Y_i^* = 1/\sqrt{1-\rho_i^2}(Y_i/\gamma_i - \rho_i X_i^*)$$

have independent errors with zero mean and common variance σ^2.

The method can be generalized in a straightforward manner to cases where one or more of the α_i's and γ_i's are equal to zero.
A user's manual [4] for an ALGOL-procedure to compute estimates in cases where the observational model is given by (7) is available from the author.

REFERENCES

[1] Numerical Methods for Unconstrained optimization, ed. by W. Murray, Academic Press London, 1972.

[2] Bradley, R.A., Gart, J.J., The asymptotic properties of M.L. estimators when sampling from associated populations. Biometrika, 19, 1962, pag. 205.

[3] Jennich, R.T., Asymptotic properties of nonlinear least squares estimators. Annals of Mathematical Statistics, 40, 1970, 633-43.

[4] Linssen, H.N., Regression with nuisance parameters, user's manual. COSOR-note R 76-17, Technological University Eindhoven, 1976.

Recent Developments in Statistics
J.R.Barra et al., editors

STABILITY THEOREMS FOR CHARACTERIZATIONS OF THE NORMAL AND OF THE DEGENERATE DISTRIBUTIONS

Eugene Lukacs

Bowling Green State University and Technical University Wien.

We need two definitions.

Definition 1. Let X and Y be two random variables and assume that the expectation $\mathcal{E}(Y)$ of Y exists. We say that Y has ε-regression on X, if $\mathcal{E}(Y \mid X) \leq \varepsilon$ almost everywhere.

Definition 2 Let F and G be two distribution functions. Their distance L(f,G) is defined by the formula

$L(F,G) = \inf \{h: F(x-h) - h \leq G(x) \leq F(x+h)+h\}$.

The following results are derived in the paper.

Theorem 1. Let X_1 and X_2 be two independently and identically distributed random variables with common distribution function F(x) and suppose the first moment of F(x) exists and that $\mathcal{E}(X_1) = 0$. Let $L_1=a_1X_1+a_2X_2$ and $L_2= b_1X_1+b_2X_2$ be two linear forms in the random variables X_1 and X_2 and suppose that L_1 has ε-regression on L_2.

We write $\alpha = a_2/a_1$, $\gamma = b_2/b_1$ and assume that either $|\alpha| < 1$ or that $|\alpha| = 1$ but $|\gamma| < 1$. Then $L(F,E) \leq C\varepsilon^{(1-2\delta)/3}$ where $0 < \delta < 1/2$ and where C is independent of ε. Here E(x) is the degenerate distribution with a single saltus at x=0.

Theorem 2 Let X_1 and X_2 be two independently and identically distributed random variables with common distribution function F(x). Suppose that the second moment of F(x) exists and that $\mathcal{E}(X_1)=0$ while $0<\mathcal{E}(X_1^2)<\infty$

Let again $L_1 = a_1X_1 + a_2X_2$ and $L_2 = b_1X_1 + b_2X_2$ be two linear forms and suppose that $a_1b_1 + a_2b_2 = 0$ while $|\gamma| = |b_2/b_1| < 1$. We assume that L_1 has ε-regression on L_2. Then there exists a normal distribution ϕ_2 with mean a_2 and variance σ_2^2 such that $L(F, \phi_2) = K(\ln\frac{1}{\varepsilon})^{-1/11}$. Here K is independent of ε.

These theorems are stability theorems for a result of C.R. Rao (Sankhya A 29, 1967, pp. 1-14).

Recent Developments in Statistics
J.R.Barra et al., editors

CLUSTER ANALYSIS AND AFFINITY OF DISTRIBUTIONS

Kameo Matusita
The Institute of Statistical Mathematics
Tokyo, Japan

The use of the affinity of distributions as a measure for evaluation of a clustering procedure is proposed and an approach to a reasonable set of clusters from those obtained by a feasible method is presented. Some examples are given.

INTRODUCTION

As is well known, cluster analysis is concerned with grouping a collection of individuals into several clusters. The definition of "cluster" depends upon circumstances. In this paper we treat the problem of clustering in the case where each individual under consideration is represented by a point in the k-dimensional (real) Euclidean space and the variables concerned are continuous.

Now, given individuals are not always a collection of samples from some populations. However, supposing that for a set of clusters the individuals in each cluster are a sample from a distribution, we want to represent the compactness or separation of the clusters by the affinity of these distributions. Of course, these distributions have to be determined from the situation. First, we choose a form of distribution, and then we estimate the parameters of the distribution for each cluster from its members. For the distributions thus obtained we calculate the affinity. The affinity of distributions is a measure for discrimination of distributions ([2],[3]). The smaller the affinity, the larger the discrepancy among the distributions. Therefore, a desirable set of clusters will be found in the class of those that minimize the affinity under some regularity conditions. However, to find such a set of clusters, we have to calculate the affinity for all fashions of partitioning the individuals (under the regularity conditions). When the number of individuals under consideration is not small, this requires a lot of time and usually seems impracticable. In this paper we want to propose using the affinity as a measure for evaluation of a clustering procedure and to present an approach to a reasonable set of clusters from those obtained by a feasible method.

In the case where we are concerned with continuous variables it is often conceived that in a cluster the individuals are more densely distributed in the middle than on the periphery. This leads us to take up Gaussian distributions as those from which the individuals are supposed to have come. According to the circumstances, we naturally take up other types of distribution than the Gaussian.

AFFINITY OF DISTRIBUTIONS

Let $F_1, F_2, \cdots, F_r$ be distributions defined in the space R, and let $p_1(x), p_2(x), \cdots, p_r(x)$ be their density functions with respect to a measure m defined in R. Then the quantity

$$\rho_r(F_1, F_2, \cdots, F_r) = \int_R (p_1(x)p_2(x)\cdots p_r(x))^{1/r}\, dm$$

is called the affinity of distributions $F_1, F_2, \cdots, F_r$.

The affinity of distributions thus defined expresses well the likeness of distributions. In other words, the affinity represents the discrepancy among distributions, and it can serve as a measure for discrimination. Concerning properties of the affinity see [1],[2],[3].

When $F_1, F_2, \cdots, F_r$ are k-dimensional Gaussian distributions:

$$F_1 = N(a_1, A_1^{-1}),\ F_2 = N(a_2, A_2^{-1}),\ \cdots,\ F_r = N(a_r, A_r^{-1})$$

where a_i are k-dimensional (column) vectors and A_i are positive definite k x k matrices, we have

$$\rho_r(F_1, F_2, \cdots, F_r) = \frac{\prod |A_i|^{1/2r}}{\left|\frac{1}{r}\sum A_i\right|^{1/2}} \exp\left[\frac{1}{2r}\left\{\left(\sum A_i a_i,\ (\sum A_i)^{-1}\sum A_i a_i\right) - \sum (A_i a_i,\ a_i)\right\}\right].$$

CLUSTERING AND AFFINITY

Suppose that we are given n individuals. For a set of clusters $C_1, C_2, \cdots, C_r$ we express the separation of the clusters by

$$\rho_r(C_1, C_2, \cdots, C_r) = \frac{\prod |U_i|^{1/2r}}{\left|\frac{1}{r}\sum U_i\right|^{1/2}} \exp \frac{1}{2r}\left[\left\{\left(\sum U_i x^{(i)},\ (\sum U_i)^{-1}\sum U_i x^{(i)}\right) - \sum (U_i x^{(i)},\ x^{(i)})\right\}\right]$$

where $x^{(i)}$ is the mean of the individuals in C_i and U_i the inverse of the variance-covariance matrix of C_i. A clustering procedure is evaluated by the value of ρ_r. For fixed r a clustering procedure which gives smaller ρ_r is considered better.

Now, we want to propose to use the affinity as a measure of goodness of partitioning for checking and modifying clusters obtained by feasible methods (for example, the k-means method, ISODATA, etc.), in order to approach a reasonable set of clusters. That is, taking up a feasible method, we calculate the value of $\rho_r(C_1, C_2, \cdots, C_r)$ for the set of $C_1, C_2, \cdots, C_r$ obtained by that method and look at the behavior of ρ_r when r increases. To make distinct the behavior of ρ_r, we can take the logarithm of ρ_r. Usually this behavior has a tendency to decrease. Then our suggestion is:

(1) to stop at the first value of r at which ρ_r has fallen down by a remarkable amount;

or

(2) to go a few steps further and consider
 (i) omitting clusters of particularly small size,
 or
 (ii) merging such clusters into other suitable clusters, for instance, into the nearest clusters in the sense of distance between centroids.

For these procedures it is recommended to set a (regularity) condition such as:

(a) $|V_i| > \varepsilon$ $(\geqq 0)$, $i = 1, 2, \cdots, r$, where $V_i = U_i^{-1}$ and ε is a preassigned constant;

(b) $\mathrm{trace}(V_i) > \varepsilon_i$, $i = 1, 2, \cdots, r$;

or

(c) the size of $C_i > n_0$, $i = 1, 2, \cdots, r$, where n_0 is a preassigned integer.

By a combination of procedure and condition such as $\{$(2)-(ii), (a)$\}$, we can reach a fairly good result. Of course, with a well-separated collection of individuals we can obtain a quite satisfactory result.

Besides ρ_r or $\log \rho_r$, we can take up $\sqrt[r]{\rho_r}$ or $\frac{1}{r}\log \rho_r$ as an average compactness per cluster.

ILLUSTRATION

The tables 1, 2, 3, 4, 5 show values of ρ_r, $\log_{10}\rho_r$ and $\frac{1}{r}\log_{10}\rho_r$ for the sets of clusters after some modification by our approach, which were originally obtained by applying the k-means method to the following collections of individuals.

(1) The collection of individuals shown in scatter diagram 1.
(2) The collection of individuals shown in scatter diagram 2.
(3) The union of the collections (1) and (2).
(4) R.A.Fisher's data of three varieties of Iris.
(5) The union of

S_1: a sample of 250 from $N\left(\begin{pmatrix}5\\7\end{pmatrix}, \begin{pmatrix}1.5 & 0\\0 & 1.5\end{pmatrix}\right)$,

S_2: a sample of 200 from $N\left(\begin{pmatrix}-3\\-7\end{pmatrix}, \begin{pmatrix}1.5 & 0\\0 & 0.3\end{pmatrix}\right)$,

S_3: a sample of 400 from $N\left(\begin{pmatrix}0\\0\end{pmatrix}, \begin{pmatrix}1 & 0\\0 & 1\end{pmatrix}\right)$,

S_4: a sample of 200 from $N\left(\begin{pmatrix}-5\\5\end{pmatrix}, \begin{pmatrix}2 & 0\\0 & 2\end{pmatrix}\right)$,

S_5: a sample of 300 from $N\left(\begin{pmatrix}-7\\-3\end{pmatrix}, \begin{pmatrix}0.5 & 0\\0 & 0.5\end{pmatrix}\right)$,

S_6: a sample of 150 from $N\left(\begin{pmatrix}7.5\\-2.5\end{pmatrix}, \begin{pmatrix}0.8 & 0\\0 & 1.5\end{pmatrix}\right)$.

Table 1 n = 50
Procedure (2)-(i), condition (a) with $\varepsilon = 0$

Step	Number of clusters	ρ_r	$\log_{10}\rho_r$	$\frac{1}{r}\log_{10}\rho_r$
2	2	0.187342	-0.727365	-0.363683
3	3	3.278309E-02	-1.48435	-0.494783
4	4	1.294278E-02	-1.88797	-0.471993
5	5	4.884808E-04	-3.31115	-0.662230
6	6	4.319601E-05	-4.36456	-0.727427
7	7	1.857549E-06	-5.73106	-0.818723
8	8	2.543142E-07	-6.59463	-0.824329
9	9	9.316324E-13	-12.0308	-1.33676
10	10	8.171960E-20	-19.0877	-1.90877

Table 2 n = 50
Procedure (2)-(i), condition (a) with $\varepsilon = 0$

Step	Number of clusters	ρ_r	$\log_{10}\rho_r$	$\frac{1}{r}\log_{10}\rho_r$
2	2	3.578260E-02	-1.44633	-0.723165
3	3	3.817347E-05	-4.41824	-1.47275
4	4	6.201851E-06	-5.20748	-1.30187
5	5	2.643545E-09	-8.57781	-1.71556
6	6	4.618550E-10	-9.33549	-1.55592
7	7	3.493725E-42	-41.4567	-5.92239
8	8	3.264694E-51	-50.4861	-6.31076
9	8	5.811513E-14	-13.2357	-1.65446
10	9	8.998442E-21	-20.0458	-2.22731

SCATTER DIAGRAM 1

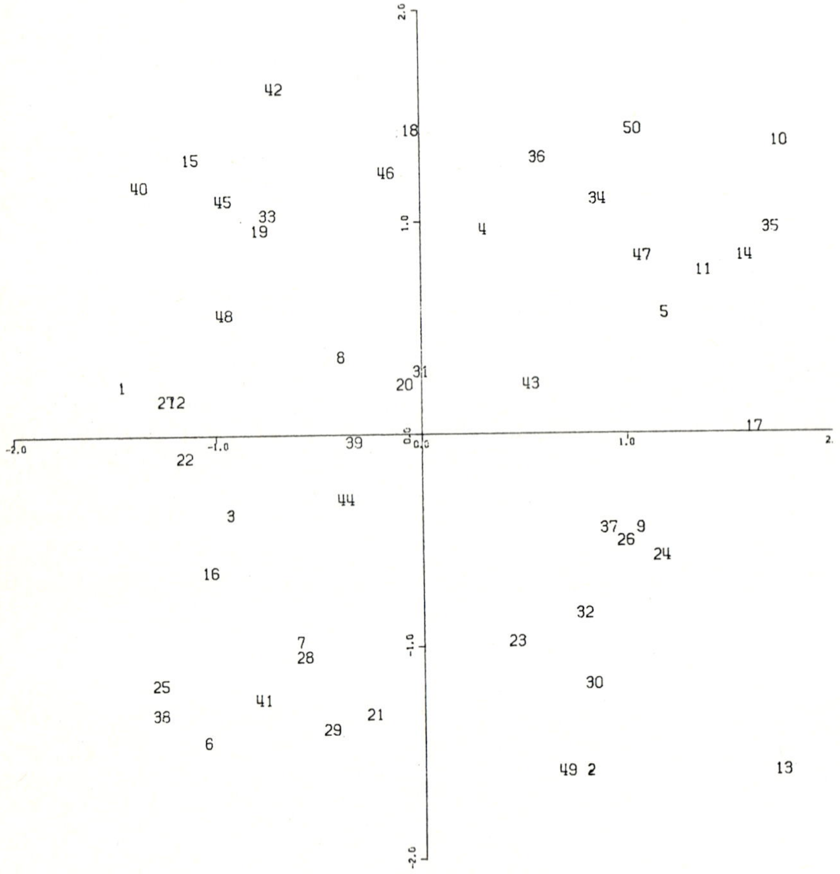

SCATTER DIAGRAM 2

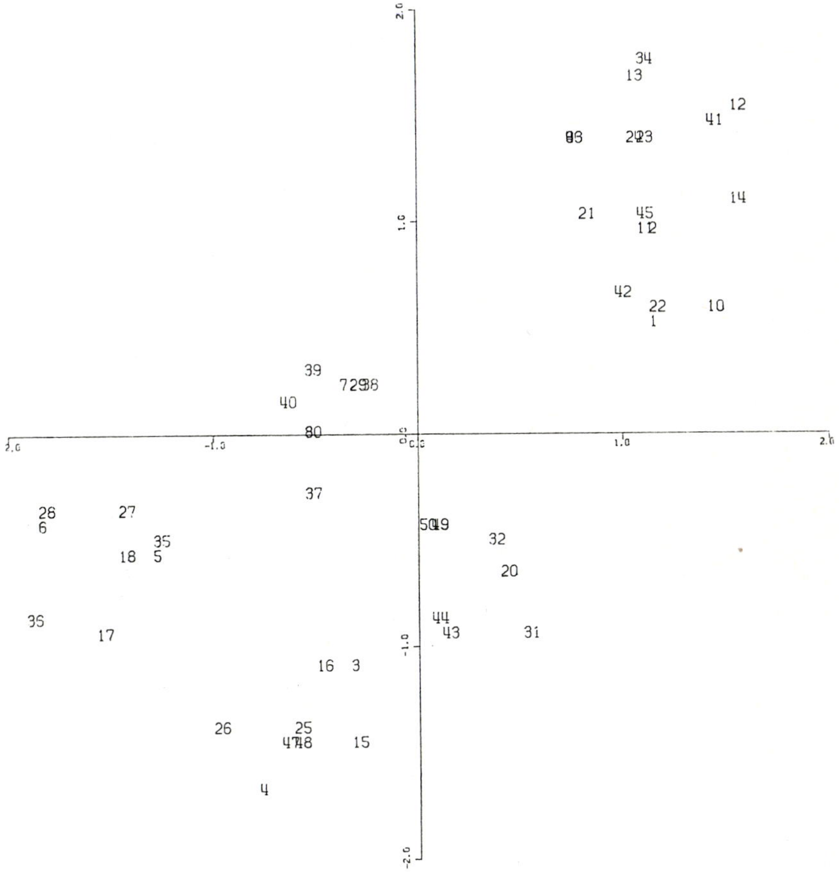

Table 3 n = 100
Procedure (2)-(i), condition (b) with $\varepsilon = 0$

Step	Number of clusters	ρ_r	$\log_{10}\rho_r$	$\frac{1}{r}\log_{10}\rho_r$
2	2	0.190047	-0.721138	-0.360564
3	3	1.523371E-02	-1.81719	-0.605730
4	4	2.830992E-03	-2.54806	-0.637015
5	5	5.026520E-04	-3.29873	-0.659746
6	6	1.038742E-05	-4.98349	-0.830582
7	7	7.647387E-07	-6.11649	-0.873784
8	8	4.573195E-07	-6.33978	-0.792473
9	9	2.074689E-10	-9.68305	-1.07589
10	10	2.677420E-09	-8.57228	-0.857228

Table 4 Fisher's data n = 150
Procedure (2)-(i), condition (a) with $\varepsilon = 0$

Step	Number of clusters	ρ_r	$\log_{10}\rho_r$	$\frac{1}{r}\log_{10}\rho_r$
2	2	1.226153E-02	-1.91146	-0.955730
3	3	7.704271E-10	-9.11327	-3.03776
4	4	1.943683E-14	-13.7114	-3.42785
5	5	1.394101E-16	-15.8557	-3.17114
6	6	1.153068E-15	-14.9381	-2.48968
7	7	4.790832E-15	-14.3196	-2.04566
8	8	3.502373E-21	-20.4556	-2.55695
9	9	7.890280E-24	-23.1029	-2.56699
10	10	2.242541E-27	-26.6492	-2.66492

Table 5 n = 1500
Procedure (2)-(i), condition (c) with $n_0 = 20$

Step	Number of clusters	ρ_r	$\log_{10}\rho_r$	$\frac{1}{r}\log_{10}\rho_r$
2	2	0.274727	-0.561099	-0.280550
3	3	4.307285E-03	-2.36580	-0.788599
4	4	3.192892E-05	-4.49582	-1.12395
5	5	3.816198E-12	-11.4184	-2.28367
6	6	6.677442E-16	-15.1754	-2.52923
7	7	7.472493E-16	-15.1265	-2.16093
8	8	8.680058E-15	-14.0615	-1.75768
9	9	1.120992E-15	-14.9504	-1.66115
10	10	1.147665E-15	-14.9402	-1.49402
11	11	1.823940E-16	-15.7390	-1.43082
12	12	2.769511E-17	-16.5576	-1.37980
13	13	7.530937E-18	-17.1231	-1.31717
14	14	5.812297E-18	-17.2356	-1.23112
15	15	1.786852E-18	-17.7479	-1.18319
16	16	1.274523E-20	-19.8946	-1.24341
17	16	3.734450E-22	-21.4278	-1.33924
18	15	5.048148E-21	-20.2969	-1.35312
20	18	3.695809E-26	-25.4323	-1.41290
22	18	1.492871E-27	-26.8260	-1.49033
24	20	1.768210E-31	-30.7525	-1.53762
26	22	5.318806E-36	-35.2742	-1.60337
28	16	9.993475E-32	-31.0003	-1.93752
30	21	3.428487E-37	-36.4649	-1.73642

Our judgment from these tables is that:

(I) for the case (1) the set of 9 clusters is reasonable;

(II) for the case (2) the sets of 5 and 6 clusters are both reasonable;

(III) for the case (3) the existence of a reasonable set of clusters is doubtful;

(IV) for the case (4) (Fisher's data) the set of 3 clusters is reasonable;

(V) for the case (5) the set of 6 clusters is reasonable.

Concerning the 6 assigned clusters in the case (5) we obtained the following result. Denote them by C_1, C_2, C_3, C_4, C_5, C_6, respectively.

	C_1	C_2	C_3	C_4
centroid:	$\begin{pmatrix} 4.98579 \\ 6.99167 \end{pmatrix}$,	$\begin{pmatrix} -2.98619 \\ -7.01352 \end{pmatrix}$,	$\begin{pmatrix} -2.246035E-02 \\ 2.605480E-02 \end{pmatrix}$,	$\begin{pmatrix} -5.00628 \\ 5.00139 \end{pmatrix}$.
size:	250	200	402	198

	C_5	C_6
centroid:	$\begin{pmatrix} -7.00346 \\ -3.01204 \end{pmatrix}$,	$\begin{pmatrix} 7.47921 \\ -2.54380 \end{pmatrix}$
size:	300	150

Between $\{S_1, S_2, \cdots, S_6\}$ and $\{C_1, C_2, \cdots, C_6\}$ we actually had the following correspondence table.

	C_1	C_2	C_3	C_4	C_5	C_6	
S_1	250						250
S_2		200					200
S_3			400				400
S_4			2	198			200
S_5					300		300
S_6						150	150
	250	200	402	198	300	150	1500

We can say this is a quite satisfactory result.

Thanks are due to N.Ohsumi, the Institute of Statistical Mathematics, for his suggestion and help in experimental work.

REFERENCES

[1] K.Matusita (1967), On the notion of affinity of several distributions and some of its applications, Ann.Inst.Statist.Math., 19, 181-192.

[2] K.Matusita (1971), Some properties of affinity and applications, Ann.Inst. Statist.Math., 23, 137-155.

[3] K.Matusita (1973), Discrimination and the affinity of distributions, Discriminant Analysis and Applications (Academic Press), 213-223.

Recent Developments in Statistics
J.R.Barra et al., editors

A GENERAL STRUCTURE FOR INFERENCE ABOUT VARIANCES AND COVARIANCES

A. O'Hagan
Department of Statistics
University of Warwick
Coventry, England

1. THE GENERAL STRUCTURE

In discussing an unknown covariance matrix there are degrees of "unknown-ness". For example in the linear model the covariance matrix of the observations is generally assumed to be almost completely known - the entire matrix Σ is described by a single parameter σ^2 through $\Sigma = \sigma^2 I$, or in generalised least squares $\Sigma = \sigma^2\Omega$ where Ω is known. It is often convenient to express greater degrees of "unknown-ness" by describing Σ in terms of a few parameters: thus in a one-way analysis of variance with random effects the covariance matrix of the observations in each group is of the form

$$\Sigma = \begin{pmatrix} \sigma^2+\sigma_t^2 & \sigma_t^2 & \cdots & \sigma_t^2 \\ \sigma_t^2 & \ddots & & \vdots \\ \vdots & & \ddots & \sigma_t^2 \\ \sigma_t^2 & \cdots & \sigma_t^2 & \sigma^2+\sigma_t^2 \end{pmatrix}, \tag{1}$$

often called an intra-class covariance matrix. Here Σ has two unknowns. However, for the analysis of models containing such covariance matrices, no general results are available and each case must be treated separately. In contrast, the situation with mean vectors is very different, for the linear model provides a general theory encompassing a very wide variety of models. The purpose of this paper is to present a class of covariance matrix parametrisations which can also be treated in a general way.

The general form is

$$\Sigma = \sum_{i=1}^{k} A_i \otimes \Sigma_i, \tag{2}$$

where (i) Σ is the p x p unknown covariance matrix,

(ii) we define the direct product $A \otimes B$ to be the partitioned matrix

$$A \otimes B = \begin{pmatrix} a_{11}B & \cdots & a_{1n}B \\ \vdots & & \vdots \\ a_{m1}B & \cdots & a_{mn}B \end{pmatrix},$$

where a_{ij} is the (i,j)th element of the m x n matrix A,

(iii) $A_1,\ldots\ldots,A_k$ are known r x r matrices, $\Sigma_1,\ldots,\Sigma_k$ are unknown q x q matrices and p = qr,

(iv) each A_i is symmetric and idempotent and

$$\sum_{i=1}^{k} A_i = I_r$$

the r x r identity matrix.

For example the form $\Sigma = \sigma^2 I$ is the case k = 1, q = 1, $A_1 = I_r$, $\Sigma_1 = \sigma^2$ of (2), and the form (1) corresponds to k = 2, q = 1, $A_1 = I_r - r^{-1} J_r$, $A_2 = r^{-1} J_r$, $\Sigma_1 = \sigma^2$, $\Sigma_2 = \sigma^2 + r\sigma_t^2$, where J_r is the r x r matrix of ones.

2. LIKELIHOOD FACTORISATION

The value of writing Σ in the form (2) arises from the following simple theorem.

Theorem: If Σ is of the form (2), then

(i) $|\Sigma| = \prod_{i=1}^{k} |\Sigma_i|^{r_i}$, where r_i is the rank of A_i;

(ii) $\Sigma^{-1} = \sum_{i=1}^{k} A_i \otimes \Sigma_i^{-1}$

(iii) Σ is positive definite if and only if every A_i is positive definite.

To save space the proof has been omitted here, but is available from the author.

Now suppose we wish to make inference about Σ, i.e. about $\Sigma_1,\ldots,\Sigma_k$. In models employing normal distributions the likelihood function will take the Wishart form

$$L(\Sigma) \propto |\Sigma|^n \exp(-\tfrac{1}{2} \operatorname{tr}\Sigma^{-1} S)$$

(if there are other parameters then typically S will be a function of them, but our own concern at the moment is with Σ).

The Theorem now shows that

$$L(\Sigma) = L(\Sigma_1,\ldots,\Sigma_k) \propto \prod_{i=1}^{k} \{|\Sigma_i|^{nr_i} \exp(-\tfrac{1}{2} \operatorname{tr} \Sigma_i^{-1} S_i)\} \quad (3)$$

where

$$\operatorname{tr}\Sigma_i^{-1} S_i = \operatorname{tr}(A_i \otimes \Sigma_i^{-1}) S,$$

so that

$$S_i = \sum_{j=1}^{r} \sum_{k=1}^{r} a_{ijk} S_{kj}$$

where a_{ijk} is the (j,k)-th element of A_i and S_{kj} is the (k,j)-th block when S is partitioned into blocks of size q x q. Equation (3) represents a very neat factorisation of the likelihood function, and the fact that each factor is itself of the Wishart form is important, because it means that we can express any of the Σ_i itself in a form such as (2) and thereby factorise the likelihood still further.

Alternatively there are two other similar factorisation classes which we could apply instead:

(a) $\Sigma = B\Sigma_1 B^T$, where B is p x p nonsingular and known so that
$|\Sigma|^n \exp(-\frac{1}{2} \operatorname{tr} \Sigma^{-1} S) \propto |\Sigma_1^{-1}|^n \exp(-\frac{1}{2} \operatorname{tr} \Sigma_1^{-1} S_1)$
where $S_1 = B^{-1} S(B^T)^{-1}$;

(b)

$$\Sigma = \begin{pmatrix} \Sigma_1 & & 0 \cdots 0 \\ & \ddots & \vdots \\ 0 & & 0 \\ \vdots & \ddots & \\ 0 \cdots 0 & & \Sigma_k \end{pmatrix}$$

where Σ_i is $p_i \times p_i$, so that
$|\Sigma|^n \exp(-\frac{1}{2} \operatorname{tr}\Sigma^{-1} S) = \prod_{i=1}^{k} \{|\Sigma_i|^n \exp(-\frac{1}{2} \operatorname{tr}\Sigma_i^{-1} S_i)\}$
where S_i is the $p_i \times p_i$ (i, i)-th block of S.

We can apply either of these two and (3) in any combinations to greatly enrich the original class of structures. For example (a) enables us to include the generalised least-squares structure $\Sigma = \sigma^2\Omega$, or to write Σ in terms of $\Sigma_i \otimes A_i$ rather than $A_i \otimes \Sigma_i$.

On their own, (a) and (b) are rather trivial so that we will concentrate on (3) in considering applications. Nevertheless, any statement appropriate to (3) is applicable to successive applications of the three decompositions in any combination.

3. APPLICATIONS

Equation (3) yields maximum likelihood estimators of the Σ_i immediately, and hence of Σ through (2). Working independently of the present author, S. Tolver Jensen obtained essentially (2) and (3) but expressed in terms of Jordan algebras, and showed that the maximum likelihood estimators are Wishart distributed. Tolver Jensen's paper was also presented at this conference; similar results were obtained by Andersson(1975).

The implications for Bayesian inference are even more interesting. The Wishart form of each factor in the likelihood suggests a very simple and tractable family of prior distributions, that giving the Σ_i independent inverse-Wishart priors. If in a given context a reasonable representation of the prior information about Σ can be obtained using a member of this family, then the posterior analysis is very simple. However, there will be situations in which such a prior distribution is unacceptable, particularly the assumption of independence. Consider the intra-class covariance form (1). With k = 2, Σ_1 and Σ_2 are the scalars $\sigma_1^2 = \sigma^2$ and $\sigma_2^2 = \sigma^2 + r\sigma_t^2$ and the suggested prior structure says that these two quantities are independent *a priori*. It would seem more appropriate in the "components of variance" context to declare σ^2 and σ_t^2 independent, and if r were small this prior could produce significantly different posterior inferences. In another context the natural parametrisation might be

$$\Sigma = \sigma_o^2 \begin{pmatrix} 1 & & \rho \cdots \rho \\ & \ddots & \vdots \\ \rho & & \rho \\ \vdots & \ddots & \\ \rho \cdots \rho & & 1 \end{pmatrix}, \qquad (4)$$

where independence between $\sigma_o^2 = \sigma^2 + \sigma_t^2$ and $\rho = \sigma_t^2 / (\sigma^2 + \sigma_t^2)$ is reasonable. Yet the suggested independence of σ_1^2 and σ_2^2 becomes more plausible if we consider the vector random variable $x = (x_1, x_2, \ldots, x_r)^r$ of which Σ is the covariance matrix, for then

$$\left.\begin{aligned} \sigma_2^2 &= \sigma^2 + r\sigma_t^2 = r \text{ Var } (r^{-1} \sum_{i=1}^{r} x_i)_1 \\ \sigma_1^2 &= \sigma^2 = \tfrac{1}{2} \text{ Var } (x_i - x_j) \quad (\text{all } i \neq j). \end{aligned}\right) \qquad (5)$$

Another cause of the independent inverse-Wishart prior being unacceptable may be the ranges of variation of the Σ_i : although the theorem states that Σ is positive definite if and only if the Σ_i are all positive definite, context may demand more stringent conditions. Thus, again in the case of the "components of variance" context it is natural to argue that σ_t^2 cannot be negative, yet for some negative σ_t^2, Σ_i will still be positive definite. The problem of avoiding negative estimates of variance components is traceable to precisely this phenomenon. It is not a serious problem in the Bayesian approach since truncated inverse-Wishart (which in this case with scalar σ_2^2, is a truncated inverse χ^2) distributions can be substituted without seriously impeding posterior analysis. (This approach is exemplified by O'Hagan (1973).)

A similar application occurs in the model discussed by Dickey *et al* (1976). Inference is required about a covariance matrix Ω about which nothing as precise as (2) is known *a priori*. An inverse-Wishart prior distribution is chosen for Ω with mean Σ, conditional on Σ. Arguments of exchangeability are then invoked to infer that Σ must have a structure like (1) or equivalently (4), and the prior specification is completed by defining a prior distribution for Σ subject to this constraint. Dickey *et al* choose independent prior distributions for σ_o^2 and ρ and argue that $\rho \geq 0$ ($\sigma_t^2 \geq 0$). Use of the Theorem yields a similar factorisation in this more complex situation giving a much more tractable posterior analysis if σ_1^2 and σ_2^2 are given independent χ^2 distributions. The justification (5) is particularly forceful in this context.

Vague prior information forms of the independent inverse-Wishart prior for (2), using zero degrees of freedom, were used for several different models by Box and Tiao (1973) who found the corresponding likelihood factorisations without discovering the general form (3).

REFERENCES

Andersson, S. (1975). Invariant normal models. Ann Statist. 3, 132-54.

Box, G. E. P. and Tiao, G. C. (1973). Bayesian Inference in Statistical Analysis. Addison-Wesley, N.Y.

Dickey, J. M., Lindley, D. V. and Press, S. J. (1976). Estimation of the dispersion matrix of a multivariate normal distribution. To appear in J. Roy. Statist. Soc. B.

O'Hagan, A. (1973). Bayes estimation of a convex quadratic. Biometrika 60, 565-71.

Recent Developments in Statistics
J.R.Barra et al., editors

PROJECTION OF MARTINGALES AND LINEAR FILTERING IN HILBERT SPACES

J.Y. OUVRARD
Université Scientifique et Médicale de Grenoble
I.R.M.A., B.P. 53
38041 - Grenoble Cédex - France

SUMMARY

We use a theorem on the projection of Hilbertian martingales [12] and a Hilbertian theorem of Girsanov type [11] to study the linear filtering problem for systems governed by evolution equations in Hilbert spaces. We show that if the observation is a function of the state through an unbounded operator and with values in a Hilbert space (under some hypotheses on the observation noise), the filtered state is obtained as the solution of a differential stochastic equation, the coefficients of which are given by a solution of an operator differential Ricatti equation. We show that the linear filtering problem can be solved only if this solution is unique. We find again, with this innovation method, results from [2] in the partial differential equation context, and from [3] and [6] in the linear hereditary equation context.

1 - DESCRIPTION OF THE SYSTEM

D, H, K, L are Hilbert spaces such that D is a dense subspace of H. The canonical injection i from D to H is continuous and dense. $(A_t)_{t\in[0,T]}$ is a family of linear operators in H, the state space, with domain containing D, such that $A_{t|D}\in \mathcal{L}(D,H)$ and $\forall h\in D$, A.h is continuous from $[0,T]$ to H. Elsewhere, let $(C_t)_{t\in[0,T]}$ be a family of linear operators from H to K, the state space of the observations, with domain containing D and such that $C_{t|D}\in \mathcal{L}(D,K)$ and that $\forall h\in D$, the application C. h is measurable and bounded from $[0,T]$ to K.

We consider a Wiener process W on L relatively to the process basis $(\Omega,\mathcal{F},(\mathcal{F}_t)_{t\in[0,T]},P)$ that satisfies the customary conditions, $(\Omega,\mathcal{F})$ being moreover a Blackwell space ; the covariance operator of W is the nuclear operator $\mathcal{W}\in \mathcal{L}_1(L)$.

Let B and D be two operator-valued functions satisfying :

$$B\in \mathcal{L}^{\infty}([0,T] \ ; \ \mathcal{L}(L,H)) \text{ and } D\in \mathcal{L}^{\infty}([0,T] \ ; \ \mathcal{L}(L,K)).$$

We make the following hypotheses about the observation noise : if we write $Q_t = D_t\mathcal{W}D_t^*$ $\quad\forall t\in[0,T]$, we suppose that :

(i) $\text{rg } C_t \subset \text{rg } Q_t \quad \forall t\in[0,T]$

(ii) $\forall h \in D$ the set $\{||Q_t^+ C_t h|| \mid t \in [0,T]\}$ is bounded in K where Q_t^+ is the pseudo-inverse (non-bounded, with domain containing rg Q_t) - for a definition see [12].

(iii) rg $D_t \mathcal{W} B_t^* \subset \mathcal{D}(Q_t^+)$ $\quad \forall t \in [0,T]$

We suppose that a Green kernel is associated with the family $(A_t)_{t\in[0,T]}$, that is a family of operators $\{G(t,s) \mid 0 \leq s \leq t \leq T,\ G(t,s) \in \mathcal{L}(H,H)\}$ such that :

(1) $G(t,t) = 1_H \quad \forall t \in [0,T]$

(2) $G(u,s) = G(u,t) o G(t,s)$ if $0 \leq s \leq t \leq u \leq T$

(3) $\forall h \in H$ $G(.,s)h$ is continuous from $[s,T]$ in H.

(4) $\forall h \in D$ the Cauchy problem

$$\begin{cases} \dfrac{dx_t}{dt} = A_t\, x_t \\ x_s = h \end{cases} \qquad s \leq t \leq T$$

has a unique solution in $L_D^2([0,T]) \cap C_H([0,T])$ given by $x_t = G(t,s)h$ (i.e. $x_t = h + \int_s^t A_u x_u du$).

We assume that the Green kernel G and B have extra properties assuring the existence and unicity of the stochastic differential equation :

$$(1.1) \quad \begin{cases} dx_t = A_t x_t dt + B_t dW_t \\ x(0) = x_0 \end{cases}$$

(see, for example, [4] and [9] for different theorems about existence and unicity). A solution of (1.1) is a process with values in H, such that any trajectory is almost everywhere (for the Lebesgue measure on $[0,T]$) in D and that satisfies (1.1).

The state x_t is subject to the observation y_t related with the state through the equation :

$$(1.2) \quad \begin{cases} dy_t = C_t x_t dt + D_t dW_t \\ y(0) = 0 \end{cases}$$

The two equations (1.1) and (1.2) constitute the filtering system.

Let $\mathcal{B}_t = \sigma(y_s \mid s \leq t)$ be the σ-field generated by the observations up to time t. We intend to establish equations giving the filtered state $E^{\mathcal{B}_t} x_t$ recursively, that is to establish the equations of the Kalman filter and this, by the innovation approach (see [1] in the finite dimensional case).

2 - MARTINGALES RELATED TO THE STATE AND OBSERVATION-INNOVATION PROCESS

According to [9] we have $E \int_0^T ||x_t||_D^2 dt < +\infty$. Then, we deduce from the real case and the separability of D and H

Lemma 2.1. : There is a $(\mathcal{B}_t)_{t\in[0,T]}$ predictable process $\hat{x}$ with values in H the trajectories of which are almost everywhere in D so that :

$$P(\hat{x}_t = E^{\mathcal{B}_t} x_t) = 1 \quad \forall t\in[0,T]$$

$\hat{x}$ satisfies :

(2.1.1) $E\int_0^T ||x_t||_D^2 dt < +\infty$

Definitions 2.2. :

1°) One defines the two processes Z_s and Z_o with values respectively in H and K by

$$\forall t\in[0,T] \quad Z_s(t) = \hat{x}_t - \int_0^t A_u \hat{x}_u du \qquad Z_o(t) = y_t - \int_0^t C_u \hat{x}_u du$$

These processes are well-defined by virtue of the hypotheses made on A and C and by lemma 2.1. Z_o is called innovation process.

2°) We call error the process e with values in H defined by :

$$e_t = x_t - \hat{x}_t \qquad \forall t\in[0,T]$$

Lemma 2.2. : The trajectories of e are almost everywhere in D and satisfy :

(2.2.1) $E^{\mathcal{B}_u} e_t = 0$ if $0 \le u \le t \le T$

(2.2.2) $E\int_0^T ||e_t||_D^2 dt < +\infty$

Proposition 2.3. : Z_s and Z_o are square integrable martingales with values in H and K respectively, with respect to the process basis $(\Omega,\mathcal{F},(\mathcal{B}_t)_{t\in[0,T]},P)$.

Proposition 2.4. : Z_o is a martingale a.s. with continuous trajectories in K ; its natural process $<Z_o>$ is given by :

(2.4.1) $<Z_o>_t = \int_0^t D_u \mathcal{W} D_u^* du \quad \forall t\in[0,T]$

Proof : We apply the Ito formula ([9]) to calculate $Z_o(t)^{\otimes 2}$.

- We do not know how to prove -and it may perhaps be false- (as in the finite-dimensional case ; see [1] f. ex.) the equality of the σ-field $\mathcal{B}_t$ and the one generated by the innovation process Z_o up to time t. However, by using an idea of [7] in the case of non-linear filtering of one-dimensional systems, we establish a representation theorem for square-integrable martingales as a stochastic integral with respect to Z_o.

- The basic tool is a Girsanov theorem ([11]). For this, we define a class $\bar{\Lambda}_T^2(Z_o ; K,H)$ of processes with values in $\mathcal{L}(K,H)$ for which we can define a stochastic integral with respect to Z_o (see [9] for constructing this integral).

Definition 2.5. : Let $\Lambda_T^2(Z_o ; K,H)$ be the set of predictable processes X with values in $\mathcal{L}(K,H)$ satisfying :

(1) $\forall k\in K$ Xk is strongly predictable with values in H.

(2) $||X||^2_{\Lambda_T^2} \equiv E\int_0^T \operatorname{tr} X_t Q_t X_t^* dt < +\infty$

$\Lambda_T^2(Z_o\ ;\ K,H)$ is provided with the pre-hilbertian semi-norm $||\cdot||^2_{\Lambda 2}$. We call Λ_T^{2*} the Hilbert space obtained by completeness with respect to this semi-norm. The space $\bar{\Lambda}_T^2(Z_o\ ;\ K,H)$ is the closure in Λ_T^{2*} of the space E of the uniformly predictable step-processes with values in $L(K,H)$. The stochastic integral with respect to Z_o of any $X\in\bar{\Lambda}_T^2(Z_o\ ;\ K,H)$ is then defined by isometry ([9]).

Representation theorem 2.6. : Any separable centered square integrable martingale M on the process basis $(\Omega,F,(B)_{t\in[0,T]},P)$ can be written $\int_0^\cdot \phi dZ_o$, where $\phi\in\bar{\Lambda}_T^2(Z_o\ ;\ K,H)$.

Proof : By using the Girsanov theorem ([11]) we obtain the representation theorem of martingales with respect to the family of σ-fields generated by a Hilbert Wiener process. Here, the hypotheses (i) and (ii) on the observation noise are used in a crucial way.

Let be φ the K-valued process defined by $\varphi_u = Q_u^+ C_u \hat{x}_u \quad \forall u\in[0,T]$.

This process is well defined by virtue of (i) and (ii) and of Lemma 2.1. It is a predictable process satisfying :

(2.6.1) $\quad E\int_0^T ||Q_u^{1/2}\ _u||_K^2 < +\infty$

Then we define the positive process α by :

(2.6.2) $\quad \alpha_t = \exp\{\int_0^t <\varphi,dZ_o>_K - \frac{1}{2}\int_0^t ||Q_u^{1/2}\varphi_u||_K^2 du\}$

and, for any $n\in N$, the stopping-time T_n :

$T_n = \mathrm{Inf}\ \{t\in[0,T]\ |\ \int_0^t ||Q_u^{1/2}\varphi_u||_K^2 du > n\}$ (we let $\mathrm{Inf}\ \phi = T$)

The sequence $(T_n)_{n\in N}$ grows to T P-a.s.

The measure $\tilde{P}_n$ with density α_{T_n} with respect to P is a probability ([11]) and the process U_n defined by :

(2.6.3) $\quad U_n(t) = Z_o(t) + \int_0^{t\wedge T_n} Q_u\varphi_u du \qquad \forall t\in[0,T]$

is a martingale with respect to $(\Omega,F,(B_t)_{t\in[0,T]},\tilde{P}_n)$; its natural process is $(\int_0^t Q_u du)_{t\in[0,T]}$.

Now, by virtue of (i) :

(2.6.4) $\quad U_n(t) = Z_o(t) + \int_0^{T_n} C_u\hat{x}_u du$

So

(2.6.5) $\quad U_n(t\wedge T_n) = y(t\wedge T_n)$

Since (Ω,F) is a Blackwell space, we have the following sequence of equalities of σ-fields :

$$B_{t\wedge T_n} = \sigma(y_{s\wedge T_n}|s \leq t) = \sigma(U_n(s\wedge T_n)|s \leq t) = C^n_{t\wedge T_n}$$

where C_t^n is the σ-field $\sigma(U_n(s)|s \leq t)$.

Then let M be a square-integrable H-valued martingale on the process basis $(\Omega,F,(B_t)_{t\in[0,T]},P)$, and let us call $\tilde{M}$ the process defined by :

(2.6.6) $\tilde{M}_t = \alpha_t^{-1} M_t \qquad \forall t \in [0,T]$

$\tilde{M}_{\cdot\wedge T_n}$ is a $(\Omega, F, (C_t^n)_{t\,[0,T]}, \tilde{P}_n)$ square-integrable martingale.

By virtue of the above-mentioned representation theorem, there is a process $\phi^n \in \bar{\Lambda}_T^2(U^n ; K,H)$ such that :

$$(2.6.7) \quad \tilde{M}_{t\wedge T_n} = \int_0^{t\wedge T_n} \phi^n dU^n$$

so, by (2.6.5) and the localisation property of the stochastic integral :

$$\tilde{M}_{t\wedge T_n} = \int_0^{t\wedge T_n} \phi^n dy = \int_0^{t\wedge T_n} \phi^n dZ_0 + \int_0^{t\wedge T_n} \phi^n C_u \hat{x}_u du$$

By using the Ito formula, we obtain :

$$(2.6.8) \quad M_{t\wedge T_n} = \int_0^t \Psi^n dZ_0 + \int_0^t \Theta_u^n du$$

where $\Psi^n \in \bar{\Lambda}_T^2(Z_0 ; K,H)$ and $\int_0^{\cdot} \Theta_u^n$ has trajectories with bounded variation in H.

These martingales being continuous, we obtain :

$$M_{t\wedge T_n} = \int_0^t \Psi^n dZ_0$$

Then we construct the process ϕ by means of the processes Ψ_n.

Corollary 2.7. : Z_s is a (B_t)-martingale with continuous trajectories and there is a process $\Psi \in \bar{\Lambda}_T^2(Z_0 ; K,H)$ such that :

$$Z_s(t) = \int_0^t \Psi dZ_0 \qquad \forall t \in [0,T]$$

Proposition 2.8. : The predictable process with bounded variation (with values in $L_1(K,H)$, the space of nuclear operators from K to H) associated with the martingales Z_0 and Z_s is written as :

$$(2.8.1) \quad \langle Z_0, Z_s \rangle_t = \int_0^t (P_u C_u^* + B_u W D_u^*) du$$

where the predictable $L_1(D,H)$-valued process is defined as follows : let $\hat{P}$ be the unique predictable process with values in the projective tensor product $D \otimes_1 H$ of the two spaces D and H such that :

$$\forall t \in [0,T] \qquad P\left[\hat{P}_t = E^{B_t^+} e_t \otimes_{D,H} e_t\right] = 1$$

P is then the process isometrically associated with $\hat{P}$ by the isometry from $D \otimes_1 H$ into $L_1(D,H)$. Later on, we shall identify these two spaces.

Proof : Apply the Ito formula.

Theorem 2.9. : Consider the martingales Z_o and Z_s on the process basis $(\Omega, \mathcal{F}, (\mathcal{B}_{t^+})_{t\in[0,T]}, P)$. We have :

$$(2.9.1) \quad \forall t \in [0,T] \qquad Z_s(t) = \int_0^t (P_v C_v^* + B_v \mathcal{W} D_v^*)(D_v \mathcal{W} D_v^*)^+ \, dZ_o$$

Proof : This results immediately from corollary 2.7. and from the representation theorems for martingales of |12|.

Corollary 2.10. : There is a continuous version of the process $\hat{x}$; it is a solution of the stochastic differential equation :

$$(2.10.1) \quad \hat{x}_t = \int_0^t A_u \hat{x}_u du + \int_0^t (P_u C_u^* + B_u \mathcal{W} D_u^*)(D_u \mathcal{W} D_u^*)^+ \, dZ_o$$

Now, we show that P is a solution of a Ricatti equation in $\mathcal{L}_1(D,H)$.

Proposition 2.11. : Almost all trajectories of the process P and the covariance (in $\mathcal{L}_1(D,H)$) of the error process e are solutions of the Ricatti equation :

$$(2.11.1) \quad \begin{cases} \dfrac{dP_t i^*}{dt} = A_t P_t^* + P_t A_t^* + B_t \mathcal{W} B_t^* - (P_t C_t^* + B_t \mathcal{W} D_t^*)(D_t \mathcal{W} D_t^*)^+ (P_t C_t^* + B_t \mathcal{W} D_t^*)^* \\ P_0 = E x_o \otimes_{D,H} x_o \end{cases}$$

where $i^* \in \mathcal{L}(H,D)$ is the adjoint of the injection i from D in H.

Proof : If $\Psi_u = (P_u C_u + B_u \mathcal{W} D_u^*) Q_u^+$, we can write, by virtue of (1.1) and (2.10.1) $e_t = \int_0^t A_u e_u du + \int_0^t B_u dW_u - \int_0^t \Psi_u dZ_o$.

Using the Ito formula, we obtain, for $\alpha > 0$:

$$(2.11.2) \quad E^{\mathcal{B}_{t^+}} (e_{t+\alpha}^{\otimes 2}{}_H - e_t^{\otimes 2}{}_H) = E^{\mathcal{B}_{t^+}} \{ \int_t^{t+\alpha} [(A_u e_u) \otimes_H e_u + e_u \otimes_H (A_u e_u)] du \}$$
$$+ E^{\mathcal{B}_{t^+}} \int_t^{t+\alpha} B_u \otimes B_u (\mathcal{W}) du - E^{\mathcal{B}_{t^+}} \int_t^{t+\alpha} \Psi_u \otimes \Psi_u (Qu) du$$

But

$$E^{\mathcal{B}_{t^+}} e_{t+\alpha}^{\otimes 2}{}_H = E^{\mathcal{B}_{t^+}} E^{\mathcal{B}_{t+\alpha}} e_{t+\alpha}^{\otimes 2}{}_H = E^{\mathcal{B}_{t^+}} P_{t+\alpha}$$

and, using the Fubini theorem and usual tensor equalities :

$$E^{\mathcal{B}_{t^+}} \int_t^{t+\alpha} (A_u e_u) \otimes_H e_u du = E^{\mathcal{B}_{t^+}} \int_t^{t+\alpha} E^{\mathcal{B}_u} (e_u \otimes_{D,H} e_u) A_u^* du$$
$$= E^{\mathcal{B}_{t^+}} \int_t^{t+\alpha} P_u A_u^* du$$

Dividing the two members of (2.11.2) by $\alpha > 0$ and taking the limits when α goes to zero, we obtain the proposition.

Remark : If the equation (2.11.1) has a unique solution, then P is deterministic and is identified with the covariance function of the error $t \to E\, e_t \otimes_{D,H} e_t$. In this case, the filtering problem is completely solved.

3 - ABOUT THE UNIQUENESS OF THE SOLUTION OF THE RICATTI EQUATION (2.11.1)

We restrict ourselves to the case where Q_t is strictly positive for all $t \in [0,T]$. Then, we remark that this equation is equivalent to :

$$(3.1) \qquad \begin{cases} \dfrac{dP_t i^*}{dt} = \mathcal{A}_t P_t^* + P_t \mathcal{A}_t^* + E_t - P_t C_t^* Q_t^{-1} C_t P_t^* \\ P(o) = P_o \in \mathcal{L}(D,H) \end{cases}$$

where we have set :

$$R_t = B_t \mathcal{W} D_t^* \qquad \mathcal{A}_t = A_t - R_t Q_t^{-1} C_t \qquad E_t = B_t \mathcal{W} B_t^* - R_t Q_t^{-1} R_t^*$$

By virtue of section II, we know that there are solutions of (3.1).

To prove the uniqueness, we introduce the "dual" control problem of the filtering problem :

$$(3.2) \qquad \begin{cases} \dfrac{di'iz_t}{dt} = -A_t' iz_t - C_t' u \\ z(t_1) = a \in H \end{cases}$$

$$(3.3) \qquad \begin{cases} J(u) = \langle iz(o), P_o i^* iz(o) \rangle_H + \displaystyle\int_0^{t_1} \left[\langle iz_t, E_t iz_t \rangle_H + \langle u_t, Q_t u_t \rangle_K \right] dt \\ \operatorname{Inf}\{J(u) \mid u \in \mathcal{L}_K^2([0,T])\} \end{cases}$$

where the "prime" designates the dual operators (for example $\mathcal{A}_t' \in \mathcal{L}(H,D')$).
A solution of (3.2) is an application $z \in \mathcal{L}_D^2([0,t_1])$ such that $i'iz$ is differentiable in D', verifying (3.2) and such that $\dfrac{di'iz}{dt} \in \mathcal{L}_{D'}^2([0,t_1])$.

We suppose first that the coefficients A, B, C, D are such that there is a unique solution for (3.2). (It is a perturbation hypothesis on A). Then, if P is any solution of the equation (3.1), J is written as :

$$(3.4) \qquad J(u) = \langle iz(t_1), P_{t_1} i^* iz(t_1) \rangle_H + \int_0^{t_1} \langle Q_t \left[Q_t^{-1} C_t P_t^* iz_t + u_t \right], Q_t^{-1} C_t P_t^* iz_t + u_t \rangle_K dt$$

The optimal control u_{op} must then satisfy :

$$u_{op}(t) = -Q_t^{-1} C_t P_t^* z_{op}(t)$$

where z_{op} is the solution of the equation :

$$(3.5) \qquad \begin{cases} \dfrac{di^* iz_t}{dt} = (-\mathcal{A}_t^* + C_t^* Q_t^{-1} C_t P_t^*) iz_t \\ iz(t_1) = a \in H \end{cases}$$

Now, we deduce from the uniqueness of the optimal control (J is strictly convex) and from the uniqueness of the homogeneous equation associated with (3.2),

(u = o), that we have, if P_1 and P_2 are two solutions of (3.1) :

$$\forall t \in [0,T] \qquad C_t P_1^*(t) = C_t P_2^*(t)$$

This is sufficient for solving the filtering problem (see equation 2.10.1).

CONCLUSION

The problem of the uniqueness of the solution for (3.1), and so of completely solving the filtering problem, is reduced to a perturbation problem of linear operators. We can show that this problem is solved positively in the two following cases :

1°) If $\forall t \quad C_t \in \mathcal{L}(H,K)$

Thus, we find again, by a different method, the results of [2], [3] etc... (Note that here we do not impose the independence of state and observation noises. We remark that, in the latter case, hypothesis (iii) is fulfilled in a trivial manner).

2°) If the Green kernel satisfies $G(t,s)H \subset D$ and $G(t,s) \in \mathcal{L}(H,D)$ and if the applications $C.$ and $Q.$ from $[0,T]$ to $\mathcal{L}(D,K)$ and K respectively are Hölder continuous.

REFERENCES

[1] Balakrishnan, A.V. (1972). SIAM J. Control, Vol. 10, n° 4, 754-766.

[2] Bensoussan, A. (1971). Filtrage optimal des systèmes linéaires. Dunod.

[3] Curtain, R.F. (1975). SIAM J. Control, Vol. 13, n° 1, 89-104.

[4] Curtain, R.F., and Falb, P. (1971). Journal of Differential Equations, 10, 412-430.

[5] Curtain, R.F., and Pritchard, A.J. (1974). Journal of Mathematical Analysis and Applications, 47, 43-57.

[6] Delfour, M.C., and Mitter, S.K. (1972). SIAM J. Control, Vol. 10, n° 2, 298-328.

[7] Fujisaki, M., Kallianpur, G., and Kunita, H. (1972). Osaka J. Math., 9, 19-40.

[8] Metivier, M. (1974). Theory of Probability and Applications, XIX, 577-606.

[9] Metivier, M., and Pistone, G. (1975). Z. Wahrscheinlichkeitstheorie Verw. Gebiete, 33, 1-18.

[10] Mitter, S.K., and Vinter, R. (1974). Internat. Symp. Control Theory Numerical Methods and Computer Systems Modelling, Iria Rocquencourt, France.

[11] Ouvrard, J.Y. (1973). Ann. Inst. Henri Poincaré, Vol. IX, n° 4, 351-368.

[12] Ouvrard, J.Y. (1975). Z. Wahrscheinlichkeitstheorie Verw Gebiete, 33, 195-208.

Recent Developments in Statistics
J.R.Barra et al., editors

CHARACTERIZATION OF THE DENSITY OF THE CONDITIONAL LAW IN THE FILTERING OF A DIFFUSION WITH BOUNDARY

E. Pardoux
CNRS *

We show, using the theory developped in [5], that the solution of a certain stochastic partial differential equation gives the density of the conditional law in the filtering problem of certain diffusion processes with inelastic boundary.

§.1. DEFINITION OF A DIFFUSION PROCESS WITH INELASTIC BOUNDARY.

We follow the work of Stroock and Varadhan [7]. Let $\mathcal{O}$ be an open subset of R^N, defined by :

$$\mathcal{O} = \{x \; ; \; \Phi(x) > 0\} \text{ , where :}$$

(1.1) $\Phi \in C_b^2(\mathbb{R}^N)$ [bounded function with continuous and bounded derivatives up to second order]

(1.2) $\Gamma = \partial\mathcal{O} = \{x \; ; \; \Phi(x) = 0\}$

(1.3) $\forall \, x \in \Gamma \; , \; |\nabla \Phi(x)| = 1$

Define $a_{ij} \in C_b(\mathbb{R}_+ \times \bar{\mathcal{O}})$, $i,j=1,\dots,N$, subject to :

(1.4) $a_{ij} = a_{ji}$

(1.5) $\dfrac{\partial a_{ij}}{\partial x_i} \in L^\infty(\mathbb{R}_+ \times \mathcal{O})$, $i,j=1,\dots,N$

---

* IRIA-LABORIA - Domaine de Voluceau - Rocquencourt - 78150 LE CHESNAY (France).

(1.6) $$\sum_{i,j=1}^{N} a_{ij}(t,x)\,\xi_i\,\xi_j \geq \alpha|\xi|^2 \quad \forall\,(t,x) \in \mathbb{R}_+ \times \mathcal{O}, \quad \alpha > 0$$

(1.7) $\gamma(t,x) = [a_{ij}(t,x)]\,\nabla\,\Phi(x)$ is a locally Lipchitz function of (t,x) on $\mathbb{R}_+ \times \Gamma$.

Define $b_i \in B(R_+ \times \mathcal{O})$, $i=1,\dots,N$ [Borel measurable and bounded functions]. And let :

$$L_t = \frac{1}{2}\sum_{i,j=1}^{N} a_{ij}(t,x)\,\frac{\partial^2}{\partial x_i \partial x_j} + \sum_{i=1}^{N} b_i(t,x)\,\frac{\partial}{\partial x_i}$$

$$\Omega_1 = C(\mathbb{R}_+ ; \mathbb{R}^N)$$

$$\mathcal{B}_1^s = \sigma\,\{\omega_1(t)\ ;\ t \geq s\}$$

Define $x_1(t,\omega_1) = \omega_1(t)$. The following results are proved in [7] :

Proposition 1.1 : $\forall\,(s,x) \in \mathbb{R}_+ \times \bar{\mathcal{O}}$, $\exists$ a unique probability measure P_{sx} on $(\Omega_1, \mathcal{B}_1^s)$, which satisfies :

(i) $P_{sx}(x_1(s) = x) = 1$

(ii) $P_{sx}(x_1(t) \in \bar{\mathcal{O}}) = 1$, $\forall\, t \geq s$

(iii) $\forall\, f \in C_0^{1,2}(\mathbb{R}_+ \times \mathbb{R}^N)$ s.t. $\{\gamma(t,x).\nabla_x f(t,x) \geq 0,\ \forall\,(t,x) \in \mathbb{R}_+ \times \Gamma\}$,

$$f(t,x_1(t)) - \int_s^t \mathbb{1}_{\mathcal{O}}(f'_\sigma + L_\sigma f)(\sigma,x_1(\sigma))\,d\sigma$$

is a P_{sx} sub-martingale.

[$C_0^{1,2}(R_+ \times R^N)$ denotes the space of real valued functions which are continuous, as well as their derivatives, up to first order in t and second order in x, with compact support in $\mathbb{R}_+ \times \mathbb{R}^N$].

∎

Proposition 1.2. : $\forall\, s \in \mathbb{R}_+$ $\exists$ a unique continuous non decreasing and non anticipating process $\xi_s(t)$, $t \geq s$, such that :

(i) $\xi_s(s) = 0$

(ii) $\forall\, f \in C_0^{1,2}(\mathbb{R}_+ \times \mathbb{R}^N)$,

$$f(t,x_1(t)) - \int_s^t \mathbb{1}_{\mathcal{O}}(f'_\sigma + L_\sigma f)(\sigma,x_1(\sigma))\,d\sigma - \int_s^t \langle\nabla_x f,\gamma\rangle\, d\xi_s(\sigma)$$

is a P_{sx} martingale.

∎

Remark 1.1. In [7], hypothesis (1.5) is not assumed to hold, and $\gamma(t,x)$ is not supposed to be given by (1.7). These hypotheses will be necessary in the sequel.

∎

§.2 A FUNDAMENTAL RESULT.

We express the semi-group associated with the diffusion defined in §.1. Let E_{sx} denqte the expectation with respect to the probability measure P_{sx}.

Theorem 2.1. $\forall\ t > s,\quad \forall\ f \in \mathbb{B}(\bar{\mathcal{O}})$.

(2.1) $$E_{sx}\ f(x_1(t)) = \bar{u}(s,x)$$

where $\bar{u}$ denotes the unique solution of equation :

(2.2) $$\begin{cases} \frac{\partial u}{\partial \sigma}(\sigma,x) + L_\sigma u(\sigma,x) = 0\ , & \forall(\sigma,x) \in\ ]0,t[\ \times \mathcal{O} \\ \gamma(\sigma,x).\ \nabla\ u(\sigma,x) = 0\ , & \forall(\sigma,x) \in\ ]0,t[\ \times \Gamma \\ u(t) = f & \end{cases}$$

Remark 2.1. Theorem 2.1 holds true without hypothesis (1.5), and also when $\gamma(t,x)$ is not given by (1.7), but satisfies the conditions given in [7].

∎

Sketch of the proof : Applying the results in [4] to equation (2.2), we get :

$$\bar{u} \in C^{0,1}\ ([0,t] \times \bar{\mathcal{O}})$$

Using a regularizing technique , we can apply Proposition 1.2 to some $\bar{u}_n$. Passing to the limit, we get (2.1).

∎

Let us rewrite equation (2.2). Define :

$$H^1(\mathcal{O}) = \{\ u \in L^2(\mathcal{O})\ ;\ \frac{\partial u}{\partial x_i} \in L^2(\mathcal{O}),\quad i=1,\dots,N\}$$

$A(\sigma) \in \mathcal{L}(H^1(\mathcal{O}),\ (H^1(\mathcal{O}))')$ as :

$$\langle\ A(\sigma)u,v\ \rangle = \tfrac{1}{2} \sum_{i,j=1}^{N} \int_{\mathcal{O}} a_{ij}(\sigma,x)\frac{\partial u}{\partial x_i}(x)\ \frac{\partial v}{\partial x_j}(x)\ dx\ + $$
$$+ \sum_{i=1}^{N} \int_{\mathcal{O}} a_i(\sigma,x)u(x)\frac{\partial v}{\partial x_i}(x)\ dx$$

where $a_i = \frac{1}{2} \sum_{j=1}^{N} \frac{\partial a_{ij}}{\partial x_j} - b_i$

Then equation (2.2) is equivalent to :

$$(2.3)\qquad \begin{cases} -\frac{d}{d\sigma} u(\sigma) + A^*(\sigma)\, u(\sigma) = 0 \\ u(t) = f \end{cases}$$

§.3 STATEMENT OF THE FILTERING PROBLEM.

a) The signal process :

Let μ be a probability measure on $\mathcal{O}$, with a density p_o with respect to Lebesgue measure. We assume :

$$(3.1)\qquad p_o \in L^2(\mathcal{O})$$

Let P_μ be the probability measure on $(\Omega_1, \mathcal{B}_1^o)$ defined by :

$$P_\mu(x_1(t) \in B) = \int_{\mathcal{O}} P_{ox}(x_1(t) \in B)\, p_o(x)\, dx$$

where B is any Borel subset of R^N.

The signal process is the canonical process $x_1(t)$ of the space $(\Omega_1, \mathcal{B}_1^o)$, equipped with the law P_μ.

b) The observation process :

Let $\Omega_2 = C(\mathbb{R}_+ ; \mathbb{R}^d)$ and $\mathcal{B}_2 = \sigma\{\omega_2(t),\ t \geq 0\}$. Let P_2 be the Wiener measure on $(\Omega_2, \mathcal{B}_2)$, and $x_2(t)$ the canonical process of Ω_2.

We suppose that the observation $Z(t)$ is given by :

$$Z(t) = \int_0^t h(s, x_1(s))ds + x_2(t)$$

where $h \in \mathbb{B}(\mathbb{R}_+ \times \mathbb{R}^N, \mathbb{R}^d)$.

Define :

$$(\Omega, \mathcal{B}, P) = (\Omega_1 \times \Omega_2, \mathcal{B}_1^o \otimes \mathcal{B}_2, P_\mu \otimes P_2)$$

$$\mathcal{F}_t = \sigma\{Z(s),\ s \leq t\}$$

Our problem is to characterize the law of $x_1(t)$, conditioned by $\mathcal{F}_t$.

We know from [1] that the innovation process :

$$w(t) = Z(t) - \int_0^t E^{\mathcal{F}_s}\, [h(s, x_1(s))]\, ds$$

is a $\mathcal{F}_t$ - P Wiener process.

Let us fix $T > 0$. From Girsanov's theorem [2], $\exists$ a probability law $\tilde{P}$ on $(\Omega,\mathcal{F}_T)$ such that $(z(t))_{t \in [0,T]}$ is a $\mathcal{F}_t$ - $\tilde{P}$ Wiener process.
In all the sequel, we will work with the probability space $(\Omega,\mathcal{F}_T,\tilde{P})$, and the filtration $(\mathcal{F}_t)_{t \in [0,T]}$.

§.4 STUDY OF A STOCHASTIC PARTIAL DIFFERENTIAL EQUATION.

Define $B(t) \in \mathcal{L}(L^2(\mathcal{O}),(L^2(\mathcal{O}))^d)$ as :

$$[B(t)u](x) = h(t,x)\ u(x)$$

We consider the stochastic PDE :

$$(4.1)\quad \begin{cases} du(t) + A(t)u(t)dt = \sum_{i=1}^{d} B_i(t)u(t)dZ_i(t) \\ u(0) = p_o. \end{cases}$$

Using the results and techniques of [5], we prove the following :

Theorem 4.1. Equation (4.1) has a unique solution :

$$u \in L^2(\Omega \times]0,T[\ ;\ H^1(\mathcal{O})) \cap L^2(\Omega\ ;\ C([0,T];\ L^2(\mathcal{O})))$$

where $u(t)$ is a previsible process with values in $L^2(\mathcal{O})$.

■

Theorem 4.2. The solution u of (4.1) satisfies :

(i) $u(t,x) \geq 0$ a.e. in $\mathcal{O}$, $\forall\ t \in [0,T]$

(ii) $\left[\sup_{t \in [0,T]} \| u(t)\|_{L^1(\mathcal{O})}\right] \in L^1(\Omega)$

(iii) $M(t) = \int_{\mathcal{O}} u(t,x)\,dx$ is a continuous square-integrable martingale, with value in $\mathbb{R}_+ - \{0\}$.

■

Equation (4.1) can be heuristically interpreted in terms of [under suitable hypotheses, this interpretation can be justified] :

$$(4.2)\quad \begin{cases} du(t,x) - \frac{1}{2}\sum_{i,j=1}^{N} \frac{\partial}{\partial x_j}(a_{ij}(t,x)\frac{\partial u}{\partial x_i}(t,x))dt - \sum_{i=1}^{N} \frac{\partial}{\partial x_i}(a_i(t,x)u(t,x))dt \\ = \sum_{i=1}^{d} h_i(t,x)u(t,x)dZ_i(t),\ (t,x) \in]0,T[\times \mathcal{O} \\ \frac{1}{2}\gamma(t,x).\nabla u(t,x) + \rho(t,x)u(t,x) = 0\ ,\quad (t,x) \in]0,T[\times \Gamma \\ u(0,x) = p_o(x) \end{cases}$$

where $\rho(t,x) = a(t,x).\ \nabla\ \Phi(x).$, $a = (a_1,\dots,a_N)$

§.5 SOLUTION OF THE FILTERING PROBLEM.

From equations (2.3) and (4.1), one easily gets :

$$(5.1)\qquad (u(t),\ \bar{u}(t)) = (p_o,\ \bar{u}(0)) + \sum_{i=1}^{d}\int_0^t (u(s),\ B_i(s)\bar{u}(s))dZ_i(s)$$

Define $\Pi_{s,t}$ as the non homogeneous two-parameter semi-group associated with $x_1(t)$:

$$[\Pi_{s,t}\ f]\ (x) = E_{sx}\ f(x_1(t))$$

Define Λ_t as the process with values in the space of measures on $\mathcal{O}$ with density $u(t,x)$:

$$\Lambda_t\ f = \int_{\mathcal{O}} u(t,x)\ f(x)\ dx$$

Then we may rewrite (5.1) as :

$$(5.2)\qquad \Lambda_t\ f = \Lambda_o(\Pi_{o,t}\ f) + \sum_{i=1}^{d}\int_0^t \Lambda_s(h_i(s)\ \Pi_{s,t}\ f)\ dZ_i(s)$$

It is easy to show, using the technique of [3] or [6], that equation (5.2) has a unique solution, which is the "unnormalised conditional law". In other words, $p(t,x) = (\int_{\mathcal{O}} u(t,x)dx)^{-1}\ u(t,x)$ is the density of the conditional law we are looking for.

Remark 5.1 : In the numerical applications as well as in the theory, equation (4.1) is easier to use than the equation satisfied by p, because it is linear. Moreover, the "noise" in equation (4.1) is the observation, i.e. what we know in the applications.

Remark 5.2 : Our results apply in particular to the case $\mathcal{O} = \mathbb{R}^N$, i.e. $x_1(t)$ is a usual diffusion process.

BIBLIOGRAPHY.

[1] M. Fujisaki - G. Kallianpur - H. Kunita, Stochastic differential equation for the non linear filtering problem. Osaka J. Math. 9 (1972), 19-40.

[2] I.V. Girsanov, On transforming a certain class of Stochastic processes by absolutely continuous substitution of measures. Theor. Prob. Appl. 5 (1960), 285-301.

[3] H. Kunita, Asymptotic behavior of the non-linear filtering errors of Markov processes. Th. Mult. Anal. 1 (1972), 365-393.

[4] Ladyzenskaja - Solonnikiv - Ural'ceva, Linear and quasi-linear Equations of Parabolic type. AMS Translation of Math. Monographs, 23 (1968).

[5] E. Pardoux, Equations aux dérivées partielles stochastiques non linéaires monotones. Thèse U. Paris XI (1975).

[6] J. Szpirglas, Sur des équations différentielles stochastiques intervenant dans le filtrage non-linéaire. This Meeting.

[7] D.W. Stroock - S.R.S. Varadhan, Diffusion processes with boundary conditions. Comm. Pure and Applied Math. 24 (1971), 147-225.

Recent Developments in Statistics
J.R.Barra et al., editors

A GOODNESS OF FIT TEST FOR FAMILIES OF RANDOM VARIABLES DEPENDING ON TWO PARAMETERS WITH CENSORED DATA (')

Fortunato Pesarin
Institute of Statistics
University of Padua (Italy)

This work concernes a goodness of fit test, based on censored data, for families of random variables depending on two parameters. The sampling distribution of the test is asymptotically invariant with regard to the parameters and the truncation level.

INTRODUCTION

In many practical problems (e.g. reliability, survival curves, etc.) there is the necessity to verify statistically, by use of censored data, a so called "functional hypothesis" concerning the observable phenomena under consideration. The functional hypothesis which we deal with can be defined in the following way: let $\mathcal{X}$ denote a random variable (r.v.) member of a specified family $\mathcal{F}$ of r.v. including all the r.v. with cumulative distribution function (c.d.f.) of the type $F(x;a,b)$, $(a,b)\in\pi$, where π is the parametric space, supposed real. A statement which implies that the observable r.v. describing the phenomena is a r.v. $\mathcal{X}\in\mathcal{F}$, is a (properly nonparametric) functional hypothesis.

In this paper we give the sufficient conditions for a consistent test with censored samples, which has an asymptotically invariant sampling distribution over π and over the truncation level, supposed unknown.

For want of space the proof of the theorem here recalled is omitted: the reader can find it in a more complete paper to appear in

---

(') This reaserch was supported by the C.N.R. grant n° 74.00544.10 .

"Metron".

PRELIMINARY REMARKS

Let us suppose given a truncation value $x_\circ$ so that we can observe the $x_\circ$-censored r.v. $(\mathscr{X}|\mathscr{X}\leq x_\circ)$, denoted by ${}_c\mathscr{X}$. This r.v. has c.d.f. ${}_cG(x) = G(x)/G(x_\circ)$ if $x\leq x_\circ$ and 1 elsewhere, $G(x)$ being the c.d.f. of the r.v. $\mathscr{X}$ (complete) under consideration. Let ${}_c\mathscr{F}$ denote the class of all the r.v. such that their c.d.f. is of the type ${}_cF(x;a,b) = F(x;a,b)/F(x_\circ;a,b)$, $(a,b)\in\pi$,if $x\leq x_\circ$ and 1 elsewhere, where F is of a type defined above.

We remark that ${}_c\mathscr{F}\supset\{(\mathscr{X}\,|\,\mathscr{X}\leq x_\circ),\mathscr{X}\in\mathscr{F}\}$.

Let us suppose drawn from the r.v. $\mathscr{X}$ (complete) a random sample of n elements whose $n_\circ$ only are observable and satisfy the relation $\mathscr{X}\leq x_\circ$; in other words we suppose that the random sample $\mathscr{X}_1,\ldots,\mathscr{X}_{n_\circ}$ is drawn from the $x_\circ$-censored r.v. ${}_c\mathscr{X}$, while the remaining $n-n_\circ$ elements are supposed not observable.

We know that, denoting by $\mathscr{X}'_1\leq\mathscr{X}'_2\leq\ldots\leq\mathscr{X}'_{n_\circ}\leq x_\circ$ the ordered random samples, the sampling c.d.f., defined by the relation:

$$ {}_cF_n(x) = \begin{cases} 0 & -\infty < x < \mathscr{X}'_1 \\ i/n_\circ & \mathscr{X}'_i \leq x < \mathscr{X}'_{i+1} \\ 1 & \mathscr{X}'_{n_\circ}\leq x<+\infty \end{cases} $$

converges almost surely, if $n_\circ$ diverges,to the true c.d.f. ${}_cG(x)$ of the $x_\circ$-censored r.v. .

We remark that ${}_cF_n$ coincides, exept for the known factor $n_\circ/n$, with the analogous F_n of the complete sample over the whole interval $-\infty<x\leq x_\circ$.

By means of the random sample we can test the null hypothesis:

$$ H_\circ = \{\ {}_c\mathscr{X}\in{}_c\mathscr{F}\ \}, $$

or equivalently :

$$ H_\circ = \{\ \exists(a,b)\in\pi\ :\ {}_cG(x) = {}_cF(x;a,b)\ \} $$

against the alternative:

$$ H_1 = \{\ {}_c\mathscr{X}\in\Theta-{}_c\mathscr{F}\ \}, $$

where the class Θ will be defined below.

CONDITIONS AND TEST

Condition 1. Let μ and φ be two functions, continuous with probability one with respect to every member of $\mathscr{F}$, such that:

$$1) \qquad \mu\{F(x;a,b)\} = a + b\,\varphi(x)\,, \qquad -\infty < x < +\infty\,,$$

i.e. there exist two functions by which we can linearize the c.d.f. of every r.v. of the family $\mathscr{F}$.

We note that μ and φ linearize also the c.d.f. of every r.v. of the class ${}_c\mathscr{F}$.

We observe that, for instance, Normal, Logistic, Gumbel, Weibull, etc. are families of r.v. whose c.d.f. can be linearized according to 1).

Condition 2. Let the members of $\mathscr{F}$ satisfy the following relations:

$$2) \quad \begin{cases} 0 < \operatorname{var}\{\varphi(\mathscr{X})\} = V(a,b) < +\infty \text{ and } V \text{ continuous over } \pi \\ 0 < \operatorname{var}\{\mu[F(\mathscr{X};a,b)]\} < +\infty\,, \ \forall\,(a,b) \in \pi\,. \end{cases}$$

We remark that also every member of ${}_c\mathscr{F}$ satisfies 2).

We now introduce the following

Definition: let Θ be the class of r.v. for which the c.d.f. $G(x)$ satisfies the relation: $\operatorname{var}\{\mu[G(\mathscr{X})]\} < +\infty$.

We note that if μ is bounded over the closed interval $[0, 1]$, then Θ is the class of all the r.v..

Condition 3. For every member of $\mathscr{F}$ let the distribution of $(\mathscr{X}-\mathscr{Y}|\mathscr{Y})$, where $\mathscr{Y} = \mu\{{}_cF_n({}_c\mathscr{X})\cdot n_0/n\}$ and $\mathscr{X} = \varphi({}_c\mathscr{X})$, be independent of $\mathscr{Y}$, i.e. constant and independent of the truncation level.

We observe that the condition 3 usually is not simple to check, but we think that for the asymptotic behavior of the test, for practical purposes, it is sufficient that this condition is approximately satisfied, in other words we think that the test is robust with regard to this condition.

Theorem. Under the above conditions let us consider the test:

$$\hat{r}^2_{n_\circ} = 1 - \frac{(1-r_n^2)\left(\sum_{i=1}^{n_\circ}\{\varphi(x_i')\}^2 - \{\sum\varphi(x_i')\}^2/n_\circ\right)}{n_\circ\, V(\hat{a}_\circ,\hat{b}_\circ)} \quad ,$$

where: i) x_i', $i=1,\dots,n_\circ$, are the ordered sampling elements ; ii) r_n^2 is the squared correlation coefficient between the observed random quantities $\{z_i = \varphi(x_i')\ ,\ y_i = \mu\{_cF_n(x_i')\cdot n_\circ/n\}$, $i=1,\dots,n_\circ$; iii) $(\hat{a}_\circ,\hat{b}_\circ)$ is the linear least square estimate of (a,b) calcu - lated by the linear relation 1) on the observed random quantities (z_i,y_i) . This test is consistent with respect to $H_\circ$ against H_1 and its sampling distribution is asymptotically invariant, under $H_\circ$, with regard to every value of the parameters $(a,b)\in\pi$ and with regard to every truncation level $F(x_\circ;a,b)$.

The proof of this theorem, and other ones recalled by it, is given in [2] . A tabulation of the test sufficient for each practical problem, relative to the Weibullian family of r.v., is given in [3] .

REFERENCES

[1] D.R. Cox- D.W. Hinkley: Theoretical Statistics; Chapman and Hall, (1974).

[2] F. Pesarin : A goodness of fit test for families of random variables depending on two parameters with censored data; application to Weibullian case. (To appear in "Metron").

[3] F. Pesarin-G. Uroda:Controllo statistico di andamento Weibulliano con dati censurati. Atti del IX Convegno dell' AICQ Roma (1976).

Recent Developments in Statistics
J.R.Barra et al., editors

ESTIMATION OF PARAMETERS IN THE A.R.-M.A. MODEL WHEN THE CHARACTERISTIC POLYNOMIAL OF THE M.A. OPERATOR HAS A UNIT ZERO

PHAM-DINH Tuan
Université Scientifique et Médicale de Grenoble
I.R.M.A., B.P. 53
38041 - Grenoble Cêdex - France

We consider the estimation of parameters in the time series model :

$$X(t) = \sum_{j=1}^{q} a_j X(t-j) + \varepsilon(t) - \varepsilon(t-1) - \sum_{j=1}^{p-1} c_j\{\varepsilon(t-j)-\varepsilon(t-j-1)\}$$

where the $\varepsilon(t)$ are independently identically distributed random variables with zero mean and variance σ^2. We compute the log likelihood function of the model, propose and justify an approximation of it and then use the latter to derive estimates of the parameters. These estimates are shown to be asymptotically normal and efficient.

1 - INTRODUCTION

The estimation of parameters in the auto-regressive moving-average (A.R.-M.A.) time series model :

$$X(t) = \sum_{j=1}^{q} a_j X(t-j) + \varepsilon(t) - \sum_{j=1}^{p} b_j \varepsilon(t-j)$$

with $\varepsilon(t)$ being identically independently distributed (i.i.d.) random variables, has been thoroughly studied (Box and Jenkins (1970), Hannan (1970), ...) under the assumption that the polynomials $1 - \Sigma a_j z^j$ and $1 - \Sigma b_j z^j$ have no zero in and on the unit circle. We consider here the estimation of parameters of this model when the polynomial $1 - \Sigma b_j z^j$ does have a unit zero, i.e. :

$$1 - \sum_{j=1}^{p} b_j z^j = (1-z)(1-\sum_{j=1}^{p-1} c_j z^j)$$

Thus we are concerned with the model :

$$X(t) = \sum_{j=1}^{q} a_j X(t-j) + \varepsilon(t) - \varepsilon(t-1) + \sum_{j=1}^{p-1} c_j\{\varepsilon(t-j)-\varepsilon(t-j-1)\} \tag{1.1}$$

Our approach to derive the estimate of the parameters in this model is the following one :

(i) We compute the exact log likelihood function of the model by evaluating the innovation :

$$\zeta_\theta(t) = X(t) - E_\theta\{X(t)|X(1),\dots,X(t-1)\}$$

θ denoting $(a_1,\dots,a_q,c_1,\dots,c_{p-1},\sigma^2)$, so that under the assumption that the observed series is Gaussian, the log likelihood function of the model for a sample of length N is :

$$(1.2)\qquad \mathcal{L}_N(\theta) = -\frac{1}{2}\sum_{t=1}^{N} \mathrm{Log}[2\pi\sigma^2(t)] - \frac{1}{2}\sum_{t=1}^{N} \zeta_\theta^2(t) / \sigma_\theta^2(t)$$

where $\sigma_\theta^2(t)$ is the variance of $\zeta_\theta(t)$.

(ii) We derive an asymptotic approximation L_N of $\mathcal{L}_N$ which satisfies the condition :

$$(A)\qquad \sup_{\theta\in C} ||\mathcal{L}_N^{(i)}(\theta) - L_N^{(i)}(\theta)|| \leq (\mathrm{Log}\ N)\ Z_N , \qquad i = 0,1,2$$

where $\mathcal{L}_N^{(i)}$ and $L_N^{(i)}$, $i = 0,1,2$ denote respectively $\mathcal{L}_N$ and L_N, and the vector of first derivatives and the matrix of second derivatives of $\mathcal{L}_N$ and that of L_N, $||\cdot||$ denotes any vector (matrix) norm, C is any compact set and $\{Z_N\}$ is some sequence of random variables bounded in the L^1-norm.

(iii) We show that the L_N function possesses all classical properties of a log-likelihood function. By (A) the same is true for $\mathcal{L}_N$. Thus there exists a consistent maximum likelihood estimate which is asymptotically normal and efficient, and a consistent approximate maximum likelihood estimate (i.e. that based on the maximization of L_N) with the same properties. Using (A), we can even show that the latter differs from the former by a term of order (Log N)/N.

2 - COMPUTATION OF THE INNOVATION

In this paragraph, we shall omit the subscript θ and we denote by $Z(t|s)$ the orthogonal projection (in the sense of L^2 scalar product) of the random variable $Z(t)$ onto the linear space spanned by $X(1), \ldots, X(s)$.

The computation of the innovation $\zeta(t) = X(t) - X(t|t-1)$ is quite technical and complicated. We give here only the general idea and refer the readers to the paper of Pham-Dinh (1977) for more details. We put :

$$\phi(t) = \varepsilon(t) - \varepsilon(t-1)$$

$$V(t) = {}^T(\phi(t)\ldots\phi(t+2-p)\ \varepsilon(t+1-p))$$

The model equation (1.1) becomes :

$$X(t) = \sum_{j=1}^{q} a_j X(t-j) - {}^Tb\ V(t-1) + \varepsilon(t)$$

where $b = {}^T(1+c_1\ldots1+c_p\ 1)$. Thus :

$$(2.1)\qquad \zeta(t) = X(t) - \sum_{j=1}^{q} a_j X(t-j) + {}^Tb\ V(t-1|t-1), \qquad t > q$$

On the other hand, using (1.1), we get

$$V(t) = B\ V(t-1) + W(t)$$

where

$$B = \begin{pmatrix} c_1 & c_2 & \cdots & c_{p-1} & 0 \\ 1 & 0 & \cdots & 0 & 0 \\ 0 & 1 & \cdots & . & . \\ . & . & \cdots & . & . \\ 0 & 0 & \cdots & 1 & 1 \end{pmatrix} \quad ; \qquad W(t) = \begin{pmatrix} 1 \\ 0 \\ . \\ . \\ . \\ 0 \end{pmatrix} [X(t) - \sum_{j=1}^{q} a_j X(t-j)]$$

The above relation permits us to derive a recurrent equation for $V(t|t)$ by a method similar to that of Kalmal's filtering. Denoting by $Q(t)$ the covariance matrix of $V(t) - V(t|t)$ and $\sigma^2(t)$ the variance of $\zeta(t)$ and using the fact that $\zeta(t) = \varepsilon(t) - {}^Tb[V(t-1)-V(t-1)|t-1]$, we get :

$$\begin{aligned} V(t|t) &= B\,V(t-1|t-1) + B\,K(t)\,\zeta(t) + W(t)\,, \qquad t > q \\ K(t) &= -Q(t-1)\,b/\sigma^2(t) \\ \sigma^2(t) &= \sigma^2 + {}^Tb\,Q(t-1)b \end{aligned}$$

Now, by (2.1), $\zeta(t) = {}^Te\,W(t) + {}^Tb\,V(t-1\ t-1)$ where $e = (1\ 1\ \ldots\ 1)$, so that :

$$(2.2) \qquad V(t|t) = B[I+K(t)\,{}^Tb]\,V(t-1|t-1) + [I+BK(t)\,{}^Te]\,W(t)\,, \qquad t > q$$

On the other hand, we have :

$$V(t) - V(t|t) = B\{V(t-1)-V(t-1|t-1)-K(t)\,\zeta(t)\}$$

where $K(t)\,\zeta(t)$ is precisely the orthogonal projection of $V(t-1) - V(t-1|t-1)$ onto the linear space spanned by $\zeta(t)$. Thus :

$$(2.3) \qquad Q(t) = B\{Q(t-1)-K(t)\,\sigma^2(t)\,{}^TK(t)\}\,{}^TB \qquad t > q$$

Using the fact that $K(t) = -Q(t-1)\,b/\sigma^2(t)$, we get $Q(t)\,{}^TB^{-1}\,Q^{-1}(t-1) = B\{I-K(t)\,{}^Tb\}$, so that by remarking that ${}^Tb = {}^TeB$, the equation (2.2) becomes :

$$V(t|t) = Q(t)\,{}^TB^{-1}\,Q^{-1}(t-1)\,\{V(t-1|t-1)+B^{-1}\,W(t)\}, \qquad t > q$$

which has the solution :

$$V(t|t) = Q(t)\,\{\sum_{j=q}^{t-1} {}^TB^{j-t}\,Q^{-1}(j)\,B^{-1}\,W(j+1) + {}^TB^{q-t}\,Q^{-1}(q)\,V(q|q)\},\ t \geq q$$

Since $\zeta(t) = {}^Te\{W(t)+BV(t-1|t-1)\}$, we get :

$$\zeta(t) = {}^Te\,B\,Q(t-1)\,\{\sum_{j=q+1}^{t} {}^TB^{j-t}\,Q^{-1}(j-1)\,B^{-1}\,W(j) + {}^TB^{q-t}\,Q^{-1}(q)\,V(q|q)\}$$

for $t > q$. Now, we can solve the recurrent equation (2.3) which gives the solution : $Q(t) = \sigma^2\,B^t\,\mu^{-1}(t)\,{}^TB^t$ where :

$$(2.4) \qquad \mu(t) = \mu(q) + \sum_{j=q+1}^{t} {}^TB^j\,e\,{}^Te\,B^j\,, \qquad t \geq q$$

Thus :

$$\zeta(t) = {}^Te\,B^t\,\mu^{-1}(t-1)\{\sum_{j=q+1}^{t} \mu(j-1)\,B^{-j}\,W(j) + \mu(q)\,B^{-q}\,V(q|q)\},\quad t \geq q$$

Using the fact that

$$\mu^{-1}(t-1)\ \mu(j-1) = I - \mu^{-1}(t-1) \sum_{k=j}^{t-1} {}^{T}B^{k}\ e^{T}\ e\ B^{k}$$

we get :

$$(2.5)\qquad \zeta(t) = \hat{\varepsilon}(t) - {}^{T}e\ B^{t}\ \mu^{-1}(t-1)\{\sum_{k=q+1}^{t-1} {}^{T}B^{k}\ e\ \varepsilon(k) - \mu(q)\ B^{-q}\ V(q|q)\},\ t \geq q$$

where :

$$(2.6)\qquad \hat{\varepsilon}(t) = \sum_{j=q+1}^{t} {}^{T}e\ B^{t-j}\ W(j)$$

The variance $\sigma^2(t)$ of $\zeta(t)$, on the other hand, is easily seen to be :

$$(2.7)\qquad \sigma^2(t) = \{1 + {}^{T}e\ B^{t}\ \mu^{-1}(t-1)\ B^{t}\ e\}\ \sigma^2$$

The term $\hat{\varepsilon}(t)$ has a simple interpretation. Indeed, it can be verified that $\hat{\varepsilon}(t)$, $t = \ldots, -1, 0, 1, \ldots$ is the solution of the recurrent equation :

$$\hat{\varepsilon}(t) - \hat{\varepsilon}(t-1) = \sum_{j=1}^{p-1} c_j\{\hat{\varepsilon}(t-j)-\hat{\varepsilon}(t-j-1)\} + X(t) - \sum_{j=1}^{q} a_j\ X(t-j),\quad t > q$$

with intitial conditions $\hat{\varepsilon}(t) = 0,\ \ t \leq q$.

3 - APPROXIMATION OF THE LOG LIKELIHOOD FUNCTION

We shall denote by Θ, the open set of R^{p+q} of values of $\theta = (a_1,\ldots,a_q,c\ ,\ldots,c_{p-1},\ \sigma^2)$ such that $\sigma^2 > 0$ and that the polynomials $1 - \Sigma\ a_j\ z^j$ and $1 - \Sigma\ c_j\ z^j$ have no common zero and no zero in the closed unit circle of the complex plane. We shall use the subscript θ, when necessary, to indicate the dependence of $\zeta(t)$, $\sigma^2(t)$, B, $V(q|q)$, $\mu(t)$, $\hat{\varepsilon}(t)$ and $\hat{\phi}(t)$ introduced in §2, on θ. We are interested in deriving an asymptotic approximation of the log likelihood function. In the Gaussian case, the exact log likelihood function for a sample of length $N > q$, is :

$$(3.1)\qquad L_N(\theta) = -\frac{1}{2} \sum_{t=q+1}^{N} \text{Log}[2\pi\sigma_\theta^2(t)] - \frac{1}{2} \sum_{t=q+1}^{N} \zeta_\theta^2(t)/\sigma^2(t)$$

$$- \frac{1}{2}\ \text{Log}[\det 2\pi\ \Gamma_q(\theta)] - \frac{1}{2}\ {}^{T}X_q\ \Gamma_q^{-1}(\theta)\ X_q$$

where Γ_q is the covariance matrix of the random vector $X_q = {}^{T}(X(1)\ldots X(q))$ and where $\zeta_\theta(t)$ and $\sigma_\theta^2(t)$ are given by (2.1)-(2.7). By these formulae, it is clear that in order to obtain an asymptotic approximation of L_N, we should study the asymptotic behaviour of ${}^{T}e\ B^{t}\ \mu^{-1}(t-1)\ {}^{T}B^{t}e$, ${}^{T}e\ B^{t}\ \mu^{-1}(t-1)\ {}^{T}B^{t}e$, $\hat{\varepsilon}(t)$ and :

$$\sum_{k=q+1}^{t-1} {}^{T}e\ B^{t}\ \mu^{-1}(t-1)\ {}^{T}B^{k}\ e\ \hat{\varepsilon}(k)$$

We find it convenient to use the following definition :

"Let $U_N(.)$ be a sequence of random functions on Θ, every sample function of which is twice continuously differentiable and $U_N^{(i)}(.)$, $i = 0,1,2$ denote the vector whose components are i-th derivatives of $U_N(.)$, we say that the sequence $U_N(.)$ is $O(\phi(N))$

in L^p norm if for any compact E of Θ, the sequences of random variables :

$$\left|\sup_{\theta \in E} \| U_N^{(i)}(\theta) \| \right| / \phi(N) \qquad N \geq 1, \quad i = 0,1,2$$

are bounded in L^p norm. In particular, the sequence of (non random) functions $U_N(.)$ is said to be $O(\phi(N))$ if the above sequences are bounded".

It is shown in Pham-Dinh (1977), that the sequences of functions $\theta \rightsquigarrow {}^Te\, B^t\, \mu^{-1}(t-1)$, $t > q$ and $\theta \rightsquigarrow {}^Te\, B^t\, \mu^{-1}(t-1)\, {}^TB^t e$, $t > q$ are O(1) and the sequences of random functions θ $\hat{\varepsilon}(t)$, $t > q$ and

$$\theta \rightsquigarrow \sum_{k=q+1}^{t-1} [{}^Te\, B^t\, \mu^{-1}(t-1)\, {}^TB^k e - t^{-1}]\, \hat{\varepsilon}(k), \quad t > q$$

are respectively O(1) and $O(t^{-1})$ in L^2 norm. The proofs of these results are quite complicated. The main idea is to block diagonalise the matrix B as follows :

$$B = P \begin{pmatrix} C & 0 \\ 0 & 1 \end{pmatrix} P^{-1}$$

where the matrix C, given by :

$$C = \begin{pmatrix} c_1 & c_2 & \dots & c_{p-2} & c_{p-1} \\ 1 & 0 & \dots & 0 & 0 \\ 0 & 1 & \dots & 0 & 0 \\ . & . & \dots & . & . \\ 0 & 0 & \dots & 1 & 0 \end{pmatrix}$$

has the eigenvalues of modulus strictly inferior to 1 for all $\theta \in \Theta$, and the matrix P is given by :

$$P = \begin{pmatrix} I & 0 \\ -{}^T\alpha & 1 \end{pmatrix} \quad , \quad {}^T\alpha = (0 \dots 0\ 1)\, (I-C)^{-1}$$

Now, it is clear that the expression ${}^Te\, B^t\, \mu^{-1}(t-1)\, {}^TB^k e$ is unchanged if B, e, $\mu(t-1)$ are replaced respectively by $\tilde{B} = P^{-1} B P$, $\tilde{e} = {}^TPe$, $\tilde{\mu}(t-1) = {}^TP\mu(t-1)P$. But :

$$\tilde{\mu}(t-1) = \tilde{\mu}(q) + \sum_{j=q+1}^{t-1} {}^T\tilde{B}^j \tilde{e}\, {}^T\tilde{e}\, \tilde{B}^j$$

$$= \tilde{\mu}(q) + \sum_{j=q+1}^{t-1} \begin{pmatrix} {}^TC^j & 0 \\ 0 & 1 \end{pmatrix} \tilde{e}\, {}^T\tilde{e} \begin{pmatrix} C^j & 0 \\ 0 & 1 \end{pmatrix}$$

Since the last component of the vector $\tilde{e}$ is 1, if we partition the matrix $\mu(t-1)$ into :

$$\tilde{\mu}(t-1) = \begin{pmatrix} \tilde{\mu}_1(t-1) & \tilde{\mu}_{12}(t-1) \\ \tilde{\mu}_{21}(t-1) & \tilde{\mu}_2(t-1) \end{pmatrix} \quad ,$$

then $\tilde{\mu}_2(t-1)$ (which is a scalar) is equal to $t-q-1+\tilde{\mu}_2(q)$ and the sequences $\tilde{\mu}_1(t-1)$, $t > q$, $\tilde{\mu}_{12}(t-A)$, $t > q$ and $\tilde{\mu}_{21}(t-1)$, $t > q$ are easily seen to be O(1).

Now let :

$$\tilde{\mu}^{-1}(t-1) = \begin{pmatrix} \{\tilde{\mu}^{-1}(t-1)\}_1 & \{\tilde{\mu}^{-1}(t-1)\}_{12} \\ \{\tilde{\mu}^{-1}(t-1)\}_{21} & \{\tilde{\mu}^{-1}(t-1)\}_2 \end{pmatrix} ,$$

then the above result implies that the sequence $\{\tilde{\mu}^{-1}(t-1)\}_1$, $t > q$ is $0(1)$, the sequences $\{\tilde{\mu}^{-1}(t-1)\}_{12}$, $t > q$, $\{\tilde{\mu}^{-1}(t-1)\}_{21}$, $t > q$ are $0(t^{-1})$ and the sequences $\{\tilde{\mu}^{-1}(t-1)\}_2 - t^{-1}$, $t > q$ are $0(t^{-2})$.

One can then deduce that :

$${}^T\tilde{e}\ \tilde{B}^t\ \tilde{\mu}^{-1}(t-1) = (0(t^{-1}) \qquad t^{-1} + 0(t^{-2}))$$

$${}^T\tilde{e}\ \tilde{B}^t\ \tilde{\mu}^{-1}(t-1)\ {}^T\tilde{B}^k\tilde{e} = t^{-1} + g(t,k)$$

where the sequence of functions $\theta \rightsquigarrow g(t,k)$ is such that :

$$\sum_{k=q+1}^{t} |g(t,k)|, \quad \sum_{k=q+1}^{t} \left|\frac{\partial}{\partial\theta_\alpha} g(t,k)\right| , \quad \sum_{k=q+1}^{t} \left|\frac{\partial^2}{\partial\theta_\alpha\partial\theta_\beta} g\ t,k)\right|$$

are bounded by K t, K being some constant.

The first half of the above result shows that the sequence $\theta \rightsquigarrow {}^Te\ B^t\ \mu^{-1}(t-1)$, $t > q$ and $\theta \rightsquigarrow {}^Te\ B^t\ \mu^{-1}(t-1)\ {}^TB^te$, $t > q$ are $0(1)$ and if the sequence $\theta \rightsquigarrow \hat{\varepsilon}(t), t > q$ is $0(1)$ in L^2 norm, then the second half of the above result shows that the sequence

$$\theta \rightsquigarrow \sum_{k=q+1}^{t-1} \{{}^Te\ B^t\ \mu^{-1}(t-1)\ {}^TB^ke - t^{-1}\}\ \hat{\varepsilon}(k) \ , \quad t > q$$

is $0(t^{-1})$ in L^2 norm. The proof that the sequence $\hat{\varepsilon}(t)$, $t > q$ is $0(1)$ in L^2 norm comes essentially from the fact that $\hat{\phi}(t) = \hat{\varepsilon}(t) - \hat{\varepsilon}(t-1)$ is the solution of the recurrent equation :

$$\hat{\phi}(t) = \sum_{j=1}^{p-1} c_i\ \hat{\phi}(t-j) + X(t) - \sum_{j=1}^{q} a_j\ X(t-j) \qquad t > q$$

and that the sum $X(1) + \ldots + X(t)$ remains bounded in L^2 norm as $t \to \infty$.

The above results suggest the approximations of $\sigma^2_\theta(t)$ by σ^2 and that of $\zeta_\theta(t)$ by :

$$\hat{\zeta}_\theta(t) = \hat{\varepsilon}_\theta(t) - \frac{1}{t} \sum_{k=q+1}^{t-1} \hat{\varepsilon}_\theta(k) \qquad t > q \tag{3.2}$$

which produce the following approximation of $\mathcal{L}_N$:

$$L_N(\theta) = -\frac{N-q}{2} \operatorname{Log}(2\pi\sigma^2) - \frac{1}{2\sigma^2} \sum_{t=q+1}^{N} \cdot\ \hat{\zeta}^2_\sigma(t) \tag{3.3}$$

We can then show (see Pham-Dinh (1977))

Theorem 1 : "Let $\mathcal{L}_N$ and L_N be given by (3.1) and (3.3), then the sequence of random functions $\mathcal{L}_N(.) - L_N(.)$, $N > q$ is $0(\operatorname{Log} N)$ in L^1 norm".

4 - ASYMPTOTIC PROPERTIES OF THE LOG LIKELIHOOD FUNCTION

We now show that the approximate (and thus the exact) log likelihood function possesses, all the classical properties of a log-likelihood function, normely :

Theorem 2 : "Let $L_N^{(i)}(.)$, $i = 1,2$ denote the vector of first derivatives and the matrix of second derivatives of L_N and θ^*, the true value of θ, then as $N \to \infty$:

(i) $N^{-1} L_N^{(1)}(\theta^*)$ tends in probability to 0.

(ii) $N^{-1} L_N^{(2)}(\theta^*)$ tends in probability to

$$K = - \begin{pmatrix} (\sigma^*)^{-2}\Gamma & \\ 0 & \frac{1}{2}(\sigma^*)^{-4} \end{pmatrix}$$

where Γ is the matrix with the general element :

$$\Gamma_{\alpha\beta} = \frac{(\sigma^*)^{-2}}{2\pi} \int_{-\pi}^{\pi} \left| \frac{\partial}{\partial\theta_\alpha} \psi_\theta(e^{i\lambda}) \frac{\partial}{\partial\theta_\beta} \psi_\theta(e^{i\lambda}) \right|_{\theta=\theta^*} |\psi_{\theta^*}(e^{i\lambda})| \quad d\lambda$$

where $\psi_\theta(z) = (1 - \Sigma\, c_j\, z^j)^{-1} (1 - \Sigma\, a_j\, z^j)$.

(iii) If the $\varepsilon(t)$ possess a finite fourth cumulant κ, then $N^{-1/2} L_N^{(1)}(\theta^*)$ tends in distribution to a gaussian random vector with zero mean and covariance matrix :

$$\begin{pmatrix} (\sigma^*)^2\Gamma & 0 \\ 0 & \frac{1}{2}(\sigma^*)^{-4} + \frac{1}{4}\kappa(\sigma^*)^{-8} \end{pmatrix}$$

Proof : A rigourous proof of this theorem is quite long, so we give here the main idea and refer the reader to the paper of the author (1977). It is easily seen that the computation of $N^{-1} L_N^{(1)}(\theta)$, $i = 1,2$ involves the expression of the type :

$$N^{-1} \sum_{t=q+1}^{N} \hat{\zeta}_\theta(t)\, \zeta_\theta^{(i)}(t) \qquad , \quad i = 0,1,2$$

$$N^{-1} \sum_{t=q+1}^{N} \hat{\zeta}_\theta^2(t) \quad \text{and} \quad N^{-1} \sigma_{t=q+1}^{N} \hat{\zeta}_\theta^{(1)}\, {}^T\hat{\zeta}_\theta^{(1)}(t)$$

where $\hat{\zeta}_\theta^{(i)}(t)$, $i = 1,2$ denotes the vector of first derivatives and the matrix of second derivatives of $\hat{\zeta}_\theta(t)$ with respect to θ. Now it can be shown that $\hat{\zeta}_\theta(t)$ can be approximated by :

$$\tilde{\varepsilon}_\theta(t) - \frac{1}{t} \sum_{k=q+1}^{t-1} \tilde{\varepsilon}_\theta(k)$$

where $\tilde{\varepsilon}_\theta(t)$ is the stationary process defined by :

$$\tilde{\varepsilon}_\theta(t) = \sum_{j=o}^{\infty} \omega_j(\theta)\, Y(t-j)$$

$$Y(t) = \sum_{j=o}^{\infty} \pi_j\, X(t-j)$$

where ω_j, $j \geq 0$, π_j, $j \geq 0$ are respectively the coefficient of the Taylor development of $(1 - \Sigma\, a_j^* z^j)^{-1} (1 - \Sigma\, c_j^* z^j)$ and of $\psi_\theta(z) = (1 - \Sigma\, c_j\, z^j)^{-1} (1 - \Sigma\, a_j\, z^j)$; the error sequence $\mu_\theta(t)$, $t > 0$ being $0(t^{-1})$ in L^2 norm. Thus, in considering the asymptotic behaviour of $N^{-1} L_N^{(i)}(\theta^*)$, $i = 1,2$ and $N^{-1/2} L_N^{(1)}(\theta^*)$ one can replace $\hat{\zeta}_{\theta^*}(t)$ and $\hat{\zeta}^{(i)}_{\theta^*}(t)$, $i = 1,2$ by :

$$\varepsilon(t) - \frac{1}{t} \sum_{k=q+1}^{t-1} \varepsilon(k)$$

$$\tilde{\varepsilon}^{(i)}_{\theta^*}(t) - \frac{1}{t} \sum_{k=q+1}^{t-1} \tilde{\varepsilon}^{(i)}_{\theta^*}(k)$$

where $\tilde{\varepsilon}^{(i)}_\theta(t)$, $i = 1,2$ denote the vector of first derivatives and the matrix of second derivatives of $\tilde{\varepsilon}_\theta(t)$ with respect to θ.

Thus we need only to show that the contribution of the last sums of the above expression to $N^{-1} L_N^{(i)}(\theta^*)$, $i = 1,2$ and to $N^{-1/2} L_N^{(1)}(\theta^*)$, tend to 0 as $N \to \infty$, the results then follow from a standard argument similar to that used in the ordinary A.R.-M.A. case.

5 - ESTIMATION OF PARAMETERS

We obtain an estimate of θ by maximizing the approximate likelihood function $L_N(\theta)$ or more frequently by resolving the equation : $L_N^{(1)}(\theta) = 0$. The existence of a approximate maximum likelihood estimate (i.e. verifying the equation $L_N^{(1)}(\theta) = 0$) can be shown by standard arguments (see for ex. Aitchison (1958) or Pham-Dinh (1975)) using theorem 2 and the following result :

Theorem 3 : "There exists a function $g : (0,\infty) \to [0,\infty]$ with $g(r) \downarrow 0$ as $r \downarrow 0$ and a sequence of random variables M_N, $N > q$ bounded in L^1 norm such that :

$$\forall\theta : ||\theta-\theta^*|| \leq r \,;\, N^{-1} || L_N^{(2)}(\theta) - L_N^{(2)}(\theta^*) || \leq g(r)\, M_N"$$

The proof of the above theorem can be found in Pham-Dinh (1977).

By theorem 1, there exist also a consistent maximum likelihood estimate $\tilde{\theta}_N$. This estimate as well as the approximate maximum likelihood estimate $\hat{\theta}_N$ are both asymptotically efficient and normal (a consequence of theorems 2 and 1). Moreover, using theorem 1, one can obtain this result :

<u>Theorem 4</u> : "<u>Let</u> $\hat{\theta}_N$ <u>and</u> $\tilde{\theta}_N$ <u>be defined as above, then the sequence</u> $N(\text{Log } N)^{-1}(\hat{\theta}_N - \tilde{\theta}_N)$ <u>is bounded in probability, in the sense that</u> :

$$\lim_{\lambda\to\infty} \limsup_{N\to\infty} P\{\|N(\text{Log } N)^{-1}(\hat{\theta}_N - \tilde{\theta}_N)\| > \lambda\} = 0"$$

The proof of this theorem is based on the mean value theorem and can be found in Pham-Dinh (1977).

In conclusion, the main results of this paper are theorems 1, 2 and 3. They ensure that we can estimate parameters by a modified Box and Jenkins method (Box & Jenkins (1970)), where the residuals are taken to be $\zeta_\theta(t)$.

REFERENCES

Aitchison, J., and Silvey, S.D. (1958). Maximum likelihood estimation of parameters subject to restraints, Ann. Math. Statist., 29, 813-828.

Box, G.P.E., and Jenkins, G.M. (1970). Time series analysis, forecasting and control, Holden day, San Francisco.

Hannan, E.J. (1970). Multiple time series, Wiley, New-York.

Pham-Dinh, T. (1975). Estimation et test dans les modèles de processus stationnaires, Thèse, Grenoble.

Pham-Dinh, T. (1977). Estimation of parameters in the A.R.-M.A. model when the characteristic polynomial of the M.A. operator has a zero on the unit circle, to appear in the Annals of Statistics.

Recent Developments in Statistics
J.R.Barra et al., editors

MARKOV-RENEWAL APPROACH TO COUNTER THEORY

M.F. Ramalhoto and D.H. Girmes
Department of Statistics and Computer Science
University College London
Gower Street, London WC1E 6BT

The $[F(.), H(.); 1]$ counter system with $F(.)$, $H(.)$ exponentially distributed will be studied in terms of Markov-Renewal theory. Most of the classical results on counters are obtained in a straightforward way. Exact and limiting distribution functions are derived for the "sojourn time the counter spends in a state 'k' in time interval $(0, t]$" for an infinitely countable number of states.

INTRODUCTION

The purpose of this paper is to show that Markov-Renewal theory is a useful mathematical tool when dealing with stochastic counter theory.

A counter is a device for transforming a sequence $\{S_i\}_{i=0}^{\infty}$ (input process) into another sequence $\{S_{n_i}\}_{i=0}^{\infty}$ (output process) . If the sequence $\{S_i\}_{i=0}^{\infty}$ represents instants of arrivals of particles (with $S_0 \equiv 0$) , then due to an inherent resolving time, which we shall call partial blocked time, the device may fail to record some of the events of $\{S_i\}_{i=0}^{\infty}$ that occur during the period of observation. The recorded events form a second sequence $\{S_{n_i}\}_{i=0}^{\infty}$ which is a subsequence of the sequence $\{S_i\}_{i=0}^{\infty}$

Points of interest are:

1. Inference from the output process about the input process.
2. Exact and limiting probability distribution functions of the state of the counter (by state of the counter we mean the number of partial blocked times present in the counter).
3. "First passage time" distribution functions.
4. "Sojourn time" problems.

A type II counter is essentially characterized by the fact that every arrival of a particle produces a non-negative random variable which we call partial blocked time and which is denoted by Y . A particle is recorded only if at its arrival no partial blocked times are present in the counter (Takács [12]).

In this paper it is assumed that in a type II counter the inter-arrival times $\{S_{i+1} - S_i\}_{i=0}^{\infty}$ for $i = 0, 1, \ldots$ (with $S_0 \equiv 0$) , are independently and identically distributed random variables with distribution function $F(x)$ given by,

$$F(x) = \begin{cases} 1 - e^{-\lambda x} & , \quad x \geq 0 \\ 0 & , \quad x < 0 \end{cases} \tag{1}$$

The sequence $\{Y_i\}_{i=0}^{\infty}$ is assumed to be a sequence of independently and identically distributed random variables, with distribution function $H(x)$ given by,

$$H(x) = \begin{cases} 1 - e^{-\mu x} & , \quad x \geq 0 \\ 0 & , \quad x < 0 \end{cases} \tag{2}$$

and independent of the sequence $\{S_i\}_{i=0}^{\infty}$; Y_0 is assumed to be zero.

This mathematical model applies to a variety of situations of which the following are examples:

1. large parking lots: where the arriving particles are the arriving vehicles and the partial blocked times are the times vehicles remain parked;

2. phenomena in the kinetics of gases: where the arriving particles are gas molecules entering a fixed region, and partial blocked times are the times spent in that region by the molecules. If the temperature is low, molecules do not interact very much and the model fits well;

3. the short-run behaviour of the size (number of families living in them) of larger cities, where the arriving particles are the families moving into the city, and the partial blocked time is the amount of time spent there by a family before it moves out;

4. telephone traffic and Geiger-Muller counters are two of the earliest applications. In general we can say that this model can be a good representation of a situation where "particles" enter unhindered into some area in the quest for service and remain there for some time, and then leave.

The assumption of Poisson input is quite acceptable for many of the relevant applications to the real world, because very often, an arrival process being examined is the superposition of a large number of uniformly sparse arrival processes.

Mathematical models like immigration - death processes, queuing systems with infinitely main servers etc. are mathematically equivalent to type II counters and they have been extensively studied by several authors (Conolly, Feller, Karlin and McGregor, Kendall, Khintchine, Takács and others) but the results are obtained by a variety of techniques and theories, usually Markov processes and infinitesimal generators. The Markov-Renewal approach helps to unify the methods of proof.

THE BASIC THEOREM FOR THE MARKOV-RENEWAL APPROACH

In this paper we are going to study the counter by looking at it every time a partial blocked time starts or ends. Let $\{T_i\}_{i=o}^{\infty}$ be the sequence of points in time at which a partial blocked time starts or ends. Let $X(t)$ be the state of the counter at time t, this is, the number of partial blocked times present in the counter at time t. Clearly $X(t)$ changes just at the points T_i's. We say that a transition (in the system) occurs whenever the state of the counter changes. If T_n is the time of the nth transition then let X_n be the state visited at time T_n. Let $M(t)$ be defined as follows:

$$M(t) = \sup\{n : T_n \leq t\} \quad .$$

Theorem 1. $\{X_n, T_n ; n = 0, 1, \dots\}$ is a Markov-Renewal process (with X_n having state space $\{0, 1, \dots\}$) induced by the semi-Markov matrix $Q(t)$, where

$$Q_{j,k}(t) = P[X_{n+1} = k , T_{n+1} - T_n \leq t | X_n = j , T_n] \text{ for } k, j\varepsilon\{0,1, \dots\}$$

is given by,

$$\begin{cases} Q_{j,j+1}(t) = \dfrac{\lambda}{j\mu + \lambda}\left[1 - e^{-(j\mu+\lambda)t}\right] & , j = 0, 1, \dots ; \quad t \geq 0 \\ Q_{j,j-1}(t) = \dfrac{j\mu}{j\mu + \lambda}\left[1 - e^{-(j\mu+\lambda)t}\right] & , j = 1, 2, \dots ; \quad t \geq 0 \\ Q_{j,k}(t) = 0 & , |k-j| > 1 ; k, j\varepsilon \{0, 1, \dots\} \ t \geq 0 \end{cases} \quad (3)$$

Proof: A transition occurs whenever a partial blocked time starts or ends. As all distributions are exponential the probability of going from state "j" to state "k" in time interval $(T_n, T_n + t]$ depends only on j, k and on t; this is,

$$P[X_{n+1} = k, T_{n+1} \leq t + T_n | X_o, \dots, X_n = j ; T_o, \dots, T_n]$$

$$= P[X_{n+1} = k, T_{n+1} \leq t + T_n | X_n = j; T_n] = Q_{j,k}(t) . \qquad t \geq 0$$

It is also easy to see that the state of the counter at time t is given by $X_{M(t)}$, where $M(t)$ denotes the total number of transitions during the time interval $(0, t]$. Because of the definition of $\{T_i\}$ it is clear that $Q_{j,k}(t) = 0$ whenever $|k-j| > 1$ for $j, k \varepsilon \{0, 1, \ldots\}$.

In particular $Q_{0,1}(t)$ is the probability of going from state "0" to state "1" in one step within time t. Obviously this probability is given by $F(t)$ and hence,

$$Q_{0,1}(t) = 1 - e^{-\lambda t}, \quad t \geq 0$$

Furthermore for $n \geq 1$ we have,

$$Q_{n,n-1}(t) = n \int_0^t [1 - H(u)]^{n-1} [1 - F(u)] \, dG(u)$$

$$= \frac{n\mu}{n\mu + \lambda} \left[1 - e^{-(n\mu+\lambda)t}\right], \qquad \text{for } t \geq 0$$

and

$$Q_{n,n+1}(t) = \int_0^t [1 - H(u)]^n \, dF(u)$$

$$= \frac{\lambda}{n\mu + \lambda} \left[1 - e^{-(n\mu+\lambda)t}\right], \qquad \text{for } t \geq 0$$

Hence the theorem is proved as the elements of the matrix $Q(t)$ satisfy the definition of a semi-Markov matrix, see Cinlar [1].

This theorem establishes the link between $[F(.), H(.) ; 1]$ system with $F(.)$ and $H(.)$ defined by (1) and (2) respectively, and Markov-Renewal theory.

Given the simple structure of $Q(t)$ the corresponding Markov-Renewal matrix $R(t)$ can be obtained by elementary techniques. The Markov-Renewal matrix $R(t)$ is given by,

$$R(t) = \sum_{n=0}^{\infty} Q^{(n)}(t)$$

where $Q^{(n)}(t)$ represents the n-fold convolution of $Q(t)$ with itself. By using Laplace-Stieltjes transforms we get,

$$\int_0^\infty e^{-st} dR(t) = r(s) = (r_{i,j}(s) \; ; \; i, j \varepsilon \{0, 1, \ldots\}) \tag{4}$$

where $r_{i,j}(s)$ represents the (i, j)th element of the matrix $r(s)$.

Furthermore,

$$r_{i,j}(s) = (j\mu + \lambda + s)\sum_{k=o}^{i}(-1)^k\binom{i}{k}[k\mu + s]^{-1}\sum_{m=o}^{\min\{k,j\}}(-1)^m\binom{k}{m}(\lambda/\mu)^{j-m}$$

$$\cdot\left[(s/\mu + 1)_{j-m}\right]^{-1}\phi(1 + j - m,\ s/\mu + k + 1 + j - m\ ;\ -\lambda/\mu) \qquad (5)$$

for $i, n\varepsilon\{0, 1, \ldots\}$; $s > 0$ and where $\phi(.\ ,\ .\ ;\ .)$ represents the confluent hypergeometric function and where $(a)_m = a(a+1)\ldots(a+m-1)$ for a real number a and non-negative integer m. For details see Ramalhoto, Girmes [7].

The matrix $R(t)$ leads to most of the relevant results on counters (and M/M/∞ systems), in a systematic and uniform way by using results of Markov-Renewal theory.

SOME RESULTS DUE TO THE UNIFYING MARKOV-RENEWAL APPROACH

1. Mean number of entrances to state "k" between two entrances to state "j":

$$\frac{\alpha_k}{\alpha_j} = \left(\frac{\lambda}{\mu}\right)^{k-j}\frac{k\mu + \lambda}{j\mu + \lambda}\frac{j!}{k!}, \quad \text{for } k, j\varepsilon\{0, 1, \ldots\}$$

where α_k, $k = 0, 1, \ldots$ are the components of the eigenvector of $Q(+\infty)$ corresponding to the eigenvalue one.

2. L - S transform of the distribution function of the output process of the counter (process formed by the recorded particles) is,

$$f_{o,o}(s) = 1 - \frac{1}{r_{o,o}(s)}$$

$$= 1 - \frac{s}{\lambda + s}\cdot\frac{1}{\sum_{j=o}^{\infty}(-1)^j\prod_{i-1}^{j}\frac{\lambda}{s + i\mu}}$$

result already achieved by Takács [12] and Pyke [6].

3. $L - S\ \{P[X(t) = k/X_o = 0]\ ;\ s\} = \frac{s}{k\mu + \lambda + s}\, r_{o,k}(s)$

see Takacs [12] and Karlin, McGregor [4].

4. L - S transform of the first passage time distribution going from state "0" to state "n"

$$f_{o,n}(s) = \frac{r_{o,s}(s)}{r_{n,n}(s)}, \quad \text{for } n = 1, 2, \ldots\ ;\ s > 0 \quad .$$

5. L - S transform of the first passage time distribution going from state "n" to state "n" :

$$f_{n,n}(s) = 1 - \frac{1}{r_{n,n}(s)} , \quad \text{for } n = 0, 1, \ldots$$

6. Let $V^{+}(t)$ be the time taken from time t , where t is fixed but arbitrary, until the next transition then

$$P[V^{+}(t) \leq y] = \sum_{k=o}^{\infty} \int_{0}^{t} R_{o,k}(du) [e^{-(k\mu+\lambda)(t-u)} - e^{-(k\mu+\lambda)(t+y-u)}]$$

$$= 1 - e^{-\int_{0}^{y} \lambda [(1-e^{-\mu t})e^{-\mu u} + 1] du}$$

SOJOURN TIME PROBLEM

We are going to discuss distributional properties of the total time the counter spends in state "k" within time interval $(0, t]$, for $k = 0, 1, 2, \ldots$. Let us consider the following stochastic process $\{X(t), t \geq 0\}$ where $X(t)$ represents the state of the counter at time t .

We partition the state space $I_o \equiv \{0, 1, 2, \ldots\}$ by,

$$I_o = \mathcal{A}_k \cup \mathcal{B}_k \quad \text{where} \quad \mathcal{B}_k = \{k\} ,$$

$$\mathcal{A}_k = \{0, 1, \ldots, k-1, k+1, \ldots\},$$

for some fixed $k = 0, 1, \ldots$.

Define a new stochastic process $\{\chi_k(t) , t \geq 0\}$ by,

$$\chi_k(t) = \begin{cases} 1 & \text{if } X(t) \varepsilon \mathcal{B}_k \\ 0 & \text{if } X(t) \varepsilon \mathcal{A}_k \end{cases}$$

and we assume $X(o) \varepsilon \mathcal{A}_k$ if $k \geq 1$.

Let us denote by A_1^k , B_1^k , A_2^k , B_2^k , ... the time spent by the counter in sets $\mathcal{A}_k$, $\mathcal{B}_k$ respectively, and let us suppose that $\{A_j^k\}_{j=1}^{\infty}$ and $\{B_j^k\}_{j=2}^{\infty}$ are non-negative independently distributed random variables. Let U_i^k denote the length of time between the ith and the $(i+1)$th first passage time from state "k" to state "k" , for $k = 0, 1, \ldots$ and $i = 1, 2, \ldots$

$$
\begin{aligned}
U_i^k &= B_i^k + A_i^k, \quad \text{for } i = 2, 3, \ldots ; k = 1, 2, \ldots \\
U_1^k &= A_1^k \qquad\quad, \quad \text{for } k = 1, 2, \ldots
\end{aligned} \tag{6}
$$

If $k = 0$, $U_i^o = B_i^o + A_i^o$, $i = 1, 2, \ldots$ $(X_o = 0)$. By definition of the random variables U_i^k we get,

$$P[A_1^k \leq x] = F_{o,k}(x) \quad , \qquad \text{for} \quad k = 1, 2, \ldots ; x \geq 0$$

$$P[U_i^k \leq x] = F_{k,k}(x) \quad , \quad \text{for } i = 2, 3, \ldots ; k = 0, 1, \ldots ; x \geq 0$$

$$P[B_i^k \leq x] = \sum_{j=o}^{\infty} Q_{k,j}(x) = Q_{k,k-1}(x) + Q_{k,k+1}(x)$$

$$= \begin{cases} 1 - e^{-(k\mu+\lambda)x} , & x \geq 0 \quad \text{for } k = 0, 1, \ldots \\ 0 & x \geq 0 \end{cases}$$

Let $B_k(t)$ denote the total sojourn time in state "k" during time interval $(0, t]$, this is,

$$\beta_k(t) = \int_0^t \chi_k(u)du$$

The Laplace-Stieltjes transforms of its moments with respect to the origin are given by,

$$L - S\{E[(\beta_k(t))^i] ; s\} = \frac{i!}{(k\mu + \lambda + s)^i} \cdot \frac{r_{o,k}(s)}{r_{k,k}(s)} F(i, b, b ; 1 - \frac{1}{r_{k,k}(s)})$$

$$\text{for } i = 1, 2, \ldots$$

where $F(. , . , . ; .)$ represents the hypergeometric function.

On the other hand if the random variables $S_A^k(n) = A_1^k + \ldots + A_n^k$ and $S_B^k(n) = B_1^k + \ldots + B_n^k$ have asymptotic distributions as $n \to \infty$ then it is reasonable to expect that $\beta_k(t)$ has an asymptotic distribution as $t \to \infty$. By the Central Limit theorem $S_A^k(n)$ and $S_B^k(n)$ have an asymptotic normal distribution if σ_A^2 and σ_B^2 are finite, (where σ_A^2 is the variance of the random variables $\{A_i\}$ (similarly for σ_B^2)). Thus if σ_U^2 (variance of the random variables $\{U_i\}$) is finite $\beta_k(t)$ is asymptotically normal with mean and variance as $t \to \infty$ given by,

$$E\{\beta_k(t)\} \approx e^{-\lambda/\mu}\frac{(\lambda/\mu)^k}{k!}\, t - \frac{1}{\mu}\frac{(\lambda/\mu)^k}{k!}\sum_{n=0}^{\infty}(-1)^n\left(\sum_{j=1}^{n+k}\frac{1}{j}\right)\frac{(\lambda/\mu)^n}{n!} \quad \text{for } k = 0, 1, \ldots$$

and

$$\text{var}\,\{\beta_k(t)\} \approx \frac{2}{\mu}\, e^{-\lambda/\mu}\frac{(\lambda/\mu)^k}{k!}\left[\sum_{n=1}^{k}(-1)^n\binom{k}{n}\frac{1}{n}\sum_{j=0}^{n}(-1)^j\binom{n}{j}\frac{(\lambda/\mu)^{k-j}}{(k-j)!}\right.$$

$$\left.\phi(k-j+1,\, k-j+1+n\, ;\, -\lambda/\mu) + a(\lambda/\mu\, ,\, k)\right] , \quad k = 0, 1, \ldots$$

where $a(\lambda/\mu, k) = \dfrac{1}{\mu}\dfrac{(\lambda/\mu)^k}{k!}\displaystyle\sum_{n=0}^{\infty}(-1)^n\left(\sum_{j=1}^{n+k}\frac{1}{j}\right)\frac{(\lambda\ \mu)^n}{n!}$

For details about the $F(. , . , . ; .)$ and $\phi(. , . ; .)$ see Erdelyi [3]; these functions are tabulated.

FINAL REMARKS

Let us consider a slightly more complicated situation where the counter is of an Albert and Nelson type; it is easy to see that the semi-Markov matrix for $0 < p \leq 1$, and $t \geq 0$ is given by,

$$Q_{j,j+1}(t\, ;p) = \frac{\lambda p}{j\mu + \lambda p}\left[1 - e^{-(j\mu+\lambda p)t}\right] , \ j = 1, 2, \ldots$$

$$Q_{j,j-1}(t\, ;p) = \frac{j\mu}{j\mu + \lambda p}\left[1 - e^{-(j\mu+\lambda p)t}\right] , \ j = 1, 2, \ldots \qquad (7)$$

$$Q_{j,k}(t\, ;p) = 0 \qquad , \ |k-j| > 1 \ ; \ k,\ j \in \{0, 1, \ldots\}$$

$$Q_{o,1}(t\, ;\, p) = 1 - e^{-\lambda t}$$

and then this system again can be treated as a Markov-Renewal process with associated semi-Markov matrix given by (7).

If the input is a Poisson process and if the partial blocked time has a general distribution function $H(x)$ and if the counter is of type II, then by Takacs [10; pp.68, 69],

$$P[a_1(t) \leq x_1, \ldots, a_k(t) \leq x_k / X(t) = k] = \prod_{i=1}^{k}\frac{\int_0^t [H(y + x_i) - H(y)]\,dy}{\int_0^t [1 - H(y)]\,dy}$$

where $a_1(t), \ldots, a_k(t)$ are the distances between t and the termination points of the partial blocked times which are present at the instant t .

Then our matrix $Q(t)$ will be for $t \geq 0$ given by,

$$Q_{o,1}(t;\, T_n) = P\left[X_{n+1} = 1,\ T_{n+1} - T_n \leq t/X_n = 0,\ T_n\right]$$

$$= 1 - e^{-\lambda t} \qquad \text{(because of Poisson input)}$$

$$Q_{j,j+1}(t;\, T_n) = \frac{\lambda}{\left[\int_0^{T_n}[1 - H(y)]\,dy\right]^j} \int_0^t e^{-\lambda u}\left[\int_0^{T_n}[1 - H(y+u)]\,dy\right]^j du$$

for $j = 0, 1, \ldots$

$$Q_{j,j-1}(t;\, T_n) = \frac{j}{\left[\int_0^{T_n}[1 - H(y)]\,dy\right]^j} \int_0^t e^{-\lambda u}\left\{\left[\int_0^{T_n}[1 - H(y+u)]\,dy\right]^{j-1} \cdot \int_0^{T_n} H_u'(y+u)\,dy\right\} du$$

$$= 1 - e^{-\lambda t}\left[\frac{\int_0^{T_n}[1 - H(y+t)]\,dy}{\int_0^{T_n}[1 - H(y)]\,dy}\right]^j -$$

$$- \frac{\lambda}{\left[\int_0^{T_n}[1 - H(y)]\,dy\right]^j} \int_0^t e^{-\lambda u}\left[\int_0^{T_n}[1 - H(y+u)]\right]^j du$$

for $j = 1, 2, \ldots$

Having the identity $\int_u^\infty [1 - H(v)]\,dv = \int_0^\infty [1 - H(v)]\,dv - \int_0^u [1 - H(v)]\,dv$, and denoting $\int_0^\infty [1 - H(v)]\,dv$ by β , (the mean partial blocked time) as $T_n \to \infty$ we get,

$$\lim_{T_n \to \infty} Q_{j,j-1}(t;\, T_n) = 1 - e^{-\lambda t}\left[1 - \frac{\int_0^t [1 - H(v)]\,dv}{\beta}\right]^j - \lambda\int_0^t e^{-\lambda u}\left[1 - \frac{\int_0^u [1 - H(v)]\,dv}{\beta}\right]^j du \quad , \quad j = 1, 2, \ldots$$

$$\lim_{T_n \to \infty} Q_{j,j+1}(t;\, T_n) = \lambda\int_0^t e^{-\lambda u}\left[1 - \frac{\int_0^u |1 - H(v)|\,dv}{\beta}\right]^j du \quad , \quad j = 0, 1, \ldots$$

$$\lim_{T_n \to \infty} Q_{j,k}(t;\, T_n) = 0, \quad |k - j| > 1 \ ; \ k, j \in \{0, 1, \ldots\} .$$

Then the stationary case can be treated as a Markov-Renewal process.

REFERENCES

[1] Cinlar, E. (1969). Markov-Renewal Theory. Adv. in Appl. Prob., 1, pp. 123-189.

[2] Conolly, B. (1975). Lecture Notes on Queuing Systems; (Ellis Harwood Ltd.).

[3] Erdelyi, Magnus, Oberhettinger and Tricomi (1953). Higher Transcendental Functions; (McGraw-Hill Book Company).

[4] Karlin, S. and McGregor, J. (1958). Many-server queuing processes ...; Pacific J.Math., 8, pp. 87-118.

[5] Kendall, D.G. (1964). Some Recent Work and Further Problems in the Theory of Queues. Th.Prob.Appl., 9, No. 1, pp. 1-13.

[6] Pyke, R. (1958). On Renewal Processes Related to Type I and Type II Counter Models. Ann.Math.Stat., 29, pp. 737-754.

[7] Ramalhoto, M.F. and Girmes, D.H. (1975). An Important Matrix in Markov-Renewal Theory (to be published).

[8] Takács, L. (1957). On Certain Sojourn Time Problems in the Theory of Stochastic Processes. Acta Math.Acad.Sci.Hung., 8, pp. 169-191.

[9] Takács, L. (1957). On Limiting Distributions Concerning a Sojourn Time Problem. Acta Math.Acad.Sci.Hung., 8, pp. 279-293.

[10] Takács, L. (1958). On a Coincidence Problem Concerning Telephone Traffic. Acta Math.Acad.Sci.Hung., 9, pp. 45-81.

[11] Takács, L. (1961). On a Coincidence Problem Concerning Particle Counters. Ann.Math.Stat., pp. 739-756.

[12] Takács, L (1962). An Introduction to Queuing Theory; (New York, Oxford University Press).

Recent Developments in Statistics
J.R.Barra et al., editors
© North-Holland Publishing Company,(1977)

M-ESTIMATORS IN ROBUST REGRESSION, A CASE STUDY

William J. J. Rey
MBLE Research Laboratory
2, av. Van Becelaere, 1170-Brussels (Belgium)

This paper is centered on a comparison of several methods of robust regression after their behaviours on a classical example. The main conclusions are

- All methods are insensitive to outliers and each method behaves differently with regard to the not-outlying observations.
- The median of the deviations is not a robust estimator of the scale.
- The method proposed by Andrews may produce aberrant results.
- Fixed point solutions are required.

INTRODUCTION

This paper belongs to the recently developing statistical body of "robustness". The general idea of the involved techniques is, by comparison with standard statistics, to limit the possible influence of any feature of the sample which could be due to the sampling operation. This may be obtained by a great variety of methods ranging from weighting of the observations to rejection of a sample part. "In a rather more general sense, a statistical procedure is described as robust if it is not very sensitive to departure from the assumptions on which it depends" [1].

After the terminology introduced by Huber [2], the M-estimators $\underline{\theta}$ are generalizations of the well known Maximum likelihood estimators. They are the set of values which, instead of maximizing the logarithmic likelihood function

$$\ln L = \sum w_i \ln f(\underline{x}_i \mid \underline{\theta}),$$

minimizes the more general form

$$M = \sum w_i \rho(\underline{x}_i \mid \underline{\theta}).$$

The standard notation is made use of, and w_i is a weighting factor set to some fixed value for each observation $\underline{x}_i$.

The regression problem we are investigating is classical and has already been considered by many authors. It is relative to the operation of a plant for the oxydation of amonia into nitric acid and can be found in [3, sec. 13. 12], [4, chap. 6], [5, chap. 5] as well as in [6].

ROBUST REGRESSION

The linear model relating dependent scalar variables y_i to independent vector variates $\underline{x}_i$

$$y_i = \underline{x}'_i \underline{\theta} + \varepsilon_i,$$

where ε_i is the error, will not be open to discussion hereinafter. We assume the model to be appropriate. In fact, our only concern is to estimate certain sets of parameters, also denoted $\underline{\theta}$, such that the regression residuals ε_i be small or, more precisely, per definition $\underline{\theta}$ is such that

$$M = \sum w_i \, \rho(\underline{x}_i \mid \underline{\theta}) \quad \text{be minimum}$$

with

$$\rho(\underline{x}_i \mid \underline{\theta}) = \rho(\varepsilon_i)$$

and

$$\varepsilon_i = y_i - \underline{x}'_i \, \underline{\theta}.$$

It will be observed that the symbol $\underline{\theta}$ is used throughout in order to avoid confusion with the ordinary $\underline{\beta}$ and its estimate $\underline{b}$. Effectively, it is a priori not clear whether the properties met in the classical regression theory are valid in the present framework. However, a few asymptotic features (large sample sizes) have been reported [7].

The selection of the function $\rho(.)$ is rather arbitrary and, therefore, various vectors $\underline{\theta}$ can be pushed forward according to the selected functions. However it appears to be common sense to require minimum conditions. We will request $\rho(.)$ to be admissible, that is to be such that any observation $\underline{x}_i$ has limited incidence on $\underline{\theta}$ evaluation.

The admissibility requirement yields to a set of conditions to fulfill almost everywhere (a.e) involving the first two derivatives of $\rho(.)$, namely $\psi(.)$ and $\phi(.)$ with

$$\phi(\varepsilon) = \psi'(\varepsilon) = \rho''(\varepsilon).$$

They are as follows [9]:

$\psi(\varepsilon_i)$ must be bounded, a.e.

$a_i = (1/\varepsilon_i)\, \psi(\varepsilon_i)$ must be bounded, a.e.

$a_i \geqslant 0$, a.e.

$\phi(\varepsilon_i)$ must be bounded, a.e.

$\phi(\varepsilon_i) \geqslant 0$, a.e.

$[\, \Sigma w_i \, a_i \, \underline{x}_i \, \underline{x}'_i \,]$ must be full rank.

$[\, \Sigma w_i \, \phi(\varepsilon_i) \, \underline{x}_i \, \underline{x}'_i \,]$ must be full rank.

This condition set can be condensed while the function $\rho(\varepsilon)$ is convex and continuous everywhere, then $\underline{\theta}$ is unique.

A natural question is whether there exists any admissible $\rho(\varepsilon)$. The answer is positive, but admissible $\rho(\varepsilon)$ cannot produce scale invariant $\underline{\theta}$. Effectively, analysis indicates that invariance can only be obtained through $\rho(\varepsilon) = |\varepsilon|^p$, which is not admissible. The lack of scale invariance has great practical importance by its implications in leading to fixed point estimators.

This fixed point feature results immediately from the next three statements where s stands for the scale factor evaluation.

Function $\rho(.)$ depends upon s.
Solution $\underline{\theta}$ depends upon $\rho(.)$.
Scale factor s depends upon $\underline{\theta}$.

It will be noted in the case study that these three dependencies are, each of them, very important even while the common definition

$$s = \text{median}\,(|\varepsilon_i|)$$

is considered for the scale factor.

CASE STUDY

Already introduced, the classical example we investigate is described by the model

$$\underset{(21x1)}{\underline{Y}} = \underset{(21x4)}{X'} \underset{(4x1)}{\underline{\theta}} + \underset{(21x1)}{\underline{E}}$$

where each row

$$y_i = \underline{x}'_i \underline{\theta} + \varepsilon_i \quad (i = 1,...,21)$$

is relative to a different observation and each set of independent parameters is four-dimensional,

$$\underline{x}_i = (1, x_{2i}, x_{3i}, x_{4i})'.$$

In this example, various techniques have revealed that four observations (i = 1, 3, 4, 21) are clearly outlying with respect to the distribution of the seventeen others which form a neat cluster. Although this does not appear in the observation coordinates, this can be ascertained [8].

Four different selections of the function

$$M = \sum w_i \ \rho(\varepsilon_i)$$

to be minimized will now be considered and for each selection some results will be reported in Table 1, namely $\underline{\theta}$ and the corresponding scale factor s. These data are correct up to the last printed digit.

Weighted least squares

With $\rho(\varepsilon) = \varepsilon^2$, the weight of the four outlying observations is gradually decreased to go smoothly from size 21 to size 17. This procedure implies prior identification of the outliers. In Table 1, Fit 1 is the ordinary least square on size 21 whereas Fit 2 corresponds with size 17.

Table 1

Fit	θ_1	θ_2	θ_3	θ_4	s	
1	- 39.920	.71564	1.2953	-.15212	1.9175	L. sq., size = 21
2	- 37.652	.79769	.57734	-.06706	1.0579	L. sq., size = 17
3	- 38.805	.82643	.64760	-.08577	1.2194	Lp, p = 1.2
4	- 38.158	.83800	.66290	-.10631	1.1330	Huber, k = 1
5	- 37.132	.81829	.51952	-.07255	.96533	Andrews, c = 1.5
6	- 37.334	.81018	.54199	-.07037	.99926	Andrews, c = 1.8
7	- 41.551	.93911	.58026	-.11295	1.4385	Andrews, c = 1.8
8	- 41.990	.93352	.61946	-.11278	1.5710	Andrews, c = 1.8

Least p-th power

Selecting the form

$$\rho(\varepsilon) = |\varepsilon|^p,$$

with p between 2 and 1, a range of solutions with more or less incidence of outlying observations is obtained by the algorithm of Rey [9]. The result for p=1.2

is reported as Fit 3.

Method of Huber [2]

In the course of a study of contaminated normal distributions, Huber has found that the function

$$\rho(\varepsilon) = \varepsilon^2, \text{ if } |\varepsilon| < ks$$
$$= ks(2|\varepsilon| - ks), \text{ otherwise}$$

exhibits some optimal properties. The scale factor s is given by

$$s = \text{median } (|\varepsilon_i|)$$

as it is throughout this paper. The result for $k = 1$ is reported as Fit 4. Let us note that the method proposed by Huber is the only admissible method among the four we are considering.

Method of Andrews [6]

A partly convex and partly concave form is selected

$$\rho(\varepsilon) = 1 - \cos(\varepsilon/cs), \text{ if } |\varepsilon| < \pi\, cs$$
$$= 2, \text{ otherwise.}$$

A result for $c = 1.5$ is reported as Fit 5. However this method is noteworthy of further discussion due to its numerous hazards. We first indicate that Fit 5 does not correspond strictly with Andrews' result [6, Table 5, last but 1 column] due to large inaccuracies in his computations. The claim that the result is the same with as without the four outlying observations is nonsense, seeing that the scale factor s on size 21 and on size 17 differ significantly. In fact, the size 21 result is a nearly fixed point solution whereas the size 17 result is not defensible. Possibly the most important weakness of this method is that it may produce aberrant results in an unnoticed way. There are discontinuities in $\underline{\theta}$ as a function of the parameter c and there may be several solutions for a given c value. Fit 6, Fit 7 and Fit 8 are three fixed point solutions obtained with $c = 1.8$, a value intermediate between the two recommended 1.5 and 2.1.

REFERENCES

1 Kendall M.G. and Buckland W.R., A dictionary of statistical terms (Int. Stat. Inst., ed.), Oliver and Boyd, Edinburgh, 3rd ed. (1971).
2 Huber P.J., Robust estimation of a location parameter, Ann. Math. Stat., 35 (1964) 73-101.
3 Brownlee K.A., Statistical theory and methodology in science and engineering, Wiley, New York, 2nd ed. (1965).
4 Draper N.R. and Smith H., Applied regression analysis, Wiley, New York (1966).
5 Daniel C. and Wood F.S., Fitting equations to data, Wiley, New York (1971).
6 Andrews D.F., A robust method for multiple linear regression, Technometrics, 16 (1974) 523-531.
7 Huber P.J., Robust regression : Asymptotics, conjectures and Monte Carlo, Ann. Stat., 1 (1973) 799-821.
8 Rey W., M-estimators in robust regression, MBLE Research Laboratory, Brussels, R329 (1976).
9 Rey W., On least p-th power methods in multiple regressions and location estimations, BIT (Nordisk Tidskr. Informationsbehandling), 15 (1975) 174-184.

Recent Developments in Statistics
J.R.Barra et al., editors

β-EXPECTATION TOLERANCE REGIONS FOR THE DIFFERENCE BETWEEN TWO RANDOM VARIABLES - STRUCTURAL APPROACH

STEFAN RINCO*

Department of Mathematics, University of New Brunswick, Fredericton, Canada

In this paper a procedure of construction of β-expectation tolerance regions for the difference between two random variables is suggested. The procedure is developed on the structural method of inference using two location-scale models. The results are applied to models with normal and exponential error variables.

1. INTRODUCTION

Statistical tolerance regions are extensively used in problems of statistical inference such as quality control, life testing and process reliability studies. The type of tolerance regions investigated in this paper can be utilized for example in quality control, when not only two variables have to be under control, but also their difference has to be under control. Two parts under control would not fit together if the difference of their measurements would be too small or too large. Even though such a situation has small probability of occurring, the proposed tolerance regions can eliminate this doubt. Since these tolerance regions depend on statistics calculated previously (for each variable alone), it is not hard to implement the added tolerance region in a real life situation.

Let $(\mathcal{X}, \mathcal{A}, P_\theta)$ be a probability space. A statistical tolerance region $Q(X)$ is a statistic defined over $\mathcal{X}$ which takes values in some σ-algebra. The coverage of the tolerance region, $C(Q)$ is the probability content of $Q(X)$. $Q(X)$ is a set function depending on the random variable X and so $C(Q)$ is also a random variable with probability measure induced by P_θ. There are several types of tolerance region $Q(X)$. Guttman [4] gives a brief description of different kinds of tolerance regions.

$Q(X)$ is a β-expectation tolerance region if the average coverage is equal to β. For a given set of data, it is natural to make any statistical inference utilizing the data. Therefore, we will look for a tolerance region so as to satisfy

$$E\{C(Q)/x\} = \beta \ , \tag{1.1}$$

AMS 1970 subject classification: Primary 62F25.

Key words and phrases: Tolerance region, β-expectation tolerance region, prediction distribution, structural distribution, location-scale model, normal distribution, exponential distribution, t distribution, Behrens-Fisher distribution.

*Research partially supported by the National Research Council of Canada.

where x represents the data, and the region Q depends on the future responses y, based on data x. It has been shown by Rinco [6] and Haq and Rinco [5] that, from a structural point of view, to obtain the tolerance region (1.1) it is enough to find a region Q as to satisfy

$$\beta = \int_Q h(y/x)\, dy \,, \tag{1.2}$$

where h(y/x) is the prediction distribution of future response y, conditional on the data x. In the structural method of inference the prediction distribution is given as

$$h(y/x) = \int_\Omega p(y/\theta)\, g(\theta/x)\, d\theta \,, \tag{1.3}$$

where $p(y/\theta)$ is a distribution of future response depending on the values of the parameter θ and $g(\theta/x)$ is the structural distribution of the parameter obtained from the given structural model.

In this paper, using the prediction distribution, β-expectation tolerance regions have been constructed for the difference between two responses from two independent location-scale models with different location and same scale parameters. The construction of the tolerance regions takes into account the internal structure of the system of responses which is described in Section 2. The results have been applied to the models with error variables having normal and exponential distribution.

2. THE LOCATION-SCALE STRUCTURAL MODEL

Consider the set of responses $x_{j\alpha}$ $(\alpha = 1, 2, \ldots, n_j;\ j = 1, 2)$ and the corresponding realized but unknown error variables $e_{j\alpha}$. $x_{j\alpha}$'s are generated from $e_{j\alpha}$ by location-scale transformations

$$x_{j\alpha} = \mu_j + \sigma e_{j\alpha}, \quad \alpha = 1, 2, \ldots, n_j;\ j = 1, 2 \tag{2.1}$$

or

$$\underset{\sim}{x}_1 = \mu_1 \underset{\sim}{1} + \sigma \underset{\sim}{e}_1$$

and (2.2)

$$\underset{\sim}{x}_2 = \mu_2 \underset{\sim}{1} + \sigma \underset{\sim}{e}_2$$

It is further assumed that $e_{j\alpha}$'s are independently and identically distributed with probability density function $f(\cdot)$. For $n_j \geq 2$, the model is a structural model and straightforward structural analysis (for references see Fraser [2]) yields the joint structural density of μ_1, μ_2, σ as

$$g(\mu_1,\mu_2,\sigma) = \psi(X)\sigma^{-(n_1+n_2+1)} \prod_{j=1}^{2} \prod_{\alpha=1}^{n_j} f\left(\frac{x_{j\alpha}-\mu_{.j}}{\sigma}\right) . \tag{2.3}$$

Now consider the future responses from the same structural model:

$$\begin{aligned} Y_1 &= \mu_1 + \sigma e_1^* \\ Y_2 &= \mu_2 + \sigma e_2^* \,, \end{aligned} \tag{2.4}$$

where $e_1{}^*$ and $e_2{}^*$ are the error variables corresponding to the future variables. Then following Fraser and Haq [3], the prediction density of Y_1 and Y_2 is obtained as

$$h(y_1, y_2 / \underset{\sim}{x}) = \int_{\mu_1}\int_{\mu_2}\int_{\sigma} \frac{1}{\sigma^2} f\left(\frac{y_1-\mu_1}{\sigma}\right) f\left(\frac{y_2-\mu_2}{\sigma}\right) g(\mu_1,\mu_2,\sigma)\, d\mu_1\, d\mu_2\, d\sigma \tag{2.5}$$

From this joint prediction distribution of Y_1 and Y_2 the prediction distribution of $Y_1 - Y_2 = Y$ may be obtained in a straightforward manner. Denote this density of Y as $h(y/x)$. Then a β-expectation tolerance region of Y is obtained by using (1.2).

3. THE NORMAL DISTRIBUTION

Let us now assume that the error variables have the normal distribution

$$f(e_{j\alpha}) = (2\pi)^{-1/2} \exp\left\{-\frac{e_{j\alpha}^2}{2}\right\}, \tag{3.1}$$

$$\alpha = 1, 2, \ldots, n_j;\ j = 1, 2.$$

Then defining the transformation variables

$$\bar{x}_j = n_j^{-1} \sum_{\alpha=1}^{n_j} x_{j\alpha}, \quad j = 1, 2 \tag{3.2}$$

and

$$s_x^2 = \sum_{j=1}^{2} \sum_{\alpha=1}^{n_j} (x_{j\alpha} - \bar{x}_j)^2 \tag{3.3}$$

and using (2.3) we obtain the structural distribution of μ_1, μ_2 and σ as

$$\begin{aligned} &g(\mu_1, \mu_2, \sigma/\underset{\sim}{x}_1, \underset{\sim}{x}_2) \\ &= (n_1 n_2)^{1/2} A_{n_1+n_2-2} (2\pi)^{-\frac{n_1+n_2}{2}} \\ &\times \exp\left\{-\frac{1}{2\sigma^2}\left[n_1(\bar{x}_1-\mu_1)^2 + n_2(\bar{x}_2-\mu_2)^2 + s_x^2\right]\right\} s_x^{n_1+n_2-2} \\ &\times \sigma^{-(n_1+n_2+1)} \end{aligned} \tag{3.4}$$

where

$$A_{n_1+n_2-2} = \frac{2\pi^{\frac{n_1+n_2-2}{2}}}{\Gamma\left(\frac{n_1+n_2-2}{2}\right)}$$

Now, by using (2.5) the joint prediction distribution of Y_1 and Y_2 is

$$h(y_1, y_2 / \underset{\sim}{x}_1, \underset{\sim}{x}_2)$$

$$= \left[\frac{n_1 n_2}{(n_1 + 1)(n_2 + 1)}\right]^{1/2} \frac{\Gamma(\frac{n_1+n_2}{2})}{\pi s_x^2 \Gamma(\frac{n_1+n_2-2}{2})}$$

$$\times \left[1 + \frac{n_1(y_1 - \bar{x}_1)^2}{(n_1 + 1)s_x^2} + \frac{n_2(y_2 - \bar{x}_2)^2}{(n_2 + 1)s_x^2}\right]^{-\frac{n_1+n_2}{2}} \qquad (3.5)$$

To obtain the prediction distribution of $Y = Y_1 - Y_2$ we use well known results on transformation of t-distribution (for reference see Cornish [1]). The prediction distribution of Y then is

$$h(y/\underset{\sim}{x}_1, \underset{\sim}{x}_2)$$

$$= \left(\frac{n_1 n_2}{n_1 + n_2 + 2n_1 n_2}\right)^{1/2} \frac{\Gamma(\frac{n_1+n_2-1}{2})}{\pi^{1/2} s_x \Gamma(\frac{n_1+n_2-2}{2})}$$

$$\times \left[1 + \frac{n_1 n_2 [y - (\bar{x}_1 - \bar{x}_2)]^2}{(n_1 + n_2 + 2n_1 n_2)\, s_x^2}\right]^{-\frac{n_1+n_2-1}{2}} \qquad (3.6)$$

That is we have, that the prediction distribution of Y is such that

$$T_{n_1+n_2-2} = \left(\frac{n_1 n_2}{n_1+n_2+2n_1 n_2}\right)^{1/2} \frac{Y - (\bar{x}_1-\bar{x}_2)}{s_x/(n_1+n_2-2)^{1/2}} \qquad (3.7)$$

has the Student t-distribution with n_1+n_2-2 degrees of freedom.

Now by (1.2) central β-expectation tolerance region is the region

$$Q = \left(\bar{x}_1 - \bar{x}_2 - K \frac{s_x}{(n_1+n_2-2)^{1/2}},\ \bar{x}_1 - \bar{x}_2 + K \frac{s_x}{(n_1+n_2-2)^{1/2}}\right], \qquad (3.8)$$

where $\bar{x}_1$ and $\bar{x}_2$ are defined by (3.2), s_x is defined by (3.3) and

$$K = (2 + n_1^{-1} + n_2^{-1})^{1/2}\, t_{n_1+n_2-2;(1-\beta)/2}, \qquad (3.9)$$

with $t_{n_1+n_2-2;(1-\beta)/2}$ being the value of t-distribution (n_1+n_2-2 degrees of freedom) exceeded with probability $(1-\beta)/2$. The other types of tolerance regions (left-hand tail or right-hand tail) can be obtained similarly.

4. THE EXPONENTIAL DISTRIBUTION

Now we will assume that the error variables have the exponential distribution

$$f(e_{j\alpha}) = \exp\{-e_{j\alpha}\}, \qquad e_{j\alpha} > 0, \qquad \alpha = 1, 2, \ldots, n_j;\ j = 1, 2 \tag{4.1}$$

The transformation variables are defined as follows:

first order statistics: $\qquad x_{1(1)}$ and $x_{2(1)}$ (4.2)

and

$$c_x = \sum_{j=1}^{2} \left[\sum_{\alpha=1}^{n_j} x_{j(\alpha)} - (n-1)x_{j(1)} \right] . \tag{4.3}$$

Then by (2.3) the structural distribution of μ_1, μ_2 and σ is

$$g(\mu_1, \mu_2, \sigma / \underset{\sim}{x}_1, \underset{\sim}{x}_2)$$
$$= \frac{n_1 n_2}{\Gamma(n_1+n_2-2)} \exp\{-\frac{1}{\sigma}[n_1(x_{1(1)}-\mu_1) + n_2(x_{2(1)}-\mu_2) + c_x]\}$$
$$\times c_x^{n_1+n_2-2}\, \sigma^{-(n_1+n_2+1)}, \quad x_{1(1)} > \mu_1, \ x_{2(1)} > \mu_2 . \tag{4.4}$$

By using (2.5) the joint prediction distribution for Y_1 and Y_2 is

$$h(y_1, y_2 / \underset{\sim}{x}_1, \underset{\sim}{x}_2) = \begin{cases} C\left[1 + \dfrac{n_1(x_{1(1)}-y_1)+n_2(x_{2(1)}-y_2)}{c_x}\right]^{-(n_1+n_2)}, & y_1 \le x_{1(1)},\ y_2 \le x_{2(1)} \\ C\left[1 + \dfrac{n_1(x_{1(1)}-y_1)+y_2-x_{2(1)}}{c_x}\right]^{-(n_1+n_2)}, & y_1 \le x_{1(1)},\ y_2 > x_{2(1)} \\ C\left[1 + \dfrac{y_1-x_{1(1)}+n_2(y_2-x_{2(1)})}{c_x}\right]^{-(n_1+n_2)}, & y_1 > x_{1(1)},\ y_2 \le x_{2(1)} \\ C\left[1 + \dfrac{y_1-x_{1(1)}+y_2-x_{2(1)}}{c_x}\right]^{-(n_1+n_2)}, & y_1 > x_{1(1)},\ y_2 > x_{2(1)}, \end{cases} \tag{4.5}$$

where

$$C = \frac{(n_1+n_2-1)(n_1+n_2-2)}{(n_1+1)(n_2+1)\, c_x} .$$

From this by straightforward calculation it can be found that the prediction distribution for $Y = Y_1 - Y_2$ is

$$h(y/\underset{\sim}{x}_1, \underset{\sim}{x}_2) = \begin{cases} \dfrac{n_1n_2(n_1+n_2-2)}{2c_x(n_1-1)(n_2+1)}\left[1+\dfrac{x_{1(1)}-x_{2(1)}-y}{c_x}\right]^{-(n_1+n_2-1)} - \dfrac{n_1n_2(n_1+n_2-2)}{c_x(n_1+n_2)(n_1^2-1)}\left[1+\dfrac{n_1(x_{1(1)}-x_{2(1)}-y)}{c_x}\right]^{-(n_1+n_2-1)} & y \le x_{1(1)} - x_{2(1)} \\ \dfrac{n_1n_2(n_1+n_2-2)}{2c_x(n_1+1)(n_2-1)}\left[1+\dfrac{y-(x_{1(1)}-x_{2(1)})}{c_x}\right]^{-(n_1+n_2-1)} - \dfrac{n_1n_2(n_1+n_2-2)}{c_x(n_1+n_2)(n_2^2-1)}\left[1+\dfrac{n_2[y-(x_{1(1)}-x_{2(1)})]}{c_x}\right]^{-(n_1+n_2-1)}, & y > x_{1(1)} - x_{2(1)} \end{cases} \tag{4.6}$$

To obtain the desired β-expectation tolerance region we have to use (1.2). Unfortunately there is no analytical solution, but for given values $x_{1(1)}$, $x_{2(1)}$ and c_x the tolerance region can be found numerically with the use of a computer.

5. ADDITIONAL REMARKS

First we would like to point out that if the β-expectation tolerance regions for the sum of two random variables are needed, the calculations are similar to those for the difference of two random variables.

Secondly, it is possible to assume the different scale parameters for two sets of response variables. In such a case the slightly different structural model is used (as a reference see Rinco [6]). The resulting prediction distribution then involve Behrens-Fisher types of distribution. For example with the normally distributed error variables the central β-expectation tolerance region for the difference is

$$Q = (\bar{x}_1 - \bar{x}_2 - d_{1-\beta}r,\ \bar{x}_1 - \bar{x}_2 + d_{1-\beta}r]\ , \tag{5.1}$$

where

$$r^2 = s_{x_1}^2 \frac{n_1+1}{n_1(n_1-1)} + s_{x_2}^2 \frac{n_2+1}{n_2(n_2-1)}\ ; \tag{5.2}$$

$$\bar{x}_j = n_j^{-1} \sum_{\alpha=1}^{n_j} x_{j\alpha}\ ,\ j = 1, 2 \tag{5.3}$$

$$s_{x_j}^2 = \sum_{\alpha=1}^{n_j} (x_{j\alpha} - \bar{x}_j)^2\ ,\ \ j = 1, 2 \tag{5.4}$$

and $d_{1-\beta}$ is the point exceeded with probability $1-\beta$ when using the Behrens-Fisher distribution with $(n_1 - 1)$ and $(n_2 - 1)$ degrees of freedom and the parameter δ, given by

$$\delta = \arctan \{ s_{x_2} \left[\frac{n_2+1}{n_2(n_2-1)} \right]^{1/2} \Big/ s_{x_1} \left[\frac{n_1+1}{n_1(n_1-1)} \right]^{1/2} \} . \qquad (5.5)$$

REFERENCES

[1] Cornish, E. A. (1954). The multivariate t-distribution associated with a set of normal sample deviates. *Australian Journal of Physics*. 7, 531-542.

[2] Fraser, D. A. S. (1968). *The Structure of Inference*. Wiley, New York.

[3] Fraser, D. A. S. and Haq, M. S. (1969). Structural probability and prediction for the multivariate model. *J. Roy. Statist. Soc. Ser. B*. 31, 317-332.

[4] Guttman, T. (1970). *Statistical Tolerance Regions: Classical and Bayesian*. Griffin, London.

[5] Haq, M. S. and Rinco, S. (1976). β-expectation tolerance regions for a generalized multivariate model with normal error variables. *J. Multivar. Anal.* 6, 414-421.

[6] Rinco, S. (1973). β-expectation tolerance regions based on the structural models. Ph.D. thesis, University of Western Ontario, London, Canada.

Recent Developments in Statistics
J.R.Barra et al., editors

ON THE EQUIVALENCE OF SOME MEASURE-VALUED STOCHASTIC DIFFERENTIAL EQUATIONS OCCURRING IN NON LINEAR FILTERING THEORY.

J. Szpirglas
Centre de Fiabilité CPM/FMI
Centre National d'Etudes des Télécommunications
196, rue de Paris. 92220 Bagneux. France.

In the markovian filtering theory, H. KUNITA [2][3] has shown that the filtering process is a solution of two equations and has solved the second one. We show a general equivalence between the two equations considered as measure-valued equations, thus solving the first one. Then, using some ideas of ZAKAI [7], we show the existence of two other equations, we call them simplified filtering equations, which are equivalent to the first ones.

I. HYPOTHESIS. FILTERING EQUATIONS.

I.1. Definition of the filtering process.

A filter is given by the statistics of the signal X, the signal-observation Z, and some initial conditions.

Let :

- T is a finite interval (OT) of R.
- The signal X is a conservative, homogeneous right continuous Markov process $(\Omega, A, (X_t)_{t\epsilon T}, (Px)_{x\epsilon S})$ with values in a separable complete metric space $(S,\mathcal{S})$. X has P_t for semi-group on B(S), set of the R^d-valued bounded Borel functions, (L,D(L)) for generator and μ for initial law.
- The observation process Z_t is given on $(\Omega, A, P\mu)$ by :

$$Z_t = \int_0^t h(Xs)ds + W(t)$$

  where h is in B(S) and W(t) is a standard Wiener process independent from X.
- The initial conditions are summarized by the σ-field F_o.
- We call $(F_t)_{t\epsilon T}$ the completion for $\mathbb{P}\mu$ of the filtration $(F_o V\sigma(Z_s s\leqslant t))_{t\epsilon T}$ where $\sigma(Z_s, s\leqslant t)$ is the σ-field generated by the observations up to time t.

We shall define the filtering process in the following theorem by a "good" version of the conditional expectation $E_\mu[f(X_t)|F_t]$.

THEOREM I-1 (cf [3]) *There exists a M(S)-valued kernel,* $\Pi_t(\omega, dy)$ *from* (S,S) *to* $(\Omega \times T, O)$ *such that :*

1°/ *For all f of* B(S), $\Pi_t(\cdot, f)$ *is an optional projection of* $f(X_t)$ *with respect to* $(F_{t^+})_{t\varepsilon T}$.

2°/ Π_t *is* $P\mu$*-right continuous for the weak topology on* M(S).

O means the F_{t+}-optional σ-field, $M(S)$ is the set of probabilities on S. Π_t is the filtering process of X with respect to $(F_t)_{t\varepsilon T}$.

1.2. Filtering equations. (cf [2] [3]).

We have the two following theorems.

THEOREM I-2 *For all f of* $\mathcal{D}(L)$ $P\mu$ *a.s.*

(1) $$\Pi_t f = \Pi_o f + \int_0^t \Pi_s(Lf)ds + \int_0^t d\nu_s\{\Pi_s(fh) - \Pi_s(f)\Pi_s(h)\}$$

where ν_t *is the innovation process :*

$\nu_t = Z_t - \int_0^t \Pi_s(h)ds$. *It is a* $(\Omega, (F_t)_{t\varepsilon T}, P\mu)$ *Wiener process.*

THEOREM I-3 *For all f of* B(S) $P\mu$ *a.s.*

(2) $$\Pi_t f = \Pi_o(P_t f) + \int_0^t dZ_s\ \{\Pi_s(hxP_{t-s}f) - \Pi_s(h)\Pi_s(P_{t-s}f)\}$$

II. EQUATIONS (1) (2) (3) (4)

II.1. Hypothesis.

Let :

- (S,S) a separable complete metric space, $B(S)$ (resp. $C_b(S)$) is the set of all bounded Borel (resp. continuous) functions on S. $M(S)$ (resp. $\tilde{M}(S)$) is the set of probabilities (resp. bounded positive measures) on S.
- P_t is a conservative semi-group on $B(S)$ with $(L,\mathcal{D}(L))$ as infinitesimal generator.
- h is in $B(S)$.

II.2. Definitions.

We call a solution of equation (1) respectively (2), (3), (4) the set

$(\Omega, A, (F_t)_{t\varepsilon T}, u_o, (u_t)_{t\varepsilon T}, \dot{B}_t, P)$ where

a) $(\Omega, A, (F_t)_{t\varepsilon T}, P)$ is a probability space with a P-complete increasing family of σ-fields.

b) u_o is a F_{o+}- measurable $M(S)$-valued random variable defined on (Ω, A, P).

c) u_t is a F_{t+}- measurable $M(S)$-(resp. $M(S)$-, $\tilde{M}(S)$-, $\tilde{M}(S)$-) valued stochastic process defined on (Ω, A, P). u_t is right continuous for the weak topology, $u(o) = u_o$.

d) B_t is a (Ω, A, F_t, P) Wiener process with values in R^d.

e) u satisfies the following relations :

For all f in $\mathcal{D}(L)$ P-a.s

(1) $u_t(f) = u_o(f) + \int_0^t u_s(Lf)ds + \int_0^t dB_s\{u_s(fh) - u_s(f)u_s(h)\}$

Respectively :

For all f in $\mathcal{B}(S)$ P-a.s

(2) $u_t(f) = u_o(P_t f) + \int_0^t dB_s\{u_s(hxP_{t-s}f) - u_s(h)u_s(P_{t-s}f)\}$

For all f in $\mathcal{D}(L)$ P-a.s

(3) $u_t(f) = u_o(f) + \int_0^t u_s(Lf)ds + \int_0^t dB_s u_s(fh)$

For all f in $\mathcal{B}(S)$ P-a.s

(4) $u_t(f) = u_o(P_t f) + \int_0^t dB_s u_s(hxP_{t-s}f)$.

Equations (3) and (4) are called the simplified filtering equations.
Shortly we shall write ; u or (u_t, B_t) is a solution of (1), (2), (3), (4).

II.3.

<u>THEOREM II-1</u> *There is equivalence between equations (1) and (2) and equations (3) and (4).*

<u>Proof</u> We make it for equations (3) and (4)

(i) For f of $\mathcal{D}(L)$, and (u_t, B_t) a solution of (3) or (4) let :

(5) $v(t, f) = u_o(P_t f) - u_o(f) + \int_0^t dB_s\{u_s(hP_{t-s}(f)) - u_s(hf)\} - \int_0^t u_s(Lf)ds$

Since $P_t f - f = \int_0^t P_s Lfds$ and because of a Fubini type theorem, we get :

(6) $v(t, f) = \int_0^t u_o(P_s Lf)ds - \int_0^t u_s(Lf)ds + \int_0^t dk \int_0^{t-k} dB_s u_s(hP_k Lf)$

(ii) a) Let u be a solution of (3) and g a resolvant of f : $g = R_\lambda f$ g is in $\mathcal{D}(L)$ and v(t, g) satisfies (6).

(7) $\int_0^{t-k} dB_s u_s(hP_k Lg) = u_{t-k}(P_k Lg) - u_o(P_k Lg) - \int_0^{t-k} u_s(P_k LLg)ds$

Reporting (7) in (6) we get : $v(t, g) = 0$

b) We know that the weak limit of λg, as λ tends to infinity, is f.

Then we get : $v(t, f) = 0$ for all f of $\mathcal{D}(L)$.

c) Since $C_b(S)$ is in the weak closure of $\mathcal{D}(L)$, $v(t, f) = 0$ for all f of $C_b(S)$.

d) A monotone class theorem (cf [2]) induces that u is a solution of equation (4).

(iii) Let u be a solution of (4). Therefore:

(8) $u_s(Lf) = u_o(P_s Lf) + \int_0^s dB_k u_k(hxP_{s-k}Lf)$

Reporting (8) in (5) we get our result, $v(t, f) = 0$ for all f of $\mathcal{D}(L)$.

III. QUASI-EQUIVALENCE OF SYSTEMS (1), (2) AND (3), (4).

THEOREM III-1 Let :

- The hypothesis of II-1
- $(\Omega, A, F_t, u_o, u_t, I_t, P)$ a solution of system (1)(2)
- $O_t = I_t + \int_o^t u_s(h)\,ds$
- $\psi_t = \exp\{\int_o^t u_s(h)\,dI_s + 1/2 \int_o^t (u_s(h))^2 ds\}$
- $\chi_t = \exp\{-\int_o^t \widehat{u_s(h)}\,dI_s - 1/2 \int_o^t (\hat{u}_s(h))^2 ds\}$

  where $\widehat{u_s(h)}$ is the predictable projection of $u_s(h)$
- $\tilde{P} = \chi_T P$
- $\tilde{u}_t = \psi_t u_t$

Then $(\Omega, A, F_t, u_o, \tilde{u}_t, O_t, \tilde{P})$ is a solution of system (3)(4).

THEOREM III-2 Let :

- The hypothesis of II-1
- $(\Omega, A, F_t, u_o, \tilde{u}_t, \tilde{P})$ a solution of the system (3)(4).

  Then the following change of variable

  $u_t = \tilde{u}_t/\tilde{u}_t(1)$ is valid

  Let : - $I_t = O_t - \int_o^t (\tilde{u}_s(h)/\tilde{u}_s(1))\,ds$

  - $\psi_t = \exp\{\int_o^t (\tilde{u}_s(h)/\tilde{u}_s(1))\,dO_s - 1/2 \int_o^t (\tilde{u}_s(h)/\tilde{u}_s(1))^2 ds$

ψ_t actually is a martingale

- $P = \psi_T \tilde{P}$

Then $(\Omega, A, F_t, u_o, u_t, I_t, P)$ is a solution of system (1)(2).

Proof We use Ito formula to the products of semi martingales $\psi_t u_t(f)$ and $\psi_t u_t(P_{u-t} f)$ for theorem III-1, $\tilde{u}_t(f)(\tilde{u}_t(1))^{-1}$ and $\tilde{u}_t(P_{u-t} f)(\tilde{u}_t(1))^{-1}$ for theorem III-2.

IV. EXISTENCE AND UNIQUENESS OF SOLUTIONS.

We use Yamada-Watanabe's method (6).

Definition IV.1. We shall say that the pathwise uniqueness holds for (2)(3) or (4) if for any two solutions (u_t, B_t) and $(u'_t\ B'_t)$ defined on a same probability space (Ω, A, F_t, P) $u_o = u'_o$ and $B_t \equiv B'_t$ imply $u_t \equiv u'_t$.

THEOREM IV.1 *The pathwise uniqueness holds for the four equations.*

THEOREM IV.2 *For all probability laws ν on $M(S)$, a solution of (1)(2) and (3)(4) exists such that u_o has ν for probability law.*

We get the following theorem :

THEOREM IV.3 *If* $(\Omega, A, F_t, \mu, u_t, B_t, P_\mu, \mu \in M(S))$ *is a solution of* (1)(2) *or* (3)(4), *it is a Markov process.*

From equations (1) or (3) we get that u_t is weakly continuous.

We have then the last theorem :

THEOREM IV.4 *If* P_t *is a semi-group of density* $q(t, x, y)$, *the solution* u *(resp.* u*) of equations* (1)(2) *(resp.* (3)(4)*) is absolutely continuous with respect to the Lebesgue measure*

$$u_t(f) = \int_S f(y) p(t, y, w)\, dy \qquad (\text{resp. } \tilde{u}_t(f) = \int_S f(y) \tilde{p}(t, y, w)\, dy)$$

We have : $p_t = \tilde{p}_t / \int_S \tilde{p}_t(y)\, dy.$

Reversely this study may be useful to identify some solutions of stochastic differential equations as conditional densities of a filtering problem [4].

REFERENCES

(1) C. DELLACHERIE et P.A. MEYER. Probabilités et potentiel, Hermann, Paris (1975).

(2) M. FUSIJAKI - G. KALLIANPUR - H. KUNITA. Stochastic differential equations for the non linear filtering problem, Osaka J. Math. 9 (1972) p. 19-40.

(3) H. KUNITA. Asymptotic behavior of the non linear filtering errors of Markov processes, J. of Multivariate Analysis. Vol. 1 n° 4 (Dec. 1971).

(4) E. PARDOUX. Caractérisation de la densité conditionnelle dans le problème du filtrage d'une diffusion réfléchie, Congrès Européen des statisticiens (1976).

(5) M. YOR. Sur les théories du filtrage et de la prédiction (à paraître).

(6) T. YAMADA and S. WATANABE. On the uniqueness of solutions of stochastic differential equations, J. of Math. of Kyoto Univ. Vol. 11 n° 1 (1971).

(7) M. ZAKAI. On the optimal filtering of diffusions processes, Z. Wahrs. Verw. Geb. 11 (1969) p. 230-243.

I am grateful to Mrs N. EL KAROUI who directed this study.

Recent Developments in Statistics
J.R.Barra et al., editors

ON CUMULATIVE SEMI-MARTINGALES

Norbert Therstappen
Lehrstuhl für Mathematik
Fachrichtung Operations Research
Technische Hochschule Aachen
Aachen, Germany

We regard cumulative processes as semi-martingales, that means we deal with these processes from the point of view of the » théorie générale « . It will be shown, that cumulative semi-martingales are special and hence they can be written as the sum of a local martingale and a previsible process of bounded variation. Applying Wald's lemma we establish these results using signed measures on the previsible space.

INTRODUCTION

The class of cumulative semi-martingales contains counting processes of renewal processes (specially Poisson processes) and renewal reward processes (specially compound Poisson processes). For applying semi-martingale theory to cumulative processes we first of all need the following general definitions.

Let us fix a probability space $(\Omega,\mathfrak{F},P)$ with a filtration $(\mathfrak{F}_t)_{t\in\mathbb{R}_+}$.

By an increasing process $(A_t)_{t\in\mathbb{R}_+}$ we mean an adapted, right continuous process with $A_o=o$ and non-decreasing paths. Let $\mathfrak{B}^+$ denote the set of increasing processes and let $\mathfrak{B} = \mathfrak{B}^+-\mathfrak{B}^+$ denote the set of right continuous processes with paths of bounded variation on every finite interval We define $\mathfrak{A}^+ = \{(A_t)\in\mathfrak{B}^+/\lim_t E(A_t)< \infty\}$ and call $\mathfrak{A} = \mathfrak{A}^+-\mathfrak{A}^+$ the set of right continuous processes with integrable variation on $\mathbb{R}_+$. Finally we use $\mathfrak{A}_{loc}=\{(V_t)\in\mathfrak{B}/ \exists$ st.t. $(T_n);T_n\uparrow\infty$ such that $(V_{t\wedge T_n})\in\mathfrak{A}\}$ and the set $\mathfrak{L}$ of local martingales.

Definition 1: A real valued stochastic process $(X_t)_{t\in\mathbb{R}_+}$ is cumulative relative to a renewal sequence $(\tau_i)_{i\in\mathbb{N}_o}$, if

(i) $(X_t)\in\mathfrak{B}$

(ii) $\{X_i^*=X^*_{\theta_i} - X^*_{\theta_{i-1}}$ $i=0,1,\ldots\}$ are independent and identically distributed. $\theta_n = \sum_1^n \tau_i$; $X_t^* = \bigvee_o^t X$

(iii) $\{X_i=X_{\theta_i}-X_{\theta_{i-1}}$ $i=0,1,\ldots\}$ are i.i.d. random variables.$\lrcorner$

A cumulative process X which is constant for $t\in[\theta_i(\omega),\theta_{i+1}(\omega)[$ is called a renewal reward process. It has the representation

$$X_t = \sum_1^{\nu_t} X_i,\ \nu_t= \sup\{n/\theta_n\leq t\}.$$

Definition 2: Let X be an adapted stochastic process.
a) X is called semi-martingale, if $X=M+V \quad M\in\mathfrak{L} \quad V\in\mathfrak{B}$
b) X is a special semi-martingale, if $X=M+V \quad M\in\mathfrak{L} \quad V\in\mathfrak{A}_{loc}$
c) X is a quasi-martingale (see Rao(1969)), if there exists a constant M, such that

$$\sup \sum_1^n E(|X_{t_i} - E(X_{t_{i+1}}|\mathfrak{F}_{t_i})|) \leq M$$

where the supremum is taken over all finite sets $t_1<t_2<\ldots<t_n, t_i\in\mathbb{R}_+$

DECOMPOSITIONS

The following theorem states the connection between cumulative processes and semi-martingales.
Theorem 3: Let X be a cumulative process, continous for $t\in]\theta_i(\omega),\theta_{i+1}(\omega)[$, then X is a special semi-martingale.
Proof: a) X is a semi-martingale; we only use a decomposition $X = \alpha - (X-\alpha), \alpha\in\mathbb{R}_+$, because every constant is a martingale, hence a local martingale and $(X-\alpha)\in\mathfrak{B}$.
b) For proving that X is special semi-martingale, we first regard a renewal reward process Y and the stopping times (θ_n). For the stopped process $(Y_{t\wedge\theta_n})$ we get

$$Y_{t\wedge\theta_n} = Y^1_{t\wedge\theta_n} - Y^2_{t\wedge\theta_n} \quad \text{with}$$

$$Y^1_{t\wedge\theta_n} = \sum_1^{\min(\nu_t,n)} X_i^+ \quad \text{and} \quad Y^2_{t\wedge\theta_n} = \sum_1^{\min(\nu_t,n)} X_i^- \Rightarrow \sup_t E|Y^1_{t\wedge\theta_n}| \leq$$

$$\leq \sup_t E\Big(\sum_1^{\min(\nu_t,n)} X_i^+\Big) \leq \sum_1^n E(X_i^+) = nE(X_i^+) < \infty \Rightarrow (Y^1_{t\wedge\theta_n}) \text{and} (Y^2_{t\wedge\theta_n}) \in\mathfrak{A}^+$$

$\Rightarrow (Y_t)\in\mathfrak{A}_{loc}$ and this means that Y is a special semi-martingale.
c) For the generalisation to cumulative processes we use the following theorem (see Meyer(1976), Yoeurp(1975)).

Let X be a semi-martingale. The following conditions are equivalent.
(i) X is a special semi-martingale.
(ii) The increasing process $(\sum_{o<s<t} \Delta s^2)^{1/2} \in\mathfrak{A}_{loc}; \Delta X_s=X_s-X_{s-}$.

Now assume X to be a cumulative and Y to be a renewal reward process. ⇒ (see (b))

$$(\sum_{o<s<t} \Delta Y^2)^{1/2} = (\sum_1^{\nu_t} X_i^2)^{1/2} = (\sum_{o\leq s\leq t} \Delta X^2)^{1/2} .$$

Hence X is a special semi-martingale. ┘
Corollary 4: (Stricker(1976)) Let X be a cumulative semi-martingale, then X can be written as the sum of a local martingale M and a previsible process A of bounded variation.
Remark 5: If X is a semi-martingale $X = X^1 - X^2$, such that $X^1\vee X^2\in\mathfrak{A}^+$, then there exists a Rao decomposition for X; this means that X can be written as the difference of two supermartingales. The proof follows immediately from Meyer (1966) Chap. VII, 2. If we assume X to be a cumulative semi-martingale, then the stopped process $(X_{t\wedge\theta_n})$ is decomposable into the difference of two nonnegative supermartingales, which means that $(X_{t\wedge\theta_n})$ is a quasi-martingale. (see Stricker (1976)). ┘
For quasi-martingales X Rao obtains a Riesz decomposition; That means X can be written in essentially one way as the sum of a martingale and a quasi-potential. The following counterexample shows

that semi-martingales have no Riesz decomposition in general.
Counterexample 6: Let X be a cumulative semi-martingale and assume X to have a Riesz decomposition $X=M+P \Rightarrow \lim_t E(X_t) = \lim_t E(M_t) = E(M_t) < \infty$. Define $Y_t = \frac{X_t}{t} \Rightarrow \lim_t E(Y_t) = o$.
That's a contradiction to a theorem of Smith(1955), which states: $\lim_t E(Y_t) = \kappa/\mu; \kappa = E(X_i), \mu = E(\tau_i)$.┘
Let X be cumulative, Y defined as before. If we assume Y to be a quasi-martingale, we can use the Riesz and Rao decompositions to prove the convergence theorems of Smith.

MEASURES ON $\mathfrak{P}$

Now we examine the existence of Pellaumail (Föllmer) measures on the previsible space $(\Omega,\mathfrak{P})$.
Theorem 7: Let ν be the counting process of a renewal process; $\nu_t = \sup\{n/\theta_n \leq t\}$; then there exists a measure μ^ν on $\mathfrak{P}$, such that

$$\mu^\nu(]\!]S,T]\!]) = E((\nu_T - \nu_S) 1_{\{S<\infty\}}) \qquad (*)$$

S,T stopping times with S<T.
Proof: Following Pellaumail(1973) we have to show, (i) $\nu_t \in L^1(\Omega,\mathfrak{F}_t,P)$ (ii) ν_t is increasing (iii) $\forall s \in \mathbb{R}_+ \Rightarrow \lim_{t \downarrow s} E(X_t - X_s) = o$.
(i) $E(\nu_t) = H(t) < \infty$. (H renewal function; see Feller(1966)). (ii) is obvious. (iii) H(t) is right continuous, because

$$H(t) = \sum_1^\infty F_n(t) \quad \text{with} \quad F_n(t) = F_{n-1}(t) * F_{\tau_1}(t) \;;\; F_o(t) = 1$$

$\sum F_n$ is convergent; $\forall\, t,\varepsilon \;\exists\; N \in \mathbb{N}$ such that $\sum_{n=N}^\infty F_n(t+1) < \frac{\varepsilon}{2}$. The F_i are right continuous $\Rightarrow F_i(t+h) - F_i(t) < \frac{\varepsilon}{2N}, h \leq \delta(t,\varepsilon)$. Let be

$$\delta = \min_{i \leq N} \delta_i(t,\varepsilon) > 0 \Rightarrow H(t+h) - H(t) = \sum_{i=1}^\infty (F_i(t+h) - F_i(t))$$

$$\leq \sum_{i=1}^{N-1} (F_i(t+h) - F_i(t)) + \frac{\varepsilon}{2} < \frac{\varepsilon}{2} + \frac{\varepsilon}{2} = \varepsilon. \;┘$$

Corollary 8: Let X be a cumulative semi-martingale, right continuous in L^1, then there exists a signed measure on $\mathfrak{P}$ such that (*) is fulfilled, if X^1 or X^2 are bounded in L^1. (see remark 5).┘
Theorem 9: Let X be a stationary renewal reward process (that means $V_t = \theta_{\nu_t+1} - t$ and X_{ν_t+1} are stationary random variables for all $t \in \mathbb{R}_+$)

$$\Rightarrow \mu^X(]\!]s,t]\!]) = \kappa \cdot \mu^\nu(]\!]s,t]\!]) \qquad s,t \in \mathbb{R}_+ \text{ with } s<t.$$

Proof: $\mu^X(]\!]s,t]\!]) = E(X_t - X_s) = E(\sum_{i=1}^{\nu_t} X_i) - E(\sum_{i=1}^{\nu_s} X_i) =$

$= E(\sum_{i=1}^{\nu_t+1} X_i) - E(X_{\nu_t+1}) - E(\sum_{i=1}^{\nu_s+1} X_i) + E(X_{\nu_s+1}) =$ (using Wald's identity (see Doob(1953)).

$= (H(t)+1)E(X_i) - E(X_{\nu_t+1}) - (H(s)+1)E(X_i) + E(X_{\nu_s+1}) =$

$= (H(t)-H(s))E(X_i) - E(X_{\nu_t+1} - X_{\nu_s+1}) = \kappa \cdot \mu^\nu(]\!]s,t]\!])$. ┘

Corollary 1o: Let $(X^n_t)_{t\in\mathbb{R}_+}$ be the stopped process of a renewal reward process X, then

$$\mu^{X^n}(]\!]s,t]\!]) = \kappa\cdot\mu^{\nu^n}(]\!]s,t]\!]) - \mu^{X^n_{\nu+1}}(]\!]s,t]\!]) \quad\text{and}$$

$$|\mu^{X^n}|(\bar{\Omega}) = |\kappa|\,|\mu^{\nu^n}|(\bar{\Omega}) - |\mu^{X^n_{\nu+1}}|(\bar{\Omega}) < \infty$$

Proof: We only have to verify the given representation, because the calculation of $|\mu^{X^n}|$ follows immediately from the theorems of Föllmer(1973) and Pellaumail(1973). Hence we have to show the applicability of the Wald lemma to the stopped process X^n. This is a consequence of a theorem of Blackwell-Girshick(1946), if we use the following terms:

$X_i = X_i\ \tau_i; f = |pr_1|; \phi_i = \sum_{j=1}^{i}(X_i \otimes \tau_i - \kappa); n(\omega) = \min(\nu_t(\omega), n)$. The left side of these equalities indicates the terminology of Blackwell-Girshick. ┘

Remark 11: On the other side we can use corollary 1o for proving that a cumulative semi-martingale is special. Thereby we start with the process (ν_t), which is special because its jumps are bounded by 1 (Stricker(1976)). Using the submartingale properties of (ν_t) it follows immediately from definition 2c that $(\nu_{t\wedge\theta_n})$ is a quasi-martingale. With the theorem of Föllmer-Pellaumail and the above corollary, it can be shown that a renewal reward process is a special semi-martingale. The generalisation to cumulative semi-martingales uses the same argument as in theorem 3. ┘

The main purpose of this representation of cumulative processes as semi-martingales is the application of the theory of stochastic integrals, which will be published in a forthcoming paper.

REFERENCES

Blackwell,D.,and Girshick,M.A.(1946). Ann.Math.Statist,17,31o.

Brown,M.,and Soloman,H.(1975). Stochastic Processes,3,3o1.

Doob,J.L.(1953). Stochastic Processes; (John Wiley and Sons, New York)

Feller.W.(1966). An Introduction to Probability Theory and its Application;Vol.II (John Wiley and Sons,New York)

Föllmer,H.,(1973). Ann.Probability,1,No.4.58o.

Meyer,P.A.,(1966).Probabilités et Potentiel; (Hermann Paris)

Meyer,P.A.,(1976). Séminaire de Probabilités,1o,245.

Neffke,H.,(1976). Beiträge zur Theorie bewerteter Erneuerungsprozesse;(Diss. Aachen)

Pellaumail,J.,(1973). Sur l'Intégrale Stochastique et la décomposition de Doob-Meyer; (Soc.Math.de France,Paris)

Rao,K.M.,(1969). Math.Scand.,24,79.

Smith,W.L.,(1955). Proc.Roy.Soc.A,232,6.

Stricker,C.,(1976). C.R.Acad.Sc.Paris A,283,375.

Yoeurp,Ch.,(1975). Contributions à la Théorie des Intégrales Stochastiques; (Diss.Strasbourg) (see "Séminaire de Probabilités 1o" too)

Recent Developments in Statistics
J.R.Barra et al., editors

ASYMPTOTIC DISTRIBUTIONS OF UNIVARIATE AND BIVARIATE M-TH EXTREMES

J.Tiago de Oliveira
Center of Statistics and Applications (INIC)
Faculty of Sciences, Lisbon

1. INTRODUCTION

In this short note we will obtain the asymptotic distributions of univariate and bivariate m-th maxima, extending some special previous results obtained by Gumbel (1935), connected with Gumbel distribution and detailed in Gumbel (1958). The methodology is somewhat different from the one of the complete paper of Smirnov (1949)(part II). The extension of those results to multivariate maxima is obvious as well as its conversion to minima results.
Let us recall that if n independent and identically distributed random variables $X_1,\ldots,X_n$ have the distribution function F(x), the distribution function of $\max(X_1,\ldots, X_n)$ is $F^n(x)$. It is well known that there exist attraction coeficients λ_n and $\delta_n(>0)$,not uniquely defined,such that,as $n \to \infty$ $F^n(\lambda_n+\delta_n x)\overset{W}{\to}L(x)$.
The reduced asymptotic distributions L(x) are of the forms:

$$\Lambda(x) = \exp(-e^{-x}) \qquad \text{Gumbel distribution}$$
$$\Phi\,\alpha(x) = 0 \quad \text{if } x<0 \qquad \text{Fréchet distribution}$$
$$= \exp(-x^{-\alpha})(\alpha>0) \quad \text{if } x\geq 0$$
$$\Psi\,\alpha(x) = \exp-(-x)^{\alpha}(\alpha>0) \quad \text{if } x<0$$
$$= 1 \quad \text{if } x\geq 0 \qquad \text{Weibull distribution.}$$

In practical applications we will aproximate $F^n(x)$ by the convenient distribution $L((x-\lambda/\delta)$ which has a location and a dispersion parameter and eventually a shape parameter (α) in the 2nd and 3rd forms.
Let us recall that the 2nd and 3rd forms can be reduced to Gumbel distribution by logarithmic transformations,a fact that can be used in the study of bivariate maxima.

2. THE ASYMPTOTIC DISTRIBUTIONS OF THE M-TH MAXIMA

Let us consider a sample of n random variables $(X_1,\ldots,X_n)$, independent and identically distributed with distribution function F(x), and let $X'_1\leq X'_2\leq \leq\ldots\leq \leq X'_{n-1}\leq X'_n$ denote the ordered sample.

The random variable X'_{n-m+1} is called the m-th maximum; for m=1 we obtain evidently $X'_n = \max(X_1,\ldots,X_n)$.
The distribution function of X'_{n-m+1} is:

$$P(X'_{n-m+1} \leq x) = ((k > n-m+1 \text{ random variables}) \leq x) = \sum_{k=n-m+1}^{n} \binom{n}{k} F(x)^k (1-F(x))^{n-k} = \sum_{0}^{m-1} \binom{n}{p} F(x)^{n-p}(1-F(x))^p.$$

Let us suppose that $F^n(\lambda_n+\delta_n x) \xrightarrow{w} L(x)$.

Then $P(X'_{n-m+1} \leq \lambda_n+\delta_n x) = \sum_0^{m-1} \binom{n}{p} F(\lambda_n+\delta_n x)^{n-p}(1-F(\lambda_n+\delta_n x))^p$ which, as $n \to \infty$ converges to $L(x) \sum_0^{m-1} \frac{1}{p!} (-\log L(x))^p$.

For the proof, let us recall that $F^n(\lambda_n+\delta_n x) \xrightarrow{w} L(x)$ is equivalent to $n(1-F(\lambda_n+\delta_n x)) \to -\log L(x)$ and that $\frac{n!}{(n-p)!} \sim n^p$ as $n \to \infty$.

If $L(x)=\Lambda(x)$ we get as asymptotic distribution $\Lambda(x)\sum_0^{m-1} \frac{1}{p!} e^{-px}$ which corresponds to the result given in Gumbel (1935) if we take $e^{-x} = me^{-y}$; if $L(x)=\Phi_\alpha(x)$ we have as asymptotic distribution $\Phi_\alpha(x) \sum_0^{m-1} \frac{1}{p!}(x)^{-p\alpha}(x>0)$ and if $L(x)=\Psi_\alpha(x)$ we obtain as asymptotic distribution $\Psi_\alpha(x) \sum_0^{m-1} \frac{1}{p!}(-x)^{p\alpha}(x<0)$.

3. THE ASYMPTOTIC DISTRIBUTION OF ℓ-TH, M-TH MAXIMA

Consider now the sample of n random pairs $(X_1,Y_1),\ldots,(X_n,Y_n)$, independent and identically distributed with distribution function $F(x,y)$ and denote now by $X'_{n-\ell+1}$ and Y'_{n-m+1} the ℓ-th maximum of the margin sample $\{X_i\}$ and the m-th maximum of the margin sample $\{Y_j\}$. Let $A(x) = F(x,+\infty)$ and $B(y) = F(+\infty,y)$ be the marginal distributions of X_i and Y_j.
The distribution function of $(X'_{n-\ell+1}, Y'_{n-m+1})$ is

$$P(X'_{n-\ell+1} \leq x, Y'_{n-m+1} \leq y) = P((i \geq n-\ell+1 \text{ of } X) \leq x, (j \geq n-m+1 \text{ of } Y) \leq y$$

$$= \sum \frac{n!}{k!(i-k)!(j-k)!\,(n+k-i-j)!} F^k(x,y) \left[A(x)-F(x,y)\right]^{i-k} \left[B(y)-F(x,y)\right]^{j-k}$$

$\times\left[1+F(x,y)-A(x)-B(y)\right]^{n+k-i-j}$ where k, in the summation, denotesthe number of random pairs (X_p,Y_p) with $X_p \leq x, Y_p \leq y$, and the summation is extended to the i,j,k, such that $i \geq n-\ell+1, j \geq n-m+1, \max(0, i+j-n) \leq k \leq \min(i,j)$.

Taking i=n-a, j=n-b and k=n-c we get $P(X'_{n-\ell+1} \leq x, Y'_{n-m+1} \leq y) =$

$$= \Sigma \frac{n!}{(n-c)!\ (c-a)!\ (c-b)!\ (a+b-c)!} F(x,y)^{n-c} [A(x)-F(x,y)]^{c-a} \times [B(y)-F(x,y)]^{c-b}$$

$[1+F(x,y)-A(x)-B(y)]^{a+b-c}$ where the summation is extended to a,b,c such that $a < \ell-1$, $b < m-1$, $\max(a,b) \leq c \leq \min(n,a+b)$. Suppose now that $F^n(\lambda_n+\delta_n x, \lambda'_n+\delta'_n y) \overset{W}{\to} L(x,y)$ or equivalently, that $n[1-F(\lambda_n+\delta_n x,\ \lambda'_n+\delta'_n y)] \to -\log L(x,y)$ which implies $n[1-A(\lambda_n+\delta_n x)] \to -\log L(x,+\infty)$ and $n[1-B(\lambda'_n+\delta'_n y)] \to -\log L(+\infty,y)$.

Then $P(X'_{n-\ell+1} \leq \lambda_n+\delta_n x,\ Y'_{n-m+1} \leq \lambda'_n+\delta'_n y) \overset{W}{\to} L(x,y)$.

$$\Sigma \frac{1}{(c-a)!\ (c-b)!\ (a+b-c)!} \left(\log \frac{L(x,+\infty)}{L(x,y)}\right)^{c-a} \left(\log \frac{L(+\infty\ ,\ y)}{L(x,y)}\right)^{c-b} \left(\log \frac{L(x,y)}{L(x,+\infty)L(+\infty,y)}\right)^{a+b-c}$$

as $\frac{n!}{(n-c)!} \sim n^c$ and $n(A(\lambda_n+\delta_n x)-F(\lambda_n+\delta_n x,\lambda'_n+\delta'_n y)) \to \log \frac{L(x,+\infty)}{L(x,y)}$

$n(B(\lambda'_n+\delta'_n y)-L(\lambda_n+\delta_n x,\ \lambda'_n+\delta'_n y)) \to \log \frac{L(+\infty,y)}{L(x,y)}$ and $n(1+F(\lambda_n+\delta_n x,\ \lambda'_n+\delta'_n y)-A(\lambda_n+\delta_n x)-B(\lambda'_n+\delta'_n y)) \to \log \frac{L(x,y)}{L(x,+\infty)\ L(+\infty,y)}$, the summation being extended to the a,b,c such that $a \leq \ell-1$, $b \leq m-1$, $\max(a,b) \leq c \leq a+b$. It should be noted that if we have asymptotic independence of the maxima ($\ell=m=1$), i.e., we have $L(x\ y)=L(x,+\infty)\ L(+\infty,y)$, all summands are null except if c=a+b, so that the asymptotic distribution is:

$$L(x,+\infty) \sum_{a=0}^{\ell-1} \frac{1}{a!} [-\log L(x,+\infty)]^a \times L(+\infty,y) \sum_{b=0}^{m-1} \frac{1}{b!} [-\log L(+\infty,y)]^b.$$

As it splits in the product of the margins, the asymptotic independence of bivariate maxima implies the asymptotic independence of the (ℓ-th, m-th) maxima, as could be expected.

It was shown that if the bivariate asymptotic distribution has Gumbel margins then the bivariate distribution is of the form $\Lambda(x,y)=\exp(-(e^{-x}+e^{-y})k(y-x))$, where the dependence function k satisfies some conditions as $\Lambda(x,y)$ is a distribution function; see Tiago de Oliveira (1962/63) and (1975). In that case the asymptotic distribution of (ℓ-th, m-th)maxima is:

$$\Lambda_{\ell m}(x,y) = \Lambda(x,y)\ \Sigma \frac{1}{(c-a)!(c-b)!(a+b-c)!} \times$$

$$\times [(e^{-x}+e^{-y})\ k\ (y-x)-e^{-x}]^{c-a} [(e^{-x}+e^{-y})k(y-x)-e^{-y}]^{c-b} \times$$

$\times [(e^{-x}+e^{-y})(1-k(y-x))]$, where the summation is extended to the a,b,c such that $a \leq e-1$, $b \leq m-1$, $\max(a,b) \leq c \leq a+b$, from which we can build models corresponding to the ones developed.

In the diagonal case, where $\Lambda(x,y) = \exp[-\max(e^{-x}, e^{-y})]$, i.e., where for the (reduced) bivariate maxima we have X=Y with probability one, we have for $x \leq y$

$\Lambda_{e,m}(x.y) = \exp(-e^{-x}) \Sigma \frac{(e^{-x} - e^{-y})^{a-b}}{(a-b)!\, b!} e^{-by}$, the summation being extended to the a,b,c such that $a \leq e-1$, $b \leq m-1$, $b \leq a$ and similarity for $y \leq x$.

$\Lambda_{e,m}(x,y)$ has no density as $\Lambda_{1,1}(x.y) = \Lambda(x,y)$ but is not degenerate as happens to the last one.

4. SAMPLE SIZE ADJUSTMENT FOR THE PRATICAL USE OF ASYMPTOTIC DISTRIBUTIONS

It is known that, in applications for large n, as $F^n(\lambda_n + \delta_n x) \overset{W}{\to} L(x)$ and $L(x)$ is continuous, the convergence is uniform, so that $F^n(x)$ is approximated by $L((x-\lambda) / \delta)$ for convenient λ and $\delta(>0)$. In many applications, e.g. Ramachandran (1974), the attention was called to the case of large fires, where the samples are not necessarily of the same size. Consequently we have different values of λ and δ and a comparison should be possible. We will obtain general relations connecting the values of λ_n and δ_n so that the pooling of data for different samples is possible. The point was first raised, so the author's knowledge, in Ramachandran papers, under very special conditions implying that the asymptotic distribution of the maximum was a Gumbel one.

The point was developed more generally by Tiago de Oliveira(1976) for this case and, here, will be extended.

It is well known that if there exist α and $\beta>0$ such that $e^{\alpha+\beta x}(1-F(x)) \to 1$ as $x \to \infty$ the asymptotic distribution is the Gumbel one. Then taking $\lambda_n = \frac{\log n - \alpha}{\beta}$, $\delta_n = \frac{1}{\beta}$ we get coefficients such that $n(1-F(\lambda_n+\delta_n x)) \to e^{-x}$. Consequently if we denote by λ_n and δ_n the attraction coefficients for a sample size n, the attraction coefficients for a sample size N are $\lambda_N = \lambda_n + \delta_n \log \frac{N}{n}$, $\delta_N = \delta_n$. If the approximation for $F^n(x)$ is $\Lambda((x-\lambda)/\delta)$, the approximation for $F^N(x)$ is then

$\Lambda((x-\lambda)/\delta + \log N/n)$.

It is also known that if $(\beta\, x)^{\alpha}(1-F(x)) \to 1$ as $x \to \infty$, the asymptotic distribution is a Fréchet one. Then, if we take $\lambda_n = 0$, $\delta_n = n^{1/2}/\beta$ we have $n(1-F(\delta_n x)) \to x^{-\alpha}$. In that case if $F^n(x)$ is approximated by $\Phi_\alpha(x/\delta)$, then $F^N(x)$ is approxited by $\Phi\, \alpha(\frac{x}{\delta} \cdot (\frac{n}{N})^{1/\alpha})$. Finally if x_0 is such that $F(x)<1$ for $x<x_0$ and $F(x_0) = 1$ and $(\beta(x_0-x)^{-\alpha}(1-F(x)) \to 1$ as $x \to x_0$, we know that the asymptotic distribution is the Weibull one. In that case we can take $\lambda_n = x_0$ and $\delta_n = n^{-1/2}/\beta$; then if $F^n(x)$ is approximated by $\Phi_\alpha((x-x_0)/\delta)$, $F^N(x)$ is approximated by $\Psi_\alpha/(\frac{x-x_0}{\delta}) \cdot (\frac{N}{n})^{1/\alpha})$.

It must be noted that the condition $e^{\alpha+\beta x}(1-F(x)) \to 1$ as $x \to +\infty$ does not apply to the normal distribution; in that case, although normal distribution is attracted to Gumbel distribution, the relations between attraction coefficients for different values of n are different and easily obtained.
For the attraction to Φ_α and Ψ_α the relations between attraction coefficients given are always asymptotically true.
The attraction coefficients for the m-th maxima in the univariate and bivariate case are, evidently, the same.
We thank Prof. L.Haan for some remarks.

REFERENCES

Gumbel, E.J. (1935). Les Valeurs Extrêmes des Distributions Statistiques, Ann.Inst.Henri Poincaré, vol.4.

Gumbel, E.J. (1958). Statistics of Extremes. (Columbia University Press).

Ramachandran, G. (1974). Extreme Value Theory and large fire losses, The Astin Bull.vl.VII.

Smirnov, N.V. (1949). Limit Distributions for the Terms of a Variational Series. Trans.Series 1, vol.11, Amer.Math. Soc..

J.Tiago de Oliveira (1962/63). Structure of Bivariate Extremes, Extensions. Est.Mat.,Est. e Econom., vol 7.

J. Tiago de Oliveira (1975). Bivariate and Multivariate Extreme Distributions. Statistical Distributions in Scientific Work, Vol.1 (D.Reidel Publ.C^{e}).

J. Tiago de Oliveira (1976). Statistical Methodology for Large Chains. (to be published in The Astin Bulletin).

SUMMARY

In this paper the general distributions of m-th maxima in the univariate case and of the (ℓ-th, m-th)maxima in the bivariate case are obtained. Finally the use of the asymptotic distributions with location and dispersion parameters as finite sample size approximations of the exact distribution is considered for the pooling of samples of different sizes.

Key-Words: m-th, maxima asymptotic distributions, sample size adjustment.

Recent Developments in Statistics
J.R.Barra et al., editors

ON OPTIMAL BLOCK DESIGNS

A. Veevers and J.M. Taylor
Department of Computational and Statistical Science
University of Liverpool
Liverpool, U.K.

In situations where practical constraints dictate the use of non-orthogonal block designs methods of deciding which of the feasible arrangements to use are necessary, a criterion of optimality based on Calinski's (1971) matrix M_0 is investigated and shown to approximate D-optimality.

1. INTRODUCTION

In an experiment with t treatments, b blocks and N observations, let r be the vector of treatment replications and k be the vector of block sizes then the $t \times b$ incidence matrix n satisfies $n1=r$ and $n'1=k$, where 1 is a vector of ones. Denoting any vector z expressed as a diagonal matrix by z^δ, with inverse $z^{-\delta}$, attention is focussed on Tocher's (1952) matrix Ω here defined by

$$\Omega^{-1} = r^\delta - nk^{-\delta} n' + \frac{rr'}{N}. \tag{1}$$

Pearce (1963) proposed a classification of non-orthogonal designs based on the pattern of Ω^{-1} and later (1970) used an infinite series expansion for Ω in the development of an expression for the efficiency of block designs. Calinski (1971) derived the same series expansion in a general iterative procedure for Ω, expressing it as

$$\Omega = r^{-\delta} + \sum_{h=1}^{\infty} M_0^h r^{-\delta} \tag{2}$$

specifically in terms of the matrix M_0, where

$$M_0 = r^{-\delta} (nk^{-\delta} n' - \frac{rr'}{N}), \tag{3}$$

which was shown to play a useful role in describing the properties of a design. A criterion of optimality based on M_0 is given in the next section and some of its properties are briefly reported.

2. THE M_0 CRITERION OF OPTIMALITY

Federov (1972) gives a comprehensive account of the theory of optimal experiments and, for example, Mitchell (1974) provides an algorithm for constructing D-optimal designs.

For given r and k in the above context a D-optimal discrete design will be defined by an incidence matrix n for which $|\Omega|$ is a minimum. Now, seeking to minimize $|\Omega|$ for a specified r is, using (2), equivalent to finding

$$\min \left| I + \sum_{h=1}^{\infty} M_0^h \right| , \qquad (4)$$

since $|r^{-\delta}|$ is constant.

Let λ_i, i = 1,2,...,t, be the eigenvalues of M_0 then, since $M_0 1 = 0$, one of them is zero and, further, the others can be shown to belong to the half open interval [0,1). Let u_i be the eigenvector corresponding to the eigenvalue λ_i then

$$\left(I + \sum_{h=1}^{H} M_0^h\right) u_i = \left(u_i + \sum_{h=1}^{H} \lambda_i^h u_i\right)$$

$$= \left(1 + \sum_{h=1}^{H} \lambda_i^h\right) u_i$$

so that, as $H \to \infty$, the eigenvalues of $(I + \sum_{h=1}^{H} M_0^h)$ approach $1/(1-\lambda_i)$. The D-optimality criterion (4) now becomes

$$\min \prod_{i=1}^{t} 1/(1-\lambda_i)$$

or

$$\max \prod_{i=1}^{t} (1-\lambda_i). \qquad (5)$$

To a first approximation

$$\prod_{i=1}^{t} (1-\lambda_i) = 1 - \sum_{i=1}^{t} \lambda_i$$

$$= 1 - \mathrm{Tr}(M_0)$$

suggesting that

$$\min \mathrm{Tr}(M_0) \qquad (6)$$

will approximate D-optimality. Higher order approximations can be made, if necessary, involving successive minimization of the trace of powers of M_0, as seen by expanding the logarithm of (5).

If designs of type O or T are feasible it can be shown that the criterion (6) is equivalent to both D and A-optimality. The intuitively appealing nature of (6) as a design criterion can be seen when it is viewed as

$$\min N.\mathrm{Tr}(M_0) = \min \sum_{i=1}^{t} \sum_{j=1}^{b} \left(n_{ij} - \frac{r_i k_j}{N}\right)^2 \frac{N}{r_i k_j} \qquad (7)$$

which has the contingency table χ^2 form and is here a quadratic measure of closeness to orthogonality.

In incomplete equal sized block situations, for example, $Tr(M_0)$ is invariant for any feasible binary design so the second or higher order approximation is used. The performance of this procedure for constructing optimal block designs, Taylor (1974), compares favourably with that reported by Mitchell (1973) for incomplete block designs.

REFERENCES

Calinski, T. (1971). On some desirable patterns in block designs. Biometrics 27, 275-92.

Federov, V.V. (1972). Theory of optimal experiments. Academic Press.

Mitchell, T.J. (1973). Computer construction of small "D-optimal" incomplete block designs. Bull. Int. Statist. Inst. 45 book 2, 199-205.

Mitchell, T.J. (1974). An algorithm for the construction of "D-optimal" experimental designs. Technometrics 16, 203-10.

Pearce, S.C. (1963). The use and classification of non-orthogonal designs. J.R. Statist. Soc. A 126. 353-69.

Pearce, S.C. (1970). The efficiency of block designs in general. Biometrika 57, 339-46.

Taylor, J.M. (1974). Ph.D. Thesis, Liverpool University.

Tocher, K.D. (1952). The design and analysis of block experiments. J.R. Statist. Soc. B 14, 45-100.

Recent Developments in Statistics
J.R.Barra et al., editors

AN INVARIANCE PRINCIPLE IN STOCHASTIC APPROXIMATION

H. Walk
Fachbereich Mathematik
Universität Essen - Gesamthochschule
4300 Essen
Bundesrepublik Deutschland

The Robbins-Monro method of stochastic approximation yields an estimation of a zero point ϑ of a regression function f according to the recursion formula

$$X_{n+1} := X_n - \frac{d}{n} Y_n \text{ with } E(Y_n | X_1,\ldots,X_n) = f(X_n) \text{ a.s.}$$

and $d \in (0,\infty)$. In the case $f : \mathbb{R}^k \to \mathbb{R}^k$ with k-dimensional random vectors X_n, Y_n the asymptotic normality of the distribution of $\sqrt{n}(X_n - \vartheta)$ - for k=1 first investigated by Chung [2] - has been proved by Sacks [11] and Fabian [4] under assumptions not including the case of an unsymmetric $k \times k$-matrix of the first partial derivatives of f at ϑ.

By use of a modified norm in $\mathbb{R}^k$ the assumptions above can be weakened in view of a not necessarily symmetric matrix. The result can be generalized for a separable real Hilbert space H, where the Fréchet derivative of f at ϑ replacing the matrix need neither be symmetric nor compact, and yields thus an answer to a problem mentioned by Schmetterer [12] , which has also been treated by Révész independently at this meeting. An application concerns a Robbins-Monro type procedure for an initial value problem written in the form f(x) = 0 with $f : L_2[0,1] \to L_2[0,1]$ defined by

$$f(x) := x(\cdot) - a_0 - \int_0^{\cdot} h(s,x(s))\, ds,$$

where $a_0 \in \mathbb{R}$, $h : [0,1] \times \mathbb{R} \to \mathbb{R}$ with $\sup_{s,y} |h_y(s,y)| < \frac{1}{2}$, continuity of h_y and only empirically available function values of h and where a Monte Carlo method is used for estimating the integral.

The result can further be generalized to the following invariance principle formulated with $\vartheta = 0$, d = 1 without loss of generality.

<u>THEOREM.</u> Let

$$X_{n+1} := X_n - \frac{1}{n} f(X_n) + \frac{1}{n} V_n \text{ , } n \in \mathbb{N}$$

with a measurable function $f : H \to H$ possessing a Fréchet derivative A at 0 such that

$$\inf\{\| \exp[u(\tfrac{I}{2} - A)]\|;\ u \in \mathbb{R}_+\} < 1 \text{ (I identity operator)}$$

or - equivalently - spectrum(A) $\subset \{\lambda \in \mathbb{C} : \operatorname{Re}\lambda > \frac{1}{2}\}$ and with H-valued random variables X_1, V_n on a probability space $(\Omega, \mathfrak{A}, P)$ such that

$$\forall_{n \in N} \quad E\|V_n\|^2 < \infty \quad , \; E\,(V_n | X_1, \ldots, X_n) = 0 \text{ a.s.}$$

Let further $S : H \rightarrow H$ be an S-operator (symmetric, positive semi-definite, compact linear operator with finite trace [9] , [8]) different from the zero operator and S^n a covariance operator of V_n given $X_1, \ldots, X_n$. Suppose

$$E\,\|\frac{1}{n} \sum_{j=1}^{n} S^j - S\,\| \rightarrow 0 \qquad (n \rightarrow \infty),$$

$$\frac{1}{n} \sum_{j=1}^{n} E\,\|V_j\|^2 \rightarrow \operatorname{trace}(S) \qquad (n \rightarrow \infty),$$

$$\forall_{r>0} \quad \frac{1}{n} \sum_{j=1}^{n} E(\,\|V_j\|^2\, \chi\{\|V_j\|^2 \geq rj\} | X_1, \ldots, X_j) \underset{P}{\rightarrow} 0 \;(n \rightarrow \infty),$$

$$X_n \rightarrow 0 \quad (n \rightarrow \infty) \text{ a.s.}$$

Then the sequence of random elements Z_n of $C_H[0,1]$ with sup-norm which are defined by

$$Z_n(t) := \frac{1}{\sqrt{n}} R_{[nt]} + (nt - [nt])\, \frac{1}{\sqrt{n}} (R_{[nt]+1} - R_{[nt]}), \; t \in [0,1],$$

with $R_n := nX_{n+1}$ converges in distribution to a Gaussian Markov process G with

$$G(t) = W(t) + (I-A) \int_{(0,1]} \exp[(\ln u)\,(A-2I)]\; W(tu)\, du \qquad (t \in [0,1])$$

$$= \int_{(0,t]} \exp\left[\,(\ln \tfrac{v}{t})\;\;(A-I)\right]\, dW(v) \qquad (t \in (0,1]),$$

where W is a Brownian motion in H with W(0)=0, EW(1)=0 and covariance operator S of W(1).

In the Theorem the assumption of a.s. convergence $X_n \rightarrow 0$ follows according to Venter [14] from the two conditions

$$\exists_{c \in \mathbb{R}_+} \quad \forall_{x \in H} \quad \|f(x)\| \leq c\|x\|,$$

$$\forall_{\varepsilon \in (0,1)} \quad \inf_{\varepsilon \leq \|x\| \leq \varepsilon^{-1}} \quad (x, f(x)) > 0.$$

In the special case $H = \mathbb{R}$ the limit distribution is also induced by the random function η with

$$\eta(0) = 0, \; \eta(t) = t^a\, \zeta\!\left(\frac{\sigma^2}{2a-1}\, t^{1-2a}\right) \qquad (0 < t \leq 1),$$

where ζ is a standard Brownian motion in $\mathbb{R}$ and (A,S) is replaced by $(a, \sigma^2) \in (\frac{1}{2}, \infty) \times (0,\infty)$. After the announcement of this talk the paper [7] of McLeish appeared giving the invariance principle in the case $H = \mathbb{R}$ with another proof under slightly more restrictive assumptions.

In the proof of the Theorem at first the case $X_1=0$, $f(x)\equiv x$ is treated which yields an invariance principle for a martingale with H-valued random variables, namely $(\sum_{j=1}^{n} V_j)$ (compare Brown [1] and Kuelbs [6] for the case of real-valued resp. independent Banach space valued summands); this together with a theorem of Skorokhod [13] and Dudley [3] on the connection between weak and a.s. convergence (compare also Pyke [10]) leads to the result in the case $X_1=0$, $f(x)\equiv Ax$; regarding the convergence behaviour of $\sqrt{n}\,(f(X_n)-AX_n)$ and using a truncation idea of Hodges and Lehmann [5] one obtains finally the general result. - A detailed proof is given in [15] .

REFERENCES

1. Brown, B.M. (1971). Martingale central limit theorems. Ann. Math. Statist. 42, 59-66.
2. Chung, K.L. (1954). On a stochastic approximation method. Ann. Math. Statist. 25, 463-483.
3. Dudley, R.M. (1968). Distances of probability measures and random variables. Ann. Math. Statist. 39, 1563-1572.
4. Fabian, V. (1968). On asymptotic normality in stochastic approximation. Ann. Math. Statist. 39, 1327-1332.
5. Hodges, J.L., Jr.; Lehmann, E.L. (1956). Two approximations to the Robbins-Monro process. Proc. Third Berkeley Symp. Math. Statist. Prob., I, 95-104.
6. Kuelbs, J. (1973). The invariance principle for Banach space valued random variables. J. Multivar. Anal. 3, 161-172.
7. McLeish, D.L. (1976). Functional and random central limit theorems for the Robbins-Monro process. J. Appl. Prob. 13, 148-154.
8. Parthasarathy, K.R. (1967). Probability Measures on Metric Spaces (Academic Press, New York).
9. Prokhorov, Yu.V. (1956). Convergence of random processes and limit theorems in probability theory, Theory Prob. Appl. 1, 157-214.
10. Pyke, R. (1969). Applications of almost surely convergent constructions of weakly convergent processes. Probability and Information Theory, 187-200 (Berlin, Springer).
11. Sacks, J. (1958). Asymptotic distribution of stochastic approximation procedures. Ann. Math. Statist. 29, 373-405.
12. Schmetterer, L. (1958). Sur l'itération stochastique. Le Calcul des Probabilités et ses Applications, Colloq. Intern. Centre Nat. Rech. Sci. 87, 55-63.
13. Skorokhod, A.V. (1956). Limit theorems for stochastic processes. Theory Prob. Appl. 1, 261-290.
14. Venter, J.H. (1966). On Dvoretzky stochastic approximation theorems. Ann. Math. Statist. 37, 1534-1544.
15. Walk, H. An invariance principle for the Robbins-Monro process in a Hilbert space (submitted for publication).

PART TWO: Selected Contributed Papers

B

Applied Statistics

Recent Developments in Statistics
J.R.Barra et al., editors

STATAL, A STATISTICAL LIBRARY

OF PROCEDURES AND PROGRAMS

J.G. Bethlehem

Department of Mathematical Statistics

Mathematical Centre

Amsterdam

A library of statistical programs and procedures is described. The most important view-points of the constructors are mentioned and illustrated with examples.

The department of Mathematical Statistics of the Mathematical Centre in Amsterdam is at the moment constructing a library (called STATAL) containing statistical procedures and programs, which can be used in the CDC-Cyber 73-26 computer of SARA (Academic Computing Centre Amsterdam).
The makers of the library are in the first place statisticians and only in the second place programmers. One can imagine that as to designing and constructing such a library the view-points are somewhat different than those of the professional software specialists. I will mention some of our most important view-points and illustrate them with some examples.

First of all we aim at two different categories of users:
1. The ALGOL-60-user can make use of the set of ALGOL-60 procedures. In general these procedures need (except for transmission of values via parameters) no input nor produce output, so that they are to a high degree independent of the compiler used. When, for instance, someone has two samples of observations and wants to have the value and the two-sided tail probability of Wilcoxon's two-sample test, he can use the procedure WILCOXONS W2, which has the following heading:

```
REAL PROCEDURE WILCOXONS W2(SAMPLE1, L1, U1, SAMPLE2, L2, U2, SHIFT, P2);
```

The procedure assumes the first sample to be contained in the array elements SAMPLE1[L1],...,SAMPLE1[U1] and the second sample in SAMPLE2[L2],..., SAMPLE2[U2]. SHIFT is an input parameter with as value the hypothetical shift in the two distributions and P2 is an output parameter which is to contain the two-sided tail probability of the test statistic under the null-hypothesis. The value of the test statistic itself is assigned to the procedure identifier WILCOXONS W2.
The user himself has to produce a piece of program which will cause the arrays to be filled with the samples and a piece of program to print the results. If sizes of the samples are m and n respectively (which are to be read by the program), then the program could look as follows:

```
BEGIN INTEGER M, N; REAL W, PROB, SHIFT;
    REAL PROCEDURE WILCOXONS W2(PARAMETERS); CODE 42401;
    INREAL(60, M); INREAL(60, N); SHIFT:= 0;
    BEGIN REAL ARRAY S1[1:M], S2[1:N];
        INARRAY(60, S1); INARRAY(60, S2);
        W:= WILCOXONS W2(S1, 1, M, S2, 1, N, SHIFT, PROB);
        OUTPUT(61, "("+5ZD.D, 3B, +D.6D")", W, PROB)
    END
END
```

The user keeps in this way much in his own hands and the routine calculations are left to the library procedures.

2. The computer user without much programming knowledge may take one of the statistical programs. These perform the complete computations from input up to and including output of a statistical problem. The programs only demand from the user knowledge about the organisation of the input data.
If, for instance, he wants to solve the problem from the previous section he can take the program STUWIL2, which contains a number of two-sample tests. The complete job for our computer would now look as follows:

```
STAT,CM77000,T30.                   (specification of amounts of
ACCOUNT,ME1F1234V.                   central memory and computing
                                     time and accountnumber)
MODLIB,STATAL1,MCSTAT,*,STUWIL2.    (retrieving, loading and
STUWIL2.                             executing the program)
*EOR
CHANNEL,63=60                       (assignment of input and output
CHANNEL,64=61                        channels)
CHANNEL,END
EXAMPLE                             (name of the job)
WILCOXON                            (test to be perfomed)
0.05                                (level of significance)
                                    (no pairchart to be printed)
PROBLEM1                            (name of the problem)
50                                  (size of the first sample)
40                                  (size of the second sample)
Observations of the first sample
Observations of the second sample
END INPUT
*EOF
```

The program performs all the necessary computations and delivers a neat readable output. The user has not much work to do but he is on the other hand stuck to the program as it is.

In constructing the library the makers emphasized among others the following points:

1. <u>User's convenience</u>.
The user must not be troubled, especially in the case of programs, with intricate input and difficult to interpret output. The makers did not consider it a drawback that as a consequence a program is sometimes a little less efficient, because manhours are still much more expensive than computer seconds.
We will illustrate this view-point with the program MULTREG for multiple regression analysis.
Suppose an experimentor has made 10 observations on a dependent variable y (yield) and two independent variables x_1 (temperature) and x_2 (pressure) and wants to test the following model:

$$E(\sqrt{\underline{y}}) = \beta_0 + \beta_1 \ln(x_1) + \beta_2 \frac{x_1}{x_2}$$

Suppose the data has the following structure:

$$\begin{matrix} x_{1,1} & x_{2,1} & y_1 \\ x_{1,2} & x_{2,2} & y_2 \\ . & . & . \\ . & . & . \\ . & . & . \\ x_{1,10} & x_{2,10} & y_{10} \end{matrix}$$

In former days he first had to rearrange his data by hand or with a separate program, to get his vector of independent values and his design matrix.

$$\sqrt{y_1} \quad \sqrt{y_2} \ \dots \ \sqrt{y_{10}} \qquad\qquad \begin{matrix} 1 & \ln(x_{1,1}) & x_{1,1} / x_{2,1} \\ 1 & \ln(x_{1,2}) & x_{1,2} / x_{2,2} \\ . & . & . \\ . & . & . \\ . & . & . \\ 1 & \ln(x_{1,10}) & x_{1,10} / x_{2,10} \end{matrix}$$

Now the data was ready for the regression program.
In our MULTREG program the following input statements are sufficient to rearrange the data and perform the regression analysis on the right model:

```
"MODEL" SQRT(YIELD) = BETAØ + BETA1 * LN(TEMPERATURE) +
                      BETA2 * TEMPERATURE / PRESSURE
"INPUT" 10 * [TEMPERATURE, PRESSURE, YIELD]
```

The model specification is in accordance with the usual statistical terminology. The user can use his own variable names which facilitates reading the output still more. The input specification describes the structure of the input data. The existence of such an input specification makes it possible that a great number of different structures are accepted by the program. The program itself makes the right correspondences between model and input specification and determines which are the variables and which the coefficients to be estimated.

2. User protection.
Especially in the case of procedures care has been taken to prevent wrong application. When a procedure call is made while a parameter has a wrong value, a diagnostic error message is produced. After this message the execution of the program is terminated to avoid further computations with wrong values. The following example illustrates this. The procedure to compute the binomial distribution function has the heading:

```
REAL PROCEDURE BIN(X, N, P);
```

where x is the number of successes, n the total number of experiments and p the probability of success in a single experiment.
One can imagine that when a value outside the unit interval is assigned to the probability p the result will be an error message. But also the call PROB:= BIN(5, 7.2, 0.5), indicating the number of experiments to be 7.2, will cause an error message and no rounding-off will take place, as is often done in other libraries.
It would lead to the following error message:

```
STATAL3 ERRORMESSAGE :
NAME OF THE PROCEDURE:  BIN
ERRORNUMBER          :  2
WRONG VALUE          :  +7.2000000000000"+000
```

In general the value of the errornumber corresponds to the number of the parameter in the procedure call.

3. Structure.
The programmers of procedures and programs give their work a clear structure. Pieces of programs which, from a technical or statistical point of view, are a logical unit, are put together in a procedure or block. Names of variables show what they are used for. To reach this the use of long variable names is not avoided.
These two elements, structure and naming, make it possible that someone who is not the programmer himself is able to understand the working of the program and make changes and improvements.

4. Precision.
Especially in the case of procedures which compute probability distributions it has been strived for a large relative precision. In general this is not necessary. In statistical tests, for instance, one is only interested to know whether the tail-probability is greater or less than 5 %.
If, however, distribution functions are used as a tool to compute complicated probabilities, for instance combined probabilities, then a small precision of a used procedure can be fatal for the ultimate result. Also in this case the slightly bigger amount of computing time is not considered disadvantageous.

The library as we intend to make it, is still in the constructing phase although a part of it is already in use.
Below follows a global survey of the final contents, but contributions from outside the Mathematical Centre can be inserted if they fit into the framework of STATAL.

Procedures:

1. Probability distributions (distribution functions, densities and inverses)
2. Test-statistics (also approximate), tail-probabilities under the null-hypothesis
3. Correlation coefficients (also approximate), tail-probabilities under the null-hypothesis
4. Jøreskog factor analysis (with user-supplied procedure to read the data)
5. Sorting & Ranking (one- and two-dimensional arrays)
6. Permutations and combinations (on frequency tables)
7. Random number generators (distributions, permutations, combinations)
8. Tables and pictures

Programs:

1. One-sample problem (Student, Wilcoxon, Robust estimations)
2. Two-sample problem (Student, Wilcoxon, Fisher, Ansari-Bradley)
3. K-sample problem (Kruskal-Wallis, Terpstra, Steel, Doornbos, Spjøtvoll)
4. Multiple regression analysis (also ridge regression)
5. Multivariate two-sample problem (Hotelling, Box)
6. Principal components analysis (including tests)
7. Factor analysis (Maximum likelihood, Jøreskog, equal residual variance)
8. Independence problem (correlation, chi-squared)
9. Hierarchical cluster analysis (7 methods)
10. Item and scale analysis (also Mokken)

REFERENCES:

Bethlehem, J.G. (ed.). (1976). STATAL reference manual. (Mathematical Centre, Amsterdam).

Gelderen, M. van. (1975). A users program for multiple regression analysis, Mathematical Centre report, SW 35/76. (Mathematical Centre, Amsterdam).

Recent Developments in Statistics
J.R.Barra et al., editors

A MIXED MODEL FOR SURVIVAL APPLIED TO BRITISH CHILDREN WITH NEUROBLASTOMA

J.F. Bithell and R.G. Upton

Childhood Cancer Research Group, Oxford
and
Department of Biomathematics, Pusey Street, Oxford

SUMMARY - Data on the survival of 487 children suffering from neuroblastoma or ganglioneuroma are analysed using a model which postulates a probability of indefinite survival and a time-independent hazard rate, both of which are related logistically to linear functions of observations on six concomitant variables. The latter include age at ascertainment, sex and four qualitative factors describing the tumour. There was a considerable degree of confounding, but it is clear that age is of fundamental importance. Various models were investigated; that finally selected is given in full and permits the prediction of survival.

Neuroblastoma is a tumour of infancy which is generally highly malignant but which also has a relatively high rate of spontaneous regression. These features make the survival prognosis of considerable importance and render unsuitable many of the models of survival used in other areas of cancer research. The prognosis appears to depend upon a number of factors, notably age at ascertainment, sex, the stage of progression of the tumour at ascertainment and its morphological type. These factors are generally confounded, however, and it is one object of this analysis to examine the inter-relationships in order to ascertain which factors appear to have an intrinsic effect on survival and which are perhaps fortuitously associated with other determinants. Another aim is to provide a parametric model which fits the data and which could consequently be used for predicting the outcome in individual cases.

In many analyses of survival, death of the individual is virtually certain to occur within a time comparable to the period of follow-up, so that, even if follow-up is not complete for all cases, an empirical survival curve (E.S.C.) declining asymptotically towards zero will be observed. Consequently, models assuming a continuing positive hazard function - such as the Weibull family - may well be appropriate. In other situations - particularly in the case of neuroblastoma - one has to consider a non-zero "asymptote" to the E.S.C. (see Fig.) corresponding to "indefinite" survival - i.e. survival for a period which is long in comparison with observed times to death from the tumour. Indeed the greatest clinical interest centres not so much on the expectation of life of a child who eventually dies as on the probability of an indefinite survival, which may be regarded as a "cure". In practice it is difficult to know what length of

FIGURE. Survival curves of children with neuroblastoma by age at anniv.

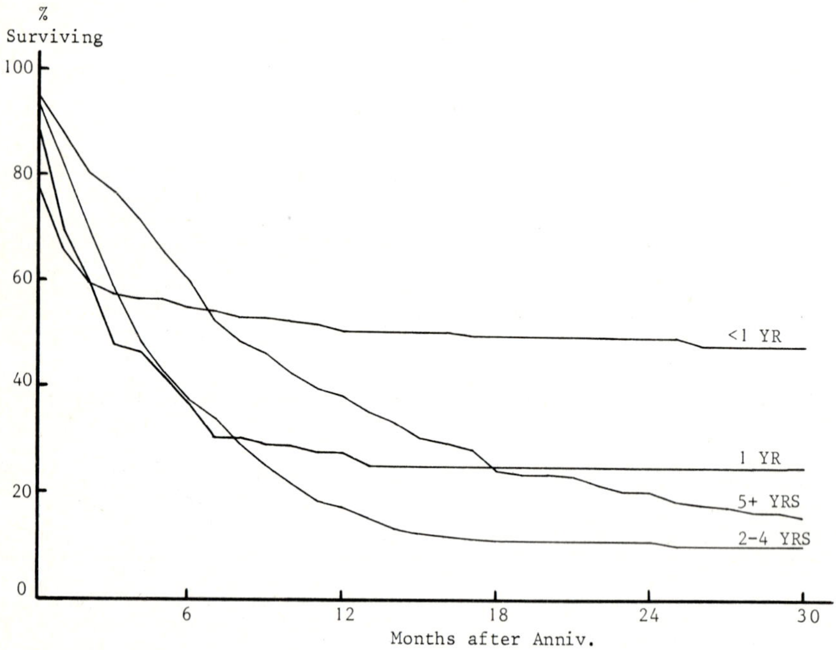

survival is equivalent to a cure because of the inherently long tails of survival distributions. Thus Gross et al (1959) consider that there is an appreciable risk of recurrence for only two years, while Collins et al (1956) have devised a formula which relates the risk period to the age of the child at ascertainment. Although the great majority of children who ultimately die have done so within three years, it seems unsatisfactory to postulate a fixed cut-off point or indeed to express a prognosis in terms of certainties. What we have aimed at is to describe by a probability model both the chance of surviving indefinitely and the hazard rate for a child ultimately dying of the disease.

The Data

Between 1962 and 1967 some 539 children with neuroblastoma or ganglioneuroma under the age of 15 years were registered in Britain. Registration records were sent to the Oxford Survey of Childhood Cancers (O.S.C.C.) whose investigators abstracted hospital and other records and conducted a follow-up survey. Some doubt attached to the histology in 52 of the cases and these have therefore been excluded from our analysis. The remaining 487 children constitute the largest series recorded in the literature to date; they have been reported on more fully by Kinnier Wilson and Draper (1974), although there are differences between the histological classifications used by these authors and by ourselves.

For each case the "anniversary date" or "anniv" is defined as the date (to the nearest month) when treatment was begun or, if there was no treatment, the date of first admission to hospital. "Anniv age" is similarly defined and, like survival time (to death or last follow-up) it is recorded to the nearest month. The great majority (84%) of all deaths recorded had occurred within one year of anniv and 98.7% within 3 years, while all surviving children were followed up to two years and all but four to three years from anniv. For these reasons it is not greatly misleading to look at crude survival rates (which are nearly equivalent to three-year survival rates) as a means of identifying the factors of greatest importance. In addition to age at anniv, which was treated as a quantitative variable, some five qualitative variables were analysed, namely sex, number of deposits, histology, site and tumour stage. The levels of the latter four factors are shown, together with their sex-specific crude survival rates, in Table 1. The number of deposits - or sites initially affected - was recorded, together with the site and stage of the largest one, since in the case of neuroblastoma it is often not possible to identify a primary site in distinction to metastases. Certain groupings of the levels shown in Table 1 were used in some of the analyses in order to economise on parameters.

Reference to Tables 1 and 2 suggests that each of the six factors selected for study could have a bearing on the survival probability. Thus increasing age, for example, clearly worsens the prognosis, at any rate throughout the first three years of life. There are also appreciable differences between the risks for tumours at different sites and with different histologies, maturing tumours offering the best and undifferentiated tumours the worst prognoses. Females appear to have a better survival rate than males, a finding which has not been statistically significant in previous series (Fortner et al, 1968). However, on allowing for age, the effect partly disappears, while the converse comparison leaves a very strong age effect for both sexes considered separately. Similar cross-tabulations throw light on other relationships, but the numbers will obviously become small in higher dimensional tables. An advantage of the parametric model described below is that it makes possible such multi-factorial comparisons.

TABLE 1. Basic levels of four qualitative variables with frequency and crude survival rates, by sex.

Variable	Males			Females			Total	
	No. Surv.	No. Cases	% Surv.	No. Surv.	No. Cases	% Surv.	No. Cases	% Surv.
Number of deposits								
1 1 deposit	38	111	34.2	41	88	46.6	199	39.7
2 2 deposits	9	67	13.4	15	67	22.4	134	17.9
3 3+ deposits	0	66	0.0	5	49	10.2	115	4.3
4 "Multiple"/NS/None	0	24	0.0	1	15	6.7	39	2.6
Histology								
1 Undifferentiated	6	76	7.9	7	53	13.2	129	10.1
2 Rosettes	8	54	14.8	12	42	28.6	96	20.8
3 Unspec.differentiation	18	87	20.7	26	77	33.8	164	26.8
4 Ganglioneuroblastoma/ Ganglioneuroma	13	19	68.4	17	26	65.4	45	66.7
5 Other/NS/None	2	32	6.3	0	21	0.0	53	3.8
Site of largest deposit								
1 No largest dep/NR	1	46	2.2	4	36	11.1	82	6.1
2 Adrenal	7	47	14.9	4	38	10.5	85	12.9
3 Liver	2	8	25.0	9	17	52.9	25	44.0
4 Abdomen (other)	6	28	21.4	4	24	16.7	52	19.2
5 Retroperitoneal	9	49	18.4	4	26	15.4	75	17.3
6 Ganglia	8	14	57.1	19	32	59.4	46	58.7
7 Thoracic/head/neck/ pelvis	13	54	24.1	17	39	43.6	93	32.3
8 Bone	1	19	5.3	1	6	16.7	25	8.0
9 Other	0	3	0.0	0	1	0.0	4	0.0
Stage of largest deposit								
1 Localised	17	44	38.6	32	57	56.1	101	48.5
2 Adherent	12	31	38.7	5	24	20.8	55	30.9
3 Invasive	8	74	10.8	12	54	22.2	128	15.6
4 Not spec.	9	73	12.3	9	48	18.8	121	14.9
5 No largest dep/NR	1	46	2.2	4	36	11.1	82	6.1
TOTAL	47	268	17.5	62	219	28.3	487	22.4

Mathematical Model

We now develop a model in which the survival characteristics of the ith child ($i = 1,2....n$) are related to two parameters, p_i and λ_i: $p_i = 1-q_i$ represents the probability that the child never dies of the disease (assuming that observation is continued indefinitely), while λ_i represents the discrete-time hazard rate - or risk per month - conditional on the certainty of his ultimate death from the disease. Because these two parameters are probabilities it seems reasonable to relate them to the concomitant information available, $\underline{X}_i$, $\underline{Y}_i$ by means of logistic functions:

$$\lambda_i = \exp(\underline{\alpha}'.\underline{X}_i)/\left(1+\exp(\underline{\alpha}'.\underline{X}_i)\right) \quad (1)$$

$$p_i = 1-q_i = \exp(\underline{\beta}'.\underline{Y}_i)/\left(1+\exp(\underline{\beta}'\ \underline{Y}_i)\right) \quad (2)$$

where $\underline{\alpha} = \{\alpha_1,\alpha_2,....\alpha_r\}'$ and $\underline{\beta} = \{\beta_1,\beta_2,\beta_3....\beta_s\}'$ are vectors of descriptive parameters and $\underline{X}_i$, $\underline{Y}_i$ are vectors of variables (possibly identical) relating to the ith child.

The model therefore associates with each combination of factors $(\underline{X},\underline{Y})$ a specific probability distribution for survival; the corresponding survival curve has a slope determined by the (constant) value of λ and declines asymptotically to a level given by p.

It is easily seen that the likelihood of the data under this model is given by

$$L = \prod_{i=1}^{n}\{p_i+q_i(1-\lambda_i)^{s_i}\}^{A_i}\{q_i\lambda_i(1-\lambda_i)^{s_i}\}^{1-A_i} \quad (3)$$

where A_i is an indicator variable such that $A_i = 1$ if the ith child was last followed-up and found to be alive s_i months after anniv and $A_i = 0$ if the ith child is known to have died s_i months after anniv.

Parametrisation

The qualitative variables were incorporated into equations (1) and (2) by means of the usual dummy variables technique, so that a t-level factor would be associated with t-1 components of $\underline{\alpha}$ (or $\underline{\beta}$) and at most one of the corresponding elements of $\underline{X}_i$ (or $\underline{Y}_i$) would be unity as opposed to zero. For this purpose the base levels of the factors were selected arbitrarily without loss of generality.

This scheme had to be modified whenever site and stage variables entered the model together since the first level of the former was logically equivalent to the fifth level of the latter. Cases in this category were associated with a single parameter, the corresponding variable value being zero for all other cases, for whom stage and site were parametrised in the usual way.

TABLE 2. Frequency and crude survival rates by age at anniv (in completed years) and by sex.

Age at Anniv. (Yrs)	Male			Female			Total	
	No. Surv.	No. Cases	% Surv.	No. Surv.	No. Cases	% Surv.	No. Cases	% Surv.
0	24	59	40.7	35	62	56.5	121	48.8
1	7	37	18.9	11	42	26.2	79	22.8
2	4	37	10.8	3	22	13.6	59	11.9
3	2	26	7.7	4	21	19.0	47	12.8
4	0	26	0.0	2	17	11.8	43	4.7
5-9	8	54	14.8	3	38	7.9	92	12.0
10+	2	29	6.9	4	17	23.5	46	13.0
TOTAL	47	268	17.5	62	219	28.3	487	22.4

Age was treated as a quantitative variable and a number of different relationships were explored, as described below. The parametrisation in each case involved assigning powers of the age at anniv as components of $\underline{X}_i$, $\underline{Y}_i$; the corresponding parameters are then analogous to regression coefficients.

Finally, constant parameters were included in both $\underline{\alpha}$ and $\underline{\beta}$. The corresponding components of $\underline{X}_i$, $\underline{Y}_i$ were of course invariably unity.

Estimation of the Parameters

Parameter estimation was by maximum likelihood, i.e. by finding solutions to the equations

$$\frac{\partial \ell}{\partial \alpha_j} = \frac{\partial \ell}{\partial \beta_k} = 0 \qquad \begin{array}{l} j = 1,2,\ldots r \\ k = 1,2,\ldots s \end{array} \tag{4}$$

where $\ell = \ln L$.

Using the relations

$$\frac{\partial \lambda_i}{\partial \alpha_j} = \lambda_i(1-\lambda_i)X_{ij}; \qquad \frac{\partial p_i}{\partial \beta_k} = p_i(1-p_i)Y_{ik} \tag{5}$$

we find, after a little manipulation, that the equations (4) become

$$\frac{\partial \ell}{\partial \alpha_j} = \sum_{i=1}^{n} X_{ij}\{\lambda_i s_i (B_i - q_i) + (1-\lambda_i)(1-A_i)\} = 0 \tag{6}$$

$$\frac{\partial \ell}{\partial \beta_k} = \sum_{i=1}^{n} Y_{ik} B_i = 0 \tag{7}$$

where
$$B_i = p_i\{A_i(p_i + q_i(1-\lambda_i)^{s_i})^{-1} - 1\} \tag{8}$$

These equations do not admit of an explicit solution but the Newton-Raphson method is generally quite straightforward for this model and converges quickly to a solution. For starting values, all parameters except the constants were set to zero (equivalent to the null hypothesis that none of the concomitant information has any effect on survival). The latter were chosen as

$$\lambda_0 = 1/(1+\bar{s}) \quad \text{or} \quad \alpha_0 = -\ln \bar{s} \tag{9}$$

where s is the mean survival of the n-m cases dying, and

$$p_0 = {}^m/n \qquad \text{or} \quad \beta_0 = \ln m - \ln(n-m) \tag{10}$$

Use of these values is equivalent to estimating the hazard rate from the mean survival of the children dying and the indefinite survival probability from the empirical proportion surviving. It can be justified as a first order approximation to a solution of equations (6) and (7) under the null hypothesis mentioned above, on the assumption that the survivors have a small probability of subsequent death.

Further differentiation of the log likelihood leads to a Hessian matrix whose negated inverse provides an estimate of the asymptotic dispersion matrix of the parameter estimates in the usual way.

Various models were fitted, increasing the number of parameters in a stepwise manner and assessing their marginal importance by means of likelihood-ratio tests. Because this investigation had to be carried out in both the p- and λ- domains, the number of models fitted was moderately large. For this reason, we give only a qualitative summary of the initial explorations; further details are given in Upton (1973).

Results

(1) Assuming a constant hazard rate, λ, and fitting factors separately for the survival rate, p, revealed a significant effect for each qualitative factor and for linear age, the latter being negatively associated with p.

TABLE 3. Results of the Final Run of the Model, Showing Parameter Estimates and their Standard Errors.

FACTOR		Basic Levels	$\hat{\alpha}$	S.E.($\hat{\alpha}$)	$\hat{\beta}$	S.E.($\hat{\beta}$)
Constant		-	-0.419	0.196	-1.869	1.010
Age:	Stage 1, Linear		-0.169	0.467	-0.769	0.309
	Stage 1, Quadratic	-	0.018	0.004	0.049	0.024
	Other stages, Linear	-	-0.303	0.058	-1.195	0.267
	Other stages, Quadratic	-	-	-	0.054	0.025
Sex:	Male	1	-	-	-0.210	0.354
No. deposits:	1 Deposit	1	-0.454	0.154	2.648	0.597
	2 Deposits	2	-0.373	0.147	1.576	0.605
Stage:	Localised	1	0.294	0.277	0.260	0.673
	Adherent	2	-	-	0.539	0.617
	No largest/NR	5	-0.058	0.158	-1.919	0.800
Site	Adrenal	2	-	-	-2.471	0.668
	Liver	3	-	-	-2.354	0.815
	Abdomen (other)	4	-	-	-1.224	0.699
	Retroperitoneal	5	-	-	-1.989	0.705
	Thoracic Etc.	7	-0.496	0.168	-0.503	0.631
	Bone	8	-0.848	0.232	-6.300	0.630
Histology	Undifferentiated	1	-	-	0.891	0.886
	Rosettes/unspec	2,3	-0.333	0.121	2.045	0.824
	Ganglioneuroblastoma/ Ganglioneuroma	4	-0.502	0.314	5.074	1.027

32 Parameters

Increase in log likelihood (over constant p, λ model) = 181.58

(2) Conversely, the factors fitted for λ with constant p gave significant associations except in the case of sex. Age was also negatively associated with λ, exemplifying the fact that a factor may imply a prognosis good for λ but bad for p.

(3) Fitting a linear age relationship for p, while keeping λ constant, reduced the apparent effect of stage and sex, though all five factors remained significant.

(4) Fitting a linear age relationship for λ with p constant reduced the effect of site (from $\chi_5^2 = 44.8$ to $\chi_5^2 = 19.2$) but the most and least favourable sites (which were bone and liver respectively) remained the same.

(5) Four different models for age were then fitted, namely

(i) Linear only;

(ii) Linear and quadratic;

(iii) Linear up to 60 months, then constant;

(iv) Separate linear effects for Stage 1 and other Stages.

When fitted to both p and λ but only with sex, the likelihood ratio χ^2 statistics were in the above order, namely 109.0(6 d.f.), 135.2(8 d.f.), 136.6(8 d.f.), 162.5(8 d.f.).

There was thus very little difference between the quadratic relationship and the linear/constant relationship and the latter was therefore eliminated from further consideration as involving an arbitrary change-over point. The idea of fitting a separate relationship for Stage 1 tumours derives from the work of Breslow and McCann (1971).

(6) The finding that both the quadratic relationships and the distinguishing of Stage 1 tumours gave an appreciable improvement in the fit led to the decision to include separate quadratic relationships in the final run, except that a quadratic term for the hazard rate was only estimated for Stage 1 tumours.

Table 3 shows the results of the final run, in which 32 parameters were estimated. These are associated with the levels indicated in column 2; all the basic levels of a given factor (as defined by Table 1) which are not shown in Table 3 are grouped and regarded as a base level relative to which the others are measured. The age parameters are related to annual rates of change. It will be recalled that high values of α and low values of β are associated with a poor prognosis.

To illustrate the application of Table 3, suppose that a male child aged 10 months at anniv had a single deposit localised in the liver with an unspecified degree of differentiation. This would give:

$$\underline{\alpha}'\underline{X} = -0.419 - 0.169\times(10/12) + 0.018\times(10/12)^2 - 0.454 + 0.294 - 0.333 = -1.04$$

$$\underline{\beta}'\underline{Y} = -1.869 - 0.769\times(10/12) + 0.049\times(10/12)^2 - 0.210 + 2.648 + 0.260 - 2.354 + 2.045 = -0.087$$

Hence predictions of the hazard rate and survival probability are:

$$\lambda^* = 0.261 \text{ per month; } p^* = 0.478$$

To predict the outcome for a child (or a group of children) for whom some of the information fitted is missing or unspecified, it would not be possible to use the model directly, without taking into account the distribution of the missing variable(s). As alternatives, a model with the appropriate subset of variables could be fitted or the expected outcome could be obtained by averaging the predicted survival curves for the index cases in the appropriate subset. This latter method was used to calculate predictions of the total numbers dying in successive months after anniv with results which agree reasonably well with the observed numbers (Upton, 1973), though it is not obvious how to obtain a formal significance test which allows for the parameter estimation.

Interpretation of the Results

If only linear age relationships are fitted, Stage 1 and other cases provide β-coefficient values of -0.16 and -0.88, so that in Stage 1 the age appears to be of much less importance. It appears from the results described in Table 3, however, that this is an oversimplification in that there are significant linear and quadratic coefficients in both relationships. Thus, if x represents age at anniv, then

$$\text{Stage 1 effect} = 0.26-0.77x+0.049x^2 \quad (11)$$

$$\text{Stage 2 effect} = 0.54-1.2x+0.054x^2 \quad (12)$$

$$\text{Other Stage effects} = \text{const}-1.2x+0.054x^2 \quad (13)$$

Equation (11) suggests that, other factors being equal, the adverse effect of age on survival is maximal at around 8 years of age, after which the prognosis starts to improve. Moreover, comparison of equations (11) and (12) shows that for very young children - up to around 8 months - the risk estimate is actually lower for Stage 2 tumours. It is only right to emphasise, however, that the estimates have appreciable standard errors and that the numbers in many groups are small: there were only 10 children with Stage 2 tumours aged less than one year, for example.

Small numbers of deposits and advanced differentiation are both associated with improved prognosis, as are certain sites, notably the liver and ganglia. The inferences from the model in these respects are in accordance with the empirical survival rates.

On the other hand, the apparent effect of sex on empirical survival rates is largely accounted for by other factors. For more girls are young, have advanced differentiation and have tumours in favourable sites. These observations do not, of course, explain why sex is associated with such other prognostic indicators, but they make it less plausible that sex is a primary factor.

Discussion

As far as is known, the model described above has not been used before. Mould and Boag (1975) describe the fit of several models to data on the survival of women with cervical carcinoma; their models certainly include a parameter like p describing the probability of indefinite survival, but they fitted curves independently in each family, rather than relating the parameters directly to concomitant information as we have done. As they point out, a particular advantage of a mixed model is that it allows the estimation of the survival curve from the histories of patients followed-up for a relatively short time; inasmuch as the great majority of our children were observed to the end of the normal risk period, we would expect that predictions of, say, two-year survival rates from our model could be similar to those that would be obtained, for example, by the general method of Breslow and McCann (1971), who were concerned solely with these rates and not with the survival curves. The latter workers used a model similar to ours, in that they related the two-year survival probability to a logistic function of the tumour stage and the age in a series of 246 children treated for neuroblastoma.

It is probable, therefore, that our model, considering jointly the indefinite survival probability and the survival time distribution may well be more useful in application to diseases with longer survival periods. In such cases it offers the possibility of estimating a "true" probability of surviving as opposed to the 3- or 5-year rates which are generally employed but which leave open to question the extent of the subsequent risk.

ACKNOWLEDGEMENTS

We wish to thank the many people who have contributed to the work of the Oxford Survey of Childhood Cancers - formerly under the direction of Dr. Alice Stewart - including the staff of borough and county borough health departments throughout Britain. Our thanks are particularly due to those who worked on the analysis of the data described herein, especially Dr. Kinnier Wilson (Marie Curie

Memorial Foundation) and Mr. Gerald Draper (Childhood Cancer Research Group). During the relevant periods, the work has been supported by grants from the U.S. Public Health Service (Grant No. CA-12208 and Contract No. FDA 72-126), the Medical Research Council (Grant No. G 964/230/C) and the Department of Health and Social Security.

REFERENCES

Breslow, N. and McCann, B. (1971). Statistical Estimation of Prognosis for Children with Neuroblastoma. Cancer Research 31, 2098-2103.

Collins, V.P., Loeffler, R.K., and Tivey, H. (1956). Observations on Growth Rates of Human Tumours. Amer. J. Roentgenology, 76, 988-1000.

Fortner, J., Nicastri, A. and Murphy, M.L. (1968). Neuroblastoma: Natural History and Results of Treating 133 Cases. Ann. Surg. 167, 1, 132-142.

Gross, R.E., Farber, S. and Martin, L.W., (1959). Neuroblastoma Sympatheticum: A Study and Report of 217 Cases. Paediatrics, 23, 6, 1179-1191.

Kinnier Wilson, L.M. and Draper, G.J. (1974). Neuroblastoma, its Natural History and Prognosis: A Study of 487 Cases. Brit. Med. J. No. 5926, 3, 301-307.

Mould, R.F. and Boag, J.W. (1975). A Test of Several Parametric Statistical Models for Estimating Success Rate in the Treatment of Carcinoma Cervix Uteri. Br. J. Cancer, 32, 529-550.

Upton, R.G., (1973). An Analysis of the Survival of Children Suffering From Neuroblastoma. M.Sc. Dissertation in the University of Oxford.

Recent Developments in Statistics
J.R.Barra et al., editors

A FLOW CHART OF MULTIVARIATE STATISTICAL METHODS

Pierre Dagnelie
Statistique et Informatique
Faculté des Sciences Agronomiques de l'Etat
Gembloux, Belgique

1. INTRODUCTION AND SUMMARY

The development of computer science has considerably increased, during the last twenty years, the possible uses of multivariate statistical methods, the number of users of these methods, and to some extent, the number and diversity of these methods themselves.

But the use of always more and more powerful computers and programs, the abundance of periodic literature, and the publication of new practical books, such as those by Morrison (1967), Cooley and Lohnes (1971), Press (1972), Bertier and Bouroche (1975), and Dagnelie (1975), have not solved the problem of choosing the best method(s) to achieve a precise aim, starting from some given data.

To make this choice easier, especially to the least qualified users, a *general flow chart* of multivariate statistical methods is presented and discussed (section 3). Beforehand, some *choice criteria* are defined (section 2), and afterwards, some *bibliographic informations* are given (section 4).

This flow chart is a tentative synthesis of a few informations given in some of the books mentioned above : "overview of multivariate methods" and "strategy consideration in multivariate research" by Cooley and Lohnes (1971) and "guide de l'utilisation des méthodes" by Bertier and Bouroche (1975). This general flow chart might possibly be completed with some specific charts.

2. CHOICE CRITERIA

The main factors that may influence the choice of multivariate statistical methods are the structure of the data matrix, the nature of the data, and the aim of the study.

The initial *data matrix* is, generally speaking, a p x n rectangular matrix, with a row assigned to each variable or attribute and a column assigned to each observed individual.

In some situations, this matrix has no a priori structure, all the variables being considered in the same way and all the individuals being considered as belonging to the same sample or to the same population : we will indicate this situation by the symbol (1, 1), as it concerns only one group of variables or attributes and one group of individuals.

In some other situations, the initial data matrix refers to two or more sets of variables (for instance one or several explanatory variables and one or several dependent variables) and to a single set of individuals : we will call this case the $(k_1, 1)$ situation.

Sometimes also, the initial data matrix is related to a single set of variables and to two or more groups of individuals : this situation will be called the $(1, k_2)$ situation.

Finally, but not very often, the data may be related both to two or more groups of variables and to two or more groups of individuals : this will be the (k_1, k_2) situation.

These four possibilities are sketched in figure 1.

As far as the *nature of the data* is concerned, a distinction may be pointed out between continuous quantitative data (results of measurements), discrete quantitative data (results of enumerations), qualitative data that may be ranked in some logical way (different levels of the same qualitative character), qualitative data that may not be ranked in some logical way, and the special case of binary data. In some problems, different kinds of data can also be found together (simultaneous study of data of various types).

Moreover, in some particular studies, of preferences for instance, the initial data matrix may be a square matrix indicating the superiority of each individual over each other. This matrix might be compared with a correlation or a similarity matrix coming from qualitative or quantitative data of type (1, 1).

The *different aims* are difficult to define and to classify in some general way, because of their diversity : description or reduction of data (ordination, hierarchisation, etc.), forecasting or estimation, explanation of the observed relations between variables, comparison of two or more groups of individuals, definition of classes within an unstructured set of individuals (classification, clustering, etc.), assignment of individuals to well-defined groups, etc.

Of these three criteria, we are considering the structure of the data matrix and the aim of the study as *the most important*. Without underestimating for instance the distinction between methods which are related to the multivariate normal distribution and methods which are distribution-free, we are noticing that the nature of the data is, from a practical point of view, a choice criteria of minor importance. We may mention for instance that some methods which are fundamentally related to continuous data, such as factor analysis, are successfully used with discrete and even qualitative data, while on the other hand, some methods which are fundamentally related to qualitative data, such as the "analyse factorielle des correspondances", are commonly used with continuous or mixed data.

3. GENERAL FLOW CHART

Figure 2 presents the first part of our general flow chart, which is related to the *first steps of the statistical analysis*.

The first question in whether the problem under consideration is, completely or partially, a multivariate one (2)*.

If the answer is positive, and after having done, if necessary, some univariate and bivariate calculations (4), one will proceed to a careful examination of the

---

* Numbers in parentheses refer to the different cells of figures 2 to 5 and of table 1.

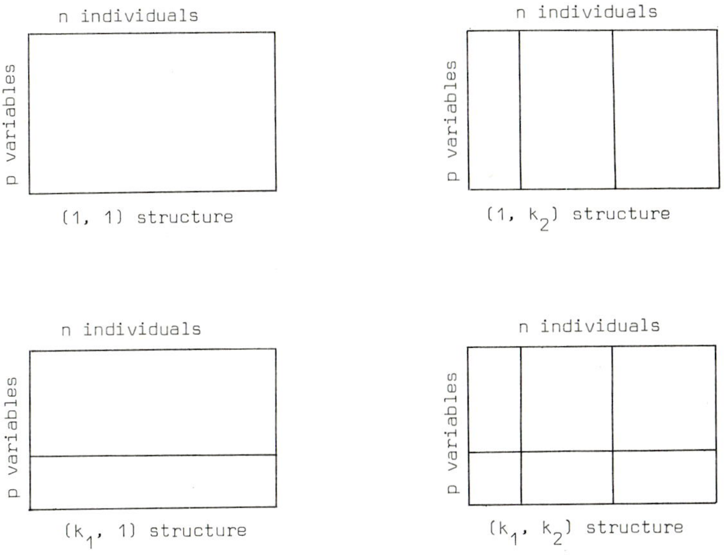

Figure 1. Schematic representation of the different structures of the initial rectangular data matrix.

initial data, for instance by considering bivariate frequency distributions (5). In situations (1, k_2) and (k_1, k_2), i.e. for two or more groups of individuals, this preliminary examination will be done separately for each group, and possibly afterwards for the whole set of individuals. If necessary, two- or multidimensional tests of normality may be performed at this stage.

This preliminary examination will usually lead at least to three questions : is any transformation of variables necessary (6), are there outliers (8), and are there missing data (10) ? According to the different answers, one will carry out the necessary transformations (7), remove the outliers (9), and estimate the missing data (11), taking again into account, if any, the presence of different groups of individuals.

Lastly, the end of figure 2 makes the distinction between the four fundamental structures that were defined in section 2, the main solution for structure (k_1, k_2) being, as far as we know, multivariate covariance analysis or similar methods (18).

Figures 3 to 5 give more informations about the three other structures.

In situation (k_1, 1), i.e. for *two or more groups of variables* and a single set of individuals (17), the first question deals with the number of groups of variables k_1 (19). If this number is larger than 2, we can only carry on studies of correlation between groups (20). If not, the question is to know whether, on the one side, the two groups should be considered at the same level, as interdependent, or on

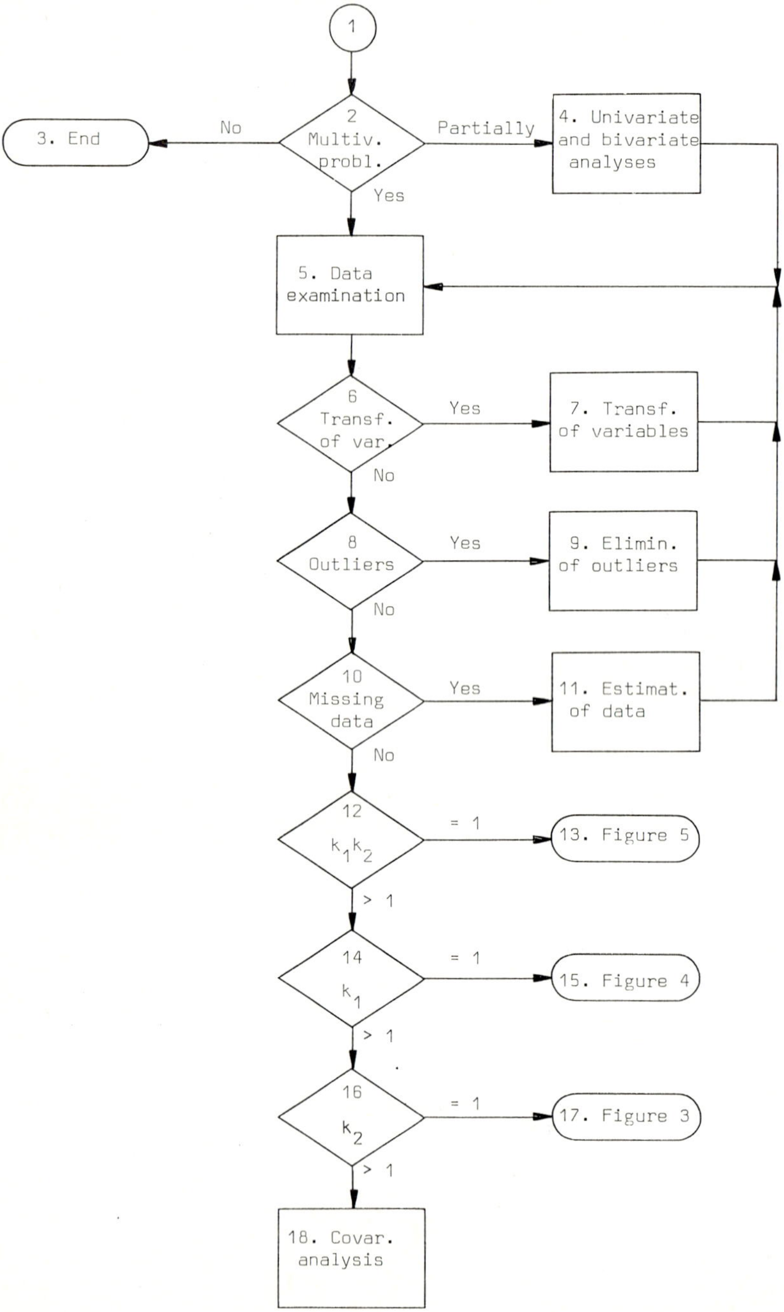

Figure 2. General flow chart (part 1).

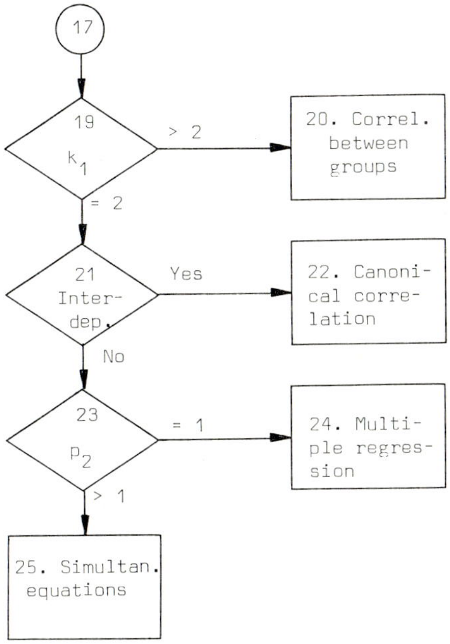

Figure 3. General flow chart (part 2) : structure (k_1, 1).

the other side, one group must be considered as explanatory and the other as dependent (21). In the first situation, we will normally turn towards canonical correlation methods (22), including multiple correlation, and in the other situation, towards multiple regression and related methods (24) or towards simultaneous regression equations (25), according to the number of dependent variables p_2 (23).

As far as situation (1, k_2) is concerned, i.e. for a single set of variables and *two or more groups of individuals* (15), we may either simply compare the different groups (26), making some tests about means (including analysis of variance), about variances and covariances, and/or about correlation coefficients (27), or consider the scattering of the different groups in a one-, two- or multidimensional space (28), by canonical analysis (29), or even assign individuals of unknown origin to one of the different groups, by discriminant analysis (30).

At last, in the apparently simplest, but not the best-defined situation where the initial data matrix has *no a priori structure*, the problem under consideration may be a simple estimation problem (31), leading to confidence regions (32), or a problem of conformity (33), leading to tests about means, variances and covariances, and/or correlation coefficients (34), or a problem of data reduction (35), bringing about for instance component analysis (36), or a problem of scattering of individuals in a one-, two-, or multidimensional space (37), which may be solved by some kind of factor analysis (38), or even a problem of classification (39).

We will also stress that, after a first analysis, it is always useful to consider the opportunity of doing some *further analysis*, based on the same principles, but taking into account other methods or other data (such as, for instance, the residuals coming from the first analysis).

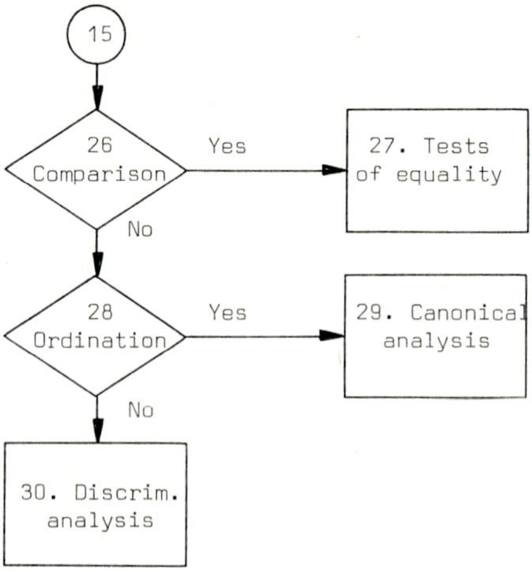

Figure 4. General flow chart (part 3) : structure $(1, k_n)$.

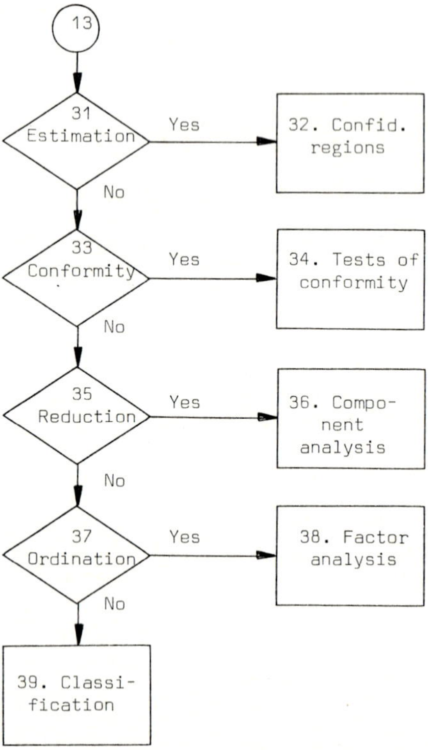

Figure 5. General flow chart (part 4) : structure (1, 1).

4. BIBLIOGRAPHIC INFORMATIONS

As a complement to our general flow chart, table 1 gives some bibliographic informations related to the main multivariate methods or groups of methods : it contains the numbers of the chapters of the five books quoted in the introduction, which are partially or completely devoted to these classes of methods.

When consulting these books, one should be especially careful as far as terminology is concerned, the vocabulary of multivariate analysis being very confusing, for instance for the words canonical, discriminant and factorial (canonical analysis, canonical correlation, canonical variate, discriminant analysis, factorial analysis, etc.).

Table 1. Numbers of the chapters of five books on multivariate analysis which are devoted to the different methods quoted in the general flow chart (figures 2 to 5).

Nos	Methods	B-B	C-L	D	M	P
7-9-11	Tests of normality, transformations of variables, detection of outliers, estimation of missing data			4,10		8
18	Analysis of covariance		11	12	5	8
20	Correlation between groups of variables			7	6	11
22	Canonical correlation, multiple correlation, partial correlation	IX	3,6,7	5,7	3,6	11
24	Multiple regression	XI	3	4,8	3	8
25	Simultaneous regression equations			6		5,8
27	Tests of equality of means, variances and covariances, and correlation coefficients, analysis of variance (analysis of dispersion)		8,12	7,10 11,12	4,5	7,8
29	Canonical analysis (factorial discriminant analysis)	X	9	13		
30	Discriminant analysis	X	10	14	4	13
32	Confidence regions			10	4	
34	Tests of conformity of means, variances and covariances, and correlation coefficients			7,10,12	3,8	7
36	Component analysis (principal component analysis)	VII	4	8	7	9
38	Factor analysis, "analyse factorielle des correspondances", scalogram analysis, latent structure analysis	III,VIII, XII,XIII	5	9	8	10
39	Classification, taxonomy, typology	IV,VI		13		15

B-B = Bertier and Bouroche (1975), C-L = Cooley and Lohnes (1971), D = Dagnelie (1975), M = Morrison (1967) and P = Press (1972).

5. REFERENCES

Bertier P. and Bouroche J.M. (1975). *Analyse des données multidimensionnelles.* Paris, Presses Universitaires de France, 270 p.

Cooley W.W. and Lohnes P.R. (1971). *Multivariate data analysis.* New York, Wiley, 364 p.

Dagnelie P. (1975). *Analyse statistique à plusieurs variables.* Gembloux, Presses Agronomiques, 362 p.

Morrison D.F. (1967). *Multivariate statistical methods.* New York, McGraw Hill, 338 p.

Press S.J. (1972). *Applied multivariate analysis.* New York, Holt, Rinehart and Winston, 521 p.

6. RESUME

Le développement de l'informatique a provoqué, durant les vingt dernières années, une extension considérable des possibilités d'utilisation des méthodes d'analyse statistique à plusieurs variables, du nombre de leurs utilisateurs et, dans une certaine mesure aussi, du nombre et de la diversité des méthodes elles-mêmes.

Mais le recours à des moyens de calcul toujours plus puissants, en matériel comme en logiciel, le foisonnement de la littérature périodique et la parution de livres nouveaux à orientation pratique, comme ceux de Morrison (1967), Cooley et Lohnes (1971), Press (1972), Bertier et Bouroche (1975) et Dagnelie (1975), n'ont pas résolu, loin s'en faut, tous les problèmes que pose le choix de la ou des méthodes les plus adéquates pour atteindre tel ou tel objectif à partir de telles ou telles données.

Pour faciliter ce choix, surtout aux utilisateurs les moins avertis, un *organigramme général* des méthodes d'analyse à plusieurs variables est présenté et commenté (paragraphe 3). Au préalable, les principaux *critères de choix* sont définis et discutés (paragraphe 2) et, ensuite, des *indications bibliographiques* sont données au sujet des diverses méthodes citées (paragraphe 4).

Cet organigramme général tente de synthétiser les quelques informations qui figurent à ce sujet dans certains des ouvrages dont il vient d'être question : "overview of multivariate methods" et "strategy considerations in multivariate research" de Cooley et Lohnes (1971) et "guide d'utilisation des méthodes" de Bertier et Bouroche (1975). Cet organigramme général pourrait éventuellement être complété par quelques organigrammes particuliers.

Recent Developments in Statistics
J.R.Barra et al., editors

A STOCHASTIC MODEL FOR THE REGULATION OF CELLULAR METABOLISM

J. DEMONGEOT
Université Scientifique et Médicale de Grenoble
I.R.M.A., B.P. 53
38041 - Grenoble Cédex - France

For the main steps of the cellular metabolism we consider a stochastic model, which is a synthesis between well-known stochastic ([1] & [2]) and deterministic ([3]) models and original remarks.

Successively we study :

a) The storage of genetical information in DNA, using the Ising model.

b) The communication channel with noise between DNA and ribosomes.

c) The enzymatic kinetics.

Using the partial models, we describe the feedback differential system which governs random numbers of reactant molecules at the time t, considered as continuous random processes.

We exactly solve the expectation system under certain hypotheses and the system of generating functions in a simple case.

Finally, the stability of the model is shown for certain criteria and we mention methods testing hypotheses concerning the randomness of the systems considered (markovian character of the genetical storage and stationarity of the probabilities for the enzyme sites to be occupied by substrate molecules).

1 - INTRODUCTION

Let us consider the steps of simplified cellular metabolism :

I - Storage of genetical information in DNA.

II - Production, by the source DNA, of two types of messages which are transmitted over the communication channel RNA-m (denoted by X) to the ribosomal decoder : the message produced by the structural gene G_1 (resp. the regulatory gene G_2) governs the building up of enzyme E (resp. repressor R).

III - Catalysis by E of the reaction, which transforms substrate S coming from the extracellular medium into a product M ; M activates R producing R_a, which blocks G_1 producing G_1^-.

This control mechanism is illustrated in the following scheme :

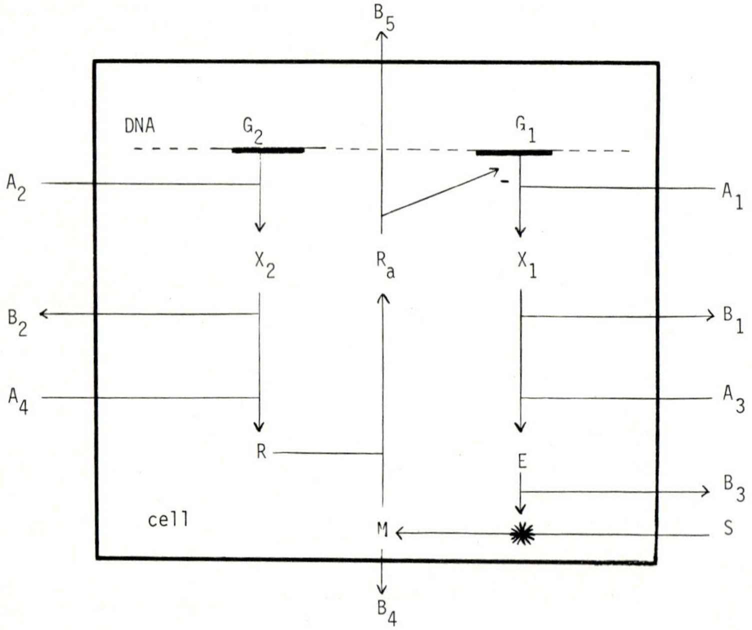

The A_i's (resp. B_i's) are the building blocks (resp. decay products) coming from (resp. going to) the extracellular medium. The scheme can be described in terms of the following chemical reactions :

I - $G_i + A_i \xrightarrow{k_o} G_i + X_i$ $\quad i = 1,2$

II - $X_i + A_{i+2} \xrightarrow{k_1} E_i + X_i$

$X_i \xrightarrow{k_2} B_i$ $\quad i = 1,2$ (où $E_1 = E$ et $E_2 = R$)

III - $E + nS \underset{d}{\overset{a}{\rightleftharpoons}} ES + (n-1)\,S \underset{d}{\overset{a}{\rightleftharpoons}} \ldots \underset{d}{\overset{a}{\rightleftharpoons}} ES_n \xrightarrow{h} E + M$

$E \xrightarrow{k_3} B_3$

$M + R \underset{k_4'}{\overset{k_4}{\rightleftharpoons}} R_a$ $\quad M \xrightarrow{k_5} B_4$

$G_1 + R_a \underset{k_6'}{\overset{k_6}{\rightleftharpoons}} G_1^-$

In the following we denote by $Y(t)$ ($Y = G_1$, E, ...) the number of reactant molecules of Y at time t in the cell.

2 - MODELS FOR THE DIFFERENT STEPS

2.1 - Step I

The alphabet coding genetical information consists of 4 letters T, A, C, G ; therefore we can represent the DNA molecule by an element of $\{0,1\}^B$, where B is the set of points of Z^2 having ordinate 0 or 1.

We suppose that the discrete random process defined by the letter succession is stationary and markovian (see [4] for justifications) ;then we can associate with this process a Gibbs measure μ on $\{0,1\}^B$, defined as in the Ising model ; we assume that this measure is translation-invariant and unique. Then the transmission has minimal mean error, because of results as the following :

Theorem 2.1. ([10] & [11]) : $(\{0,1\}^B, \mu, \mathcal{T})$, where $\mathcal{T}$ is the group of horizontal translations of B, is a Bernoulli scheme ; let H denote its entropy, then, for every memoriless channel with capacity $C \geq H$, there exists an invertible code, such that the transmission rate is H.

2.2 - Step II (from [5] & [6])

We obtain X by transcription from an element of $\{0,1\}^F$, where F is the finite set of the points of B having abscissae between n_F and m_F ; let $P_i(t)$ denote the probability that the points having abscissa n_F+i are reached by transcription after time t ; it is proved in [5] that:if $|F| = 2n$, then $\frac{dP_n(t)}{dt} = \dot{P}_n(t) = \frac{e^{-\nu t}(\nu t)^{n-1}}{n-2!\ t} = J_n(t)$,

where ν is the forward rate of transcription.

Let us denote by $Q_m(t)$ the probability that a protein having m aminoacids is built during time t ; it is proved in [6] that, if $G_1(t)$ is stationary : $Q_m(t) = C(m)$, where C depends only on the forward rate of construction.

2.3 - Step III (cf. [4])

If we assume, as in [12], that the probability that one molecule of S gets bound to complex ES_i during the interval $(t, t+dt)$, is equal to $aS(t)\ ES_i(t)\ dt$ $(i = 1,\dots,n)$ and if we assume also that at most one fixation can occur in this interval, defining :

$$P_{0,m}(t) = P(\{S(t) = m\})$$

$$P_{i,m}(t) = P(\{ES_i(t) = m\}), \quad i = 1,\dots,n$$

$$P_{n+1,m}(t) = P(\{M(t) = m\}),$$

we obtain a differential system (S_0), which governs the $P_{j,m}(t)$'s with initial conditions : $P_{0,S}(0) = P_{i,0}(0) = 1,\ \forall i \geq 1.$

Now, let us remark that :

$$\dot{\mathbb{E}}(ES_i(t)) = \sum_m m\, \dot{P}_{i,m}(t) \quad \text{(where } \mathbb{E}(Y) \text{ denotes expectation of } Y\text{)}$$

If we suppose S constant, we can deduce from (S_0) the system (S_1), which governs expectations :

$$\begin{bmatrix} \dot{\mathbb{E}}(ES_1(t)) \\ \vdots \\ \dot{\mathbb{E}}(ES_i(t)) \\ \vdots \\ \dot{\mathbb{E}}(M(t)) \end{bmatrix} = M \begin{bmatrix} \mathbb{E}(ES_1(t)) \\ \vdots \\ \mathbb{E}(ES_i(t)) \\ \vdots \\ \mathbb{E}(M(t)) \end{bmatrix} - \begin{bmatrix} naSE(0) \\ 0 \\ \vdots \\ \vdots \\ 0 \end{bmatrix}$$

with initial conditions : $\mathbb{E}(M(0)) = \mathbb{E}(ES_i(0)) = 0$, $\forall i \geq 1$. M is a tridiagonal-row matrix of order n+1, which is a function of S, n, of the association constant a, of the dissociation constant d (we assume that a and d are the same for all the n sites of E) ; we have :

$$M = \begin{bmatrix} -(2n-1)\,aS-d & naS+2d & -naS & \cdots & & -naS & 0 \\ (n-1)\,aS & -(n-2)\,aS-2d & 3d & 0 & \cdots & 0 & 0 \\ 0 & \ddots & \ddots & \ddots & & & \\ \vdots & & & & & & \\ 0 & \cdots & (n+1-i)\,aS & -(n-i)\,aS-id & (i+1)\,d & \cdots & 0 \\ \vdots & & & \ddots & \ddots & \ddots & \\ 0 & \cdots & \cdots & \cdots & aS & -(nd+h) & 0 \\ 0 & \cdots & \cdots & \cdots & \cdots & h & 0 \end{bmatrix}$$

Multiplying the two equation members of (S_0) by s^m and summing over m, we obtain the differential system (S_2) which governs the generating functions $\Psi_j(t,s)$ of the studied processes. Let (S_1') denote the system (S_1) without its last row ; then we have :

Lemma 2.3.1. : (S_1) has one and only one solution, denoted by $\{\mu_i(t)\}_{i=1}^{n+1}$ and, if d and $||\mathbb{E}(E(t))-E(0)||_\infty$ are sufficiently small, (S_1') is asymptotically stable and, if we denote by $\{\mu_i\}_{i=1}^{n}$ the limit of the solution of (S_1'), as t increases infinitely, $\mu_{n+1}(t)$ is asymptotically equivalent to $h\mu_n t$.

Now, if we assume that :

$\forall t \in R_+$, $\forall i$, $\mathbb{E}(ES_i(t) \mid ES_{i-1}(t)) = \mathbb{E}(ES_i(t) \mid ES_{i+1}(t)) = \mu_i(t)$, then we have for (S_2), if we define $\Psi_j'(t,s) = \dfrac{\partial \Psi_j(t,s)}{\partial s}$:

$$\dot{\Psi}_i(t,s) = ((i+1)d\mu_{i+1}(t) + (n+1-i)aS\mu_{i-1}(t))(s-1)\Psi_i(t,s)$$
$$+ (id+(n-i)aS)(1-s)\ \Psi_i'(t,s) \quad \text{avec } \mu_0(t) = S,\ \forall i = 1,\dots,n-1$$

$$\dot{\Psi}_n(t,s) = aS\mu_{n-1}(t)(s-1)\Psi_n(t,s) + (nd+h)(1-s)\Psi_n'(t,s)$$

$$\dot{\Psi}_{n+1}(t,s) = h(s-1)\mu_n(t)\Psi_n(t,s)$$

Then we have :

<u>Lemma 2.3.2.</u> : If d and $||\mathbb{E}(E(t))-E(0)\ ||_\infty$ are sufficiently small, (S_2) is asymptotically stable and its solution has a limit, denoted by $\{\Psi_i\}_{i=1}^n$, such that Ψ_i is poissonian with parameter μ_i ; finally, we have contiguity in law between M(t) and the process Y(t) which has the expectation $h\mu_n t$ and the generating function defined by : $\varphi(t,s) = 1 + h(s-1)\mu_n\ e^{\mu_n(s-1)}\ t.$

2.4 - <u>Remark</u>

We have assumed that the diffusion of metabolites in the cell or through its membrane is negligible.

3 - <u>GENERAL MODEL</u>

As in 2.3, if we define $P_{Y,m}(t) = P(\{Y(t) = m\})$, for $Y = G_1$, E, ..., we obtain the system (T_0) which governs the $P_{Y,m}(t)$'s, the system (T_1) which governs the expectations $\mathbb{E}(Y(t))$ and the system (T_2) which governs the generating functions $\Psi_Y(t,s)$. For example, let us write the equation for $P_{G_1,m}(t)$:

$$\dot{P}_{G_1,m}(t) = k_6\mathbb{E}(R_a(t))\ ((m+1)P_{G_1,m+1}(t) - mP_{G_1,m}(t)) + k_6'\ \mathbb{E}(G_1^-(t))\ (P_{G_1,m-1}(t) -$$
$$P_{G_1,m}(t)) + k_oA_1((m+1)P_{G_1,m+1}(t) - mP_{G_1,m}(t)) + k_o \int_0^t J_{n_1}(\tau)\ \dot{\mathbb{E}}(G_1(t-\tau))d\tau$$
$$(P_{G_1,m-1}(t) - P_{G_1,m}(t)),$$

if the length of G_1 is n_1 and if we assume that A_1 is constant. The last term on the right-hand side is relative to molecules whose transcription ends during the interval (t, t+dt) ; as in 2.3, we have assumed that :

$$\forall t,\ \tau \in R_+,\ \tau \leq t,\ \mathbb{E}(G_1(t-\tau)\ |\ G_1(t)) = \mathbb{E}(G_1(t-\tau)).$$

In the same way, if $(1-\eta)$ denotes the probability of the transcription error for a letter of G_1 calculated as in 2.1, we have :

$$\dot{P}_{X_1,m}(t) = k_o(1-\eta)^{n_1} \int_0^t J_{n_1}(\tau)\ \dot{\mathbb{E}}(G_1(t-\tau))\ d\tau(P_{X_1,m-1}(t) - P_{X_1,m}(t))$$

$$+ k_2((m+1)\ P_{X_1,m+1}(t) - mP_{X_1,m}(t)) + k_1A_3((m+1)\ P_{X_1,m+1}(t) - mP_{X_1,m}(t))$$

$$+ k_1 \int_{t-t_o}^t \dot{Q}_{p_1}(\tau)\ \dot{\mathbb{E}}(X_1(t-\tau))\ d\tau(P_{X_1,m-1}(t) - P_{X_1,m}(t)),$$

if E has p_1 amino-acids and if we assume A_3 to be constant, and where we have defined : $t_o = \inf\{t\ ;\ Q_{p_1}(t) = 1\}$.

We resolve (T_1) under the following assumptions : the number of molecules of G_i (resp. X_i) $(i = 1,2)$, which participate in the transcription (resp. building), is negligible with respect to the number of free molecules of G_i (resp. X_i) at any time t ; therefore we have :

Theorem 3.1. : Under the simplifying assumptions above and under the hypothesis of lemma 2.3.1., (T_1) is asymptotically stable, and, if we interpret $\mathbb{E}(G_1(t))$ as the input of the feedback system below :

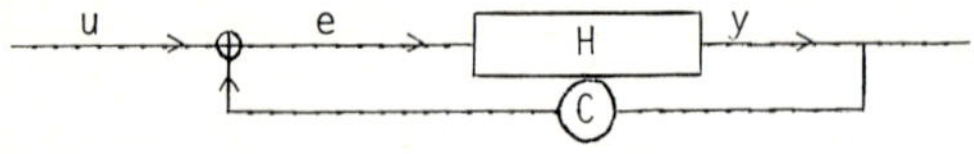

, with

$$y(t) = \int_0^t H(t,\tau)\ e(\tau)\ d\tau \quad , \quad e(t) = u(t) - C(t)\ y(t), \quad \text{where } u = \mathbb{E}(G_1) \text{ and}$$

$$y = \begin{bmatrix} \mathbb{E}(X_1) \\ \mathbb{E}(X_2) \\ \mathbb{E}(E) \\ \mathbb{E}(M) \\ \mathbb{E}(G_2) \\ \mathbb{E}(R) \end{bmatrix}$$

, we have :

the feedback system is $L^1(R_+)$-stable, i.e. there exists a constant k, such that :

$\forall u \in L^1(R_+),\ ||e||_{L^1} \leq k\ ||u||_{L^1}$.

Finally, as in 2.3, (T_2) is asymptotically stable and the limit of its solution is poissonian.

4 - METHODS TESTING ASSUMPTIONS OF THE MODEL

4.1 - Markovian character of the process on $\{0,1\}^B$

We cannot observe the letters of G, but only the amino-acids of E ; we can prove that the markovian character persists after transmission. In [7], testing the frequency of occurrence of a configuration of order 2, the assumption of spatial independence of components of the process defined by succession of amino-acids in E is not rejected ; we can improve this test, using configurations of higher order (cf. [8]). Let us remark that the spatial independence for E implies that the process on $\{0,1\}^B$ is markovian of order 2, because each amino-acid of E is the transcription of a subset F of length 2.

4.2 - Stationarity of the occupation probabilities of the enzyme sites

We have assumed in 2.3, that the association constant a is the same for all the n sites of E. Let us observe that the occupation probability of any site i of E is equal to :

$$\theta_i = \frac{aS}{d+aS}, \quad i = 1,\ldots,n$$

A test between hypothesis "θ_i constant for all the sites" and, for example, hypothesis "$\theta_1 < \theta_2 \leq \cdots \leq \theta_n$" (in the allosteric case) can be performed by the comparison of the quantity F(t) to 2 adequate numbers a(t) and b(t) :

$$F(t) = \frac{\theta_1^{\sum_{i=2}^{n} iES_i(t)} \, (1-\theta_1)^{\sum_{i=1}^{n} (n-i)ES_i(t)}}{\theta_2^{\sum_{i=2}^{n} (i-1)ES_i(t)} \, (1-\theta_2)^{\sum_{i=1}^{n} (n-i)ES_i(t)}}$$

This test is an immediate adaptation of the classical sequential tests (cf. [9]). Let us note that we observe practically the limit of $ES_i(t)$, as t increases infinitely, but the test is still valid, because of its asymptotic expression.

REFERENCES

[1] Mc Donald, C.T., Gibbs, J.H. & Pipkin, A.C. (1968). Kinetics of biopolymerisation on nucleic acid templates, Biopolymers, 6, 1-25.

[2] Mc Donald, C.T. & Gibbs, J.H. (1969). Concerning the kinetics of polypeptide synthesis on polyribosomes, Biopolymers, 7, 707-725.

[3] Babloyantz, A. & Nicolis, G. (1972). Chemical instabilities and multiple steady state transitions in Monod-Jacob type models, J. Theor. Biol., 34, 185-192.

[4] Demongeot, J. (1975). Thèse, Grenoble.

[5] Goel, N.S. & Richter-Dyn, N. (1974). Stochastic models in biology, Acad. Press.

[6] Gibbs, J.H. (1969). The kinetics of biopolymerisation on nucleic acid templates Biopolymers, 7, 707-725.

[7] Jorre, R.P. (1974). A statistical analysis of the evolution of proteins and of the genetic code, Ph. D. Thesis, Reading.

[8] Barra, J.R. (1971). Contrôle statistique des nombres "au hasard", Séminaire de Statistique, Grenoble.

[9] Wald, A. (1947). Sequential analysis, Wiley.

|10| Di Liberto, F., Gallavotti, G. & Russo, L. (1973). Markov processes, Bernoulli schemes and Ising model, Com. Math. Phys., 33, 259-282.

|11| Billingsley, P. (1965). Ergodic theory and information, Wiley.

|12| McQuarrie, D. (1967). Stochastic approach to chemical kinetics, J. Appl. Prob., 4, 413-478.

Recent Developments in Statistics
J.R.Barra et al., editors

STATISTICAL DISCRIMINATION IN SEISMOLOGY

Eva Elvers
National Defence Research Institute
S-104 50 Stockholm
Sweden

There are several seismological methods to discriminate between earthquakes and underground nuclear explosions, and an outline of them is given. One method, which is based on a model for magnitudes from shortperiod body waves and longperiod surface waves, is studied in detail. The two alternative statistical distributions involve the event size as a form of nuisance parameter, and the principle of choice of discriminant is discussed, as well as related empirical procedures. All variables are not always obtained, and a method is judged not only by its capability at best, but also by its applicability, i.e. for how weak events it generally can be used. The case is studied, where upper bounds of missing values are available. Such bounds can be utilized in the discrimination procedure, increasing the applicability.

INTRODUCTION

Signals received at a seismological station can be used to discriminate between earthquakes and underground nuclear explosions. These two kinds of seismic events are of different character, but there are variations within the groups with respect to the generation of signals, and these are moreover influenced along the travel path. When looking for discrimination rules it is usually not the whole signal that one has tried to model, but instead different quantities have been selected, which seem characteristic either empirically or for physical reasons.

From a statistical point of view, when the set of variables is given, the problem is one of discrimination between two multivariate populations. It contains, of course, the usual questions of model choice, classification rule etc., but there are some points which are at least unusual in other applications.

1. One type of error, to classify an earthquake as an explosion, is generally regarded as considerably more serious than the other, and one then works with a preset value of its probability. To look for a good rule with this restriction is an approach similar to hypothesis testing, (although the alternatives are formulated as to which population a new observation belongs) instead of a minimization of some "total" error rate or misclassification cost.

2. The sampling of events is not controlled, and the number of observations is not always large, especially not for explosions. This is strengthened by the splitting into more homogeneous regions, which is sometimes needed.

3. The way events are generated makes it reasonable to consider earthquake samples as representative for future observations, but this is less valid for explosions.

4. Some variables depend on the size of the event, which can be viewed as a form of nuisance parameter of the model.

5. Missing data is rather frequent. This can happen irrespectively of the variable itself, but it is often because the signal is too weak to measure. Sometimes an upper bound of the variable can be provided; a partial information which can be utilized.

6. The classification probabilities are generally functions of the event-size, and it is important to know for how weak events the methods are applicable.

7. A large amount of data is obtained at a seismic station, which necessitates some simple rule to sort out most events, and it seems reasonable to have several stages of classification due to computational and collecting efforts.

8. The possible variables are many, they are of different types (qualitative as well as quantitative), and it is difficult to include some of them in a classification rule of scores, for example the local time of occurrence.

These points contain problems, several of which are statistically challenging as such. The description below is a mixture of seismological models and methods and of general considerations in statistical terms. The last chapter, which is intended to give a general view, is by necessity brief, while the three first ones are somewhat more detailed, dealing with the points 3-6 in particular.

DISCRIMINATION BETWEEN TWO LINEAR RELATIONSHIPS

Model

The probably most used method to discriminate between earthquakes and underground nuclear explosions is the so-called $m_b(M_s)$-method. The two variables m_b and M_s are magnitudes of shortperiod body waves and longperiod surface waves, respectively, obtained at a seismic station. They are scattered around two lines, one for each kind of event, which are fairly parallel. The discriminant is taken to be a linear combination of the magnitudes, and it is based on simple fits or computed from more or less specified models. Some different methods in use will be given, but first a rather theoretical discussion.

Ericsson (1971a) gave the following model for explosions, where the magnitudes are now denoted by letters S.

$$S_1 = a_1 + b_1K + \epsilon_1$$
$$S_2 = a_2 + b_2K + \epsilon_2 \qquad (1)$$

The a_i's and b_i's are parameters (depending on the location of the event and the recording station), K is the logarithm of the yield (energy) of the explosion, and the ϵ_i's are "errors". These are assumed to follow a bivariate normal distribution with mean values zero and with known coefficient of correlation and ratio between the standard deviations.

The same type of model can be used for earthquakes too. Ericsson (1971b) investigated the linear relationship, its parallelism with the explosion correspondence, and the distribution of a suitably chosen linear combination of the magnitudes with respect to normality. He found all this reasonable, but more firm conclusions were not permitted then, due to the limited number of observations. Ericsson did not give an explicit definition of the size-parameter K, but more recent research confirms the linear form in (1) with the seismic moment in the place of the yield, Aki (1972) and Chinnery and North (1975). (The two size-parameters will be further discussed below.) A second subscript (j) is added to distinguish between the two sets of parameters, with the populations numbered by one for earthquakes and two for explosions.

The mean values ξ_{ij} of the magnitudes thus satisfy either of the two relationships

$$\xi_{2j} = \lambda_j + \varkappa_j \xi_{1j}, \text{ where}$$
$$\varkappa_j = b_{2j} / b_{1j} \text{ and } \lambda_j = a_{2j} - \varkappa_j a_{1j} \quad (j = 1,2) \qquad (2)$$

The slopes $\varkappa_j$ are assumed to be equal ($\varkappa$), while the intercepts λ_j are different, with a greater value for explosions, for which there are geophysical reasons.

The model (1) can be extended to include several seismic stations, which measure the same event. The magnitude subscript i then runs up to I, say, which usually is twice the number of stations, but not necessarily, as some stations only provide one type of magnitude. It will be assumed below that all ε_{ij}'s are uncorrelated, and that all standard deviations are equal for each kind of event, with the two values σ_1 and σ_2.

"Optimal" discriminant — general

When the two alternative distributions of the chosen discriminatory variables are completely specified the discriminant is formed by the ratio of the frequency functions. Its "cut-off level" is usually obtained either from prior probabilities for the two populations or from these in combination with costs of misclassifications. With the present principle the level is, however, determined by the preset value of the probability of the serious error. When the alternatives are not simple some other method is needed, as here, where each size-parameter K_j specifies the location of the mean values on their line. Different approaches are possible, and there are two main lines depending on whether K_j is regarded as a fixed number or as a realization of some stochastic process. This corresponds to regarding the relationship as functional and structural respectively, in an often used terminology.

Functional approach

When K_j is regarded as a number it is a nuisance (or incidental) parameter, and one method to form discriminant is by taking likelihood ratio. The K_j which maximizes the frequency function for the observed S_i's is simply found, Elvers (1975b), and the obtained maximum is proportional to the frequency function of $(Y_1, \ldots, Y_{I-1})$, where

$$Y_i = S_i - \frac{b_{ij}}{b_{Ij}} S_I \qquad (3)$$

The discrimination problem is thus reduced to one between two multivariate normal distributions, $N_p(\mu_j, \Sigma_j)$ in short notation. The dimension p is (I-1), and the parameters μ_j and Σ_j are easily found from (3). The covariance matrices Σ_j are proportional, so the best linear discriminant has the following form

$$D = \left[\Sigma_1^{-1}(\mu_2 - \mu_1)\right]^T Y \quad \text{or} \quad D = f^T S, \text{ where}$$

$$f = (A - (A \cdot B)B)/\sigma_1^2 \text{ with the vectors given by} \tag{4}$$

$$A_i = a_{i2} - a_{i1}, \; B_i = b_{ij}/\sqrt{\sum_{i=1}^{I} b_{ij}^2} \quad i = 1,\ldots,I$$

The Σ_j's are equal when $\sigma_1 = \sigma_2$, and the above discriminant is then overall best. There are several other arguments leading to the same discriminant. Some are presented by Rao (1966) together with further attractive properties, and this discrimination model has been studied also by Burnaby (1966).

D has a univariate normal distribution with the alternative mean values ζ_j and variances ω_j^2, say. $(\zeta_2 - \zeta_1)$ equals ω_1^2, and so does ω_2^2 when $\sigma_1 = \sigma_2$, whereby this quantity can be regarded as a Mahalanobis' distance. It gives a measure of the discrimination capability, and also of the relative importance of the different magnitudes. If the i:th magnitude is omitted, for example, ω_1^2 is reduced by

$$\left\{A_i - (A \cdot B)B_i\right\}^2 / \left\{(1 - B_i^2)\,\sigma_1^2\right\} \tag{5}$$

from the total quantity $\left\{(A \cdot A) - (A \cdot B)^2\right\}/\sigma_1^2$

Structural approach

The above classification rule has error probabilities which do not depend on the event sizes K_j. (The situation is not so simple in practice, due to missing data, for example, as will be seen below.) This may seem desirable, and it can be used as a defining property. Another aspect is that the K_j's are restricted to a finite range and that the frequency of sizes is not constant. If the K_j's are taken to be random variables, with some distribution, the "total" frequency functions of magnitudes can be computed for each kind of event. The second main line to construct discriminant is to base it on these two alternative distributions. The approach is reasonable for earthquakes (which will be described), but not for explosions, as one cannot tell the future distribution of logyields K_2. There is thus a logical difficulty as the alternatives are of different character. It would, of course, be possible to use some other kind of argument, as importance of correct classification, but this deviates from the principle adopted.

One can view both the sizes of earthquakes and their occurrence as stochastic, but there is a dependence between events. If one considers time periods, it is, however, reasonable to assume both an intensity of occurrence and a size-distribution. The latter is traditionally described in terms of magnitudes, which are considered to follow an exponential distribution (region-dependent). With the present model it is more attractive to regard the underlying measure of size, K_1, as a random variable. This is also in line with more recent work, as Wyss (1973), Aki (1972), and Chinnery and North (1975), (of which the two last were mentioned when describing the model). Assuming that the random errors ϵ_{ij} are independent of K_1, one can then compute the distribution of the S_i's from the model (1). The probability to misclassify the "next" earthquake is, however, not defined, because weak events are not detected. One could make a restriction to detected earthquakes, but the "rate" of misclassifications is perhaps more interesting. Adding a model for the varying detection capability, as was done by Elvers (1975a) to whom is referred for a full description, the rate can be expressed in terms of parameters, all of which are estimable. This will be touched upon again below.

MAGNITUDE DISCRIMINANTS, BASED ON SAMPLES

General statistical considerations

The model parameters are, of course, not known in reality, as assumed above, but the information available is contained in samples. The discriminant can then be constructed in various ways, for example by simply "plugging in" parameter estimates in the optimal rule, by considering total alternative likelihood functions in the likelihood ratio method, and by Bayesian posterior arguments, leading to among other logistic and predictive discriminants.

The discrimination capability can be described by the error probabilities (α_1, α_2). Each discriminant has a curve of possible (α_1, α_2)-values between $(0,1)$ and $(1,0)$, and the "cut-off" level chooses a point from this curve. An empirical discriminant will generally have a curve inferior to the optimal. When it is based on random samples, the curve will be random too. The true probabilities, which depend not only on the rule but also on unknown parameters, are unknown.

The choice of level is not trivial. The principle is that α_1 should equal a preset value, but this cannot be guaranteed. One can aim at the given value, for example by using the expected value of the asymptotic expansion (with respect to large samples) of α_1, as described by Anderson (1973) for the model $N_p(\mu_j, \Sigma)$. One can also choose it with "marginal", with regard to the variation over samples of α_1.

Different methods to construct a discriminant can be compared through the true error probabilities. Their expected values, the so-called unconditional probabilities, is one possible measure, but the variation of the conditional values should be considered too. This is not easily done, and neither does it seem to be constructively useful. When the alternatives are composite, as here when the K_j's are regarded as incidental, there are two more problems.

One problem is that even if the discriminant searched for has alternative distributions independent of the size-parameters, this will not be true for the empirical discriminant. The error probabilities will thus depend on the K_j's. These are bounded from both above and below (for detection reasons). This means that there will be a rectangle of possible outcomes in the (α_1, α_2)-plane for each rule. It is not so easy to make an appropriate choice of level, and one arrives then also at the second problem, which concerns estimation. The number of unknown parameters increases with the number of observations. If the maximum likelihood method is used the ordinary large-sample properties do thus not apply. Much study has been devoted to the behaviour of estimators for a linear functional relationship, for example Robertson (1974), Dolby (1976), and Anderson with discussants (1976). It is clear already intuitively that it is desirable that the spread of mean values along the line is great in comparison with the measurement errors. It is important to observe that the ordinary estimator of σ_j^2 is not consistent, and to correct for its underestimation.

Some different magnitude discriminants

It has been said that the $m_b(M_s)$-method is much used and that the linear discriminant is often obtained by a simple fit to a plot of data. Sometimes a method is presented, and a few of the arguments will be repeated.

When the discriminant is based on parameter estimates, these can be obtained from either population or from both. Weichert and Basham (1973) find none of the alternatives uniquely attractive, and they follow Ericsson (1971b) in choosing the explosions only, partly also because there is a smaller scatter within this population. Bungum and Tjøstheim (1976) do the same. The earthquake population is chosen by Elvers (1975b) and by Dahlman and Israelson (1976), who in their work emphasize the standpoint that only the samples of earthquakes can be considered as representative for future events.

The use of the normal distribution is criticized by Weichert and Basham (1973). They estimate the "trend line" of the magnitudes non-parametrically with reference to Bartlett (1949), (using M_s for ranking). Then they construct the discriminant from the explosion population, compute the discriminant values for the observations, and plot these. They find that the normal distribution is approximately valid, but, coming to the main point, they demonstrate the danger of extrapolating the tails according to these normal distributions in that it may give quite misleading estimates of the error probabilities.

Another approach, which is more data-oriented, is to use principal components. It is one of the suggestions of Burnaby (1966), and inherent in the recent work by Tjøstheim (1976). If the model (1) is true, the first axis will generally not be parallel to the mean value lines, but the methods are not really comparable.

The above refers to a single station. With several stations the magnitudes are "corrected" as to be comparable. Simple arithmetic means are usually taken for each kind of magnitude (M_s and m_b), possibly with the exclusion of those which deviate considerably. Another approach is taken in the study by Basham et al. (1976) where the model (1) is extended to the simultaneous distribution of I magnitudes as already presented.

A level of the discriminant is usually not given in seismological work, but the emphasis is put on the separation and on the applicability, which will now be discussed.

MISSING VALUES AND APPLICABILITY

Two magnitudes

Discriminants have been described above by their error probabilities, but under the restriction that all variables are available. In this chapter the effect of a special form of missing data will be considered. The discussion is made in terms of magnitudes, but the results are valid also for other variables. Often, when a magnitude is missing, an upper bound of its value can be given, from the noise level. Otherwise expressed, there is a detection threshold below which the magnitude is not obtained. The upper bound can be utilized to improve the discrimination procedure, and this will also be shown. The formulations were given by Elvers (1974).

A single station is studied in this section, and the parameters of the model are assumed to be known, for simplicity. To begin with, the size-parameters are regarded as incidental and the thresholds are assumed to be fixed numbers t_i. The random variables S_1 and S_2, given by (1), are censorized from below at t_1 and t_2 respectively, but their distribution is also truncated, because an event with both magnitudes below their thresholds remains undetected. A discriminant can still be constructed, although it seems necessary to refrain from a classification in certain cases. A reasonable rule is to use the discriminant (4) when possible and to classify an event as explosion (earthquake) when the single obtained magnitude $S_2(S_1)$ is above some level. The most interesting classification probabilities are α_1 and β_2, say, to classify an explosion correctly. They can be written as follows, where the subscript j indicates which set of parameters to use.

$$P_j\{(s_1,s_2);\ s_1 \geq t_1,\ s_2 \geq t_2,\ s_2 - \varkappa s_1 \geq C_1\} + P_j\{(s_1,s_2);\ s_1 < t_1,\ s_2 \geq t_2,\ s_2 \geq C_2\} \quad (6)$$

The two terms can be expressed by the univariate and bivariate standardized normal distribution, Φ.

The first term shows the capability of the $m_b(M_s)$-method in its original form. It is in each case a function of the size-parameter K_j which increases with K_j from zero to the asymptotic value $\Phi(-(C_1-\lambda_j)/\sigma_j\sqrt{1+\varkappa^2})$. The method can be said to be applicable for those event sizes for which the probability is not far from the corresponding asymptotic value.

The utilization of the upper bound comes in the second term, and there is then a second level C_2 to choose, in accordance with the preset α_1-value, which acts as an upper bound. An obvious choice is $C_2 = C_1 + \varkappa t_1$, with C_1 as above, which is equivalent to substituting S_1 when not detected by its upper bound t_1 in the discriminant $(S_2 - \varkappa S_1)$. This is called identification by negative evidence. Although simple, the method has the effect of extending the applicability to weaker events. If C_2 is increased, the function (6) will get a maximum, and C_1 will have to be decreased so that the maximum is kept at α_1 for earthquakes. The asymptotic value is then decreased, which is generally not desirable for explosions, but some further principle is needed to find the "optimal" combination of levels.

The classification rule has been obtained with the functional approach and its performance has thus been measured by classification probabilities for single events. One can, however, take up again the structural model, which it is reasonable to formulate for earthquakes, and compute the rate of misclassified earthquakes, as a kind of summarizing measure. The rate is computed from the regions in (6) with K_1 having an exponential distribution (the starting point of which is small enough in comparison with the detection capability to be without influence) and with regard to the intensity of occurrence of earthquakes. The second term gives a positive contribution for all C_2, but the increase in the rate (or rather the modification of levels needed to keep it unchanged) may seem negligible compared with the extended applicability for explosions.

Elvers (1974, 1975a) applies these methods to data, which refer to events in US and USSR respectively, measured at the Hagfors Observatory in Sweden. In the first case, for example, $\varkappa = 0.6$, $\lambda_1 = 2.4$, $\lambda_2 = 3.5$, $\sigma_1 = \sigma_2 = 0.3$, $a_{11} = a_{12} = 2$, $b_{11} = b_{12} = 1$, $t_1 = 4.0$, $t_2 = 4.8$. If α_1 is permitted to be 1 per mil only, the asymptotic value of β_2 is found to be 0.52. For $K_2 = 2$, which is the value making the mean ξ_{12} equal to t_1, β_2 is only 0.17. The inclusion of identification by negative evidence increases this value as much as to 0.44. A visual inspection of the β_2-curves shows that the applicability is lowered about half a magnitude for this set of parameters, which is a considerable improvement in seismological terms. The second set of data shows similar results, and here also the rate of misclassified earthquakes is studied, confirming the above statement on applicability.

The detection thresholds have so far been assumed to be fixed, but in fact they vary. This does not present any logical difficulties, as long as they are measurable, because the classification rules are easily modified. It has, however, the effect that the classification probabilities for an event of given size vary. This is intuitively clear, and it means, of course, also that the applicability of the methods is not constant.

More than two magnitudes

The analysis above can be extended to more than two magnitudes, but there is a difference inasmuch that a "real" discriminant can be constructed when at least two magnitudes have been obtained. The likelihood ratio does not give an explicit discriminant, and here only a special case will be considered. It has its origin in the study by Basham et al. (1976), and it has been treated by Elvers (1975b). Let there be only one magnitude, S_I, with threshold, and assume that the others are always measured. The discriminant (4) is an obvious choice when all magnitudes are available. When S_I is not obtained, there are two simple possibilities, namely (4) based on the remaining magnitudes and the negative evidence technique.

If the first choice is taken, and if f_I is negative (which it usually will be for M_s-magnitudes, which have the greater practical interest here), the two classification probabilities α_1 and β_2 (which now equals $(1-\alpha_2)$) will have horizontal asymptotes for small and great K_j-values, with a minimum between. The levels of the discriminants are then chosen so that the asymptotic values of α_2 are equal to the preset value. The second possibility mentioned gives, of course, zero as asymptotic value of α_1 and β_2 for weak events, but it also has advantages. One can try to combine these two rules, a simple way being to classify an event as explosion as soon as either of them does. It is not clear analytically whether this will imply a maximum of α_1, but numerical examples have given maxima, although little above the permitted value. There is again a considerable improvement in β_2, due to the simple negative evidence technique, but the curve forms also suggest that further gains are possible.

SELECTION OF VARIABLES AND PROCEDURE

General statistical considerations

At a first glance it may seem desirable to base a discriminant on many variables, but it is not so when the model contains unknown parameters and the samples are given. This has been demonstrated for both continuous and discrete variables, usually in terms of detoriation in expected error probabilities. The model $N_p(\mu_j, \Sigma)$ is generally chosen in the first case. It has the advantage that the separation between the two populations is determined by a single quantity, the Mahalanobis' distance $\Delta^2 = (\mu_2-\mu_1)^T \Sigma^{-1}(\mu_2-\mu_1)$. Numerical results are given for example by Dunn (1971) for different values of Δ^2, the number p of classification variables, and the sample sizes. The error probabilities are sensitive to p, so the variables have to be chosen with care. When the variables are discrete there is no such established model. Here many have pointed out that the achieved reduction in the number of parameters to estimate often motivates the use of a less accurate model when constructing the classification rule.

When a discriminant is based on discrete variables, for example by forming a linear combination, the exact weightings are not essential, but only the ranking of outcomes. There is a positive probability that an empirical procedure is equivalent to the optimal, and the rate of convergence towards optimality is of different type. Discrete and continuous variables are thus not easily compared. They can be combined, although not much attention has been paid to such procedures in the literature, as pointed out by Krzanowski (1975). The behaviour is then of the continuous type. For relationship-variables there are further properties to consider, as described above.

It is not only the theoretical capability and its loss due to estimation, which are of interest. Frequency of missing data is also important, and the effort of collection and computation. Here stagewise procedures are valuable. One such method has been investigated by Zielezny and Dunn (1975).

Methods in use

There are many variables which can be used to discriminate between earthquakes and underground nuclear explosions. Only magnitudes have been mentioned so far, but the two important concepts capability and applicability for a variable have been treated. A further important property when judging a variable is its sensitivity to the location of the event, and also to the depth.

When one of the magnitudes at a station is missing, it is usually M_s, and it is especially the shortperiod body wave which has been searched for characteristics. Dahlman and Israelson (1976) say in their review that most of the used variables can be associated with the wave form complexity or with the frequency content, and that the greater separation between earthquakes and explosions seems to be found in the second group. It is a drawback of the shortperiod discriminants that they are highly dependent on both location and depth, Israelson (1975). This means that great data-bases are needed to enable an appropriate discriminant to be found. There are many other suggestions of variables, for example similar characteristics from other waves and autoregressive parameters, see further Dahlman and Israelson (1976).

The classification of an event is generally made in several stages. The first considers the epicenter, whereby a great part of the events can be sorted out, as they with great confidence have occurred in regions where the possibility of an explosion can be excluded. The second step is of similar kind, with regard to depth. The elimination of deep earthquakes has the positive effect of decreasing the scatter within the population. There is even a systematic improvement for magnitudes, because deep earthquakes tend to give more explosion-like magnitudes (in a way not easy to model). Again one has, of course, to consider the uncertainty in estimates.

The discriminatory variables are often used in stages, also after these two steps of reduction. The $m_b(M_s)$-method is usually taken alone first, because of its high capability and its rather good robustness towards regional variations. If it cannot be applied, some shortperiod discriminant is chosen. When measurements from several stations are available, they are usually combined into a single criterion, which was described for magnitudes above. Different discriminants have also been combined, for example linearly and by taking maximum. The choice is generally not motivated by a model, but data-bases are used for construction within some pre-selected class of discriminants and for evaluation. One example is Bungum and Tjøstheim (1976), where two variables from a signal are used to create a new one, after which two such variables are combined linearly. Recently pattern recognition methods have also been applied, Tjøstheim (1976). It is notable that most studies are data-oriented, and much attention is given to spurious events, which do not follow the expected behaviour.

Another set of variables, which is not yet in use, but likely to turn out helpful, is of the presence/absence type. The so-called secondary phases from surface reflected waves, for example, can, when they are identified, be used as indicative of deep earthquakes, Yamamoto (1974). The classification will probably be made through a thus improved estimate of the depth together with further variables of the same binary type. There are other qualitative variables, which are considered in the discrimination procedure now, but so far their use has not been formalized.

Another trend is work on increased exchange of data and on a global network. Great data-bases increase the understanding of seismic phenomena, which lead to improved methods, especially if measurements from several suitably spaced stations can be combined, and the capability will then also be extended to weaker events.

REFERENCES

Aki, K. (1972) Scaling Law of Earthquake Source Time-Function, Geophys. J. R. astr. Soc. 31, 3-25.

Anderson, T.W. (1973) Asymptotic Evaluation of the Probabilities of Misclassification by Linear Discriminant Functions, in Discriminant Analysis and Applications, ed. T. Cacoullos, Academic Press.

Anderson, T.W. (1976) Estimation of Linear Functional Relationships: Approximate Distributions and Connections with Simultaneous Equations in Econometrics, J.R. Statist. Soc. B, 38, 1-36, with discussion.

Bartlett, M.S. (1949) Fitting a Straight Line when Both Variables are Subject to Error, Biometrics 5, 207-212.

Basham, P.W., Dahlman, O., Elvers, E., Israelson, H., and Slunga, R. (1976) Definition and Identification of Seismic Events in the US in 1972, FOA Report to appear.

Bungum, H. and Tjøstheim, D. (1976) Discrimination between Eurasian Earthquakes and Underground Explosions Using the m_b:M_s Method and Short-Period Autoregressive Parameters, Geophys. J. R. astr. Soc. 45, 371-392.

Burnaby, T.P. (1966) Growth-Invariant Discriminant Functions and Generalized Distances, Biometrics 22, 96-110.

Chinnery, M.A. and North, R.G. (1975) The Frequency of Very Large Earthquakes, Science 190, 1197-1198.

Dahlman, O. and Israelson, I. (1976) Monitoring Underground Nuclear Explosions. To be published.

Dolby, G.R. (1976) The Ultrastructural Relation: A Synthesis of the Functional and Structural Relations, Biometrika 63, 39-50.

Dunn, O.J. (1971) Some Expected Values for Probabilities of Correct Classification in Discriminant Analysis, Technometrics 13, 345-353.

Elvers, E. (1974) Seismic Event Identification by Negative Evidence, Bull. Seism. Soc. Am. 64, 1671-1683.

Elvers, E. (1975a) Seismic Event Identification by the $m_b(M_s)$-Method, FOA Report C 20070-T1. Research Institute of National Defence, Stockholm.

Elvers, E. (1975b) A Discrimination Problem from Seismology, FOA Report C 20080-T1. Research Institute of National Defence, Stockholm.

Ericsson, U. (1971a) A Linear Model for the Yield Dependence of Magnitudes Measured by a Seismographic Network, Geophys. J. R. astr. Soc. 25, 49-70.

Ericsson, U. (1971b) Event Identification by m(M) Observations from Networks, FOA Report C 4480-A1, Research Institute of National Defence, Stockholm.

Israelson, H. (1975) Short Period Identification, in Exploitation of Seismograph Networks, ed. K.G. Beauchamp, Nordhoff, Leiden, The Netherlands.

Krzanowski, W.J. (1975) Discrimination and Classification Using Both Binary and Continuous Variables, J. Am. Statist. Ass. 70, 782-790.

Rao, C.R. (1966) Discriminant Function between Composite Hypothesis and Related Problems, Biometrika 53, 339-345.

Robertson, C.A. (1974) Large-Sample Theory for the Linear Structural Relation, Biometrika 61, 353-359.

Tjøstheim, D. (1976) Improved Seismic Discrimination Using Pattern Recognition. Presented at 7th Nordic Seminar on Dection Seismology, Copenhagen, May 1976.

Weichert, D.H. and Basham, P.W. (1973) Deterrence and False Alarms in Seismic Discrimination, Bull. Seism. Soc. Am. 63, 1119-1133.

Wyss, M. (1973) Towards a Physical Understanding of the Earthquake Frequency Distribution, Geophys. J. R. astr. Soc. 31, 341-359.

Yamamoto, M. (1974) Estimation of Focal Depth by pP and sP Phases, FOA Report C 20027-T1, Research Institute of National Defence, Stockholm.

Zielezny, M. and Dunn, O.J. (1975) Cost Evaluation of a Two-Stage Classification Procedure, Biometrics 31, 37-47.

Recent Developments in Statistics
J.R.Barra et al., editors

A FISSION MODEL FOR YEAST CELLS

J. Gani
CSIRO Division of Mathematics and Statistics
PO Box 1965
Canberra City
ACT 2601
Australia

This paper considers a branching process which describes the reproduction of yeast cells. As time $t \to \infty$, it is proved that the proportions of cells having 0,1,2,..., offspring approach the ratios $2^{-1}, 2^{-2}, 2^{-3}, \ldots$.

1. Introduction

After reproduction, a yeast cell carries a scar where its daughter cell has budded from it. The number of scars on the mother cell is thus equal to the number of its offspring. In 1975, James Barnett of the University of East Anglia obtained some empirical data on the numbers of scars for certain varieties of yeast cells; to fit these, some theoretical fission models seemed necessary. These were developed by Gani and Saunders (1976); we describe here the simplest of them, a multitype branching process of the Bienaymé-Galton-Watson (BGW) type.

For this model, let us assume that in each generation, the probabilities of fission, inactivity, and death for each of the cells are p,q, and r respectively $(p+q+r=1)$. Let $N_0(t), N_1(t), N_2(t), \ldots,$ be the numbers of cells having 0,1,2,..., offspring at generation t $(t=0,1,2,\ldots)$. Starting from the probability generating function of the multitype BGW process, one can readily show that the means $n_0(t), n_1(t), n_2(t), \ldots,$ of the cell numbers satisfy the equations

$$\begin{aligned} n_0(t+1) &= (p+q)n_0(t) + p\Sigma_{j=1}^{\infty} n_j(t) \\ n_1(t+1) &= pn_0(t) + qn_1(t) \\ n_2(t+1) &= pn_1(t) + qn_2(t) \\ &\cdot\ \cdot\ \cdot\ \cdot\ \cdot\ \cdot\ \cdot \end{aligned} \tag{1.1}$$

The solution of these, subject to the initial conditions $N_0(0) = x_0$, $N_1(0) = x_1$, $N_2(0) = x_2, \ldots,$ will allow us to formulate a hypothesis about the proportions of mean numbers $n_0(t), n_1(t), n_2(t), \ldots,$ when the number t of generations becomes large.

2. Values of mean numbers

The method of solution of equations (1.1) is simple; one considers first

$$n_0(t+1) = (p+q)n_0(t) + p\Sigma_{j=1}^{\infty} n_j(t)$$
$$\Sigma_{j=1}^{\infty} n_j(t+1) = pn_0(t) + (p+q)\Sigma_{j=1}^{\infty} n_j(t) \tag{2.1}$$

where the last equation is the sum of all equations in (1.1) starting from the second. We obtain directly that

$$n_0(t) = (2p+q)^t \tfrac{1}{2}x + q^t(x_0 - \tfrac{1}{2}x)$$
$$\Sigma_{j=1}^{\infty} n_j(t) = (2p+q)^t \tfrac{1}{2}x - q^t(x_0 - \tfrac{1}{2}x) \tag{2.2}$$

where $x = \Sigma_{j=0}^{\infty} x_j$.

Once $n_0(t)$ is known, we can then solve the equation

$$n_1(t+1) - qn_1(t) = pn_0(t) \tag{2.3}$$

to derive the result

$$n_1(t) = (2p+q)^t \frac{x}{4} + q^t \{(x_1 - \frac{x}{4}) + t\frac{p}{q}(x_0 - \tfrac{1}{2}x)\}. \tag{2.4}$$

Using the same method, one obtains for each $n_k(t)$, $k = 1,2,\ldots,$ the value

$$n_k(t) = (2p+q)^t \frac{x}{2^{k+1}} + q^t \{\Sigma_{j=0}^{k}(x_{k-j} - \frac{x}{2^{k+1-j}})(\frac{p}{q})^j \binom{t}{j}\} \tag{2.5}$$

We note that $(2p+q)$ is the larger of the two eigenvalues; when t is large, we thus see that

$$n_k(t) \sim (2p+q)^t \frac{x}{2^{k+1}}. \tag{2.6}$$

This means that if $n(t) = \Sigma_{j=0}^{\infty} n_j(t) = (2p+q)^t x$ is the mean number of all cells in the population,

$$\lim_{t\to\infty} \frac{n_k(t)}{n(t)} = 2^{-k-1} . \tag{2.7}$$

We can then readily obtain that the proportions of cells with $0,1,2,\ldots,$ offspring tend, for large t, to the ratios 2^{-1}, 2^{-2}, 2^{-3}, ..., independently of the values of p,q, and r.

3. Observed data and theoretical model

We set out below a simplified table of the empirical data on cell numbers observed by James Barnett, together with the theoretical means based on the result (2.7).

Table 1

Observed and expected numbers of cells with k = 0,1,2,..., offspring

Yeast*	k =	0	1	2	3	4	$k \geq 5$	χ_5^2
1	o_k	399	193	117	47	23	14	8.3
	e_k	396.5	198.3	99.1	49.6	24.8	24.8	
2	o_k	287	149	98	43	16	11	11.9*
	e_k	302	151	75.5	37.8	18.9	18.9	
3	o_k	685	267	213	88	57	32	34.5**
	e_k	671	335.5	167.8	83.9	41.9	41.9	

*Yeast 1 = Baker's yeast; Yeast 2 = Saccharomyces Cerevisiae (exponential growth phase); Yeast 3 = Saccharomyces Cerevisiae (stationary phase).

o_k, e_k = observed, expected number of cells with k offspring.

To test our fission model, we use the χ^2 test of hypothesis. The value of χ^2 in the case of baker's yeast is 8.3; this is not significant at the 5% level. In the case of Yeast 1 the model seems entirely adequate. For Yeast 2 the value 11.9 of χ^2 is significant at the 5% level; the BGW model does not seem entirely appropriate. In the case of Yeast 3, χ^2 has the value 34.5 which is significant at the 1% level; here the model is clearly not satisfactory.

Should one then abandon the hypothesis of a multitype BGW process? Some recent work of Gani and Saunders (1977) indicates that a slight modification to the probabilities of fission, inactivity and death may lead to a more appropriate model. This consists in assuming that the relevant probabilities p_0, q_0 and r_0 for cells without offspring (k=0), are different from the probabilities p, q, and r for all other cells (k=1,2,...). In this case, as $t \to \infty$, the ratios of means $n_k(t)/n(t)$ of the cells tend to the values

$$\frac{\rho-p-q}{\rho-p-q+p_0}\ ,\ \frac{\rho-p-q}{\rho-p-q+p_0}\left(\frac{p_0}{\rho-q}\right)\ \frac{\rho-p-q}{\rho-p-q+p_0}\left(\frac{p_0}{\rho-q}\right)\left(\frac{p}{\rho-q}\right)\ ,\ldots, \qquad (3.1)$$

where $\rho = \frac{1}{2}\{p_0+p+q_0+q+\ [(p_0+p+q_0+q)^2 - 4(p_0q+q_0p+q_0q)]^{\frac{1}{2}}\}$.

After estimating the relevant probabilities, one is led to a closer agreement between Barnett's empirical data and the multitype BGW model for yeast cell fission. However, some oscillations in the ratios of the means $n_k(t)/n(t)$ during the growth phase of the cell population still remain to be accounted for.

4. References

Gani, J. and Saunders, I.W. (1976). On the parity of individuals in a branching process. J. Appl. Prob. 13, 219-230.

Gani, J. and Saunders, I.W. (1977). Fitting a model to the growth of yeast colonies. Biometrics 33 (to appear).

Recent Developments in Statistics
J.R.Barra et al., editors

PORTABLE GENERATORS FOR THE RANDOM VARIABLES USUAL IN RELIABILITY SIMULATION

J. ARTHUR GREENWOOD

Institute of Marine and Atmospheric Sciences, City University of New York

To guarantee that an exponential generator (coded in standard FORTRAN) yield the same output, to within rounding, on any computer, it is convenient to prescribe, first, that the uniform random number generator used to drive the exponential generator shall be machine-independent; second, that all IF statements be computed on integer variables (thus bypassing uncertainty whether a difference in rounding affects the branch taken). A generator is given fulfilling these conditions: the logarithm of a uniform random number is calculated by a 5-term approximation to the inverse hyperbolic tangent; on the average 256/255 random integers are used for each exponential number generated. Normally distributed random numbers are computed by the polar method of Box & Muller, using a 4-term approximation to the circular cotangent; on the average 511/510 random integers are used for each normal number generated.

Generators for random variables from the lognormal and Weibull distributions afford no difficulties in mathematical analysis, but require the detection of overflow and underflow. For completeness, a gamma generator is given: the code is machine-independent, but essential use is made of IF statements computed on real variables.

1. INTRODUCTION

This paper provides algorithms, in machine-independent FORTRAN, which generate pseudo-random numbers with the distributions usual in reliability simulation. Even when the algorithms are machine-independent, their output can be machine-dependent. The algorithms below achieve independence by two devices: (1) they employ a portable generator for random integers, in which intermediate results are never permitted to exceed 31 bits plus sign; (2) all arithmetic expressions embedded in IF statements are of type INTEGER, so that the execution of the IFs is unaffected by rounding.

The portable integer generator IRAN31 has been modified from Greenwood (1976). IRAN31 produces integers uniformly distributed on $[0,2^{31}-1]$; the integers are almost independent in blocks of 46; serial correlations vanish out to lag 131.

Let n_i, $i = 0, \ldots, 119$, be integers as close as convenient to

2^{27}, such that $12n_i+5$ and $24n_i+11$ are both prime; a supply of such integers has been found by sieving. Then all quadratic non-residues of $24n_i+11$, except -1, are primitive roots; in particular, 2 is a primitive root of $24n_i+11$. Set $q_i = 24n_i+12-2^{31}$. Then the recurrence $X_{i,j+1} = \text{Mod}(2(X_{ij}+q_i-1), 24n_i+11)-q_i+1$ has cycle length $24n_i+10$; a preliminary rejection reduces this cycle to $24n_{i-1}+11$. The generator IRAN31 consists in running the 119 recurrences cyclically, and excluding all negative values of X_{ij} from the output to any external program. This double use of rejection as a step in a uniform random number generator is a heroic application of a device due to Miller & Prentice (1968). IRAN31 has cycle length

$$2^{31}\prod(24n_{i-1}+11)\sum(24n_{i-1}+11)^{-1} \simeq 119\cdot 3^{118}\cdot 2^{3451} > 10^{1097}.$$

The array JB contains the 120 q_i. If the initial contents of array JA are in the range [-1073658098,+2147483647] then the output of the generator and the future contents of array JA are machine-independent.

IRAN31 as printed is not self-initializing: the calling program should insert a random integer (in the range stated) into each of the 120 locations of array JB. Subroutine INIT may be used for this insertion; the device of expanding initial values from a short seed is taken from Bright & Enison (1976).

2. EXPONENTIAL DISTRIBUTION

Marsaglia (1961) has expressed an exponential random variable as the sum of a truncated exponential variable on the range $[0,\alpha)$ and a geometric-progression variable on the lattice $\{n\alpha\}$. In theoretical work it is customary to take $\alpha = 1$; Sibuya (1962) exposed the advantage of taking $\alpha = \log_e 2$. Greenwood (1974) showed how to apply this decomposition to yield an exponential distribution with a tail extending as far as desired; the choice $\alpha = \log_e 1000$ in that paper is inconvenient for binary computers.

The generator REXP expresses an exponential random variable Z as the sum $Z = X_1+X_2+X_3$, where

$$\Pr\{X_1=8i\log_e 2\} = 2^{-8i}(1-2^{-8}), \quad i = 0,1,2,\ldots;$$

$$\Pr\{X_2=i\log_e 2\} = 2^{-i-1}(1-2^{-8})^{-1}, \quad i = 0,1,\ldots,7;$$

$$\Pr\{X_3\leq x\} = \begin{cases} 0, & x < 0; \\ 2(1-e^{-x}), & 0 \leq x \leq \log_e 2; \\ 1, & x > \log_e 2. \end{cases}$$

X_1 and X_2 are effectively computed by coin-tossing. $X_3 = -\log_e U$, where U is a uniform random number on $[\frac{1}{2},1)$ and the logarithm is computed by the transformation $\log_e U = 2 \text{ ar } \tanh((U-1)/(U+1))$, where the approximation

$$2 \text{ ar } \tanh y \simeq y(2.000000042+.6666591253y^2+.4003750857y^4+ +.2781697041y^6+.287037037y^8)$$

is good to 2^{-24} on the range $|y| < 1/3$. The tail of the exponential distribution is accurate out to $Z = 255.07$.

The device of computing a logarithm as an inverse hyperbolic tangent is probably as old as Wolfram and was revived by Hastings (1955, pp. 107–109); the countersuggestion that ar tanh „can be computed most economically as a natural logarithm" (Kogbetliantz 1960, p. 30) is never good computing practice and can be fatal in floating-point work. The alternative computation

$$-\log_e U = \tfrac{1}{2}\log_e 2 - 2 \text{ ar } \tanh((U-2^{-\frac{1}{2}})/(U+2^{-\frac{1}{2}}))$$

saves two multiplications and one addition by confining y to the range $|y| < 3-\surd 8$; but it effectively computes a fixed-point value of X_3. The resultant granularity of X_3 is acceptable when exponential random numbers are required as such or as a step in generating normally distributed numbers; but, when exponential numbers are transformed into numbers with the Gumbel distribution, it emerges as unpleasantly coarse granularity in the range $(4 \log_e 2, 12 \log_e 2)$ of the Gumbel distribution.

3. NORMAL DISTRIBUTION

Given a good exponential generator, it is natural to build the normal generator upon it; this dictates the method of Box & Muller (1958). The sine and cosine occurring must be computed directly. The FORTRAN code required to compute the sine and cosine by rejection, keeping the IF statement machine-independent, say:

```
      DOUBLE PRECISION U(2),US(2),UUS
10    DO 11 IA = 1,2
        U(IA) = (IRAN31(JA)/32)*2-67108863
        U(IA) = U(IA)/67108864D0
        US(IA) = U(IA)**2
11    CONTINUE
      UUS = US(1)+US(2)
      IF (UUS .GT. 1D0) GO TO 10
      VS = UUS
      V1 = (SNGL(US(1))-SNGL(US(2)))/VS
      V2 = 2.*SNGL(U(1))*SNGL(U(2))/VS
```

requires double-precision arithmetic to yield a single-precision result, and is particularly awkward on a UNIVAC 1100, where the format of a single-precision number differs from that of the left half of a double-precision number.

Tocher (1963, p. 34) observes that it suffices to compute the sine and cosine on the interval $[0,\frac{1}{2}\pi)$. Since the sine and cosine can be interchanged before computing X_1 and X_2, the interval $[0,\frac{1}{4}\pi)$ suffices: V_1 and V_2 are computed from U, a uniform random number on $[0,1)$, by the sequence

$$Y = \tan(\pi U/8) \simeq U/(2.54679091-.1308997701U^2- -.0013453775U^4-.0000203746U^6);$$

$$V_1 = (1-Y^2)/(1+Y^2),\ V_2 = 2Y/(1+Y^2);$$

then, after affixing signs to V_1 and V_2 and interchanging them if necessary, $V_1 = V_1\surd(2Z)$, $X_2 = V_2\surd(2Z)$, where Z is an exponential random number computed by the method of Section 2. The tails of the normal distribution are accurate out to $X = \pm 15.97$.

Note that subroutine RNORM produces normal deviates in pairs; the use of an indicator variable to deliver them one at a time (Ahrens & Dieter 1972) is inconvenient in ISO FORTRAN. The user's program should employ both X_1 and X_2 (they are independent); otherwise, the program will execute twice the necessary number of calls to RNORM.

4. OTHER DISTRIBUTIONS

4.1 Uniform distribution. Function subprograms RUNIF1 and RUNIF2 are included for completeness; the experienced simulator will combine integer-to-real conversion with other operations. RUNIF1 suffices for tutorial purposes; RUNIF2 solves exactly the following not very practical problem: To generate a pseudo-random sequence of the 2^{23} fractions $2^{-24}(2i-1)$, $1 \le i \le 2^{23}$, the sequence being completely independent of the machine on which run.

4.2 Lognormal distribution. When e^X would overflow or underflow, subroutine RGALT returns X; when the parameters are outside reasonable ranges, an error indicator is returned. Note that RGALT produces lognormal numbers in pairs.

4.3 Gumbel distribution. Function subprogram RGUMB generates numbers from the standardized distribution of extreme values

$$\Pr\{Y \le y\} = \exp(-e^{-y}),\ -\infty < y < +\infty.$$

Compute an exponential random number Z_1. If $\exp(-Z_1) < 1-2^{-12}$, set $Y = -\log_e Z_1$. Otherwise, compute a new exponential number Z_2; then $Y = -\log_e\{-\log_e(1-2^{-12}\exp(-Z_2))\} \simeq 12 \log_e 2 + Z_2 - 2^{-13}\exp(-Z_2)$; the approximation of Hastings (1955, sheet 57) is used for $\exp(-Z_2)$.

4.4 Weibull distribution. Notations for this distribution vary. Subroutine RWEIB generates numbers X from the distribution

$$\Pr\{X \le x | b,p\} = 1 - \exp(-(x/b)^p), \quad x,b,p > 0,$$

so that $X = b \exp(-Y/p)$, where Y is a random number from the Gumbel distribution. When the exponential would overflow or underflow, $\log_e b - Y/p$ is returned; when $|p|$ is outside reasonable range, or $b \le 0$, an error indicator is returned. Note that $p < 0$ is lawful and generates a random number with a Cauchy-type extremal distribution.

4.5 Gamma distribution. Function subprogram RGAMMA is adapted from Greenwood (1974). The code is machine-independent; the numbers generated are not. Specifically, the IFs in statements 16 and 24 of RGAMMA are applied to the results of floating-point computation. The code printed below seeks to isolate the effects of these IFs by always calling other generators in sets of two uniform numbers and one exponential number (for $p < 1$) or one normal and one exponential number (for $p > 1$). Further attempts to reduce machine dependence yield diminishing returns: whether p is calculated by the calling program or read in as a decimal fraction, it will in general not be exactly representable as a binary fraction; it will receive different binary representations in different machines; and necessarily there will be marginal values of RGAMMA that one machine delivers and another machine rejects.

5. FORTRAN PROGRAMS

The programs below will run unmodified on machines accommodating integers of 31 or more bits plus sign. In oral discussion at Grenoble and Berlin in September of 1976 it was pointed out that ICL 1900 computers use integers of 23 bits plus sign; a FORTRAN floating-point number occupies two words, with 39 bits allotted to the mantissa. These programs can be transferred to those machines (at some cost in running time) by declaring the function IRAN31 to be REAL and using real arithmetic therein. (The arithmetic operations in IRAN31 are additions and subtractions; no truncation or overflow problems will occur.)

(N.B. The sign ⁎ is an asterisk)

```
      SUBROUTINE INIT (JA,JB)
C TO EXPAND A 4-WORD SEED IN JB INTO A 120-WORD ARRAY IN JA
      DOUBLE PRECISION DA(4)
      INTEGER JA(120),JB(4)
      DO 10 IA = 1,4
        DA(IA) = ABS(FLOAT(JB(IA)))
10    CONTINUE
      DA(1) = DA(1)+DA(4)+1D0
      KA = 9
      KC = 1
      KD = 4
11    KA = KA+1
```

```
      KC = KC+1
      IF (KC .EQ. 5) KC = 1
      KD = KD+1
      IF (KD .EQ. 5) KD = 1
      DA(KC) = DA(KC)+DA(KD)
      IF (DA(KC).LE. 3D11) GO TO 11
      DA(KC) = DA(KC)/FLOAT(KA)
      DA(KC) = DA(KC)-DMOD(DA(KC),1D0)
      JA(1) = DMOD(DA(KC),1D2)
      DO 13 IA = 2,120
        KA = KA+1
        KC = KC+1
        IF (KC .EQ. 5) KC = 1
        KD = KD+1
        IF (KD .EQ. 5) KD = 1
        DA(KC) = DA(KC)+DA(KD)
        IF (DA(KC).LE. 3D11) GO TO 12
        DA(KC) = DA(KC)/FLOAT(KA)
        DA(KC) = DA(KC)-DMOD(DA(KC),1D0)
12      JA(IA) = DMOD(DA(KC),3D9)-1D9
13    CONTINUE
      RETURN
      END
      INTEGER FUNCTION IRAN31(JA)
C 31-BIT GENERATOR
      INTEGER JA(120),JB(120)
      DATA JB(1),JB(2),JB(3),JB(4),JB(5),JB(6),JB(7),JB(8),JB(9),
     $JB(10),JB(11),JB(12),JB(13),JB(14),JB(15),JB(16),JB(17),
     $JB(18),JB(19),JB(20),JB(21),JB(22),JB(23),JB(24),JB(25),
     $JB(26),JB(27),JB(28),JB(29),JB(30),JB(31),JB(32),JB(33),
     $JB(34),JB(35),JB(36),JB(37),JB(38),JB(39),JB(40),JB(41),
     $JB(42),JB(43),JB(44),JB(45),JB(46),JB(47),JB(48),JB(49),
     $JB(50),JB(51),JB(52),JB(53),JB(54),JB(55),JB(56),JB(57),
     $JB(58),JB(59),JB(60)/
     $1073657236,1073658100,1073659900,1073660452,1073661052,
     $1073662756,1073663236,1073663356,1073664940,1073665420,
     $1073666140,1073666380,1073666452,1073667556,1073668996,
     $1073671540,1073674900,1073678212,1073678356,1073678692,
     $1073679196,1073679316,1073686252,1073687980,1073690476,
     $1073692756,1073697172,1073697340,1073697796,1073697916,
     $1073698972,1073699380,1073700820,1073701732,1073702836,
     $1073704612,1073709772,1073710612,1073712412,1073713876,
     $1073714836,1073717956,1073718340,1073723740,1073726812,
     $1073727460,1073728780,1073729380,1073729476,1073730436,
     $1073733292,1073734036,1073736436,1073737780,1073738572,
     $1073739052,1073747572,1073747956,1073748460,1073749132/
      DATA JB(61),JB(62),JB(63),JB(64),JB(65),JB(66),JB(67),JB(68),
     $JB(69),JB(70),JB(71),JB(72),JB(73),JB(74),JB(75),JB(76),
     $JB(77),JB(78),JB(79),JB(80),JB(81),JB(82),JB(83),JB(84),
     $JB(85),JB(86),JB(87),JB(88),JB(89),JB(90),JB(91),JB(92),
     $JB(93),JB(94),JB(95),JB(96),JB(97),JB(98),JB(99),JB(100),
     $JB(101),JB(102),JB(103),JB(104),JB(105),JB(106),JB(107),
     $JB(108),JB(109),JB(110),JB(111),JB(112),JB(113),JB(114),
     $JB(115),JB(116),JB(117),JB(118),JB(119),JB(120)/
     $1073751220,1073753980,1073755876,1073756380,1073756932,
     $1073757436,1073758660,1073759116,1073762572,1073765452,
     $1073766076,1073766676,1073768932,1073770420,1073772940,
     $1073776492,1073777092,1073778220,1073779012,1073782972,
     $1073783092,1073783716,1073785660,1073785876,1073788612,
     $1073790796,1073792452,1073793436,1073794516,1073792692,
```

```
     $1073797180,1073798020,1073798980,1073800372,1073800516,
     $1073801500,1073804356,1073806156,1073806996,1073807332,
     $1073807740,1073808916,1073808952,1073809972,1073810956,
     $1073811292,1073812372,1073812996,1073813596,1073814340,
     $1073815276,1073816860,1073817916,1073818180,1073818972,
     $1073819356,1073820100,1073820940,1073822212,1073823532/
      KB = IABS(JA(1))
10    IF (KB .GE. KA) KB = 1
      KB = KB+1
11    IF (JA(KB)) 13,15,12
12    IRAN31 = JA(KB)-2147483647
      JA(KB) = JA(KB)+IRAN31-1
      IF (JA(KB).GE. 0) GO TO 16
      IRAN31 = JA(KB)+JB(KB)
      IF (IRAN31 .GT. 0) GO TO 14
      JA(KB) = IRAN31+2147483647
      GO TO 16
13    IRAN31 = JA(KB)+JB(KB)
      JA(KB) = JA(KB)+IRAN31-1
      IF (JA(KB).GE. 0) GO TO 16
14    IF (JA(KB)+JB(KB-1).GT. 0) GO TO 10
      GO TO 11
15    JA(KB) = JB(KB)-1
16    IRAN31 = JA(KB)
      JA(1) = KB
      RETURN
      END

      SUBROUTINE PARAM (PA)
C TO INITIALIZE PARAMETERS IN RGAMMA
      REAL PA(7)
C SEE FUNCTION RGAMMA FOR CONTENTS OF ARRAY PA
      PA(2) = PA(1)
      IF (PA(1).GT. 0.) GO TO 10
      PA(7) = -1.
      RETURN
10    IF (PA(1).GE. 8E-9) GO TO 11
      PA(7) = -2.
      RETURN
11    IF (PA(1).LE. 1E12) GO TO 12
      PA(7) = -3.
      RETURN
12    PA(7) = 0.
      IF (PA(1).GT. 1.) GO TO 13
      PA(3) = 1./PA(1)
      PA(4) = 1./(1.-PA(1))
      RETURN
13    PA(3) = PA(1)-1./3.
      PA(4) = SQRT(PA(1))/3.
      AG = PA(1)*(1.-1.73205081*PA(4))**3
C 1.73205081 = SQRT(3)
      PA(5) = PA(3)*ALOG(AG)-AG+.5*(PA(4)-1.73205081)**2
      PA(6) = 1E8
      RETURN
      END

      REAL FUNCTION REXP(JX)
C EXPONENTIAL GENERATOR
      INTEGER JX(120)
      DATA BA,BB/5.545177444,.6931471806/
C BA = LOGE(256), BB = LOGE(2)
      REXP = 0.
```

```
10    KA = IRAN31(JX)
C IRAN31 IS A 31-BIT GENERATOR
      IF (KA-8388608) 11,12,13
C 8388608 = 2**23
11    REXP = REXP+BA
      GO TO 10
12    REXP = REXP+BA
      RETURN
13    IF (KA-1073741824) 14,15,16
C 1073741824 = 2**30
14    REXP = REXP+BB
      KA = KA+KA
      GO TO 13
15    REXP = REXP+BB
      RETURN
16    KA = 2147483647-KA
      BC = KA+1
      BC = BC/(BC+(65536.*65536.))
C 65536.*65536. = 2**32
      BD = BC**2
      REXP = REXP+((((.287037037*BD+.2781697041)*BD+.4003750857)*BD+
     $        .666591253)*BD+2.000000042)*BC
      RETURN
      END

      SUBROUTINE RGALT (AM,SI,Y1,Y2,OV1,OV2,JX)
C LOGNORMAL GENERATOR
C Y1 AND Y2 ARE INDEPENDENT LOGNORMAL RANDOM NUMBERS
C LOGE(Y) IS NORMALLY DISTRIBUTED WITH MEAN AM AND
C   STANDARD DEVIATION SI
C IF Y OVERFLOWS, IT IS SET TO -1. IF Y UNDERFLOWS, IT IS SET TO 0.
C IF Y OVERFLOWS OR UNDERFLOWS, OV CONTAINS LOGE(Y)
C ERROR MESSAGES
C   Y1 = -2. AM OR SI IS TOO LARGE. Y WILL OFTEN OVER- OR UNDERFLOW
C   Y1 = -3. SI IS TOO SMALL TO COMPUTE A USEFUL LOGNORMAL VARIABLE
C             IN SINGLE PRECISION
      INTEGER JX(120)
      IF (ABS(AM)+ABS(SI).LT. 88) GO TO 10
      Y1 = -2.
      RETURN
10    IF (ABS(SI).GT. 1E-6*(1.+ABS(AM))) GO TO 11
      Y1 = -3.
      RETURN
11    CALL RNORM (X1,X2,JX)
C RNORM IS A NORMAL GENERATOR
      X1 = AM+SI*X1
      IF (ABS(X1).GE. 88) GO TO 12
      Y1 = EXP(X1)
      GO TO 20
12    Y1 = 0.
      IF (X1 .GT. 0.) Y1 = -1.
      OV1 = X1
20    X2 = AM+SI*X2
      IF (ABS(X2).GE. 88.) GO TO 22
      Y2 = EXP(X2)
      RETURN
22    Y2 = 0.
      IF (X2 .GT. 0.) Y2 = -1.
      OV2 = X2
      RETURN
      END
```

```
      REAL FUNCTION RGAMMA(PA,JX)
C GAMMA GENERATOR
      REAL PA(7)
      INTEGER JX(120)
C PA(1), EXPONENT P IN DENSITY X**(P-1)*EXP(-X). PA(2), ECHO FOR P
C PA(3),PA(4),PA(5), FUNCTIONS OF P
C PA(6), STORAGE FOR A RANDOM NORMAL DEVIATE
C N.B. CALLING PROGRAM SHOULD CONTAIN THE STATEMENT
C     DATA PA(2),PA(6),PA(7)/1.,1E9,0./
C ERROR MESSAGES. PA(7) = 0., SUCCESSFUL COMPLETION
C                         -1., P .LE. 0.
C                         -2., P .LT. 8E-9
C                         -3., P .GT. 1E12
      IF (PA(1).NE. 5) GO TO 11
      IF (PA(6).LT. 1E7) GO TO 10
      CALL RNORM (AA,PA(6),JX)
C RNORM IS A NORMAL GENERATOR
      RGAMMA = .5*AA**2
      GO TO 12
10    RGAMMA = .5*PA(6)**2
      PA(6) = 1E8
      GO TO 12
11    IF (PA(1).NE. 1.) GO TO 13
      RGAMMA = REXP(JX)
C REXP IS AN EXPONENTIAL GENERATOR
12    IF (PA(7).LT. 0.) PA(2) = PA(1)
      PA(7) = 0.
      RETURN
13    IF (PA(1).NE. PA(2)) CALL PARAM (PA)
C PARAM IS AN INITIALIZATION ROUTINE
      IF (PA(7).EQ. 0.) GO TO 14
      RGAMMA = 0.
      RETURN
14    IF (PA(1).GT. 1.) GO TO 20
15    RGAMMA = (FLOAT(IRAN31(JX))/(32768.*65536.))**PA(3)
C IRAN31 IS A 31-BIT GENERATOR. 32768.*65536. = 2**31
      KA = IRAN31(JX)
      AB = REXP(JX)
      IF (RGAMMA .EQ. 0) RETURN
      AA = (FLOAT(KA)/(32768.*65536.))**PA(4)+RGAMMA
16    IF (AA .GT. 1.) GO TO 15
      RGAMMA = AB*RGAMMA/AA
      RETURN
20    IF (PA(6).LT. 1E7) GO TO 21
      CALL RNORM (AA,PA(6),JX)
      GO TO 22
21    AA = PA(6)
      PA(6) = 1E8
22    AB = REXP(JX)
      AC = 1.+PA(4)*(AA-PA(4))
      IF (AC .LE. 0.) GO TO 20
      AC = PA(1)*AC**3
24    IF (AC .GT. AB-PA(5)+PA(3)*ALOG(AC)+.5*AA**2) GO TO 20
      RGAMMA = AC
      RETURN
      END
```

```
      REAL FUNCTION RGUMB(JX)
C GUMBEL GENERATOR
      INTEGER JX(120)
      RE = REXP(JX)
C REXP IS AN EXPONENTIAL GENERATOR
      IF (RE .GE. 3E-4) GO TO 10
      KA = JX(1)
      IF (JX(KA).GE. 2146959360) GO TO 11
10    RGUMB = -ALOG(RE)
      RETURN
C THE BRANCH BELOW IS TAKEN WITH FREQUENCY 1/2**12
11    RE = REXP(JX)
      IF (RE .LT. 8.317766167)
     $  RE = RE-(1./8192.)/
     $       (((.0038278*RE+.0292732)*RE+.2502713)*RE+1.)**4
      RGUMB = RE+8.317766167
C 8.317766167 = LOGE(2**12)
      RETURN
      END

      SUBROUTINE RNORM (X1,X2,JX)
C NORMAL GENERATOR
      INTEGER JX(120)
      AA = REXP(JX)
C REXP IS AN EXPONENTIAL GENERATOR
      BA = KA
      BA = BA/(32768.*65536.)
C 32768.*65536. = 2**31
      BB = BA*BA
       BC = ((-2.073463969E-5*BB-1.345377537E-3)*BB-.1308997701)*BB
     $      +2.546479091
      BD = BC*BC
      BE = SQRT(AA+AA)/(BB+BD)
      BA = 2.*BA*BC*BE
      BB = (BB-BD)*BE
      KA = MOD(KA,8)
      IF (KA .LT. 4) GO TO 10
      BA = -BA
      KA = KA-4
10    IF (KA .LT. 2) GO TO 11
      BB = -BB
      KA = KA-2
11    IF (KA .EQ. 0) GO TO 12
      X1 = BB
      X2 = BA
      RETURN
12    X1 = BA
      X2 = BB
      RETURN
      END

      REAL FUNCTION RUNIF1(JX)
C UNIFORM GENERATOR
      INTEGER JX(120)
      RUNIF1 = FLOAT(IRAN31(JX))/(32768.*65536.)
C IRAN31 IS A 31-BIT GENERATOR. 32768.*65536. = 2**31
      RETURN
      END
```

```
      REAL FUNCTION RUNIF2(JX)
C DEAD FLAT UNIFORM GENERATOR, TAKING ON THE VALUES 1/2**24,
C   3/2**24, ..., (2**24-1)/2**24 EACH WITH FREQUENCY 1/2**23
      INTEGER JX(120)
      RUNIF2 = FLOAT((IRAN31(JX)/256)*2+1)/(4096.*4096.)
C IRAN31 IS A 31-BIT GENERATOR. 4096.*4096. = 2**24
      RETURN
      END

      SUBROUTINE RWEIB (B,P,X,OV,JX)
C WEIBULL GENERATOR
C IF P .GT. 0, X HAS C.D.F. 1 - EXP(-(X/B)**P)
C IF P .LT. 0, X HAS C.D.F.     EXP(-(X/B)**P)
C IF X OVERFLOWS, IT IS SET TO -1. IF X UNDERFLOWS, IT IS SET TO 0.
C IF X OVERFLOWS OR UNDERFLOWS, OV CONTAINS LOGE(X)
C ERROR MESSAGES
C   X = -2. B .LE. 0.
C   X = -3. P IS SO SMALL THAT X WILL OFTEN OVERFLOW OR UNDERFLOW
C   X = -4. P IS TOO LARGE TO COMPUTE A USEFUL WEIBULL VARIABLE IN
C       SINGLE PRECISION
      INTEGER JX(120)
      IF (B .GT. 0.) GO TO 10
      X = -2.
      RETURN
10    IF (ABS(P).GE. .02) GO TO 11
      X = -3.
      RETURN
11    IF (ABS(P).LE. 1E6) GO TO 12
      X = -4.
      RETURN
12    Y = -RGUMB(JX)/P+ALOG(B)
C RGUMB IS A GUMBEL GENERATOR
      IF (ABS(Y).GE. 88.) GO TO 13
      X = EXP(Y)
      RETURN
13    X = 0.
      IF (Y .GT. 0.) X = -1.
      OV = Y
      RETURN
      END
```

REFERENCES

Ahrens, J. H.; Dieter, U. (1972) Computer methods for sampling from the exponential and normal distributions. *Comm. ACM 15:* 873–882.

Box, G. E. P.; Muller, M. E. (1958) A note on the generation of random normal deviates. *Ann.Math.Statist. 29:* 610–611.

Bright, H. S.; Enison, Richard L. (1976) Cryptography using modular software elements. *AFIPS Natl.Comput.Conf.Expo.Conf.Proc. 45:* 113–123.

Greenwood, J. Arthur (1974) A fast generator for gamma-distributed random variables. *In* G. Bruckmann, F. Ferschl, L. Schmetterer, eds., *Compstat 1974: Proceedings in computational statistics.* Wien: Physica Verlag, pp. 19–27.

Greenwood, J. Arthur (1976) A fast machine-independent long-period generator for 31-bit pseudo-random numbers. *In* J. Gordesch, P. Naeve, eds., *Compstat 1976: Proceedings in computational statistics.* Wien: Physica Verlag, pp. 30–36.

Hastings, Cecil (1955) *Approximations for digital computers.* Princeton (New Jersey): Princeton Univ. Press.

Kogbetliantz, E. G. (1961) Generation of elementary functions. *In* A. Ralston, H. S. Wilf, eds., *Mathematical methods for digital computers.* New York: John Wiley & Sons, pp. pp. 7–35.

Marsaglia, George (1961) Generating exponential random variables. *Ann.Math.Statist. 32:* 899–900.

Miller, J. C. P.; Prentice, M. J. (1968) Additive congruential pseudo-random number generators. *Computer J. 11:* 341–346.

Sibuya, M. (1962) Exponential and other random number generators. *Ann.Inst.Statist.Math. 13:* 231–237.

Tocher, K. D. (1963) *The art of simulation.* London: English Universities Press.

Author's address: J. Arthur Greenwood
Box A 39 Grand Central
New York NY 10017 USA

Recent Developments in Statistics
J.R.Barra et al., editors

STATISTICAL PROPERTIES OF THE CANKERED LEAF SURFACE OF A PLANT POPULATION AFFECTED WITH A MYCOSE

E. Jolivet
Laboratoire de Biométrie
INRA - CNRZ
78350 Jouy-en-Josas (France)

INTRODUCTION

The problem we shall treat takes its origin in the studies of F. RAPILLY on the glume blotch, which is a corn disease due to rain dispersed mushrooms. We shall expose a part of the model constructed to describe the behaviour of an epidemy and to compare this behaviour on various varieties of corn. All the statistical and probabilistic results we present are well known, but the field of application seems to be rather new.

A MODEL OF EPIDEMY

The disease evolves through spores transported on the sound leaf surface by the splashing of the rain drops on the plants. The spores germinate on the leaf surface and the mushrooms of a new generation appear after a latent period and grow as rather circular blotches : the symptoms. These mushrooms in turn give spores which are rain dispersed and germinate. The cycle runs on until the crop, according to the process we can schematize as follows :

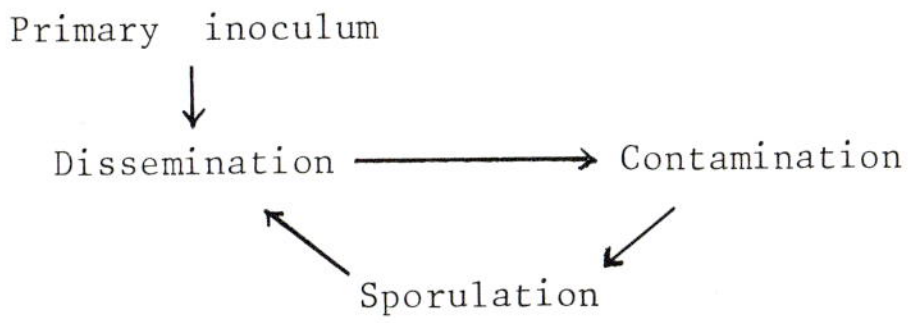

At instant t, it seems correct to consider the mushrooms of one generation as circles with the same radius and with centers distributed as an homogeneous Poisson process restricted to the measurable closed part of R^2 formed by the leaf surface of the plants. The union of these circles is called cankered surface. The area of this surface is the important "output" of the model, as it enables us to evaluate the phase of the disease.

To simplify the presentation, we shall consider the disease due to an unique generation of mushrooms, but the results are easily extended to the (real) case of superposition of many generations.

At a fixed time, the cankered surface is the part of the leaf surface $F \subset R^2$ formed by the union of circles of radius r whose centers are distributed as an homogeneous Poisson process M whose density is m. We are concerned by $\lambda(S)$, where λ is the Lebesgue measure in R^2. $\lambda(S)$ is a random variable and the Robbins formula enables us to calculate all its moments.

$$E\left[\lambda(S)^n\right] = \int_{F^n} P(x_1 \in S, x_2 \in S, \ldots, x_n \in S)\, d\lambda(x_1) \ldots d\lambda(x_n)$$

In our case, for example, the expectation of $\lambda(S)$ is :

$$E\left[\lambda(S)\right] = \lambda(F)(1 - \exp(-ms))$$

if edge effects are neglected.

But the second order moment is more complicated, depending particularly upon the distribution of distances of two points independently, identically and uniformly distributed on F.

Then, we are able to calculate the expectation of $\lambda(S)$, but we don't know its law of distribution and the second order moment is rather difficult to compute. Through an asymptotic study, we can nevertheless manage this information.

ASYMPTOTIC STUDY

Leaf surface model

To study the asymptotic properties of $\lambda(S)$, we shall briefly indicate a possible construction of a leaf surface model. The set of leaves is represented by a domain process in R^2 as defined by Miles (4), each domain being the convex polygonal hull of an n-configuration (n fixed). Miles (3) indicates a possible parametrization of such a configuration. We suppose that any two domains are formed from two distinct configurations with the same shape parameters, with scale parameters independently and identically distributed on the segment $[b, B]$ $(0 < b < B)$ and with orientation and position parameters chosen to ensure that the two domains are disjoint. From now on, we suppose we have a realization of this domain process with a countable infinity of domains. We number these domains F_i, $i = 1, 2, \ldots, n, \ldots$

Thus the leaf surface we examine from now on is

$$F = \bigcup_{i=1}^{\infty} F_i$$

We suppose that the probability space $(\Omega', \mathcal{B}, Q)$ on which the domain process is constructed is independent of $(\Omega, \mathcal{A}, P)$ on which M, the Poisson process of circle centers is constructed.

We note S_i the part of the cankered surface on F_i and we neglect edge effects on each F_i.

<u>Almost sure convergence and convergence in law of the cankered proportion of the leaf area.</u>

We define the random sequence

$$Y_n = \frac{\sum_{i=1}^{n} \lambda(S_i)}{\sum_{i=1}^{n} \lambda(F_i)}$$

We are able to calculate

$$E\left[\sum_{i=1}^{n} \lambda(S_i)\right] = \int_{\Omega} \sum_{i=1}^{n} \lambda(S_i)\ P(d\omega) = \sum_{i=1}^{n} \lambda(F_i)(1 - \exp(-ms))$$

The sequence Y_n depends upon $\omega \in \Omega$ and upon $\omega' \in \Omega'$. But we fixed ω' so Y_n is a random variable on Ω conditioned by ω'. The area of F_i is bounded, whatever i, $(a < \lambda(F_i) < A)$ because its scale parameter ℓ_i belongs to $[b, B]$.

We put $Z_i = \lambda(S_i) - (1 - \exp(-ms))\ \lambda(S_i)$.

Then we have the relation

$$U_n^{-1} \sum_{i=1}^{n} Z_i = Y_n - (1 - \exp(-ms))$$

where $U_n = \sum_{i=1}^{n} \lambda(F_i)$.

$\{Z_i(\omega)\}_{i \in N}$ is a sequence of P-independent random variables : $\lambda(S_i)$ and $\lambda(S_i)$ are constructed from points of a Poisson process distributed on disjoint parts of R^2.

On the other hand, we can prove ([1], [5]) that

$$\mathrm{var}\left[\lambda(S_i)\right] = \left[\lambda(F_i)\right]^2 \int_0^{\Delta_i} \exp(-2\ ms)\left[\exp(m\sigma(\delta)) - 1\right] \phi_i(\delta)\ d\delta$$

where Δ_i is the biggest distance between two points which are independently, identically and uniformly distributed on F_i.

$\sigma(\delta)$ is the common part of two circles of radius r whose centers are distant by δ. We have the relation $\sigma(\delta) \leq s$, where s is the area of a circle of radius r.

We can overestimate

$$\int_0^{\Delta_i} \exp(-2\ ms)\left[\exp(m\sigma(\delta)) - 1\right] \phi_i(\delta)\ d\delta$$

by $\exp(-ms)\left[1 - \exp(-ms)\right]$ which is lower than 1.

Then the positive series $\sum_{i=1}^{n} U_i^{-2} E(Z_i^2)$ is overestimated by the series $A^2 a^{-2} \sum_{i=1} i^{-2}$ and thus converges. This property and the independence property of the Z_i involve the convergence of the random sequence $U_n^{-1} \sum_{i=1}^{n} Z_i$ to 0, P-almost surely and for the quadratic norm when n increases to infinity.

This result shows that if we consider a large and homogeneous enough cultivated area, the cankered blotched leaf area proportion is near equal to its mathematical expectation. Under these requirements the result of the model is a good evaluation of the cankered surface area at time t.

Studying the convergence in law of the variable

$$T_n \left[\text{var } T_n\right]^{-1/2}$$

where $T_n = \sum_{i=1}^{n} Z_i$

we can improve these results.

We observe that $Z_i(\omega)$ fulfils the Lindeberg condition (2). Indeed

i) Z_i is a centered random variable

ii) $\sum_{i=1}^{\infty} E(Z_i^2)$ is infinite.

If we consider the realization $(F_i)_{i \in N}$ of the domain process, the scale parameters ℓ_i of each domain are independently identically distributed according to the probability law $\mathcal{L}$.

It follows that $\{\text{var } \lambda(S_i)\}_{i \in N}$ is a random sequence of variables distributed according to the law image of $\mathcal{L}$ by the application $\ell_i \to \text{var}[\lambda(S_i)]$ and then is a sequence of independently identically distributed random variables. If we note $E_{\mathcal{L}}$ the expectation with respect to $\mathcal{L}$, $E_{\mathcal{L}}[\text{var } (\lambda(S_i))]$ is finite, $\text{var}[\lambda(S_i)]$ being bounded as it was proved.

Let $T_n = \sum_{i=1}^{n} Z_i$, then we have $\text{var}[T_n] = \sum_{i=1}^{n} \text{var}[\lambda(S_i)]$ and $\text{var}(T_n)$ is $\mathcal{L}$-almost surely equivalent to $n\, E\, [\text{var}[\lambda(S_i)]]$. Except when M is the null process on R^2, $\text{var}[\lambda(S_i)]$ is a strictly positive real number as the variance of a positive, on Ω constructed random variable whose law range is not reduced to a point. Then the expectation with respect to $\mathcal{L}$ of $\text{var}[\lambda(S_i)]$ is strictly positive and $\sum_{i=1}^{\infty} E(Z_i^2)$ is infinite $\mathcal{L}$-almost surely.

iii) The ratio

$$\frac{\sum_{i=1}^{n} \int_{\{|Z_i| > \varepsilon \sqrt{var(T_n)}\}} Z_i(\omega)^2 \, P(d\omega)}{\sum_{i=1}^{n} \int_{\Omega} Z_i(\omega)^2 \, P(d\omega)}$$

tends to 0 when n increases indefinitely whatever ε, $\mathcal{L}$-almost surely.

Indeed, $Z_i = \lambda(S_i) - (1 - \exp(-ms))\,\lambda(F_i)$ is bounded :

$$- (1 - \exp(-ms))\,A \leq Z_i \leq A - (1 - \exp(-ms))\,a$$
$$|Z_i| < A$$

And then, whatever ε, there is an integer $n_0(\varepsilon)$ such that for all n bigger than $n_0(\varepsilon)$, $[var(T_n)]^{1/2}$ is bigger than $A\,\varepsilon^{-1}$ $\mathcal{L}$-almost surely, and it is clear that

$$\frac{\sum_{i=1}^{n} \int_{\{|Z_i| > A\}} Z_i(\omega)^2 \, P(d\omega)}{\sum_{i=1}^{n} \int_{\Omega} Z_i(\omega)^2 \, P(d\omega)} \text{ is null}$$

These three conditions imply the low convergence of the law of the random variable

$$x_n(\omega) = T_n(\omega) \left[var(T_n(\omega))\right]^{-1/2}$$

to the reduced normal law $\mathcal{L}$-almost surely.

Let

$$f_n = \sum_{i=1}^{n} \lambda(F_i)$$

$$v_n = \sum_{i=1}^{n} var\left[\lambda(S_i)\right]$$

$$x_n = \frac{Y_n - f_n(1 - \exp(-ms))}{v_n}$$

then

$$P\{|x_n| \leq \alpha\} = P\left\{\left|\frac{Y_n}{f_n} - (1 - \exp(-ms))\right| \leq \frac{\alpha v_n}{f_n}\right\}$$

tend to

$$\Phi(\alpha) = \frac{2}{\sqrt{2\pi}} \int_0^{\alpha} E^{-t2/2}\, dt$$

Now, if we put $v^2 = E\left[\mathrm{var}(\lambda(S_1))\right]$

$$f = E\left[\lambda(F_1)\right]$$

when n tends to infinity, $\frac{v_n}{f_n}$ is $\mathcal{L}$-almost surely equivalent to $v\, f^{-1}\, n^{-1/2}$ and the event

$$\{|x_n| \leq \alpha\}$$

has the same measure as the event

$$\{\sqrt{n}\ \left|\frac{Y_n}{f_n} - (1 - \exp(-ms))\right| \leq \alpha\ v\ f^{-1}\}$$

To obtain approximate confidence intervals of the ratio $\frac{Y_n}{f_n}$, it is sufficient to solve the equation in ε :

$$\Phi(\varepsilon\ f\ v^{-1}) = P$$

with probability P, when n tends to infinity, the ratio

$$\frac{Y_n}{f_n} = \frac{\sum_{i=1}^{n} \lambda(S_i)}{\sum_{i=1}^{n} \lambda(F_i)}$$

belongs to an interval we can approximate by

$$\left[1 - \exp(-ms) - \frac{\varepsilon}{\sqrt{n}},\ 1 - \exp(-ms) + \frac{\varepsilon}{\sqrt{n}}\right]$$

This last result gives us a better knowledge of the behaviour of $\lambda(S)$. It is easy to compute $E(Y_n)$; furthermore we know the convergence speed of $\frac{Y_n}{f_n}$ to its expectation and we are able to give approximate confidence intervals of this quantity. We note that, in the corn example, it is meaningful to use asymptotic results in practice as soon as we consider a cultivated surface of about ten square meters.

REFERENCES

[1] Kendall, M. G. and Moran, P. A. P. (1963). Geometrical probability (Griffin, London)

[2] Krickeberg, L. (1963).Wahrscheinlichkeitstheorie(Teubner, Stuttgart)

[3] Miles, R. E. (1970). On the homogeneous Poisson point process Math. Biosc. 6, 85 - 127

[4] Miles, R. E. (1974). On the elimination of edge effects in planar sampling in stochastic geometry, Harding and Kendall ed. (J. Wiley and Sons, London)

[5] Rapilly, F. and Jolivet, E. (1976). Construction d'un modèle (Episept) permettant la simulation d'une épidémie de Septoria Nodordum Berk sur blé ; Rev. Stat. Appl. 24, n° 3, 31 - 60

[6] Rapilly F. (1976). Essai de modélisation d'une épidémie de Septoriose due à Septoria Nodorum Berk sur blé : recherche de critères de résistance horizontale ; thèse d'état, Université Paris XI

Recent Developments in Statistics
J.R.Barra et al., editors

CHOOSING PARAMETERS FOR

CONGRUENTIAL RANDOM NUMBERS GENERATORS

Yoshiharu KURITA

Laboratoire National de Recherche de Métrologie de Tokyo

et

IRIA - LABORIA

1 - INTRODUCTION

An obvious characteristic of today's digital computers is their deterministic behaviour. For this reason, when a computer is used to generate a sequence of numbers to represent the realization of a random variable, one speaks of pseudo-random numbers in order to point out the determinism of the algorithms which have been used to generate those numbers. Moreover, the definition of the randomness of a sequence exists only for infinite sequences, but no proper definition is available for finite sequences, as is always the case for sequences generated by a computer. We can only say that a sequence of numbers is pseudo-random if it is a finite sequence in which obvious regularities are difficult to find.

Several techniques have been proposed to generate pseudo-random numbers, and the most widely used are based on congruential methods. Another approach, based on shift registers which generate Tausworthe sequences [6] has been recently developed . The major drawback of this technique is that it has not yet been thoroughly tested. This is not the case for congruential generators and an important literature is available on this subject.

We present in this paper a methodology to select the parameters of a multiplicative congruential random generator and we illustrate our approach with numerical examples.

2 - MULTIPLICATIVE CONGRUENTIAL RANDOM GENERATORS

We shall restrict our study to the multiplicative random generator presented by Lehmer in 1951 . The nth random number generated is given by the recursion

(1) $$x_n = ax_{n-1} \underline{\text{mod}}\ m$$

It can be noted that three parameters are required to use equation (1): x_0, m and a. The choice of x_0 is not critical (but x_0 must be odd) and in practice x_0 is selected from the results of statistical tests performed by the user in the conditions of the experiments. On the other hand, one expects the period of the sequence of random numbers to be as large as possible. This constraint restricts the choice of m among integers having particular properties.

2.1- Determination of a maximum period

From (1), it is obvious that the period of the sequence cannot exceed M.

The parameter m will thus be chosen as large as possible, and will usually be the largest integer which the registers of the computer can represent. However, it does not imply that the period T of the generator is large.

From (1), we have

$$x_n = a^n x_0 \ \underline{\text{mod}}\ m,$$

and T is given by

$$T = \min\ (\lambda \mid a^\lambda = 1 \ \underline{\text{mod}}\ m).$$

By definition, we say that a is a primitive root of m if and only if

$$T = m-1 .$$

If m is selected so that it has primitive roots and a is one of them, a maximum period T = m-1 is guaranteed. In order to satisfy the conditions of existence of primitive roots m must have the following form

(2) $$m = p^s \text{ or } m = 2p^s$$

where p is a prime number greater than 2 and s is a positive integer In order to compute rapidly the congruence modulo m on a digital computer [4] one selects

(3) $$m = 2^r - 1 .$$

Given (2) and (3), m is a Mersenne or Fermat number [5]. If the computer has a 32 bits register we shall only consider :

$$m = 2^{31} - 1$$

Otherwise prime numbers of the following form may be considered

$$m = 2^\alpha \pm 2^\beta \pm 1 , \quad (\alpha, \beta \in N^+)$$

In this case the congruence modulo m can also be simply computed. Among the acceptable values of m, we have for example

$$m = 2^{31} - 1,\ 2^{35} - 31,\ 2^{47} - 127, \text{ etc...}$$

Given m it remains to select a among the $\varphi(m-1)$ primitive roots of m where φ is the Euler function*. For instance, if $m = 2^{31} - 1$ then $\varphi(2^{31} - 2) = 534\ 600\ 000$. However, all primitive roots of m do not ensure good statistical properties to the generated sequence. We present in the next section the statistical criteria which can be used to select a satisfaction multiplier a.

---

* $\varphi(n)$ is the number of integers less than n prime relatively to n.

2.2 - The first-order autocorrelation

It is desirable for the first-order correlation on a whole period to be as small as possible. If so, there will be a weak relationship between two consecutive numbers in the generated sequence. The avantage of this criterion is that it can be easily computed when the period of the generator is maximum.

The first-order correlation is defined as :

$$C = \frac{m \sum_{n=1}^{m-1} x_n (ax_n \underline{\text{mod}}\ m) - \left(\sum_{n=1}^{m-1} x_n\right)^2}{m \sum_{n=1}^{m-1} x_n^2 - \left(\sum_{n=1}^{m-1} x_n\right)^2}$$

and if the sequence $\{x_n\}$ has the maximum period $(m-1)$, C can be computed by a generalized Dedekind summation [2]. We present in figure 1 the values of C which have been obtained for $m = 2^{31} - 1$ and $a = q, q+1, \dots, q+50$ with $q = 2^9, 2^{10}, \sqrt{m}, \dots$ $\dots, m-\sqrt{m}$

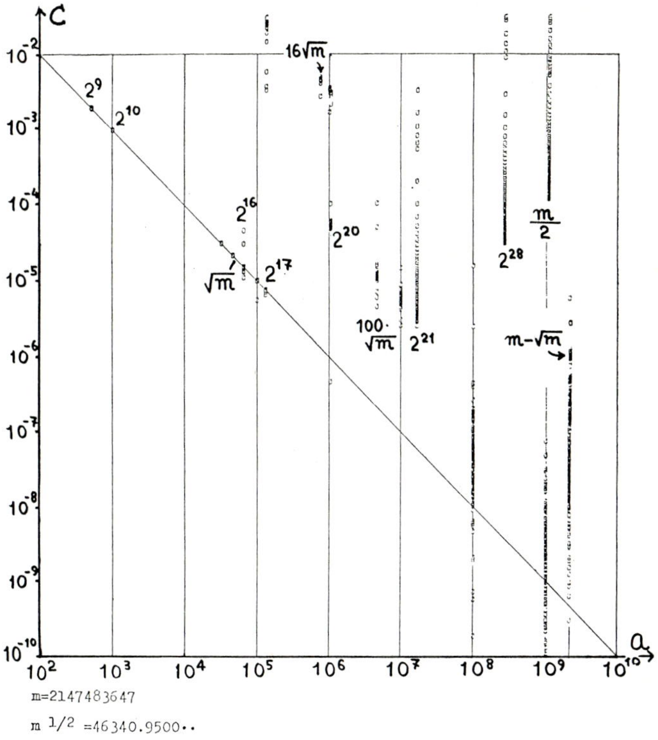

Figure 1: Autocorrelation on a whole period

It can be observed on figure 1 that if $a/m \ll 1$, then $C \simeq 1/a$. Otherwise, when a is close to m, C can take large values, but even in this case it is clear that a small C can also be found.

It should be noted that C is computed on the complete period of the generator, and that its value does not guarantee the statistical properties of a sub-sequence which is extracted from it.

2.3 - Spectral analysis of the generated numbers

We now propose a stronger test known as "hyperplane structure". The purpose of this test is to study the repetitions of sub-sequences of length r, $r = 1, 2, \ldots, m-1$ in the sequence of the generated numbers. Two approaches can be taken, a geometrical one [3] and an analytical one using a spectral analysis method [1].

2.3.1 - The geometrical representation

Let $\{x_n\}$ be the sequence defined by

$$x_n = a\, x_{n-1} \ \underline{\mathrm{mod}}\ m,$$

where a and m are chosen so that all the points of $\{x_n\}$ belong to one cycle, and π_i, $i = 1, 2, \ldots$ the sub-sequences of $\{x_n\}$ of length r defined by

$$\pi_i = \{x_i, x_{i+1}, \ldots, x_{i+r-1}\}.$$

The geometrical approach consists in considering the π_i as points with coordinates $(x_i, x_{i+1}, \ldots, x_{i+r-1})$ in an r-dimension space. These points form a lattice of parallel and equidistant hyperplanes as shown in figure 2 for r=2. The generator will be the more regular the greater the minimal number ν_r of hyperplanes. It can be shown that ν_r is given by

$$\nu_r = \min(s_1^2 + s_2^2 + \ldots + s_r^2)^{1/2}$$

where s_i, $i = 1, 2, \ldots, r$ satisfies

$$s_1 + a s_2 + \ldots + a^{r-1} s_r = 0 \ \underline{\mathrm{mod}}\ m$$

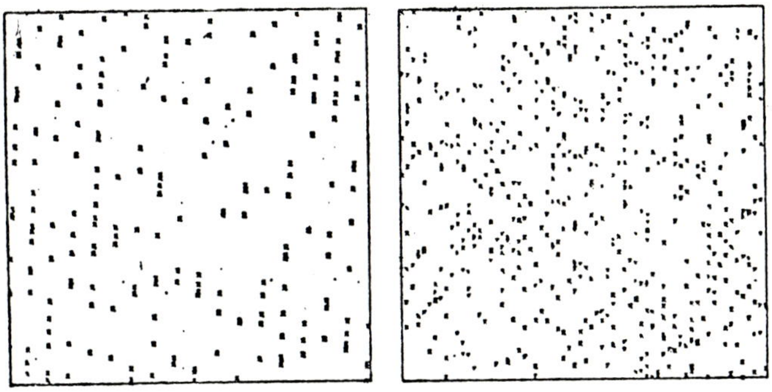

$x_n = (2^4+1)\, x_{n-1} \ \underline{\text{mod}}\ 2^{30}$ (220 points) $x_n = (2^6+1)\, x_{n-1} \ \text{mod}\ 2^{30}$ (565 points)

Figure 2 : Hyperplane structures for r=2

2.3.2 - The spectral analysis representation

This approach has been developed in [1]. Let us consider a sequence $\{x_n\}$ such that all its points belong to one cycle, and let $F(t_1, t_2,\dots,t_r)$ be the limiting density of the number of occurrences of the sub-sequences $(t_1, t_2,\dots,t_r)$ among all the sub-sequences of length t. We have

$$F(t_1, t_2,\dots,t_r) = \lim_{N\to\infty} \frac{1}{N} \sum_{k=0}^{N-1} \delta_{x_k,t_1}\, \delta_{x_{k+1},t_2} \cdots \delta_{x_{k+t-1},t_r}$$

where $\delta_{i,j} = 1$ if i=j, 0 otherwise.

If the sequence $\{x_n\}$ is uniform, we must have

$$F(t_1, t_2,\dots,t_r) = 1/m^r,$$

and in this case, the sequence $\{x_n\}$ is said to be uniform of order r. In order to analyse this property we can compute the Fourier transform of $F(t_1, t_2,\dots,t_r)$. We have

$$(4) \qquad f(s_1, s_2,\dots,s_r) = \delta[\Phi(a)/m],$$

where

$$\delta(z) = 1 \text{ if } z \in N, \text{ 0 otherwise,}$$

and

$$\Phi(a) = s_1 + as_2 + \dots + a^{r-1} s_r .$$

If $\{x_n\}$ is uniform of order r, then :

$$f(s_1, s_2, \ldots, s_r) = \begin{cases} 1 \text{ if } s_1 = s_2 \equiv s_r \ \underline{\text{mod}}\ m, \\ 0 \text{ otherwise} \end{cases}$$

An indication of the irregularity of $\{x_n\}$ will be obtained by looking at the vectors $(s_1, s_2, \ldots, s_r) \neq 0$ for which $f(s_1, s_2, \ldots, s_r) \neq 0$.

Each number $f(x_1, x_2, \ldots, x_2)/m^r$ is the amplitude of the wave the length of which is $L_r = 1/(s_1^2 + s_2^2 + \ldots + s_r^2)^{1/2}$ in the Fourier transform of $F(t_1, t_2, \ldots, t_r)$.

This wave will arise more perturbations if its wave length is long and so, we shall try to minimize its "frequency" $\nu_r = 1/L_r$. In the multiplicative sequence $x_n = ax_{n-1}$ mod m, the hypothesis "$f \neq 0$"and the relation (4) imply : $\phi(a) = 0$ mod m, and so :

$$\nu_r = \min(s_1^2 + s_2^2 + \ldots + s_r^2)^{1/2}$$

with :

$$s_1 + as_2 + \ldots + a^{n-1} s_r = 0 \ \underline{\text{mod}}\ m$$

Using those two remarks, this test will be made by computing the ν_r's with an algorithm given in [2] : a will then be chosen as giving the largest $\nu \rightarrow r$ However, ν_r cannot be arbitrarily large because of Hermite's theorem concerning the minimum of a positive definite quadratic form which implies :

$$\nu_r < \gamma_r \ m^{1/r} \quad \text{where } \gamma_r \text{ is a constant [7].}$$

An important consequence of this inequality is that a congruential generator cannot be made arbitrarily good by choosing adequate parameters.

In the next paragraph, we shall present examples of computations of a maximizing ν_r.

3 - NUMERICAL RESULTS

According to the results on autocorrelation presented in § 2.2, the parameter a can be chosen in the following interval, when $2^{31} - 1$:

$$2\,100\,000\,000 \leq a \leq 2\,100\,000\,000 + 30\,206$$

There are 7 440 primitive roots in this interval, and each of them has been examined by "screening" them in the following way :

1) Take the 1st primitive root ;

2) for r=2 to 7

```
begin  if ν_r < ν̄_r  then go to 4,
       else
end ;
```

3) Compute ν_8 and adopt this primitive root ;

4) Take the following primitive root and go to 2).

The $\bar{\nu}_r$'s$(r = 1,\dots,7)$ represent the fineness of the screen. They have been chosen between 60 °/o and 75 °/o of the upper bound $\gamma_2\, m^{1/r}$. By taking $\bar{\nu}_2 = 30\ 000$, $\bar{\nu}_3 = 1\ 086$, $\bar{\nu}_4 = 192$, $\bar{\nu}_5 = 60$, $\bar{\nu}_6 = 30$ and $\bar{\nu} = 19$, 11 primitive roots have been kept by screening the 7 440 possible values. They are shown in Table 1 with the values of the ν_r's and the corresponding autocorrelation coefficient.

	ν_2	ν_3	ν_4	ν_5	ν_6	ν_7	C
							$x10^{-8}$
2100004546	33492.3	1313.9	212.3	64.0	30.9	19.5	-2.8
2100005341	43487.0	1201.5	205.7	65.2	31.9	19.3	1.1
2100008090	41396.6	1223.1	197.8	60.8	31.7	19.6	-1.3
2100009038	39752.6	1100.5	194.1	68.9	34.3	19.5	-5.1
2100009088	30328.3	1155.3	206.9	65.0	31.0	19.5	-0.5
2100009592	34471.2	1334.2	195.3	69.0	33.1	19.9	-2.2
2100014651	31168.0	1279.3	209.2	64.0	35.2	20.3	-4.7
2100016018	35193.2	1154.9	223.4	70.3	34.0	19.3	8.8
2100017008	44643.0	1196.5	194.2	68.5	31.8	21.2	-1.4
2100019167	41742.2	1166.4	203.6	79.2	33.2	19.2	-0.8
2100023033	31118.3	1164.5	209.7	65.2	37.3	20.5	-3.8
UPPER LIM.	(49796.6)	(1448.2)	(256.0)	(90.5)	(46.4)	(29.0)	

Table 1 : Possible values of the parameters a for the sequence $x_n = a.\,x_{n-1}$ mod $2^{31}-1$, which maximize ν_r, $r = 1,\dots,7$.

The Table 2 presents similar results under the supplementary constraint on the form of a : $a = 2^\alpha \pm 2^\beta \pm 1$

This form allows fast computations for the product $a.\ x_{n-1}$.

m	a	ν_2	ν_3	ν_4	ν_5	C
	$2^{36}-2^{12}$	8.4×10^6	4.7×10^4	2.3×10^3	5.6×10^2	-4.4×10^{-11}
$m=2^{47}-2^7+1$	$2^{44}-2^{28}+1$	8.9×10^6	4.9×10^4	2.1×10^3	5.5×10^2	-5.8×10^{-11}
	borne sup	1.2×10^7	5.8×10^4	4.1×10^3	8.3×10^2	
	$2^{27}-2^{18}-1$	1.7×10^5	2.1×10^3	3.1×10^2	1.1×10^2	2.6×10^{-10}
$m=2^{35}-2^5+1$	$2^{30}-2^{18}-1$	1.3×10^5	2.9×10^3	3.4×10^2	1.1×10^2	-5.4×10^{-9}
	borne sup	1.9×10^9	3.6×10^3	5.1×10^2	1.5×10^2	
	$2^{22}-2^{14}+4$	3.1×10^4	1.2×10^3	1.8×10^2	6.6×10^2	1.3×10^{-7}
$m=2^{31}-1$	$2^{26}-2^{16}+2$	3.4×10^4	1.2×10^3	1.8×10^2	5.1×10^2	-5.4×10^{-9}
	borne sup	4.9×10^4	1.4×10^3	2.6×10^2	9.0×10^2	

Table 2 : Possible values of the parameter $a = 2^\alpha \pm 2^\beta \pm 1$, for the sequences $x_n = a.x_{n-1}$ mod m, for three values of m.

5 - CONCLUSION

In this paper a method for choosing the multiplicative parameter of a congruential random numbers generator has been presented. The method is based on two tests : the 1st one allows to determine an interval of the parameter insuring a low first-order autocorrelation : the values in this interval are then screened and those that are kept are such that the sub-sequences of sizes 2, 3,... they generate are distributed as uniformly as possible. These tests are illustrated by numerical results for different values of the parameters.

It remains to verify if the quality of such generators improves the results of experiments in which they are used, in simulation experiments for instance. A simulated M/M/1 queue is being analysed at LABORIA, and the first results show an actual improvement. Other experiments would be necessary and this a wide and yet unexplored field of research.

REFERENCES

[1] COVEYOU, R.R., Mac PHERSON, R.D. - "Fourier analysis of uniform random numbers" - J.A.C.M. - Vol. 14 - 1967.

[2] KNUTH, D.E. - "The art of computer programming" - Vol. 2 - Seminumerical Algorithms - Adison Wesley - 1969.

[3] MARSAGLIA, G. - "Random number fall mainly on the planes" - Proc. N.A.S - Vol. 61 - 1968.

[4] PAYNE, W.H. - "Coding the Lehmer pseudo-random number generator" - C.A.C.M. - Vol. 12 - 1969.

[5] SIERPINSKI, W. "Elementary theory of numbers" - Porska Academia Nauk - 1964.

[6] TAUSWORTHE, R.C. - "Random number generated by linear recurrence modulo two" - Math. Comp. - Vol. 19 - 1965.

[7] CASSLES, J.W.S. - "An introduction to the geometry of numbers" - Springer - Berlin - 1959.

Recent Developments in Statistics
J.R.Barra et al., editors

A DATA BASE SYSTEM FOR SCIENTIFIC PURPOSES.

Leif S. Mortensen and Mogens E. Larsen.

Department for data processing
Herlev Hospital
Copenhagen
Denmark

A data base system (SCIBAS) created in a hospital environment for scientific purposes, especially for statistical analysis, is presented.

OBJECT

The aim has been to make a special purpose system capable of storing and using data structures, which can be utilized in statistical analysis. The system should furthermore be provided with instructions being powerful tools for controlling analysis. As far as possible, SCIBAS should control the analysis automatically by utilizing parameters stored in the system and describing the characteristics of the data. Finally, it was wished to avoid producing a lot of nearly identical programs by making analysis modules and other data processing modules oriented towards their specific functions and fully data independent.

DESCRIPTION

Data is always stored in two dimensional matrices, variables in one dimension and cases in the other. The matrices are transposed which means they are stored physically variable by variable in the data base disk area. Each variable is packed with optimal density. The single data element is identified in the data matrix by its sequence number in the variable.

To every data matrix there belongs a list, also stored in the data base, containing characteristics for each of its variables (variable definitions). They include among other things a type indication for controlling the different functions. They comprise the most essential step for totally data independent modules.

Furthermore each data matrix has four auxiliary matrices, and some superior information including project name, sizes of matrices, physical adresses etc.

The components belonging to different data matrices are stored in arbitrary order in the data base disk area.

Apart from the index that appears indirectly from the physical order of the cases in a matrix, it is possible to generate and store secondary indexes each based on one or several variables.

Pointers from cases in one data matrix to cases in another data matrix can be generated and utilized to gain access to logically connected data.

Apart from this, attempts have not yet been made to extend SCIBAS further with characteristics of a relational data base even if this seems possible.

CONTROL

One task for SCIBAS is generally defined by specification of the following four pieces of information:

1. The identification of the data matrix that has to be processed.
2. A filter that defines which cases are to be used in the actual task (Example: Sex = female and Age > 40).
3. One or several specifications of actual variables.
4. A specification of the function to be carried out.

Each of the four pieces may degenerate. In the simplest instance, a task specification degenerates into specification of one parameter (a function code).

Today, SCIBAS is controlled by (virtual) punched cards with control parameters and without mnemonic aids. It is up to the user to keep track of the meaning of the parameters in different invocations of the functions. This is the source of many errors. The reason that SCIBAS, in spite of this, can be used is due to the fact that most of the information necessary for the solution of each task is stored in the system in advance.

Programs for interactive specification of tasks for the system are under preparation.

Functions in SCIBAS that can be controlled by the user are, of course, also accessible from programs. This makes it possible to let the analysis programs store the results from a series of analysis as data in the data base. Consequently, it is feasible to organize a half or fully automated multistep analysis in which one step operates on results from the previous steps.

EXPERIENCE

SCIBAS has been in daily use for one year even though all of the planned facilities are not yet implemented in the system.

We have been able to satisfy a number of needs that we otherwise would have been forced to give up. Data matrices from about 30 projects have been processed until now.

Each new project is different from all the preceding projects, nevertheless, three main types can be distinguished:

a. Medical research with fairly small data matrices, typically 100 variables and 50-500 cases. The most common need for statistical analysis is non parametric hypothesis testing and hypothesis generation. Less frequently, there is a need for multivariate analysis and model building.

b. Administrative production sum up, typically with 25 variables and 100,000 cases. Primitive statistical methods are usually sufficient.

c. Clinical production control. Here is a need for sequential statistical analysis on sets of small data matrices, describing the trend of disease for a small patient population.

SCIBAS has not yet the facilities necessary for the latter type, but they will be achieved later. SCIBAS is planned for this purpose on the basis of experience with this kind of projects using a less powerful program system on a small machine. The intention is to let the single steps of the sequential analysis start automatically, either periodically or when supplementary data arrive.

SCIBAS is programmed for IBM system/370, mainly in PL/I. The programming costs comprising about two man years are adequately paid back because of its data independency.

There is an initial cost starting every new project caused by definition of variables and specification of other characteristics. With a data matrix in the system the single task is cheaper, both in preparation and in use of hardware resources compared with cases where traditional programming and storing principles are applied.

Use of SCIBAS may be quite expensive when only limited analysis is to be performed, but if there is an extensive need for analysis, SCIBAS is a useful tool for reducing costs and for making complicated data processing possible.

Statistical functions and data management modules in SCIBAS are developed with support from the Danish Medical Research Council, from the IBM Research Foundation, the Technical University of Denmark and from the Danish Hospital Foundation for Medical Research. Region of Copenhagen, the Faroe Islands and Greenland.

Recent Developments in Statistics
J.R.Barra et al., editors

STRONG DRAININGS AND SAMPLING PROBLEMS

by H. RAYNAUD
UNIVERSITE SCIENTIFIQUE ET MEDICALE DE GRENOBLE - IRMA
BP . 53 - 38041 GRENOBLE CEDEX

INTRODUCTION

A. We can sketch a model for how a chemist works out linear macro-molecules in this way : he puts in a bath tub small balls of different colours that he calls monomers. He fixes to the draining apparatus an "exciting" device and opens the draining. The first monomer presenting itself at the hole gets "excited". Among the other monomers, the "excited" one looks for a "fitting" second monomer which, when found, gets "excited" in its turn, takes the place of the first one at the draining hole, while the first monomer is drained out of the bath tub.

The way for a monomer to know if another one can be fitting consists in looking at a certain given compatibility graph indicating which successions through the draining are permitted and which are not.

Different problems arise naturally from this model ; for example :

(1) Under which conditions the polymer chain that goes out of the bath tub will have a similar constitution all along the "draining process" ?

(2) Under which conditions the polymer chain will use "almost all" the monomers present in the bath tub at the beginning of the process ?

B. Suppose you observe a large (but finite) population of bacterias of type A_i, $i \in I$ in a microscope preparation including a proportion p_i of bacterias of type A_i, $\forall i \in I$. Suppose you follow a rule of the type : after the observation of a bacteria of type A_i, observe the next bacteria under the microscope among those of type A_j, $j \in \Gamma(i)$, where Γ is an application of I into $P(I)$. Under which conditions this procedure will give you, after a certain number of observations, proportion of bacterias observed comparable to the p_i's - or to any other wanted quota ?

C. Suppose a multi-processor computer, in which each processor works independently from the others, but calls for a "bus" for all his finished jobs, waiting in a line to be taken by the bus to another processor. The bus can take only one job at a time.

Let us suppose that the bus can "decide" which processor it will satisfy ; suppose it will "decide" it according to a rule indicating which processors are allowed to be served next, depending only on the last one served ; the choice between the processors at one instant will be made at random with equal probability between the waiting jobs in the permitted processors. A certain frequency of jobs issued from processors being known, which rule of succession will allow the processors to be reasonably satisfied - i.e. no queue in the processors being divergent but all almost surely finite ?

D. All those problems can be enounced as draining problems, the solution of which being very often a little paradoxical. The reason of the paradoxes can be found in this very simple example :

Consider an infinite population of bacterias of types A and B, in equal proportions, and suppose that we follow the very simple succession rule shown on the graph of Fig. 1.

Fig. 1

Let us draw at random a sample of the infinite population according to this rule ; let n be the number of items in the sample, n_A (resp. n_B) the number of bacterias A (resp. B) in the sample. Everything happens like in a Markov chain, the transition matrix of which being

$$P = \begin{bmatrix} 1/2 & 0 \\ 1/2 & 1 \end{bmatrix}$$

It is easy to see that, surprisingly, $P^n \rightarrow \begin{bmatrix} 0 & 0 \\ 1 & 1 \end{bmatrix}$ i.e. the system takes state A infinitely more often than state B - which will be neglected !

In order to study the question, we have introduced the concepts of strong drainings and stable drainings.

Some theorems on drainings possibly useful for statistical questions are going to be explained in the following. Their proofs being generally very technical, only one simple example has been completely treated in order to allow the reader to imagine the types of proofs of the other theorems (which are all published in papers such as theses, etc...).

1 - DEFINITIONS

For the sake of simplicity, we borrow from J.G. Dion [1976] a definition for strong drainings which, if it is not as general as possible, offers however a very good efficiency for proofs. A slightly more general one can be found in H. Raynaud [1968].

Let G be a graph with p vertices, oriented or not, with or without loops ; let $v_1,\ldots,v_p$ be its vertices, A(G) the set of its arcs (or edges). Let us associate with each vertex v_i of G a positive integer N_i $i = 1,2,\ldots,p$. Let us denote

$$\vec{N} = (N_1,N_2,\ldots,N_p) \text{ and } N = N_1+N_2+\ldots+N_p$$

The couple $(G,\vec{N})$ will be called weighted graph.

With any weighted graph $(G,\vec{N})$ we associate an urn containing N balls among which N_1 balls of colour $1,\ldots,N_i$ balls of colour $i,\ldots,N_p$ balls of colour p, urn from which we draw the balls without replacement according to the following algorithm :

- The first ball to be drawn is chosen at random with equal probability among the balls.
- If in urn U there remain n_k balls of colour k, $k = 1,2,\ldots,p$, and if a ball of colour i has just been drawn, the next ball to be drawn (without replacement) will be of colour j with probability zero if $(v_i,v_j) \notin A(G)$ and with probability

$$\frac{n_j}{n_{j_1}+\ldots+n_{j_1}}$$

if $(v_i, v_j) \in A(G)$ and if $\{(v_i, v_{j_t}) : t = 1,2,\ldots,l\}$ runs over the set of all the arcs in G issuing from v_i. This is repeated until the probability to draw a ball of colour j is zero for all $j = 1,2,\ldots,p$.

Let ν be the number of balls drawn when the drawing stops. ν is a random variable, the distribution of which being well defined as soon as $(G,\vec{N})$ is given.

Now let N increase infinitely while $\frac{N_i}{N}$ remains equal to a strictly positive limit λ_i, $i = 1,2,\ldots,p$ with $\lambda_1+\lambda_2+\ldots+\lambda_p = 1$. We shall say that $(G,\vec{N})$ is a draining if $\nu \xrightarrow{ps} \infty$ when $N \to \infty$, and that $(G,\vec{N})$ is a strong draining if $\frac{\nu}{N} \xrightarrow{P} 1$ when $N \to \infty$ in other words :

$$\forall\ \alpha,\beta > 0,\ \exists\ N_o \text{ such that } N \geq N_o \Rightarrow P\{\frac{\nu}{N} > 1-\alpha\} > 1-\beta$$

$(G,\vec{N})$ is a stable draining if whatever be the place of a sample of successive balls taken in the drawing, the probability of the occurrences of the different sample sequences depends only on their length and nature.

2 - STRONG DRAININGS WITH TWO KINDS OF BALLS

We shall repeat here only the results - proofs can be found in J.P. Diet and H. Raynaud [1977]

 for any N ;

 for any N_1, N_2 ;

 for $\frac{N_1}{N} = \frac{1}{2}$

are strong drainings. They are, of course stable drainings.

 is never a strong draining. Balls of type one are drawn out of the urn always quicker than balls N_2 when N_1 and N_2 are of the same order of magnitude. G. Kuntz has programmed for us the following situation : one thousand balls of types 1 and 2 are drawn. Consider Fig. 2: along the Y-axis is the number of balls of type 2 remaining in the urn when the drawing stops. Along the X-axis we have the proportion of balls of type N_1 in the urn at the beginning of the drawing.

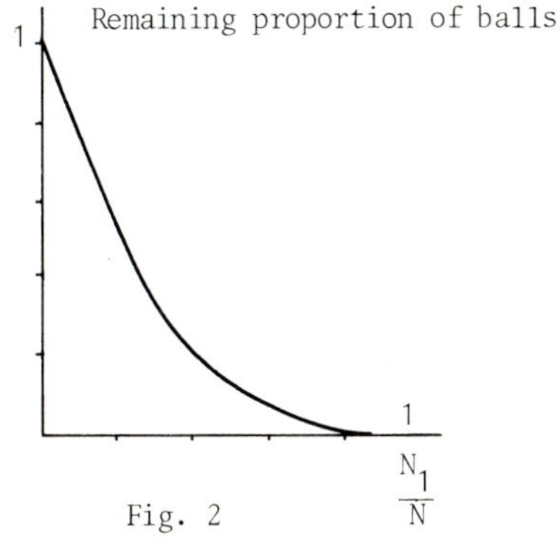

Fig. 2

Clearly 

will not be a stable draining, exactly like the example of the introduction suggests. However, we have not been able to find a good analytic approximation for the proportion of balls of type 2 remaining in the urn at the end of a drawing when one knows the proportion of balls at the beginning of the drawing A rather loose approximation can be found in the pre-quoted paper by Diet and Raynaud.

3 - STRONG DRAININGS WITH 3 KINDS OF BALLS

Some strong drainings with three kinds of balls can be identified as stable strong drainings, but there exist strong drainings with three types of balls which are not stable.

Of course 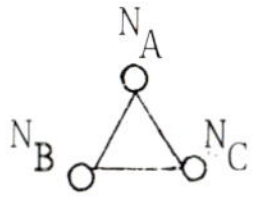

is a strong draining, whatever are N_A, N_B and N_C.

F. PETRY and myself established in 1971 the following theorem. The proof contained one error which was corrected in J.G. Dion's thesis.

THEOREM - The weighted graph $(K_3\ ;. N/3,\ N/3,\ N/3)$ is a stable strong draining.

Proof : Let A,B,C be the three vertices of K_3, the corresponding balls being called A-balls, B-balls, C-balls.

We shall say that A is the nth ball in the drawing if the nth ball to be drawn is a an A-ball. We denote this event by (A,n). We want to show that

$$\forall \alpha,\beta > 0, \quad \exists N_0 \text{ such that } N \geq N_0 \Rightarrow P\left\{\frac{\nu}{N} > 1-\alpha\right\} > 1-\beta$$

where ν is the number of balls drawn out of the urn when the drawing stops. To do this, using symmetry properties in K_3, we shall see that the proportion of balls of the three kinds drawn among the k first ones (under the hypothesis $k \leq \nu$), converges in probability towards the proportion of balls of the three different kinds at the beginning of the drawing. Then it will be easy to derive the wanted result. More precisely, K_3 being point-symmetrical, and its weights being equal, it is clear that

$$P\{(L_1,n)\} = P\{(L_2,n)\} \text{ for any } L_1, L_2 \in \{A,B,C\}$$

Hence $$P\{(L_1,n) \mid \nu \geq n\} = \frac{1}{3} \ \forall L_1 \in V \text{ and } P\{(L_1,n)\} \to \frac{1}{3} \ \forall L_1 \in V, \forall n, N\to\infty$$

as ν increases necessarily infinitely.

Let $X_{A,k} = \frac{1}{k} \sum_{i=1}^{k} I_{(A,i)}$ where $I_{(A,i)}$ is the indicator of the event (A,i).

$$E(X_{A,k}) = \frac{1}{k} \sum_{i=1}^{k} P(A,i) \to \frac{1}{3} \quad \text{when } N \to \infty$$

Let us consider

$$E(X_{Ak}^2) = \frac{1}{k^2} \sum_{i,j=1}^{n} P\{(A,i) \cap (A,j)\}$$

$$= \frac{1}{k^2} [2 \sum_{1\leq i\leq j\leq k} P\{(A,i) \cap (A,j)\} + \sum_{1=1}^{k} P(A,1)]$$

$$= \frac{2}{k^2} \sum_{i=1}^{k-1} \sum_{j=i+1}^{k} P\{(A,i)\cap(A,j)\} + \frac{1}{k^2} \sum_{1=1}^{k} P(A,1)$$

Let us consider now the sums

$$\frac{2}{k^2} \sum_{i=1}^{k-1} \sum_{j=i+1}^{k} P\{(A,i)\cap(A,j)\} \qquad (1)$$

that we can write in the form

$$\frac{2}{k^2} \sum_{i=1}^{k-1} \sum_{j=i+1}^{k} P(A,i)\, P\{(A,j) \mid (A,i)\} \qquad (2)$$

Denoting by $[\![x]\!]$ the integer part of x and by Ln the neperian logarithm of n, consider, instead of (2), the sum

$$\frac{2}{k^2} \sum_{i=1}^{k-1-[\![Lk]\!]} \sum_{j=i+1+[\![LLk]\!]}^{k} P(A,1)P\{(A,j) \mid (A,i)\} \qquad (3)$$

In this sum we neglect $[\![Lk]\!]\ [\![LLk]\!]$ terms. These terms being bounded by one, their contribution to the value of (1) has limit zero when k tends to infinity.

(3) is of course equal to $\frac{2}{3k^2} \sum_{i=1}^{k-1-[\![Lk]\!]} \sum_{j=i+1\,[\![LLk]\!]}^{k} P\{(A,j)\,|\,(A,i)\}$ (4).

Consider now

$$\sum_{j=i+1+[\![LLk]\!]}^{k} P\{(A,j)\,|\,(A,i)\},$$

i being at most equal to $k-1-[\![Lk]\!]$, this sum contains a number of terms at least of the order of magnitude of $[\![Lk]\!]$ when k tends to infinity. But how much does the fact that the ith ball drawn is an A-ball affect the probabilities of ulterior events (A,j) as soon as j-i becomes superior to LLk ? As we shall see, by a quantity which tends to zero with $\frac{1}{k}$: when j goes from i+1 to $i+1+[\![LLk]\!]$, the proportions of the different kinds of balls in the urn vary at most by about $\frac{LLk}{Lk}$; and for $k \geq i$ let the vector of a priori probabilities

$$\vec{V}_k = (P\{(A,k)\,|\,(A,i)\},\ P\{(B,k)\,|\,(A,i)\},\ P\{(C,k)\,|\,(A,i)\})$$

It is clear that we can write $\vec{V}_{k+1} = \vec{V}_k P_k$ and $\vec{V}_j = \vec{V}_i P_i P_{i+1} \dots P_1 \dots P_{j-1}$ where P_1 denotes the matrix of a priori transition probabilities which converge to

$$\begin{vmatrix} 0 & 1/2 & 1/2 \\ 1/2 & 0 & 1/2 \\ 1/2 & 1/2 & 0 \end{vmatrix}$$

when N increases infinitely.

As the transition probabilities depend only on the proportions of balls inside the urn, P_1 differs from P_i only by quantities of the order of $\frac{LLk}{Lk}$ at most. P_1 can then be written $P_i[I+\varepsilon_1]$ when ε_1 is a matrix, the terms of which are at most of the order of $\frac{LLk}{Lk}$.

And

$$\vec{V}_j = \vec{V}_i\ P_i^{j-i} \prod_{1=1}^{j-1} [I+\varepsilon_1] = \vec{V}_i\ P_i^{j-i}[I+\varepsilon]$$

where ε is a matrix which tends towards zero with $\frac{1}{k}$ for

$$\prod_{1=0}^{[\![LLk]\!]} \left(1+\frac{LLk}{Lk}\right) \simeq 1+\frac{(LLk)^2}{Lk} \xrightarrow[k\to\infty]{} 1$$

P_i being a stochastic matrix $P_i^n \to R$, matrix with all coefficients equal to $\frac{1}{3}$.

Hence $P\{(A,j)\,|\,(A,i)\}$, for j at least equal to $i+1+[\![LLk]\!]$, converges to $\frac{1}{3}$ when k increases infinitely and

$$\sum_{j=i+1+[\![LLk]\!]}^{k} P\{(A,j)\,|\,(A,i)\} \approx \frac{k-[\![LLk]\!]-i}{3}$$

for all i fixed when k tends to infinity - which implies for the sum (4) that

$$\frac{2}{3k^2}\sum_{i=1}^{k-1-[\![Lk]\!]}\ \sum_{j=i+1+[\![LLk]\!]}^{k} P\{(A,j)\,|\,(A,i)\} \simeq \frac{1}{9}\,\frac{k(k-1)}{k^2} \text{ when } k\to\infty.$$

Hence

$$E(X^2_{Ak}) = \frac{1}{9}\,\frac{k(k-1)}{k^2} + \frac{1}{3k} = \frac{1}{9} - \frac{1}{9k} + \frac{1}{3k}$$

and

$$Var(X_{A,k}) = \frac{1}{3k} - \frac{1}{9k} = \frac{2}{9k}$$

which tends to zero with $\frac{1}{k}$. By this $X_{A,k} \xrightarrow{P} \frac{1}{3}$ when $k \to \infty$, $k \le \nu$, and it is the same for $X_{B,k}$ and $X_{C,k}$.

This means that the proportion of balls of each kind drawn among the k first ones tends in probability towards $\frac{1}{3}$ when k tends to infinity, $k \le \nu$. More precisely

$$\forall \varepsilon\ ,\eta > 0,\ \exists L(\varepsilon,\eta) \text{ such that}$$

$$k \ge L(\varepsilon,\eta) \Rightarrow P\{\sup_i |X_{ik} - \tfrac{1}{3}| > \varepsilon \,|\, k \le \nu\} < \eta \qquad (5)$$

Using this property for successive larger and larger values of k, we shall obtain the wanted result.

It is clear that $\nu \ge \frac{2N}{3}$. From (5), if one knows that $\nu \ge k_o$ for all kinds of balls the probability that at most $\frac{k_o}{3} + k_o\,\varepsilon$ balls of this sort be among the k_o first balls to be drawn out is greater or equal to $1-\eta$. And, if this event happens, it is clear that one can still draw, after the k_o first balls $\frac{2N}{3} - 2(\frac{k_o}{3} + k_o\varepsilon)$ balls for the same reason as previously. This means that :

$\forall\varepsilon,\eta > 0$, $\forall N$ large enough, $\exists L(\varepsilon,\eta)$ such that

$$k_o \ge L(\varepsilon,\eta) \Rightarrow P\{\nu \ge k_o + \frac{2N}{3} - 2(\frac{k_o}{3} + k_o\varepsilon) \,|\, \nu \ge k_o\} > 1-\eta \qquad (6)$$

this property, by induction, and using the fact that $P\{\nu \ge \frac{2N}{3}\} = 1$ implies that

$$\forall\varepsilon,\eta > 0,\ \forall k \in \mathbb{N},\ \exists N_o(=4L(\varepsilon,\eta)) \text{ such that}$$

$$N \ge N_o \Rightarrow P\{\nu \ge \frac{2N}{3}\sum_{i=0}^{k}(\tfrac{1}{3} - 2\varepsilon)^i\} \ge (1-\eta)^k \qquad (7).$$

As a matter of fact, suppose that $\frac{2N}{3} \ge \frac{8}{3}L(\varepsilon,\eta)$; (7) is true for k = 0 since $P\{\nu \ge \frac{2N}{3}\} = 1$. Let us suppose that (7) is true for k, and let us prove that it is true for k+1.

In (6) let $k_o = \frac{2N}{3} \sum_{i=0}^{k} (\frac{1}{3} - 2\varepsilon)^i$

$$k_o \geq \frac{8}{3} L(\varepsilon,\eta) \sum_{i=0}^{k} (\frac{1}{3}-2\varepsilon)^i = \frac{8}{3} L(\varepsilon,\eta) \frac{1-(\frac{1}{3}-2\varepsilon)^{k+1}}{\frac{2}{3}+2\varepsilon}$$

$$\geq 2L(\varepsilon,\eta)\ [1-(\frac{1}{3}-2\varepsilon)^{k+1}] \text{ for } \varepsilon \text{ small enough}$$

$$\geq 2L(\varepsilon,\eta)[1-(\frac{1}{3})^{k+1}]$$

$$\geq L(\varepsilon,\eta)$$

Thus we obtain

$$P\{\nu \geq \frac{2N}{3}+\Big[\frac{2N}{3} \sum_{i=0}^{k} (\frac{1}{3}-2\varepsilon)^i\Big](\frac{1}{3}-2\varepsilon) \mid \nu \geq \frac{2N}{3} \sum_{i=0}^{k} (\frac{1}{3}-2\varepsilon)^i\} \geq 1-\eta$$

i.e. $P\{ \nu \geq \frac{2N}{3} \sum_{i=0}^{k+1} (\frac{1}{3}-2\varepsilon)^i \mid \nu \geq \frac{2N}{3} \sum_{i=0}^{k} (\frac{1}{3}-2\varepsilon)^i\} \geq 1-\eta$

Hence

$$P\{\nu \geq \frac{2N}{3} \sum_{i=0}^{k+1} (\frac{1}{3}-2\varepsilon)^i\}$$

$$= P\{\nu \geq \frac{2N}{3} \sum_{i=0}^{k+1} (\frac{1}{3}-2\varepsilon)^i \mid \nu \geq \frac{2N}{3} \sum_{i=0}^{k} (\frac{1}{3}-2\varepsilon)^i\} P\{ \nu \geq \frac{2N}{3} \sum_{i=0}^{k} (\frac{1}{3}-2\varepsilon)^i\}$$

$$\geq (1-\eta) P\{\nu \geq \frac{2N}{3} \sum_{i=0}^{k} (\frac{1}{3}-2\varepsilon)^i\} \geq (1-\eta)^{k+1}$$

according to the induction hypothesis.

Remarking that $\frac{2N}{3} \sum_{i=0}^{\infty} (\frac{1}{3})^i = N$, it is sufficient now to choose ε,η and k conveniently to show that

$$\forall\ \alpha,\beta\ ,\ \exists N_1 \text{ such that } N \geq N_1 \Rightarrow P\{\frac{\nu}{N} > 1-\alpha\} > 1-\beta$$

First, choose k and ε such that $\frac{2}{3} \sum_{i=0}^{k} (\frac{1}{3}-2\varepsilon)^i > 1-\beta$, then η such that $(1-\eta)^k > 1-\beta$, and, finally $L(\varepsilon,\eta)$ satisfying the conditions of property (6). Then, taking N_1 equal to N_o, as defined by property (7), we obtain

$$P\{\frac{\nu}{N} > 1-\alpha\} \geq P\{\frac{\nu}{N} \geq \frac{2}{3} \sum_{i=0}^{k} (\frac{1}{3}-2\varepsilon)^i\} \quad \text{from the choice of } \varepsilon \text{ and } \eta$$

$$= P\{\nu \geq \frac{2N}{3} \sum_{i=0}^{k} (\frac{1}{3}-2\varepsilon)^i\} \geq (1-\eta)^k \text{ from (7)}$$

$$\geq 1-\beta \quad \text{from the choice of } \eta\ .$$

Hence $(K_3\ ;\ \frac{N}{3}, \frac{N}{3}, \frac{N}{3})$ is a strong draining.

XXX

What about other drainings with three kinds of balls ?

Nothing is said about 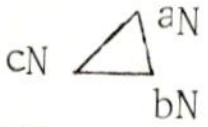

, a,b,c being three different coefficients.

It is clear that 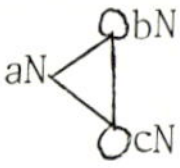

is never a strong draining. And it is not very difficult to see that :

THEOREM 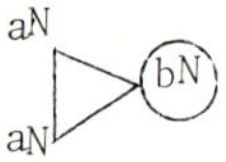

is a strong - but not stable- draining.

Proof : Let us say that the associated urn contains aN balls A, aN balls B and bN balls C.

Consider a drawing of the balls out of the associated urn and extract from it the part S composed only of sequences of balls A or B. Two such sequences are separated in the drawing by one or many consecutive balls C. From S, extract again the only sequences counting an odd number of balls. Let $s_1 \ldots s_i \ldots s_n$ be those latter sequences extracted.

1°) Suppose that n is not infinitely large with N : it means that the number of balls A and the number of balls B drawn at the end of the drawing would be of the same order of magnitude compared to N -and the draining would be strong.

2°) Let us suppose now that n increases infinitely together with N. Let us consider the variable X_i equal to +1 if the sequence s_i begins by an A-ball, otherwise equal to -1.

It is clear that $P\{X_i=1\} = P\{X_i=-1\} = \frac{1}{2}$ and $E\frac{\Sigma X_i}{n} = E\frac{\Sigma X_i}{N} = 0$

Let us consider now

$$E\left(\frac{\Sigma X_i}{n} - E\frac{\Sigma X_i}{n}\right)^2 = E\left(\frac{\Sigma X_i}{n}\right)^2 = \frac{1}{n} + \frac{2}{n^2}\sum_{i<j} X_i X_j$$

If the sequence s_i has one A-ball in excess and if one has no other information on the drawing, this implies for s_j a probability superior to $\frac{1}{2}$ of having one B-ball in excess.

Hence

$$EX_iX_j = P\{\{x_i=1\}\{x_j=1\}\}+P\{\{x_i=-1\}\{x_j=-1\}\}-P\{\{X_i=1\}\{X_j=-1\}\}-P\{\{x_i=-1\}\{x_j=1\}\}$$

Thus $EX_iX_j < 0$, from the previous remark.

The quantity $\mathrm{Var}\left(\frac{\sum_1^n X_i}{n}\right)$ is, by the way, inferior to $\frac{1}{n}$ and the standard deviation of $\sum_1^n X_i$ is inferior to n. At the end of the drawing, the difference between the proportion of A-balls and the proportion of B-balls drawn from the urn converges toward zero in probability. In any case, the considered draining is strong.

XXX

4) SOME MORE THEOREMS

Strong draining theory had not been presented at any statisticians meeting until this conference. This is why we have thought it useful to summarize some of the recent results and conjectures in the theory.

THEOREM (J.G. Dion [1976])

Let G be a strongly connected graph with p vertices denoted $\{1,2,\dots,p\}$.

If $(G,\vec{N})$ is a draining, then we call associated Markov chain, the Markov chain in which the state at step n is the kind of the $n^{\underline{th}}$ ball drawn from the urn, the drawing being made with replacement.

If $\vec{N} = (N_1,N_2,\dots,N_p)$ is such that $N_i = \alpha_i N$ for $i = 1,2,\dots,p$ where $\vec{\alpha} = (\alpha_1,\alpha_2,\dots,\alpha_p)$ is the only probability vector such that $\vec{\alpha} P = \vec{\alpha}$, P being the probability transition matrix of the associated Markov chain, then $(G,\vec{N})$ is a strong draining.

Definition

We say that we have made an elementary expansion of a weighted graph $(G,\vec{N})$ if we consider the graph $(G',\vec{N}')$ with p+k vertices (p being the number of vertices in G) constructed like this :

Let $V_1,V_2,\dots,V_p$ be the vertices of G. $V_1',\dots,V_p',\dots,V_{p+k}'$ are the vertices of G'

$\vec{N}' = (N_1\ N_2,\dots,N_{p-1},\ N_p',\ N_{p+1}',\dots,N_{p+k}')$

and $N_{p+1}' \geq 0$, $i = 0,1,2,\dots,k$ $\qquad N_p'+N_{p+1}' +\dots+ N_{p+k}' = N_p$

For all $i \leq p$, $j \leq p$; $(V_i',V_j') \in A(G') \iff (V_i,V_j) \in A(G)$

For all $i < P$, $j < p, p+1,\dots,p+k$; $(V_i',V_j') \in A(G') \iff (V_i,V_p) \in A(G)$ and

$(V_j',V_i') \in A(G') \iff (V_p,V_i) \in A(G)$.

If V_p has a loop, the subgraph generated by $V_p',\dots,V_{p+k}'$ in G' is the complete subgraph with loops ; and if V_p has no loop, the subgraph generated by $V_p',\dots,V_{p+k}'$ is without any arcs.

An expansion will be a product of elementary expansions.

THEOREM (J.G. Dion [1976])

Let G be a graph with p vertices $V_1, V_2, .., V_p$, $\vec{N}' = (N'_1, N'_2, ..., N'_p)$ an integer finite fixed and positive vector $(G', \vec{M}')$ a weighted graph with p+k vertices resulting from an expansion of $(G, \vec{N}')$.

If for all $\vec{N} = (N_1, N_2, ..., N_p)$, the ratios $\frac{N_i}{N} \to \frac{N'_i}{N'}$ for all i when $N \to \infty$, and for all $\vec{M} = (M_1, M_2, .. M_{p+k})$, the ratios $\frac{M_i}{M} \to \frac{M'_i}{M'}$ for all i when $M = N \to \infty$, then the weighted graph $(G, \vec{N})$ is a strong draining if the weighted graph $(G', \vec{M})$ is a strong draining.

Open problems :

a) Characterize strong drainings !

b) If it is not easy, begin by the following one which looks likely

Is 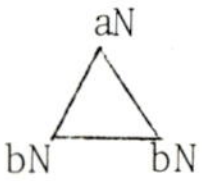

$a > \frac{1}{3}$, $a+2b = 1$ a strong draining ?

BIBLIOGRAPHY

DIET J.P., and RAYNAUD H. (1977)
Les vidanges fortes les plus simples Discrete Math., (to appear)

DION J.G. (1976)
De plus en plus fortes !...ou les avatars d'une conjecture sur les vidanges.
Thèse de 3ème cycle. Université Scientifique et Médicale de Grenoble.

PETRY F. (1971)
Résolution d'une conjoncture issue de la théorie des graphes aléatoires.
Thèse de Doctorat de 3ème Cycle. Faculté des Sciences de Paris.

RAYNAUD H. (1968)
Sur les graphes aléatoires.
Ann. Inst. Henri Poincaré, section B. Vol. IV, n° 4 (1968) pp. 255-329.

Recent Developments in Statistics
J.R.Barra et al., editors

ESTIMATING INPUT DENSITIES FOR OPERATING SYSTEM MODELS

Anne SCHROEDER

IRIA-LABORIA
BP 105
78150 LE CHESNAY
FRANCE

1. INTRODUCTION

Because of their complexity, the study of operating systems makes frequently use of modeling techniques. Given an "input" information, a model is a tool which gives a corresponding "output" information. This output is supposed to approximate the actual output of the modelised system.

Once a model is built, it is important to feed it with an input information as realistic as possible. In paragraph 2, we shall see that this required information has to be a probability density function for a wide class of models. To find such densities, real samples have to be observed and the densities deduced from them : this is a problem of statistical estimation. In paragraph 3, we shall present an estimation technique for a family of densities (mixture densities) that often appear in models. Then, some numerical results will be given in the fourth paragraph.

2 - AN ESTIMATION PROBLEM IN MODELING

2.1 - Operating systems modeling

Modeling is often used to study operating systems and their performances. It allows to analyse the behaviour of a system when changing some of its characteristics without having to realize actual experiments, or to predict the performances of a system during its design.

For instance, if the modelised system is a time-sharing system, the input information can refer to the users (requirements, multiprogramming level,...) or to the devices (service times of a disc or a drum,...), while the output may be a response time or utilisation rates of those devices. Program behaviour in some given environment may also be modelised : the input then is a full reference string of a program or a more synthetical information on its requirements in time, space, I/O,... while its real execution time may be the observed output.

We shall consider two types of models :

- Analytical models (Mun 75) : In those models, input and output information are the probability distribution functions (pdf) of some parameters. In the first example above, the input can be the pdf of the service time of the CPU and the output the pdf of the response-time of the system. These models deduce the output from the input by using formulae, i.e. by describing the system with equations.

- Simulation models (Ler 76) deal with actual samples of parameters as input or output information (e.g. reference strings, sample of arrival times of users in the system...). Then, the behaviour of the system is described logically step-by-step.

2.2 - Input Information

In analytical models, the input is a pdf the type of which depends on the equations involved. Up to now, the most adequate models for operating systems are queueing networks models. Those models require mixtures of exponential or Erlangian densities as input (Cox 55, Bas 75). There are also approximate models based on diffusion processes (Kob 74, Gel 75) which theoretically accept any kind of density ; in this case, a density estimate is needed. To make the model efficient, this estimate has to be defined by a small number of parameters -which is not the case of usual density estimates . So, unless some usual density can easily be fitted to the sample, we propose to approximate the observed density by a Gaussian mixture.

In simulation models, the input is a sample. However, an actual observed sample may be expensive to manipulate (reference strings), or, on the contrary, too small to permit several experiments. Then, it is necessary to know how to generate as many samples as needed, which have the same properties as the initial observed sample. For this purpose, we also propose to estimate an approximate density of the actual sample as a Gaussian mixture. It is then easy to generate other samples drawn from this distribution.

So the problem of giving adequate input to operating system models may quite often be stated in terms of estimating a mixture of several densities of a given type.

3 - MIXTURES ESTIMATION

3.1 - The problem

Given an observed sample E, the statistical problem is to estimate the unknown parameters of a probability density of the following form :

(1) $$f(x) = \sum_{1 \leq i \leq k} c_i f_{\lambda_i}(x) \qquad (x \in R)$$

where $(f_\lambda / \lambda \in \mathcal{L} \subset R^s)$ is a given family of densities depending on an unknown parameter λ (which may be a vector). Knowing the number k of components in the mixture, there are k+sk-1 parameters to estimate : the λ_i's and the c_i's ; c_i represents the proportion of the ith density f_{λ_i} in the mixture, and we have the equality : $\sum_{1 \leq i \leq k} c_i = 1$.

The problem of estimating this kind of density is a common problem in statistics. There are many techniques to estimate the unknown parameters of univariate Gaussian mixtures. There are few methods allowing to deal theoretically with other kinds of distributions and in practice they are only used in the Gaussian case. We shall present a quite general method which can be applied to any kind of mixtures (Erlangian or exponential for instance, or multivariate, etc...). It is also valid for any number of components.

We shall give here a variant of this method which maximizes a likelihood criterion, but other criteria may also be optimized within the same general frame. The most general case is presented in (Sch 76a).

3.2 - The algorithm

It is an iterative algorithm which simultaneously searches a partition of the sample and densities fit to the classes of this partition. This idea of a simultaneous search of a partition and of characteristic features of the classes has first been used in clustering in the so-called Dynamic Clusters method (Did 72). It has then be extended to other characteristic features (Did 74).

One example of these extensions is the mixture estimation method which we present. There probability densities are taken as characteristic features of the classes.

Given the sample E, the number k of components and the family $(f_\lambda / \lambda \in \mathcal{L})$, the method gives the $(\lambda_i / 1 \leq i \leq k)$ and a partition $(P_i / 1 \leq i \leq k)$ of E in k classes, such that the likelihoods of the P_i's, considered as samples of the distributions f_{λ_i} are as good as possible. The c_i's are then estimated as the empirical frequency of the ith component P_i, i.e. : $c_i = \mathrm{Card}(P_i)/\mathrm{Card}(E)$.

The results thus give a theoretical density maximizing a likelihood criterion which can be compared to the empirical density by goodness-of-fit tests.

More precisely, the algorithm consists in building a sequence $(L^{(n)}, P^{(n)})_{n \geq 1}$ such that :

- for all $n \geq 1$, $W(L^{(n)}, P^{(n)}) \geq W(L^{(n-1)}, P^{(n-1)})$ where W is the following likelihood criterion :

$$W(L^{(n)}, P^{(n)}) = \prod_{i=1,k} \prod_{x \in P_i^{(n)}} f_{\lambda_i^{(n)}}(x)$$

Thus, W is the product of the likelihoods of the k sub-samples $P_i^{(n)}$ for the corresponding densities $f_{\lambda_i^{(n)}}$.

This algorithm is thoroughly described and its convergence properties proved in (Sch 76a).

Let us just note that the sequence $(L^{(n)}, P^{(n)})$ converges towards a solution (L*, P*) which locally optimizes the likelihood criterion W. Different choices for $(L^{(0)}, P^{(0)})$ will lead to different solutions among which will be chosen the one giving the best value for the criterion.

The particular cases of univariate Gaussian, multivariate Gaussian and Erlangian densities may be found in (Sch 76a) while details on the hyperexponential case are in (Sch 76b).

4 - APPLICATIONS

4.1 - Approximation by Gaussian mixtures of samples of CPU times to execute different commands of an operating system.

Measurements have been taken on the CP-CMS system of the University of Grenoble (France) to analyse the CPU time required to execute some commands, such as the editing command, or a FORTRAN or PL/1 compilation. Such service times are used as input data in operating system models. All detail on the sampling and on the analysis of those measurements can be found in (Ler 73) and (Ler 75).

Here is an example : a sample of 670 execution times of the ASSEMBLER command. Since we did not know a priori the number k of Gaussian densities required to get a good approximation, we made several runs of the algorithm taking successively k equal to 2,3,...,10. The number of 6 components has finally be chosen as giving the best global fitness ot the estimated and empirical distributions.

Results :

$$f(x) = \frac{1}{\sqrt{2\Pi}} \sum_{1 \leq i \leq 6} c_i \frac{1}{\sigma_i} \exp - \left[\frac{(x-\mu_i)^2}{2\sigma_i^2} \right]$$

where the values of the parameters are :

i	1	2	3	4	5	6
c_i	.26	.07	.11	.09	.15	.32
μ_i	10.30	67.58	40.33	31.39	23.66	3.64
σ_i	3.68	25.12	2.98	1.59	2.73	0.33

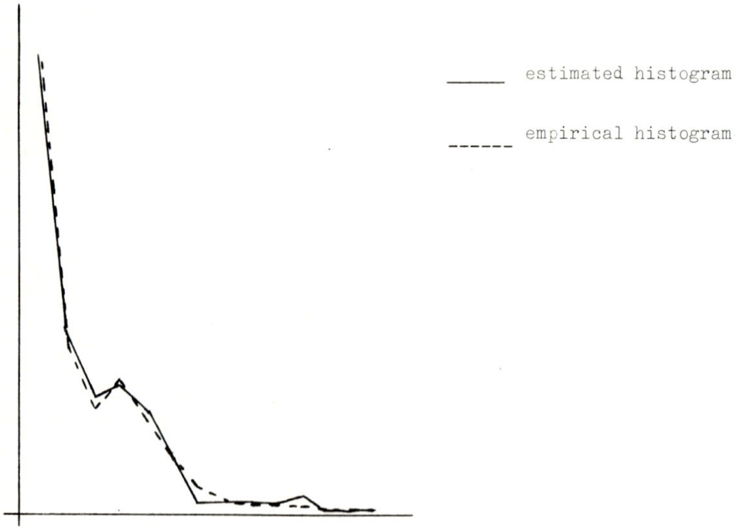

Figure 1

4.2 -Estimation of Gamma mixtures on those same samples.

In analytical models with queues, input data are often linear combinations of exponential or Erlangian densities. (see § 4.3 for an application to the exponential case).

Erlangian densities are of the following form :

$$f_{\alpha\beta\gamma}(x) = \begin{cases} \dfrac{\alpha^\beta}{\Gamma(\beta)} (x-\gamma)^{\beta-1} \exp\left[-\alpha(x-\gamma)\right] & \text{if } x \geq \gamma \\ 0 & \text{if } x < \gamma \end{cases}$$

with $\alpha \in R^+$, $\beta \in N$, and $\gamma \in R$.

Gamma densities are the same, except that the parameter β can take any real value greater than one : $\beta \in [1, \infty[$.

Though models explicitly involve Erlangian densities, our estimation method deals with Gamma densities from which approximate Erlangian ones may be deduced.

As an example, let us take a sample of 969 execution times of the FORTRAN compiler. The number of three components in the mixture has been determined in the same way as in § 4.1.

Results :

$$f(x) = \sum_{1 \leq i \leq 3} c_i \, 1_{[\gamma_i, \infty[}(x) \, \frac{\alpha_i^{\beta_i}}{\Gamma(\beta_i)} (x-\gamma_i)^{\beta_i - 1} \exp[-\alpha_i(x-\gamma_i)]$$

where

$$1_{[\gamma_i, \infty[}(x) = \begin{cases} 1 & \text{if } x \geq \gamma_i \\ 0 & \text{else} \end{cases}$$

with :

i	c_i	i	β_i	γ_i
1	.30	.1	1.0	.9
2	.41	1.6	1.3	.9
3	.29	.5	1.5	4.3

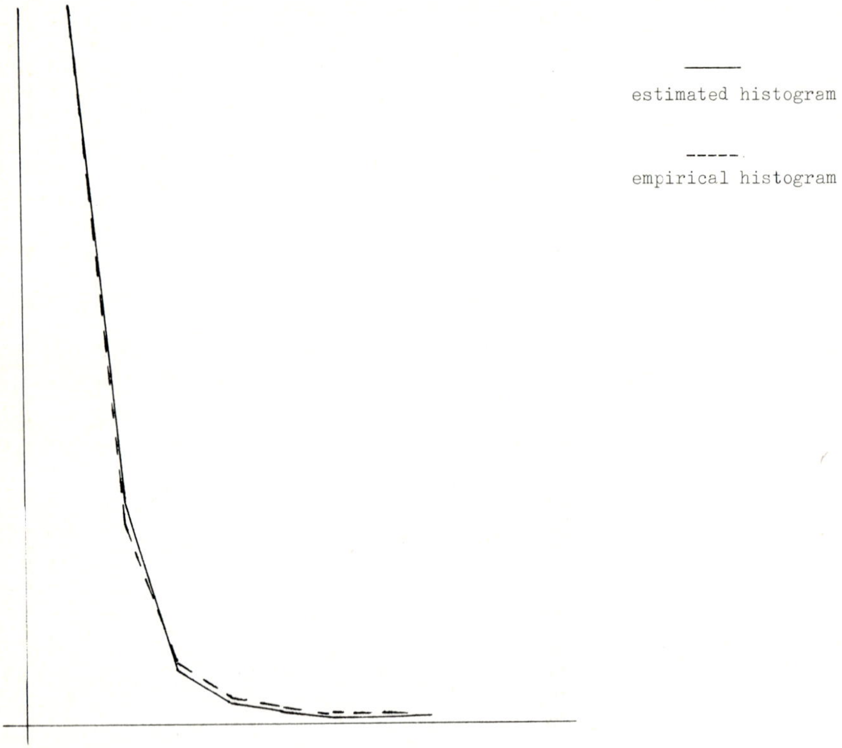

Figure 2

4.3 - Estimation of hyperexponential densities of samples of virtual life times of a program between successive page faults.

These samples have been obtained by interpreting reference strings of programs running on a CII - 10070 of the University of Rennes (France). All results on this work may be found in (Bur 76).

Distributions of such life times of a program may be used as input data in models of program behaviour. Wishing to have hyperexponential densities, we used the same method as above for a mixture of two exponential densities.

Example : a sample of 7382 life times between successive page faults during the execution of a FORTRAN compiler (60 pages) for a memory size of 20 pages, a page size of 512 words and a LRU replacement algorithm.

The hyperexponential density obtained is :

$$f(x) = 10^{-3}(9.92\ e^{-0.016x} + 0.38\ e^{-0.001x})$$

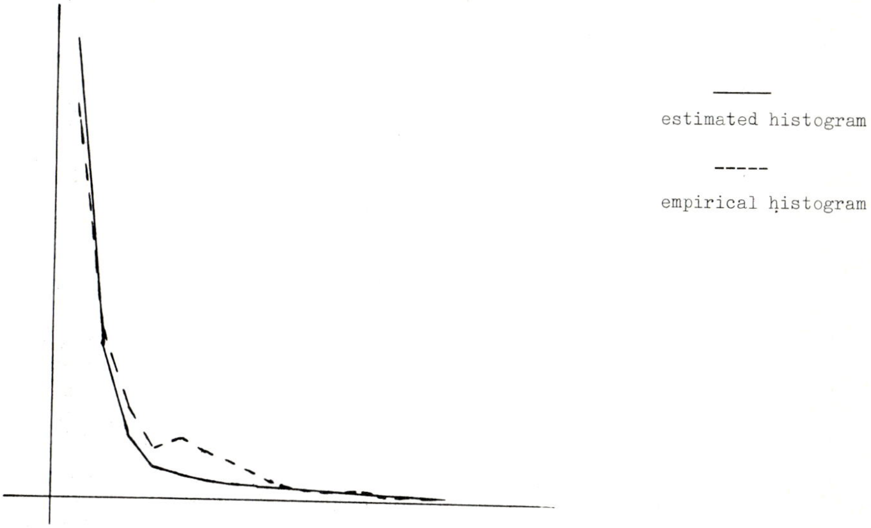

Figure 3

5. CONCLUDING REMARKS.

We have presented a tool which can help to improve the validity of different kinds of models. Our purpose was not, in any way, to legitimate the use of modeling, but only to help in bringing the abstraction of a model closer to the reality.

From a different point of view, characterising empirical distributions as mixtures of densities can also be used in a purpose of description : their interpretation may for instance reveal the presence of distinct behaviours in the sample.

Though the above applications were in the Computer Science area, the techniques presented can be put to use in any field using modeling.

ACKNOWLEDGEMENTS.

The measurements and the analyses presented as applications are due to Jacques Leroudier and Patrice Burgevin ; may they find here our most friendly thanks.

REFERENCES

[Bas 75] F. BASKETT ; K.M. CHANDY ; R.R. MUNTZ ; F. G. PALACIOS - Open,closed and mixed Networks of Queues with Different Classes of Customers JACM, Vol 22, n°2 - 1975.

[Bur 76] P. BURGEVIN ; J. LEROUDIER - Characteristics and Models of Program Behavior - Annual Conference of the ACM, Houston (Texas) - October 1976.

[Cox 55] D.R. COX - A Use of Complex Probabilities in the Theory of Stochastic Processes - Proc. Camb. Phil. Soc., 51 , 1955.

[Did 72] E. DIDAY - Nouvelles méthodes et nouveaux concepts en classsification automatique et reconnaissance des formes - Thèse, Paris VI, 1972.

[Did 74] E. DIDAY ; A. SCHROEDER ; Y. OK - The Dynamic Clusters Method in Pattern Recognition - IFIP Congress, Stockholm - 1974.

[Gel 75] E. GELENBE - On approximate Computer System Models - JACM, Vol. 22, n°2 - 1975.

[Kob 74] H. KOBAYASHI -Application of the Diffusion Approximation to Queueing Networks : Equilibrium Queue Distribution - JACM, Vol. 21, n°2 - April 1974.

[Ler 73] J. LEROUDIER - Analyse d'un système à partage de ressources - RAIRO, série bleue - Octobre 1973.

[Ler 75] J. LEROUDIER ; A. SCHROEDER - A Statistical Approach to the Estimation of Service Times Distributions in Operating Systems Modeling - ICS75, North-Holland publ. - 1975.

[Ler 76] J. LEROUDIER ; M. PARENT - Discrete Event Simulation Modelling of Computer Systems for Performance Evaluation - Rapport de Recherche IRIA - LABORIA - 1976.

[Mun 75] - R.R. MUNTZ - Analytic Modeling of Interactive Systems - Proceedings of the IEEE - June 1975.

[Sch 76a] A. SCHROEDER - Analyse d'un mélange de distributions de probabilité de même type - Revue de Statistique Appliquée, Vol. XXIV, n°1- 1976.

[Sch 76b] A. SCHROEDER - Estimation de distributions d'entrée dans les modèles de Systèmes Informatiques - in "La statistique : outil d'analyse des Systèmes Informatiques", Journées organisées par le Chapitre Français de l'ACM - Juin 1976.

PART TWO: Selected Contributed Papers

C

Data Analysis and Connected Topics

Recent Developments in Statistics
J.R.Barra et al., editors

TALLIS' CLASSIFICATION MODEL FOR THREE GROUPS

A.I. ALBERT
Department of Probability and Statistics
University of Liege
Liege, Belgium

Tallis' general classification model is envisaged in the case of three populations and applied to biochemical data from patients with liver diseases. ML estimators are not considered here. Attention was only focused on the estimators obtained by extending the method of Hannan and Tate to the case of a multivariate normal distribution when one variable is trichotomized.

INTRODUCTION

The problem of classification of an individual into one of several populations has always been a fascinating one to the statistician. Just remember the question asked by the archaeologist about the Highdown Skull, or more recently the problems of medical diagnosis. The very first problem was that of classifying an individual into one of two normally distributed populations with different means, $\ddot{\mu}_1 \neq \ddot{\mu}_2$, but common dispersion matrix $\ddot{\Sigma}$, giving rise to the well-known linear discriminant function, Fisher (1936).

When the covariance matrices are different, $\ddot{\Sigma}_1 \neq \ddot{\Sigma}_2$, the rule becomes more difficult and is a quadratic function of the variates, Smith (1946). Lachenbruch (1975) is now looking at the problem of classification when the two populations have common mean, $\ddot{\mu}_1 = \ddot{\mu}_2$, but different covariance matrices, $\ddot{\Sigma}_1 \neq \ddot{\Sigma}_2$.

Normality plays an important role in the classification models proposed, and Lachenbruch et al. (1973) have recently shown how tedious the problem gets, when non-normal distributions are envisaged. When more than two populations are involved, the solutions proposed are not completely satisfactory.

TALLIS' CLASSIFICATION MODEL

Tallis et al. (1975) proposed recently a general classification model, which was applied in the particular case of response to adrenalectomy in women with breast cancer.

Let Π be a population characterized by a random variable X_o assumed to be normally distributed N(0,1) and by a p-components random vector $\ddot{X}' = (X_1 \ldots X_p)$ with multinormal distribution $N(\ddot{\mu}, \ddot{\Sigma})$. Moreover suppose that the joint distribution of X_o and $\ddot{X}$ is also multinormal $N(\ddot{\nu},\ddot{V})$, where

$$\ddot{\nu}' = (0,\ddot{\mu}'),\ \ddot{V} = \begin{bmatrix} 1 & \ddot{\sigma}' \\ \ddot{\sigma} & \ddot{\Sigma} \end{bmatrix}.$$

The model assumes that X_o defines in the $(p+1)$-dimensional space two populations Π_1 and Π_2, as follows :

$$\Pi_1 = \{(X_o,\ddot{X}) : X_o \leq a\}$$
$$\Pi_2 = \{(X_o,\ddot{X}) : X_o > a\}$$

The problem is then to classify any new individual into Π_1 or Π_2 on the mere basis of the $\ddot{X}$-measurements, X_o being unknown.
The authors obtained the $\ddot{X}$-conditional probabilities for an individual to belong to Π_1 and Π_2.
We suggest here a method to extend the model to a three-groups situation and give an application in the field of liver diseases.

EXTENSION OF TALLIS' MODEL

Under the same assumptions as Tallis et al. (1975) about the distributions of X_o and $\ddot{X}$, suppose now that X_o defines in the $(p+1)$-dimensional space the three mutually exclusive populations,

$$\Pi_1 = \{(X_o,\ddot{X}) : x_o \leq a\}$$
$$\Pi_2 = \{(X_o,\ddot{X}) : a < X_o \leq b\}$$
$$\Pi_3 = \{(X_o,\ddot{X}) : b < X_o\}$$

where $a < b$.
Given the $(X_1 \ldots X_p)$-measurements, what are the individual's probabilities to belong to Π_1, Π_2 and Π_3 , X_o being unknown?
Conditionally to $\ddot{X}$, X_o is a normal variable with mean and variance respectively equal to :

$$E(X_o|\ddot{X}) = \ddot{\sigma}'\ \ddot{\Sigma}^{-1}(\ddot{X}-\ddot{\mu})$$

$$Var(X_o|\ddot{X}) = 1 - \ddot{\sigma}'\ \ddot{\Sigma}^{-1}\ \ddot{\sigma} = 1 - R^2,$$

where R^2 is the multiple correlation coefficient of X_o with $X_1 \ldots X_p$. It follows immediately, that the three required probabilities are :

$$P_1(R,u) = F_G\left(\frac{a - Ru}{\sqrt{1 - R^2}}\right)$$

$$P_2(R,u) = F_G\left(\frac{b - Ru}{\sqrt{1 - R^2}}\right) - F_G\left(\frac{a - Ru}{\sqrt{1 - R^2}}\right)$$

$$P_3(R,u) = 1 - F_G\left(\frac{b - Ru}{\sqrt{1 - R^2}}\right),$$

where $u = \ddot{\sigma}' \ddot{\Sigma}^{-1}(\ddot{X} - \ddot{\mu})/R$ is a standardized normal variable $N(0,1)$ and $F_G(.)$ is the standard normal cumulative distribution function.
For all u and $R > 0$, the following relation holds :

$$P_1(R,u) + P_2(R,u) + P_3(R,u) = 1$$

As in the two-groups situation, given R, the three probability curves can be plotted as function of u as shown in Fig. 1.
Knowing the $(X_1 \ldots X_p)$-measurements of an individual and therefore the value of the linear criteria u, it is possible to obtain valuable information about this individual's chances to belong to any of the three populations.

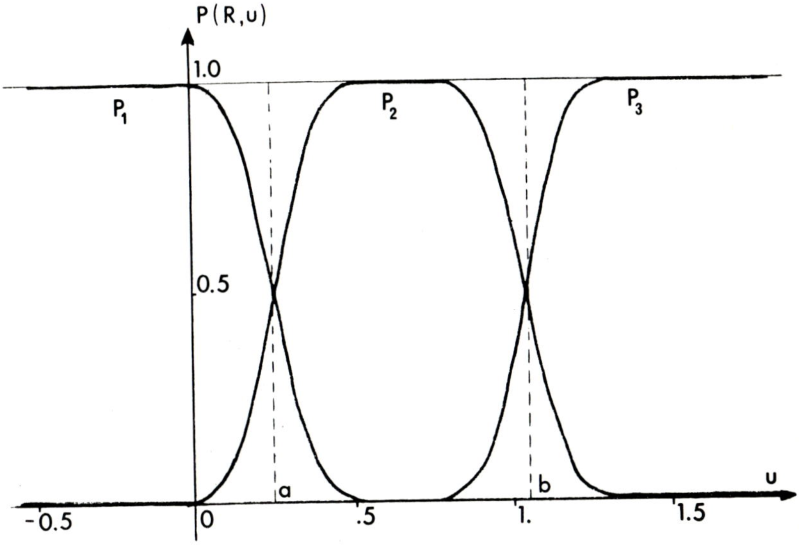

Figure 1. PROBABILITY CURVES

a=0.2533 $\hat{R}$=0.9952
b=1.0364

ESTIMATION OF THE PARAMETERS OF THE MODEL

The model, previously defined, involves $(p+1)(p+4)/2$ unknown parameters, i.e. $\ddot{\mu}$, $\ddot{\Sigma}$, $\ddot{\sigma}$, a and b, which in a practical situation have to be estimated from a random sample of size n, drawn from Π.
Maximum likelihood estimators of $\ddot{\sigma}$, a and b have not been explored. The present estimators are obtained by extending the method of Hannan and Tate (1965) to the case of a multivariate normal distribution when one variable is trichotomized.
Define a random variable Z, such that

$$\begin{aligned} Z &= -1 \text{ if } X_0 \leq a \\ &= 0 \text{ if } a < X_0 \leq b \\ &= +1 \text{ if } b < X_0 \end{aligned}$$

(It can easily be shown that the final estimators do not depend on the values taken by Z, provided these are equidistant.)
Let $\ddot{\underset{\sim}{\sigma}} = \mathrm{cov}(Z,\ddot{X})$ be the covariances for the $(Z,\ddot{X})$-distribution. The relation between $\ddot{\underset{\sim}{\sigma}}$ and $\ddot{\sigma}$ is straightforward,

$$\ddot{\underset{\sim}{\sigma}} = [f_G(a) + f_G(b)]\ddot{\sigma} \quad (1)$$

where $f_G(.)$ is the standard normal probability density function.

Any random sample of size n drawn from Π can thus be written in the form, $\{(Z_i,\ddot{X}_i),\ i=1,n\}$, where $Z_i = -1, 0$ or $+1$ if the individual belongs to Π_1, Π_2 or Π_3.

It follows immediately that the estimators of a and b are,

$$\hat{a} = Q_G(\frac{n_1}{n})$$

$$\hat{b} = Q_G(1 - \frac{n_3}{n}),$$

where $Q_G(.)$ is the inverse of the standard normal cumulative distribution function.

The vector of arithmetic means $\bar{\ddot{X}}$ and the sample covariance matrix $\ddot{S}$ will be taken as the usual ML estimators of $\ddot{\mu}$ and $\ddot{\Sigma}$, based on the $\ddot{X}_i$ observations.

If we let $L(\bar{\ddot{X}})$ be the following linear compound of the arithmetic means of $\ddot{X}$ in the three populations, i.e.

$$L(\bar{\ddot{X}}) = -n_1(n_2+2n_3)\bar{\ddot{X}}_{(1)} + n_2(n_1-n_3)\bar{\ddot{X}}_{(2)} + n_3(2n_1+n_2)\bar{\ddot{X}}_{(3)} \ ,$$

then,

$$\hat{\ddot{\underset{\sim}{\sigma}}} = n^{-2}\ L(\bar{\ddot{X}})$$

Relation (1) gives finally the estimator for $\ddot{\sigma}$,

$$\hat{\ddot{\sigma}} = \{n^2\ [f_G(\hat{a}) + f_G(\hat{b})]\}^{-1}\ L(\bar{\ddot{X}})$$

Estimators of the multiple correlation coefficient $R = (\ddot{\sigma}'\ \ddot{\Sigma}^{-1}\ \ddot{\sigma})^{\frac{1}{2}}$ and of the correlation vector $\ddot{\rho} = \mathrm{corr}(X_0,\ddot{X})$ follow immediately

$$\hat{R} = \{n^2\ [f_G(\hat{a}) + f_G(\hat{b})]\}^{-1}\ [L(\bar{\ddot{X}})'\ \ddot{S}^{-1}\ L(\bar{\ddot{X}})]^{\frac{1}{2}}$$

$$\hat{\ddot{\rho}} = \{n^2\ [f_G(\hat{a}) + f_G(\hat{b})]\}^{-1}\ L(\bar{\ddot{X}})\ \ddot{D}_{\ddot{S}}^{-\frac{1}{2}} \ ,$$

where $\ddot{D}_{\ddot{S}}^{-\frac{1}{2}}$ is the diagonal matrix, whose element $(\ddot{D}_{\ddot{S}}^{-\frac{1}{2}})_{ii} = (\ddot{S}_{ii})^{-\frac{1}{2}}$.

It is clear that, except for $R = 0$, these estimators are continuous functions of sample moments, differentiable in the neighbourhood of their expectations and, following a theorem by Cramér (1946), have asymptotically normal distribution.

Remark : it sometimes happens that one has some knowledge of the relative proportions (p_1, p_2, p_3) of the three populations in Π, and therefore of a and b. It follows that, $a = Q_G(p_1)$ and $b = Q_G(1-p_3)$, and if we let,

$$\mathcal{L}(\bar{\ddot{X}}) = -p_1(p_2+2p_3)\bar{\ddot{X}}_{(1)} + p_2(p_1-p_3)\bar{\ddot{X}}_{(2)} + p_3(2p_1+p_2)\bar{\ddot{X}}_{(3)}, \text{ then}$$

$$\hat{\ddot{\sigma}} = [f_G(a) + f_G(b)]^{-1} \mathcal{L}(\bar{\ddot{X}})$$

$$\hat{R} = [f_G(a) + f_G(b)]^{-1} [\mathcal{L}(\bar{\ddot{X}})' \ddot{S}^{-1} \mathcal{L}(\bar{\ddot{X}})]^{\frac{1}{2}}$$

MEDICAL APPLICATION

Plomteux et al. (1975) showed that the biochemical differenciation between medical jaundice, surgical jaundice without metastases and secondary liver cancer is of greatest importance in the diagnosis of liver diseases. The usual classification model, assuming multinormality of the three populations and common dispersion matrix, does not always give satisfaction, as the underlying assumptions are generally not satisfied. Therefore the above three groups model was proposed as an alternative to the latter. Moreover the assumption of the existence of a variable X_o, closely associated with the state of the liver and taking increasing values as one goes from the first to the third group, is plausible and a priori not senseless.

The statistical material consists of 110 patients, divided as follows :

56 medical jaundices
28 surgical jaundices without metastases
26 secondary liver cancers

A biological profile of 15 tests of the liver function was performed on each subject, as shown in Table I.
The relative proportions of the three groups in the mixture were assumed to be known in the ratio 0.60 : 0.25 : 0.15, thus giving $a = 0.2533$ and $b = 1.0364$. In order to satisfy normality assumptions, nearly all variables were transformed by a logarithmic (ln) or a square root (sqrt) transformation.
The most important results are summarized in Table I, which gives the arithmetic means of each test in the three groups, the estimated covariances and correlations of X_o with $X_1 \ldots X_p$ and the coefficients of the linear criteria u. The multiple correlation coefficient was found to be equal to 0.9952, a very significant value, and the three probability curves were plotted as shown in Fig. I.
The linear criteria u and the associated $\ddot{X}$-conditional probabilities were recalculated for each patient in the sample.
If P denotes the individual's probability to belong to his own group obtained by the criteria u, Table II gives for each group the number of patients, for which $P > 0.5$, $0.2 < P \leq 0.5$ and $P \leq 0.2$. This shows that nearly 87.3% of the patients are correctly classified by the linear function u and that 10% are definitely misclassified. We finally conclude that Tallis' classification model for three groups works quite satisfactory in this case and that calculation of a simple linear criteria for a patient can give valuable information about his illness.

TABLE I : Statistical results

Biochemical tests with unities		Transf.	$\bar{\ddot{X}}_{(1)}$	$\bar{\ddot{X}}_{(2)}$	$\bar{\ddot{X}}_{(3)}$	$\hat{\ddot{\sigma}}$	$\hat{\ddot{\rho}}$	$\hat{\ddot{\sigma}}'\ddot{S}^{-1}/\hat{R}$
1-Total Bil.	mg/l	ln	3.86	2.99	2.17	-0.7533	-0.5694	-0.0912
2-Alk.Phosph.	MUI	ln	5.35	5.71	5.79	0.2210	0.2755	0.2072
3-5.Nucleo.	MUI	ln	3.62	3.61	4.07	0.1563	0.1792	-0.1235
4-Thymol test	UML	ln	1.82	0.33	-0.55	-1.1029	-0.8726	-0.3192
5-Albumin	g/l	-	38.46	39.16	32.26	-2.0503	-0.2544	-0.0154
6-α_1-Globulin	g/l	sqrt	1.57	1.61	1.84	0.0974	0.3916	-0.1622
7-α_2-Globulin	g/l	sqrt	2.30	2.70	3.01	0.3214	0.7425	0.7375
8-γ-Globulin	g/l	ln	2.76	2.54	2.41	-0.1628	-0.4552	-0.3015
9-IGA	mg%	ln	5.69	5.73	5.71	0.0165	0.0340	-0.0455
10-IGG	mg%	ln	7.42	7.21	7.07	-0.1629	-0.3977	0.1089
11-IGM	mg%	ln	5.50	5.03	4.77	-0.3406	-0.5424	-0.1300
12-LDH iso-V	MUI	sqrt	6.99	3.19	7.57	-0.4837	-0.0911	0.0433
13-SGOT	MUI	ln	5.43	3.75	3.92	-0.8376	-0.6424	-0.0443
14-SGPT	MUI	ln	5.73	3.93	3.72	-1.0332	-0.7001	-0.1267
15-OCT	MUI	ln	6.19	4.71	5.12	-0.6462	-0.4540	-0.1058

$\hat{R} = 0.9952$

TABLE II : Performance of the method

	$0.5 < P$	$0.2 < P \leq 0.5$	$P \leq 0.2$	
Group 1	55	0	1	56
Group 2	23	2	3	28
Group 3	18	1	7	26
	96	3	11	110
	87.3%	2.7%	10%	

REFERENCES

Cramér, H. (1946), Mathematical Methods of Statistics. (Princeton University Press).

Fisher, R.A. (1936). The use of multiple measurements in taxonomic problems. Ann. Eugenics 7, 179-188.

Hannan, J.F., and Tate, R.F. (1965). Estimation of the parameters for a multivariate normal distribution when one variable is dichotomized. Biometrika, 52, 664-668.

Lachenbruch, P.A. (1975). Zero-mean difference discrimination and the absolute linear discriminant function. Biometrika, 62, 397-401.

Lachenbruch, P.A., Sneeringer, C., and Revo, L.T. (1973). Robustness of the linear and quadratric discriminant functions to certain types of non-normality. Comm. in Statist. 1(1), 39-56.

Plomteux, G., Toulet, J., Albert, A.I., and Amrani, N. (1975). Traitement statistique des données biochimiques par la méthode d'analyse discriminante. Sélection des variables biochimiques discriminantes. Ann. Biol. Clin., 33, 411-422.

Smith, C.A.B. (1946). Some examples of discrimination. Ann. Eugenics, 13, 272-282.

Tallis, G.M., Leppard, P. and Sarfaty, G. (1975). A general classification model with specific application to response to adrenalectomy in women with breast cancer. Computers and biomedical Research, 8, 1-17.

Recent Developments in Statistics
J.R.Barra et al., editors

CLUSTERING TIME VARYING DATA

Antonio Bellacicco
Istituto di Calcolo delle Probabilità e Istituto di Statistica
Facoltà di Scienze Statistiche-Università di Roma

The aim of this paper is to enlarge the usual domain of cluster analysis to the case of samples from a multivariate stationary stochastic process and to discuss the consequences of a linear compression of the data, both in the time domain and in the frequency domain, for some basic assumptions in the area of cluster analysis.

CUBIC DATA

The following case is very common in data analysis: a set of N units $S=\{S_1, S_2, \ldots, S_N\}$, a set of M characters or predicates $X=\{X_1, X_2, \ldots, X_M\}$ and a set $T=\{t_1, t_2, \ldots, t_T\}$ of instants of time.

We can arrange the whole set of data in a cubic matrix (S,X,T) whose generic element x_{ijh} for $i=1,2,\ldots,N$, $j=1,2,\ldots,M$ and $h=1,2,\ldots,T$ can be called a cubic data because of its three indices i, j, h. In this paper we are interested in the enlargement of the usual domain of application of cluster analysis techniques, that is to pass from plane matrices, that is the usual matrices, to cubic matrices like (S,X,T). Moreover we will consider the dimension T as a time dimension and the set T as the set of the first T natural numbers with the usual mathematical properties of this set. Finally we will consider both the case of T, very short as a sequence, namely $T \leq 50$, and the case of T, long enough to allow the use of spectral techniques.

To summarise this paper, we will treat the following topics:

i) multiple time series modelling, that is optimal linear combination of M time series which is able to embody the most relevant features of each time series;

ii) spectral analysis of the "latent" time series Y, optimal linear combination of the M time series $\{X_1, X_2, \ldots, X_M\}$ and determination of a "generative" law $\mathcal{L}(Y)$, of autoregressive type;

iii) cluster analysis of the plane matrix whose generic element y_{ih} is just the determination under the law $\mathcal{L}(Y)$ on the element S_i of the character Y at the time t_h.

As a matter of fact, the analysis of cubic data can be inbedded in the general framework of the so called "data compression" problem, which can be considered actually, the main problem in data analysis. An interesting survey on this topic can be considered Bouroche (1972).

In the case of plane data, that is data with two indices, the data compression problem can be treated in the mathematical framework of normed linear spaces. As is well known, the compression is obtained through an optimal linear combination of the original system of variables by maximizing the variance "explained" of the linear combination itself. Unfortunately, we do not have at our disposal a general mathematical theory for cubic data as we have for plane data. Actually, we have at our disposal both data compression techniques for plane data matrices, like (S,X), namely the principal components techniques, and data compression tecniques for plane matrices like (X,T), that is multivariate spectral analysis and autoregressive vector schemes.

In order to simplify the subsequent discussion let us introduce the following notation. Let $(S,X;\ t_h)$ be the plane matrix, namely the so called object-predicate table, at time t_h, which can be represented by the matrix

$$\begin{array}{c|c} & X_1, X_2, \ldots, X_j, \ldots, X_M \\ \hline S_1 & \vdots \\ S_2 & \vdots \\ \vdots & \vdots \\ S_i & \cdots\cdots\cdots\, x_{ij}(h) \\ \vdots & \\ S_N & \end{array}$$

whose generic element $x_{ij}(h)$ represents the determination of variable X_j on the unit S_i at the instant t_h. By $(S,T;\ X_j)$ we denote the matrix, in short O-T table

$$\begin{array}{c|c} & t_1, t_2, \ldots, t_h, \ldots, t_T \\ \hline S_1 & \vdots \\ S_2 & \vdots \\ \vdots & \vdots \\ S_i & \cdots\cdots\cdots\, x_{ih}(j) \\ \vdots & \\ S_N & \end{array}$$

whose generic element represents the determination of the variable X_j at the instant t_h on the unit S_i and the whole table represents the T determinations of the variable X_j on the set S. Let us call the matrix $(S,T;X_j)$ an "object-time" table, for j= 1,2,..., M. The last case will represent the determination on a given unit S_i of the set of determinations of the variables $\{X_1, X_2, \ldots, X_M\}$ along T instants of time in the following matrix, which will be called "variable-time" table.

$$\begin{array}{c|c} & t_1, t_2, \ldots, t_h, \ldots, t_T \\ \hline X_1 & \vdots \\ X_2 & \vdots \\ \vdots & \vdots \\ X_j & \cdots\cdots\cdots\, x_{jh}(i) \\ \vdots & \\ X_M & \end{array}$$

We will denote by $(X,T;S_i)$ the X-T table which can be obtained as a projection of the cubic matrix (S,X,T) over a plane cutting the S axis at S_i. In the same way it is possible to obtain the plane matrices $(S,X;t_h)$ and $(S,T;X_j)$ from the cubic matrix (S,X,T).

CLUSTER ANALYSIS ON PLANE DATA

Cluster analysis deals with the following problem: on a given object-predicate table like $(S,X;t_h)$ we introduce a partition of the units of S such that the variability inside each class of the partition is minimum and at the same time the distance among the classes themselves is maximum. We do not discuss in this paper the mathematical foundation of cluster analysis, that is the formal theory of optimal partitioning, which was actually treated from a pure topological point of view in a previous paper by Bellacicco & Labella,(1976). We simply remark in this paper that cluster analysis deals mainly with plane data, namely with object-predicate tables like $(S,X;t_h)$ for a given h. Moreover we have to remark from a theoretical point of view that we can still consider both the hierarchical clustering procedures and the clustering procedures with partially overlapping classes, as special cases of optimal partitioning problems and therefore we will not consider them at all in this paper. On the other side we will consider the case of "object-time" tables as the aim of this paper.

Actually, cluster analysis on cubic matrices has to deal with data with three indices and therefore the main effort for a cluster analyst is to reduce the set of indices, through a suitable compression of the information, just "smashing" the cubic matrix on a plane matrix like the object-predicate table. Unfortunately, we cannot consider the three dimensions of the cubic matrix just in the same way. We have to recall that the T axis has some intrinsic properties like the ordering property of the natural numbers which cannot be found in the other dimensions. This property cannot be lost and therefore the compression has to take into account this fact. On the other side we can consider the units belonging to the set S, and in the same way the variables belonging to X, as elements of a generic set without any definite mathematical structure. As a final remark, we consider the fact that we can save other relevant features in the cubic matrix (S,X,T). In any case, we will develop in this paper the case previously mentioned, that is we will compress the cubic matrix in an "object-time" table which will save the ordering property of the set T.

In the next paragraph we will consider the mathematical structure of the set T.

THE T DIMENSION

As a matter of fact, the T dimension plays a central role in the analysis of cubic data, and it is necessary to take into account its mathematical structure. Actually, the elements belonging to the S and to the X are labelled by a set of indices like "1,2,...", which simply denote the elements itself. The T dimension has the usual meaning of the time dimension, which means at least a way of ordering. For each instant t_h we can consider an actual "cluster structure", that is an optimal partition of S on the corresponding object-predicate table $(S,X;t_h)$ and therefore to cluster time varying object-predicate tables means to analyse the "evolution" of the units of S; that is the transitions from one class to another.

The lack of any "transition rule", that is any law which "explains" the evolution of each unit of S through the classes of the partitions in the long run, compels us to look for a new way of dealing with the T dimension. Actually, a statistical rule can be assumed, that is, it is possible to consider time varying cluster structures as a markovian process, provided it is reasonable to assume the "existence" of the classes and moreover the possibility of identifying the classes themselves.

The previous assumptions are not so easy to accept from the epistemological point of view and therefore it is more convenient to look for other approaches.

A simple way of dealing with the T dimension is just to consider each S_i, for i=1,

2,...,N, as a path in the space $(X,T;S_i)$ and to cluster directly just the paths themselves(°).

As we will see in the coming paragraphs it is possible to introduce indices of similarity between paths, by distinguishing between the case of T "not too long" and the case of T "long enough" as we will see later.

In case T "not too long", T is simply represented by the mathematical structure of the first k natural numbers, that is 1,2,...,k which behave just as a pure ordering structure, that is for each S_i we can consider the path $\{x_1,x_2,\ldots,x_k\}$ through the mapping

$$S_i : X \longrightarrow T$$

where $T = \{1,2,\ldots,k\}$. It is easy to see that the only relevant feature of each path which is possible to detect, when k is not too large, is actually the so called <u>trend</u>, that is the fundamental behaviour of the path in the long run.

When T is "long enough" it is possible to represent the set of all the possible paths for each S_i by the structure of Hilbert space. In other words, Parzen has shown that it is possible to "represent" in the mathematical sense, the random function $\{X(t),\ t \in T\}$, for each S_i by the Hilbert space structure if and only if the inner product can be represented by the so called <u>convariance kernel</u>, that is

$$< X(s),X(t)>_H = R(s,t) = E\left[X(s),X(t)\right]$$

where X(s) and X(t), for $s,t \in T$, are elements of the Hilbert space H, the brackets <.,.> represent the inner product in H, R(s,t) is the well known covariance operator and E is the expected value operator,(1961).The Hilbert space H, as is well known, is a complete inner product vector space and the T dimension has to be considered, from a theoretical point of view, as a continuous set which can be represented by the set of real numbers. We will see later how to take advantage of this representation in order to compare all the paths which are sequences of elements of the space H. The way to do this will be therefore the actual construction of the covariance kernel which will be accomplished by the crosspectra operator, which is an hermitian operator.

THE SET T IS DISCRETE AND "NOT TOO LONG"

As pointed out earlier in this paper, a clustering procedure is based on the choice of an index of variability among the units of the set S. As a matter of fact, there are two types of indices of variability, that is the indices based on the average of the "distances" between each unit of a given cluster and his centroid and the indices which are based on the average of the "distances", or "similarities" between couples of units in each cluster. We will not discuss the concept of distance and the concept of similarity, because these concepts are basic concepts in cluster analysis. We only remark that generally speaking each index of distance, or each index of similarity, is computed variable by variable, that is independently in each column of the object-predicate table; see Bock (1974).

In the previous paragraph we discussed the problem of the T dimension just from a pure mathematical point of view without considering any relation to cluster analysis. We remark now that dealing with an object-time table we can consider the choice of an index of variability as the choice of a suitable weighting of the columns;

(°) From the practical point of view we will denote by X both the set of variables $\{X_1,X_2,\ldots,X_M\}$ and one single variable which is supposed to represent the whole set.

the weights will represent the temporal importance of the O-T table. Actually, the most recent informations are more or less important following a subjective choice of the researcher. Therefore the importance of the information can be measured through a suitable weighting of the corresponding columns, which has to be introduced in the formula of the indices of variability. In the case of T discrete and represented by the first k natural numbers, 1,2,...,k, for k ≤ 50, there is a "natural" weighting, as it is possible to show, from Bellacicco (1975), that is

$$(1-1/k),\ (1-2/k),\ (1-3/k),\ldots,(1-h/k),\ldots,(1-k/k) \qquad (1)$$

which gives a lot of weight to the past and just nothing to the present information. A weight like (1) was introduced by us just for measuring the trend of a short sequence through the formula

$$B(k,0)= 1 - \frac{2 \sum_{h=1}^{k}(k-h)x_h}{k\ (S - \bar{x})} \qquad (2)$$

where k is the lenght of the sequence and 0 is the actual minumum element of the sequence itself, $\{x_i\}$, S is the sum $\Sigma\ x_h$ and $\bar{x}$ is the mean.

As a binary index of similarity between two time sequences we can consider a suitable application of this formula, just considering the sequence

$$x_1= \left|y_{i1}-y_{r1}\right|,\ x_2= \left|y_{i2}-y_{r2}\right|,\ldots,\ x_k= \left|y_{ik}-y_{rk}\right|$$

where by y_{ih} and y_{rh} we denote two generic elements of the object-time table, at time t_h, corresponding to the rows S_i and S_r. The index of similarity between two sequences becomes

$$B(S_i,\ S_r;\ k)= \left|1 - \frac{2 \sum_{h=1}^{k} (k-h)x_h}{(k-1)\Sigma\ x_h}\right| \qquad (3)$$

The index of trend presence in a sequence ranges from -1 to +1, while $B(S_i,S_r;k)$ ranges only from 0 to 1; 0 occurs when $y_{ih}= y_{rh}$, for every h= 1,2,...,k and 1 occurs when $y_{ih}= y_{r,k-h+1}$ or $y_{i,k-h+1}= y_{ih}$, that is when the two sequences are symmetrically covariant. As a name for $B(S_i,S_r;k)$ we propose <u>cotrend index</u> for two time series. As mentioned earlier, we do not discuss in this paper any sampling assumption; in any case from a pure descriptive point of view, we can generalize both and the way of measuring the simple distances

$$d_{ir}^{(j)}= |y_{ij} - y_{rj}|$$

We do not go further on this topic, just limiting ourselves to the previous remarks. We only note that still in this case the treatement of an object-time table implies the choice of a weighting procedure.

THE SET T IS DISCRETE AND "LONG ENOUGH"

In the case in which the set T of instances is "long enough" to allow the use of spectral techniques we can try to compress the cubic matrix (S,X,T) in order to obtain an object-time table (S,T; Y) where Y is an "optimal" linear combination of the set of variables $\{X_1,X_2,\ldots X_M\}$ which is able to save the most relevant features of each variable. As it is well known, in case the T dimension is not at all considered, that is all the information is instantaneous, it is possible to compress the set of variables, X, through an optimal linear combination of the variables themselves which maximize the explained variance and therefore the variance is the relevant information to save.

On the other side, in the case in which T has to be considered, we can compress the variables by considering them as finite realizations of a stationary multivariate stochastic process. Therefore, the relevant features to save by the compression are the cycle structures of the M time series. It is easy to see that in case T "not too long" the only relevant feature of each time series is the trend, while in case T "long enough" it is possible to consider other hidden cycles than the trend, which has to be removed by a suitable filtering procedure. We will see later that the cycle structure of a time series is just a variance structure and in some way it is possible to consider principal component techniques as valid techniques for getting the desired compression.

Let $\underline{X}^T = \{X_1(t), X_2(t), \ldots, X_M(t)\}$ a set of M finite realizations, of lenght T, from a stationary multivariate stochastic process X, where stationary has to be understood in the covariance sense, and let $\underline{V}(u) = E\left[(X(t)-\underline{m})(X(t+u)-\underline{m}')\right]$ the crosscovariance matrix

$$\begin{bmatrix} V_{11}(u), & V_{12}, \ldots, & V_{1M}(u) \\ \vdots & \vdots & \vdots \\ V_{M1}(u), & V_{M2}, \ldots, & V_{MM}(u) \end{bmatrix} \quad (4)$$

where $\underline{m}$ is the vector of M means, which are supposed time invariant, the symbol " ' " is the transposition operator for vectors, E means expected value, "u" is the "lag", $V_{jj}(u)$ is the autocovariance of the time series $X_j(t)$ and $V_{rs}(u)$ is the crosscovariance between $X_r(t)$ and $X_s(t)$.

Matrix $\underline{V}(u)$ is not a symmetric matrix, while it is true that $\underline{V}'(u) = \underline{V}(-u)$, as it is easy to see, by Jenkins & Watts (1969).

Let us consider now the well known Fourier transforms which relate the autocovariance and crosscovariance matrix to their spectral representation, namely

$$f_{rs}(\lambda) = \int_{-\infty}^{+\infty} V_{rs}(u)\, e^{-i\, 2\pi\lambda u}\, du \quad \text{for } i = \sqrt{-1}$$

$$V_{rs}(u) = \int_{-\infty}^{+\infty} f_{rs}(\lambda)\, e^{i\, 2\pi\lambda u}\, d\lambda$$

where $f_{rs}(\lambda)$ is the spectral function at the frequency λ.

The matrix $F(\lambda) = \{f_{rs}(\lambda)\}$ for r,s= 1,2,...,M, is called the autospectra and crosspectra matrix; $F(\lambda)$ is actually an hermitian positive semidefinite matrix.

Let us consider now the linear combination

$$Y(t) = \theta_1^\circ X_1(t) + \theta_2^\circ X_2(t) + \ldots + \theta_M^\circ X_M(t) \quad (5)$$

where θ_j°, j=1,2,...,M, is a set of M complex coefficients, which have to be obtained in a way such that the explained variance of Y(t), namely $\underline{V}_{yy}$ is maximum. The solution to this problem can be still obtained through principal component techniques, after suitable modifications which take into account the presence of the tem poral dimension T.

Let

$$\underline{V}_{yy}(u) = E\left[(Y(t) - \underline{m}_y)(Y(t+u) - \underline{m}_y)^\circ\right]$$

$$= E\left[(\theta^{\circ\prime}(X(t)-\underline{m})\ (X(t+u) - \underline{m})^{\circ\prime}\ \theta\right] \qquad (6)$$

$$= E\left[\theta^{\circ\prime}\ \underline{V}\ (u)\ \theta\right]$$

where by $\theta^{\circ\prime}$ we denote the transpose of the complex vector θ°.

We can consider now the Fourier transform of the previous quadratic from (6), just obtaining

$$F_{yy}(\lambda) = \theta^{\circ\prime}\ F(\lambda)\ \theta \qquad (7)$$

Actually, $F_{yy}(\lambda)$ is a true variance for each λ namely the variance of a linear combination. The problem to face is therefore the choice of the complex elements of the vector θ° which maximizes $F_{yy}(\lambda)$, for every λ.

It is useful to observe that θ° does not depend on λ and on the other side for each λ we can estimate the contribution of each $X_s(t)$, for $s= 1,2,\dots,M$, to the general variance.

We can write therefore,

$$\max F_{yy}(\lambda) = \max \theta^{\circ\prime}\ F(\lambda)\theta$$

under the usual normalization condition $\theta^{\circ\prime}\ \theta= 1$. It is known that $F(\lambda)$ is an unitary hermitian operator which possesses a set of non negative eigenvalues, which can be ordered according to their magnitude. Therefore we can select the largest eigenvalue and its corresponding eigenvector, which is actually the vector θ°.

We have indeed

$$\emptyset\ (\theta^\circ) = \frac{\partial}{\partial\theta}\left[F_{yy}(\lambda) + a_1\ (1 - \theta^{\circ\prime}\theta)\right] \qquad (8)$$

where a_1 is the Lagrange multiplier, ∂ is the partial derivative operator and $\underline{I}$ is the unit vector. Letting $\emptyset(\theta^\circ)= 0$, we get a_1 as the first eigenvalue and $\theta^\circ(a_1)$ the corresponding eigenvector which is actually the vector of the coefficient in the linear equation (5). We can easily see that a_1 is actually the variance var $Y(t)$ to maximize. We have indeed

$$a(\lambda)= \theta^{\circ\prime}\ F(\lambda)\theta= F_{yy}(\lambda)= \text{var } Y(t)$$

for each frequency λ, and therefore for every λ we can construct the linear combination of the random functions $\{X(t),\ t\varepsilon T\}$, which maximizes the variance.

As a matter of fact, we can use the power spectrum $F_{yy}(\lambda)$ either directly for each S_i as a summarised description of the set of time series associated with S_i or indirectly for the construction in the time domain of a "law" of autoregressive type, $\mathcal{L}\,|Y(t)|$, that is

$$Y(t)= b_1 Y(t-1) + b_2 Y(t-2) + \ldots + Z(t) \tag{9}$$

where $(b_1, b_2, \ldots)$ are the coefficients of the autoregressive scheme, and $Z(t)$ is a pure random process with definite statistical properties. It is easy to see that the presence of a "lag" in (9) corresponds to the presence of a frequency in the power spectrum $F_{yy}(\lambda)$ and viceversa. Therefore, the amount of information carried by the spectral representation $F_{yy}(\lambda)$ and by the equation (9) is exactly the same. In any case we can assume the existence of a stochastic law which approximates in the statistical sense the data at disposal. We do not discuss further the technical details both of the statistical estimation of the spectrum $F_{yy}(\lambda)$ and of the statistical estimation of the coefficients of the equation (9). All these topics are already well developed in Hillinger (1969) and Jenkins & Watts (1969).

We simply recall that at each frequency λ we have actually pooled the contribution to the power spectrum of $Y(t)$ of the spectral densities of the random functions $\{X(t),\ t\varepsilon T\}$, and this contibution is a variance contribution. It is easy to show that the spectral component at the frequency λ of each random function $X_j(t),\ t\varepsilon T$, for $j= 1,2,\ldots,M$ is just a variance, which is actually the weight of the corresponding oscillating component of the process $X_j(t)$ itself.

CLUSTERING O-T TABLES UNDER A LAW, $\mathcal{L}$.

In the previous paragraphs we tried to set up a way of taking advantage of the dimension T in cluster analysis. Actually, we can deal with two types of object-time tables that is $(S, F_{yy}(\lambda))$ and $(S, \mathcal{L}|Y(t)|)$, where by $\mathcal{L}|Y(t)|$ we mean the autoregressive scheme which can be estimated by the power spectrum $F_{yy}(\lambda)$.

In case we have to do with tables like $(S, F_{yy}(\lambda))$, the main problem is to introduce a suitable metric which can measure the distance or at least the similarity between the elements of S, that is the rows of the matrix. As a matter of fact, the so called χ^2 metric is the best way to handle the table. We have to compare the weight of the hidden oscillations of each time series and therefore the χ^2 metric is able to take into account the relative relevance of each column in the table. A methodological approach fully developed on the basis of the χ^2 metric is due to Benzécri who considers the table as a column (row) vector space; Benzécri (1973).

The case we are interested in here is the case of an object-time table like $(S, \mathcal{L}|Y(t)|)$ where the columns $Y(t)$ are generated by a law, namely an autoregressive scheme.

In the previous paragraph 4 we noted that the absence of any transition rule does not allow the use of a sequential cluster analysis. On the other side, in case we dispose of a a rule like a law $\mathcal{L}$ it is reasonable to consider time varying cluster stuctures. In other words, for each instant of time t_h there is a cluster structure, that is an optimal partition of the units of S, and the set of cluster structures are connected by the law $\mathcal{L}$. As a consequence, from a pure theoretical point of view, it is enough to consider only a cluster structure at a given instant of time t_h just because it is possible to "prolong " the cluster structure to the subsequent instants, by $\mathcal{L}$. In this case we can speak of "cluster forecasting" under a statistical law $\mathcal{L}$.

The main problem is the following: Can the sequential cluster structure under $\mathcal{L}$ still be considered under the label "optimal", just in the same way as we spoke of "optimal partition" for the instantaneous cluster structure?

Actually, it is not too difficult to answer the previous question. We have indeed a substitution of object-time tables: $<$ the original table (S,T; X) by the artificial table $(S,T,\mathcal{L}|Y|)>$ where by X we simply denote a specified variable which can represent a set of variables and by $\mathcal{L}|Y|$ we denote the theoretical variable Y under the law $\mathcal{L}$. The similarity between the objects $S_i \in S$ is therefore reduced to the similarity between the laws themselves $\mathcal{L}_i|Y|$. In other words, it is enough to consider a cluster structure at the last instant of time t° because the determination of the variable Y at the end of the sequence indexed by the set T, namely $\{y_{i1}, y_{i2}, \ldots, y_{it°}\}$, for $i = 1, 2, \ldots, N$, is a representative by $\mathcal{L}_i$ of the whole sequence. In short, there is a unique cluster structure, which is the cluster structure at the instant t°, which is the last instant of observation, and that cluster structure is obviously an optimal cluster structure, in the sense of optimal partition of the set S.

As a sketch of the clustering procedure outlined in the previous pages we will summarize briefly a clustering algorithm on a cubic matrix:

i. let (S,X,T) be a cubic matrix;
ii. define for each S_i a law $\mathcal{L}_i|Y(t)|$ such that we obtain a matrix $(S, \mathcal{L}|Y(t)|)$

$$
\begin{array}{c|cccccc}
 & Y(t_1), & Y(t_2), & \ldots, & Y(t_h), & \ldots, & Y(t_T) \\
\hline
S_1 & & & & \vdots & & \\
S_2 & & & & \vdots & & \\
\vdots & & & & \vdots & & \\
S_i & \cdots\cdots\cdots\cdots\cdots\cdots\cdot & y_{ih} & & & & \\
\vdots & & & & & & \\
S_N & & & & & & \\
\end{array}
$$

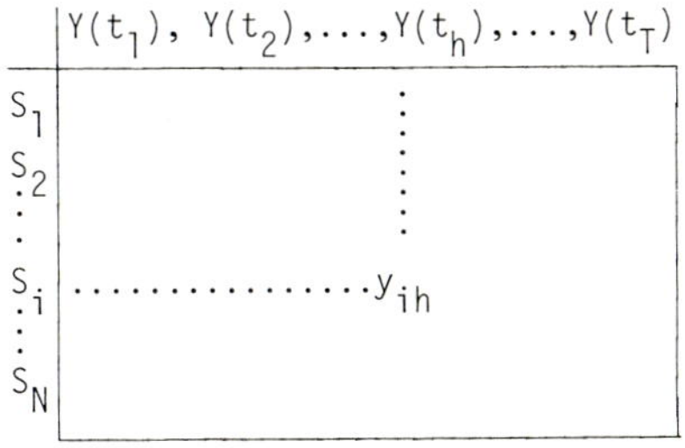

iii. let t° be the instant of time for which the model $\mathcal{L}|Y(t)|$ is identified; actually t° can be an instant of forecasting, namely t° > T or t° can be the last instant of time, namely t° = T, where T is the maximum span of time at disposal;
iv. let $\pi(t°)$ a cluster structure at t°, namely a partition of S, under $\mathcal{L}$, at t°;
v. the family $\{\pi(t), t \in T\}$ is a cluster process by the law $\mathcal{L}(Y)$.

REFERENCES

Bellacicco A. (1975). Metron - Roma, Vol. XXXIII, n.1-2.
Bellacicco A., Labella A. (1976). Rendiconti di matematica, Roma, Vol. 9, VI, 2.
Benzécri J.P. (1973). L'Analyse des Données.(Dunod, Paris).
Bock H.H. (1974). Automatische Klassifikation. (Vandenhoek & Ruprect, Gottingen).
Bouroche J.M. (1972) Metra Internationale, Note de Travail, n. 174.
Hillinger C. (1969) Mathematical Systems Theory and Economics (H.W. Kuhn & G.P. Szegö editors, Springer-Verlag, Berlin).
Jenkins G.M., Watts D.S. (1969) Spectral Analysis and its Applications (Holden-Day, S. Francisco).
Parzen E. (1961). Annals of Mathematical Statistics, 32.

Recent Developments in Statistics
J.R.Barra et al., editors

SOME METHODS OF QUALITATIVE DATA ANALYSIS.

J.M. BOUROCHE
C O R E F

G. SAPORTA
I.U.T. PARIS V

M. TENENHAUS
C E S A

Most methods of data analysis have been conceived for numerical data (factor analysis, canonical analysis...), sometimes for both numerical and nominal data (analysis of variance and covariance, discriminant analysis). Other methods (multidimensional scaling, preference analysis, isotonic regression) enable the treatment of ordinal data but in a very restrictive and particular context.

Various statisticians have already proposed some methods for a mix of nominal and ordinal data e.g: Benzecri [1], Bock [3], Bouroche, Saporta and Tenenhaus [4], Carroll [5], Kruskal [8] [9], Lebart [10], de Leeuw [11], Masson [12], Nishisato [13], Pagès [14], Saporta [15], Tenenhaus [18], Young [19].

In the C.O.R.E.F-D.G.R.S.T. n° 75 07 0230 project the authors of this paper have attempted to achieve the following purposes :

- to complete existing syntheses
- to propose new methods
- to apply methods of analysis of qualitative data on real data to make sure of the value of results.

We present here a first synthesis of this project.

1.- DATA AND PROBLEMS

Before analysing a set of data, two questions arise :

- What is the nature of the data ?
- What are the problems to solve ?

The choice of a method derives from the answers to these questions.

1.1. Nature of data.

A variable is nominal if the set of its categories is finite and has no ordinal structure.

A variable is ordinal if the set of its categories is finite with an ordinal structure.

A variable is numerical if it takes its values in $\mathbb{R}$.

Moreover F.W. Young [19] defines the concept of "underlying (or generating) process" which can be either discrete or continuous.

So we have six types of measurements : for instance a variable will be continuous-ordinal if the ordered set of its categories represents a continuous underlying process.

In a first stage, we have to determine exactly the nature of each variable. Even if a variable generally belongs to only one class, it may be interesting to modify arbitrarily its nature.

For instance we may consider a discrete ordinal variable as either discrete nominal (we neglect the ordinal structure) or continuous ordinal or discrete-ordinal.

1.2. Problems.

There are two kinds of problems. Either we have few information about the set of data and we are looking for a description, or we need a prediction of one (or more) specified variable through the others.

In the first case we shall use factor analysis or clustering techniques, in the second case methods derived from least squares and canonical analysis.

2.- METHODS

According to the nature of variables and to the problem, different methods are available. Before quoting a few of them in § 2.2., we present briefly the principle of optimal scaling.

2.1. Quantification of qualitative data.

Roughly speaking, the question is to allot a numerical value to each category of a discrete nominal or ordinal variable. If the variable is ordinal we require that the order of the numerical values represent the order of the categories.

Let E be the set of individuals, X a nominal variable. X is a mapping from E to $\mathfrak{X}$, set of categories. Let δ be a mapping from $\mathfrak{X}$ to $\mathbb{R}$. The variable $\delta_\circ X$ is then numerical : it is a quantification (scaling) of X. As there is an infinite number of scalings, the choice of an optimal scaling is actually function of a criterion related to the method (cf. § 3).

If the variable X is continuous nominal or continuous ordinal, an interval of $\mathbb{R}$ corresponds to a category of X and the scaled observations are required to fall in the interval but have not necessarily to be equal. If X is ordinal, we require that the order of intervals corresponds to the order of categories, (cf. F.W. Young [19]).

2.2. Choice of methods.

We shortly quote some methods and their authors. For further details see the reference list or the final C.O.R.E.F report.

2.2.1. Descriptive methods.

We deal only with methods of principal component analysis type.

a). All variables are nominal.
See Bock [3], Bouroche, Saporta, Tenenhaus [4], Lebart [10] , Nishisato [13].

b). Nominal and numerical variables
see Tenenhaus [18] and § 3.

c). Ordinal variables
see Kruskal and Shepard [9].

d). Mix of ordinal, nominal, numerical variables
see Young, de Leeuw, Takane [20], de Leeuw [11].

2.2.2. Predictive methods.

The method proposed by Young, de Leeuw and Takane [21] is the only one available for the general case. Let us point out two particular cases :

a). One nominal dependent variable, all predictors nominal : Carroll [5], Masson [12], Saporta [15].

b). One ordinal dependent variable, all predictors nominal : Kruskal [8].

Let us emphasize on the great flexibility of the Young, de Leeuw and Takane [21] approach : their MORALS/CORALS algorithm allows the treatment of any mix of variables in terms of regression analysis or canonical analysis.

More modestly, we shall now present two methods based upon optimal scaling of nominal data : a predictive method (DISQUAL) and a descriptive one (PRINQUAL).

3.- TWO NEW METHODS.

3.1. DISQUAL : a method and a program for stepwise discriminant analysis of nominal variables [17].

p nominal variables with $m_1, m_2, \dots m_p$ categories are measured on n individuals and we attempt to predict the categories of an outside nominal variable with only k of these p predictors $(k < p)$.

The medhod consists of two relatively separate parts : on the one hand the selection of the predictors, on the other hand the discrimination performed by means of the selected predictors.

3.1.1. Stepwise selection with Escoufier's operators.

According to Escoufier [6] and Pagès [14] we associate to each of the $p+1$ nominal variables the orthogonal projector P_i on the subspace of $\mathbb{R}^n$ of zero-mean variables spanned by the indicator variables of its categories This is the subspace of zero-mean discrete numerical variables obtained by scaling the nominal variable.

Projectors belong to the subset of the vector space of symetrical operators in which we define an inner product and a norm by

$$<P_i \; ; \; P_j> \; = \; \text{Trace}(P_i \, P_j)$$

$$||P_i||^2 = \text{Trace} \; P_i^2.$$

In the case of nominal variables we know that $\text{Trace}\; P_i \, P_j = \Phi^2$ where Φ^2 is the K-Pearson measure of dependance between variables i and j (canonical or correspondence analysis of two nominal variables).

Furthermore $\text{Trace}\; P_i^2 = \text{Trace}\; P_i = m_{i-1}$ and the cosine of the angle between two operators associated to two nominal variables is nothing else than the Tschuprow coefficient

$$T_{ij} = \frac{\phi^2}{\sqrt{(m_{i-1})\ (m_{j-1})}}$$

which is thus equivalent to a correlation coefficient between nominal variables if we identify a variable to its associated projector.

According to the usual geometry of correlation, we may then define the partial Tschuprow coefficient between nominal variables. For three variables i,j,k we have

$$T_{ij/k} = \frac{T_{ij} - T_{ik}\, T_{jk}}{\sqrt{(1-T_{ik}^2)\ (1-T_{jk}^2)}} \quad \text{and so on.}$$

Then stepwise selection method is :

- The first predictor is the one which maximizes Tschuprow's coefficient with the dependent variable

- The second predictor is the one which maximizes the partial Tschuprow coefficient with the dependent variable, given the first predictor.

3.1.2. Discriminant analysis.

k predictors being selected we have now an array of data of the following kind :

$$(A||X_1\ |X_2\ | \ldots\ |X_k)$$

where X_i (and A) are $n \times m_i$ matrices where the columns are the indicators of the categories. The rank of $X = (X_1|X_2\ \ldots\ |X_k)$ is inferior or equal to

$$(\sum_{i=1}^{k} m_i - k),$$

since the sum of the columns of each X_i is the vector $\underline{1}$.

An ordinary discriminant analysis being impossible here, because X'X is not regular, we substitute to X a new matrix of nearly equivalent numerical variables : these new variables are the Guttman principal components of scale of the k nominal variables (which are also the components of Carroll's generalized canonical analysis [4] or of correspondence analysis of X).
Among the $\sum_{i=1}^{k} m_i - k$ principal components, we retain only those which have a sufficient discriminating power.

A discriminant factor analysis performed on the selected components gives the discriminating scores of the nominal variables which we directly use for the classification technique if the dependent variable is binary (Fisher's function) ; otherwise we use a **classical procedure** based on the distances to the centroids.

3.2. PRINQUAL : a method and a program of principal component analysis of a set of nominal and numerical variables [18].

We are looking for a scaling of the nominal variables in order to get the best principal component analysis with factors in the sense of a maximum explained variance.

3.2.1. Method

Let E be the set of individuals, $D_1, D_2, \dots D_k$ nominal variables, $X_1, X_2, \dots X_\ell$ numerical variables. Let $\delta_i \circ D_i$ be the scaled variables. We know that the first m principal components of $\delta_i \circ D_i$ (supposed known) and X_i realize

$$\operatorname*{Max}_{Z_1, Z_2, \dots Z_m} \sum_{i=1}^{k} \sum_{j=1}^{m} \mathrm{cor}^2(\delta_i \circ D_i, Z_j) + \sum_{i=1}^{\ell} \sum_{j=1}^{m} \mathrm{cor}^2(X_i\ Z_j)$$

where the Z_j are uncorrelated.

Thus the principle of the method consists in obtaining the optimal scaling δ_i by maximizing the previous expression both on δ_i and Z_j.

3.2.2. Algorithm.

The algorithm is iterative and maximizes the criterion alternatively on the δ_i and on the Z_j.

The initial solution $Z_j^{(o)}$ is optimally chosen by using generalized canonical analysis [18].

At step t we get the $\delta_i^{(t)}$ by

$$\operatorname*{Max}_{\delta_1, \delta_2, \dots \delta_k} \Sigma\,\Sigma\ \mathrm{cor}^2\,(\delta_i \circ D_i\ ,\ Z_j^{(t-1)}) + \Sigma\,\Sigma\ \mathrm{cor}^2\,(X_i\ ,\ Z_j^{(t-1)})$$

Let $\lambda(t)$ be the value of this maximum.

We have now the $Z_j^{(t)}$ by

$$\operatorname*{Max}_{Z_1, Z_2, \dots Z_m} \Sigma\,\Sigma\ \mathrm{cor}^2\,(\delta_i^{(t)} \circ D_i\ ,\ Z_j) + \Sigma\,\Sigma\ \mathrm{cor}^2\,(X_i, Z_j)$$

Let $\mu(t)$ be the value of this maximum.

It is possible to show that :

$$\lambda(t) \leq \mu(t) \leq \lambda(t+1) \leq k+\ell$$

Thus the algorithm converges and

$$L = \lim_{t \to \infty} \lambda(t) : \lim_{t \to \infty} \mu(t)$$

Let

$$\delta_i^{*} = \lim \delta_i^{(t)}$$

$$Z_j^* = \lim Z_j^{(t)}$$

with $\delta_i \circ D_i$ and Z_j of zero-mean and unit-variance.

3.3. Principal component analysis of the $\delta_i^* \circ D_i$ and X_i

The Z_j^* are the first m principal components.

The part of explained variance is $\frac{L}{k+\ell}$.

We may represent observations and variables as usual in principal conponent analysis.

The categories are represented in $\mathbb{R}^m$ as follows

To the category k of variable D_i we associate the vector :

$$M_{i,k} = \frac{1}{|D_i^{-1}(k)|} \sum_{e \in D_i^{-1}(k)} \begin{pmatrix} Z_1^*(e) \\ Z_2^*(e) \\ \vdots \\ Z_m^*(e) \end{pmatrix}$$

REFERENCES

[1] Benzecri, J.P. (1974). L'analyse des données, Tomes I and II.

[2] Bertier P. & Bouroche J.M. (1975). L'analyse des données multidimensionnelles. (P.U.F.).

[3] Bock, D. (1960). Methods and applications of optimal scaling-Psychometric Laboratory, Report n° 26, University of North Carolina.

[4] Bouroche, J.M., Saporta, G., & Tenenhaus (1975). Generalized canonical analysis of qualitative data. U.S. Japan Seminar on multidimensional scaling and related methods, San Diego.

[5] Carroll, J.D. (1969). Categorical conjoint measurement, Ann Arbor, Michigan, Meeting of Mathematical Psychology.

[6] Escoufier, Y. Le traitement des variables vectorielles, Biometrics 29, p. 751.

[7] Hayashi, C. (1950). On the quantification of qualitative data from the mathematical-statistical point of view. Annals of the Institute of Statistical Mathematics, 2, 35-47.

[8] Kruskal, J.B. (1965). Analysis of Factorial Experiments by Estimating Monotone Transformation of the data, JRSS, Serie B, 27.

[9] Kruskal, J.B. and Shepard, R. (1974). A non metric variety of linear factor analysis, Psychometrika, 39, 123-157.

[10] Lebart, L. (1973). Recherche sur la description automatique des données socio-économiques, Rapport de recherche, CREDOC.

[11] De Leeuw, J. (1976). HOMALS, Spring meeting of the Psychometric Society, Murray Hill.

[12] Masson, M. (1974). Thèse d'Etat, Université de Paris VI.

[13] Nishisato, A.P.S. (1975). Non linear programming approach to optimal scaling of partially ordered categories, Psychometrika, 40, 522-548.

[14] Pagès, J.P. (1974). A propos des opérateurs d'Escoufier, Séminaires de l'I R I A.

[15] Saporta, G. (1975). Liaison entre plusieurs ensembles de variables et codage des données qualitatives, Thèse de 3ème Cycle, Paris VI.

Recent Developments in Statistics
J.R.Barra et al., editors
© North-Holland Publishing Company,(1977)

A LINEAR MODEL FOR PREDICTION IN MULTIDIMENSIONAL CONTINGENCY TABLES -THE PROBLEM OF EMPTY CELLS-

A. CARLIER
Laboratoire de Statistique[(1)]
Université Paul Sabatier
Toulouse - FRANCE

This study is concerned with the prediction of a value for a specified qualitative variable (the predictand) given the values of p other qualitative variables (the predictors) on the same individual. A related concern is with the graphical analysis of the relations between the predictand and the predictors. We consider the case where the associated (p+1) - way contingency table has a great number of cells with respect to the size n of the observed sample (section 6).

As a consequence, most observed cell frequencies are zero or close to zero. Therefore prediction based on direct estimation of conditional probabilities (i.e. the observed frequencies) is impossible or meaningless. We cope with this problem making assumptions of non additive interactions (like in analysis of variance). Under these assumptions, the conditional probability may be expressed linearly (section 3).

A specific linear regression (section 5-2) gives these conditional probabilities (using only some marginal tables of the contingency table). Furthermore, a multiple discriminant analysis of indicators coincides with a correspondence analysis on the whole contingency table (sections 2 and 5-1). We give some properties of possible resulting representations (section 5-1). The case where the conditions are relaxed to some extent is also considered (section 5). A condition of stability of the method under sampling is stated (section 6).

1. THE CANONICAL ANALYSIS

Let H_1 and H_2 be two closed subspaces of a separable real Hilbert space H. We know (see [3]) that the canonical analysis (C.A.) of H_1 and H_2 is obtained through the spectral analysis of P_1+P_2-I, or of $P_1 \circ P_2$, where P_i is the orthogonal projection onto H_i. When this analysis is compact (i.e. when $P_1 \circ P_2$ is a compact operator), the analysis is also given by $(\{\rho_i\}_{i\in I}, \{f_i\}_{i\in I}, \{g_i\}_{i\in I})$ where I is a finite or countable set, $\{\rho_i^2\}_{i\in I}$ is the non-increasing full sequence of the eigenvalues of $P_1 \circ P_{2|H_1}$ (resp. $P_2 \circ P_{1|H_2}$), $\{f_i\}_{i\in I}$ (resp. $\{g_i\}_{i\in I}$) is the associated orthonormal sequence of eigenvectors of $P_1 \circ P_{2|H_1}$ (resp. $P_2 \circ P_{1|H_2}$).

Let (X,Y) be a couple of random variables (r.v.) defined on $(\Omega, \mathcal{A}, P)$ and valued in $(E \times E_0, \mathcal{E} \otimes \mathcal{E}_0)$ and let $\mathcal{B}$ and $\mathcal{C}$ be the complete sub σ-fields of $\mathcal{A}$ induced by X and Y respectively. The C.A. of the closed subspaces $L^2(\mathcal{B})$ (set of the $\mathcal{B}$-measurable r.v.'s for which $|f|^2$ is P-integrable) and $L^2(\mathcal{C})$ of $L^2=L^2(\Omega, \mathcal{A}, P)$ is also called non-linear C.A. of X and Y or C.A. of the σ-fields $\mathcal{B}$ and $\mathcal{C}$. It is obtained through the spectral analysis of $E^{\mathcal{B}} \circ E^{\mathcal{C}}{}_{|L^2(\mathcal{B})}$.

(1) E.R.A. - C.N.R.S. N°591

If E and E_o are finite, so are $\mathcal{B}$ and $\mathcal{C}$. Thus the dimensions of $L^2(\mathcal{B})$ and $L^2(\mathcal{C})$ are finite. $\mathcal{B}$ (respectively $\mathcal{C}$) can be generated from a partition $(B_1,\dots,B_p)$ (resp. $(C_1,\dots,C_q)$) of Ω, and $\{1_{B_i}/i=1,\dots,p\}$ is a basis of $L^2(\mathcal{B})$. The projections $E^{\mathcal{B}}$ and $E^{\mathcal{C}}$ are compact and therefore the C.A. is compact. Considering statistical variables (i.e. defined on a finite set Ω), we obtain the correspondence analysis (Corr. A.) of (X,Y).

Let E (resp. E_o) be $\mathbb{R}^p$ (resp. $\mathbb{R}^q$) and its Borel σ-fields. Let us assume that each component X_i or Y_j of X and Y belongs to $L^2(\Omega,\mathcal{A},P)$. We call linear C.A. of X and Y, the C.A. of $F_X = \text{vect}\{1_\Omega,X_1,\dots,X_p\}$ (subspace spanned by X_i and 1_Ω) and of $F_Y = \text{vect}\{1_\Omega,Y_1,\dots,Y_q\}$. This C.A. is compact. For a finite set Ω, we retrieve the usual C.A.of X and Y ([2]). When $q=1$, we find the linear regression of Y on X. If $E = \mathbb{R}^p$ and if E_o is finite, we may study the proximities between the linear combinations of the X_i and those of the indicators of Y by considering the C.A. of F_X and $L^2(\mathcal{C})$ in L^2. We thus obtain, in the case of statistical variables (Ω is finite), the discriminant analysis of Y by X (see [2]).

2. APPROXIMATION OF A CORRESPONDENCE ANALYSIS

In the sequel, let (X,X_o) be a couple of r.v.'s defined on $(\Omega,\mathcal{A},P)$ and valued in $(E\times E_o, \mathcal{E}\otimes\mathcal{E}_o)$ such that $X=(X_1,\dots,X_p)$, $E = E_1\times\dots\times E_p$, each X_i being valued in the finite set E_i $(i=0,\dots,p)$. Let $\mathcal{B}$ (resp. $\mathcal{B}_i$) be the sub σ-field of $\mathcal{A}$ induced by X (resp. X_i) and $\mathcal{E}$ (resp. $\mathcal{E}_i$) the σ-field of all sets in E (resp. E_i). If $I=\{1,\dots,p\}$, we notice that $\mathcal{B}$ is also the σ-field generated by $\bigcup_{i\in I}\mathcal{B}_i$.
The Corr. A. of (X,X_o) is given through the spectral analysis of $E^{\mathcal{B}_o}\circ E^{\mathcal{B}}{}_{|L^2(\mathcal{B}_o)}$.
Substituting in the C.A. of $L^2(\mathcal{B})$ and $L^2(\mathcal{B}_o)$ a subspace V of $L^2(\mathcal{B})$ (with finite dimension and hence closed) for $L^2(\mathcal{B})$, we get an approximation of this Corr. A. . The C.A. of V and $L^2(\mathcal{B}_o)$ (or discriminant analysis of X_o by the r.v.'s spanning V) which approximate the C.A. of $L^2(\mathcal{B})$ and $L^2(\mathcal{B}_o)$, is given by spectral analysis of $E^{\mathcal{B}_o}\circ\pi_{V|L^2(\mathcal{B}_o)}$, where π_V denotes the orthogonal projection from L^2 on V.

We can show that these C.A. are the same if and only if :

$$\pi_{V|L^2(\mathcal{B}_o)} = E^{\mathcal{B}}{}_{|L^2(\mathcal{B}_o)} \qquad (1)$$

Since V is contained in $L^2(\mathcal{B})$ (1) means that :

$$\forall f\in L^2(\mathcal{B}_o) \qquad E^{\mathcal{B}}(f)\in V$$

In other words, since any f of $L^2(\mathcal{B}_o)$ is a linear combination of event indicators A of $\mathcal{B}_o$, and since $E^{\mathcal{B}}(1_A) = P^{\mathcal{B}}(A)$:

$$\forall A\in\mathcal{B}_o \qquad P^{\mathcal{B}}(A)\in V \qquad (2)$$

We are now led to consider the choice of the subspace V of $L^2(\mathcal{B})$.

$L^2(\mathcal{B})$ is spanned by the event indicators B of $\mathcal{B}$.
We assume here that the subspace V is spanned by the event indicators of a subclass $\mathcal{B}'$ of $\mathcal{B}$ such that Ω belongs to $\mathcal{B}'$ or $\mathcal{B}'$ contains a partition of Ω.

If the r.v.'s X_i have only one-order additive interactions with respect to X_o (see § 3.) then we choose :

$$V = \sum_{i=1}^{p} L^2(\mathcal{B}_i) \qquad (3)$$ (Note that $\mathcal{B}' = \bigcup_{i\in I}\mathcal{B}_i$ imply $V = \sum_{i\in I} L^2(\mathcal{B}_i)$)

In the case of more general interactions, we choose a collection $\mathcal{J}$ of subsets of I and define $V = \sum_{J \in \mathcal{J}} L^2(\mathcal{B}_J)$ where for any J of $\mathcal{J}$, $\mathcal{B}_J$ denotes the sub-σ-field generated by $\bigcup_{i \in J} \mathcal{B}_i$.

Remark 1 : If $\mathcal{B}' = \bigcup_{J \in \mathcal{J}} \mathcal{B}_J$, then $V = \sum_{J \in \mathcal{J}} L^2(\mathcal{B}_J)$.

Remark 2 : In all cases, we can use the definition (3) of V, substituting for the X_i the r.v. $X_i' = 1_B$ $(B \in \mathcal{B}')$. Hence :
$V = \sum_{i=1}^{\text{card}\,\mathcal{B}'} L^2(\mathcal{B}_i')$ where $\mathcal{B}_i'$ denotes the σ-field induced by X_i'.

It is useful to choose a $\mathcal{B}'$ such that the corresponding X_i' are linearly independent. Thus dim V = card $\mathcal{B}'$ = p.

In the sequel we shall always assume this to be the case, unless otherwise stated.

3. THE HYPOTHESIS OF NO ADDITIVE INTERACTIONS

Let $V = \sum_{i \in I} L^2(\mathcal{B}_i) = \text{vect}\,[1_B \mid B \in \mathcal{B}']$. The condition (2) may be written as follows :
For any event A in $\mathcal{B}_o$ and any event B' in $\mathcal{B}'$, there exists a scalar $\lambda_{B'}^A$ such that

$$P^{\mathcal{B}}(A) = \sum_{B' \in \mathcal{B}'} \lambda_{B'}^A \, 1_{B'}$$

and then, for every event B in $\mathcal{B}$ such that P(B) > 0

$$P^B(A) = \sum_{B' \in \mathcal{B}'} \lambda_{B'}^A \, P^B(B') \qquad (4)$$

or

$$P(B \cap A) = \sum_{B' \in \mathcal{B}'} \lambda_{B'}^A \, P(B \cap B') \qquad (4')$$

Let P_i (resp. P_X) be the probability (pr.) induced by X_i (resp. X) on $\mathcal{E}_i$ (resp. $\mathcal{E}$).

Let $\mathcal{B}' = \{[X_i = e_i] \mid e_i \in E_i, P_i(\{e_i\}) > 0, i \in I\}$. It is easily seen that the indicators of events of $\mathcal{B}'$ span $\sum_{i \in I} L^2(\mathcal{B}_i)$, some of them being in this case linearly dependent. $\mathcal{E}'$ being the image of $\mathcal{B}'$ by X (i.e. here : $e \in \mathcal{E}'$ when e is a non negligible element of one of the E_i), (4) can be expressed as follows :

For all $e_o \in E_o$ and for all $e_i \in \mathcal{E}'$, there exists a scalar $\lambda_{e_i}^{e_o}$ such that, if $e = (e_i)_{i \in I}$ with $P_X(\{e\}) > 0$, then: $P^{[X=e]}[X_o = e_o] = \sum_{i \in I} \lambda_{e_i}^{e_o}$ (5)

In other terms, the probability of e_o given e is a sum of effects associated with the levels of the r.v.'s X_i. So we can also call this condition "*hypothesis of additivity of the effects*".

Definition 1 : *The r.v.'s $(X_1, \ldots, X_p)$ are said to have no greater-than-one-order additive interaction with respect to X_o when, for any A in $\mathcal{B}_o$:*

$$P^{\mathcal{B}}(A) \in \sum_{i \in I} L^2(\mathcal{B}_i)$$

Example : Let I be {1,2} and let P_X be a strictly positive probability. (X_1, X_2) have only one-order additive interaction with respect to X_o, if there exists a set of scalars $\{\lambda_{e_i}^{e_o} \mid e_o \in E_o, e_i \in E_i, i \in I\}$ such that for any $e = (e_1, e_2) \in E_1 \times E_2$, and for any $e_o \in E_o$:

$$P^{[X=e]}[X_o=e_o] = \lambda^{e_o}_{e_1} + \lambda^{e_o}_{e_2}$$

Using an orthogonal decomposition of $e \to P^{[X=e]}[X_o=e_o]$ in the Hilbert space $L^2(E, \mathcal{E}, P_1 \otimes P_2)$ (i.e. according to the metric induced by the product probability $P_1 \otimes P_2$) :

$$P^{[X=e]}[X_o=e_o] = P_o(e_o) + \phi_{e_o}(e_1) + \psi_{e_o}(e_2)$$

where $\phi_{e_o}(e_1) = \sum_{e_2 \in E_2} P_2(e_2) \, [P^{[X=(e_1,e_2)]}[X_o=e_o] - P_o(e_o)]$

$$\psi_{e_o}(e_2) = \sum_{e_1 \in E_1} P_1(e_1) \, [P^{[X=(e_1,e_2)]}[X_o=e_o] - P_o(e_o)]$$

In this formula, $P_o(e_o)$ is the average effect of X on e_o, $\phi_{e_o}(e_1)$ the differential effect due to e_1, and $\psi_{e_o}(e_2)$ the differential effect due to e_2. If the hypothesis of one-order interaction does not hold, we might define second-order interaction between X_1 and X_2 with respect to X_o as follows :

$$P^{[X=e]}[X_o=e_o] - P_o(e_o) - \phi_{e_o}(e_1) - \psi_{e_o}(e_2)$$

and we are mainly interested in knowing whether the above expression vanishes or not.

4. COMPARISON WITH THE "LEAST MEAN SQUARES METHOD"

The criterion associated with the method studied in this paper is maximized for the most correlated r.v.'s of V and $L^2(\mathcal{B}_o)$.

Apart from the solution ($\rho_o=1$, $u_o=1_\Omega$, $v_o=1_\Omega$), the first non trivial solution is obtained for the pair of centered variables of $L^2(\mathcal{B}_o)$ and V with maximal correlation. Let (r,v,u) be this solution and $L^2_o(\mathcal{B}_i)$ the centered subspace of $L^2(\mathcal{B}_i)$. Then

$$v \in L^2_o(\mathcal{B}_o), \quad u \in \sum_{i \in I} L^2_o(\mathcal{B}_i) \qquad r = \frac{\langle u,v \rangle}{||u|| \; ||v||} \text{ is maximum, and } ||u|| = ||v|| = 1$$

Setting $u = \sum_{i \in I} u_i$, where $u_i \in L^2_o(\mathcal{B}_i)$, the correlation may be explicitly written as follows :

$$r = \frac{\sum_{i \in I} \langle v, u_i \rangle}{||v|| \sqrt{\sum_{(i,i') \in I^2} \langle u_i, u_{i'} \rangle}}$$

Let Z be the r.v. (X,X_o) and P_Z the probability induced by Z on $\mathcal{E} \otimes \mathcal{E}_o$. Consider now the image of this C.A. by Z (see [3]), or equivalently, substitute for each solution (r,v,u), obtained as above, the triplet (r,g,f) such that :

$$g \in L^2_o(\mathcal{E}_o) \quad \text{and} \quad v = g \circ X_o,$$

$$f \in \sum_{i \in I} L^2_o(\mathcal{E}_i) \quad \text{and} \quad u = f \circ X,$$

where $L^2_o(\mathcal{E}_i)$ is the centered subspace of $L^2(E_i, \mathcal{E}_i, P_i)$, $i=0,\ldots,p$.

Letting $f = \sum_{i \in I} f_i$ $(f_i \in L_0^2(\mathcal{E}_i))$ and $\alpha_i = \|f_i\|$, the criterion can then be written

$$r(g,f) = \frac{\sum_{i \in I} \alpha_i \, r(g,f_i)}{\sqrt{\sum_{(i,i') \in I^2} \alpha_i \alpha_{i'} \, r(f_i,f_{i'})}} \qquad (6)$$

the metric being that of $L^2(\mathcal{E} \otimes \mathcal{E}_0)$.

Consider now the "least mean squares method" (see [6] and [5]).
Let $Z_i = (X_i, X_0)$ and $\Delta = \bigcup_{i \in I} E_i$, let P_{Z_i} be the pr. induced by Z_i on $\mathcal{E}_i \otimes \mathcal{E}_0$. Let P_{Δ,E_0} be the pr. defined on $S(\Delta x E_0)$ such that its restrictions on any $\mathcal{E}_i \otimes \mathcal{E}_0$ are $1/p \; P_{Z_i}$. The "least mean squares method" consists in making the Corr. A. of P_{Δ,E_0} on $(\Delta x \; E_0, S(\Delta x \; E_0))$

The first solution (ρ,g,f) is such that, $g \in L_0^2(\mathcal{E}_0)$ and $f_i \in L_0^2(\mathcal{E}_i)$,
$\rho^2 = 1/p \sum_{i \in I} r^2(g,f_i)$ is maximum
and f is the variable defined on Δ whose restriction to E_i is f_i (see [5]).

If the variables X_i , $i \in I$, are mutually uncorrelated, the two criteria are equivalent (see [6]). In this case, the maximum of (6) is reached for

$$\|f_i\|^2 = \frac{r^2(g,f_i)}{\sum_{i \in I} r^2(g,f_i)}$$

under the restriction that $\sum_{i \in I} \|f_i\|^2 = 1$.
From a computing stand point, the proposed method is more expensive than the "least mean squares method". However it is more flexible (freedom in the choice of V) and it takes into account the interactions between the explicative variables. It is also less sensitive to multicollinearity since it tends to extract the variable of $L^2(\mathcal{E}_0)$ most correlated with the greatest number of the least correlated variables.

5. APPLICATIONS

5.1. Approximation of a correspondence analysis

Here Ω is $\{\omega_1,\ldots,\omega_n\}$ and $\mathcal{A}$ is $S(\Omega)$, the class of all sets in Ω. $\{1_B \mid B \in \mathcal{B}'\}$ is a basis of V. p denotes the dimension of V and the cardinality of $\mathcal{B}'$. Any individual ω_j can be represented in $\mathbb{R}^p$ by the vector $x(\omega_j)$ whose coordinates on the canonical basis are the values of the indicators at ω_j. For simplicity, let 1_{e_0} denote $1_{[X_0=e_0]}$. Analogously, if card $E_0=q$, any individual ω_j can be represented in $\mathbb{R}^q$ by the vector $y(\omega_j)$:

$$x(\omega_j) = \begin{bmatrix} \vdots \\ 1_B(\omega_j) \\ \vdots \end{bmatrix}_{B \in \mathcal{B}'} \qquad y(\omega_j) = \begin{bmatrix} \vdots \\ 1_{e_0}(\omega_j) \\ \vdots \end{bmatrix}_{e_0 \in E_0}$$

P_0 is assumed to be strictly positive on $\mathcal{E}_0$.
Let X (resp. Y) denote the n × p (resp. n × q) matrix whose columns are the $x(\omega_j)$ (resp. $y(\omega_j)$).

Let D_p (resp. D_{p_o}) denote the $n \times n$ (resp. $q \times q$) diagonal matrix whose element (j,j) is the probability $P(\omega_j)$ (resp. $P_o(e_o)$) :

$$D_p = \begin{bmatrix} \ddots & & 0 \\ & P(\omega_j) & \\ 0 & & \ddots \end{bmatrix}_{\omega_j \in \Omega} \qquad D_{p_o} = \begin{bmatrix} \ddots & & 0 \\ & P_o(e_o) & \\ 0 & & \ddots \end{bmatrix}_{e_o \in E_o}$$

Then we get the following duality scheme (cf. [2]) where the metric of $\mathbb{R}^n$ is associated with the matrix D_p in the canonical basis ($\mathbb{R}^n$ is thus identified with $L^2(\Omega, \mathcal{Q}, P)$, and $\cdot^*$ denotes the dual space of $\cdot$):

$$\begin{array}{ccccc} \mathbb{R}^q & \xleftarrow{Y} & \mathbb{R}^{n*} & \xrightarrow{X} & \mathbb{R}^p \\ D_{p_o} \Big\uparrow & & D_p \Big\uparrow (XD_p{}^tX) & (XD_p{}^tX)^{-1} \Big\downarrow & \\ \mathbb{R}^{q*} & \xrightarrow{{}^tY} & \mathbb{R}^n & \xleftarrow{{}^tX} & \mathbb{R}^{p*} \end{array}$$

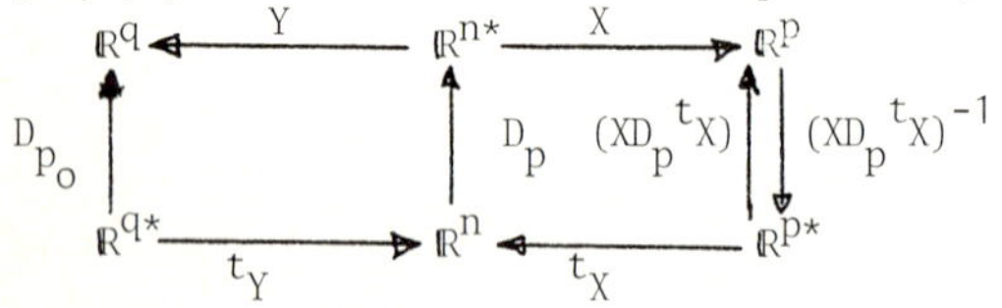

The metric on $\mathbb{R}^p$ is Mahalanobis's type and has for associated matrix $(XD_p{}^tX)^{-1}$ in the canonical basis.
If S_1 (resp. S_2) denotes the orthogonal projection from $\mathbb{R}^n$ into $\text{Im}^t Y = L^2(\mathcal{B}_o)$ (resp. $V = \text{Im}^t X$), then :
$S_1 = {}^tY D_{p_o}^{-1} Y D_p$, $S_2 = {}^tX(XD_p{}^tX)^{-1} X D_p$ and $D_{p_o} = Y D_p {}^tY$

Remark : S_1 (resp. S_2) replaces here $E^{\mathcal{B}_o}$ (resp. π_V) as defined in section 2.

As indicated above, this C.A. is also a discriminant analysis. The spectral analysis of $E^{\mathcal{B}_o} \circ P_{V|L^2(\mathcal{B}_o)}$ reduces to a simple diagonalization of $S_1 \circ S_2$ or $S_2 \circ S_1$.
Apart from the trivial eigenvector corresponding to the eigenvalue 1, the remaining eigenvectors ξ_ℓ, $\ell=1,\ldots,r$, of $S_2 \circ S_1$ are called the linear discriminators. They are the images of the discriminant factors b_ℓ of $\mathbb{R}^{p*}$ by ${}^tX$.
$b_\ell \in \mathbb{R}^p$ is associated in $\mathbb{R}^p$ with the discriminant axis spanned by the vector u_ℓ such that : $\forall x \in \mathbb{R}^p$ $<u_\ell, x>_{\mathbb{R}^p} = b_\ell(x)$.

The k-dimensional subspace spanned by $u_1,\ldots,u_k$ is called the k-dimensional discriminant subspace. According to the criterion maximizing the ratio of the among-group variation to the pooled within-group variation, the k dimensional discriminant subspace is, among all k-dimensional subspaces, that which best discriminates the groups $[X_o = e_o]$, $e_o \in E_o$.
On the discriminant subspaces, we may represent :
. individuals ω_j (their coordinates on the discriminant axes are given by the values of the corresponding factor at $x(\omega_j)$)
. for any e_o of E_o (resp. $\forall i \in I$ and $\forall B \in \mathcal{B}_i$) mean vectors g^{e_o} (resp. g^B) of the vectors which represent, in the same subspace, the individuals ω_j such that $X_o(\omega_j) = e_o$ (resp. $X_i(\omega_j) \in B$).
Such an analysis is an approximation of the Corr. A. of (X, X_o). Furthermore under (2) the two analyses coincide. In this case, the representation of any discriminant subspace is one of the standard "simultaneous representations" of Corr. A. (see [1]).
The levels e_o of the variable X_o are then represented by the g^{e_o} (weighted mean vector of the vectors associated with the levels of X with weights $P^{[X_o=e_o]}_{[X=e]}$). The representation of the level e of X, $X^{-1}(e) \neq \phi$, is given by that of any individual belonging to $X^{-1}(e)$.

5.2. Approximation of the probability given $\mathcal{B}$

The value of π_V on the indicator 1_{e_o} of E_o is an approximation of $P^{\mathcal{B}}[X_o=e_o]$. It is given by the linear regression of 1_{e_o} on the r.v.'s which span the subspace V. Indeed :

$$\pi_V(1_{e_o}) = \pi_V \circ E^{\mathcal{B}}(1_{e_o}) = \pi_V[P^{\mathcal{B}}[X_o=e_o]]$$

Under condition (2), this approximation and the conditional pr. coincide. According to the notations of section 5.1., π_V becomes S_2. The duality scheme of 5.1. is still applicable. For any e_o of E_o, the variable $S_2(1_{e_o})$ defines the linear fonctional d^{e_o} on $\mathbb{R}^p$ whose image by ${}^tX$ is $S_2(1_{e_o})$. As a result :

$S_2(1_{e_o}) = {}^tX(XD_p{}^tX)^{-1}XD_p1_{e_o}$. Since $g^{e_o} = \frac{1}{P_o(\{e_o\})} X D_p 1_{e_o}$, then

$S_2(1_{e_o}) = P_o(\{e_o\})\,{}^tX(XD_p{}^tX)^{-1}g^{e_o}$ and $d^{e_o} = P_o(\{e_o\})(X D_p {}^tX)^{-1} g^{e_o}$

The image of $x(\omega_j)$ by d^{e_o} is then :

$$d^{e_o}(x(\omega_j)) = \langle x(\omega_j), g^{e_o}\rangle_{\mathbb{R}^p}$$

Moreover, for any i of I and for any B of $\mathcal{B}_i$, $E^{\mathcal{B}_i} \circ S_2 = E^{\mathcal{B}_i}$ and $d^{e_o}(g^B) = E^B(S_2(1_{e_o}))$, g^B being defined as in section 5.1.

Thus

$$d^{e_o}(g^B) = P^B[X_o=e_o] \qquad (7)$$

If we also assume that condition (2) holds, (7) extends to any non-negligible B of $\mathcal{B}$. In particular when $B = [X = e]$, $e \in E$, x is constant over [X=e] and for each $j=1,\dots,n$, $x(\omega_j) = g^{[X=X(\omega_j)]}$ and :

$$d^{e_o}(x(\omega_j)) = P^{[X=X(\omega_j)]}[X_o=e_o] \qquad (8)$$

Note that (8) is equal to (4) with $A = [X_o = e_o]$ and where d^{e_o} has coordinates $\lambda_A^{B'}$, $B' \in \mathcal{B}'$, in the dual basis of $\mathbb{R}^{p*}$.

We can take the d^{e_o} as discriminant linear functions (d.l.f.) associated with each group $[X_o = e_o]$ in Ω.

In the case of two groups, we find (up to an additive and a scale factor) Fisher's d.l.f.. For any pair (e_o, e_o') of elements of E_o, the hyperplane of $\mathbb{R}^p$, whose equation is $(d^{e_o}-d^{e_o'})x = 0$, separates $\mathbb{R}^p$ in regions R_{e_o} associated with each level of E_o. Each region R_{e_o} contains all mean vectors g^B so that $B \in \bigcup_{i\in I} \mathcal{B}_i$ and P^B is maximized on $[X_o=e_o]$.

If (2) holds this results also extends to all B of $\mathcal{B}$.

6. A CONDITION OF STABILITY

Let us consider a case where the variable (X,X_o) has not been observed everywhere on Ω, but only on a subset Ω' in Ω. To simplify, let us assume that Ω is finite, and that $\mathcal{Q}$ is the class of all sets in Ω. It is a standard model for statistical variables observed on a sample. The linear regression of 1_{e_o} on V is also the C.A. of vect$[1_{e_o}]$ and V. Then in the C.A. of $L^2(\mathcal{B}_o)$ and V (resp. vect$[1_{e_o}]$ and V), we use only the restriction of the metric of $L^2(\mathcal{B} \otimes \mathcal{B}_o)$ on $L^2(\mathcal{B}_o)+V$ (resp. on the subspace vect$[1_{e_o}]+V$ of $L^2(\mathcal{B}_o)+V$). The restrictions depend only on

$$\langle 1_B, 1_{B'}\rangle \text{ where } B \in \mathcal{B}' \cup \mathcal{B}_o \quad , \quad B' \in \mathcal{B}' \cup \mathcal{B}_o$$

(resp. $B \in \mathcal{B}' \cup \{A\}$, $B' \in \mathcal{B}' \cup \{A\}$ with $A = X_o^{-1}(e_o)$).

In other words, they depend only on the values of P on the subclass $\mathcal{F}$ of $\mathcal{B}\otimes\mathcal{B}_o$,

$$\mathcal{F} = \{B \cap B' \mid B \in \mathcal{B}' \cup \mathcal{B}_o \ , \ B' \in \mathcal{B}' \cup \mathcal{B}_o\}$$

That is, the image of this C.A. by Z depends on the values of P_Z on $\mathcal{F}' = Z(\mathcal{F})$ (since $\mathcal{F} \subset \mathcal{B}\otimes\mathcal{B}_o$), or on the two-order marginal pr. of P_Z(since $\mathcal{B}' = \bigcup_{i\in I}\mathcal{B}_i$).

We now define :
- $\mathcal{Q}'$ as the class of all sets in Ω' ,
- P' as the probability defined on $\mathcal{Q}'$ by : $\forall A \in \mathcal{Q}' \quad P'(A) = P^{\Omega'}(A)$

(if P assigns equal probability to all ω in Ω , then P' does in Ω')
- $Z_{\Omega'}$ and $1'_{e_o}$ as the respective restrictions of Z and 1_{e_o} on Ω'.

The probabilities induced by Z from P and by $Z_{\Omega'}$ from P' coincide on $\mathcal{F}'$ under the condition

$$\forall A \in \mathcal{F} \quad , \ P^{\Omega'}(A) = P(A) \qquad (9)$$

(i.e. Ω' is independent of all events of $\mathcal{F}$).
Furthermore, let $L^2(\mathcal{B}_o)'$ and V' be the concepts defined by substituting in the definition of $L^2(\mathcal{B}_o)$ and V, for $(\Omega,\mathcal{Q},P)$ the pr. space $(\Omega',\mathcal{Q}',P')$.

Therefore, the image by Z of the C.A. of $L^2(\mathcal{B}_o)$ and V (resp. vect$[1_{e_o}]$ and V) and the image by $Z_{\Omega'}$ of the C.A. of $L^2(\mathcal{B}_o)'$and V' (resp. vect$[1'_{e_o}]$ and V') are the same.

In conclusion, we can say that if we substitute $(\Omega,\mathcal{Q},P)$ by $(\Omega',\mathcal{Q}',P')$ as indicated above, the analysis will not be modified under (9).

Consequently, if we want (9) to hold, we are led to choose a $\mathcal{B}'$ with reasonable cardinality. This implies that we assume no high-order additive interactions and/or that we reduce the number of explicative variables with respect to card Ω'. We thus eliminate the fluctuations due to the choice of Ω' in Ω. In the other hand, to ensure a good approximation, condition (2) must hold. We are then led to choose V large enough. With p not too small and card Ω' large enough with respect to p, we can hope to approach simultaneously the two conditions.
In applying this result in section 5, under condition (9), all the levels e of X so that $P_X(e) > 0$ can be represented on the discriminant subspaces. Similarly, the linear regression obtained from $(\Omega',\mathcal{Q}',P')$ will allow us to approximate all conditional pr.

$$P^{[X=e]}[X_o=e_o]$$

so that $P_X(e) > 0$.

Note : This condition, $P_X(e) > 0$, does not imply that e has been observed.

Assuming (2) and (9), these results agree with those which could be obtained through the Corr. A. of (X,X_o) and through the regression analysis of 1_{e_o} on V.

These results allow us to approximate the conditional pr. in the empty cells (or the nearly empty cells) of the contingency table, provided that the hypothesis

$$P_X(e) > 0$$

is not contradictory to condition (9). In other words, choose $\mathcal{B}'$ not too big so that, if $A = X^{-1}(e)$, all the F of $\mathcal{F}$ containing A satisfy $P'(F\cap\Omega') \neq 0$.

7 - COMPUTATIONAL PROBLEMS.

Both approximations (Corr. A and conditional pr.) require the computation of the p x p inner product matrix of the indicators 1_B, $B\in\mathcal{B}'$, and its inverse. The Corr. A. approximation additionally requires the diagonalization of a qxq matrix (q=card E_o).

At the present stage, a programme generates, by a process of elimination, a basis of $V = \sum_{J \in \mathcal{J}} L^2(\mathcal{B}_J)$. It computes both approximations and gives the habitual representations. Then linked programmes give other estimations of the conditional pr.. These are based on the method of Fix and Hodges (see [4]) and on the usual method of density estimation under k-dimensional normal hypothesis.

Comparisons between the obtained percentages of well-classified can then be made (provided that an individual is classified in its most probable group).

As an example, a medical problem has been treated and seems to give interesting results. At the present time, applications covering this problem are still being developed.

REFERENCES

[1] BENZECRI J.P. (1973) L'analyse des données - Dunod - Tome II

[2]CAILLIEZ F. et PAGES J.P. (1976) Introduction à l'analyse des données. S.M.A.S.H

[3]DAUXOIS J. et POUSSE A. (1976) Les analyses factorielles en Calcul des Probabilités et statistiques : essai d'étude synthétique. Thèse Univ.P.Sabatier TOULOUSE

[4]COLLOMB G. (1976) Estimation non paramétrique de la densité par la méthode du noyau. Thèse de Docteur-Ingénieur. Université Paul Sabatier - TOULOUSE

[5]LECLERC A. (1976) Une étude de la relation entre une variable qualitative et un groupe de variables qualitatives. Int. Stat. Rev. N°2 - Vol.44 [241-248]

[6]SAPORTA G. (1975) Liaison entre plusieurs ensembles de variables et codages de données qualitatives. Thèse de 3ème Cycle - Université Paris VI

Recent Developments in Statistics
J.R.Barra et al., editors

LEAST SQUARES ANALYSIS OF ORDINAL DATA

G. DROUET D'AUBIGNY
Université de St Etienne - Département de Mathématiques
23, rue du Docteur Paul Michelon
42000 - Saint Etienne - France

An isotonic correlation index is presented which is estimated on finite samples by Kruskal's Stressform 2 [11]. A property of strong consistency of this estimator is applied to the comparison of least squares data analysis methods.

Our results are exemplified on multidimensional scaling methods. Further applications are quoted.

1 - INTRODUCTION

Numerous methodological proposals have characterized the last decade in data analysis, owing to the spread of computing ability. Here we are interested in comparing them from the point of view of measurement. As a matter of fact, it is often difficult to determine the actual kind of information contained in the data, when applying these methods to the behavioral and the social sciences. Young and De Leeuw [16] give this problem an empirical answer : they perform several analyses of the data set, under varying measurement assumptions ; the appropriate level being, in their mind, the highest one that yields the best fit.

Such an expansive method is shown to be needless if we are not chiefly interested in the determination of the measurement level of the observations. Although the result is exemplified on multidimensional scaling methods, it remains valid for all methods one can express in our probabilistic formulation [7, 8] ; individual differences and preferences analyses are well-known examples of such methods.

2 - DEFINITION AND PROPERTIES OF THE ISOTONIC CORRELATION INDEX

The analysis of the dependence of a real valued random variable Y on a random element X, defined on the same probability space $(\Omega,\mathcal{A},P)$ is a regression problem. X is generally valued in a measurable space $(E,\mathcal{E})$, where E is a real linear space. When the order relation $\mathcal{R}$ on E is the unique significant information on the observed phenomena, we are only interested in the elements U in $\mathcal{E}$, the indicator $1_U(.)$ of which is an isotonic function from $(E,\mathcal{R})$ into $(\{0,1\}, \geq)$. The class $\mathcal{U}$ of such parts U of E is a unitary sub-σ-lattice of $\mathcal{E}$.

Definition 1 [2] : A class $\mathcal{U}$ of parts of E is a σ-lattice if it is closed under countable union and intersection. It is unitary if it contains $\emptyset$ and E. An extended real valued function m on E is $\mathcal{U}$-measurable (i.e. $m \in \mathcal{M}(\mathcal{U})$) if $m^{-1}([x,+\infty[)$ belongs to $\mathcal{U}$ for every real number x.

Then m belongs to $\mathcal{M}(\mathcal{U})$ if and only if m is isotonic from $(E,\mathcal{R})$ into $(\overline{R}, \geq)$ and $\mathcal{E}$-measurable.

Definition 2 : $(E,\mathcal{U})$ is called an Ordinal-Measurable space and every function

$$X : (\Omega,\mathcal{A},P) \to (E,\mathcal{U})$$

such that $\mathcal{B} = X^{-1}(\mathcal{U})$ is a part of $\mathcal{A}$, is an ordinal random variable.

$\mathcal{B}$ is a unitary sub-σ-lattice of $\mathcal{A}$ and one can construct the conditional expectation of Y given the σ-lattice $\mathcal{B}$ in terms of the Radon-Nikodym derivative [2]. Since we are interested in least squares data analysis, we shall assume that all the encountered random variables belong to $L^2_R(\mathcal{A}) = L^2_R(\Omega,\mathcal{A},P)$.

Theorem 1 [2] : Let $L^2_R(\mathcal{B})$ be the set of real valued $\mathcal{B}$-measurable random variables :

$$L^2_R(\mathcal{B}) = L^2_R(\mathcal{A}) \cap \{m_o X \ ; \ m \in M(\mathcal{U})\},$$

$L^2_R(\mathcal{B})$ is a closed convex cone and a σ-complete lattice in $L^2_R(\mathcal{A})$.

While the algebraic structure of $L^2_R(\mathcal{B})$ gives a theoretical foundation to Guttman's inverse-rank-image algorithm, its geometrical structure warrants the uniqueness of the orthogonal projection of any element Y in $L^2_R(\mathcal{A})$ on $L^2_R(\mathcal{B})$, as the closest point in $L^2_R(\mathcal{B})$ to Y.

Definition 3 : Let $E(Y/\mathcal{B})$ be the unique closest point of $L^2_R(\mathcal{B})$ to Y. $E(Y/\mathcal{B})$ is called the conditional expectation of Y given the σ-lattice $\mathcal{B}$. It satisfies the relation

$$\int_\Omega |Y-E(Y/\mathcal{B})|^2 \, dP \leq \int_\Omega |Y-m_o X|^2 \, dP$$

where $E(Y/\mathcal{B})$ is identified with one of its elements.

Another consequence of the geometrical structure of $L^2_R(\mathcal{B})$ is that the variance decomposition formula remains true, when $\mathcal{B}$ is but a σ-lattice :

$$\mathrm{Var}(Y) = \mathrm{Var}(E(Y/\mathcal{B})) + E(|Y-E(Y/\mathcal{B})|^2)$$

So the analogy with the correlation ratio leads us to

Definition 4 [7] : The isotonic correlation index is

$$\nu^2_{Y/X} = \rho^2(Y,E(Y/\mathcal{B})) = 1 - \frac{E(|Y-E(Y/\mathcal{B})|^2)}{\mathrm{Var}(Y)}$$

where $\rho^2(.,.)$ is the Pearson-Bravais coefficient of linear correlation.

This asymmetric index undergoes a proportional reduction in error interpretation, in the sense of Blalock [1], since :

$$\nu^2_{Y/X} = \frac{\mathrm{Var}(E(Y/\mathcal{B}))}{\mathrm{Var}(Y)}$$

and one can easily ascertain that it is valued in [0,1]. Moreover we have proved

Proposition 1 :

i) $\nu^2_{Y/X} = \nu^2_{Y/U}$ if and only if $U = m_o X$ where m belongs to $\mathcal{M}(U)$.

ii) $\nu^2_{Y/X} = 0$ if and only if a non degenerate element of $L^2_R(\mathcal{B})$ has a negative linear correlation with Y. Then E(Y) belongs to $E(Y/\mathcal{B})$.

iii) If there exists a non degenerate element in $L^2_R(\mathcal{B})$ having a positive linear correlation with Y, then $E(Y/\mathcal{B})$ satisfies
$\nu^2_{Y/X} = \mathrm{Max}(\rho^2(Y,V)\ ;\ V \in L^2_R(\mathcal{B}))$.

Elements of proof :

i) is a consequence of the definition. ii) and iii) find a direct demonstration in [7], but more simply, one can observe that in our case, they express the convex duality theorem.

Remark :

ii) and iii) show that $\nu^2_{Y/X}$ is an index of positive dependence in the sense of Lehmann.

Since $\mathcal{B}$ is just a σ-lattice, the conditional expectation $E(Y/\mathcal{B})$ is no more a linear operator. Consequently, an evaluation of $\nu^2_{Y/X}$ reveals itself impossible without calculating $E(Y/\mathcal{B})$, as usual with nonlinear regression methods.

When we are given a weighted sample $\{((x_i,y_i),p_i)\ ;\ i = 1,\dots,p\}$ from (X,Y), $\nu^2_{Y/X}$ is estimated by

$$\hat{\nu}^2_{Y/X} = 1 - S^2_{Y/X} \qquad \text{where} \qquad S^2_{Y/X} = \mathrm{Inf}(S^2_{Y/X}(m)\ ;\ m \in \mathcal{M}(U))$$

$$S^2_{Y/X}(m) = \frac{\sum_{i=1}^{p} p_i |y_i - m(x_i)|^2}{\sum_{i=1}^{p} p_i |y_i - y_.|^2} = \frac{||Y-m(X)||^2_{D_p}}{||Y-y_.J||^2_{D_p}}$$

and $S^2_{Y/X}$ is evaluated by the pool-adjacent-violator algorithm ([2], [10]).

Notations :

$||X||^2_{D_p}$ is the Euclidean norm of the vector X in R^p provided with the metric of weights associated with the matrix

$$D_p = \begin{bmatrix} p_1 & & & 0 \\ & p_2 & & \\ 0 & & \ddots & \\ & & & p_p \end{bmatrix} \quad , \quad p_i > 0,\ \sum_{i=1}^{p} p_i = 1$$

$J \in R^p$, ${}^tJ = (1,1,\ldots,1)$.

$y_. = {}^tY \, D_p \, J$.

Then we can state the following

Proposition 2 [12] : If the conditional expectation of Y given the σ-lattice $\mathcal{B}$ is an affine function of X, then $\hat{\nu}^2_{Y/X}$ converges almost everywhere towards $\rho^2(X,Y)$.

Elements of proof :

Prop. 1 iii shows that when $E(Y/\mathcal{B})$ converges, its limit is the best linear predictor of Y, under our hypothesis. So the proof consists in the demonstration of that convergence almost everywhere.

Remark :

The use and interpretation of the index $\hat{\nu}^2_{Y/X}$ are founded upon the fact that the empirical coefficient of linear correlation, $r^2_{X,Y}$, is also a strongly consistent estimator of $\rho^2(X,Y)$. Then for n high enough, the following inequality holds :

$$P(|\hat{\nu}^2_{Y/X} - r^2_{X,Y}| \geq \varepsilon) < \varepsilon \qquad (\forall \varepsilon, \ \varepsilon \in R^+ \setminus \{0\})$$

So we can use $\hat{\nu}^2_{Y/X}$ as a biaised estimator of $\rho^2(X,Y)$ which tends to overestimate the true value of $r^2_{X,Y}$, since we have

$$r^2_{X,Y} \leq \hat{\nu}^2_{Y/X}.$$

3 - APPLICATION TO THE LEAST SQUARES ANALYSIS OF ORDINAL DATA

The experiment in social sciences often leads to one of the following three types of measurement scales :

- The weakest one is the ordinal scale, in which the information contained in the data set Δ is preserved by any one-to-one isotonic transformation :

$$T = m_o \Delta.$$

- When any affine transformation on the data set

$$T = a\Delta + c, \ a \in R^+ \setminus \{0\}, \ c \in R,$$

gives the same information on the observed phenomena, we call this an interval scale.

- Finally, when the information contained in the data is invariant under any linear transformation

$$T = a\Delta, \ a \in R^+ \setminus \{0\},$$

the observations are made on a ratio scale.

While the latter two levels give some importance to the numerical value of the data (thus they lead to Metric methods), in the ordinal case the order relations induced by the collected numbers are solely significant (thus they lead to Nonmetric methods).

Our comparison of these two kinds of methods is illustrated on the multidimensional scaling problem.

3.1 - The multidimensional scaling problem

We are given a triplet (I,Δ,P), where the data set Δ (respectively P) consists of dissimilarities measured on (resp. weights assigned to) a finite set I of objects (items, stimuli, brands, products, show business stars, ...). We note

$$\Delta = (\delta_{ij})_{i,j\in I}, \quad P = (p_{ij})_{i,j\in I}$$

We want to represent I in a normed linear space V, the norm of which is noted $||\text{-}||_V$

$$X : I \to V$$

such that

- the dimensionality r of B be minimal,

- the fit between the data Δ and the reconstructed interpoint distances D be the best.

$$D = (d_{ij})_{i,j\in I}, \quad d_{ij} = ||X(i) - X(j)||_V$$

If r is known, we must solve a regression problem which can be written in least squares terms :

$$(P_1) \quad ||D-T||^2_{D_p} = \text{MIN !}$$

where D_p is the metric of weights in R^p, $p = \frac{n(n-1)}{2}$ if card(I) = n.

If Δ is measured on an ordinal scale, the problem (P_1) is solved by the Kruskal algorithm [11], which computes

$$(1) \quad d(D(I),C^\circ) = \text{MIN}\left(\text{MIN}\left(\frac{||D-T||^2_{D_p}}{||D-d_{.}J||^2_{D_p}} \; ; \; T \in C^\circ\right) \; ; \; D \in D(I)\right)$$

where

$$C^\circ = \{T \; ; \; T \in R^p \; ; \; \Sigma T \geq 0\}, \quad \Sigma_{pp} = \begin{bmatrix} 1 & & & & \\ -1 & 1 & & 0 & \\ & -1 & 1 & & \\ 0 & & \ddots & \ddots & \\ & & & -1 & 1 \end{bmatrix}$$

is the set of the Disparities $T = m_o\Delta$, m being an isotonic function. D(I) is the set of the interpoint distances compatible with the norm defined on V. This is a nonpolyhedral cone, the convexity of which is warranted only if W is an Euclidean space [7].

3.2 - Comparison with the factorial analysis of the triplet (I,Δ,P)

Our attention was drawn to these problems by the increasing use of ordinal multi-dimensional scaling as a more robust (in an indefinite way!) substitute for the usual factor analysis of a triplet (I,Δ,P), developed by Torgerson [15]. This is a misleading opinion for several reasons that we are going to explain, after having briefly recalled the problem.

Let us call Δ_J the straight line of constants in R^n, and π the D_n-orthogonal projector on the hyperplan Δ_J, D_n-orthogonal to Δ_J.

$\pi = (I_n - D_n\, J\,{}^tJ)$ where I_n is the identity matrix in R^n and J is the vector of Δ_J, the coordinates of which equal 1.

The Torgerson double centering method associates to Δ the pseudo-scalar product matrix

$$W = -\frac{1}{2}\,\pi\Delta^{(2)}\pi\ , \qquad \Delta^{(2)} = (\delta^2_{il})_{i,l\in I}.$$

Remark :

i) W is a matrix of association coefficients in the sense of Gower [9].

ii) W is called a pseudo-scalar product matrix, because it can be semi-definite. This problem is let aside here ; if interested, refer to [3] or [7] for a discussion of this important problem.

The configuration X is then built by interpreting the operator W_oD_n in the usual duality diagram of principal components analysis [3].

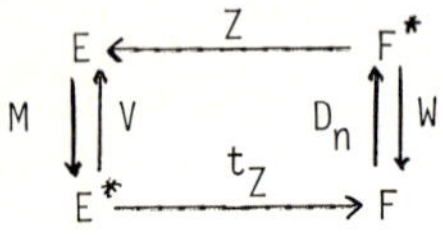

$$W = {}^tZ_oM_oZ$$
$$V = Z_oD_n{}_o{}^tZ$$
$$M = V^{-1}$$

where $Z = Y_o\pi$ is associated with a dummy data set Y of unobserved latent random variables, on which the proximities of elements in I are measured. This farfetched interpretation of the data leads to an unappropriate criterium for judging the quality of the representation X in V since the usual index is

$$\text{INDEX} = \frac{I_V}{I_g}$$

which is the ratio of inertias of points $\dot{Y}(i)$ respectively calculated in relation to the affine manifold D_n-orthogonal to V in R^n, and to the centroid g. A straightforward calculation shows that

$$I_g = \frac{1}{2}\,\delta^2_{..} = \frac{1}{2}\,\Sigma\{p_ip_j\,{}^{2}_{ij}\ /\ (i,j)\in I\mathrm{x}I\}$$

$$I_V = \frac{1}{2}\,d^2_{..} = \frac{1}{2}\,\Sigma\{p_ip_jd^2_{ij}\ /\ (i,j)\in I\mathrm{x}I\}$$

with

$$d_{ij} = ||X(i)-X(j)||_V$$

so

$$\text{INDEX} = \frac{d^2_{..}}{\delta^2_{..}}$$

which is a rather rough index of fit ! But a more important statistical inadequacy of this method appears in terms of invariance. Any accurate method in data analysis must respect the invariance level of the data. While the principal components analysis is invariant under reflexions and translations, W and $\Delta^{(2)}$ are equivalent maximal L(p)-invariant statistics [13], where L(p) is the group of affine transformations (i.e. translations and non singular linear transformations).

The affine invariant method closest to principal components analysis, proposed by Obenchaïn [14], minimizes

$$|||\Delta^{(2)}-D^{(2)}|||^2_{I_n} = 4|||W_oD_n-{}^tXX|||^2_{I_n} + 2({}^tJe)^2 + 2n\,{}^tee \ ,$$

where $|||W_oD_n-{}^tXX|||^2_{I_n}$ is the criterium minimized in principal components analysis,

and e is the vector in R^n that contains the discrepancies in the squared distances of the n points from their respective centroïds in the configurations Y and X. So the affine invariant method puts more emphasis upon minimizing the discrepancies between distances (explained in e) than between angles (fitted by principal components analysis).

Moreover, since

$$|||\Delta^{(2)}-D^{(2)}|||^2_{D_n} = \Sigma\{p_ip_j(\delta_{ij}+d_{ij})^2.(\delta_{ij}-d_{ij})^2 \ / \ (i,j)\in I\text{x}I\}$$

such criteria give more emphasis to high distances than De Leeuw's approach in which we minimize

$$|||\Delta-D|||^2_{D_n} = \Sigma\{p_ip_j(\delta_{ij}-d_{ij})^2 \ / \ (i,j)\in I\text{x}I\}$$

although this method remains affine invariant in the sense of Obenchaïn.

4 - EXTENSIONS OF THESE RESULTS

Isotonic regression methods spread to the analysis of the dependence of a random vector Y, element of $L^2_Rm(\Omega,\mathcal{A},P)$, on an ordinal random variable X, defined on the same probability space $(\Omega,\mathcal{A},P)$. This necessitates a new assumption :

We assert the existence of a real linear functional $<.,w>$, defined on $L^2_Rm(\Omega,\mathcal{A},P)$. Hence the random variable

$$Z : (\Omega,\mathcal{A},P) \longrightarrow (R,\mathcal{B}_R)$$

$$\omega \sim \longrightarrow Z(\omega) = <Y(\omega),w>$$

is an element of $L^2_R(\Omega,\mathcal{A},P)$.

The purpose of the ordinal linear model is to build the best predictor of <.,w> in a least squares sense. This only requires to know the isotonic regression of Z on X.

Therefore, the consequence of proposition 2 remains true for any method one can formulate in terms of the ordinal linear model, the first formulation of which is due to De Leeuw [5]. The interested reader is referred to [7] and [8] for a formalisation of individual differences and preferences analyses in this framework. De Leeuw shows how to express polynomial regression and analysis of variance in the ordinal case. We are working too on the formulation of optimal scaling, in this context.

5 - REFERENCES

[1] Blalock, H.M. (1974). Measurement in the social sciences : Theory & strategies, Aldine Publishing Company, Chicago.

[2] Barlow, R.E. & aliis (1972). Statistical inference under order restrictions, Wiley, New York.

[3] Caillez, F. & aliis (1976). Introduction à l'analyse des données, S.M.A.S.H. 9, rue Duban, 75016 Paris, France.

[4] Cooper, L.G. (1972). A new solution to the additive constant problem in metric MDS, Psychometrika, Vol. 37, n° 3, 311-322.

[5] De Leeuw, J. (1968). Non metric linear model. Mimeograph RN00369, Department of data theory, Leiden, The Netherlands.

[6] De Leeuw, J. (1976). Application of convex analysis to multidimensional scaling, Invited paper European Meeting of Statisticians, Grenoble, France.

[7] Drouet d'Aubigny, G. (1975). Description statistique des données ordinales : analyse multidimensionnelle, Thèse 3ème cycle, Grenoble, France.

[8] Drouet d'Aubigny, G. (1976). Description multidimensionnelle des données ordinales. Séminaire I.R.I.A. (to appear).

[9] Gower, J.C. (1966). Some distance properties of latent roots & vector methods used in multivariate analysis, Biometrika, n° 53, Vol. 3, 325-338.

[10] Guttman, L. (1968). A general non metric technique for finding the smallest coordinate space for a configuration of points. Psychometrika, Vol. 33, n° 4, 469-506.

[11] Kruskal, J.B. (1964). Multidimensional scaling by optimizing goodness of fit to a non metric hypothesis. Psychometrika, Vol. 29, 1-27 and 115-129.

[12] Mayer, L.S. (1973). Estimating a correlation coefficient when one variable is not directly observed, J.A.S.A., n° 68, 420-421.

[13] Obenchaïn, R.L. (1971). Multivariate procedures invariant under linear transformations, Annals of Math. Stat., Vol. 42, n° 5, 1569-1578.

[14] Obenchaïn, R.L. (1973). Affine multidimensional scaling. Mimeograph Bell Telephone Laboratories. Murray Hill, New Jersey, U.S.A.

[15] Torgerson, W.S. (1958). Theory and methods of scaling. Wiley, New York.

[16] Young, W. & De Leeuw, J. (1975). Non metric individual differences multidimensional scaling. Psychometric Laboratory Report, n° 146, Univ. of North Carolina, U.S.A.

Recent Developments in Statistics
J.R.Barra et al., editors

EIGENVALUES AND EIGENVECTORS IN HIERARCHICAL CLASSIFICATION

M. Gondran
Direction des Etudes et Recherches EDF
1, av. du Général de Gaulle
92140 CLAMART

This paper shows that any level of a hierarchical classification which is based on an ultrametric distance corresponds to an eigenvalue of the distance matrix. Consequently, there is a one-to-one correspondence between the eigenvectors which generate the semi-module corresponding to such an eigenvalue, and the types of the associated level.

The above interpretation of the levels and types of a hierarchical classification leads to a unified theory of both, factorial analysis and classification.

1. INTRODUCTION

The two conventional approaches to data analysis, factorial analysis and classification appear, at first sight, to be incompatible.

The factorial analysis seems to be concerned with quantity, whereas the classification is seemingly interested in quality.

In this paper, we try to show that these approaches are very closely related. Indeed, it will be shown that the search for a hierarchical classification based on an ultrametric distance is reduced to a search for the eigenvalues and the eigenvectors of the dissimilarity matrix in a suitable algebraic structure (semi-module on a semi-ring, see GONDRAN[1], GONDRAN and MINOUX[2]).

Thus, the difference between the two approaches lies actually in the algebraic structure chosen for data processing.

In paragraph 2, we introduce the algebraic structure subjacent to an ultrametric distance. It is the semi-ring $(S, \oplus, *)$ where $S = R^{+} \cup \{+\infty\}$ and where the operations $\oplus$ and $*$ correspond to the operations "min" and "max". We show then the relation between the subdominant ultrametric distance and the algebraic structure, and recall the relations connecting this ultrametric distance to the minimum spanning tree and to the indexed hierarchical classifications.

In paragraph 3, we state the basic theorem relating the eigenvalues and eigenvectors of the dissimilarity matrix to the levels and the different types of an hierarchical classification.

Thus, any level of a hierarchical classification based on the subdominant ultrametric distance corresponds to an eigenvalue of the dissimilarity matrix. Consequently, there exists a one-to-one correspondence between the eigenvectors, which generate the semi-module corresponding to such an eigenvalue, and the different classes of the associated level.

Actually, this theorem has a wider interpretation, for, to any indexed hierarchy corresponds an ultrametric distance (see BENZECRI[13]). We can then draw the conclusion that there is a one-to-one correspondence between the levels and types of any hierarchical classification, and the eigenvalues and eigenvectors of the distance matrix of the corresponding ultrametric distance.

Finally, in paragraph 4, we give an example illustrating the previous results.

2. THE PROPERTIES OF THE CLASSIFICATIONS RELATED TO AN ULTRAMETRIC DISTANCE

2.1. The subjacent algebraic structure

Given a set of n objects, we define, between any two of them i, j, a dissimilarity index $a_{ij} \in R^+$, which may not satisfy the distance axioms.

The matrix $A = (a_{ij})$ can be considered as the generalized incidence matrix of a symmetrical graph G, whose edges will be valued by the a_{ij}.

The subdominant ultrametric distance can then be defined in the following manner (see ROUX [10]) :

$$\delta(i,j) = \min_{\Pi \in C_{ij}} \left[\max(a_{i_1 i_2}, a_{i_2 i_3}, \ldots, a_{i_r, i_{r+1}}) \right] \qquad (2.1)$$

where

- . C_{ij} represents the set of the pathes from i to j.
- . Π represents the path $(i_1, i_2, i_3, \ldots, i_r, i_{r+1})$ with $i_1 = i$, $i_{r+1} = j$.

Consider the semi-ring $(S, \oplus, *)$ where $S = R^+ \cup \{+\infty\}$ and where the operations "addition" and "multiplication" correspond to the operations "min" and "max" ($\oplus$ = min, $*$ = max).

$\oplus$ admits the neutral element $\varepsilon = +\infty$, called the zero element and $*$ admits as neutral element $e = 0$, called the unit.

Furthermore, ε is absorbent for $*$.

The addition and the multiplication for the n order square matrices with elements in S will be defined by the laws $\oplus$ and $*$. The set of the matrices $M_n(S)$, defined in this way, then admits the same structure as S with the matrix :

$$\Sigma = \begin{bmatrix} \varepsilon & \cdots & \varepsilon \\ \vdots & & \vdots \\ \varepsilon & \cdots & \varepsilon \end{bmatrix}$$

as zero element, and with the matrix :

$$E = \begin{bmatrix} e & & \varepsilon \\ & \ddots & \\ \varepsilon & & e \end{bmatrix}$$

as unit.

$\oplus$ and $*$ will also denote the operations induced on $M_n(S)$.

It can then be shown (see CARRE [3], GONDRAN [4] [5]) that the matrix of the subdominant ultrametric distance is :

$$A^* = A \oplus A^2 \oplus \ldots \oplus A^{n-1} = A \oplus A^2 \oplus \ldots \oplus A^n = \ldots = A^{n-1} \qquad (2.2)$$

and verifies the equations :

$$A^* = AA^* \oplus A = A^*A \oplus A \tag{2.3}$$

even here, since $A^* = A^* \oplus E$,

$$A^* = AA^* = A^*A = A^*A^*. \tag{2.4}$$

A^* can then be calculated from (2.3) by the GAUSS method (see CARRE [3], GONDRAN [4][5]).

2.2. The minimum spanning tree

Considering a_{ij} as the lower capacity of the edge (i,j), $\delta(i,j)$ represents *the capacity of the minimum capacity path between i and j.*

Consider then the minimum spanning tree of the graph with the length a_{ij}. The following property can be stated (see KALABA [6], HU [7]) : *a minimum capacity path between the vertices i and j is obtained when considering the path connecting i to j in the minimum spanning tree.*

So, the search for a minimum spanning tree enables the subdominant ultrametric distance between all the objects to be obtained.

We are going to draw some very important conclusions from the preceding property.

a) With a tree having n - 1 edges, the matrix A^* contains at most n distinct coefficients (the n - 1 lengths of the minimum spanning tree edges, and the unit element e).

If all the coefficients contained in the matrice A are distinct (except for the $a_{ii} = e = 0$ and for those equal to $\varepsilon = +\infty$), then the matrix A^* has exactly n distinct coefficients.

b) According to the KRUSKAL algorithm [8], the minimum spanning tree, and consequently also the subdominant ultrametric distance, depend uniquely on the "arrangement" between the dissimilarity indexes a_{ij} (see the JOHNSON algorithms [11] and the LERMAN algorithms [12]).

c) Let the p ($p \leqslant n$) elements of A^* be arranged :

$$\lambda_1 \geqslant \lambda_2 \geqslant \ldots \geqslant \lambda_k \geqslant \ldots \quad \lambda_p = e = 0.$$

Then, the classification tree of the hierarchical classification associated with the subdominant ultrametric distance will have p levels, with the λ_k being the indexes for these levels.

The classification tree will be obtained by connecting, at each stage, the adjacent vertices of the minimum spanning tree taken in the order of the increasing λ_k.

The λ_k level classification corresponds to the connected components of the minimum spanning tree from which the edges greater than λ_k have been removed (cf. [14]).

In [9], we present a simple and efficient algorithm founded on the preceding remarks.

3. EIGENVALUES AND EIGENVECTORS IN TYPOLOGY

A vector $V \in S^n$ is an *eigenvector* of the matrix A in the semi-ring S, if $V \neq \varepsilon$ and if there exists $\mu \in S$, called the *eigenvalue*, such that :

$$AV = \mu V \tag{3.1}$$

where μV corresponds to the component vector $\mu * v_i$.

Lemma 1 - *For every* $\mu \in S$ *the vector* $\begin{pmatrix}\mu\\ \vdots\\ \mu\end{pmatrix}$ *is an eigenvector for A.*

Proof :

$$a_{ij} * \mu \geq \mu \quad , \text{ hence } \sum_{j=1}^{n} a_{ij}\, \mu \geq \mu$$

and since :

$$a_{ii} = 0, \quad \sum_{j=1}^{n} a_{ij}\, \mu = \mu, \text{ we have}$$

$$A \begin{pmatrix}\mu\\ \vdots\\ \mu\end{pmatrix} = \mu \begin{pmatrix}\mu\\ \vdots\\ \mu\end{pmatrix} . \quad \square$$

Consequently, all the values of S are eigenvalues.

Let $\mathcal{V}_\mu$ be the set of the eigenvectors corresponding to the eigenvalue μ.

Lemma 2 - *For any* $V \in \mathcal{V}_\mu$ *, we have :*

$$V = \mu V = AV = A^* V . \tag{3.2}$$

Proof : (3.2) implies for any i :

$$\sum_j a_{ij}\, v_j = \mu v_i$$

it is : $v_i = a_{ii} v_i \geq \mu v_i$, hence $v_i = \mu v_i$.

$V = AV$ then implies : $V = A^2 V = A^3 V = \ldots = A^* V$. $\square$

Assume that :

$$V_\mu^i = \mu (A^*)^i . \tag{3.3}$$

For μ given, the vector V_μ^i can be associated with any object i.

Lemma 3 - *Two objects i and j belong to the same type of level* μ *of the classification tree, if and only if their associated vectors* V_μ^i *and* V_μ^j *are equal.*

Proof : Immediate consequence, considering the connexion between the minimum spanning tree and the matrix A^*.

Lemma 4 - $\mathcal{V}_\mu$ *is identical with the semi-module generated by* V_μ^i

Proof :

(i) It will be shown at first that any element of the semi-module is a vector of $\mathcal{V}_\mu$.

Any element W of the semi-module generated by the V^i_μ is of the form :

$$W = \sum_{i=1}^{n} x_i \, V^i_\mu .$$

Since, according to (2.4), $(A^*)^i = A(A^*)^i$, and because of $\mu^2 = \mu$, we have :

$$A\,V^i_\mu = \mu A(A^*)^i = \mu(A^*)^i = \mu\mu(A^*)^i = \mu V^i_\mu \qquad (3.4)$$

it follows that :

$$AW = \mu W.$$

(ii) It will be shown now that any vector of $\mathcal{V}_\mu$ is generated by the V^i_μ.

Immediate, for, according to (3.2), $V = A^* \mu V$

it is :

$$V = \sum_{i=1}^{n} v_i \, V^i_\mu . \quad \square \qquad (3.5)$$

Lemma 5 - *A vector* V^i_μ *cannot be generated by any combination of other vectors.*

Proof :

Assume that $V^i_\mu = \Sigma\, \gamma_k\, W^k$ with $W^k \in \mathcal{V}_\mu$.

(i) Since the i component of V^i_μ is μ, there exists a k' such that $(\gamma^{k'} W^{k'})_i = \mu$, which involves $\gamma^{k'} \leqslant \mu$.

Since $W^{k'} \in \mathcal{V}_\mu$, it follows from the lemma 2 that $\gamma^{k'} W^{k'} = W^{k'}$; hence $V^i_\mu \leqslant W^{k'}$.

(ii) According to (3.5), we have :

$$W^{k'} = \sum_i (W^{k'})_i \, V^i_\mu$$

and since $(W^{k'})_i = \mu$, we derive :

$$W^{k'} \leqslant \mu \; V^i_\mu = V^i_\mu .$$

From the preceding inequalities we derive $W^{k'} = V^i_\mu$.

THEOREME : *Any level of a hierarchical classification based on an ultrametric distance corresponds to an eigenvalue of the distance matrix. There exists then a unique base of the semi-module of the corresponding eigenvectors; each of these eigenvectors defines a type of the hierarchical classification of this level.*

Note : The eigenvalues and the eigenvectors of A^* are identical with those of matrix A. This allows a wider interpretation of the previous theorem.

Indeed, since to any indexed hierarchy corresponds an ultrametric distance (see Benzecri [13]), we can draw the conclusion that there exists a one-to-one correspondence between the levels and classes of any hierarchical classification, on the one hand, and the eigenvalues and eigenvectors of the distance matrix of the corresponding ultrametric distance, on the other.

4. EXAMPLE

Consider the matrix of the following dissimilarities :

$A =$

	a	*b*	*c*	*d*	*e*	*f*	*g*	*h*	*i*
a	0								
b	7	0							
c	5	2	0						
d	8	10	7	0					
e	10	9	11	8	0				
f	8	9	10	4	9	0			
g	10	10	9	11	5	10	0		
h	12	11	9	11	1	9	6	0	
i	10	9	9	10	6	7	3	6	0

The minimum spanning tree is then :

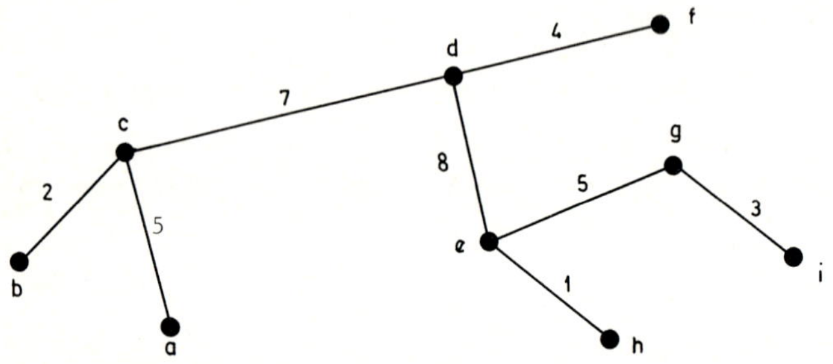

The classes of level 5 are then the following connected components :

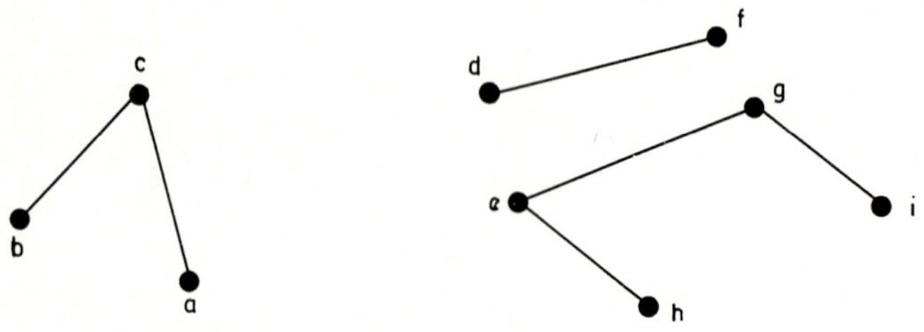

and the system of the associated eigenvectors is :

$$\begin{bmatrix} 5 \\ 5 \\ 5 \\ 7 \\ 8 \\ 7 \\ 8 \\ 8 \\ 8 \end{bmatrix} \text{ for } abc, \qquad \begin{bmatrix} 7 \\ 7 \\ 7 \\ 5 \\ 8 \\ 5 \\ 8 \\ 8 \\ 8 \end{bmatrix} \text{ for } df, \qquad \begin{bmatrix} 8 \\ 8 \\ 8 \\ 8 \\ 5 \\ 8 \\ 5 \\ 5 \\ 5 \end{bmatrix} \text{ for } eghi.$$

REFERENCES

1. M. GONDRAN - "Théorème de PERRON-FROBENIUS dans les semi-anneaux". Note EDF. (to appear).

2. M. GONDRAN et M. MINOUX - "Valeurs propres et vecteurs propres en théorie des graphes". Note EDF. HI 1941/02, 1975. To appear in *Actes du Colloque CNRS : Problèmes combinatoires et théorie des graphes*, Juillet 1976.

3. B.A. CARRE - "An algebra for network routing problems". *J. Inst. Maths Applics* - 7 - (1971) - p. 273-294.

4. M. GONDRAN - "Algèbre linéaire et cheminement dans un graphe". *Rev. Française Inf. Recherche Opér.* - 9ème année - janvier 1975 - V 1 - p. 81-103.

5. M. GONDRAN -"Algèbre des chemins et algorithmes". *Bulletin des Etudes et Recherches EDF*, série C, n° 2, 1975. See also "Path Algebra and Algorithms" *Combinatorial Programming : Methods and applications* - D. Reidel Publishing Co. - Dordrecht - Hollande - 1975 (B. Roy editor).

6. R. KALABA - "On some communication network problems" - Ch. 21 in *Combinatorial Analysis*, Proc. Sympos. Appl. Math - American Mathematical Society - RHODE ISLAND - 1960.

7. T.C. HU - "The maximum capacity route problem" - *Ops. Res.* 9 - 1961 - p. 898-900.

8. J.B. KRUSKAL Jr. - "On the shortest spanning subtree of a graph and the traveling salesman problem" - *Proc. Amer. Math. Soc.* n° 7 - 1956 - p. 40-50.

9 P. COLLOMB et M. GONDRAN - "Un algorithme efficace pour l'arbre de classification". Note EDF HI 1857/02 du 2 juin 1975. To appear *Rev. Française Inf. Recherche Opér.*, Janvier 1977, V 1.

10. M. ROUX - "Un algorithme pour construire une hiérarchie particulière". Thèse de 3ème cycle (L.S.M. I.S.U.P.) 1968.

11. S.C. JOHNSON - "Hierarchical clustering schemes", Psychometrica, 32, p. 241-243, 1967.

12. I.C. LERMAN - "Les bases de la classification automatique", Gauthier-Villars, Paris 1970.

13. J.P. BENZECRI - "L'analyse des données" Tome 1 : Taxinomie, Dunod Paris 1974.

14. C3E - "Analyse des données multidimensionnelles, tome III, 1972.

Recent Developments in Statistics
J.R.Barra et al., editors

AN APPLICATION OF COMBINATORIAL THEORY TO HIERARCHICAL CLASSIFICATION

Bruno Leclerc
Centre de Mathématique Sociale
Ecole des Hautes Etudes
en Sciences Sociales
54 Bd Raspail 75270 Paris 06
France

INTRODUCTION

We consider the problem of constructing a hierarchical classification of a finite set X on which is defined a dissimilarity coefficient (DC) $d : X \times X \to \mathbb{R}^+$, such that $d(x,x) = 0$ and $d(x,y) = d(y,x)$, for every $x,y \in X$.

Johnson [5] and Benzecri [1] have shown that an ultrametric DC, such that for every $x,y,z \in X$, $d(x,z) \leq \max(d(x,y), d(y,z))$, provides a hierarchical classification. Thus, one may consider the problem of assigning an ultrametric DC to each DC. The simplest solution is the subdominant ultrametric (SDU) of d ; that is, the unique ultrametric r such that $r \leq d$ and for every ultrametric $r' \leq d$, $r' \leq r$. This solution, equivalent to that of single linkage cluster analysis (SLCA) satisfies some interesting axioms of classification. But it has inconveniences; the main one being the "chaining" effect (see, for example, Jardine and Sibson [4]).

In part 1, we give some general definitions and results from combinatorial theory, and recall the relation between SDU's and minimum spanning trees (MST). In part 2, we improve a theorem on chains of MSTs, related to the chaining effect, and, in part 3, another result giving an optimality property of the SDU.

1. COMBINATORIAL DEFINITIONS AND RESULTS

1.1. Bottleneck theorem.

Let E be a finite set, Φ a family of subsets of E such that $I \in \Phi$, $J \subset I \Rightarrow J \not\subset \Phi$. We note $\tau(\Phi)$ the family of all subsets of E whose intersection with any element of Φ is not empty, and minimal (for inclusion) with this property. Edmonds and Fulkerson [2] gave two general results on Φ and $\tau(\Phi)$.

Theorem 1. $\tau(\tau(\Phi)) = \Phi$

Theorem 2 (Bottleneck theorem). For every function $f : E \to H$, where H is a totally ordered set, $\min_{I \in \Phi} \max_{i \in I} f(i) = \max_{J \in \tau(\Phi)} \min_{j \in J} f(j)$.

1.2. Definitions from graph theory.

A graph G is a doublet (X,E), where X is the finite set of vertices of G and E is a set of pairs (subsets of cardinality two) of vertices, the edges of G. A subgraph $G' = (Y,F)$ of G is a graph such that $Y \subseteq X$ and $F \subseteq E$.

A chain of G between two vertices $x,y \in X$ is a subset of E of the form $\{\{x,z_1\}, \{z_1,z_2\}, \ldots, \{z_k,y\}\}$. If $x = y$, the chain is a circuit. If, except for x and y, $x, z_1, z_2, \ldots, z_k, y$ are all distinct, the chain (circuit) is elementary.

We note $\mathcal{C}(x,y)$ the set of all elementary chains between x and y. A connected component of G is a maximal subgraph (Y,F) such that for every $x,y \in F$, there is a chain between x and y. If G is itself a connected component, G is a connected graph.

A spanning tree T of a connected graph G is a subset of E such that the subgraph (X,T) is connected and has no circuits. T is a spanning tree if and only if there is in T a unique chain, noted $T(x,y)$, between any two vertices x and y. We note $\mathcal{T}$ the set of all spanning trees of G.

A cutset ω of G is a minimal subset of E such that $(X,E-\omega)$ has one more connected component than G. We note Ω the set of all cutsets of G and, when x,y are in the same connected component of G, $\Omega(x,y)$ the subset of all $\omega \in \Omega$ such that x,y are in different connected components of $(X,E-\omega)$.

We shall use the following classical properties of trees, chains and cutsets :

(i) $\mathcal{T} = \tau(\Omega)$ and $\mathcal{C}(x,y) = \tau(\Omega(x,y))$.

(ii) Let $t \in T \in \mathcal{T}$. Then, there exists a unique cutset $\omega(T,t)$ such that $T \cap \omega(T,t) = t$. For every $e \in \omega(T,t)$, $T - \{t\} \cup \{e\} \in \mathcal{T}$.

(iii) Let $T \in \mathcal{T}$ and $e = \{x,y\} \in E - T$. Then, $T \cup \{e\}$ contains a unique circuit equal to $T(x,y) \cup \{e\}$ and, for every $t \in T(x,y)$, $T - \{t\} \cup \{e\} \in \mathcal{T}$.

1.3. Minimum spanning trees and subdominant ultrametric.

We consider now (and hereafter) the complete graph $K_X = (X,E)$, where E is the set of all pairs of elements of X, valued by the DC d here thought of as an application of E into $\mathbb{R}^+$ (this is equivalent to the preceding definition of a DC). Obviously, K_X is a connected graph.

A minimum spanning tree (MST) is then a spanning tree T of K_X minimizing $\sum_{e \in T} d(e)$.

A complete mathematical study of MST has be done by Rosenstiehl [7]. Gower and Ross [3] have related SLCA to MST and given an important bibliography about the applications of MST.

If T is a MST, the SDU r of d is given by $r(x,y) = \max \{d(e) \;/\; e \in T(x,y)\}$, for any $x,y \in X$.

2. CHARACTERIZATION OF THE CHAINS OF MST

We define the minimax chains (of K_X valued by d) between $x,y \in X$ as the chains C satisfying $\max \{d(e) \;/\; e \in C\} = \min_{C' \in \mathcal{C}(x,y)} \max \{d(e) \;/\; e \in C'\}$, the minimax edges as the edges e such that {e} is a minimax chain, and the totally minimax (or t-minimax) chains as the chains C such that every chain $C' \subseteq C$ is a minimax chain.

Let T be a MST and $t \in T(x,y)$ such that $d(t) = \max \{d(e) \;/\; e \in T(x,y)\}$. The cutset $\omega(T,t)$ is an element of $\Omega(x,y)$. If there exists $e \in \omega(T,t)$ with $d(e) < d(t)$, then, by (ii) above, $T - \{t\} \cup \{e\} \in \mathcal{T}$ and T is not a MST. Thus $d(t) = \min \{d(e) \;/\; e \in \omega(T,t)\}$. By (i) and the bottleneck theorem, we have :

<u>Theorem 3.</u> $\min_{C \in \mathcal{C}(x,y)} \max_{e \in C} d(e) = \max_{e \in T(x,y)} d(e) = d(t) = \min_{e \in \omega(T,t)} d(e) = \max_{\omega \in \Omega(x,y)} \min_{e \in \omega} d(e)$

Therefore, any chain included in T is minimax (Kalaba, [6]). Moreover, $\omega(T,t)$ is a "maximin" cutset. This point will be studied further on.

Suppose now that $e = \{x,y\}$ is a minimax edge. Then $d(e) = d(t)$ and $T - \{t\} \cup \{e\}$ is, by (iii), a MST. So any minimax edge is element of a MST (Roux, [8]). The complete result is :

Theorem 4. A chain C is t-minimax if and only if C is included in a MST.

The only part of the theorem not yet proved is that any t-minimax chain C, with $|C| > 1$, is included in a MST. Suppose this to be true for $|C| \leq k-1$ and let $C = \{e_1, \ldots, e_k\}$ be a t-minimax chain with $e_i = \{x_{i-1}, x_i\}$, $i = 1, \ldots, k$.

By the induction hypothesis, there exists a MST T such that $\{e_2, \ldots, e_k\} \subset T$. If $e_1 \notin T$, consider the minimax chain $T(x_o, x_1)$. Notice that this chain is of the form $\{e_2, \ldots, e_\ell, e'_1, \ldots, e'_m\}$, where $\ell \leq k$ ($\{e_2, \ldots, e_\ell\}$ can be empty), $m \leq |X| - k$ and $e'_i \notin C$, for $i = 1, \ldots, m$.

The chains $C' = \{e_1, \ldots, e_\ell\}$ and $C'' = \{e'_1, \ldots, e'_m\}$ are both minimax chains between x_o and x_ℓ, by $C' \subset C$ and $C'' \subset T$. As $d(e_1) = \max\{d(e) / e \in T(x_o, x_1)\}$ and $\max\{d(e) / e \in C\} = \max\{d(e) / e \in C'\}$, there exists $e \in T(x_o, x_1) - C$ with $d(e) = d(e_1)$ and, by (iii) above, C is included in the MST $T - \{e\} \cup \{e_1\}$. Thus the theorem is improved.

We will not give here the demonstration of the following result :

Theorem 5. A tree T is a MST if and only if every chain $C \subset T$ is minimax.

The theorem 4 makes, it seems, that the chains of the MSTs correspond to an intermediarity between elements of X : a t-minimax chain C between x and y may be thought of as a step by step similarity between x and y, for any chain 'C' between x and y contains an edge no smaller than those of C, this being also true for any subchain of C. Although x and y may be very dissimilar, the chain C provides an optimal path between x and y by successively smallest possible dissimilarities. This point of view is quite different from that of classification.

3. AN OPTIMALITY PROPERTY OF THE SDU

For every pair Y_1, Y_2 of disjoint subsets of X, let $D(Y_1,Y_2)$ be the interclasses dissimilarity of the SLCA, defined by $D(Y_1,Y_2) = \min\{d(x_1,x_2) / x_1 \in Y_1, x_2 \in Y_2\}$. A partitioning of X into two classes X_1 and X_2 such that $X_1 \cap X_2 = \emptyset$, $X_1 \cup X_2 = X$, maximizing $D(X_1,X_2)$ is given by the sets of vertices of the two connected components of the subgraph $(X, E - \omega)$ of K_X, where ω is a cutset maximizing $\min\{d(e) / e \in \omega\}$. By the bottleneck theorem, we have, for any MST T and $t \in T$ such that $d(t) = \max\{d(e) / e \in T\}$:

Theorem 6. $\min_{U \in \mathcal{T}} \max_{e \in U} d(e) = \max_{e \in T} d(e) = d(t) = \min_{e \in \omega(T,t)} d(e) = \max_{\omega \in \Omega} \min_{e \in \omega} d(e)$

So, an optimal partitioning corresponds to the cutset $\omega(T,t)$. Let $x_1 \in X_1$, $x_2 \in X_2$. By $t \in T(x_1,x_2)$, we have $r(x_1,x_2) = d(t) = \min\{d(e) / e \in \omega(T,t)\} = D(X_1,X_2)$.

Consider now the complete graphs $K_{X_1} = (X_1,E_1)$ and $K_{X_2} = (X_2,E_2)$, valued by the restrictions of d to E_1 and E_2. It is not difficult to see that $E_i \cap T$ is a MST of K_{X_i}, $i = 1,2$. Then, theorem 6 can be applied to K_{X_1} (resp. K_{X_2}) for partitioning X_1, (resp. X_2) into two classes X_{11} and X_{12}, (resp. X_{21} and X_{22}), and so on. The optimal hierarchical classification obtained is that provided by the SDU.

It is interesting to notice that SLCA is usually described as an agglomerative cluster method. We have shown above that it may also be thought of as a divisive cluster method. We know of no other method for obtaining a hierarchical classification optimal for both division and agglomeration.

REFERENCES

[1] Benzecri, J.P., Problemes et méthodes de la taxinomie (1965), in : BENZECRI, J.P. et coll., L'analyse des données. 1. La taxinomie, Paris, Dunod, 1973.
[2] Edmonds, J., Fulkerson, D.R., Bottleneck extrema, J. of Combinatorial Theory 8 (1970) 299-306.
[3] Gower, J.C., Ross, G.J.S., Minimum spanning tree and single linkage cluster analysis, Applied Statistics 18 (1969) 54-64.
[4] Kalaba, R., On some communication network problems, ch.21 in : Comb. analysis, Proc. Symp. Appl. Math. 10 Math.Soc. 1960.
[5] Jardine, N., Sibson, R., Mathematical Taxonomy, New-York, Wiley 1972.
[6] Johson, S.C., Hierarchical clustering schemes, Psychometrika 32 n°3 (1967), 241-254.
[7] Rosenstiehl, P., L'arbre minimum d'un graphe, in : Theorie des graphes, Rome, 1966, Paris, Dunod and New-York, Gordon and Breach 1967.
[8] Roux, M., Notes sur l'arbre de longueur minima, Rev. Stat. Appl. 23, n°2 (1975).

Recent Developments in Statistics
J.R.Barra et al., editors

FORMAL ANALYSIS OF A GENERAL NOTION OF PROXIMITY BETWEEN VARIABLES

I.C. LERMAN
Université de RENNES
U.E.R. Mathématiques et Informatique
Boite postale 25 A
35031 - RENNES Cédex
FRANCE

INTRODUCTION

We have worked out a set of algorithms that make possible the arrangement of the relationships between variables (retained by the Expert to describe the population under study), into classes and subclasses according to a given structure. We were thus brought to study the nature of the variables as they appear in the Humanities and Natural Sciences ; and to make clear the unique steps leading to the definition of a measure of proximity between the variables of the same type. We have thus refound some known measures and discovered some of the new ones.

Though the ideas developped here intervene crucially in the definition of our classification algorithms, there will be no mention of the latter in this paper.

TYPES OF THE DESCRIPTIVE VARIABLES

J. Hájek and Z. Sidak in their renowned book "Theory of Rank Tests" state that "The whole theory of rank tests is based on the essential assumption that all the observed random variables are governed by continuous distributions". In our approach to the Analysis of Data we insist on a mathematical representation that respects the poverty of scale of the descriptive variables to be treated ; then, the validity of the preceding assumption cannot be justified. In fact we distinguish between two principal types of descriptive variables : the one that can be represented by a subset of the set E of objects or by a weighting on E ; and the other whose representation is a subset of ExE or a weighting on ExE.

This attribute of description (i.e. feature), represented by the subset of objects possessing it, and the numerical variable which defines a weighting on E may be put into the first category. The following belong to the second : the ranking variable defining a total order o on E that we represent by its graph

$$R(o) = \{(x,y) \in ExE \mid x < y \text{ for } o\},$$

the descriptive character with a completely ordered set of values defining a total preorder W on E represented by the subset

$$R(W) = \sum_{i<j} E_i x E_j \text{ (set operation } \textstyle\sum)$$

of ExE where E_i is the ith class of the preorder ; the descriptive character the set of values of which is without any structure, defining a partition π on E represented by $R(\pi)$ in the set F of unordered object pairs, where

$$R(\pi) = \left\{ \{x,y\} \in F / \exists\ j,\ 1 \leqslant j \leqslant k,\ x \in E_j \text{ and } y \in E_j \right\}$$

and $\pi = \{E_1, E_2 \ldots . E_j, \ldots . E_k\}$;

and the "weighting" variable on ExE which can be represented by a square matrix

$$\{\mu_{xy} / (x,y) \in ExE\}$$

where μ_{xy} is a weighting attached to the couple (x,y).

GENERAL EXPRESSION OF THE MEASURE IN THE "DISCRETE" CASE

If (α,β) is a couple of variables of a same type (i.e. defining the same type of structure on E) represented by a couple $(R(\alpha), R(\beta))$ of subsets of E in case of descriptive attributes and subsets of ExE in case of the variables of second category, we introduce the "rough" index of proximity :

$$s = \text{card}\ (R(\alpha) \cap R(\beta)) \tag{1}$$

With α (respectively β) is associated the set A (resp. B) of structures of the same type and having the same cardinal characteristics as α (resp. β). Thus if α is a total preorder on E whose "composition" is $u = (n_1, n_2, \ldots, n_k)$, where n_i is the cardinal of the ith class, A will be the set of all the total preorders of the same composition u on E. In the different cases considered above, we have proved in a unique way that the distribution of the random variable S_α = Card $(R(\alpha) \cap R(\beta'))$ is the same as that of S_β = Card $(R(\alpha') \cap R(\beta))$ where α' (resp. β') is a random element in A (resp. B) having a uniform probability distribution. We give the precise expression of the moments of such distribution and its asymptotic normality. (cf. [8] and [9]).

If $\mu_{\alpha\beta}$ and $\sigma^2_{\alpha\beta}$ denote the mean and the variance of the r.v. S_α (resp. S_β), we adopt the following index

$$Q_{\alpha\beta} = (s - \mu_{\alpha\beta}) / \sigma_{\alpha\beta} \tag{2}$$

Denoting by N the above mentioned hypothesis of non link, the final index referring to a probability scale that we shall consider can be written as

$$P(\alpha,\beta) = P_r^N\ \{S < s\} \tag{3}$$

where S is one of the two identically distributed r.v's. S_α, S_β. In other words, the bigger the value of s compared to the hypothesis N, the higher the degree of resemblance between α and β.

The formula (3) is obtained from (2) by using the following relation which is a good approximation

$$P(\alpha,\beta) = \phi\ [Q(\alpha,\beta)] \tag{4}$$

where ϕ is the distribution function of the N(0,1) variate. The following results are obtained :

If (α,β) is a couple (a,b) of features, then Q(a,b) is nothing else than K. Pearson's coefficient of association.

In the case of a couple (o,o') of rankings (i.e. totally and strictly ordinal), Q(o,o') is M.G. Kendall's τ where the denominator is replaced by the standard deviation $\sigma_{oo'}$ (the denominator of this index (τ) is the maximum of the absolute value of the numerator).

In the case of comparing a couple (W,W') of the total preorders, a completely new index is obtained. It is shown that the index proposed by M.G. Kendall in this case is biased in the sense that the expected value of the associated r.v. in the non link hypothesis N is different from zero. In fact M.G. Kendall has defined his index by simply assigning, in the algorithm of computation of τ, the value of an ordinal function to different objects belonging to the same class of the preorder (treatment of ties).

In the same way, a new index is obtained in the case of comparing two partitions, π and π', of E defined by a couple of descriptive characters. This index is essentially different from the χ^2 associated to the contingency table where the two partitions are crossed.

The experimental work of comparing the two statistics has already been undertaken.

GENERALISATION OF THE COMPARISON OF TWO FEATURES TO THE COMPARISON OF TWO WEIGHTINGS ON E

If E is coded by a set of subscripts $I = \{1, 2, ..., i, ..., n\}$, it is easy to see that the rough index of proximity between two features a and b can be put in the from

$$s = \sum_{i \in I} \alpha_i \beta_i$$

where α (resp. β) is the indicating function of the subset E_a (resp. E_b), and that the two r.v.'s S_a and S_b are the same as

$$\sum_{i \in I} \alpha_i \beta_{\sigma(i)} \quad \text{and} \quad \sum_{i \in I} \alpha_{\sigma(i)} \beta_i$$

where $(\sigma(1), ..., \sigma(1), ..., \sigma(n))$ is a random permutation of $(1, 2, ..., i, ...,n)$; in other words, where σ is a random element in the set $\mathfrak{S}_n$, provided with a uniform measure of probability, of all permutations on $(1, 2, ... n)$.

Therefore, the rough index of proximity between two weightings on E which is the generalization of the rough index of proximity between two features :

$s = \sum \alpha_i \beta_i$ where α_i (resp. β_i) is, now, the value assigned to the object coded as i by the first (resp. second) variable. The two r.v.'s associated are

$$\sum_{i \in I} \alpha_i \beta_{\sigma(i)} \quad \text{and} \quad \sum_{i \in I} \alpha_{\sigma(i)} \beta_{(i)}$$

where $(\sigma(1), ..., \sigma(i), ..., \sigma(n))$ is a random permutation of $(1, 2, ..., i, ...,n)$. It is known that the famous theorem of Wald, Wolfowitz and Noether studies the asymptotic behaviour of this kind of distribution.

GENERALIZATION TO TWO WEIGHTINGS ON ExE

In the same way we extend the notion of proximity between two discrete variables of the same type (represented by the subsets of ExE) - as the ones defining the total orders, total preorders or the partitions on E - to the comparison of two weightings on ExE of the form

$$\{\mu_{ij} \ / \ (i,j) \in I^{[2]}\} \quad \text{and} \quad \{\nu_{ij} \ / \ (i,j) \in I^{[2]}\}$$

with $I^2 = (I \times I - \Delta)$, Δ being the diagonal of $I \times I$, where we suppose that for all (i,j) belonging to $I^{[2]}$, $r_{ij} \in \mathbb{R}^+$ (resp. $\nu_{ij} \in \mathbb{R}^+$)

It is natural to consider the rough index

$$s = \sum_{(i,j) \in I^{[2]}} r_{ij}\, \nu_{ij} \tag{5}$$

and to associate the dual variables :

$$S = \sum_{(i,j) \in I^{[2]}} r_{ij}\, \nu_{\sigma(i)\sigma(j)} \quad \text{and} \quad T = \sum_{(i,j) \in I^{[2]}} r_{\sigma(i)\sigma(j)}\, \nu_{ij} \tag{6}$$

shown to have the same distribution, where σ is defined as before.

It can in fact be shown that the different cases of comparison of discrete variables considered above appear formally as particular cases of the present situation.

Inspired by an old paper of H.E. Daniels, G. Lecalvé got the idea of this extension which led to a new measure. Nevertheless we have resumed this approach in a more precise way, especially in the calculation of moments to better justify the normal approach of the distribution of S (resp. T).

In fact the moments of S are compared to those of the "linear" permutation statistic

$$U = \sum_{(i,j) \in I^{[2]}} \xi_{ij}\, \eta_{\tau(i,j)} \tag{7}$$

where τ is a random permutation of $m = n(n-1)$ elements of $I^{[2]}$.

To start with, we notice that for fixed (i,j) the distribution of $\eta_{\sigma(i)\sigma(j)}$ is the same as that of $\eta_{\tau(i,j)}$ and hence the mean of S is equal to the mean of U.

For the comparison we use

$$X_{ij} = \xi_{ij} / (m_2(\xi))^{1/2}, \quad Y_{ij} = \eta_{ij} / (m_2(\eta))^{1/2} \tag{8}$$

where $m_2(\xi)$ (resp. $m_2(\eta)$) is the second moment of the distribution

$$\{\xi_{ij} / (i,j) \in I^{[2]}\} \quad (\text{resp. } \{\eta_{ij} / (i,j) \in I^{[2]}\})$$

so that

$$\frac{1}{m} \sum_{I^{[2]}} X_{ij}^2 = \frac{1}{m} \sum_{I^{[2]}} Y_{ij}^2 = 1 \tag{8'}$$

The rth moments of the two r.v.'s.

$$U_1 = m^{-1/2} \sum_{I^{[2]}} X_{ij} \, Y_{\tau(i,j)} \quad \text{and} \quad S_1 = m^{-1/2} \sum_{I^{[2]}} X_{ij} \, Y_{\sigma(i)\sigma(j)} \tag{9}$$

can be written, respectively, as

$$m^{-r/2} \sum c(r\ ;\ e_1, e_2, \ldots, e_k) \sum_{I_2^{(k)}} x_{q_1}^{e_1} \ldots x_{q_k}^{e_k} \frac{1}{m!} \sum_{\mathfrak{S}_m} y_{\tau(i_1,j_1)}^{e_1} \ldots\ldots\ldots \; y_{\tau(i_k j_k)}^{e_k} \tag{10}$$

and

$$m^{-1/2} \sum c(r\ ;\ e, \ldots, e_k) \sum_{I_2^{(k)}} x_{q_1}^{e_1} \ldots x_{q_k}^{e_k} \frac{1}{n!} \sum_{\sigma \in \mathfrak{S}_n} y_{\sigma(i_1)\sigma(j_1)}^{e_1} \ldots\ldots\ldots\ldots \; y_{\sigma(i_k)\sigma(j_k)}^{e_k}$$

where the first $\sum$, in both expressions, extends over all partitions of r into k parts ; $r = e_1 + \ldots + e_k$, and

$$c(r\ ;\ e_1 \ldots e_k) = \frac{r!}{e_1! \ldots e_k!\ s_1! \ldots s_h!} \tag{11}$$

If there are exactly h distinct integers e_j, the first one occurring s_1 times, etc ..., the last one s_h times.

Obviously, $k = s_1 + s_2 + \ldots + s_h$

The second $\sum$ denotes the summation over all the mutually distinct k-tuples of the elements of $I^{[2]}$. There are m(m-1) ... (m-k+1) such k-tuples in $I_2^{(k)}$. This sum is split up according to a partition of $I_2^{(k)}$, each of its classes $G_k^{(c)}$ being characterized by a configuration (c) of the k-tuple, which specifies the positions where an object of I is repeated. If d is the number of distinct subscripts used in the configuration (c), we have

$$n(c) = \text{card}\ (G_k^{(c)}) = n(n-1) \ldots (n-d+1) \tag{12}$$

Hence (10) can be put as

$$m^{-1/2} \sum c(\,r;\ e_1 \ldots e_k) \sum_{(c)} n(c)\ \overline{X}_c\ (e_1 \ldots e_k)\ \overline{Y}'_k\ (e_1 \ldots e_k)$$

and

$$m^{-r/2} \sum c(\,r;\ e_1 \ldots e_k) \sum_{(c)} n(c)\ \overline{X}_c\ (e_1 \ldots e_k)\ \overline{Y}_c\ (e_1 \ldots e_k) \tag{13}$$

$$\text{where } \overline{X}_c\,(e_1 \dots e_k) = \frac{1}{\mathrm{Card}(G_k^{(c)})} \sum_{G_k^{(c)}} X_{q_1}^{e_1} \dots X_{q_k}^{e_k}$$

$$\text{and } \overline{Y}'_k\,(e_1 \dots e_k) = \frac{1}{m(m-1)\dots(m-k+1)} \sum_{I_2^{(k)}} Y_{q_1}^{e_1} \dots Y_{q_k}^{e_k} \tag{14}$$

relatively to a sequence of the pairs of tables

$((X_{ij}^{(n)}), (Y_{ij}^{(n)}))$, such that $\overline{X}(2) = \overline{Y}(2) = 1$ (condition (8')).

The convergence of $\overline{X}_c\,(e_1 \dots e_k)$, $\overline{Y}_c\,(e_1 \dots e_k)$ and $\overline{Y}'_k\,(e_1 \dots e_k)$ towards a finite limit, is ensured by the following condition :

$$\mathcal{E}_o\,(Z_{\tau(q_1)} + \dots + Z_{\tau(q_k)})^r = O(1), \text{ for all } r \geqslant 3,\ k \leqslant r \tag{15}$$

for each of the sequences $(X_{ij}^{(n)})$ and $(Y_{ij}^{(n)})$, where $\mathcal{E}_o$ denotes the operator $(\frac{1}{m!}\sum_{\tau \in \mathfrak{S}_m})$.

n(c) is maximum for d = 2k. In the expression (13) of the rth moment, the part of the sums corresponding to k < r/2 tends to 0 as n tends to infinity. As a result, the non-zero part of each of the sums corresponds to $k \geqslant r/2$. Besides, the part of each of the sums corresponding to the configuration other than the (0) - configuration in which no element is repeated in different pairs q_j belonging to the k-tuple $(q_1, \dots, q_k)$, tends to be negligible compared to the sum corresponding to the (0) - configuration. In fact

$$\mathrm{card}\,(G_k^{(o)}) = n(n-1)\dots(n-2k+1)$$

$$\text{and } \mathrm{card}\,(G_k^{(o)}) / \mathrm{card}\,(I_2^{(k)}) \to 1 \text{ as } n \to \infty \tag{16}$$

So we just have to compare

$$m^{-r/2} \sum_{\{Pr,k \mid k \geqslant r/2\}} c(r\,;\,e_1 \dots e_k)\, n(0)\, \overline{X}_o\,(e_1 \dots e_k)\, \overline{Y}_o\,(e_1 \dots e_k)$$

with

$$m^{-r/2} \sum_{\{Pr,k \mid k > r/2\}} c(r\,;\,e_1 \dots e_k)\, n(0)\, \overline{X}_o\,(e_1 \dots e_k)\, \overline{Y}'_k\,(e_1 \dots e_k) \tag{17}$$

where Pr,k is the set of partitions of r into k parts (splitting of the integer r into the shares $e_1, \dots, e_k$).

Similarly, we may write

$$\overline{Y}'_k\,(e_1 \dots e_k) = \sum_{(c)} \frac{\mathrm{card}\ G_k^{(c)}}{\mathrm{card}\,(I_2^{(k)})}\, \overline{Y}_c\,(e_1, \dots, e_k) \tag{18}$$

which tends towards $\overline{Y}_0 (e_1, \ldots, e_k)$.

Finally, if μ_r and ν_r are the rth moments about zero of the r.v.'s S and U respectively, we have the following theorem :

<u>Theorem</u> : Under the condition (15) the ratio μ_r/ν_r tends towards 1 as n tends to infinity.

As a result the distribution of S may be compared to that of U which is asymptotically normal by virtue of the theorem of Wald and Wolfowitz. This is a general indication to be confirmed by an experimental analysis made possible by the computer. There are two ways of doing this. First, by generating the different sets $G_k^{(c)}$; the exact values, or very close approximations (by calculating the sums in (13) over the set $\{P_{r,k} | k \geq r/2\}$, mentioned above) of μ and ν, can be determined. It is easier to do so in the case of two partitions or total preorders, where $x_{q_1}^{e_1} \ldots x_{q_k}^{e_k} = x_{q_1} \ldots x_{q_k} = 0$ or 1 (resp. $y_{q_1}^{e_1} \ldots y_{q_k}^{e_k} = y_{q_1} \ldots y_{q_k} = 0$ or 1).

The second way is to simulate the distribution of S from a large size random sample of permutations σ. Thus Mr. Bigorgne (post graduate student) has shown, by a computational analysis, the form of the distribution of Card $(R(\pi_0)\ R(\pi'))$ where π' is random, in the particular case of partitions, each one having two classes. The tendency towards the normal distribution appears more and more clearly as the difference between the sizes of two classes increases. If there are more classes in a partition this tendency might be stronger.

CONCLUSION

Although the random element considered is a permutation σ, the presentation of M.G. Kendall's τ, generally done in non parametric statistics, does not seem to be well integrated with the permutation statistics. We are content with assuring that the statistic corresponding to τ is not linear, with respect to the one of the form

$\sum_i \xi_i \eta_{\sigma(i)}$, subject to a very minute analysis (Wald, Wolpowitz, Noether, Hájek).

Between S considered by us and the latter, is situated the statistic studied by Mr. Motoo (1957) which can be presented as

$$V = \sum_{1 \leq i \leq n} \gamma_{i\sigma(i)} \tag{19}$$

Such a statistic makes it possible to compare two weightings on ExE with respect to the permutation of rows or (exclusively) the columns of the square table representing the weighting. In fact, in this situation we are led to the statistic

$$W = \sum_{i \leq j \leq n} \sum_{1 \leq i \leq n} \xi_{ij} \eta_{i\sigma(j)} \tag{20}$$

where we have put

$$\gamma_{j\ \sigma(j)} = \sum_{1 \leq i \leq n} \xi_{ij} \eta_{i\sigma(j)} \tag{21}$$

On the other hand, the study of the statistics attached to the contingency tables (which occur when we have to cross two partitions or total preorders) appear in the literature of non parametric statistics as quite separated from the permutational statistics. Now we have seen how our measure combines the two aspects in a very natural way.

We must insist that relatively to such tables our point of view remains essentially different from that developped in a series of articles by L.A. Goodman and W.A. Kruskal :

Let A (resp. B) be the set indexing the rows (resp. columns) of the contingency table, and let

$$\{\pi_{ab} \ / \ (a,b) \quad A \times B\} \qquad (22)$$

be the table of proportions or theoretical probabilities for the entire or hypothetical population. The above mentioned authors, after some intuitive and somewhat arbitrary considerations, propose different measures which are functions of π_{ab}. Each of them is supposed to measure a certain type of relationship. In the centre of this work is the asymptotic behaviour of the sampling distribution of estimators of each of those measures.

The distribution of the permutation statistics is studied in favour of the development of the non parametric theory of the statistical tests where the hypothesis of independence or non-link is tested. But in our approach to the data in Humanities there is always a link, however subtle it may be, which can be "measured" from what is observed. However, the hypothesis of no link plays a crucial role, because it enables to establish the scale of reference for the evaluation of the link. These steps have led us to a uniform approach at the different stages of organisation of the links between different variables in the form of a tree , the most significant nodes of which can be recognized from similar considerations.

BIBLIOGRAPHY

[1] H.E. Daniels .- "The Relation Between Measures of Correlation in the Universe of Sample Permutations", Biometrika, vol 33, (1944).

[2] L.A. Goodman & W.H. Kruskal .- "Measures of Association for Cross Classifications", J.A.S.A., 49 (Déc. 1954), 732-64.

[3] L.A. Goodman & W.H. Kruskal .- "Measures of Association for cross Classifications", Approximate Sampling Theory", J.A.S.A., 58 (June 1963), 310-64.

[4] D.A.S. Fraser .- "Nonparametric Methods in Statistics", John Wiley, New York, (1957).

[5] J. Hájek & Z. Sidak .- "Theory of Rank Tests", Academic Press, New York and London, (1967).

[6] M.G. Kendall .- "Rank Correlation Methods", Charles Griffin, London, (fourth edition, 1970).

[7] G. Lecalvé .- "Sur un Indice de Similarité entre Variables" (Personal communication)) (Univ. Rennes II, Nov. 1975).

[8] I.C. Lerman .- "Etude Distributionnelle de Statistiques de Proximité entre Structures Finies de même type ; Application à la Classification Automatique". Cahiers du B.U.R.O. n° 19, Paris, (1973). (50 p.).

[9] I.C. Lerman .- "Etude Statistique de la Notion de Ressemblance" in "Cours sur la Reconnaissance et Classification des Structures Finies en Analyse des Données", chap. 2, Laboratoire de Statistique, Univ. de RENNES I (1974-75).

[10] M. Motoo .- "On the Hoeffding's Combinatorial Central Limit Theorem", Ann. Inst. Stat. Math. 8, (1957), 145-154.

[11] E. Noether .- "On a Theorem by Wald and Wolfowitz", Ann. Math. Stat. 20, (1949), 455-458.

[12] M.L. Puri & P.K. Sen .- "Non Parametric Methods in Multivariate Analysis", John Wiley, New York, (1971).

[13] A. Wald & J. Wolfowitz .- "Statistical Tests Based on Permutations of the Observations", Ann. Math. Stat., vol. 15, (1944).

Recent Developments in Statistics
J.R.Barra et al., editors

IDENTIFICATION RULES BASED ON PARTIAL INFORMATION ON THE PARAMETERS

F. Streit
University of Geneva
Geneva, Switzerland

In this paper the problem of constructing identification rules for situations involving two populations with partially known distributions is discussed from the decision theoretic point of view. In particular the case in which both populations are normally distributed with the same unknown covariance matrix Σ and different known mean vectors $\underset{\sim}{\mu}_1$ and $\underset{\sim}{\mu}_2$ is considered. For this problem of identification a decision rule is presented which satisfies several different criteria of optimality.

1. On the construction of identification rules

The basic problem to be considered is the following:
Let g_1 and g_2 be two populations with corresponding k-dimensional random vectors $\underset{\sim}{X}_1$ and $\underset{\sim}{X}_2$ and corresponding distribution functions F_1 and F_2 $[F_1 \neq F_2]$. $\underset{\sim}{x}$ denotes a realization of a random vector $\underset{\sim}{X}$. It is assumed that either $\underset{\sim}{X} \sim F_1$ or $\underset{\sim}{X} \sim F_2$. Thus we assume that $\underset{\sim}{x}$ is an observation of a sample unit taken either from g_1 or from g_2. The problem consists in constructing reasonable statistical rules for deciding whether the sample unit is an element of g_1 or of g_2.

In practical applications the distribution functions F_1 and F_2 are seldom completely known. Often they are only known up to a parameter θ_1 and θ_2, i.e. we may only assert that the distribution function F_1 of $\underset{\sim}{X}_1$ is an element of the class $\mathfrak{F}_1 = \{F_1(\cdot:\theta_1);\ \theta_1 \in \Theta_1\}$, that the distribution function F_2 of $\underset{\sim}{X}_2$ is an element of the class $\mathfrak{F}_2 = \{F_2(\cdot:\theta_2);\ \theta_2 \in \Theta_2\}$ and that $F_1 \neq F_2$. Under such conditions it is customary to estimate the quantities θ_1 and θ_2 based on a training sample $\underset{\sim}{x}^{(1)},\ldots,\underset{\sim}{x}^{(n_1)}$ from g_1 and a training sample $\underset{\sim}{x}^{(n_1+1)},\ldots,\underset{\sim}{x}^{(n_1+n_2)}$ from g_2. This leads to the following concept of identification rule which is relevant when only partial information on the distributions is available.

Definition:

Any function ϕ defined on $R^{(n_1+n_2+1)\cdot k}$ taking its values in [0, 1] is called identification rule for the comparison of $\mathcal{F}_1$ and $\mathcal{F}_2$.

Remark:

The values $\phi((\tilde{x}'; \tilde{x}^{(1)'}, \ldots, \tilde{x}^{(n_1+n_2)'})')$ are interpreted as the conditional probability of assigning the sample unit to g_1 given the observation $\tilde{X} = \tilde{x}$ and the measurements $\tilde{X}^{(i)} = \tilde{x}^{(i)}$ $[i = 1, \ldots, n_1+n_2]$ obtained from a realization of the training samples.*)

By Φ we designate the class of all identification rules. Of course it is crucial to know how to choose from Φ decision rules with desirable statistical properties. Some of the criteria, which may reasonably be applied for selecting an identification rule under these circumstances are listed below. They are expressed in using the following notation and abbreviations:

$n: = n_1 + n_2$ represents the size of the pooled training samples;

$\tilde{D}: = (\tilde{X}'; \tilde{X}^{(1)'}, \ldots, \tilde{X}^{(n)'})'$; $\tilde{d}: = (\tilde{x}'; \tilde{x}^{(1)'}, \ldots, \tilde{x}^{(n)'})'$,

i.e. $\tilde{d}$ represents the observed data collected from the training samples and the sample unit to be classified;

$\Lambda(\theta_1, \theta_2 : i, \tilde{d})$ represents the likelihood function of the parameters θ_1 and θ_2 conditional on $\tilde{D} = \tilde{d}$ and $\tilde{X} \sim F_i$ $[i = 1, 2]$.

Similarly $\Lambda(\theta_1, \theta_2 : \tilde{x}^{(1)}, \ldots, \tilde{x}^{(n)})$ denotes the likelihood function of θ_1 and θ_2 given a realization $\tilde{x}^{(1)}, \ldots, \tilde{x}^{(n)}$ of the training samples and $\Lambda(\theta_i : i, \tilde{x})$ the likelihood function of θ_i given that the identification index I equals i and that $\tilde{X} = \tilde{x}$, where I = i is equivalent to $\tilde{X} \sim F_i$. The distribution functions are assumed to be sufficiently regular, so that these concepts are well defined;

by C_i we denote the cost of a misclassification of a sample unit to g_i $[i = 1, 2]$ - it is assumed that C_i is a constant depending only on i - ;

by $G(\cdot : i; \theta_1, \theta_2)$ we designate the joint distribution function of $\tilde{D}$ conditional on I = i $[i = 1, 2]$;

P represents a prior distribution of the unknown quantities θ_1, θ_2 and I ;

we write $\bar{\phi}: = 1 - \phi$ for the probability of assigning the sample unit to g_2;

---

*) The sign ' is used to indicate the transposition of a matrix.

and we designate the risks of a misclassification by

$$r_{(\theta_1,\ \theta_2)}(2|1:\phi) := C_2 \int_{R^{(n+1)k}} \bar{\phi}(\underset{\sim}{d})\ dG(\underset{\sim}{d}: 1;\ \theta_1, \theta_2)$$

and $$r_{(\theta_1,\ \theta_2)}(1|2:\phi) := C_1 \int_{R^{(n+1)k}} \phi(\underset{\sim}{d})\ dG(\underset{\sim}{d}: 2;\ \theta_1, \theta_2)\ .$$

Some criteria of optimality for identification rules (partial information on the distributions available):

a.) For each $(\theta_1, \theta_2) \in \Theta_1 \times \Theta_2$ let $\psi_{(\theta_1,\ \theta_2)}$ denote an identification rule (fulfilling some recognized optimality criterion) for the comparison of $F_1(\cdot: \theta_1)$ and $F_2(\cdot: \theta_2)$, when θ_1 and θ_2 are known. Furthermore let T_1 and T_2 be estimators of θ_1 and θ_2 taking their values t_1 and t_2 in Θ_1 and Θ_2 respectively. Then the identification rule $\hat{\phi} := \psi_{(t_1,\ t_2)}$ is called the plug-in-rule for comparing $\mathcal{F}_1$ and $\mathcal{F}_2$ based on ψ, T_1 and T_2.

b.) Let $T_1^{(i)}$ and $T_2^{(i)}$ be the maximum likelihood estimators of θ_1 and θ_2 calculated from $\underset{\sim}{D}$ under the condition $I = i$ and write $\Lambda_i(\underset{\sim}{d})$ for $\Lambda(t_1^{(i)}, t_2^{(i)}: i, \underset{\sim}{d})$ $[i = 1, 2]$. An identification rule ϕ_M with the property that

$\Lambda_1(\underset{\sim}{d}) > \Lambda_2(\underset{\sim}{d})$ implies $\phi_M(\underset{\sim}{d}) = 1$ and

$\Lambda_1(\underset{\sim}{d}) < \Lambda_1(\underset{\sim}{d})$ implies $\phi_M(\underset{\sim}{d}) = 0$

is called maximum likelihood rule for comparing $\mathcal{F}_1$ and $\mathcal{F}_2$ (Anderson (1966, pp. 141-142)).

Remark:

Das Gupta (1973, p. 93) introduces a somewhat modified version of the maximum likelihood rule which is based on the conditional distributions of $\underset{\sim}{X}$ given $\underset{\sim}{X}^{(1)}, \ldots, \underset{\sim}{X}^{(n)}$ for $I = 1$ and for $I = 2$ and defined as follows:

b*.) Let $$\lambda_i(\underset{\sim}{d}) := \sup_{\substack{\theta_1 \in \Theta_1 \\ \theta_2 \in \Theta_2}} \left\{\frac{\Lambda\ (\theta_i : i, \underset{\sim}{x})}{\Lambda\ (\theta_1, \theta_2: \underset{\sim}{x}^{(1)}, \ldots, \underset{\sim}{x}^{(n)})}\right\} \quad [i = 1, 2]$$

An identification rule ϕ_C with the property that

$\lambda_1(\underset{\sim}{d}) > \lambda_2(\underset{\sim}{d})$ implies $\phi_C(\underset{\sim}{d}) = 1$ and

$\lambda_1(\underset{\sim}{d}) < \lambda_2(\underset{\sim}{d})$ implies $\phi_C(\underset{\sim}{d}) = 0$

is called conditional maximum likelihood rule for comparing $\mathcal{F}_1$ and $\mathcal{F}_2$.

In general criterion b.) and criterion b*.) do not lead to the same

identification rules as may be seen from the following example:
Let $\mathcal{F}_i := \{Be(q_i) ; \frac{1}{4} \leq q_i \leq \frac{3}{4}\}$ [i = 1, 2] , where Be (q) characterizes the distribution function of the 0 - 1 - distribution with $Pr(X = 0) = q$ and $Pr(X = 1) = 1-q$. For the identification problem comparing $\mathcal{F}_1$ and $\mathcal{F}_2$ based on a training sample of sample size 1 from g_1 and a training sample of sample size 1 from g_2 the realizations $X = 0$, $X^{(1)} = 0$, $X^{(2)} = 1$ yield the following values of the relevant statistics:

$$\Lambda_1(\underset{\sim}{d}) = \frac{27}{64}, \quad \Lambda_2(\underset{\sim}{d}) = \frac{3}{16}, \quad \lambda_1(\underset{\sim}{d}) = 4, \quad \lambda_2(\underset{\sim}{d}) = 12 .$$

For the indicated observed values a maximum likelihood rule assigns the sample unit to g_1 because the inequality $\Lambda_1(\underset{\sim}{d}) > \Lambda_2(\underset{\sim}{d})$ holds; a conditional maximum likelihood rule however assigns the same sample unit to g_2, because the inequality $\lambda_1(\underset{\sim}{d}) < \lambda_2(\underset{\sim}{d})$ holds. In this example the two methods for the construction of identification rules do therefore not generate equivalent decision rules.

c.) An identification rule ϕ_B which minimizes the expression

$$r(P,\phi) := \int_{(\theta_1,\theta_2)\in\Theta_1\times\Theta_2} r_{(\theta_1,\theta_2)}(2|1:\phi) \; dP(1;\theta_1,\theta_2)$$

$$+ \int_{(\theta_1,\theta_2)\in\Theta_1\times\Theta_2} r_{(\theta_1,\theta_2)}(1|2:\phi) \; dP(2;\theta_1,\theta_2)$$

among all $\phi \in \Phi$ is called <u>Bayes rule</u> for comparing $\mathcal{F}_1$ and $\mathcal{F}_2$ with respect to the prior distribution P. Here $dP(i;\theta_1,\theta_2)$ denotes the joint probability element of $I = i$, θ_1 and θ_2 [i = 1, 2] induced by P.

d.) Each identification rule ϕ_a with the property, that it does not exist a better identification rule than ϕ_a - i.e. there does not exist a $\phi \in \Phi$ with $r_{(\theta_1,\theta_2)}(2|1:\phi) \leq r_{(\theta_1,\theta_2)}(2|1:\phi_a)$ and $r_{(\theta_1,\theta_2)}(1|2:\phi) \leq r_{(\theta_1,\theta_2)}(1|2:\phi_a)$ for all $(\theta_1,\theta_2)\in\Theta_1\times\Theta_2$, where at least one of these inequalities is a strict inequality - is called an <u>admissible identification rule</u> for comparing $\mathcal{F}_1$ and $\mathcal{F}_2$.

2. On a particular identification rule based on partial information on the parameters

In section 1 a few criteria for the construction of identification rules have been presented. Of course the choice of the criterion which is used to select an element of Φ has to depend entirely on the particularities of the specific identification problem under investigation, but an identification rule may be considered as being especially valuable if it simultaneously fulfills several criteria of optimality. In this section we consider the problem of discrimination between two normally distributed populations with different known mean vectors and common unknown covariance matrix. Various other identification problems involving two normally distributed populations and partial information on their parameters are investigated in the literature (for a prominent example see John (1961); further references may be found in Das Gupta (1973), chapter 5). In many applications it seems however reasonable to assume that the populations differ only in their mean vectors, that the location parameters are known, and that the other parameters can only be estimated. It will be shown that for this identification problem there exists a decision rule satisfying the optimality criteria a.) - d.) listed in section 1.

In the sequel the following notation will be used:
By $N_k(\underset{\sim}{\mu},\Sigma)$ we denote the k-variate normal distribution with mean vector $\underset{\sim}{\mu}$ and covariance matrix Σ, by $n_k(\underset{\sim}{x} : \underset{\sim}{\mu}, \Sigma)$ its density function, by $N_k(\underset{\sim}{x} : \underset{\sim}{\mu}, \Sigma)$ its distribution function; by $W_k(n, \Sigma)$ we denote the k-dimensional (central) Wishart distribution with n degrees of freedom and scale matrix Σ and by $w_k(A : n, \Sigma)$ its density function;
given a quadratic matrix Ψ, we express by '$\Psi > 0$' that Ψ is positive definite;
the sign $\leftrightarrow$ is used to indicate the equivalence of two statements.

Theorem

Let ϕ^* be the identification rule characterized by the relations

$$\phi^*(\underset{\sim}{d}) = 1 \quad \text{for} \quad (\underset{\sim}{x}-\underset{\sim}{\mu}_1)'\,\bar{S}_n^{-1}(\underset{\sim}{x}-\underset{\sim}{\mu}_1) < (\underset{\sim}{x}-\underset{\sim}{\mu}_2)'\,\bar{S}_n^{-1}(\underset{\sim}{x}-\underset{\sim}{\mu}_2)$$

$$\text{and} \quad \phi^*(\underset{\sim}{d}) = 0 \quad \text{for} \quad (\underset{\sim}{x}-\underset{\sim}{\mu}_1)'\,\bar{S}_n^{-1}(\underset{\sim}{x}-\underset{\sim}{\mu}_1) > (\underset{\sim}{x}-\underset{\sim}{\mu}_2)'\,\bar{S}_n^{-1}(\underset{\sim}{x}-\underset{\sim}{\mu}_2)$$

for the problem of comparing $\mathcal{F}_1 := \{N_k(\cdot:\underset{\sim}{\mu}_1, \Sigma_1);\ \Sigma_1 > 0\}$ and $\mathcal{F}_2 := \{N_k(\cdot:\underset{\sim}{\mu}_2, \Sigma_2);\ \Sigma_2 > 0\}$ under the restrictions that $\underset{\sim}{\mu}_1$ and $\underset{\sim}{\mu}_2$

are known, that $\underset{\sim}{\mu}_1 \neq \underset{\sim}{\mu}_2$, that $\Sigma_1 = \Sigma_2 = \Sigma$ and that Σ is unknown; here $\bar{S}_n$ denotes the realization of the estimator

$$S_n := \frac{1}{n} \left(\sum_{i=1}^{n_1} (\underset{\sim}{X}^{(i)} - \underset{\sim}{\mu}_1)(\underset{\sim}{X}^{(i)} - \underset{\sim}{\mu}_1)' + \sum_{i=n_1+1}^{n} (\underset{\sim}{X}^{(i)} - \underset{\sim}{\mu}_2)(\underset{\sim}{X}^{(i)} - \underset{\sim}{\mu}_2)' \right)$$

for Σ and it is calculated from the observed values $\underset{\sim}{x}^{(1)}, \ldots, \underset{\sim}{x}^{(n_1)}$ and $\underset{\sim}{x}^{(n_1+1)}, \ldots, \underset{\sim}{x}^{(n)}$ of the training samples from g_1 and from g_2 $[n \geq k]$. The following statements hold for ϕ^* :

α.) ϕ^* is a maximum likelihood rule for comparing $\mathcal{F}_1$ and $\mathcal{F}_2$.

β.) ϕ^* is a Bayes rule for comparing $\mathcal{F}_1$ and $\mathcal{F}_2$.

γ.) ϕ^* is an admissible rule for comparing $\mathcal{F}_1$ and $\mathcal{F}_2$.

Remarks:

1. Note that ϕ^* is the plug-in-rule based on the commonly used decision rule appropriate for the corresponding identification problem with known Σ (Das Gupta (1973, p.98)) and based on the estimator S_n.

2. For the identification problem stated in the theorem, criterion b.) does not generate the same class of decision rules as criterion b^*.). The application of the principle of conditional maximum likelihood rule leads to a less convincing solution of the problem. This and the example presented in section 1 suggest that criterion b^*.) is perhaps not a suitable principle for the selection of identification rules.

Proof of the theorem

The theorem refers to an identification problem which involves the common unknown quantity $\theta_1 = \theta_2 = \theta \in \Theta$ with $\theta = \Sigma$ and $\Theta = \{\Sigma : \Sigma > 0\}$.

Derivation of the statement α.)

It is easy to verify that

$$\hat{\Sigma}^{(i)} = (n+1)^{-1} (nS_n + (\underset{\sim}{X} - \underset{\sim}{\mu}_i)(\underset{\sim}{X} - \underset{\sim}{\mu}_i)')$$

and

$$\Lambda_i(\underset{\sim}{D}) = (2\pi)^{-\frac{n+1}{2} \cdot k} (|\hat{\Sigma}^{(i)}|)^{-\frac{n+1}{2}} e^{-\frac{(n+1)k}{2}} \qquad [i = 1, 2] .$$

A maximum likelihood rule ϕ_M for comparing $\mathcal{F}_1$ and $\mathcal{F}_2$ is therefore specified by the following properties

$$\phi_M(\underset{\sim}{d}) = 1 \quad \text{if } \Lambda_1(\underset{\sim}{d}) > \Lambda_2(\underset{\sim}{d})$$

$$\leftrightarrow \; |n\bar{S}_n + (\underset{\sim}{x} - \underset{\sim}{\mu}_2)(\underset{\sim}{x} - \underset{\sim}{\mu}_2)'| > |n\bar{S}_n + (\underset{\sim}{x} - \underset{\sim}{\mu}_1)(\underset{\sim}{x} - \underset{\sim}{\mu}_1)'|$$

$$\leftrightarrow \; (\underset{\sim}{x} - \underset{\sim}{\mu}_2)' \bar{S}_n^{-1} (\underset{\sim}{x} - \underset{\sim}{\mu}_2) > (\underset{\sim}{x} - \underset{\sim}{\mu}_1)' \bar{S}_n^{-1} (\underset{\sim}{x} - \underset{\sim}{\mu}_1)$$

and

$$\phi_M(\underset{\sim}{d}) = 0 \quad \text{if } \Lambda_2(\underset{\sim}{d}) < \Lambda_2(\underset{\sim}{d})$$

$$\leftrightarrow \quad |n\bar{S}_n + (\underset{\sim}{x}-\underset{\sim}{\mu}_2)(\underset{\sim}{x}-\underset{\sim}{\mu}_2)'| < |n\bar{S}_n + (\underset{\sim}{x}-\underset{\sim}{\mu}_1)(\underset{\sim}{x}-\underset{\sim}{\mu}_1)'|$$

$$\leftrightarrow \quad (\underset{\sim}{x}-\underset{\sim}{\mu}_2)'\,\bar{S}_n^{-1}(\underset{\sim}{x}-\underset{\sim}{\mu}_2) < (\underset{\sim}{x}-\underset{\sim}{\mu}_1)'\,\bar{S}_n^{-1}(\underset{\sim}{x}-\underset{\sim}{\mu}_1) .$$

This shows that ϕ^* is a maximum likelihood rule.

<u>Derivation of the statement β.)</u>

We show that ϕ^* is a Bayes rule ϕ_B for comparing $\mathcal{F}_1$ and $\mathcal{F}_2$ with respect to the prior distribution which assigns equal probability to any of the two admissible values of I and for which the distribution of Σ^{-1} is given by the vague prior distribution and is independent of I (The concept of vague prior distribution is for instance explained by Press (1972, p. 72).). This amounts to choose the distribution function P_0 of the unknown quantities I and Σ^{-1} which is specified by the relation

$$P_0(I,\Sigma^{-1}) = P^{(1)}(I) \cdot P^{(2)}(\Sigma^{-1})$$

with $dP^{(1)}(1) = dP^{(1)}(2) = \frac{1}{2}$ and $dP^{(1)}(i) = \Pr(I=i) \; [i=1,2]$

and $dP^{(2)}(\Sigma^{-1}) \doteqdot (|\Sigma^{-1}|)^{-\frac{k+1}{2}} \, d\Sigma^{-1}$

where $\doteqdot$ means 'proportional to' .

For this choice of P and for equal cost of misclassification, i.e. for $C_1 = C_2 = C$, the Bayes risk of a decision rule ϕ depending on $\underset{\sim}{d}$ only through the values of the sufficient statistics $\underset{\sim}{x}$ and $A_n = n\bar{S}_n$ *) is given by

$$r(P_0,\phi) \doteqdot \frac{1}{2} \cdot \Big(\int_{\Sigma>0} \int_{R^k} \int_{A_n>0} \bar{\phi}(\underset{\sim}{x},A_n)\, w_k(A_n:n,\Sigma)\, n_k(\underset{\sim}{x}:\underset{\sim}{\mu}_1,\Sigma)\,dA_n\, d\underset{\sim}{x}$$
$$(|\Sigma^{-1}|)^{-\frac{k+1}{2}}\, d\Sigma^{-1} \Big) \cdot C$$

$$+ \frac{1}{2} \cdot \Big(\int_{\Sigma>0} \int_{R^k} \int_{A_n>0} \phi(\underset{\sim}{x}, A_n)\, w_k(A_n:n,\Sigma)\, n_k(\underset{\sim}{x}:\underset{\sim}{\mu}_2,\Sigma)\,dA_n\, d\underset{\sim}{x}$$
$$(|\Sigma^{-1}|)^{-\frac{k+1}{2}}\, d\Sigma^{-1} \Big) \cdot C$$

A Bayes rule ϕ_B minimizes $r(P_0,\phi)$. This is equivalent with minimizing the expression

---

*) It is adequate to consider only such identification rules, since they form an essentially complete class in Φ (see Ferguson (1969, p. 120)).

$$\frac{1}{2}\, C \int_{R^k} \int_{A_n>0} \phi(\underset{\sim}{x}, A_n) \quad J(A_n, \underset{\sim}{x}) \quad dA_n\, d\underset{\sim}{x}$$

with $J(A_n,\underset{\sim}{x}) := \int_{\Sigma>0} w_k\,(A_n:n,\Sigma)\; n_k(\underset{\sim}{x}:\underset{\sim}{\mu}_2,\Sigma)\,(|\Sigma^{-1}|)^{-\frac{k+1}{2}}\, d\Sigma^{-1}$

$$- \int_{\Sigma>0} w_k\,(A_n:n,\Sigma)\; n_k(\underset{\sim}{x}:\underset{\sim}{\mu}_1,\Sigma)\,(|\Sigma^{-1}|)^{-\frac{k+1}{2}}\, d\Sigma^{-1}$$

Thus

$\phi_B(\underset{\sim}{d}) = 1$ if $J(A_n, \underset{\sim}{x}) < 0$

and

$\phi_B(\underset{\sim}{d}) = 0$ if $J(A_n, \underset{\sim}{x}) > 0$.

Taking into account that $w_k(A_n:n,\Sigma) = c_{k,n}(|A_n|)^{\frac{n-k-1}{2}} (|\Sigma^{-1}|)^{\frac{n}{2}}$ $\exp\,[-\frac{1}{2}\,\mathrm{tr}\,(\Sigma^{-1} A_n)]$ where $\mathrm{tr}(\Psi)$ denotes the trace of a quadratic matrix Ψ and $c_{k,n}$ is a normalizing constant, we find that

$$\phi_B(\underset{\sim}{d}) = 1 \text{ if } \int_{\Sigma>0} (|\Sigma^{-1}|)^{\frac{n-k}{2}} \exp\,[-\frac{1}{2}\,\mathrm{tr}(\Sigma^{-1}T_n^{(2)})]d\Sigma^{-1} < \int_{\Sigma>0} (|\Sigma^{-1}|)^{\frac{n-k}{2}} \exp\,[-\frac{1}{2}\,\mathrm{tr}(\Sigma^{-1}T_n^{(1)})]\, d\Sigma^{-1}$$

and $\phi_B(\underset{\sim}{d}) = 0$ if the reverse inequality holds.

Here $T_n^{(i)} := A_n+(\underset{\sim}{x}-\underset{\sim}{\mu}_i)(\underset{\sim}{x}-\underset{\sim}{\mu}_i)'$ [i=1, 2] . Since $T_n^{(i)}>0$ [i=1, 2] with probability 1, $\mathrm{tr}(\Sigma^{-1}T_n^{(i)}) = \mathrm{tr}(T_n^{(i)}\Sigma^{-1})$ and

$\int_{\Sigma^{-1}>0} w_k(\Sigma^{-1}: n+1, T_n^{(i)})\, d\Sigma^{-1} = 1$ we may express this situation equivalently by the relations

$$\phi_B(\underset{\sim}{d}) = 1 \text{ if } (|T_n^{(2)}|)^{-\frac{n+1}{2}} < (|T_n^{(1)}|)^{-\frac{n+1}{2}}$$

$$\leftrightarrow (\underset{\sim}{x}-\underset{\sim}{\mu}_1)'\,\bar{S}_n^{-1}(\underset{\sim}{x}-\underset{\sim}{\mu}_1) < (\underset{\sim}{x}-\underset{\sim}{\mu}_2)'\,\bar{S}_n^{-1}(\underset{\sim}{x}-\underset{\sim}{\mu}_2)$$

and

$$\phi_B(\underset{\sim}{d}) = 0 \text{ if } (|T_n^{(2)}|)^{-\frac{n+1}{2}} > (|T_n^{(1)}|)^{-\frac{n+1}{2}}$$

$$\leftrightarrow (\underset{\sim}{x}-\underset{\sim}{\mu}_1)'\,\bar{S}_n^{-1}(\underset{\sim}{x}-\underset{\sim}{\mu}_1) > (\underset{\sim}{x}-\underset{\sim}{\mu}_2)'\,\bar{S}_n^{-1}(\underset{\sim}{x}-\underset{\sim}{\mu}_2)$$

This shows that ϕ^* is indeed a Bayes rule with respect to the prior distribution specified by P_o.

Derivation of the statement γ.)

We have already verified that ϕ^* is a Bayes rule (statement β.)). According to the general theory of statistical decisions (see Ferguson (1969, p. 60)) ϕ^* is an admissible identification rule if it is unique up to equivalence. We therefore have to check, whether the relations

$$\Pi_i: = \Pr((\underset{\sim}{X}-\underset{\sim}{\mu}_1)' S_n^{-1}(\underset{\sim}{X}-\underset{\sim}{\mu}_1) = (\underset{\sim}{X}-\underset{\sim}{\mu}_2)' S_n^{-1}(\underset{\sim}{X}-\underset{\sim}{\mu}_2) : i, \Sigma) = 0 \quad [i=1, 2]$$

hold. An alternative representation of Π_i is

$$\Pi_i = \Pr((\underset{\sim}{\mu}_2-\underset{\sim}{\mu}_1)' S_n^{-1}(\underset{\sim}{X}-\frac{\underset{\sim}{\mu}_1+\underset{\sim}{\mu}_2}{2}) = 0 : i, \Sigma)$$

$$= \int_{\bar{S}_n>0} \Pr(D(S_n) = 0 | S_n = \bar{S}_n : i, \Sigma) \cdot w_k(\bar{S}_n : n, \frac{1}{n}\Sigma)\, d\bar{S}_n$$

with $D(\bar{S}_n) := (\underset{\sim}{\mu}_2-\underset{\sim}{\mu}_1)' \bar{S}_n^{-1} (\underset{\sim}{X}-\frac{\underset{\sim}{\mu}_1+\underset{\sim}{\mu}_2}{2})$.

From $\underset{\sim}{X} \sim N_k(\underset{\sim}{\mu}_i, \Sigma)$ for $I = i$, it follows that $D(\bar{S}_n)$ has a univariate normal distribution with the positive variance $(\underset{\sim}{\mu}_2-\underset{\sim}{\mu}_1)' \bar{S}_n^{-1}\Sigma \bar{S}_n^{-1}(\underset{\sim}{\mu}_2-\underset{\sim}{\mu}_1)$. Thus

$$\Pr(D(S_n) = 0 | S_n = \bar{S}_n : i, \Sigma) = 0 \quad \text{for } i=1, 2 \text{ and } \bar{S}_n > 0$$

and this implies

$$\Pi_i = 0 \quad [i = 1, 2] \quad .$$

This shows that ϕ^* is indeed an admissible identification rule for comparing $\mathcal{F}_1$ and $\mathcal{F}_2$.

An abstract of some work on the properties of the identification rule ϕ^*, in particular on its behaviour for $n \to \infty$, may be found in Hug & Streit (1976). An additional justification for its use is presented in Enis & Geisser (1970).

References

Anderson, T.W. (1966). An Introduction to Multivariate Statistical Analysis. (J. Wiley, New York).

Das Gupta, S. (1973). Theories and Methods in Classification: A Review. In 'Discriminant Analysis and Applications' (Cacoullos, T., Editor; Academic Press, New York), pp. 77-137.

Enis, P. & Geisser, S. (1970). Sample Discriminants which Minimize Posterior Squared Error Loss. S.Afr.Statist.J. 4, 85-93.

Ferguson, T.S. (1969). Mathematical Statistics. A Decision Theoretic Approach. (Academic Press, NewYork).

Hug, G. & Streit, F. (1976). Linear Discriminant Analysis when $\underset{\sim}{\mu}_1$ and $\underset{\sim}{\mu}_2$ Are Known and Σ Is Unknown. Bulletin IMS, 5, 207.

John, S. (1961). Errors in Discrimination. Ann.Math.Statist. 32, 1125-1144.

Press, S.J. (1972). Applied Multivariate Analysis. (Holt, Rinehart and Winston, New York).

AUTHOR INDEX